T0236346

More information about this series at http://www.springer.com/series/7911

Marinos Themistocleous · Vincenzo Morabito (Eds.)

Information Systems

14th European, Mediterranean,
and Middle Eastern Conference, EMCIS 2017
Coimbra, Portugal, September 7–8, 2017
Proceedings

 Springer

Editors

Marinos Themistocleous
Department of Digital Systems
University of Piraeus
Piraeus
Greece

Vincenzo Morabito
Department of Management and Technology
Bocconi University
Milan
Italy

ISSN 1865-1348   ISSN 1865-1356 (electronic)
Lecture Notes in Business Information Processing
ISBN 978-3-319-65929-9   ISBN 978-3-319-65930-5 (eBook)
DOI 10.1007/978-3-319-65930-5

Library of Congress Control Number: 2017949520

Printed on acid-free paper

This Springer imprint is published by Springer Nature
The registered company is Springer International Publishing AG
The registered company address is: Gewerbestrasse 11, 6330 Cham, Switzerland

Preface

The European, Mediterranean, and Middle Eastern Conference on Information Systems (EMCIS) is an annual research event addressing the IS discipline from a regional as well as a global perspective. EMCIS has successfully helped to bring together researchers from around the world to freely exchange innovative ideas in a friendly atmosphere. EMCIS was founded in 2004 by Brunel University, London, UK and it is now an annual event. This year's event was organized by the University of Coimbra and the University of Lisbon, Portugal. Many respected collaborations have been established between different local universities across the destinations chosen each year and EMCIS still proves to attract many further partnerships.

EMCIS is one of the premier conferences in Europe and the Middle Eastern region for Information Systems academics and professionals, covering technical, organizational, business, and social issues in the application of Information Technology. EMCIS is dedicated to the definition and establishment of Information Systems as a discipline of high impact for the methodological community and IS professionals, focusing on approaches that facilitate the identification of innovative research of significant relevance to the IS discipline following sound research methodologies that lead to results of measurable impact.

EMCIS 2017 received 106 papers from 32 countries from all continents. The papers were submitted through the easychair.com online system and were sent for double blind review. The reviewers were either members of the conference committee or external. Papers submitted by the track chairs were reviewed by one conference co-chair and one member of the executive committee. The papers of the conference chairs and conference program chairs were reviewed by one senior external reviewer and one senior member of the executive committee. Overall, 53 papers were accepted by EMCIS 2017; 37 of them as full papers and another 16 as short papers submitted to one of the following tracks:

- Big Data and Semantic Web
- Cloud Computing
- Digital Services, Social Media, and Digital Collaboration
- e-Government
- Enterprise Systems
- Blockchain and Fintech
- Information Systems Security and Information Privacy Protection
- Healthcare Information Systems
- Management and Organizational Issues in Information Systems
- IT Governance

The papers were accepted for their theoretical and practical excellence and for the promising results they present. We hope that the readers will find the papers interesting and we are open for a productive discussion that will improve the body of knowledge on the field of Information Systems.

July 2017   Marinos Themistocleous
                                                                      Vincenzo Morabito

Acknowledgments

We express our gratitude to all those who contributed to the successful organization of EMCIS 2017. We would like to thank the authors for their contributions, the reviewers for their valuable comments, the track chairs for their efforts in selecting the best papers for their track. Also, we are grateful to the conference chairs, Paulo Rupino da Cunha and Miguel Mira da Silva, for their wide-ranging organizational involvement in the conference. We thank the members of the committee for their continuous support and our volunteers, who helped with the organization of this conference. Last but not least, we express our gratitude to Ralf Gerstner and Alfred Hofmann from Springer, without whom these proceedings would not have come into being.

Conference Organization

Conference Co-chairs

Paulo Rupino Cunha	University of Coimbra, Portugal
Miguel Mira da Silva	University of Lisbon, Portugal

Conference Executive Committee

Vincenzo Morabito	Bocconi University, Italy (Program Chair)
Marinos Themistocleous	University of Piraeus, Greece (Program Chair)
Gianluigi Viscusi	École Polytechnique Fédérale de Lausanne (EPFL), Switzerland (Publications Chair)
Muhammad Kamal	Brunel University, UK (Public Relations Chair)
Leonidas Katelaris	University of Piraeus, Greece (Website Coordinator)

International Committee

Janice Sipior	Villanova University, USA
Piotr Soja	Cracow University of Economics, Poland
Stanisław Wrycza	University of Gdansk, Poland
Celina M. Olszak	University of Economics in Katowice, Poland
Peter Love	Curtin University, Australia
Gail Corbitt	California State University, USA
Vishanth Weerakkody	Brunel University, UK
Lasse Berntzen	Buskerud and Vestfold University College, Norway
Kamel Ghorab	Alhosn University, UAE
Marijn Janssen	Delft University of Technology, The Netherlands
Inas Ezz	Sadat Academy for Management Sciences - SAMS, Egypt
Ibrahim Osman	American University of Beirut, Lebanon
Przemysław Lech	University of Gdansk, Poland
Małgorzata Pańkowska	University of Economics in Katowice, Poland
Euripidis N. Loukis	University of the Aegean, Greece
Mariusz Grabowski	Cracow University of Economics, Poland
Aggeliki Tsohou	Ionian University, Greece
Paweł Wołoszyn	Cracow University of Economics, Poland
Sofiane Tebboune	Manchester Metropolitan University, UK
Fletcher Glancy	Miami University, USA
Aurelio Ravarini	Universitario Carlo Cattaneo, Italy
Wafi Al-Karaghouli	Brunel University, UK
Ricardo Jimenes Peris	Universidad Politécnica de Madrid (UPM), Spain
Federico Pigni	Grenoble Ecole de Management, France

Paulo Henrique de Souza Bermejo	Universidade Federal de Lavras, Brazil
May Seitanidi	University of Kent, UK
Sevgi Özkan	Middle East Technical University, Turkey
Demosthenis Kyriazis	University of Piraeus, Greece
Karim Al-Yafi	Qatar University, Qatar
Manar Abu Talib	Zayed University, UAE
Alan Serrano	Brunel University, UK
Steve Jones	Conwy County Borough, UK
Tillal Eldabi	Ahlia University, Bahrain
Carsten Brockmann	Capgemini, Germany
Ella Kolkowska	Örebro University, Sweden
Grażyna Paliwoda-Pękosz	Cracow University of Economics, Poland
Janusz Stal	Cracow University of Economics, Poland

Contents

Digital Services, Social Media and Digital Collaboration

e-Government

Healthcare Information Systems

Information Systems Security and Information Privacy Protection

IT Governance

Management and Organizational Issues in Information Systems

Big Data and Semantic Web

Efficient Big Data Modelling and Organization for Hadoop Hive-Based Data Warehouses

Eduarda Costa[iD], Carlos Costa[⊠][iD], and Maribel Yasmina Santos[iD]

ALGORITMI Research Centre,
University of Minho, 4800 058 Guimarães, Portugal
{eduardacosta, carlos.costa, maribel}@dsi.uminho.pt

Abstract. The amount of data has increased exponentially as a consequence of the availability of new data sources and the advances in data collection and storage. This data explosion was accompanied by the popularization of the Big Data term, addressing large volumes of data, with several degrees of complexity, often without structure and organization, which cannot be processed or analyzed using traditional processes or tools. Moving towards Big Data Warehouses (BDWs) brings new problems and implies the adoption of new logical data models and tools to query them. Hive is a DW system for Big Data contexts that organizes the data into tables, partitions and buckets. Several studies have been conducted to understand ways of optimizing its performance in data storage and processing, but few of them explore whether the way data is structured has any influence on how quickly Hive responds to queries. This paper investigates the role of data organization and modelling in the processing times of BDWs implemented in Hive, benchmarking multidimensional star schemas and fully denormalized tables with different Scale Factors (SFs), and analyzing the impact of adequate data partitioning in these two data modelling strategies.

Keywords: Big Data · Data Warehousing · Hive · Modelling · Partitioning

1 Introduction

The advancements in Information and Communications Technology (ICT) contributed to the ever-increasing volume of data being currently generated in our daily activities [1]. Organizations are currently drowning in data, leading to severe data storage and processing difficulties when using traditional technologies [2]. This phenomenon is known as Big Data, mainly defined as data with high volume and variety (e.g., different data types, structures and sources), flowing at different velocities (e.g., batch, interactive and streaming) [3, 4], and aims to solve the problems of traditional data storage and processing technologies (e.g., centralized relational databases), since they provide high performance, scalability and fault-tolerance [5], generally at significant lower costs than traditional enterprise-grade technologies [6]. Hadoop is one of the main open source technologies in Big Data environments. It is divided into two main components: the Hadoop Distributed File System (HDFS), which is Hadoop's scalable and schema-less storage layer [7], capable of processing huge amounts of unstructured data on a cluster of commodity hardware [8]; and YARN, frequently defined as a sort of

© Springer International Publishing AG 2017
M. Themistocleous and V. Morabito (Eds.): EMCIS 2017, LNBIP 299, pp. 3–16, 2017.
DOI: 10.1007/978-3-319-65930-5_1

operative system for Hadoop, assuring that batch (e.g., MapReduce, Hive), interactive (e.g., Hive, Tez, Spark) and streaming (e.g., Spark Streaming, Kafka) applications use the cluster's resources as efficiently as possible, by adequately managing the resources in the cluster [9, 10].

Since DWs have long been a fundamental enterprise asset to support decision making [11], practitioners are looking into new ways of modernizing their current Data Warehousing installations [12, 13]. Hadoop is seen as one of the potential candidates for such endeavor. However, due to the fact that there is a lack of methodological approaches for DW design on Hadoop, practitioners often apply their current knowledge on traditional DWs, i.e., deploying star or snowflake schema DWs on Hive (the propeller of the SQL-on-Hadoop movement) [14]. Despite current efforts to accommodate this type of data processing on Hive (e.g., cost based optimizer, map joins) [15], Hadoop was originally conceived to store and process huge amounts of data in a sequential fashion. Therefore, there are questions that remain fairly unanswered by the scientific community and practitioners: Is a multidimensional DW a suitable design pattern for Data Warehousing in Hive? Are these schemas more efficient than fully denormalized tables? Do Hive's partitions have a significant effect in the execution times of typical Online Analytical Processing (OLAP) queries?

In order to help the community planning their BDWs and contribute to certain methodological aspects of data modelling and organization, this paper attempts to provide answers to these three questions, by benchmarking a Hive DW based on the Star Schema Benchmark (SSB) [16], using different SFs and two query engines, in order to provide observations based on different SQL-on-Hadoop systems: Hive on Tez [15], which can be considered the more efficient and stable query execution engine currently available for Hive; and Engine-X (real name omitted due to licensing limitations), which is a SQL-on-Hadoop system that can execute queries on Hive tables to provide low latency query execution for Business Intelligence (BI) applications, since Hive on Tez's latency still remains relatively high [17]. Furthermore, this paper also addresses relevant guidelines regarding Hive's data organization capabilities, such as data partitioning, which can considerably increase the performance of Hive DWs. Practitioners can use the insights provided by this paper to build their modern Data Warehousing infrastructure on Hadoop, or to migrate from a DW based on a Relational Database Management System (RDBMS).

This document is structured as follows: Sect. 2 presents scientific contributions related to this work; Sect. 3 discusses the materials and methods used in this research process; Sect. 4 presents the results obtained in the benchmark; Sect. 5 contains a discussion regarding the results and their usefulness for an adequate DW design on Hadoop (Hive), concluding with some remarks about this work.

2   Related Work

Today's data volume and data structure appear to be a major problem challenging the processing power of traditional DWs, since the inherent rules/strategies for relational data models can be less effective and efficient for patterns extracted from text, images, videos or sensor data, for example [6]. Data is no longer centralized and limited to the

Online Transaction Processing (OLTP) systems of the organizations, being now highly distributed, with different structures, and growing at an exponential rate. Therefore, the BDW differs substantially from the traditional DW, since its schema must be based on new logical models that are more flexible than relational ones [18]. The BDW implies new features and changes, such as highly distributed data processing capabilities; scalability at low cost; ability to analyze large volumes of data without creating samples; processing and visualization of data at the right time to improve the decision-making process; integration of diverse data structures from internal or external data sources; support of extreme processing workloads [19, 20].

In this new context, the data schema can change over time according to the storage or analytical requirements, being important to consider an adequate data model, as it ensures that the analytical needs are properly considered, allowing different analytical perspectives on data [21, 22].

Hive is a widely used DW, adopted by many organizations to manage and process large volumes of data, and was created by Facebook as a way to improve Hadoop query capabilities that were very limiting and not very productive [14]. This DW software for Big Data contexts organizes the data into tables (each table corresponding to a HDFS directory), partitions (sub-directories of the table directory) and buckets (segments of files in HDFS), and provides a SQL-based query language called HiveQL [14, 22]. Inside Facebook, it is heavily used for reporting, ad hoc querying and analysis [8], and according to [14], the main advantages include the simplicity in the implementation of ad hoc analysis and the ability to provide data processing services at a fraction of the cost of a more traditional storage infrastructure.

Total denormalization of data can be a way to improve query performance, as [23] demonstrates by comparing a relational DW with a fully denormalized DW using Greenplum, a Massively Parallel Processing (MPP) DW system. Other works such as [22] propose specific rules for structuring a data model on Hive by transforming a multidimensional data model (commonly used for traditional DWs) into a tabular data model, allowing data analysis in Big Data Warehousing environments. One of these rules includes a suggestion for the identification of Hive partitions and buckets, mentioning that is important to study the balance between the cardinality of the attributes and their distribution, also following Hive's official documentation [24].

There are also some works discussing the implementation of BDWs using NoSQL databases, despite the fact that they are mainly designed to scale OLTP applications with random access patterns [25], instead of fast sequential access patterns. Examples of such works can be highlighted: [26] studies the implementation of a DW based on a document-oriented NoSQL database; and, [27] discusses a set of rules to transform a multidimensional DW in column-oriented and document-oriented NoSQL data models.

SQL-on-Hadoop systems have been pointed as the *de facto* solution for Big Data Warehousing. Although the list of SQL-on-Hadoop systems is fairly extensive, this work points systems such as Hive [14]; Presto [28]; Impala [29]; Spark SQL [30]; and Drill [31], which were already evaluated, in works that compare and discuss their performance [17, 30, 32–34].

There is, however, a significant absence in the literature about the way data should be modeled in Hive, and how the definition of partitions and buckets can be optimized, as these can significantly improve the performance of BDWs. This work seeks to fulfill this scientific gap by benchmarking multidimensional star schemas and fully denormalized tables implemented in several Hive DWs that use different SFs (sizes), providing a clear overview of the impact of the adopted data models in the system efficiency. Moreover, the impact of data partitions is also analyzed for both data modelling strategies. This is of major relevance to both researchers and practitioners related to the topic of Big Data Warehousing, since it can foster future research and support design patterns for DWs in Big Data environments, exposing reproducible and comparable results.

3   Materials and Methods

Since this paper discusses some best practices for Big Data modelling and organization in Hive DWs, the guidelines and considerations here provided must be adequately validated, in order to produce strongly-supported and replicable results. Consequently, a benchmark was conducted to evaluate the performance of a Hive DW in different scenarios. This section describes the materials and methods used in this research.

3.1   Infrastructure

The infrastructure used in this work consists of a Hadoop cluster with 5 nodes, configured as 1 HDFS NameNode (YARN ResourceManager) and 4 HDFS DataNodes (YARN NodeManagers). The hardware used in each node includes: (i) 1 Intel core i5, quad core, with a clock speed ranging between 3.1 GHz and 3.3 GHz; (ii) 32 GB of 1333 MHz DDR3 Random Access Memory (RAM), with 24 GB available for query processing; (iii) 1 Samsung 850 EVO 500 GB Solid State Drive (SSD) with up to 540 MB/s read speed and up to 520 MB/s write speed; (iv) 1 gigabit Ethernet card connected through Cat5e Ethernet cables and a gigabit Ethernet switch. The operative system installed in every node is CentOS 7 with an XFS file system. The Hadoop distribution being used is the Hortonworks Data Platform (HDP) 2.6.0 with the default configurations, apart from the HDFS replication factor, which was set to 2. Besides Hadoop itself, an Engine-X master is also installed on the NameNode, as well as 4 Engine-X workers on the 4 remaining DataNodes. All Engine-X's configurations are left to their defaults, except the memory configuration, which was set to use 24 GB of the 32 GB available in each worker (same as the memory available for YARN applications in each DataNode/NodeManager).

3.2   Datasets and Queries

This work uses the SSB [16] and, therefore, it considers both the SSB dataset, which is a traditional sales DW modelled according to the multidimensional structures (stars) presented in [11], and the thirteen SSB queries to analyze its performance for typical OLAP workloads. As this paper is interested in evaluating the performance of different data modelling and organization strategies for Hive DWs, despite the significant advancements both in Hive (on Tez) and Engine-X to adequately process traditional star schemas (e.g., cost based optimizers, map joins) [15], denormalized models are also evaluated, since these structures have been typically preferred in the Hadoop ecosystem. In fact, as previously discussed in Sect. 2, recent approaches for Big Data Warehousing are already focusing on fully denormalized structures to support analytical tasks [22, 23, 35]. Therefore, both the original SSB relational tables and a fully denormalized table were implemented in the Hive DW, in order to evaluate their performance in Big Data environments. The data is stored as Hive tables using the ORC format and compressed using ZLIB. All the thirteen SSB queries were also adapted to the fully denormalized table, providing the same results as the original SSB dimension and fact tables.

3.3 Test Scenarios

As previously mentioned, this paper aims to analyze the impact of different data models and different ways to organize the data in the BDW, so the tests to be performed will consist of: (i) Scenario A: Measuring the variation in the processing time using the star model vs. the denormalized model; (ii) Scenario B: Measuring the variation in the processing time using data partitions, applied both to the star model and to the denormalized model. The insights provided by this paper also take into consideration two SQL-on-Hadoop systems, to verify if the results are comparable among different query engines. In this benchmark, Engine-X and Hive (on Tez) are used. Furthermore, since one of the objectives of this paper is to understand the impact of different data organization strategies, and being partitioning an inherent feature of Hive, it is important to understand how it behaves when partitions are created. Figure 1 presents the test scenarios implemented in this work, namely scenario A and B. In order to achieve more rigorous results, several scripts were coded for executing each query four times. These scripts were adapted according to the SQL-on-Hadoop system (Hive or Engine-X), the applied data model (star or denormalized) and the organization strategy (with or without partitions). All Hive scripts are available on GitHub (https://github.com/epilif1017a/bigdatabenchmarks).

Moreover, it is important to highlight that for Engine-X, the queries were executed using two different joins strategies: distributed joins and broadcast joins. The

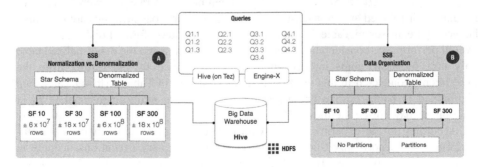

Fig. 1. Test scenarios.

distributed joins strategy can handle larger join operations but is typically slower. Broadcast joins can be substantially faster, but require that the right side of the join fits in a fraction of the memory available in each node. Such distinction is not made for Hive, since cost based optimization, map joins and other related configurations work by default in HDP 2.6.0, which automatically assure the best join strategies according to the cluster's configuration.

4 Results

This section presents the results achieved in the test scenarios discussed previously, not only comparing the performance of a Hive DW modeled using the star schema approach [11] with one modelled using a fully denormalized table, but also analyzing the performance impact of adequate data partitioning strategies in Hive. The results depicted in this section are relevant to highlight several factors that practitioners must take into consideration when designing BDWs.

4.1 Scenario A: Star Schema vs. Fully Denormalized Table

Relational DWs have long been the backbone of many decision support systems in organizations. In contrast, with the wide acceptance of Hadoop and Hive, denormalized structures became more common, allowing the reading and writing of data to large and contiguous sections of disk drives, optimizing I/O performance [36]. During a significant period of time, Hadoop and the processing of huge amounts of denormalized data (often unstructured) became synonymous. However, with the constant improvements made in Hadoop and related projects, and with organizations increasingly demanding adequate support for relational structures in Big Data environments, there was a wave of efforts to improve the performance of SQL-on-Hadoop systems, in order to efficiently support traditional BI applications based on multidimensional DWs. Nevertheless, there is one question that still remains fairly unanswered by the scientific community: "What are the performance advantages/disadvantages of structuring Hadoop-based DWs according to traditional and relational rules?".

Table 1 illustrates the results for different scaling factors, SF = 10, SF = 30, SF = 100 and SF = 300 workloads, both for Hive and Engine-X query engines. As previously discussed in Subsect. 3.3, two different join strategies were tested for Engine-X: distributed and broadcast. Processing times for broadcast joins are presented between parentheses in Table 1, and since they always outperformed distributed joins, overall performance considerations between the star schema and the denormalized table only take broadcast joins into account. As demonstrated by the results from the experiments conducted in this work, which are presented in Table 1 (lower values are highlighted in bold), despite being feasible to implement multidimensional DWs in Hadoop, it may not be the most efficient approach.

According to the experimentations with different SFs, the star schema only outperformed the denormalized table in 14 out of the 104 query executions, namely in Hive's SF = 10 workload (Q1.1, Q1.2, Q1.3 and Q2.2); in Hive's SF = 30 workload

Table 1. SSB execution times (in seconds): star schema (SS); denormalized table (DT).

Queries	SF = 10		SF = 30		SF = 100		SF = 300	
	SS	DT	SS	DT	SS	DT	SS	DT
Hive								
Q1.1	**19**	20	24	**23**	28	**23**	**44**	59
Q1.2	**20**	21	24	24	29	**22**	**43**	62
Q1.3	**19**	21	**23**	24	29	**22**	**43**	60
Q2.1	24	**21**	33	**26**	70	**45**	543	**95**
Q2.2	**21**	27	**33**	36	**58**	72	538	**181**
Q2.3	23	**20**	33	**24**	56	**37**	528	**81**
Q3.1	26	**22**	35	**28**	58	**47**	643	**112**
Q3.2	24	**22**	32	**28**	54	**46**	677	**108**
Q3.3	21	21	35	**25**	228	**37**	665	**84**
Q3.4	22	22	35	**25**	241	**37**	673	**94**
Q4.1	27	**22**	38	**28**	105	**51**	225	**121**
Q4.2	26	**23**	45	**28**	71	**31**	141	**122**
Q4.3	25	**22**	36	**28**	68	**32**	113	119
Total	297	**−4%**	426	**−19%**	1095	**−54%**	4876	**−73%**
Engine-X								
Q1.1	5 (4)	**2**	6 (5)	5	16 (12)	**5**	45 **(33)**	40
Q1.2	5 (4)	**3**	5 (5)	5	13 (13)	**3**	37 **(33)**	43
Q1.3	5 (5)	**3**	5 (5)	5	13 (13)	**3**	37 **(33)**	44
Q2.1	12 (5)	**3**	28 (7)	**4**	88 (20)	17	257 (56)	**37**
Q2.2	11 (5)	**2**	27 (7)	**4**	87 (19)	15	253 (55)	**35**
Q2.3	12 (5)	**2**	26 (7)	**4**	87 (19)	15	253 (52)	**34**
Q3.1	10 (5)	**3**	22 (8)	**4**	70 (29)	12	206 (82)	**36**
Q3.2	9 (5)	**3**	19 (5)	**4**	60 (17)	14	175 (52)	**39**
Q3.3	8 (5)	**2**	18 (6)	**4**	58 (15)	12	166 (42)	**33**
Q3.4	9 (5)	**3**	18 (6)	5	57 (15)	**9**	168 (43)	**42**
Q4.1	16 (6)	**3**	40 (13)	5	133 (46)	17	359 (125)	**46**
Q4.2	12 (6)	**3**	27 (8)	5	89 (26)	**8**	258 (69)	**45**
Q4.3	12 (5)	**3**	26 (8)	5	87 (21)	**8**	254 (60)	**42**
Total	65	**−46%**	90	**−34%**	265	**−48%**	735	**−30%**

(Q1.3 and Q2.2); in Hive's SF = 100 workload (Q2.2); in Hive's SF = 300 workload (Q1.1, Q1.2, Q1.3 and Q4.3); and, in Engine-X's SF = 300 workload (Q1.1, Q1.2 and Q1.3). If one takes a closer look at the pattern of Q1 in the SSB, it can be concluded that it should favor the star schema. Considering that the star schema fact table is roughly 3 times smaller than the denormalized table, and since Q1 only joins the fact table with the date dimension, the smaller size should compensate for the overhead of performing a single join operation. However, this does not happen in Hive's SF = 100 workload, where the denormalized table outperformed the star schema for all Q1 queries.

Moreover, in Hive's SF = 10, SF = 30 and SF = 100 workloads, Q2.2 also performs better for the star schema. After analyzing the particularities of Q2.2, one can conclude that Hive's query execution engine (Tez) may have demonstrated some performance degradation when performing string range comparisons in larger amounts of data. In this case, Q2.2 contained the following predicate: *"p_brand1 between 'MFGR#2221' and 'MFGR#2228'"*. However, this trend is not transposed to the SF = 300 workload, in which Q2.2 is significantly slower for the star schema when compared to the denormalized table.

Regarding the remaining 90 query executions, the denormalized table outperformed the star schema 84 times, and both achieved the same result 6 times. In Hive's workloads, the minimum overall performance advantage for the denormalized table was 4% (SF = 10) and the maximum overall advantage was 73% (SF = 300), which means that, in the best scenario, the denormalized table was able to complete the workload 73% faster than the star schema. With the increase of the dataset size, also increases the gain in performance, clearly benefiting Big Data scenarios. This performance difference is more noteworthy than in Engine-X's workloads, wherein the denormalized table tends to perform between 30% and 48% faster than the star schema. These results demonstrate that while it is feasible to implement multidimensional DWs in Hive, this approach is not always optimal when analyzing query execution times, since it is often outperformed by a purely denormalized structure.

If one takes a closer look at Engine-X's SF = 300 workload, at a first glance, it might seem that the performance advantage of the denormalized table faded out when compared to previous workloads. However, the SF = 300 is the first workload in which the total size of the denormalized table does not fit in the total amount of memory available for querying in the cluster (96 GB), containing around 139 GB of data stored in ORC files, while the entire star schema DW for the same SF contains approximately 51 GB of data in the same format. Given the fact that one is comparing a DW that is 45% larger than the total amount of memory with a DW that totally fits in memory, and it is still 30% faster on average, this highlights the fact that denormalized structures bring more benefits in terms of pure performance when processing queries over large datasets.

There are certain queries in which the star schema's execution times although slower, are fairly comparable with the denormalized table's execution times, but there is also a significant number of queries in which the star schema is more than 50% slower (sometimes more than 100% slower). Even with Engine-X's broadcast joins strategy, results do not favor the multidimensional approach for DWs in Big Data environments. Such phenomenon is also typically aggravated with an increased SF (specially for Hive), highlighting the data volume bottleneck in multidimensional BDWs. Consequently, whether one uses HiveQL or a more interactive SQL-on-Hadoop system like Engine-X's to query a Hadoop-based DW, it can be stated that a denormalized structure is able to typically outperform a multidimensional approach, with less potential bottlenecks when the volume of data increases.

Another troubling factor for multidimensional DWs on Hadoop is the fact that their performance, according to this benchmark, is only comparable to the performance of a denormalized table when efficient techniques such as broadcast joins are applied. With the distributed joins strategy in Engine-X or with Hive (on Tez), the star schema performance is often alarming and query execution times are significantly high. However, there is one relatively important caveat when employing broadcast joins: the broadcasted table (right side of the join) must be small enough to fit in the memory of each node (technically, in a small fraction of the memory available for queries), because if the broadcasted input is too large, "out of memory" errors can occur, due to the lack of memory to process all inputs. This suits to several dimension tables, as they are typically small, but for DWs implementing type 2 slowly changing dimensions [11], for example, this join strategy may not be feasible [23]. Not only does the star schema present certain memory requirements, but also, in this benchmark, the star schema tends to show a significant CPU overhead when compared to the denormalized modelling approach. On average, in Engine-X's SF = 300 workload, the star schema uses 143% more CPU time, despite being on average 43% slower than the denormalized table. Consequently, as Fig. 2 demonstrates, higher CPU usage can be considered as another disadvantage of the star schema approach for DWs in Big Data environments.

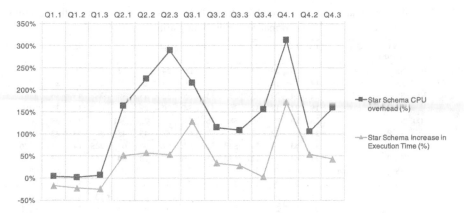

Fig. 2. Star schema and denormalized table CPU time and execution time comparison.

4.2   Scenario B: Data Organization (Hive Partitions)

One of the biggest problems of a DW, either traditional or in Big Data environments, is how to partition the data. Data partitioning refers to splitting data into separate physical units that can be treated independently. Thus, proper partitioning can bring benefits to a DW in terms of loading, accessing, storing and monitoring data [37]. According to [24], the attribute(s) on which partitioning will be applied should have low cardinality, in order to avoid the creation of a high number of subdirectories on HDFS.

Usually, Hive analyzes the entire table to answer a query and that may compromise its performance. For huge volumes of data, partitioning can dramatically improve queries performance. However, according to [36], this only happens if the partition schemes reflect common filters to the data, being usually defined with the attribute that appears more often in the "where" conditions of the queries. However, there is a question that remains: "What is the true impact of partitioning on processing time?".

Table 2 presents the results for SF = 10, SF = 30, SF = 100 and SF = 300 workloads for a star schema and for a denormalized table, respectively, both for Hive and Engine-X. Considering the results obtained in Sect. 4.1, the tests for Engine-X using the star schema will only contemplate the broadcast join strategy.

Analyzing the results, one can conclude that there is a significant improvement when using data partitioning, being few the cases where tables with no partitions behave better. Considering the percentages of decrease/increase in times obtained when using partitions, in all scenarios there is a decrease in queries execution time. In Hive's workloads, the minimum overall performance advantage was 2% and the maximum advantage was 63%, which means that, in the best scenario (star schema SF = 300), the partitioned table was able to complete the workload 63% faster than the table without partitions. Engine-X is generally faster than Hive and in its workloads the maximum advantage was 46%, in the denormalized table SF = 300 workload.

As mentioned before, usually, a table should be partitioned by the attribute that appears more often in the "where" condition of the queries. Considering the queries

Table 2. SSB execution times (in seconds): star schema (SS); star schema with partitions (SS-P); denormalized table (DT); denormalized table with partitions (DT-P).

	SF	Data schema	Q	1.1	1.2	1.3	2.1	2.2	2.3	3.1	3.2	3.3	3.4	4.1	4.2	4.3	
Star schema	10	SS	Hive	19	20	19	**24**	**21**	**23**	26	24	**21**	**22**	**27**	26	25	297
		SS-P		**14**	20	**14**	25	25	24	26	24	24	23	29	**23**	**21**	−2%
	30	SS		24	24	23	33	**33**	33	35	32	35	35	**38**	45	36	426
		SS-P		**19**	24	**18**	**32**	36	**31**	**34**	**30**	**29**	**28**	43	**32**	**28**	−10%
	100	SS		28	**29**	29	70	**58**	56	**58**	54	228	241	105	71	68	1095
		SS-P		**21**	30	**22**	**62**	72	**52**	69	**52**	**45**	**45**	**94**	**53**	**42**	−40%
	300	SS		44	**43**	43	543	**538**	528	643	677	665	673	**225**	141	113	4876
		SS-P		**27**	44	**25**	**127**	676	**95**	**136**	**97**	**79**	**84**	254	**102**	**73**	−63%
	10	SS	Engine-X	4	4	5	5	5	5	5	5	5	5	6	6	5	65
		SS-P		**2**	**3**	**2**	**4**	**4**	**4**	**4**	**3**	**3**	**3**	6	**3**	**3**	−32%
	30	SS		5	5	5	7	7	7	8	**5**	6	**6**	13	8	8	90
		SS-P		**2**	5	**2**	9	8	7	9	6	6	7	15	**6**	**5**	−3%
	100	SS		12	13	13	**20**	19	19	29	17	15	**15**	**46**	26	21	265
		SS-P		**3**	13	**3**	24	22	21	**28**	17	15	16	48	**15**	**12**	−11%
	300	SS		33	33	33	**56**	**55**	52	82	52	42	**43**	**125**	69	60	735
		SS-P		**7**	33	**6**	69	61	56	**79**	**48**	42	45	132	**41**	**32**	−11%
Denormalized	10	DT	Hive	20	21	21	21	27	**20**	22	22	**21**	22	**22**	23	22	284
		DT-P		**18**	**19**	**17**	21	27	21	**21**	**21**	22	**21**	23	**20**	**20**	−5%
	30	DT		23	24	24	26	**36**	24	28	28	25	25	28	28	28	347
		DT-P		**20**	**21**	**20**	**25**	37	**24**	**25**	**26**	**23**	**24**	**27**	**21**	**22**	−9%
	100	DT		23	22	22	45	72	37	47	46	37	37	51	31	32	502
		DT-P		**21**	**20**	**20**	**39**	67	**32**	**40**	**40**	**31**	**32**	43	**26**	**27**	−13%
	300	DT		59	62	60	95	181	81	112	108	84	94	121	122	119	1298
		DT-P		**25**	**26**	**24**	**80**	159	**57**	**84**	**85**	**54**	80	102	**43**	**43**	−34%
	10	DT	Engine-X	2	3	3	3	2	2	3	3	2	3	3	3	3	35
		DT-P		**1**	**1**	**1**	2	2	2	**2**	**2**	2	2	**2**	1	2	−37%
	30	DT		5	5	5	4	4	4	4	4	4	5	5	5	5	59
		DT-P		**2**	**2**	**2**	4	4	4	4	4	**3**	**4**	5	**2**	**3**	−27%
	100	DT		5	3	3	17	15	15	12	14	12	9	17	8	8	138
		DT-P		**3**	3	**1**	**11**	9	9	**10**	9	**8**	**7**	13	4	5	−33%
	300	DT		40	43	44	37	35	34	36	39	33	42	46	45	42	516
		DT-P		**7**	**11**	**7**	**30**	**26**	**24**	**28**	**31**	**24**	**28**	**39**	**12**	**12**	−46%

used in this work, most of them were filtered by year, namely Q1.1, Q1,3, Q3.1, Q3.2, Q3.3, Q4.2 and Q4.3. Therefore, the attribute "year" was used to create the partitions and, as can be seen in Table 2, generally these were the queries that were more influenced by data partitioning. For almost all scenarios, these queries obtained better results after partitioning the data. In the cases where additional time was needed due to partitioning, it was not very significant, only surpassing the 10 s' mark in one of the cases (on Hive).

However, it is important to highlight that the creation of partitions, although it optimized certain queries, can be a disadvantage for other queries that do not consider the filters by the attribute used for partitioning. Q2.2 and Q4.1 are two of the most affected queries, both for Hive and Engine-X, since the increase can reach the 138 and 29 s, respectively. Q3.1, Q3.2 and Q3.3 are the queries that not always suffer a decrease by the application of partitions, despite having the filter. After analyzing the "where" conditions of these queries, it is important to highlight that although they are filtered by year, this filter only excludes the partition year of 1998, so to answer these queries, both Hive and Engine-X must search the other 6 partitions (6 folders), thus justifying that in some cases it may take more time or not be affected by the defined data organization. This outperformance in a context with no partitions, is also mainly verified in lower SFs, since when the volume of data is smaller the advantages of not having to search the entire dataset is pointless, because it does not contain a significant amount of data. In contrast, for higher SFs, not having to scan the entire dataset means not having to process a considerable amount of data.

Other queries showing a decrease in execution time, even without considering the filter in their "where" conditions, may be related to a rearrangement of the queries optimizer used in each tool, or to particularities of the queries that execute faster when data is organized by folders. This should be further evaluated in future work. The results here obtained also consolidate the results presented in the previous section, since it is clear the performance advantage of the denormalized table when it is also partitioned. Thus, the best scenario (the one with the lowest processing time), is the scenario considering a partitioned and denormalized table, using Engine-X for query execution.

5 Discussion and Conclusions

This work presented an evaluation and discussion of adequate data modelling and organization strategies for Hadoop (Hive) DWs in Big Data environments. The SSB benchmark was used to evaluate the performance of a star schema DW and a fully denormalized DW, with and without data partitioning. Four SFs were used, namely: SF = 10; SF = 30; SF = 100; and SF = 300. Two SQL-on-Hadoop systems (Hive on Tez and Engine-X) were used to query the DW and to observe if the insights provided in this paper are reproducible in more than one Big Data querying engine.

Regarding the comparison between DWs built using star schemas and DWs built using fully denormalized tables, this paper concludes that Hadoop-based DWs benefit from using a fully denormalized data modelling strategy, since the results achieved in the benchmark showed a significant performance advantage over the multidimensional DW design strategy throughout all scaling factors. Despite being feasible to implement

DWs on Hive using the star schema, this may not be the most efficient design pattern, despite saving a significant amount of storage, since the SSB fully denormalized dataset was 3 times bigger than the original dataset in a multidimensional format. Storage space, which is becoming increasingly cheaper, is the price to pay in BDWs that use fully denormalized tables, but the advantages can be sufficient to overcome this disadvantage, including the following, according to this benchmark: faster query execution (very often more than 50% faster); less memory requirements regarding specific join strategies (broadcast joins); and less intensive CPU usage.

Consequently, despite the innumerous efforts in several SQL-on-Hadoop systems to adequately support queries over multidimensional structures (e.g., adequate cost based optimizers, map joins), it can be stated that Hadoop-based DWs still favor fully denormalized structures that do not rely on join operations to answer OLAP queries. Such approach avoids the cost of performing join operations in Big Data environments. These results corroborate previous studies (please see Sect. 2 for more details) arguing that denormalized tables showed better performance than star schemas when using specific MPP Data Warehousing systems.

Regarding the different data organization strategies, this paper also concludes that there is a clear advantage in partitioning data, since the results obtained in the benchmark showed a significant decrease in query execution time when Hive tables are adequately partitioned. Thus, the results presented in this paper reinforce the potential benefit of creating data partitions easier to process, since, depending on the query being executed, these techniques can drastically reduce processing time. It also proves that, once the queries are known in advance, using the attribute that appears more often as a filter for partitioning is effectively one of the best partitioning strategies, consolidating previous studies.

For future work, one aims to continue studying the impact of other data partitioning strategies, as well as the impact of adequate bucketing strategies in Hive DWs, in order to complement the insights depicted in this paper. Moreover, one also aims to propose structured guidelines for the automatic creation of materialized views in Big Data Warehousing environments.

Acknowledgements. This work is supported by COMPETE: POCI-01-0145- FEDER-007043 and FCT – *Fundação para a Ciência e Tecnologia* within the Project Scope: UID/CEC/00319/ 2013, and funded by the SusCity project, MITP-TB/CS/0026/2013, and by the Portugal Incentive System for Research and Technological Development, Project in co-promotion nº. 002814/2015 (iFACTORY 2015–2018).

References

1. Dumbill, E.: Making sense of big data. Big Data **1**, 1–2 (2013). doi:10.1089/big.2012.1503
2. Chen, M., Mao, S., Liu, Y.: Big data: a survey. Mob. Netw. Appl. **19**, 171–209 (2014). doi:10.1007/s11036-013-0489-0
3. Ward, J.S., Barker, A.: Undefined by data: a survey of big data definitions. arXiv:1309.5821 [cs. DB] (2013)

4. NBD-PWG: NIST Big Data Interoperability Framework: Volume 6, Reference Architecture. National Institute of Standards and Technology (2015)
5. Philip Chen, C.L., Zhang, C.-Y.: Data-intensive applications, challenges, techniques and technologies: a survey on Big Data. Inf. Sci. **275**, 314–347 (2014). doi:10.1016/j.ins.2014.01.015
6. Krishnan, K.: Data Warehousing in the Age of Big Data. Morgan Kaufmann Publishers Inc., San Francisco (2013)
7. Shvachko, K., Kuang, H., Radia, S., Chansler, R.: The hadoop distributed file system. In: 2010 IEEE 26th Symposium on Mass Storage Systems and Technologies (MSST), pp. 1–10 (2010)
8. Thusoo, A., Shao, Z., Anthony, S., Borthakur, D., Jain, N., Sen Sarma, J., Murthy, R., Liu, H.: Data warehousing and analytics infrastructure at Facebook. In: Proceedings of the 2010 ACM SIGMOD International Conference on Management of Data, pp. 1013–1020. ACM, New York (2010)
9. Apache Hadoop: Welcome to Apache Hadoop. https://hadoop.apache.org/
10. Vavilapalli, V.K., Murthy, A.C., Douglas, C., Agarwal, S., Konar, M., Evans, R., Graves, T., Lowe, J., Shah, H., Seth, S., Saha, B., Curino, C., O'Malley, O., Radia, S., Reed, B., Baldeschwieler, E.: Apache hadoop YARN: yet another resource negotiator. In: Proceedings of the 4th Annual Symposium on Cloud Computing, pp. 5:1–5:16. ACM, New York (2013)
11. Kimball, R., Ross, M.: The Data Warehouse Toolkit: The Definitive Guide to Dimensional Modeling. Wiley, Hoboken (2013)
12. Russom, P.: Evolving data warehouse architectures in the age of big data. The Data Warehouse Institute (2014)
13. Russom, P.: Data warehouse modernization in the age of big data analytics. The Data Warehouse Institute (2016)
14. Thusoo, A., Sarma, J.S., Jain, N., Shao, Z., Chakka, P., Zhang, N., Antony, S., Liu, H., Murthy, R.: Hive-a petabyte scale data warehouse using hadoop. In: IEEE 26th International Conference on Data Engineering (ICDE), pp. 996–1005. IEEE (2010)
15. Huai, Y., Chauhan, A., Gates, A., Hagleitner, G., Hanson, E.N., O'Malley, O., Pandey, J., Yuan, Y., Lee, R., Zhang, X.: Major technical advancements in apache Hive. In: Proceedings of the 2014 ACM SIGMOD International Conference on Management of Data, pp. 1235–1246. ACM, New York (2014)
16. O'Neil, P.E., O'Neil, E.J., Chen, X.: The star schema benchmark (SSB) (2009)
17. Floratou, A., Minhas, U.F., Özcan, F.: SQL-on-hadoop: full circle back to shared-nothing database architectures. Proc. VLDB Endow. **7**, 1295–1306 (2014). doi:10.14778/2732977.2733002
18. Tria, F.D., Lefons, E., Tangorra, F.: Design process for big data warehouses. In: 2014 International Conference on Data Science and Advanced Analytics (DSAA), pp. 512–518 (2014)
19. Goss, R.G., Veeramuthu, K.: Heading towards big data building a better data warehouse for more data, more speed, and more users. In: 2013 24th Annual SEMI Advanced Semiconductor Manufacturing Conference (ASMC), pp. 220–225. IEEE (2013)
20. Mohanty, S., Jagadeesh, M., Srivatsa, H.: Big Data Imperatives: Enterprise: Big Data Warehouse, BI Implementations and Analytics. Apress, New York City (2013)
21. Santos, M.Y., Costa, C.: Data models in NoSQL databases for big data contexts. In: Tan, Y., Shi, Y. (eds.) DMBD 2016. LNCS, vol. 9714, pp. 1–11. Springer, Cham (2016). doi:10.1007/978-3-319-40973-3_48
22. Santos, M.Y., Costa, C.: Data warehousing in big data: from multidimensional to tabular data models. In: Ninth International C* Conference on Computer Science & Software Engineering (C3S2E), pp. 51–60. ICPS (ACM) (2016)

23. Jukic, N., Jukic, B., Sharma, A., Nestorov, S., Korallus Arnold, B.: Expediting analytical databases with columnar approach. Decis. Support Syst. **95**, 61–81 (2017). doi:10.1016/j. dss.2016.12.002
24. Apache Hive: Apache Hive Documentation - Apache Software Foundation. https://cwiki. apache.org/confluence/display/Hive/Home
25. Cattell, R.: Scalable SQL and NoSQL data stores. ACM SIGMOD Rec. **39**, 12–27 (2011). doi:10.1145/1978915.1978919
26. Chevalier, M., Malki, M.E., Kopliku, A., Teste, O., Tournier, R.: Document-oriented models for data warehouses - NoSQL document-oriented for data warehouses. Presented at the 18th International Conference on Enterprise Information Systems 2 March (2017)
27. Yangui, R., Nabli, A., Gargouri, F.: Automatic transformation of data warehouse schema to NoSQL data base. Procedia Comput Sci. **96**, 255–264 (2016). doi:10.1016/j.procs.2016.08. 138
28. Presto: Presto | Distributed SQL Query Engine for Big Data. https://prestodb.io/
29. Kornacker, M., Behm, A., Bittorf, V., Bobrovytsky, T., Choi, A., Erickson, J., Grund, M., Hecht, D., Jacobs, M., Joshi, I., Kuff, L., Kumar, D., Leblang, A., Li, N., Robinson, H., Rorke, D., Rus, S., Russell, J., Tsirogiannis, D., Wanderman-milne, S., Yoder, M.: Impala: a modern, open-source SQL engine for hadoop. In: Proceedings of the CIDR 2015, California, USA (2015)
30. Armbrust, M., Xin, R.S., Lian, C., Huai, Y., Liu, D., Bradley, J.K., Meng, X., Kaftan, T., Franklin, M.J., Ghodsi, A., et al.: Spark SQL: relational data processing in spark. In: Proceedings of the 2015 ACM SIGMOD International Conference on Management of Data, pp. 1383–1394. ACM (2015)
31. Hausenblas, M., Nadeau, J.: Apache Drill: interactive ad-hoc analysis at scale. Big Data **1**, 100–104 (2013)
32. Chen, Y., Qin, X., Bian, H., Chen, J., Dong, Z., Du, X., Gao, Y., Liu, D., Lu, J., Zhang, H.: A study of SQL-on-hadoop systems. In: Zhan, J., Han, R., Weng, C. (eds.) BPOE 2014. LNCS, vol. 8807, pp. 154–166. Springer, Cham (2014). doi:10.1007/978-3-319-13021-7_12
33. Kornacker, M., Behm, A., Bittorf, V., Bobrovytsky, T., Choi, A., Erickson, J., Grund, M., Hecht, D., Jacobs, M., Joshi, I., Kuff, L., Kumar, D., Leblang, A., Li, N., Robinson, H., Rorke, D., Rus, S., Russell, J., Tsirogiannis, D., Wanderman-milne, S., Yoder, M.: Impala: a modern, open-source SQL engine for hadoop. In: Proceedings of the CIDR 2015, California, USA (2015)
34. Santos, M.Y., Costa, C., Galvão, J., Andrade, C., Martinho, B., Lima, F.V., Costa, E.: Evaluating SQL-on-hadoop for big data warehousing on not-so-good hardware. In: Proceedings of International Database Engineering & Applications Symposium (IDEAS 2017), Bristol, United Kingdom (2017)
35. Santos, M.Y., Martinho, B., Costa, C.: Modelling and implementing big data warehouses for decision support. J. Manag. Anal. **4**, 111–129 (2017)
36. Capriolo, E., Wampler, D., Rutherglen, J.: Programming Hive. O'Reilly Media Inc., Sebastopol (2012)
37. Inmon, W.H.: Building the Data Warehouse. Wiley, Hoboken (2005)

Managing Modular Ontology Evolution Under Big Data Integration

Hanen Abbes$^{(\boxtimes)}$ and Faiez Gargouri

MIRACL Laboratory, Higher Institute of Computer Science and Multimedia,
Sfax University, Sfax, Tunisia
abbes.hanen@gmail.com, faiez.gargouri@isims.usf.tn

Abstract. Big Data integration frameworks provide unified view of the data available from heterogeneous data sources. These data sources are continuously evolving, forcing systems that integrate them to adapt their global schema after each change. This gets more challenging when aiming to maintain the global schema always reflecting data sources content. To cope with such complexity, in this paper we describe evolution scenarios and manage modular ontology evolution within Big Data integration framework in an a priori way according to changes performed against the data sources.

Keywords: Ontology evolution · Big Data integration · Data source evolution · Modular ontology

1 Introduction

According to [1], "Big Data can be defined as data that exceed the processing capacity of conventional database systems. This implies that the data count is too large, and/or data values change too fast, and/or it does not follow the rules of conventional database management systems". Big Data are characterized along three important dimensions, namely volume, variety and velocity [2, 3] known as 3Vs.

Despite this complexity, users usually look for a unified view of the data available from heterogeneous data sources. Consequently, several Big Data integration systems were proposed. Nevertheless, they do not cope with the evolutionary aspect of Big Data sources. Indeed, maintaining an integrated view over such evolving and heterogeneous set of data sources is a challenging problem which current systems fail to address.

Regardless of the great amount of work done in ontology-based Big Data integration, an important problem that most of the systems are likely to ignore is that ontologies are living artifacts and are subject to change and evolution as well. Ontologies are frequently changed to reflect the new knowledge that is acquired. The problem that occurs is the following: when data sources change, the mappings may become invalid and should be updated. Ontology evolution is defined as the "timely adaptation of an ontology to the arisen changes and the consistent management of these changes" [4].

In this paper, we address the problem of ontology evolution under Big Data integration. We argue that data sources changes should be considered when designing ontology-based Big Data integration systems. A distinctive solution would be to update

© Springer International Publishing AG 2017
M. Themistocleous and V. Morabito (Eds.): EMCIS 2017, LNBIP 299, pp. 17–28, 2017.
DOI: 10.1007/978-3-319-65930-5_2

the mappings and then regenerate the dependent ontologies each time a data source evolves. We propose an a priori method to correct incoherencies caused by a change and rely on ontology learning and ontology merging tools to manage the evolution process.

This paper is organized as follows. Section 2 exposes our research context. In the third section, we describe scenarios that drive to evolve the ontology in a Big Data integration context. Our approach to manage modular ontology evolution in Big Data integration is detailed in Sect. 4. We start by describing the evolution process, then we give examples of coherence constraints that we respect to evolve the ontology. Inspired by previous research work, we adapt existing notions to our needs and provide an illustrative example. Section 5 examines related work. A discussion comparing our research to previous ones encloses this section. Finally Sect. 6 draws conclusions and suggests further research.

2 Research Context

This work joins within the scope of a Big Data integration approach [5] such that the departure corpus is formed by Big Data, whereas the target schema is an OWL[1] (Ontology Web Language) ontology. We are interested particularly to OWL-DL since it supports the maximum expressiveness while retaining computational completeness and decidability.

The original approach aims to build an ontology for Big Data integration where Big Data are seen as data from many sources having different formats, each source contains a very big amount of data and grows and evolves independently from the other ones [5]. This approach is based on three main steps (Fig. 1):

- *Wrapping data sources to MongoDB[2] databases*: the content of each data source is converted to a MongoDB database,
- *Mapping MongoDB databases to ontology modules*: each MongoDB database is mapped to an OWL ontology module by means of transformation rules [6, 7]. The first phase is the creation of the ontology skeleton. It consists of defining ontology classes and detecting subsumption relationships between them. The second phase is to learn concepts properties (dataTypeProperties and objectProperties). Individuals are identified in the third phase. In the fourth phase, class axioms (equivalence and disjoining), property axioms (inverseOf) and constraints (cardinality constraints, value constraints) are deduced. Finally, in the fifth phase, the ontology is enriched with classes' definition operators (union, intersection, complement).
- *Merging ontology modules to get a global one*: the modules obtained in the previous step are merged together in order to get a global ontology [8]. Our algorithm is based on three main actions. The first action is to detect overlaps between the two modules to be merged. The second action is to compute similarities between concepts belonging to the two ontology modules and the third action is to update the reference ontology module with concepts, attributes, as well as relationships from

---

[1] https://www.w3.org/TR/owl-features/.

[2] http://www.mongodb.org/.

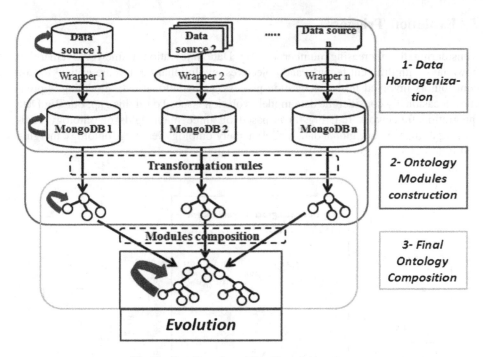

Fig. 1. Ontology-based Big Data integration.

the input ontology module. To measure the similarity between two concepts, one from the input ontology module and the other from the reference ontology module, we combine syntactic matching and semantic matching. The syntactic matching compares strings characterizing concept names as well as concept attributes whereas the semantic matching is based on relationships similarity. To compute the syntactic matching, we apply a distance function over a pair of strings. We adopted the Levenshtein distance [9] which returns the number of character changes needed to transform one string into another (LEV). The smaller this dissimilarity is (i.e. the less character changes needed), the more similar are the strings.

The methodology followed to build the global ontology follows a modular conceptualization since the beginning of the ontology development cycle where an ontology module represents a point of view covered by a data source containing data about the modeled domain.

Considering that Big Data are dynamic by nature, they are exposed to different updates. These updates must be sent up to the global ontology and evolution issues have to be dealt with. Indeed, new sources of data may appear and data about a domain may change according to different manners. On the one hand, new data may appear, and this leads to establish new concepts. On the other hand, some data may become obsolete, and so, some concepts must be removed from the global ontology. Besides, modeling a domain may necessitate concepts redefinition. New concepts that are more specific than the pre-existent ones can be defined if the domain needs to be more precise, or more abstract if we want to simplify the domain and facilitate its comprehension.

3   Evolution Triggers

Considering the increasing number of Big Data integration frameworks emerging nowadays with different features and objectives, evolutionary issues are critical for the contributors involved in the integration process. Conventional data modeling approaches occasionally consider the data model evolution issue. To fill this gap, our Big Data integration framework (c.f. Fig. 1) is based on three main fields of ontology engineering to manage our global data model, namely ontology learning, ontology merging and ontology evolution (Fig. 2).

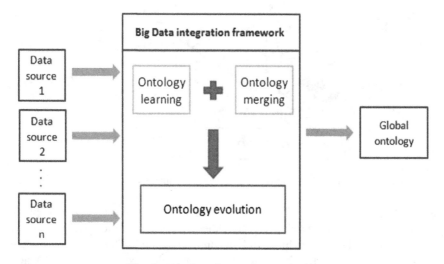

Fig. 2. Big Data integration evolution.

Accordingly, the derived data model represents a shared data model for the various data sources it integrates and should be always up to date in compliance with the changes met by the data sources. However, to deal with the evolution of data models over time, three scenarios may trigger the evolution process as depicted in Fig. 3.

We consider that ontology modules corresponding to each data source are stored separately in addition to the global ontology for subsequent use.

– *Scenario 1* (Introducing a new data source): A new data source may be added to the framework. It is required to acquire its corresponding ontology module by means of ontology learning according to the process defined in Fig. 1 (wrapping the data source to a MongoDB database then mapping the MongoDB database to an ontology module) [6, 7]. The learned knowledge of the new data source is represented in a separate ontology module, i.e. Module (New), and then the similarities between Module (New) and the global ontology are calculated [8]. At the end, Module (New) is integrated into the global ontology according to the ontology merging results.

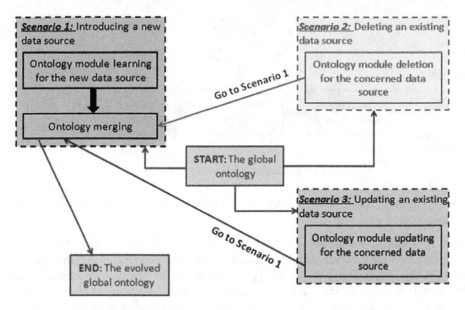

Fig. 3. Evolutionary scenarios for Big Data integration.

- *Scenario 2* (Deleting an existing data source): A current data source may leave the framework in case the data it contains become obsolete or due to reasons coming from the outside of the integration framework. Consequently, the corresponding ontology module (Module (i)) is deleted from the modules database and the other ontology modules have to be re-merged to constitute the new version of the global ontology.
- *Scenario 3* (Updating an existing data source): When an existing data source undergoes a modification of one of its entities (addition, deletion, renaming or modification), the corresponding ontology module must be updated accordingly and the global ontology as a consequence. We consider two levels of evolution: inter-modular evolution and intra-modular evolution. The inter-modular level concerns the global ontology update and is managed by the merging process. After evolving the corresponding ontology module, the ontology merging process has to be re-iterated to produce the new global ontology. The intra-modular evolution concerns the update of the concerned module itself. Detailed characterization of the intra-modular evolution is described in the next section.

4   Intra-modular Evolution Approach

4.1   Evolution Process

To carry out the ontology module evolution task, we propose an a priori process to address changes composed of three main steps as depicted in Fig. 4. For each change that occurs in the database, the corresponding mapping must be updated, since the

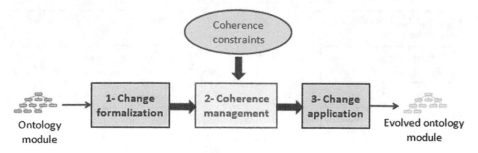

Fig. 4. Ontology evolution process.

ontology module has to always reflect the structure of the database and subsequently the data source.

Step 1: Change Formalization. This step consists of representing the change expressed over the database entities according to their correspondents into the ontology module. We define four types of changes similarly to the changes undergone in the database level, namely insertion, deletion, renaming and updating. Thus, the list of changes is given in Table 1. The list of changes is performed in compliance with the transformation rules presented in [6, 7].

Table 1. Changes performed against the ontology module analogously to those performed against the database.

	Database level	Ontology module level
Insertion	-Collection	-Insert concept
	-Document into collection	-Insert individual
	-Basic field into document	-Insert dataTypeProperty
	-DBList into document	-Insert cardinality constraint
	-Embedded document into document	-Insert objectProperty
	-Reference with DBRef into document	-Insert objectProperty
	-Parent reference into document	-Insert hierarchy relationship
Deletion	-Collection	-Delete class
	-Document from collection	-Delete individual
	-Basic field from document	-Delete dataTypeProperty
	-DBList from document	-Delete cardinality constraint **OR** -Update cardinality constraint
	-Embedded document from document	-Delete objectProperty
	-Reference with DBRef from document	-Delete objectProperty
	-Parent reference from document	- Update hierarchy
Renaming	-Collection	-Rename class
	-Basic field into document	-Rename dataTypeProperty
	-Embedded document into document	-Rename objectProperty
	-Reference with DBRef into document	-Rename objectProperty
Updating	-Document	-Update individual
	-DBList	-Update cardinality constraint
	-Parent reference into document	-Update hierarchy

Step 2: Coherence Management. This step consists of studying the impact of the selected change on the ontology module coherence. To do this, we reused the notion of "change kit" of [10, 11] to maintain a priori coherence constraints defined over the ontology module. Coherence constraints are discussed in Sect. 4.2 and the specification of change kit is given in Sect. 4.3.

Step 3: Change Application. In this step, changes are implemented over the ontology module. Hence, we obtain the evolved ontology module which already respects coherence constraints.

4.2 Coherence Constraints

We classify the coherence constraints that must be respected into three main categories as in [4]. Structural constraints represent constraints imposed by the ontology representation language which is OWL in our context. Logical constraints refer to checking the semantic correctness of the ontological entities. User-defined constraints describe specific user requirements. We give examples of these constraints.

OWL Language Constraints

- Isolated classes are not allowed
- A class which is a sub-class of another class must be defined in the ontology
- Each objectProperty relationship must link two classes that are defined in the ontology
- Each individual must be linked to a class that is defined in the ontology

Logical Constraints

- A class can not be disjoint with its super-class
- Two disjoint classes can not have common sub-classes

User's Requirements

- Redundant information is not allowed
- Redundant links between two directly linked classes are not allowed
- The resulting ontology must respect the conventional requirements of the transformation rules described in [6].

4.3 Change Kit

A kit of change associates to each change the definition of its pre-conditions, its role, the mandatory additional changes, the optional additional changes and post-conditions. The specification of a kit of change is as follows:

- Pre-conditions: a set of predications that must be checked and controlled before applying the change
- Role: Description of the change

- Required additional changes: a set of changes that are necessarily attached to the current change to avoid coherence constraints violation
- Optional additional changes: a set of changes that may extend the current change
- Post-conditions: a set of predications that must be verified after the application of the change

4.4   Illustrative Example

As an example, we suppose that a modification in data source leads to insert an imbedded document in the corresponding MongoDB database. We describe the effect of inserting an embedded document (document B) into an existing document (document A) in the MongoDB database. According to the correspondences described in Table 1, this insertion leads to update the corresponding ontology module by inserting an objectProperty. The kit of change associated to the change "insert objectProperty" is as follows (Table 2).

The global evolution process monitors the following steps. The corresponding ontology module is updated according to the intra-modular ontology evolution process described in Fig. 4 and the change kit specified in Table 2 while respecting coherence constraints. Then the global ontology is updated according to the merging process performed against the updated module and the other previously existing modules that are related to the other data sources. The evolved global ontology is consequently structurally and logically consistent according to the researches developed in [6–8].

Table 2. Example of a change kit

Kit of change "insert objectProperty"	
Role	The kit of change "insert objectProperty" serves to link two classes A and B with an objectProperty relationship
Pre-conditions	-The class corresponding to the document A exists already in the ontology module -The class corresponding to document B doesn't exist in the ontology module -The objectProperty relationship doesn't exist in the ontology module
Required additional changes	-Insert class corresponding to document B
Optional additional changes	–
Post-conditions	-The class corresponding to the document A belongs to the ontology module -The class corresponding to the document B belongs to the ontology module -The name of the objectProperty is the concatenation of the word "has" and the name of the class corresponding to document B -The domain of the inserted objectProperty is the class corresponding to document A -The range of the inserted objectProperty is the class corresponding to document B

5   Related Work

To situate our research in the area of ontology evolution, we focus on a priori evolution approaches and concentrate on recent solutions addressing ontology evolution in Big Data integration frameworks.

In [10], a preventive approach that manages the inconsistencies generated by each change is described and a set of rules that must be maintained during the evolution of an ontology is defined. Authors define kits of changes to a priori manage the inconsistencies generated by each change. They rely on the UML specifications to take into account the maximum of evolution cases, independently of the ontology representation language and consider all conceptual relationships supported by the UML language, such as n-ary relationships, but do not define explicitly the type of coherence which is considered.

Authors in [11] present an evolution process composed of three main steps. The first step consists of presenting all the possible changes for the naRyQ ontological and terminological resource (OTR) evolution to the ontology engineer, from which he chooses the ones to be applied. The second step consists of preserving the coherence constraints (CC-coherence) of naRyQ during its evolution. To do this, authors adapted the notion of kit of changes of [10] to their needs, thus an additional set of changes is added automatically to maintain a priori the CC-coherence of the OTR before the application of the requested changes. In the third and last step, requested and additional changes are applied to the OTR.

Authors in [12] focused on addressing the need to reflect the evolution of ontologies used as global schemata onto the underlying data integration systems. They consider that when ontologies evolve, the changes should necessarily be rendered and used by the pre-existing data integration systems. They propose to answer query in data integration systems under evolving ontologies without mappings redefinition. This is ensured by rewriting queries among ontology versions and then sending them to the underlying data integration systems to be answered. Initially, the changes among ontology versions are automatically detected and described using a high level language of changes. These changes are then interpreted as sound global-as-view (GAV) mappings, used to produce equivalent rewritings among ontology versions.

In [13], authors present an approach that enables to integrate situational data coming from external providers, and to facilitate the co-evolution of data and analytical processes preserving backward compatibility. They introduce the Big Data Integration ontology that allows the isolation of analytical queries and applications from the technological details of the sources and accommodates syntactic evolution from the sources. Its goal is to model and integrate, in a machine-readable format, semi-structured data while preserving data independence regardless of the source formats or schema. The introduced ontology incorporates two layers to provide to the analysts an integrated and format-agnostic view of the sources. The global level provides a unified schema to query and relevant metadata about the attributes, while the source level deals with the physical details of each data source. This structure is exploited to handle the evolution of source schema via semi-automated transformations on the ontology upon service releases. Aided by semi-automatic techniques, a data steward is responsible for, first incorporating to the source level the triple-based representation of the schema of newly incoming events (*Ei*) produced by APIs, and second make such data available for data analysts to query (*Qi*) by creating mappings from the source level to the global level. To semi-automatically adapt the BDI ontology to such evolution, authors present an algorithm to aid the data steward to enrich the ontology upon new releases to shield analytical processes, implemented on top of the global level, so that they do not crash upon new API version releases. This aims to adapt the source level to schema evolution in the events, so that the global level is not affected.

The following table summarizes advantages and drawbacks of these works (Table 3).

Table 3. Comparison between ontology evolution approaches.

Approaches	Advantages	Drawbacks
[10]	Anticipatory approach that takes into account the maximum of evolution cases	Does not define the type of coherence which is considered
[11]	Anticipatory approach that is based on a clear definition of ontology coherence	Does not propagate changes to the related artifacts
[12]	The proposed architecture can be placed on top of any traditional ontology-based data integration system, enabling ontology evolution	Does not consider local schema evolution, thus the ontology used as a global schema may contain inconsistencies
[13]	The proposed method handles schema evolution using a metadata-driven approach in the context of Big Data integration	Focuses only on enrichment and does not cover other change types (deletion, renaming, updating)

From these perspectives, we notice the following remarks against the studied approaches.

Authors of [10, 11], although they develop preventive ontology evolution approaches, the latters do not fit evolution in data integration frameworks. On the other hand, evolution approaches developed in [12, 13] in the context of Big Data integration are not preventive and suffers from some limitations.

Authors of [13] aim to aid the data steward to only enrich the ontology upon new releases and do not consider other aspects of evolution such as deletion and modification. Moreover, they address the evolution locally in the source level and do not propagate changes to the global level. Conversely, we consider all evolution types namely insertion, deletion, update and renaming and we propagate changes to the global ontology by means of merging re-iteration.

Authors of [12] are interested to ontology evolution in data integration like us. But, while they focus on propagating changes from the ontological level to the data sources level, we make the opposite and try to communicate changes over the data sources to the global ontology.

The notion of change kit was initially proposed by [10] but relying on UML specification, authors of [11] adapted it to the specificities of the OWL language, and in our work, we adapted it to the context of ontology-based Big Data integration founded on data sources evolution.

Our approach has three main advantages. It firstly covers the entire ontology evolution cycle and manages incoherencies that are likely to occur in a priori manner, secondly relies on the use of background releases i.e. previously developed components [6–8] to potentially decrease, or even eliminate, user involvement, and finally fits all evolution scenarios.

6   Conclusions and Future Work

Ontologies need to be updated across their life cycle to reflect new requirements and must remain coherent. We are interested in this work to the ontology evolution in the context of Big Data integration. The majority of existing works about ontology-based Big Data integration ignores evolution issues.

When an ontology is used as a component of an advanced information system, its evolution is a complex process and raises several challenges such as the formal representation of ontology changes, the verification of ontology consistency when applying the ontology changes, and the propagation of these changes to the ontology related artifact. We discussed related work relative to a priori ontology evolution and Big Data integration evolution. We presented evolution scenarios in the context of Big Data integration and proposed a solution to deal with modular ontology evolution while considering changes performed against the data sources.

There are many interesting future directions. A prominent one is to explore how to manage change history and to record changes performed against the ontology. As a short-term goal, we plan to integrate such functionality to our approach. Other avenue of research would be to propose an approach to enhance the change propagation step to the global ontology.

References

1. Gupta, R., Gupta, H., Mohania, M.: Cloud computing and big data analytics: what is new from databases perspective? In: Srinivasa, S., Bhatnagar, V. (eds.) BDA 2012. LNCS, vol. 7678, pp. 42–61. Springer, Heidelberg (2012). doi:10.1007/978-3-642-35542-4_5
2. Zikopoulos, P., Eaton, C.: Understanding Big Data: Analytics for Enterprise Class Hadoop and Streaming Data. McGraw–Hill/Osborne Media, New York City (2011)
3. Boden, C., Karnstedt, M., Fernandez, M., Markl, V.: Large-scale social-media analytics on stratosphere. In: Proceedings of the 22nd International Conference on World Wide Web Companion, pp. 257–260 (2013)
4. Haase, P., Stojanovic, L.: Consistent evolution of OWL ontologies. In: Gómez-Pérez, A., Euzenat, J. (eds.) ESWC 2005. LNCS, vol. 3532, pp. 182–197. Springer, Heidelberg (2005). doi:10.1007/11431053_13

5. Abbes, H., Gargouri, F.: Big data integration: a MongoDB database and modular ontologies based approach. Procedia Comput. Sci. **96**, 446–455 (2016)
6. Abbes, H., Boukettaya, S., Gargouri, F.: Learning ontology from Big Data through MongoDB database. In: Proceedings of IEEE/ACS 12th International Conference of Computer Systems and Applications, pp. 1–7 (2015)
7. Abbes, H., Gargouri, F.: M2Onto: an approach and a tool to learn OWL ontology from MongoDB database. In: Madureira, A.M., Abraham, A., Gamboa, D., Novais, P. (eds.) ISDA 2016. AISC, vol. 557, pp. 612–621. Springer, Cham (2017). doi:10.1007/978-3-319-53480-0_60
8. Abbes, H., Gargouri, F.: Structure based modular ontologies composition. In: 2016 IEEE/ACS 13th International Conference of Computer Systems and Applications (AICCSA), Agadir, Morocco (2016)
9. Levenshtein, V.I.: Binary codes capable of correcting deletions, insertions, and reversals. Sov. Phys. Dokl. **10**(8), 707–710 (1966)
10. Jaziri, W., Sassi, N., Gargouri, F.: Approach and tool to evolve ontology and maintain its coherence. Int. J. Metadata Semant. Ontol. **5**(2), 151–166 (2010)
11. Touhami, R., Buche, P., Dibie, J., Ibanescu, L.: Ontology evolution for experimental data in food. In: Garoufallou, E., Hartley, R.J., Gaitanou, P. (eds.) MTSR 2015. CCIS, vol. 544, pp. 393–404. Springer, Cham (2015). doi:10.1007/978-3-319-24129-6_34
12. Kondylakis, H., Plexousakis, D.: Ontology evolution without tears. Web Semant.: Sci. Serv. Agents World Wide Web **19**, 42–58 (2013)
13. Nadal, S., Romero, O., Abelló, A., Vassiliadis, P., Vansummeren, S.: An integration-oriented ontology to govern evolution in big data ecosystems. In: Proceedings of the EDBT/ICDT 2017 Joint Conference. Published in the Workshop OLAP (2017)

A Context Aware Notification Architecture
Based on Distributed Focused Crawling
in the Big Data Era

Mehmet Ali Akyol$^{(\boxtimes)}$, Mert Onuralp Gökalp, Kerem Kayabay,
P. Erhan Eren, and Altan Koçyiğit

Informatics Institute, Middle East Technical University, Ankara, Turkey
{aliakyol,gmert,kayabay,ereren,kocyigit}@metu.edu.tr

Abstract. The amount of data created in various sources over the Web is tremendously increasing. Trying to keep track of relevant sources is an increasingly time-consuming task. The traditional way of accessing information over the Web is pull-based. Users need to query data sources in certain time intervals where an important piece of information can be lately recognized or even missed completely. Technologies including RSS help users to get push-based notifications from websites. Discovering the relevant information without a notification overload is still not possible with existing technologies. Despite some promising efforts in push-based architectures to solve this problem, they fall short to meet the requirements in the big data era. In this study, by leveraging the latest advancements in distributed computing and big data analytics technologies, we use a focused crawling approach to propose a context aware notification architecture for people to find desired information at its most valuable state.

Keywords: Big data · Stream processing · Distributed focused crawling · Context aware notifications · Distributed complex event processing

1 Introduction

The rapid growth of the data sources such as the Web and social media causes a continuous accumulation of information. Scaling up the performance of crawling engines is becoming more of a challenge in the big data era. Search engines like Google, Bing, and DuckDuckGo focus on crawling the entire Web and provide results to individuals when they request information. In spite of the search engines' significant efforts with highly powerful hardware and software, crawling the entire Web is almost impossible in today's world. Moreover, search engines may still fail to provide the latest information on websites because of their web page ranking mechanisms. Web page rankings on search engines depend on lots of issues, and freshness is only one of many factors in page ranking algorithms.

Most of the websites that search engines spend time, money, and effort to crawl are unlikely to constitute value for individuals or businesses. Therefore, focused crawling is an appropriate approach in order to find the web pages that are likely to be relevant to a predefined topic [1]. In this approach, specific categories, TLD (top-level-domains) or

© Springer International Publishing AG 2017
M. Themistocleous and V. Morabito (Eds.): EMCIS 2017, LNBIP 299, pp. 29–39, 2017.
DOI: 10.1007/978-3-319-65930-5_3

topics are crawled. While generic web crawlers may fail to perceive the differences between relevant topics and related topics in returned results, focused crawling enable us to search for a specific topic continuously and discover any new information on the predefined topic [2]. However, accessing the information is still pull-based, so users keep querying the topic for new information. A push-based notification method is necessary for users to access the latest information on web pages without intervention. Moreover, if the information loses its value over time, there are costs associated with missing or lately realizing new information for people and businesses. The push-based notification method eliminates the process of continuously querying the same topic so that users get notifications as soon as the new information becomes available.

In addition to reaching information on a push-based manner, it is also important to provide new information to users in the proper context, and this is associated with the situation of a person, place, or object [3]. Contexts are categorized into two main categories, which are external and internal. Sensors give us data on location, temperature, light, sound, and air pressure and these measure external contexts. On the other hand, user preferences and interactions specify internal contexts. Users' tasks, goals, emotional state, or workflows can be examples of internal contexts [4]. The users' preferences or behavior may change dynamically, for example; users may like to have new information about their hobbies at home or in weekends, but they want to receive work related information on weekdays or while they are at work. At this point, the utilization of context is particularly important. Context aware systems are able to adapt their features according to the context without explicit user action. Context awareness can increase usability and effectiveness of systems by considering environmental context. However, analyzing dynamically changing contexts together with focused crawling requires processing fast and voluminous data flowing from a variety of context sources, which are characteristics of big data. Therefore, the architecture for context aware systems and focused crawling engines should be highly scalable to support big data processing.

In this study, a distributed architectural framework is proposed to crawl web pages on a specified topic, and to send notifications to people according to their context preferences via different notification channels; email, SMS, mobile push notifications and chat-bot messages. This enables individuals and businesses to reach the latest valuable information in predefined topics and in proper contexts. The proposed architecture also supports the state-of-the-art big data processing, distributed messaging queues, and cloud computing technologies in order to handle the volume, velocity, and variety characteristics of the data from crawled web pages.

The rest of the paper is organized as follows: in Sect. 2, we review the background and literature of focused crawlers and context aware systems. We introduce the system architecture of our proposed framework in Sect. 3. In the last part, we conclude our proposed research with discussion and future works.

2   Related Works

Focused Crawling and Context Aware Notification systems are not new paradigms in the literature, but they need to be transformed to provide a solution for growing data volume and velocity in the big data era. This section reviews the recent focused crawling and context aware notification systems in the literature.

In the literature, there are numerous studies related to focused crawling. In order to evaluate their extensibility to provide a solution in the big data era, these studies are analyzed in two main categories, depending on whether their architecture support distributed environment settings or not. Most of the studies [1, 5, 6] related to focused crawling propose a centralized single node solution in which their performance is limited due to processing capabilities of a single server. Thus, these solutions do not provide a viable solution in the big data era.

Several distributed solutions Mercator [7], Ubicrawler [8], BUbiNG [9–11] for focused crawling are proposed in the literature. Although they serve different purposes and use cases, they commonly present a scalable focused crawling system architecture to distribute processing components across several nodes. For example, TwitterEcho [12], proposes a distributed focused crawling architecture for Twitter, they aim to collect data from the Twitter and make it available for researchers. Another study [13], proposes a focused crawler for social networks which tracks the specified Twitter topic, using MapReduce programming model to benefit from its inherent support for distributed computations. Mercator is also one of the most notable web crawling architectures in the literature, also used by AltaVista search engine. It is a scalable and extensible web crawler whose components can be distributed across several computing units. However, there is a centralized module in the Mercator's architecture to manage all the crawling process which constitutes a performance issue that affects the overall system.

There are push-based solutions that people can use in order to get notifications when certain events occur in the web content. Users can use RSS (Rich Site Summary) [14] to subscribe to changes in the content of websites that support this technology. Google Alerts [15] can notify users of new articles upon subscription. The main limitation of these technologies is that they require users to subscribe to changes in the websites that support some particular functionalities. Alert Notification System [16] tackles this problem by adding an automated intelligent agent in the communication between the server and the user. This agent detects changes like data about an object becomes part of a web page, a new property is associated with a given object in a web page, and a new value is set to a property of an object in a web page.

Most of the existing solutions do not consider context information when sending push notifications. If users are not given the choice to filter which information to receive in certain contexts, they prefer not to receive any notifications except for the very important ones. This may lead to missing a very important piece of information which is only valuable for a limited amount of time or while the user is in a certain location or situation. In context aware systems, reminders and notifications have a crucial role, as notifying people at the right time in the right channel can increase productivity and efficiency. There are many examples of context aware notification systems in the literature. In the study by Katsiri [17], registered users who are interested in a particular activity are notified. In another study [18], machine learning algorithms are used to manage the incoming notifications. Even though these examples showcase a number of context aware notification use cases, in our framework, we are dealing with a completely different use case, delivering the right information in different channels like email, SMS, mobile push notifications and chatbot messages depending on users' preferences like time and location.

In the big data era, we see distributed focused crawling and context awareness as crucial components of information retrieval. Even though there are distributed architectures in literature, they lack the needs of big data era in which there are a lot of social networks, blogs, and websites generating an immense amount of data. Different from the works in the literature that tackle this problem, a new approach can be utilized, with the help of emerging stream processing engines providing machine learning and Complex Event Processing (CEP) capabilities. Context awareness of current solutions cannot meet user expectations. Notifications systems should consider more contextual information to be used as part of big data solutions. Users should be able to define complex filters when setting up notification preferences. In the next section, we propose a new contextual framework to introduce a more comprehensive notification system in the big data context to enable people to get relevant information in the proper context.

3   Conceptual Framework

In this section, we explain how the list of websites are determined, how websites are crawled and turned into streams of raw texts, how both website and contextual data are distributed and processed inside our framework, and how context aware notifications are delivered to users. The architecture of the proposed conceptual framework is delineated in Fig. 1. It consists of the following modules:

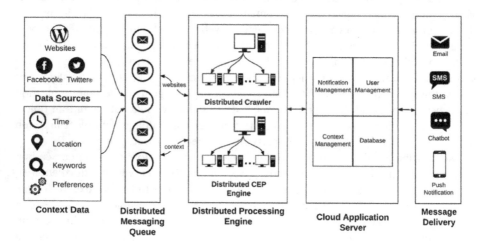

Fig. 1. Architecture of conceptual framework

- Data Sources collect data from the Web and social networks.
- Context Data is collected from users and it represents dynamically changing context.
- Cloud Application Server manages users and notifications. It also processes context data for push notifications.

- Distributed Messaging Queue builds real-time data pipeline for data sources and context data.
- Inside the Distributed Processing Engine, Distributed Crawler employs focused crawling, and Distributed CEP Engine detects patterns according to the context data.
- Message Delivery module notifies users about new information seamlessly via different channels.

3.1 Data Sources

Data Sources module collects websites from users' social network accounts and bookmarks. These websites depend on user interests that are related to their work or hobbies. Due to their heterogeneous nature, these data sources transmit data in disparate formats. Heterogeneity of data sources is an important challenge that needs to be tackled in the framework.

The links shared by users on social networks indicate users' interests and hobbies. Therefore, it is important to feed the framework with gathered links shared on social networks. Social networks allow us to collect these links through their Application Programming Interfaces (APIs). While extracting links from social networks, it is necessary to preprocess the links before feeding them into the framework. Most of the links shared on these networks are shortened by services like bit.ly [19] or goo.gle [20]. In order to use the real links in crawling, the shortened links should be resolved as full website links and this is done recursively following the HTTP redirects until there is no redirect left.

Apart from social networks, users can directly provide websites using a browser bookmark lets application or a mobile application. It is also necessary to allow users to edit or remove the links before they are fed into the framework.

3.2 Context Data

In this framework, notifications are delivered to users only if a certain predefined context is obtained. The context data collected from users is very crucial to send notifications in the most appropriate time and/or location. A mobile application is used to gather context data from users. Context data can be either rules or continuous streams. User defined preferences such as digest or real-time notifications and user defined keywords are some examples of rules. Data obtained from smartphone features such as location and time is fed into the framework as continuous data streams. External context data can also be obtained from other IoT devices to gather heart rate, weather, air pressure, and so on.

3.3 Cloud Application Server

The Cloud Application Server module consists of four submodules; User Management, Context Management, Notification Management, and Database.

User Management submodule has an important role in terms of user's interaction with the system. Users can define seed URLs within the web application. This module also handles user membership.

Context Management submodule provides users the ability to manage their context preferences. Users can define rules regarding the context preferences depending on how and when they want to receive notifications. The user defined rules are forwarded to Distributed CEP Engine. Therefore, it is crucial for CEP module to detect the relevant events depending on user interests.

Notification Management submodule manages the notifications to be sent to users depending on their preferences like time, location, and channel. Notifications can be delivered to users in real-time as well as in daily or weekly digests via channels such as email, SMS, mobile push notifications and chat-bot messages.

Database submodule holds user settings, website lists, notification preferences, and context rules. Each of these data sources has its own characteristics, complexities and ranges from unstructured to highly structured data in various formats. As a result of the wide variety of data sources, a NoSQL [21] database management system can be a solution to store these data. There are many NoSQL solutions such as HBase [22] and Cassandra [23] that use HDFS [24] which is a distributed file system designed to run on top of Hadoop ecosystem.

The application server should be fault tolerant and scalable. The main challenge in a cloud application is the orchestration of the distributed nodes in a cluster and the management resources such as CPU, RAM and disk space. There are frameworks for managing clusters and orchestrating distributed infrastructure. One of the most notable frameworks is DC/OS (Data Center Operating System). DC/OS abstracts set of clusters into a single computer, pooling distributed workloads, simplifying rollout and operations [25]. It is built on Apache Mesos [26], which abstracts CPU, memory, storage and other computational resources away from machines, enabling fault-tolerant and distributed systems to easily be built and run. A micro-services based solution using Docker [27] can be easily deployed using DC/OS for application server modules to run on the cloud.

3.4   Distributed Messaging Queue

Collecting data from the external environment is a vital function in this architecture, and it affects the overall performance of the system. Therefore, Distributed Messaging Queue module, which moves data to Distributed Processing Engine, enables scaling up the performance when necessary. Utilizing a distributed message queue also provides a resilient and fault tolerant data distribution method across the framework. This module is introduced as a layer between the external environment and the Distributed Processing Engine to create a loosely coupled architecture with no direct connections between the components. This enables us to extend and modify the system modules independently, by simply ensuring they adhere to the same interface requirements.

In this architecture, the Distributed Messaging Queue creates a data pipeline between data sources and Distributed Processing Engine. Fundamentally, one topic within the message queue carries seed URLs into the Distributed Crawling Engine while another topic carries context related data such as time, location, keywords and preferences to Distributed CEP Engine.

There are many publicly available solutions to be used as a Distributed Messaging Queue such as Apache Kafka [28], Apache Flume [29], RabbitMQ [30], and Simple Queue Service [31] provided by Amazon Web Services (AWS).

Apache Kafka is one of the notable distributed messaging queue frameworks. It is horizontally scalable and fault-tolerant. It is being used by many leading service companies including but not limited to LinkedIn, Twitter, Spotify, Netflix and PayPal [28].

Apache Flume is another alternative to distributed messaging queue frameworks. It supports distributed computing and its architecture is highly adaptable. Flume can be easily integrated with Apache HBase, HDFS, Apache Spark [32], Apache Storm [33], and Apache Flink [34].

RabbitMQ allows us to define distributed messaging pipelines. It is an open source message broker that can be deployed in distributed and federated configurations. It supports multiple messaging protocols such as AMQP [35], STOMP [36] and MQTT [37]. Even though it is not a messaging protocol, RabbitMQ can transmit messages over HTTP.

Besides open source solutions, Amazon provides a paid message queuing service called as Simple Queue Service (SQS). SQS is deployed on the cloud and it is a managed service. Its scalability issues are handled by AWS [31].

3.5 Distributed Processing Engine

The proposed framework should be able to crawl millions of websites and detect user-defined context patterns on streaming data coming from many users to deliver messages in a timely and context aware manner. This is a significant amount of processing load and it cannot be handled by a single node solution. Therefore, there is a need to constitute a scalable solution, which should be able to support stream processing to detect complex context events, and it should support batch processing to crawl the Web.

With an ever-increasing amount of big data solutions, the task of selecting a processing framework for big data can be difficult. Each big data platform has its own advantages, disadvantages, and overlapping uses. Therefore, the proposed framework may be designed to run on multiple big data platforms such as Apache Storm [38], Apache Spark, and Apache Flink.

Apache Spark, Apache Storm, and Apache Flink are the most notable distributed processing frameworks that are widely used in big data applications. Choosing the most suitable engine depends on the specific characteristics of crawling websites and detecting contexts. In the proposed architecture, we divide the distributed processing engine into two main sub-modules, which are Distributed Crawler and Distributed CEP Engine.

Distributed Crawler is responsible for crawling the websites that are gathered in the Data Sources module. In order to reduce the time to crawl these websites and handle many users, a distributed infrastructure is proposed. Crawling operations require browsing websites in a batch-processing manner. Apache Spark and Apache Flink are the most notable big data tools that support batch processing. Apache Spark is specifically designed as a batch-processing framework but it also handles stream processing by micro batches. In this approach, it processes the data with specific time intervals. Apache Flink is another tool that supports batch processing, yet its kernel is designed for streaming applications. Apache Flink represents batch data as a finite sequence of streams and runs batch applications on a streaming system.

Most of the focused crawling methods [39] in the literature are based on machine learning approaches [40]. Apache Spark and Apache Flink have machine learning libraries in their own stack. Spark MLlib and FlinkML provide distributed implementations of commonly used machine learning algorithms and utilities. Therefore, the provided architecture allows users to implement their own crawling method and improve the performance of the machine learning models incorporating big data tools.

Distributed CEP Engine is specially designed to detect patterns on streaming context data which flows into the system continuously. CEP [41, 42] is defined as the analyzing and filtering of the streaming data coming from multiple sources to detect and react to events in a timely manner. The main goal of this sub-module is delivering the results of crawling to users in a context aware manner by processing the streaming data to detect pre-defined patterns on streaming context events. Distributed CEP can be a solution for real-time processing of big data streams originating from a diverse set of contexts.

One of the most notable streaming processing engines is Apache Storm. It is a distributed real-time stream processing engine which can be used to detect user contexts in real-time. Storm does not have a built-in complex event processing library to define context patterns easily. However, there are solutions [43] to use Storm as a CEP engine. Moreover, it is also possible to utilize open source business rule engines, Drools [44] and Esper [45] to make Storm support CEP processing. Another alternative for stream processing platform is Apache Flink. Unlike Apache Storm, Apache Flink has a built-in complex event processing feature to define user patterns easily.

In order to improve the context awareness of the system, a machine learning model can also be implemented. Apache SAMOA [46] is a machine learning framework that contains a programming abstraction for distributed streaming machine learning algorithms. It also supports deploying or developing machine learning algorithms on Apache Storm and Apache Flink.

3.6 Message Delivery

In this architecture, the results of the crawling engine can be distributed to users through different notification channels like email, SMS, mobile push notifications and chat-bot transactional messages, depending on users' preferences. The Message Delivery Module can be expanded by adding new notifications channels according to the needs.

4   Conclusion

Search for information is a fundamental daily activity for people. Today's operating environment where big data is now changing the rules of competition pushes businesses to access new data faster than ever before in order to stay operational. Some information may be valuable for a limited time. Accessing that information as soon as it becomes available and preferably before competitors do so becomes critical. On the other hand, some information is only valuable in a certain context. This context can depend on time, location, type of the information, the scope of information, or the combination of these parameters. Furthermore, the end user may wish to receive information via different delivery channels depending on the contextual information.

While traditional pull-based methods for accessing information fail to satisfy the requirements of the current competitive business environment, existing push-based solutions fall short on multiple aspects in this new era. Some of these tools are only usable if the content provider implements the underlying functionality. Focused crawling is used to tackle this problem. Tools that do focused crawling do not have a distributed architecture to meet the scalability requirements of big data. Others do not consider context comprehensive enough to allow users to define intended websites, precise situations, and desired delivery channels to retrieve notifications. When users cannot define the exact situation to receive notifications, the notifications become distractions. To avoid unnecessary distractions, users tend to ignore all notifications in which case users have to retrieve information with pull-based methods.

In this paper, we propose a conceptual framework to overcome the shortcomings of push-based information retrieval systems. In this framework, users can define the exact context in which they prefer to receive information. Users can define parameters like source, type, and scope of information. They can prefer when, where, and how to receive the information. Notifications become valuable prompts, instead of distractions. Additionally, in order for focused crawling to satisfy the requirements of big data characteristics, our framework takes advantage of latest developments in the big data analytics domain and particularly in stream processing and complex event processing systems. Built on top of open source stream processing engines, our distributed crawling engine can satisfy real-time processing and delivery requirements. We are planning to implement a prototype of the proposed conceptual framework to realize various real-world use cases using this framework as a future work. This framework is expected to be enhanced by the introduction of machine learning algorithms in order to match web pages not only with user queries but also with content classification and text mining algorithms.

References

1. Chakrabarti, S., Berg, M., Dom, B.: Focused crawling: a new approach to topic-specific web resource discovery. Comput. Netw. **31**, 1623–1640 (1999)
2. Gaur, R., Sharma, D.K.: Review of ontology based focused crawling approaches. ICSCTET 2014 – International Conference Soft Computing Techniques for Engineering and Technology (2016)

3. Dey, A.K.: Understanding and using context. Pers. Ubiquit. Comput. (2001). http://dl.acm.org/citation.cfm?id=59357
4. Baldauf, M.: A survey on context-aware systems. Inf. Syst. **2**(4) 2007
5. Cho, J., Garcia-Molina, H., Page, L.: Reprint of: efficient crawling through URL ordering. Comput. Netw. **56**(18), 3849–3858 (2012)
6. Diligenti, M., Coetzee, F., Lawrence, S., Giles, C.L., Gori, M.: Focused crawling using context graphs. In: Proceedings of the 26th VLDB Conference, pp. 527–534 (2000)
7. Heydon, A., Najork, M.: Mercator: a scalable, extensible web crawler. World Wide Web **2**, 219–229 (1999)
8. Boldi, P., Codenotti, B., Santini, M., Vigna, S.: UbiCrawler: a scalable fully distributed web crawler. Softw. - Pract. Exp. **34**(8), 711–726 (2004)
9. Boldi, P., Marino, A., Santini, M., Vigna, S.: BUbiNG: massive crawling for the masses. In: International Conference on World Wide Web - WWW 2014 Companion, no. Ga 288956, pp. 227–228 (2014)
10. Yan, H., Wang, J., Li, X., Guo, L.: Architectural design and evaluation of an efficient web-crawling system. In: Proceedings - 15th International Parallel and Distributed Processing Symposium, IPDPS 2001, vol. 60, pp. 1824–1831 (2001)
11. Shkapenyuk, V.: Design and implementation of a high-performance distributed web crawler. Vladislav Shkapenyuk Torsten Suel, Department of Computer and Information Science. Technical report TR-CIS-2001-03, Design and Implementation of a High-Performance Distributed Web Crawle (2001)
12. Boanjak, M., Oliveira, E., Martins, J., Mendes Rodrigues, E., Sarmento, L.: TwitterEcho. In: Proceedings of 21st International Conference on Companion World Wide Web - WWW 2012 Companion, p. 1233 (2012)
13. Yakushev, A.V., Boukhanovsky, A.V., Sloot, P.M.A.: Topic crawler for social networks monitoring. Commun. Comput. Inf. Sci. **394**, 214–227 (2013)
14. RSS 2.0 Specification. http://blogs.law.harvard.edu/tech/rss
15. Sia, K.C., Cho, J., Cho, H.-K.: Efficient monitoring algorithm for fast news alerts, pp. 1–12
16. Gusev, M., Ristov, S., Gushev, P., Velkoski, G.: Alert notification as a new model of internet-based transactions. In: 2014 22nd Telecommunications Forum, TELFOR 2014 - Proceedings of Papers (2015)
17. Katsiri, E.: A context-aware notification service
18. Corno, F., De Russis, L., Montanaro, T.: A context and user aware smart notification system. In: IEEE World Forum Internet Things, WF-IoT 2015 - Proceedings, pp. 645–651 (2016)
19. Bitly. https://bitly.com/
20. Google URL Shortener. https://goo.gl
21. Cattell, R.: Scalable SQL and NoSQL data stores. ACM SIGMOD Rec. **39**(4), 12 (2011)
22. Apache HBase. https://hbase.apache.org/
23. Apache Cassandra. http://cassandra.apache.org/
24. Shvachko, K.: The hadoop distributed file system. In: IEEE 26th Symposium Mass Storage Systems and Technologies, pp. 1–10 (2010)
25. DC/OS. https://dcos.io/
26. Apache Mesos. http://mesos.apache.org/
27. Docker. https://www.docker.com/
28. Apache Kafka. https://kafka.apache.org/. Accessed 20 Feb 2017
29. Apache Flume. https://flume.apache.org/
30. RabbitMQ. https://www.rabbitmq.com/
31. Amazon Simple Queue Service (SQS). https://aws.amazon.com/sqs/
32. Apache Spark. http://spark.apache.org/
33. Apache Storm. http://storm.apache.org/

34. Apache Flink. https://flink.apache.org/. Accessed 20 Feb 2017
35. AMQP. https://www.amqp.org/
36. STOMP. http://stomp.github.io/
37. MQTT. http://mqtt.org/
38. Toshniwal, A., et al.: Storm@twitter. In: Proceedings of 2014 ACM SIGMOD International Conference on Management of Data - SIGMOD 2014, pp. 147–156 (2014)
39. Batsakis, S., Petrakis, E.G.M., Milios, E.: Improving the performance of focused web crawlers. Data Knowl. Eng. **68**(10), 1001–1013 (2009)
40. Pant, G., Srinivasan, P.: Learning to crawl: comparing classification schemes. ACM Trans. Inf. Syst. **23**(4), 430–462 (2005)
41. Luckham, D.: The power of events: an introduction to complex event processing in distributed enterprise systems. In: Bassiliades, N., Governatori, G., Paschke, A. (eds.) RuleML 2008. LNCS, vol. 5321, p. 3. Springer, Heidelberg (2008). doi:10.1007/978-3-540-88808-6_2
42. Etzion, O., Niblett, P.: Event Processing in Action (2010). ISBN 9781935182214
43. Gokalp, M.O., Kocyigit, A., Eren, P.E.: A cloud based architecture for distributed real time processing of continuous queries. In: Proceedings - 41st Euromicro Conference Software Engineering and Advanced Applications, SEAA 2015, pp. 459–462 (2015)
44. Drools - Business Rules Management System. https://www.drools.org/
45. Esper. http://www.espertech.com/esper/
46. Apache SAMOA. https://samoa.apache.org/

ETL Based Framework for NoSQL Warehousing

Rania Yangui[1]([✉]), Ahlem Nabli[2], and Faïez Gargouri[1]

[1] Institute of Computer Science and Multimedia,
Sfax University, BP 1030 Sfax, Tunisia
`yangui.rania@gmail.com`, `faiez.gargouri@isimsf.rnu.tn`
[2] Faculty of Sciences, Sfax University, BP 1171 Sfax, Tunisia
`ahlem.nabli@fsegs.rnu.tn`

Abstract. Over the last few years, NoSQL systems are gaining strong popularity and a number of decision makers are using it to implement their warehouses.

Building the ETL process is one of the important tasks of creating NoSQL warehouse. Traditional ETL tools require the structure of the target system to be known at advance. As NoSQL databases are schema-free, this increases the need for extending the existing ETL tool in order to be able to designing schema while integrating data. In spite of the importance of ETL processes in the NoSQL warehousing, little researches have been done in this area due to its complexity.

In this paper, we propose an ETL-based platform for transforming a multi-dimensional conceptual model into document-oriented one. We model the transformation rules using the Business Process Modeling Notation (BPMN). The resulting warehouse was evaluated in term of "Write Request Latency" and "Read Request Latency" using TPC-DS benchmark.

Keywords: Extract transform and load · Business Process Modeling Notation · NoSQL · Data warehouse · Transformation rules

1 Introduction

The explosion of social network has lead to the generation of massive volumes of user-generated data that has given birth to a novel area of research, namely data warehousing from social network. By integrating external data from social network (SN) with a company's data warehouse (DW), decision-makers can better anticipate changes in customer behavior, strengthen supply chains, improve the effectiveness of marketing campaigns, and enhance business continuity. Nevertheless, current warehousing methodologies with relational databases cannot be successfully applied for storing and analyzing such data. As result, many new technologies have emerged such as NoSQL warehousing in order to increase the performance and the availability of services. NoSQL, also known as "Not Only SQL" is a term often used to describe a class of non-relational and schema-less databases that scale horizontally to very large data sets.

M. Themistocleous and V. Morabito (Eds.): EMCIS 2017, LNBIP 299, pp. 40–53, 2017.
DOI: 10.1007/978-3-319-65930-5_4

As argued by many researchers, building the ETL process is the biggest tasks of building a warehouse. In fact, it is complex, time consuming, and uses most of data warehouse projects implementation efforts, costs, and resources [1]. What makes this process even more challenging is the schema-less feature of NoSQL databases. The schema-less nature of the document-oriented database means that the designer can store documents in any shape but the notion of schema itself does not disappear from the model. Therefore, the mapping to NoSQL data storage means a move from explicitly defined data structures to implicit ones. These databases assign the responsibility to maintain the schema to the developer. Consequently, the creation of a schema occurs while inserting data at ETL level.

In the literature, there were similar researches for loading data to NoSQL databases. However, those have considered only the extraction and loading tasks. The transformation operations that reflect the implicit target schema are not addressed.

This paper provides a general implementation strategy for NoSQL data warehousing. In this context, we explain the need of a NoSQL data warehouse over traditional warehouse and we propose rules for implementing a DW under document-oriented system. These rules are implemented using java routines integrated with Talend "Talend Open Source for Big Data". In our proposal, we suggest modeling the ETL process using BPMN that allows covering the deficit of communication between the design and implementation of such process. The obtained NoSQL data warehouse was evaluated in terms of "Write Request Latency" and "Read Request Latency" using TPC-DS benchmark.

This paper is organized as follows. Section 2 represents a literature survey. Section 3 introduces the generic approach. Section 4 presents a rule-based approach for generating NoSQL Data Warehouse. Section 5 addresses the ETL process. Section 6 represents the evaluation tasks. Section 7 concludes the paper and draws future research directions.

2 Literature Survey

In this section, we review two main themes of literature related to the contributions made in this paper. Firstly, we present works that treat the implementation of data warehouses under NoSQL databases. Secondly, we discuss previous research in modeling ETL process using BPMN.

2.1 NoSQL Data Warehousing

NoSQL systems have shown their advantages over relational systems in terms of flexibility and the ability to handle massive data. The authors in [2] explained the need of a NoSQL data warehouse over traditional warehouse. They recognized that traditional Data warehouse designed over MOLAP or ROLAP are not always easy to leverage the opportunities of storing big data. In fact, growth in data volumes, number of data sources and type of data are a major area of concern for the organization. Consequently, they proposed to use NoSQL Database for implementing a data warehouse.

In the literature, many researchers have proposed approaches for the migration from relational databases to NoSQL ones. However, few works have focused on the transformation of the multidimensional conceptual model into NoSQL logical one. For example, the author in [3] showed how to build and store multidimensional data on top of big data systems and how to extend typical OLAP operators and parallel OLAP query processing on distributed key-value data store systems. However, the main goal of this paper is to propose a benchmark.

Moreover, in [4, 5], the authors proposed three approaches which allow big data warehouses to be implemented under the column oriented NoSQL model but without giving the formalization for the modeling process. Each approach differs in terms of structure and the attribute types used when mapping the conceptual model into logical model is performed.

Furthermore, the author in [6, 7] tried to define a logical model for NoSQL data stores (oriented columns and oriented documents). The authors proposed a set of rules to map star schemas into two NoSQL models: column-oriented (using HBase) and document oriented (using "MongoDB").

Recently, the authors in [8] presented a set of rules for the transformation of multidimensional data models into Hive tables, making available data at different levels of detail. These several levels are suited for answering different queries, depending on the analytical needs.

Similarly, in [9], the authors discussed the possibilities to create data warehouse solutions by using NoSQL database management systems. The main challenge is to find a good balance between characteristics of classical data warehouses using relational database management systems and opportunities offered by NoSQL database management systems. The paper described processes of creation and production of data warehouse using a NoSQL data mart and outlined requirements for technology necessary for such processes. The research is based on practical experience when implementing NoSQL data marts with MongoDB and Clusterpoint DB.

As NoSQL data stores are becoming increasingly popular in application development, the above presented works have shown that it is possible to implement DW under NoSQL storage. These systems are attractive for developers due to their ability to handle large volumes of data, as well as data with a high degree of structural variety. However, it appears that the majority of researchers use HBase for implementing a NoSQL DW. This is justified by the resemblance between the logic model HBase and that of relational databases, particularly in terms of concepts of tables and rows. Besides, there is an interesting concept that has never been used in the proposed transformation rules, namely embedded documents and super columns.

In the other hand, the majority of authors neglected the ETL process, which is an essential stage in the construction of data warehouse. They did not explicitly define the different function of the ETL process. However, as NoSQL databases claim to be schema-less, it need careful migration due to the implicit schema in any code that accesses the data. As a result, it requires more programming to obtain needed DW.

2.2 BPMN Based ETL Process Modeling

ETL is often a complex combination of process and technology that consumes a significant portion of the data warehouse development efforts and requires the skills of business analysts, database designers, and application developers. A solid, well-designed, and documented ETL system is necessary for the success of a data warehouse project.

As the ETL process is critical in building and maintaining the DW systems, some researchers tried to represent it using BPMN language. In this context, the authors in [10] presented a method to guide BPMN specifications in the definition of conceptual models of ETL systems.

Similarly, the authors in [11] provided an independent platform for the conceptual modeling of an ETL process based on BPMN. Using the same BPMN objects presented by [12], the authors proposed a correspondence between the ETL process and the needs of decision-makers to easily identify which data are necessary and how include them in the DW.

Thereafter, in [13], the authors defined specific conceptual model that takes into account of the capture of evolutionary data, the change of dimensions, the treatment of the substitutions keys and the data quality.

Later, the authors in [14, 15] presented a two-levels approach for the construction of web warehouse by means of BPMN. The first level is focused on helping the user to configure the web data sources and the desired data quality characteristics. The second level uses the defined configuration to generate the warehouse.

As first initiative, the authors in [16] identified and implemented a framework for NoSQL Databases. In the Methodology, it is realized that it would take a lot more time and effort to create an enterprise level application for the NoSQL ETL Framework.

Notice that, the majority of works proposed ETL frameworks that work well with relational systems, only the work of [16] takes into consideration the ETL process for NoSQL warehousing. However, this work does not detail and model the proposed process. Moreover, many software vendors, including "IBM", "Informatica", "Talend", and "Pentaho", provide ETL software tools allowing migration to NoSQL systems. However, the transformations are simple and several NoSQL concepts such as "embedded document" or "super column" are not explored.

Based on the above discussion, there is a strong need for a significant ETL-based platform that allows the implementation of DW under NoSQL system. Representing the ETL process by means of BPMN notations is also recommended.

3 ETL-Based Framework for NoSQL Warehousing

In the absence of a clear approach that allows the implementation of data warehouses under NoSQL system, we propose in this section a two-level approach for data warehouse building under document-oriented system (Fig. 1).

The first level relates to the definition of the transformation rules for logical NoSQL schema definition. In fact, OLAP servers that base on relational data storage solutions

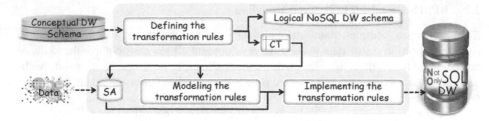

Fig. 1. Approach overview.

cannot process big volumes of data. In this level, we need to save traceability of the transformation rules on a table called Correspondence Table (CT).

Data are extracted from data sources and loaded into the staging area. The second level consists on using the resulting CT for modeling and implementing the proposed rules. As NoSQL databases are schema-less, to create a NoSQL DW, we are required to write a program that relies on some form of implicit schema. The implicit schema is a set of assumptions about the data's structure in the code that manipulate the data [17]. That mean, the creation of a schema occurs when inserting data at ETL level. In this level, we implement the defined transformation rules that reflect the implicit schema using java routines.

4 NoSQL Warehouse Logical Schema Definition

Given that a well-designed DW requires a well planned logical design, all updates and versions of a DW lead to a revision of the logical design. Generally, the mapping from the conceptual to the logical model is made according to three approaches: ROLAP (Relational-OLAP), MOLAP (Multidimensional-OLAP) and HOLAP (Hybrid-OLAP) [18]. All these models are inadequate when dealing with large amount of data which need scalable and flexible systems. As an alternative, NoSQL systems begin to grow.

4.1 The Choice of the Adequate NoSQL System

In the literature, four types of NoSQL data stores exist [19]: Key-value Stores, Document Stores, Columnar Stores and Graph Stores. In previous work, we implemented two data warehouses using Cassandra as a columns-oriented system and "MongoDB" as documents-oriented system. These warehouses were evaluated in terms of "Write Request Latency" and "Read request Latency" using "TPC-DS" benchmark. We noted that the DW built under Cassandra is faster than the DW built under "MongoDB" (1000000 rows in 203.23 s for Cassandra while 1000000 rows in 437.36 s for "MongoDB"). This is argued by the fact that Cassandra uses less memory space and it is known for effective data compression (due to column redundancy).

Subsequently, we tested the behavior of the systems against a set of queries. The results showed that the documents-oriented NoSQL data warehouse is more efficient in terms of interrogation. This is justified by the fact that data in the column-oriented systems are not available in the same place. Subsequently, we choose to implement our DW under document-oriented systems and specially MongoDB.

4.2 Transformation Rules: Hierarchical Transformation to Document-Oriented Model

Recall that a data warehouse schema consists of fact with measures, as well as a set of dimensions with attributes, we map the dimensions according to its attributes and the facts according to its measures.

The hierarchical transformation uses different collections for storing facts and dimensions, and uses the simple documents for representing measure and the composed attributes for representing dimension attributes while explaining hierarchies.

Rule 1 represents the transformation of a fact and its measures to the document-oriented model.

Rule 1: *Fact/Measures Transformation.*

Each fact $F \in MS^{Fact}$ is transformed to a document DC (DC^N, DC^{Att}) where:
- The name of the document is the name of the fact / $DC^N \leftarrow F^N$;
- Each measure $M \in F$ is transformed to a simple attribute $SA \in DC^{Att}$ / $SA^N \leftarrow M^N$;
- Each identifier of a related dimensions is transformed to a simple attribute $SA \in DC^{Att}$ / $SA^N \leftarrow D^{id}$.

Moreover, Rule 2 mentions the hierarchical transformation of a dimension and its attributes (Strong and Weak) to the document-oriented model.

Rule 2: *Dimension/Parameters Transformation*

Each dimension $D(D^N; D^{Att}; D^{Hier}) \in MS^{Dim}$ is transformed to a document $DC(DC^N; DC^{Att})$ where:
- The name of dimension D is equivalent to the name of the document / $DC^N \leftarrow D^N$;
- Each hierarchy $H(H^N, H^P, H^A)$ is transformed to a composed attribute $CA \in DCatt$ where:
  · The name of the composed attribute is the hierarchy name / $CA^N \leftarrow H^N$
  · The values of composed attributes are the simple attributes that represent the weak and the strong attributes / $CA^{Val} \leftarrow H^P \cup H^A$.

Based on the above defined rules (Rule 1 and Rule 2), we deduce the hierarchical transformation of a multidimensional schema (MS) as mentioned by Rule 3.

> **Rule 3:** *Multidimensional Schema Transformation.*
> Each multidimensional schema MS (MS^N, MS^{Fact}, MS^{Dim}, Func) is transformed to a documents collection DCC (DCC^N; DCC^{Val}) where:
> - The name of the collection is the name of the MS / $DCC^N \Leftarrow MS^N$;
> - Each fact $F \in MS^{Fact}$ is transformed to a document DC (DC^N, DC^{Att}) where the name of the document is the name of the fact / $DC^N \Leftarrow FN$ (Rule 1);
> - Each dimension $D(D^N; D^{Att}; D^{Hier}) \in MS^{Dim}$ is transformed to a document $DC(DC^N; DC^{Att})$ (Rule 2).

Figure 2 shows an example of transforming an excerpt of TCP-DS multidimensional schema.

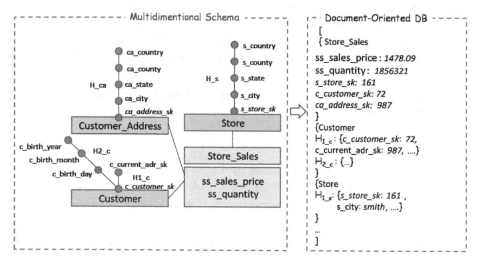

Fig. 2. MS transformation example (TCP-DS example).

In this level, the correspondence table is generated to keep trace of different transformations. Table 1 presents an excerpt of the generated correspondence table (CT).

As output of this level, we have the correspondence table with full documentation of all transformation operations. This table will be used for modeling and implementing the transformation rules within ETL. In fact, to feed the NoSQL DW, data must be identified and extracted from the source. Consequently, the data must be transformed and verified before being loaded into "MongoDB". Moreover, loading data require the structure of the target to be known at advances. As NoSQL DBs are schema-less, this increases the need for extending the existing ETL tool in order to be able to designing data warehouse while integrating data. ETL tool should be adapted with the constant

Table 1. Correspondence table excerpt.

Object source	Type	Operation	Target	Type data
Store_Sales	Fact	Fact Transf	Store_Sales	Document
ss_sales_price	Measure	Fact Transf	ss_sales_price	Simple Attribute
Customer	Dimension	Dimension Tranf	Customer	Document
H1_c	Hierarchy	Dimension Transf	H1_c	Composed Attribute
c_customer_sk	Week Attribute	Dimension Transf	c_customer_sk	Simple Attribute
c_current_adr_sk	Strong Attribute	Dimension Transf	c_current_adr_sk	Simple Attribute

changes, to produce and to modify executable code quickly. This process will be addressed in the next section.

5   Extract, Transform and Load (ETL) Process

Business Intelligence (BI) solutions are very important as they require the implementation and the design of complex ETL process. In fact, ETL plays an important role in data warehousing architecture since these ETL processes move the data from transactional or sources systems to data staging areas and from staging areas into the data warehouse. ETL is often a complex combination of process and technology that consumes a significant portion of the data warehouse development efforts and requires the skills of business analysts, database designers, and application developers. New applications, such as, NoSQL data warehousing, require agile and flexible tools. In fact, the free-schema feature of these DBs imposes new requirements on the ETL process implementation and maintenance.

To facilitate and minimize the complexity of ETL process, we propose to model it using BPMN (Business Process Modeling Notation) language. BPMN notation seems to be a good choice since it can cover a deficit of communication that often occurs between the design and implementation of business process.

5.1   BPMN Modeling of ETL Process

In the last years, several modeling methodologies for ETL scenarios have been proposed, covering both conceptual and logical level. The chosen model is based on the Business Process Modeling Notation (BPMN), a standard for specifying business processes. BPMN provides a conceptual and implementation-independent specification of such processes, which hides technical details and allows users and designers to focus on essential characteristics of such processes. An advantage of BPMN is that it helps in graphically portraying also complex processes. Another benefit is the empowerment of business process workflow with execution details, thus providing a mapping between the graphics of the notation and the underlying constructs of execution languages.

Several works have dealt with ETL process modeling using BPMN such as [12, 13]. In this section, we are based on the above works to model ETL process using BPMN (Fig. 3).

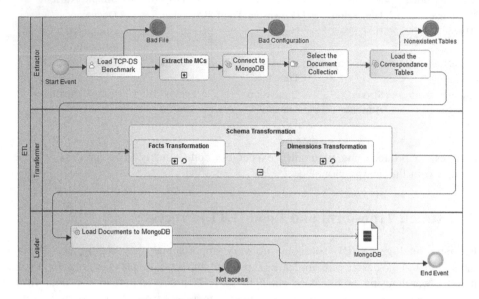

Fig. 3. BPMN modeling of ETL process.

The BPMN model presented in Fig. 3 is composed of one pool which encloses three lanes.

The first lane (Extractor) models the ETL extraction step. This step is responsible for extracting data from the source systems (TCP-DS Benchmark) and the configuration of the target system ("MongoDB"). The Extraction component must solve the problem of format heterogeneity, since data can be found in a variety of formats. Each data source has its distinct set of characteristics that need to be managed in order to effectively extract data for the ETL process. During the loading of the sources, there may occur "Throw" events, which will be fired whenever they are some inconsistent files.

The Second lane (Transformer) models the ETL transformation step. This step tends to make some conforming on the incoming data to obtain the needed system. This process includes schema transformation and data integration. It defines the granularity of the target schema. All transformation rules are implemented at this step using java routines. The Transformer lane has one sub-process which is composed of two others, representing the fact transformation and the dimension transformation.

Iteratively, the first sub-process handle all the records contained in the fact table and, one by one, invokes other two sub-processes that is in charge to select the measures and the identifiers of related dimensions (Fig. 4). Once guaranteed the integrity of the transformation, the process updates the documents.

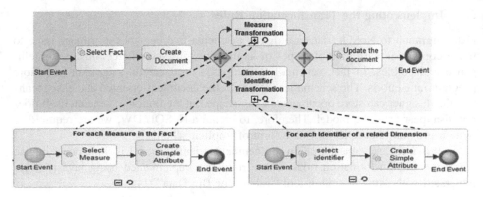

Fig. 4. BPMN representation of fact transformation.

The second sub-process contained in the Transformer lane is Dimension Transformer (Fig. 5). Iteratively, this sub-process handle all the records contained in the dimension table and, one by one, invokes another sub-processes that is in charge to create a composed attribute for each hierarchy. This sub-process contains, in turn, two others sub process that describes the week and the strong attributes transformations.

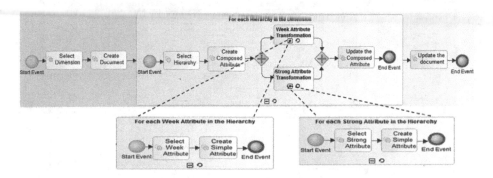

Fig. 5. BPMN representation of dimension transformation.

The third lane (Loader) models the ETL loader step. The Loader mechanism loads data into the target of an ETL process such as documents in "MongoDB". Loading data to the target NoSQL DB is the final ETL step. In this step, extracted and transformed data is written into the document-oriented structures actually accessed by the end users and application systems.

It can be seen that, most tasks in the presented ETL process are automatic of type service task, and only a few remain of type user or manual task. This is due to the focus of the process in implementing the transformation rules in order to automatically generate the NoSQL DW.

5.2   Implementing the Transformation Rules

Data migration to NoSQL database that is a schema-free becomes a primary issue to many companies with multiple types of applications in e-commerce, business intelligence and politics. In fact, schema-less databases need careful migration and more programming efforts. The schema-less nature of the document-oriented database means that the designer can store documents in any shape but the notion of schema itself does not disappear from the model. Therefore, to create a NoSQL DW, we are required to write a program that relies on some form of implicit schema.

In this work, schema migration to document-oriented model is done according to the proposed transformation rules. These rules are implemented using java routines integrated with data integration tool "Talend for Big Data" (Fig. 6).

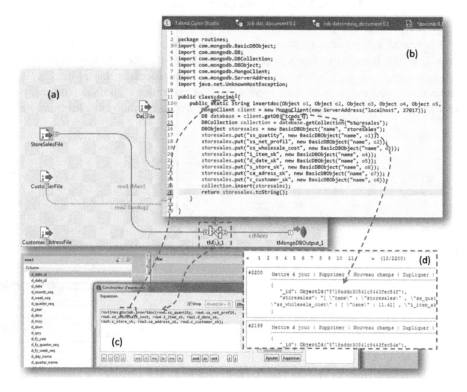

Fig. 6. Example of routine implemented with Talend.

Figure 6 shows four main interfaces: (a) the created Job, (b) an excerpt of the implemented java routine, (c) the expression editor, and (d) an excerpt of the resulting "MongoDB" database.

A Job is a graphical design, of one or more components connected together such as tFileInputDelimited (CustomerFile, StoreSalesFile, etc.), tMap and FileOutputDelimited ("MongoDB"). Otherwise, a routine is a complex Java code, generally used to optimize data processing and improve Job capacities. In this work, the routine is used for implementing the transformation rules. The implemented routine is then called and edited in the expression editor. This editor provides visual interface to write any function or transformation in a handy dedicated view.

6 Evaluation

For implementing the document-oriented DW, our choice is oriented towards MongoDB. Like any other NoSQL database, MongoDB can quickly handle large volumes of data and offers the ability to create flexible schema. Our choice is justified by the fact that MongoDB provides a rich document oriented structure with dynamic queries.

In our context, we use TPC-DS benchmark to load the NoSQL DW. As TPC-DS is a logical model, we generated the conceptual schema, sand then we used our proposed rules to obtain the corresponding NoSQL models.

In order to evaluate the implemented NoSQL DW, we chose to use two metrics: "Write Request Latency" and "Read request Latency". The first metric has the purpose to test the speed of the system during the data loading stage. As for the second metric, it evaluates the system's ability to respond quickly to user requests. Regarding requests, we chose to use two. The first category is simple. As for the second query, it is more and consists on using some operators (Table 2).

Table 2. Request description.

Request	SQL Langage	MongoDB Langage
Q1	Select * From date where d_year BETWEEN '1990' and '2000';	> db.date.find({"d_year" : {"$gte" : 1990, "$lte" : 2000} })
Q2	SELECT SUM(ss_quantity) FROM store_sales WHERE d_moy IN('2415022', '2415023') GROUP BY d_date_sk;	> db.store_sales.group({ > "key": {"d_date_sk": true}, > "initial": {" sumquantity ": 0}, > "reduce": function(obj, prev) {prev.sumquantity = prev.sumquantity + obj.ss_quantity ; }, > "cond": { "d_date_sk": { "$in": [2415022, 2415023]} } > });

Table 3 describes the system behavior against requests.

As can be seen from the above results, NoSQL data warehouse can respond to increasing queries without performance degradation (0.35 for Q1 and 0.46 for Q2). A NoSQL data warehouse can rapidly adapt to growth in data volume and query intensity without degrading performance (0.21 for 400000 records while 0.46 for 1000000 records using the same query Q2). Moreover, the NoSQL data warehouse can

Table 3. Obtained results.

Number records	Request	Returned values number	WRL (s)	RRL (s)
400000	Q1	363	0.20	0.19
	Q2	2415022: 326 2415023: 129		0.21
800000	Q1	1245	0.38	0.24
	Q2	2415022: 793 2415023: 513		0.32
1000000	Q1	3652 ·	0.43	0.35
	Q2	2415022 875 2415023:614		0.46

quickly adapt to changes in data structure and content without requiring any schema redesign, additional data migration, or new data storage structures.

Additionally, the schema-less nature of the NoSQL system eliminates the need of transformation of schema between various data sources and the data warehouse. This transformation was just a waste of time in traditional data warehouse.

7 Conclusion

Because of the wide development of the social media, a huge amount of data is now continuously available to decision support.

Since the relational systems are lack of scaling and inefficient of handling big data it's vital to extract transform and loading the data from traditional data warehouses to NoSQL ones. Some approaches have been introduced to handle this problem however the ETL process, which is an essential stage in the construction of data warehouse, was neglected. ETL processes are very important problem in the current research of data warehousing. The schema-free feature of NoSQL systems increases the need for extending the existing ETL tool in order to be able to designing schema while integrating data.

In this paper, we proposed an ETL-based framework for NoSQL warehousing. In this context, we proposed rules for transforming a multidimensional conceptual schema into document-oriented system. As NoSQL databases are schema-less, the creation of a schema occurs while inserting data at ETL level. We used the BPMN for modeling the ETL process then we implemented the defined rules as routines that reflect the implicit schema. Experiments are carried out using the benchmark TPC-DS. The obtained results showed that a NoSQL data warehouse can rapidly adapt to growth in data volume and query intensity without degrading performance.

In the future work to this paper, we aim implementing OLAP operators (D-Roll up, D-Slice, D-Dice) with the phases of the invisible join technique for a document-oriented data warehouse.

References

1. Favre, C., Bentayeb, F., Boussaid, O., Darmont, J., Gavin, G., Harbi, N., Kabachi, N., Loudcher, S.: Les entrepots de donnees pour les nuls .ou pas. 2eme Atelier d'aide à la Decision a tous les Etages (2013)
2. Chandwani, G.: NoSQL data-warehouse. Int. J. Innovative Res. Comput. Commun. Eng. **4**, 96–104 (2016)
3. Zhao, H., Ye, X.: A practice of TPC-DS multidimensional implementation on NoSQL database systems. In: Nambiar, R., Poess, M. (eds.) TPCTC 2013. LNCS, vol. 8391, pp. 93–108. Springer, Cham (2014). doi:10.1007/978-3-319-04936-6_7
4. Dehdouh, K., Boussaid, O., Bentayeb, F.: Columnar NoSQL star schema benchmark. In: Ait Ameur, Y., Bellatreche, L., Papadopoulos, G.A. (eds.) MEDI 2014. LNCS, vol. 8748, pp. 281–288. Springer, Cham (2014). doi:10.1007/978-3-319-11587-0_26
5. Dehdouh, K., Boussaid, O., Bentayeb, F.: Using the column oriented NoSQL model for implementing big data warehouses. In: Proceedings of the 21st International Conference on Parallel and Distributed Processing Techniques and Applications, pp. 469–475 (2015)
6. Chevalier, M., El Malki, M., Kopliku, A., Teste, O., Tournier, R.: Implementing multidimensional data warehouses into NoSQL. In: Proceedings of the 17th International Conference on Enterprise Information Systems, pp. 172–183 (2015)
7. Chevalier, M., El Malki, M., Kopliku, A., Teste, O., Tournier, R.: Document-oriented models for data warehouses. In: Proceedings of the 18th International Conference on Enterprise Information Systems, pp. 142–149 (2016)
8. Santos, M.Y., Martinho, B., Costa, C.: Modelling and implementing big data warehouses for decision support. J. Manag. Anal. **4**, 1–19 (2017)
9. Bicevska, Z., Oditis, I.: Towards NoSQL-based data warehouse solutions. Proc. Comput. Sci. **104**, 104–111 (2017)
10. Wilkinson, K., Simitsis, A., Castellanos, M., Dayal, U.: Leveraging business process models for ETL design. In: Parsons, J., Saeki, M., Shoval, P., Woo, C., Wand, Y. (eds.) ER 2010. LNCS, vol. 6412, pp. 15–30. Springer, Heidelberg (2010). doi:10.1007/978-3-642-16373-9_2
11. El Akkaoui, Z., Mazon, J., Vaisman, A., Zimanyi, E.: Defining ETL worfklows using BPMN and BPEL. In: Data Warehousing and OLAP, pp. 41–48 (2009)
12. El Akkaoui, Z., Mazón, J.-N., Vaisman, A., Zimányi, E.: BPMN-based conceptual modeling of ETL processes. In: Cuzzocrea, A., Dayal, U. (eds.) DaWaK 2012. LNCS, vol. 7448, pp. 1–14. Springer, Heidelberg (2012). doi:10.1007/978-3-642-32584-7_1
13. Oliveira, B., Belo, O.: BPMN patterns for ETL conceptual modelling and validation. In: Chen, L., Felfernig, A., Liu, J., Raś, Z.W. (eds.) ISMIS 2012. LNCS, vol. 7661, pp. 445–454. Springer, Heidelberg (2012). doi:10.1007/978-3-642-34624-8_50
14. Delgado, A., Marotta, A., González, L.: Towards the construction of quality-aware web warehouses with BPMN 2.0 business processes. In: RCIS, pp. 1–6 (2014)
15. Marotta, A., Delgado, A.: Data quality management in web warehouses using BPM. In: ICIQ, pp. 18–27 (2016)
16. Sahiet, D., Asanka, P.D.: ETL framework design for NoSQL databases in dataware housing. Int. J. Res. Comput. Appl. Rob. **3**, 67–75 (2015)
17. Sadalage, P.J., Fowler, M.: NoSQL Distilled: A Brief Guide to the Emerging World of Polyglot Persistence. Pearson Education, London (2012)
18. Chaudhuri, S., Dayal, U., Ganti, V.: Database technology for decision support systems. IEEE Comput. Soc. **34**, 48–55 (2002)
19. Sharma, V., Dave, M.: SQL and NoSQL databases. Int. J. Adv. Res. Comput. Sci. Softw. Eng. **2**, 20–27 (2012)

An User Interest Ontology Based on Trusted Friends Preferences for Personalized Recommendation

Mohamed Frikha$^{(\boxtimes)}$, Mohamed Mhiri, and Faiez Gargouri

MIRACL Laboratory, University of Sfax, Sfax, Tunisia
med.frikha@gmail.com, med.mhiri@gmail.com,
faiez.gargouri@isimsf.rnu.tn

Abstract. The importance of personalized recommender systems has recently increased and its role in providing better experiences to different users has been demonstrated. The quality of the recommendation can be guaranteed based on the help of user interpersonal interests in a social network. Information obtained about users and their friends makes it unnecessary to look for similar users and to measure their rating similarity. In our work, we have developed a trusted friend's calculation method for determining social trusted friends by analyzing user's profile on Facebook. We have also represented the user's model as an ontology that takes into consideration all trusted friends' preferences and the degree of trust between friends. Afterwards, we have used this ontology in a semantic tourism recommender system as an e-tourism tool able to recommend items based on the users' preferences and their trusted friends' preferences.

Keywords: Preference representation · Semantic web mining · Social network · Tourism recommender systems · Ontology

1 Introduction

The quick development of online social networks and its spread around the world has created a new place of interaction among users by helping them share their knowledge, experiences, and opinions. Online social networks can also have an impact on people's behavior in terms of traveling and purchasing products. In fact, a social network includes a set of people or groups of people who have some common interests and social relationships to interact together. It is through these interactions and relationships that we can delve into the user preferences and extract his/her interests to improve recommendations. For example, in real everyday life, when we think of buying a particular and unfamiliar product, we most often tend to seek immediate advice from some of our friends who have come across this product or experienced it. We, similarly, tend to accept and use friends' recommendations because we trust them. Therefore, integrating social networks in recommender systems can result in more accurate recommendations [1]. Generally speaking, those networks are represented by the graphs' theory and the matrix technology. Since social networks are developed by several different kinds of relationships, it remains impossible for the graphs' edges and the numerical values to

© Springer International Publishing AG 2017
M. Themistocleous and V. Morabito (Eds.): EMCIS 2017, LNBIP 299, pp. 54–67, 2017.
DOI: 10.1007/978-3-319-65930-5_5

explain all the semantic relationships. Having mentioned this problem, we propose, as a solution, a method that makes it possible to represent a social network based on ontology. Using an ontology-based method could allow us to describe all the semantic relationships and the interactions in a social network. The importance of the use of ontologies to represent the types of actors and the relationships in a social network for the purpose of semantically visualizing the databases has been demonstrated by [2]. Based on the fact that ontologies can be used to semantically describe social relationships and interactions and that social networks can generate and suggest several varied recommendations with reference to the user needs.

Traditional methods of recommendation rely on keywords to model user's interests and information requirements [3]. Due to the vagueness of keywords, the user interest model cannot represent user's interests and the relationships between different items of the target domain accurately. Besides there are cold-start and data sparsity problems [4] resulting in poor performance. Many studies have used graph representation to design user interest model. Their work has, however, been restricted to using it in traditional recommendation methods. According to the literature [5], the user model has several types of representations: a set of words, a matrix of weighted keywords, the use of taxonomy [6] or by an ontological representation of the profile through the use of ontologies. In our work, we will build an ontology to represent user preferences and his/her interests. Our proposal relies on a combination of two approaches to improve the recommendation process. The first is the trusted friends' preference to reduce the sparsity problem when we do not have many information about the ego user. The second is the ontology based user interest to represent the semantics of these interest.

The rest of the paper is organized as follows. Section 2 presents related works in tourism recommender systems. Then, we explain the importance of Ontologies in Representation of User Interest and we present preview work on ontology-based recommendations. After that, we present an overview of trust in recommender systems. Section 3 describes our method for calculating trusted friends on social networks. Section 4 presents the user interest ontology-based on trusted friends' preferences. While Sect. 5, we present the *SMART-TMT* recommender system development for the Tunisian medical tourism domain and in Sect. 6, we conclude the paper and present our future work.

2 Related Work

2.1 Recommender Systems in Tourism Domain

Tourism recommender systems's main purpose is to map the user preferences onto the leisure resources and tourist activities of the city [7]. To be able to achieve this, tourism recommender systems need some initial data explicitly provided by the user. These data are especially useful when automatically inferring the user preferences through explicit or implicit feedback. In fact, the majority of tourism recommender systems are hybrid approaches that seek to combine basic recommendation techniques such as demographic, content-based and collaborative filtering techniques [8].

Very little work that designed medical tourism recommender system has been conducted. [9] designed a recommender system for medical tourism by using the technology of the semantic web to model the domain knowledge. They used content-based recommendation techniques to elaborate recommendation for users. However, content-based methods rely only on explicit item descriptions. Such descriptions may be difficult to obtain for items like ideas or opinions. Most of the tourism recommender systems that build personalized information use the classical recommendation techniques. Recently, however, social networks have become of particular importance in the tourism industry [10]. When planning their journeys, travelers, in addition to the personal experience and opinion of relatives and friends, refer to social networks during and after their trips, thus generating interest in viewers that can become travelers themselves [11]. In our tourism recommender system, we give an important role to the social network to make recommendation based on the user's preferences as well as his/her friends' preferences.

2.2   Ontologies Based Representation of User Interest

To process the data by the recommendation system, we basically need three elements, which are: user, item and rating. In most algorithms, these three elements are represented by a matrix. In this matrix, the rows indicate users, columns indicate items, and matrix entries indicate ratings. An example of user × item matrix is given in Fig. 1 with three users and four items along with their ratings. The items may have various variables in the implementation, such as gender, age for users, price, year for items, and time for ratings.

Alternatively, we can use graph representation of user × item matrix. In graph representation, users and items are represented with nodes and the weighted edges relate users with items [12]. An example of graph representation is given in Fig. 2, which shows the same information as Fig. 1.

Ontology can be considered as one type of graph representation and is used to represent users and their related rating of items in many recommender systems. [13] are among the first to make use of ontology to build user model and provide personalized document access. Domain ontologies have been employed to organize documents.

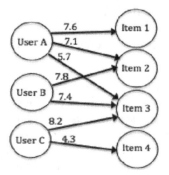

	Item 1	Item2	Item3	Item4
User A	7.6	7.1	5.7	-
User B	-	7.8	7.4	-
User C	-	-	8.2	4.3

Fig. 1. User × item matrix

Fig. 2. User × item graph

Based on a user's surfing history, the user model is built and the personalized document access is performed by referring to the interest degree of each document's concept. The semantics of concept relations and ontology structure, however, is not taken into consideration during the calculation of the user's concept interest degree.

[14] propose a user interest model based on user interest ontology. In this work, recommendation process is presented by using the ontological user interest model. The biggest shortcoming of this method is that it does not use the advantage of social network in the recommendation process. [14] use only collaborative recommendation in the process of recommendation, but collaborative recommender systems fail to help in cold-start situations, as they cannot discover similar user behavior because there is no enough previously logged behavior data upon which to base any correlations.

[15] describes an adaptive recommender system which adopts a semantic approach to assist the user both in the travel planning phase and, on-site, during the trip. [15] propose a general architecture of an adaptive recommender system that uses ontology to assist users in the travelling. In this work, extraction of the user interest is not done automatically from the social network, but users are required to answer a set of questions about personal interests and to provide detailed information regarding their requirements.

2.3 Trust Based Recommendation

Trust can play an important role across different disciplines and is an important feature of our everyday lives. Additionally, trust is a feature associated with people in the real world and with users in social media [16, 17]. In recommender systems, trust is defined with reference to the other users' ability to provide valuable recommendations [18]. The trust value can be binary or in real numbers (i.e., in the range of [0; 1]). Binary trust is the simplest way of expressing trust. Two users can choose to trust or not trust each other. A more in-depth method is a continuous trust model, which gives real values to the trust relations. In both binary trust and continuous models, 0 and 1 means no trust and full trust, respectively [19].

In recommender systems, trust is classified into explicit and implicit. Explicit trust emphasizes information explicitly stated by users. Although research on explicit trust-based recommender systems has been proposed [20, 21] and its effectiveness has been tested, the system is found to have many caveats. In spite of the fact that specific trust values are possible in real systems, publicly available datasets such as FilmTrust [21] only contain trust links without real values due to the concern of privacy. The indifferent and binary trust will prevent achieving better performance.

[22] uses the subjective logic in his attempt to define a trust model taking into account the local, collective and global trust. This method does not take into consideration the temporal factor in measuring trust between users. although a simple opinion in subjective logic has a probabilistic structure based on the multiple interactions between users, it is not time sensitive [23]. This means that all interactions have the same importance. We assume that interactions are not perceived the same way over time. To put it differently, some interactions are more important than others when computing an opinion [24].

[25] suggested a general model to measure trust among social network users. They have also explained how a general model is implemented using a specific social network as a medium. In their work, however, [25], also, have not taken into consideration the temporal factor in measuring trust between users. In the real world, the value of an interaction between two users is different according to its time of apparition [23]. In other words, many trust models were proposed in many social web applications but most of these methods do not take into consideration the time of an interaction between two users in the social network. In our recent work, we have provided a method to calculate trusted friends for an ego user with an implicit trust metrics [26]. Implicit trust is generally inferred from user behaviors and trust relationships among social friends. Our method takes into consideration the temporal factor of interactions that exists between social network friends. In this work, we will apply this method to determine trusted friends and then to extract these preferences.

3   Trusted Friends Calculation on Social Network

Research has demonstrated the effectiveness of trust in recommender systems. Lack of explicit trust information in most systems emphasizes the need to propose some trust metric to infer implicit trust from the user and his/her friends' interactions. Our goal is to determine the list of friends trusted by a $user_x$. We begin, first, by all the friends who have made interactions or shared information with the user in a very specific period of time. Interactions among users in a social network are of different types. They can be comments or mentions "Like" on objects in the profile (from the user or from friends). We apply some trust metric to calculate the value of trust between $user_x$ and each of his/her friends. Our goal is to demonstrate how an ego's social activities and his/her friends can be used to calculate a "level of trust" between two friends. Finally, we choose friends who have the highest level of trust. The extraction of the trusted friends' list of a user is performed to determine the preferences and interests of each friend. Indeed, these interests can be useful to know the interests of the *ego user*.

A *social activity is* a social interaction between two directly connected social network friends at a period d. An example of social activity can be the joint tagging of two friends on the same photo. Different social networks have different types of social activities among users. To calculate a "level of trust" between two friends in a social network, we have classified all social activities among these friends taking into consideration the temporal factor [26]. In our work, we take into consideration the video interaction between friends in Facebook (friends tagged in the same video, friends who liked or commented on a video).

To determine the list of friends near a user, we begin, first of all, by all the friends who have made interactions or shared information with the user in a very specific period of time. Thus, interactions are of different types. They can be comments or mentions "Like" on objects in the profile (from the user or from friends). Then, we calculate the number of occurrences of each friend in these interactions. Finally, we choose friends who have a number of occurrences beyond a predetermined threshold.

However, the extraction of the trusted friends' list of a user is performed to determine the preferences and interests of each friend. These preferences will be represent in the user interest ontology. In the next section we present the user interest ontology based on trusted friend's preferences.

4 User Interest Ontology Based on Trusted Friends Preferences

In this section we present an ontology based on trust to represent the user's preferences and their trusted friends preferences. We will extract these preference from the user's profile in the social network. The user profile contains information about the user, such as name, age, friends; and about his/her preferences, such as items that the user liked in the past. Indeed, the user profile contains information about the users, their preferences, behavior in the network and interactions with friends. In social networks based recommender systems, the user profile is an excellent source of information to retrieve interests of the user and to determine the trusted friends, so it is very important that this profile be treated appropriately.

Our method for extracting user interests consists of several steps. After giving the user's answers and obtaining the initial profile, we collect user data from his/her social profile and behaviors in the network. Then, we analyze the user data to extract the interests of the user. Finally, we present all these data in the form of ontology. The ontology's main concepts are the user's interests and relationships. The conceptual and semantic relationships between the user's interests are determined in the data analysis step.

4.1 User Questions Answers

When the user accesses to our Website, the system will propose questions to know about his/her health interests in order to create the initial profile and to take explicit information about his/her health state. The information provided are about the country of origin of the user, the location of the accommodation, the type of accommodation, an initial estimation of the budget, and the travel period. With these answers, our system will know the initial user profile of this user to represent it in the user interest ontology. Then the user will connect to his/her social network (Facebook) and the system collects data from the user profile and completes the construction of user interest ontology from user's preferences and his/her trust friends' preferences.

4.2 User Data Collection

The user is identified in a social network by his/her traits and behaviors (personal specifications), and his/her interactions with the various system services. In our approach, data generated by the user are used primarily to explore their interests. These data can be of different formats: photos, links pages, texts, etc. In our work, we treat the textual form, which presents the major form of data to identify user interests. In this context, the content generated by the user can be regarded as explicit and implicit data. First, the explicit data are the data about users which are explicitly declared and which represent user information and preferences. The user information may be either static or dynamic. Static user information are mainly general information such as name, first name, sex, date of birth, etc. while dynamic user information relate to marital status, level of education, place of residence, occupation, email address, phone number, nickname, photo, etc. Then the implicit data are usually discovered through actions provided by the user or by its behavior such as comments, quotes "like", publications, events, etc. and also deduced from explicit data contained in the ontology.

4.3   Data Analyzing and User Interest Extraction

In this step, we analyze the preferences and behaviors of the user (shares, comments, events, etc...). Thus, in this phase, we try to filter the raw data and display them in a representation in order to further process them. Indeed, these textual data are not structured and written in a formal language, which makes their use very difficult in their raw state.

- **User preferences analyzing**

We started by analyzing the preferences of the user, which are presented by a set of XML files extracted from the social network. These XML files are obtained with the Facebook API Open Graph that gives us the authorization to accede to the user profile. First, we begin by removing all empty concepts. Then, we construct a matrix of occurrence that can indicate the number of occurrences of each concept. We afterwards, determine the semantic relations between the concepts using "*WordNet*" which covers most common English words, and a *base of terms* containing the common concepts that do not belong to "*WordNet*". We use a calculating similarity between concepts when the concept does not belong to "*WordNet*" or to the *base of terms.*

After extracting the various concepts used by the user, we will try to determine the semantic relationships that exist between these concepts. To do this, we use the language tool "*WordNet*" to determine if there are relationships between the concepts of the matrix. In many cases, we can find concepts that do not exist in "*WordNet*". To remedy this, we use a *base of terms* that contains commonly used concepts and acronyms in the medical tourism domain. When the concept does not exist in the *base of terms*, we use a function that allows the calculation of structural similarity to check if the concept is similar to another concept in the matrix or in the *base of terms.*

We have built a base for the frequent terms taking into consideration the terms that do not exist in "*WordNet*" along with the abbreviations and special characters of the tourism domain. Sometimes, we find terms that do not exist neither in "WordNet" nor in the base. We, accordingly, use a structural similarity measurement between each of these terms and the other matrix terms.

- **User behavior analyzing**

The user provides a set of actions that enrich our knowledge about these interests. These actions can be comments (on photos, videos, articles, etc.), publications (such as photo or video sharing), and events of various interests or "I like" signs on the pages or groups. However, we begin by extracting sentences of each document (comments, status, etc.). Then, we apply the same method in the user preference analysis to extract concepts and relations between them. In this step, we use the TF-IDF method to measure the weight (density) of each term in a document. The TF-IDF algorithm is based on term weighting and has the advantage of being easily used with statistics and generic techniques. In our approach, to obtain the interests of the user, we remove from the matrix constructed in the previous step the concepts with a lower number of occurrences of a predetermined threshold. For this, we only get the most interesting concepts for the user.

4.4   Trusted Friends Preferences Determination

After we determine the list of friends trusted by the ego user, we begin by extracting the preference of every friend. We collect the trusted friend data from his/her social profile and behaviors in the network. Then, we analyze the user data to extract the preference of this user. These preferences are extracted by the common component determination method using the Principal Component Analysis procedure [27]. It is a procedure for identifying a smaller number of uncorrelated variables; "principal components" from a larger set of data. We can, accordingly, get the raw data of principal component analysis from the interest degree based on TF-IDF. Indeed, these interests can be useful to know the interests of our user. When we represented a friend's preference as a class in the *user interest ontology*, we indicated the name of the trusted friend and the degree of trust of an ego user relative to another user (level of trust between the ego user and a friend) as two datatypes properties in this OWL class.

4.5   Concept Weight in the User Interest Ontology

A key factor in this ontology is the representation of semantic relations obtained in the step of determining the types of relationships between concepts. In other words, the ontology of user interests consists of a set of concepts and a set of connections using a logic axioms language. Each item of interest in our ontology has a degree of interest (weight) to express the importance of this concept for the user. The degree of interest of a user relative to an item in the ontology is calculated using the number of occurrences and the year duration of the concept from its last apparition in the user's interactions. This weight represents the degree of interest of the concept to the user. Figure 3 show the user interest ontology-based on trusted friend's preferences representation.

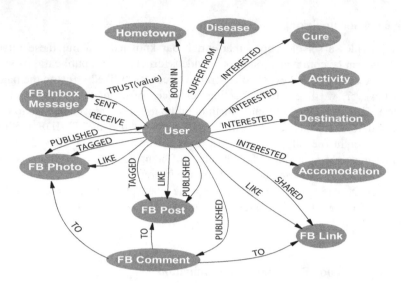

Fig. 3. Partial view of the user interest ontology

5 Social Semantic Recommender System for Tunisian Medical Tourism

We have created a Java-based prototype named Social seMAntic Recommender sys-Tem for Tunisian Medical Tourism (*SMART-TMT*) that used an e-tourism system to recommend items (hotels, restaurants, monuments, and travels) personalized for each user. The system retrieves the most interesting items to users in the Tunisian medical tourism domain. Our application runs an intelligent algorithm [28] to know the tastes and needs of each user and to accordingly recommend customized items. *SMART-TMT* is an e-tourism recommender system for personalized advertising of Tunisian medical tourism, which encapsulated knowledge-based recommender systems to help user find the best advice for his/her health and promote medical tourism in Tunisia. Medical tourism is a domain rich of data, that is why, it is very important to create a referential model that represents the medical tourism in Tunisia (SPA centers and cures, clinics, hotels, thalassotherapy activities, surgical acts for medical tourism, etc.). In our research team, we have created a Tunisian Medical Tourism ontology (*TMT ontology*) that represents all concepts and relations of the medical tourism in Tunisia [29]. The e-tourism ontology provides a way to achieve integration and interoperability through the use of shared vocabulary and meanings for terms with respect to other terms. When the user accesses to the *SMART-TMT*, at first, he/she can connect with his/her Facebook account to pick up the user's preferences and his/her trusted friends' preferences for the creation of the *user interest ontology*. Then, the system will ask him/her to complete a form and answer some questions to know his/her health situation. The main objective is to obtain as much information about his/her health interest as possible with the smallest number of questions. Afterwards, our semantic recommender system will compute the most adequate activity as well as the tourism provider service by investigating into the

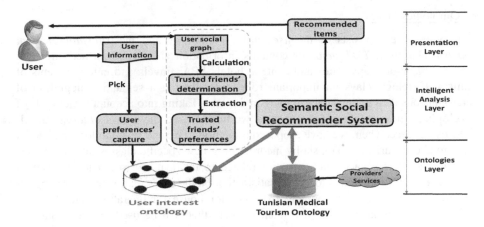

Fig. 4. Functional architecture of *SMART-TMT* system

TMT ontology taking into consideration all information represented in the *user interest ontology*. Figure 4 further illustrates the functioning of our system.

- Presentation layer

The purpose behind a presentation layer is to allow the interaction between the user and the system and to present the recommended items to the user. This layer picks the user's profile when the user answers to some health interest questions and connects to his/her account, gathers data and displays information through the *SMART-TMT*. It acts as the interface for the user to access the system. By using our Tunisian Medical Tourism interface, it enables the user to visit items which are recommended to him/her by the system. The presentation layer mainly consists of Graphical User Interface and the results of the recommendation system.

- Intelligent analysis layer

Intelligent analysis layer is made up of an intelligent analysis module and a social recommendation system. Intelligent analysis module is composed of two modules of user's preferences capture and trust friends' preferences capture. User preference is collected from user's profile, and friends' preference is collected from the user social graph presented by Facebook after the determination of the trust friends. Before capturing the user preferences, we must first analyze the entire user profile, the personalized friends' list; the News Feed; the relations and the "likes" of the user and their trusted friends. All these preferences will be represented in a *user interest ontology* (detailed in the next section). Then the social recommendation system generates recommendations with the help of the *user interest ontology*. By using social recommendation algorithm, our system looks for similar items to this user's interest ontology in the *Tunisian Medical Tourism Ontology*. Our intelligent recommendation system picks out information in agreement with the user's needs from the *TMT Ontology* and produces recommendation set. In this layer, the system can offer a dynamic guidance for the user's choice when he/she desires to travel to Tunisia for medical tourism.

- Ontologies layer

Ontologies layer consists of the *user interest ontology* and the *Tunisian Medical Tourism Ontology*. *TMT ontology* contains information about medical activities and centers, thalassotherapy cures and centers, hotels SPA, travels and entertainments of Tunisia. Ontology plays an important role in facilitating a semantic integration of heterogeneous data in the Tunisian tourism domain. Taking into account the caveats of developing a new ontology from scratch, we have reused the *Travel ontology* developed by *Protégé*. Then, we have proposed a method to extend this ontology and adapt it for medical tourism in Tunisia by incorporating new specific classes into the *Travel ontology* about the places and activities related to medical tourism. The aim behind these types of information is to ensure satisfaction on the users' part when searching for information about medical tourism. It is the storehouse for operation data. In other words, ontologies layer stores and manages data that will be used in our social recommender system.

Presentation layer, intelligent analysis layer and ontologies layer are interconnected to accomplish the function of providing personalized recommendation to the user.

After the user connects to his/her Facebook account, the system will guide the user following four steps. In the first step, the user will precise the medical kind that he/she searches. When user chooses a specific kind (eg. Plastic Esthetic, Dental Cure, Balneo Cure, Thalasso Cure), the system will display some activities in this chosen medical kind. Then the user can indicate the importance of every activity for him/her.

Then, in the second step, the user will set his/her demographic and travel information with the form presented in the next figure. These data include information about the country and town of the user, the location of the accommodation, the kind of the accommodation, the travel budget, and the period of the medical travel.

After that, when the user clicks on the "*Recommend me*" button, we will move to the next step, the system creates the user interest ontology based on these explicit information answered by the user and from the social network profile by adding the user preferences and trusted friends' preferences. Before capturing the user preference, we must first analyze the entire user profile, the personalized friends' list; the relations and the "likes" of the users and their friends. All these preferences will be represented in a user interest ontology [30]. Then the social recommender system generates recommendations with the help of the user interest ontology. By using social recommendation algorithm [28], our system will search for similar items to the user preferences in the *TMT ontology*. These preferences are important to decide which activities to recommend to the user. Recommendations are selected from *TMT ontology* as the most similar to the user interest. That is way some similarity measurement [31] are used for ontologies matching. Our semantic tourism recommender system will calculate similarity between the *user interest ontology* and the *TMT ontology* and will suggest the most similar items to the user. The *SMART-TMT* system will search the most adequate medical activities for the user taking into consideration all information represented in the user interest ontology. Figure 5 shows a screenshot of an example of recommendation presented for the user.

The system can suggest different types of recommendations (eg. Hotel, Clinic, Treatment Center) based on the information completed in the first and second steps. In

Fig. 5. An example of hotels recommendation

Fig. 5, for example, the system has suggested only hotels and treatment centers because the user has chosen *"Thalasso Cure"* in the first step. In this case, the user can choose a hotel among the hotels suggested and then move to the last step to get more details on this item (price, activities, address, position on the map) and book in this hotel if s/he is interested. Otherwise, if the user does not like the recommendations suggested, s/he can select a hotel and click on *"other similar activity"* so that the system will suggest hotels that are similar to the selected one. The user can also select other hotels that are near by the selected one by clicking on *"other activity around"*. Generally, the user can select any recommended activity by clicking on *"more details"* and the system will give information about it. Once the user makes his decision about one of the suggested activities, s/he can reserve it for his/her medical travel to Tunisia.

6   Conclusion and Future Work

In this work, we have demonstrated that a social network-based Web application can be a useful option for the implementation of a social tourism recommender system. Thanks to the tested user feedback, the *SMART-TMT* system has improved during the last year. One of the main aspects that were improved is the interface, always giving information about the whole process of recommendation, taking into consideration the user connection to their Facebook accounts and taking into consideration the trust friends' preferences in the recommended activities process. In this research, we have integrated a user interest ontology in a semantic social recommender system to deal with the lack of semantic information in personalized recommender systems in the tourism domain. We have implemented the system to evaluate the quality of recommendation and proven its importance in improving the medical tourism domain. In our future work, we are going to take into consideration the updating of the user interest ontology to encompass user's interest and trusted friends' preferences changes.

References

1. Rathod, A., Indiramma, M.: A survey of personalized recommendation system with user interest in social network. Int. J. Comput. Sci. Inf. Technol. **6**(1), 413–415 (2015)
2. Correa, C.D., Ma, K.-L.: Visualizing social networks. In: Aggarwal, C. (ed.) Social Network Data Analytics, pp. 307–326. Springer, Boston (2011). doi:10.1007/978-1-4419-8462-3_11
3. Tan, A.-H., Teo, C.: Learning user profiles for personalized information dissemination. In: The 1998 IEEE International Joint Conference on Neural Networks Proceedings, IEEE World Congress on Computational Intelligence, pp. 183–188. IEEE (2010)
4. Sarwar, B., Karypis, G., Konstan, J., Riedl, J.: Item-based collaborative filtering recommendation algorithms. In: Proceedings of the 10th International Conference on World Wide Web, pp. 285–295. ACM (2001)
5. Gauch, S., Speretta, M., Chandramouli, A., Micarelli, A.: User profiles for personalized information access. In: Brusilovsky, P., Kobsa, A., Nejdl, W. (eds.) The Adaptive Web. LNCS, vol. 4321, pp. 54–89. Springer, Heidelberg (2007). doi:10.1007/978-3-540-72079-9_2
6. Sulieman, D.: Towards semantic-social recommender systems. Cergy Pontoise (2014)
7. Staab, S., Werthner, H., Ricci, F., Zipf, A., Gretzel, U., Fesenmaier, D.R., Paris, C., Knoblock, C.: Intelligent systems for tourism. IEEE Intell. Syst. **17**, 53–64 (2002)
8. Borràs, J., Moreno, A., Valls, A.: Intelligent tourism recommender systems: a survey. Expert Syst. Appl. **41**, 7370–7389 (2014)
9. Dewabharata, A., Chou, S.-Y., Samopa, F.: A design of semantic-based recommender system for medical tourism. Jurnal Sistem Informasi **4**(3), 143–156 (2012)
10. Mislove, A., Marcon, M., Gummadi, K.P., Druschel, P., Bhattacharjee, B.: Measurement and analysis of online social networks. In: Proceedings of the 7th ACM SIGCOMM Conference on Internet Measurement, pp. 29–42. ACM (2007)
11. White, L.: Facebook, friends and photos: a snapshot into social networking. In: Tourism Informatics: Visual Travel Recommender Systems, Social Communities, and User Interface Design, pp. 115–129 (2009)
12. Ozsoy, M.G., Polat, F.: Trust based recommendation systems. In: Proceedings of the 2013 IEEE/ACM International Conference on Advances in Social Networks Analysis and Mining, pp. 1267–1274. ACM (2013)
13. Pretschner, A., Gauch, S.: Ontology based personalized search. In: 11th IEEE International Conference on Tools with Artificial Intelligence, Proceedings, pp. 391–398. IEEE (1999)
14. Zhenglian, S., Jun, Y., Haisong, C., Jiaojiao, Z.: Improving the performance of personalized recommendation with ontological user interest model. In: 2011 Seventh International Conference on Computational Intelligence and Security (CIS), pp. 1141–1145. IEEE (2011)
15. Ferraro, P., Lo Re, G.: Designing ontology-driven recommender systems for tourism. In: Gaglio, S., Lo Re, G. (eds.) Advances onto the Internet of Things. AISC, vol. 260, pp. 339–352. Springer, Cham (2014). doi:10.1007/978-3-319-03992-3_24
16. Lewis, J.D., Weigert, A.: Trust as a social reality. Soc. Forces **63**, 967–985 (1985)

17. Mayer, R.C., Davis, J.H., Schoorman, F.D.: An integrative model of organizational trust. Acad. Manag. Rev. **20**, 709–734 (1995)
18. Guo, G., Zhang, J., Thalmann, D., Basu, A., Yorke-Smith, N.: From ratings to trust: an empirical study of implicit trust in recommender systems. In: Proceedings of the 29th Annual ACM Symposium on Applied Computing, pp. 248–253. ACM (2014)
19. Gupta, S., Nagpal, S.: Trust aware recommender systems: a survey on implicit trust generation techniques. Int. J. Comput. Sci. Inf. Technol. **6**(4), 3594–3599 (2015). ISSN 0975–9646
20. Guo, G., Zhang, J., Thalmann, D.: A simple but effective method to incorporate trusted neighbors in recommender systems. In: Masthoff, J., Mobasher, B., Desmarais, M.C., Nkambou, R. (eds.) UMAP 2012. LNCS, vol. 7379, pp. 114–125. Springer, Heidelberg (2012). doi:10.1007/978-3-642-31454-4_10
21. Golbeck, J., Hendler, J.: Filmtrust: movie recommendations using trust in web-based social networks. In: Proceedings of the IEEE Consumer Communications and Networking Conference, pp. 282–286. Citeseer (2006)
22. Haydar, C.: Les systèmes de recommandation à base de confiance. Université de Lorraine (2014)
23. Haydar, C., Boyer, A., Roussanaly, A.: Time-aware trust model for recommender systems. In: International Symposium on Web AlGorithms (2015)
24. Campos, P.G., Díez, F., Cantador, I.: Time-aware recommender systems: a comprehensive survey and analysis of existing evaluation protocols. User Model. User-Adap. Inter. **24**, 67–119 (2014)
25. Podobnik, V., Striga, D., Jandras, A., Lovrek, I.: How to calculate trust between social network users? In: 2012 20th International Conference on Software, Telecommunications and Computer Networks (SoftCOM), pp. 1–6. IEEE (2012)
26. Frikha, M., Mhiri, M., Zarai, M., Gargouri, F.: Time-sensitive trust calculation between social network friends for personalized recommendation. In: Proceedings of the 18th Annual International Conference on Electronic Commerce: e-Commerce in Smart connected World, p. 36. ACM (2016)
27. Wang, X., Yin, F.L., Yang, T.R., Liu, J.B.: The research on broadcast television user interest model based on principal component analysis. In: 2015 IEEE 12th International Conference on Ubiquitous Intelligence and Computing, 2015 IEEE 12th International Conference on Autonomic and Trusted Computing, 2015 IEEE 15th International Conference on Scalable Computing and Communications and Its Associated Workshops (UIC-ATC-ScalCom), pp. 578–581. IEEE (2015)
28. Frikha, M., Mhiri, M., Gargouri, F.: A semantic social recommender system using ontologies based approach for Tunisian tourism. Adv. Distrib. Comput. Artif. Intell. J. (ADCAIJ) **4**, 90–106 (2015)
29. Frikha, M., Mhiri, M., Zarai, M., Gargouri, F.: Using TMT ontology in trust based medical tourism recommender system. In: International Conference on Computer Systems and Applications (AICCSA) (2016)
30. Frikha, M., Mhiri, M., Gargouri, F.: Toward a user interest ontology to improve social network-based recommender system. In: Sobecki, J., Boonjing, V., Chittayasothorn, S. (eds.) Advanced Approaches to Intelligent Information and Database Systems. SCI, vol. 551, pp. 255–264. Springer, Cham (2014). doi:10.1007/978-3-319-05503-9_25
31. Frikha, M., Mhiri, M., Gargouri, F.: Extraction of semantic relationships starting from similarity measurements. In: International Conference on Enterprise Information Systems (ICEIS), pp. 602–606 (2007)

A Tool for Managing the Evolution of Enterprise Architecture Meta-model and Models

Tiago Rechau[(⊠)], Nuno Silva, Miguel Mira da Silva, and Pedro Sousa

Instituto Superior Tecnico, Av. Rovisco Pais 1, 1049-001 Lisboa, Portugal
{tiago.rechau,nuno.miguel,mms,pedro.manuel.sousa}@tecnico.ulisboa.pt

Abstract. Enterprise Architecture advocates the use of models to support decisions about organizational changes aligned with the organization's strategy and vision. However, few existing tools support change impact analysis and automated model updates. This paper proposes a software tool, that supports co-evolution of the enterprise architecture meta-model and conforming model as well as prior change impact analysis of the model. The proposal focuses on the architecture and user interface of the software tool that is now being developed.

Keywords: Enterprise architecture · EA tool · EA meta-model evolution · Co-evolution

1 Introduction

Organizations, whether operating in public or private sectors, are exposed to frequent changes in their environment driven by technology advancement, demanding customers, aggressive competitors, growing dependency on information technology as well as regulatory changes. In order to remain competitive, organizations need to adapt to swiftly changing their business strategy and/or their business processes. To do so, organizations need to have a holistic view over the impact of such changes.

In recent years, enterprise architecture (EA) has gain relevance for its holistic management of information systems in an organization. Fundamentally, EA describes a coherent whole of principles, methods, and models that are used in the design and realization of an enterprise's organizational structure, business processes, information systems, and infrastructure [11]. EA is used as a means to design and communicate the desired organizational changes according to the business strategy, and to implement these changes across the operational structures, processes, and systems of the organization's business and IT domains [18].

A considerable part of EA is model-based, in the sense that diagrammatic descriptions of the systems and their environment constitute the core of the approach. Therefore, EA models are used to represent the enterprise's domains. All EA models must be compliant with the EA meta-model that specifies the elements and relationships used to structure and represent the enterprise in all its domains [6].

M. Themistocleous and V. Morabito (Eds.): EMCIS 2017, LNBIP 299, pp. 68–81, 2017.
DOI: 10.1007/978-3-319-65930-5_6

Co-evolution denotes the process of adapting models as a consequence of meta-model evolution [15,16]. As the organization changes, so must its architecture. Therefore, to keep updated architectural views of the organization is necessary to co-evolve the EA meta-model and respective models.

Existing EA software tools do not fully address the co-evolution of the EA meta-model and models, thus making the task of maintaining a consistent and updated EA repository strenuous and error-prone. For instance, if one seeks to employ evolution-specific scenarios to the EA meta-model focusing on the organization's goals, the information captured by the EA models at that time becomes inconsistent with the new EA meta-model. Without an automated support, the effort of updating and resolving all model inconsistencies manually is unsustainable and impractical when dealing with complex changes.

Furthermore, once the evolution scenarios are specified, information relative to the impact of changes in the EA models should be regarded. EA analysis, being the application of property assessment criteria on EA models, helps to understand which architectural scenarios are most appropriate towards realizing the organization's strategy. For instance, one might see fit to assess semantic inconsistencies of a specific evolution scenario and a criterion for assessment of this property could be "If altering a relationship between two EA meta-model elements creates a semantic inconsistency in a single view, then the scenario is impractical". If concerned with the impact of manual effort in updating the architectural views, a criterion for assessment of this property could be "If more manual refactoring than expected is required in updating all views, the evolution scenario should be disregarded as viable". Criteria and properties such as these may be extracted from academic literature or from empirical measurements.

This paper presents a software tool for co-evolution and changes impact analysis of the EA meta-model and models. The tool supports the creation of evolution scenarios and generates quantitative assessments of such scenarios. Assessments can be of various quality attributes, such as dependency impact, performance, semantic inconsistencies and manual refactoring effort.

The outline of this paper is as follows. Section 2 specifies the research problem. Section 3 presents a state-of-the-art review on the problem domain. Then, Sect. 4 describes the architecture, design, and usage of the tool. Section 5 presents a discussion about the usage of the too and Sect. 6 concludes the paper with a summary and themes for future work.

2   Research Problem

Managing EA change requires awareness over what has been changed and the compliance of those changes with the rules in which EA models can be updated. Hence, robust change management processes and procedures are essential in maintaining the architecture of the enterprise [10].

IT projects define and implement the strategies that guide the enterprise in its evolution [24]. An issue that typically compromises the success rate of IT projects is the complexity that EA models may bring. EA models explicit traceability

across multi-layered elements representing the various levels of the enterprise
[24] reinforces their tendency to focus on holistic views of the enterprise, rather
than on the subset of artifacts relevant for a given project, thus making the task
of reading and updating these models more complex [22].

Furthermore, many of today's enterprises struggle with transformation management from a baseline architecture to a target architecture via intermediary planned architectures [7]. Several causes were identified by Aier et al. [5]
regarding this issue, such as missing practical methodologies for architecture
road-mapping, inadequate representation of the concept of time in architectural
models and insufficient tool support for architecture planning.

More than just providing EA model visualizations, EA software tools are used
when making strategic decisions concerning the enterprise [20]. Despite the multitude of available tools on the market (Troux, Aris, Enterprise Architect, System
Architect, ABACUS, etc.), few support change impact analysis and automated
model updates while maintaining a well-integrated EA repository containing the
past, present, and future states of the architecture [13,14,19].

3   Related Work

This section presents a literature review of research domain. Section 3.1 outlines
research made on the topic of EA evolution. Then, Sect. 3.2 highlights existing
methods for meta-model evolution from both the fields of Model Driven Engineering (MDE) and Enterprise Architecture (EA). Finally, Sect. 3.3 describes
some approaches on EA model change visualization.

3.1   EA Evolution

Evolving an EA is an iterative and incremental process. The transition architecture is a state of the EA between the baseline and target architectures. An
enterprise might experience more than one transition architectures during the
evolution process. A road-map is the abstracted plan for the business or technology change, typically operating across various disciplines and over multiple
years [4]. Architecture road-maps are used to describe the transformation path,
over a certain period of time, from the baseline architecture to the desired target architecture. Thus, a time-line view is necessary for describing or visualizing
the architecture road-map, thus showing the required activities needed to be
performed to realize the target architecture.

Besides EA road-maps, other approaches have focused specifically on updating the EA meta-model and respective models. Roth and Matthes [19] proposed
a four-layered conceptual design of an interactive visualization to drill down and
analyze model differences in both the EA meta-model and EA models.

Sousa et al. consider the evolution of EA models as a means to address
change within the organization's domain [22]. In their approach, an enterprise is
perceived as a graph G of artifacts and their relationships that can be represented
and altered by two fundamental types of the enterprise type space:

- **Blueprint:** "whose instances contain references to other artefacts. A given *artefact* is represented on a given blueprint if graph G holds as a relation between them";
- **Project:** "whose instances contain references to artefacts related with the project".

A state of existence of all artefacts is also defined for all artefacts representing the EA other than *Blueprint* as one of the following states:

- **Conceived:** "if it is only related with *blueprints*";
- **Gestation:** "if it is related with alive *projects* and is not related with any other artefact other than *blueprints*";
- **Alive:** "if it is related with other artefacts in the alive state. This means that it may act upon other artefacts in conceived, gestation or alive states";
- **Dead:** "if it is no longer in the alive state".

Sousa et al. consider the relevance of IT project and life-cycle as two important concepts concerning EA evolution. Nonetheless, their approach lacks a more detailed rationale on how the life-cycle of EA model elements changes, i.e., which transformations were applied to the elements.

3.2 Meta-model Evolution

In MDE, meta-models create a semantic and symbolic manner that has to be respected when representing the models. Sometimes models can not be modeled based on the actual meta-model, thus having the need to co-evolve together with the meta-model in order to maintain consistency. Some changes in meta-model are additive changes, and this away, they don't break the respective models, but other, introduce incompatibilities and cross-version inconsistencies, therefore invalidating models. Therefore, models must be adapted to the new meta-model through migrations, as stated by Mantz et al. [12].

Therefore, for Mantz et al. [12], co-evolution of meta-model and respective models has to respect the following requirements:

1. Migrated models must belong to the evolved modeling language. This property is usually called soundness. It subdivides into well-typedness and well-definedness with regard to language constraints. A migrated model is well-typed if all its elements are typed over the evolved meta-model. Moreover, all well-definedness rules of the evolved language have to be satisfied.
2. All models of the original modeling language can be migrated to the evolved language meaning that the migration is viable. This property is usually referred to as completeness.
3. Model migration should be specified on a high abstraction level. This means that a model migration is either automatically deduced from its meta-model evolution or specified using a high-level language.
4. The specification of model migrations is reusable (see also [1]). In particular, equivalent migration steps are specified only once.

5. General strategies for model co-evolution are formulated independently of a specific meta-modeling approach.

In order to perform co-evolution of meta-model and respective models, a process and a set of migration rules need to be defined. The following sections describe a process applied to Model Driven Development (MDD) by Gruschko et al. [8] and a set of migration rules defined for EA by Silva et al. [21].

Model Driven Development Meta-model Evolution. Gruschko et al. [8] describe an approach to address the model migration problem, implementing transformations to migrate models when meta-models are changed. Their approach is based on the Meta Object Facility (MOF) meta-modeling architecture. MOF introduce a four layer division, where each layer represents a different level of abstraction. The first layer, M3, represents the meta-meta-models, that are used to define M2, the second layer. M2 describes instances models that are instances of M3. M2 models represent meta-model. The third layer, M1, describes instances of M2 and finally, M0 represents instances of M1.

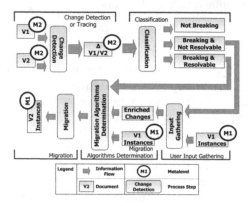

Fig. 1. Example of M1 migration when changes in M2 happens (reproduced from [8])

The focus of their approach is the migration of M1 models when M2 changes and is defined as a 5 step method (see Fig. 1):

1. Change detection or tracking;
2. Classification;
3. User input gathering;
4. Algorithms determination;
5. Migration.

The first step (*Change detection or tracking*) is comprised of two options. The first option is the direct comparison between two different versions of M2 models, in which the output will be the differences between them. On the other hand, the second option is tracing the differences, i.e., taking an older version of the M2 model and a tracing of change operations as input. This trace of changes operations represent the operations that allow the older version of an M2 model to get to the desired M2 model version.

The second step, *Classification*, refers to the classification of the detect changes in step 1. The changes can be classified into 3 groups:

1. Not breaking changes;
2. Breaking and resolvable changes;
3. Breaking and unresolvable changes.

Gruschko argues that the first group is not relevant since these changes do not change the instances of M2 models (M1), thus no migration is needed. The second group is the group of changes that can be propagated to M1 models without user input. Finally, the third group is a group of changes that although they can be propagated to M1 models, some user input to execute the propagation is needed.

The third step (*User input gathering*) execution depends on the type of changes. If there isn't any change of "Breaking and unresolvable changes" type, this step is not required.

The four step, *Algorithms Determination*, will generate changes necessary to do the migration, i.e., generate a list of necessary steps to turn the old M1 models into new M1 models aligned with the new version of the M2 model.

The fifth and last step is the *migration* of the M1 model itself.

Enterprise Architecture Meta-model and Model Co-evolution. Silva et al. [21] proposed, based on Wachsmuth research [23], a set of migration rules applied to the EA meta-model and respective models:

- Co-construction
  - Introduce EA Property
- Co-destruction
  - Eliminate EA Class
  - Eliminate EA Property
- Co-refactoring
  - Change EA Object Class
  - Change EA Property Type
  - Move EA Property
  - Inline EA Class
  - Association to EA Class
  - EA Class To Association

Every migration rule corresponds to a modification on the EA meta-model that needs to be propagated to the respective models. To do this propagation, every transformation has some properties, parameters, pre-conditions, post-conditions, and statements. The parameters are all the information needed to propagate the transformation. The pre-conditions are conditions that need to be met before the propagation is made. The post-conditions are the conditions that need to be met to guarantee that the propagation was well done. The statements are the required propagation steps.

When applied, the migration rules follow Sousa et al. life-cycle approach in which all artifacts defining both the EA meta-model and respective models had a life-cycle property composed of four states: `gestating`, `alive`, `dead`, and `retired`. So, when applying any co-destruction transformation no artifacts are removed, instead their life-cycle state changes to either `dead` or `retired`. This principle allows for a history of changes made to the EA meta-model and its models.

3.3   EA Model Changes Visualization

Buckl et al. identified the challenge in the visualization of the development of business support provided by the application landscape over time [7]. In their approach both a Gantt chart inspired graphical viewpoint for supporting EA transformation documentation and a conceptual model explaining the information demands that need to be satisfied with the creation of road-map plans were introduced. Their approach allows the analysis and visualization of the business application's life-cycle given an IT project. Nonetheless, there is no evidence of the viewpoint's expressiveness in representing both the drivers motivating each transformation as well as the nature of each transformation.

Ross et al. emphasize on the transformation procedure also from a visual perspective [17,18]. A high-level maturity model for EAs was proposed providing starting points for the design of the transformation process to enhance the used EA management process. However, neither an information model nor visualizations supporting the transformation process are discussed in their approach.

Roth and Matthes [19] proposed visualization is based on 4 layers. In the first layer, changes to the meta-model are shown in a graph where:

- New class is shown in green;
- Altered class is shown in orange;
- Delete class shown in red;

The changes on names of class tint the differences on the name with green to the added parts, and red to the deleted parts. Changes in relations between classes are also shown and follow the same logic, green to new relations, red to deleted relations and changed ones with orange.

On the second layer an overview of objects (meta-model instances) following the previous logic is showed. New instances in green, changed ones in orange and deleted ones in red. Besides the visualization is possible to filter and zoom in/out to facilitate the visualization.

The third layer shows the instance neighborhood, i.e., the neighbors of a chosen object, but focusing on relations of Objects instead of on attributes, because the expert states that links (instances of relationships) between objects are far more interesting for an analysis than changes of attributes [19]. The color code is the same as on other layers.

On the fourth and last layer, the user can choose an object and view the different versions of that object, thus comparing the different versions with the original one.

4 Research Proposal

This section describes the proposed under development software tool. The following requirements were identified from a set of interviews with practitioners:

1. Necessity of saving the motivation of a change;
2. Migration of the models;
3. Describe the impact of changes;
4. Rollback of changes made to the meta-model;

By considering the requirements above together with existing literature, the tool aims at achieving three main objectives:

1. EA practitioner support on EA co-evolution by means of a simple and interactive EA meta-model editor and visualizer;
2. EA practitioner support on EA co-evolution by providing change impact analysis features of specific model changes;
3. EA practitioner support on EA co-evolution by automated model migration.

4.1 Architecture of the Tool

In accordance with the objectives stated above, the tool's architecture must be grounded in a set of concepts and relationships that define the evolution aspects of an EA. Figure 2 illustrates the conceptual model describing the architecture of the proposed tool.

The remainder of this section presents the different elements of the conceptual model. An IT Project acts as the enabler of EA evolution by applying a set of Transformations that transform the "AS-IS" state of an EA description into a "TO-BE" state. Each transformation can be decomposed into a set of Changes, each one altering the Life-Cycle of an EA Description Element (either concept, property or relationship).

Howbeit these concepts expressing the process of EA change, the motivation and interested parties in the evolution process must also be considered. Therefore, the concepts of Driver and Stakeholder are introduced. A Driver "represents an external or internal condition that motivates an organization to define its goals and implement the changes necessary to achieve them" [1]. One or more Stakeholders, defined as in ISO 42010:2011 [9], can have one or more drivers.

Also, the tool's co-evolution process is based on the process defined by Gruschko et al. [8], although in this case, the authors consider only a baseline version of the meta-model since all changes made to the meta-model are traced. The tool will support the model migration rules proposed by Silva et al. [21].

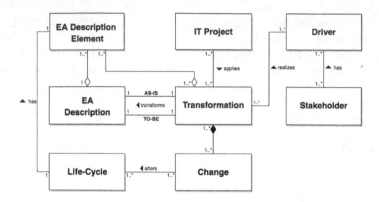

Fig. 2. Conceptual model of the tool

4.2   Design of the Tool

Tool Workflow. The design of the tool is based on a 7-step workflow as Fig. 3 illustrates.

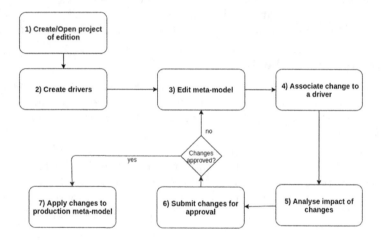

Fig. 3. Tool usage work-flow

Create/Open Project of Edition. First, the user creates or opens a saved project for meta-model edition. Only inside the project scope is possible to edit the EA meta-model. This concept of project is mapped into the concept *IT Project* of the conceptual model of the tool.

Create Driver. In order to edit the meta-model, from both retrieved practitioners and researchers data, the reason behind a specific change must exist. To do this the concept of *Driver* has been created and associated with one or more changes made to the EA meta-model.

Edit Meta-model. Editing the meta-model is mostly applying a set of predefined transformations to the meta-model. The edition can be done in an interactive way where the user can interact with the graphical representation of the meta-model.

Associate Transformation to Driver. A transformation to driver association must also exist, explaining the motivation behind a meta-model change. The association could be made while adding a transformation to the meta-model or after the transformation has been added.

Analyse Impact of Transformation. In this step, the user is presented with analysis-specific features regarding the change impact of a meta-model change to the respective models. The analysis is based on indicators of incoherences and problems with relations between classes, supporting the user in the decision-making task of which evolution scenarios would be preferred. The analysis can be specific to a list of changes, transformations, drivers or projects.

Submit Changes for Approval. The co-evolution must be approved before updating the organization's EA repository. Therefore, the tool also allows the user to submit the evolution scenario for approval.

Apply Changes to the EA Repository. After the changes have been approved they must be implemented, i.e., be applied to the current version of the EA repository. In this step, the user selects an approved project and applies the editions made to the EA repository. In this step the modifications are also associated with life-cycle dates, i.e., if the user removes a class, on February 23, 2017, and applies that change to the EA repository, the life-cycle state changes to *Dead* dating February 23, 2017. Is also at this step that the migration of the models is automatically done following the migration rules defined by Silva et al. [21].

Tool Structure. In order to support the process specified above, the tool's (see Fig. 4) interface is structured in three items: a tree-like data container with the associated drivers and transformations, EA meta-model visualizer and the transformation window.

The user can see which drivers and transformations were applied to the meta-model and also analyze the change impact of applying the transformations (and changes) to both the meta-model and models. This feature allows the user to filter for transformations that he or she wishes to see. Also, the future state of the EA meta-model, i.e., the new meta-model version after applying the changes can be seen on the meta-model visualizer with colors representing what changes were applied to a specific meta-model concept or relation type following Roth and Matthes approach [19].

Besides the graphical representation of the meta-model, presented by the meta-model visualizer, one can choose how he or she wants to view the meta-model, i.e., view "AS-IS" and "TO-BE" side by side, or as an integrated view

Metamodel Editor

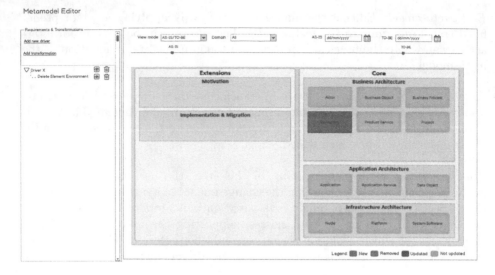

Fig. 4. Tool's interface mock-up

in which both the "AS-IS" and "TO-BE" are incorporated into a single view. Another view option is the *Domain*, an option that works as a filter in which the user can choose, from the existent EA meta-model domains, the type of architecture he wants to focus the visualization on, thus, allowing either a holistic or domain-specific view of the EA meta-model.

An AS-IS/TO-BE date options, as well as the AS-IS/TO-BE bar, work as a time navigation feature, allowing the user to navigate to a previous version of the EA meta-model and observe which changes were made between the two chosen dates.

The transformation screen allows the user to (1) fulfill the properties of a transformation, (2) apply the transformation to the EA meta-model, and (3) observe the change impact of such transformation on the models. To view the properties screen the user has two options:

1. An interactive view in which the user goes to the meta-model visualizer and right-clicks an element or architectural domain to identify the applicable transformations;
2. Via the *Add transformation* button option.

The first option allows the user to interact with the model and also filters all the possible transformations. The second one is more conventional, where the user has to fill in both the element and the transformation to be applied to that element. Both options navigate the user to the same screen where he or she has to fill some properties (for example, the association with a driver or creation of a new one at that moment, as well as seeing the change impact of that transformation).

Concerning the change impact analysis, it covers quantitative aspects as number/percentage of instances impacted, number/percentage of relations between classes lost in a depth chosen by the user.

5 Discussion

To better understand the tool's usage, the following use case was considered. Company ABC is a large company that relies on proper and updated use of EA as means of supporting the alignment between its business operations and IT infrastructure. In last years ABC had its EA supported by ArchiMate 2.1 meta-model [1], but with the appearance of ArchiMate 3.0, ABC wants to migrate its EA, in order to have its model valid with the latest version of ArchiMate.

In order to make this update of the meta-model, ABC needs to know what makes an Archimate 2.1 model invalid in ArchiMate 3.0 [2] and know how to change the meta-model. By following the information provided by The Open Group [3], the next transformations are required:

1. Rename "used by" relationships to "serving";
2. Change relationship of type assignment of an application component to a business process or function, this may be replaced by a realization relationship from the application component to the business process or function;
3. If there is an assignment of a location to another element, this may be replaced by an aggregation;
4. If other relationships between two elements in a model are no longer permitted according to with the new meta-model change the relationship with an association.

Taking the above into account and to ensure meta-model and model conformance, ABC needs to edit its meta-model and migrate the model. The interaction's workflow would be the following:

1. Create a project of edition of the repository that ABC wants to change;
2. Create a driver, for example, "Migration to ArchiMate 3.0";
3. Edit the meta-model doing the before mentioned transformations associating them with the driver created before;
4. After all transformations are done apply the changes to the repository.

Once the desired change is applied all models are migrated, making ABC's EA model valid according to ArchiMate 3.0.

6 Conclusion and Future Work

In this paper, a tool for managing the co-evolution of the EA meta-model and models was presented. Change impact analysis of each evolution scenario is also an important feature of the tool in order to support decision-making when choosing the most appropriate scenario according to the organization's goals. To fulfill its purpose the tool consists of two separate parts: one defining the co-evolution scenario and another for performing analysis over the created scenario. Both parts of the tool support the co-evolution process comprised of the creation and analysis of co-evolving both the EA meta-model and models as described in Sect. 4.

Being an ongoing research, a final and fully functional version of the tool is currently being developed. The authors consider as future effort the possibility of implementing an interactive help feature to offer the user proper guidance and more visual refinements regarding the change impact analysis, specifically in the changes to the existing EA repository data and views.

References

1. Archimate® 2.1 specification. http://pubs.opengroup.org/architecture/archimate2-doc/. Accessed 20 May 2017
2. Archimate® 3.0 specification. http://pubs.opengroup.org/architecture/archimate3-doc/. Accessed 20 May 2017
3. Changes from archimate 2.1 to archimate 3.0 (informative). http://pubs.opengroup.org/architecture/archimate3-doc/apdxe.html. Accessed 20 May 2017
4. Togaf version 9.1. http://pubs.opengroup.org/architecture/togaf9-doc/arch. Accessed 14 Nov 2016
5. Aier, S., Gleichauf, B., Saat, J., Winter, R.: Complexity levels of representing dynamics in EA planning. In: Albani, A., Barjis, J., Dietz, J.L.G. (eds.) CIAO!/EOMAS-2009. LNBIP, vol. 34, pp. 55–69. Springer, Heidelberg (2009). doi:10.1007/978-3-642-01915-9_5
6. Buckl, S., Ernst, A.M., Lankes, J., Schneider, K., Schweda, C.M.: A Pattern Based Approach for Constructing Enterprise Architecture Management Information Models, p. 65 (2007)
7. Buckl, S., Ernst, A.M., Matthes, F., Schweda, C.M.: Visual roadmaps for managed enterprise architecture evolution. In: 10th ACIS International Conference on Software Engineering, Artificial Intelligences, Networking and Parallel/Distributed Computing, pp. 352–357. IEEE (2009)
8. Gruschko, B., Kolovos, D., Paige, R.: Towards synchronizing models with evolving metamodels. In: Proceedings of the International Workshop on Model-Driven Software Evolution (2007)
9. ISO/IEC/IEEE: Systems and software engineering - architecture description. ISO/IEC/IEEE 42010:2011(E), pp. 1–46 (2011)
10. Kaisler, S.H., Armour, F., Valivullah, M.: Enterprise architecting: critical problems. In: Proceedings of the 38th Annual Hawaii International Conference on System Sciences, p. 224b. IEEE (2005)
11. Lankhorst, M.: Enterprise Architecture at Work: Modelling, Communication and Analysis. The Enterprise Engineering Series. Springer, Heidelberg (2013). doi:10.1007/978-3-642-29651-2
12. Mantz, F., Taentzer, G., Lamo, Y., Wolter, U.: Co-evolving meta-models and their instance models: a formal approach based on graph transformation. Sci. Comput. Program. **104**, 2–43 (2015)
13. Matthes, F., Hauder, M., Katinszky, N.: Enterprise Architecture Management Tool Survey 2014 (2014)

14. Matthes, F., Buckl, S., Leitel, J., Schweda, C.M.: Enterprise Architecture Management Tool Survey 2008, vol. 19 (2008)
15. Mens, T., Wermelinger, M., Ducasse, S., Demeyer, S., Hirschfeld, R., Jazayeri, M.: Challenges in software evolution. In: Eighth International Workshop on Principles of Software Evolution, pp. 13–22. IEEE (2005)
16. Rose, L.M., Kolovos, D.S., Paige, R.F., Polack, F.A.: Enhanced automation for managing model and metamodel inconsistency. In: 24th IEEE/ACM International Conference on Automated Software Engineering, pp. 545–549. IEEE (2009)
17. Ross, J.W.: Creating a strategic it architecture competency: Learning in stages. MIS Q. Exec. **2**(1), 31–43 (2003)
18. Ross, J.W., Weill, P., Robertson, D.C.: Enterprise Architecture as Strategy: Creating a Foundation for Business Execution. Harvard Business Press, Brighton (2006)
19. Roth, S., Matthes, F.: Visualizing differences of enterprise architecture models. In: Proceedings International Workshop on Comparison and Versioning of Software Models at Software Engineering (2014)
20. Schaub, M., Matthes, F., Roth, S.: Towards a conceptual framework for interactive enterprise architecture management visualizations. In: Modellierung, pp. 75–90. Citeseer (2012)
21. Silva, N., Ferreira, F., Sousa, P., da Silva, M.M.: Automating the migration of enterprise architecture models. Int. J. Inf. Syst. Model. Des. (IJISMD) **7**(2), 72–90 (2016)
22. Sousa, P., Lima, J., Sampaio, A., Pereira, C.: An approach for creating and managing enterprise blueprints: a case for IT blueprints. In: Albani, A., Barjis, J., Dietz, J.L.G. (eds.) CIAO!/EOMAS -2009. LNBIP, vol. 34, pp. 70–84. Springer, Heidelberg (2009). doi:10.1007/978-3-642-01915-9 6
23. Wachsmuth, G.: Metamodel adaptation and model co-adaptation. In: Ernst, E. (ed.) ECOOP 2007. LNCS, vol. 4609, pp. 600–624. Springer, Heidelberg (2007). doi:10.1007/978-3-540-73589-2_28
24. Wegmann, A.: The systemic enterprise architecture methodology (SEAM). Business and IT alignment for competitiveness. In: International Conference on Enterprise Information Systems (ICEIS), pp. 483–490 (2003)

A Comparative Study in Data Mining: Clustering and Classification Capabilities

Argyro Mavrogiorgou$^{(\boxtimes)}$, Athanasios Kiourtis,
Dimosthenis Kyriazis, and Marinos Themistocleous

Department of Digital Systems, University of Piraeus, Piraeus, Greece
{margy, kiourtis, dimos, mthemist}@unipi.gr

Abstract. The ICT evolution has driven on the creation of a capable society, in providing new kinds and type of information. The gathered information is stored continuously, meaning that a great amount of databases has to be created. The problem that arises is whether there is a global manner of managing and gaining knowledge out of the rising variety and volumes of data. Many efforts have been developed for addressing the emerging challenges of data mining based on statistics and machine learning techniques that can significantly boost the ability to analyze data. In this paper, a detailed study on the data mining field takes place, followed by a comparative study between clustering and classification techniques, resulting that the integration of clustering and classification techniques can provide more accurate results than a simple classification technique that classifies datasets with priorly known attributes and classes.

Keywords: Data mining · Classification · Clustering

1 Introduction

The evolution of information and telecommunication systems, has driven on the creation of a capable society, in providing new kinds and type of information. The gathered information is stored continuously, and as a result a great amount of databases needs to be created. The aforementioned fact is a modern phenomenon, which is observed as a need from the simplest to more complex issues of the people's daily lives. Therefore, the problem that arises is whether there is a manner of managing all these tons of data which are continuously up-to-date, or not. Moreover, in combination with this, it seems to be very difficult to derive the necessary information from all this data. All these important issues have driven the computer science into the creation of the data mining field, which includes a series of techniques based on various algorithms, in order to produce and classify useful insights and information for future or even immediate use.

It is an undeniable fact, that all the produced data is multiplied each year, with the disadvantage that the useful part of this data is lost and reduced, between the data's big volume. The field of data mining and knowledge discovery not only from databases but also from other various data sources, expands rapidly in the computer science field. More specifically, data mining has become necessary because of the need for the creation of techniques and tools, which will aid in the analysis and interpretation of the

© Springer International Publishing AG 2017
M. Themistocleous and V. Morabito (Eds.): EMCIS 2017, LNBIP 299, pp. 82–96, 2017.
DOI: 10.1007/978-3-319-65930-5_7

quickly increasing data. This data explosion is the result of a continuous use of computers and information systems.

For that reason, in this paper we are going to describe all the methodologies, techniques and algorithms which are daily used in order to extract and predict useful either structured or unstructured data insights. It is worth mentioning that the challenge we have to face has to do with today's data, which as mentioned before, is continuously increasing day-by-day. In particular, by taking into consideration this challenge, we will investigate and study various ways for classifying as well as clustering this data, with the same speed as its continuous development.

This paper is organized as follows. Section 2 describes the study of the state of the art and the related work regarding data mining, clustering, and classification. Section 3 compares the clustering and classification techniques according to a specific dataset, while Sect. 4 is addressing the challenges of data mining, analyzing our conclusions and future plans.

2 Related Work

2.1 Data Mining

The capabilities of both generating and collecting data have been increasing rapidly. Millions of databases have been used in business management, government administration, scientific and engineering data management, and many other applications. It is noted that the number of such databases keeps growing rapidly because of the availability of powerful and affordable database systems [1, 2]. This explosive growth in data and databases has generated an urgent need for new techniques and tools that can intelligently and automatically transform the processed data into useful information and knowledge. One of these techniques, called data mining, has become a research area with increasing importance [3–5]. In more details, data mining is an essential step in the knowledge discovery in databases (KDD) process that produces useful patterns or models from data (Fig. 1) [1]. However, the terms of KDD and data mining are different. KDD refers to the overall process of discovering useful knowledge from data,

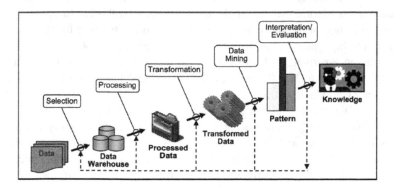

Fig. 1. Data mining and the KDD process

while data mining refers to ability to discover new patterns from a wealth of data in databases by focusing on the algorithms to extract useful knowledge [1]. Therefore, data mining is a process of discovering patterns and extracting useful information from large datasets combined with techniques of artificial intelligence, machine learning, statistics, and database systems, using either supervised or unsupervised techniques [6, 7], such as pattern recognition, clustering, association, regression, as well as classification [7].

The overall goal of a data mining process is to extract information from a database/dataset and transform it into an understandable structure for further usage. Mining information and knowledge from large databases has been recognized by many people as a key research topic in database systems and machine learning, and by many industrial companies as an important area with an opportunity of major revenues [8–11]. The discovered knowledge can be applied to information management, query processing, decision-making, process control, and numerous other applications. Furthermore, several emerging applications in information providing services, such as online services and World Wide Web, also require the existence of various data mining techniques to better understand users' behaviors, meliorate the provided services, and increase the businesses' opportunities.

As a result, one important characteristic of the data mining process is that it can use data that has been gathered for many years and transform it in valuable knowledge. The larger the database gets the better and the more accurate the knowledge becomes. An important observation is that when we refer to the use of large amounts of information in data mining, we are not always referring to usual large Database Management Systems (DBMS), which can reach terabytes of data [12, 13]. On the contrary, due to the fact that currently there exist numerous different types of databases, data mining can be applicable to numerous types of information repositories, implementing in each case the required algorithms and approaches. For example, data mining can be put into usage in different kinds of databases, such as relational, object-oriented, transaction, and NoSQL databases, in data warehouses, in unstructured and semi-structured repositories such as the World Wide Web, in advanced databases such as spatial, multimedia, time-series, as well as textual databases, and even in flat files. Some of these examples are being described more comprehensively below [14, 15, 30–33]:

- *Flat Files*: Flat files have been largely used as the data source of information used by data mining algorithms, especially at the research level. Flat files are simple data files either in text or in binary format, having a known structure for the data mining algorithm to be applied. Their format consists of rows and columns, assuming that every item in a particular file consists of the same data. One common example of this type of files is a CSV (Comma Separated Values) file, as well as a spreadsheet such as a XLS (MS Excel).
- *Relational Databases*: Relational databases consist of a set of tables containing either values of entity attributes, or values of attributes from entity relationships. Tables have columns and rows, where columns represent attributes and rows represent tuples. Thus, the data can be related to other data in the same table or other tables that have to be correctly managed, can be joined with one or more tables. There have been created many different query languages that are used in the

relational databases, of whom the most commonly used is Structured Query Language (SQL), which allows the retrieval and manipulation of the data stored in the tables, as well as the calculation of aggregated functions such as average, sum, min, max, etc. Relational databases include Oracle, MS SQLServer, IBM DB2, MySQL, SQLite and PostgreSQL among others.

- *Object-oriented Databases*: In object-oriented databases, the object and its data or attributes are being seen as one, and being accessed through pointers rather than being stored in relational table models. Object-oriented databases consist of diverse structures and are quite extensible. These databases were mainly designed to work closely with programs built with Object-oriented programming languages, thereby almost making the data and the program operate as one. With these databases, applications can treat the data as native code. There is little commercial implementation of these databases as it is still developing, however some examples include the IBM DB4o and the DTS/S1 from Obsidian Dynamics.

- *Transaction Databases*: A transaction database is a set of records representing transactions, each with a timestamp, an identifier and a set of items. Associated with the transaction files, transaction databases could also be descriptive for different items' data. Since relational databases do not allow nested tables, transactions are usually stored either in flat files or in two (2) normalized transactional tables, one for the transactions and one for the transaction items. A transaction database is usually hugely important to organizations as they include the customers' database, the personal database and the inventory database.

- *NoSQL Databases*: NoSQL databases are non-relational and largely distributed databases that enable rapid, ad-hoc organization and analysis of extremely high-volume, distributed data types. They describe an approach to database design that implements a key-value store, a document store, a column store or a graph format for data. NoSQL contrasts to databases that adhere to SQL's relational methods, where data are placed in tables, and data schema are carefully designed before the database is built. Examples of NoSQL databases include Cassandra, Hypertable, Accumulo, MongoDB, and Neo4 J.

- *Data Warehouses*: Data warehouses are repositories of data that are being collected from multiple (often heterogeneous) data sources, allowing different kinds of data analysis, which will be used as a whole under the same unified schema. Therefore, a data warehouse gives the option to analyze data from different sources that co-exist in the same location. To facilitate decision-making and multi-dimensional views, data warehouses are usually modeled by a multi-dimensional data structure. Common examples of data warehouses are IBM InfoSphere DataStage, SAP Sybase IQ, and WhereScape.

- *Multimedia Databases*: Multimedia databases include video, images, audio, and text media. They can be stored on extended object-relational or object-oriented databases, or simply on a file system. Multimedia is characterized by its high dimensionality, which makes data mining even more challenging. Data mining from multimedia repositories may require computer vision, computer graphics, image interpretation, as well as natural language processing methodologies.

- *Spatial Databases*: Spatial databases are databases that, in addition to usual data, store geographical information like maps, and global or regional positioning.

- *Time-Series Databases*: Time-series databases contain time related data such as stock market data or logged activities. These databases usually have a continuous flow of new data coming in, which sometimes causes the need for a challenging real-time analysis. Data mining in such databases commonly includes the study of trends and correlations between evolutions of different variables, as well as the prediction of trends and movements.
- *World Wide Web*: The World Wide Web is the most heterogeneous and dynamic repository available. A very large number of authors and publishers are continuously contributing to its growth and metamorphosis, and a massive number of users are accessing its resources daily. Data in the World Wide Web is organized in inter-connected documents, which can be text, audio, video, raw data, and even applications. Conceptually, the World Wide Web is comprised of three (3) major components: (i) the content of the Web, which encompasses documents available, (ii) the structure of the Web, which covers the hyperlinks and the relationships between documents, and (iii) the usage of the Web, which describes how and when the resources are accessed. Data mining in the World Wide Web, or web mining, tries to address all these issues and is often divided into web content mining, web structure mining and web usage mining.

2.2 Data Mining Methods

There are various data mining methods available in carrying out knowledge extraction from large databases. These could be classified into two (2) main categories: "Descriptive" and "Predictive". Thus, the two (2) high-level primary goals of data mining in practice tend to be prediction and description [10, 13].

As for the prediction, it involves using some variables or fields in the database to predict unknown or future values of other variables of interest. For that reason, prediction is concerned with the creation of models that are capable of producing prediction results when applied to unseen, future cases [7]. Classification and regression are the most frequent types of tasks that are applied in predictive data mining.

As for the description, it focuses on finding human-interpretable patterns describing the data. Therefore, description is concerned with explanatory models that summarize data for the purpose of inference [7]. Summarization and visualization of databases are the main applications of descriptive data mining. The usefulness of this concept is that it enables one to generalize the dataset from multiple levels of abstraction, which facilitates the examination of the general behavior of the data, since it is impossible to deduce that from a large database.

Although the boundaries between prediction and description are not sharp (some of the predictive models can be descriptive to the degree that they are understandable, and vice versa), the distinction is useful for understanding the overall discovery goal [5, 6, 8]. The relative importance of prediction and description for particular data-mining applications can vary considerably. Thus, the goals of prediction and description can be achieved using a variety of existing data mining methods that describe the type of mining and data recovery operation [10–15]. These methods are summarized below:

- *Classification* is a well-known data mining operation and it has been studied in machine learning community for a long time. Its aim is to classify cases into different classes, based on common properties (i.e. attributes) among a set of objects in a database. After the construction of the classification model, it is used to predict classes of new data that are going to be inserted in the database.
- *Clustering* is a common descriptive task where one seeks to identify a finite set of categories or clusters to describe the data. It is the task of grouping a set of objects in such a way that objects belonging into the same group (i.e. cluster) will have more similarities among them, than those that belong into different clusters. Clustering can be achieved via various algorithms that differ significantly in their notion of what constitutes a cluster and how to efficiently find them. Popular notions of clusters include groups with small distances among the cluster members, dense areas of the data space, intervals or particular statistical distributions.
- *Regression* is a data mining function that predicts a number. Usually a regression task begins with a dataset whose target values are known, where the regression algorithm estimates the value of the target as a function of the predictors for each case in the build data. Afterwards, these relationships between predictors and target values are summarized in a model, which can then be applied to a different dataset in which the target values are unknown.
- *Summarization* involves methods for finding a compact description for a subset of data. It is a key data mining concept which involves techniques for finding a compact description of a dataset. Simple summarization methods such as tabulating the mean and standard deviations are often applied for data analysis, data visualization and automated report generation. Summarization can be viewed as compressing a given set of transactions into a smaller set of patterns, while retaining the maximum possible information.
- *Dependency Modeling* consists of finding a model that describes significant dependencies between variables. Dependency models exist at two levels: the structural level of the model specifies (often in graphic form) whose variables are locally dependent on each other, and the quantitative level of the model specifies the strengths of the dependencies using some numeric scale.
- *Change and Deviation Detection* focuses on discovering the most significant changes in the data from previously measured or normative values.

2.3 Clustering

Clustering can be considered as the most important and most commonly used unsupervised learning technique, where a set of patterns, usually vectors in a multi-dimensional space, are grouped into clusters in such a way that patterns in the same cluster are similar in some sense, and patterns in different clusters are dissimilar in the same sense. It is a data mining technique used to place data elements into related groups without prior knowledge of the group definitions. However, a large number of clustering definitions can be found in the literature, with a variety of complexities. A loose definition of clustering could be the process of organizing objects into groups whose members are similar in some way [16–18]. A cluster is therefore a collection of objects that are "similar" between them and are "dissimilar" to the objects belonging to other clusters.

Generally, clustering analysis helps construct meaningful partitioning of a large set of objects based on a "divide and conquer" methodology which decomposes a large-scale system into smaller components to simplify design and implementation. As a data mining task, data clustering identifies clusters, or densely populated regions according to a distance measurement, in a large, multidimensional dataset. Given a large set of multidimensional data points, the data space is usually not uniformly occupied by the data points [18, 19]. Data clustering identifies the sparse and the crowded places, hence discovering the overall distribution patterns of the dataset. In order to decide what constitutes a good clustering, it can be stated that there is no absolute "best" criterion which would be independent of the final aim of the clustering. Consequently, it is the user which must supply this criterion in such a way that the result of the clustering will suit his needs.

There are many different ways to express and formulate the clustering problem, and as a consequence, the obtained results and its interpretations, strongly depend on the way the clustering problem was originally formulated [20, 21]. For example, the clusters or groups that are identified may be exclusive, so that every instance belongs in only one group. Alternatively, they may be overlapping, meaning that one instance may fall into several clusters. On the other hand, they may be probabilistic, whereby an instance belongs to each group depending on a certain assigned probability. Alternatively, they may be hierarchical, such as that there is a crude division of the instances into groups at a high level that is further refined into a finer level. Thus, it is very difficult to provide a categorization of clustering methods, as these categories may overlap, and a method may have features from several categories [22, 23].

In general, the major clustering methods can be classified into the following categories:

- *Partitional*: Given a set of n objects, a partitional method constructs k partitions of the data, where each partition represents a cluster and k $\leq$ n. It divides the data into k groups such that each group must contain at least one object. In other words, partitional methods conduct one-level partitioning on datasets. The basic partitional methods typically adopt exclusive cluster separation, where each object must belong to exactly one group. The general criterion of a good partitioning is that objects in the same cluster are "close" or related to each other, whereas objects in different clusters are "far apart" or very different.
- *Hierarchical*: A hierarchical method creates a hierarchical decomposition of the given set of data objects. A hierarchical method can be classified as being either agglomerative or divisive, based on how the hierarchical decomposition is formed. As for the agglomerative approach, also called the bottom-up approach, it starts with each object, forming a separate group. It successively merges the objects or groups close to one another, until all the groups are merged. As for the divisive approach, also called the top-down approach, it starts with all the objects in the same cluster. In each successive iteration, a cluster is split into smaller clusters, until eventually each object belongs to one cluster.

- *Density-based*: Density-based methods' idea has to do with the continuous growth of a given cluster as long as the density (number of objects or data points) in the "neighborhood" exceeds some threshold. Such methods can be used to filter out noise or outliers and discover clusters of arbitrary shape. Typically, density-based methods can divide a set of objects into multiple exclusive clusters, or a hierarchy of clusters, and do not consider fuzzy clusters.
- *Grid-based*: Grid-based methods quantize the object space into a finite number of cells that form a grid structure, where all the clustering operations are performed on the grid structure. Grid-based methods offer a fast processing time, being independent of the number of data objects, and being only dependent of the number of cells in each dimension in the quantized space. It must be mentioned that they can be integrated with other clustering methods.

Clustering algorithms can be categorized based on their cluster model, as listed above. There are possibly over 100 published clustering algorithms, however due to the fact that not all of them are capable of providing models for their clusters, they cannot be easily categorized. Some of them integrate the ideas of several clustering methods, and it becomes sometimes difficult to classify a given algorithm as uniquely belonging to only one clustering method category. Furthermore, some applications may have clustering criteria that require the integration of several clustering techniques. There is no objectively "correct" clustering algorithm, but the most appropriate clustering algorithm for a particular problem often needs to be chosen experimentally, unless there is a mathematical reason to prefer one cluster model over another. It should be noted that an algorithm that is designed for one kind of model has no chance on a dataset that contains a radically different kind of model. However, all the clustering algorithms should satisfy the requirements of: (i) achieving scalability, interpretability and usability, (ii) dealing with different types of attributes, (iii) discovering clusters with arbitrary shape, (iv) dealing with noises and outliers, and (v) offering high dimensionality. In Table 1, the most commonly used algorithms of each clustering method are stated:

Table 1. Clustering algorithms

Type of algorithm	Algorithms
Partitional	k-Means, k-Medoids, PAM, CLARA, CLARANS
Hierarchical	CURE, CHAMELEON, SLINK, CLINK, BIRCH
Density-based	DBSCAN, SNN, OPTICS, EnDBSCAN, DENCLUE
Grid-based	STING, CLIQUE, PROCLUS, ORCLUS

It should be noted that clustering algorithms can be applied in various different fields:

- *Marketing*, for finding groups of customers with similar behavior given a large database of customer data, containing their preferences and past buying records.
- *Biology*, for classifying plants and animals given their features.
- *Libraries*, for classifying books and books' ordering.

- *Insurance*, for identifying groups of motor insurance policy holders with a high average claim cost, or for identifying frauds.
- *City-Planning*, for identifying groups of houses according to their house type, value and geographical location.
- *World Wide Web*, for document classification and for grouping weblog data, in order to discover groups of similar access patterns.

2.4 Classification

Classification can be considered as the most important and most commonly used supervised learning technique, where objects with common properties are grouped into classes, producing a classification scheme over a set of data objects [24–26]. In more details, classification is the process which finds the common properties among a set of objects in a database/dataset and classifies them into different classes, according to a classification model. To construct such a classification model, a sample database E is treated as the training dataset, in which each tuple consists of the same set of multiple features as the tuples in a large database W, whilst each tuple has a known class identity (i.e. label) associated with it [8, 10]. The objective of the classification is firstly to analyze the training data and secondly develop an accurate description or a model for each class using the features available in the data. Such class descriptions are then used either to classify future test data in the database W or to develop a better description (i.e. classification rules) for each class in the database.

In other words, via classification we are able to predict a certain outcome based on a given input. In order to predict the outcome, the existing classification algorithms process a training dataset containing a set of attributes and the respective outcome, and try to discover relationships between the attributes that would make it possible to predict the outcome. Then, the algorithms are given a dataset not seen before, called prediction set, which contains the same set of attributes, except for the prediction attribute which is not yet known. Finally, the algorithms analyze the given input and produce a prediction. The prediction accuracy defines how "good" the algorithm is [27]. Thus, the goal of classification is to accurately predict the target class for each case in the data [28, 29]. In order to mate this prediction, classification follows a two-step process: (i) The *Model Construction* step, which describes a set of predetermined classes, where each tuple/sample is assumed to belong to a predefined class, as determined by the class label attribute. The set of tuples used for model construction is called training dataset, and (ii) The *Mode Usage* step, which is used for classifying future or unknown objects, by estimating the accuracy of the model and comparing the known label of test sample with the classified result from the model. Thus, the test dataset takes place, which is independent of the training dataset.

In general, the major classification methods can be classified into the following categories [34–38]:

- *Statistical*: Statistical methods are generally characterized by having an explicit underlying probability model, which provides a probability of being in each class rather than simply a classification. Thus, statistical methods follow a procedure in which individual items are placed into groups based on quantitative information on one or more characteristics inherent in the items (referred to as traits, variables, characters, dimensions, etc.).

- *Decision Trees:* A decision tree is a classification model that decides the target value (dependent variable) of a new sample based on various attribute values of the available data. The internal nodes of a decision tree denote the different attributes, the branches between the nodes denote the possible values that these attributes may have in the observed samples, whereas the terminal nodes denote the final value (i.e. classification) of the dependent variable.
- *Rule-based*: Rule-based methods help to define a set of rules to be used in making decisions based on training and testing datasets. Thus, rule-based methods follow an iterative process, by firstly generating a rule that covers a subset of the training datasets and then removing all examples covered by the rule from the test dataset. This process is repeated iteratively until there are no examples left to cover.
- *Neural Networks*: Neural network methods can handle problems with many parameters, being able to classify objects well, even when the distribution of objects in the N-dimensional parameter space is very complex. Therefore, these methods process one record at a time and "learn" by comparing their classification of the record (which as the beginning is largely arbitrary) with the known actual classification of the record. Errors from the initial classification of the first records are fed back into the network and used to modify the networks algorithm the second time around. This continues for many, many iterations.
- *Support Vector Machines*: Support Vector Machines are supervised learning methods used mainly for classification. They can make use of certain kernels in order to transform the problem, such that we can apply linear classification techniques to non-linear data. Applying the kernel equations arranges the data instances in such a way within the multi-dimensional space, that there is a hyper-plane that separates data instances of one kind from those of another. One important thing to note about Support Vector Machines is that the data to be separated needs to be binary.

Classification is one of the tasks most frequently carried out by the so-called Intelligent Systems. Therefore, a large number of techniques have been developed based on Artificial Intelligence, Perceptron-based techniques and Statistics (Bayesian Networks, Instance-based techniques) [28]. As the "no free lunch" theorem suggests, there is no technique that has been proven to offer the best solution to all classification or prediction problems. The selection of the classification model is critical as well as difficult, especially if there is little prior knowledge about the nature of the problem. Another problem stems from the fact that the classification process is unpredictable and quite often nondeterministic, which means that the appropriateness of the choice made cannot be immediately justified. This can be achieved after going through the entire classification process and a feedback on the performance is established. A new iteration begins again when the feedback indicates poor or unsatisfactory performance, until an acceptable level of confidence is established in the performance feedback. However, independently of the nature of the problem and the used classification algorithms, all the classification algorithms should satisfy the requirements of: (i) predicting accuracy,

(ii) achieving speed, scalability, and robustness, and (iii) offering comprehensibility, simplicity, interpretability, and quality. In Table 2, the most commonly used classification algorithms are outlined:

Table 2. Classification algorithms

Type of algorithm	Algorithms
Statistical	Naïve Bayes, k-NN, ALLOC80, SMART, CASTLE
Decision trees	ID3, C4.5, C5.0, CART, IndCART, Cal5, Bayes Tree
Rule-based	FOIL, AQ, CN2, RIPPER, PRISM
Neural networks	Kohonen, LVQ, RBF, DIPOL92
Support vector machines	SVM, LS-SVM

It should be noted that classification algorithms can be used in various different fields:

- *Computer Vision*, for medical imaging or medical imaging analysis, video tracking or optical character recognition.
- *Drug Discovery*, for developing different kinds of drugs and toxicogenomic.
- *Geostatistics*, for analyzing statistics and focusing on the geographic information systems.
- *Search Engines*, for searching quickly information over the World Wide Web.
- *Speech Recognition*, for translating spoken words into texts and finding the identity of who is speaking.
- *Document Classification*, for assigning documents to one or more classes and categories.
- *Pattern Recognition*, for recognizing patterns and regularities in data and discovering previously unknown patterns.

3 A Comparison Between Clustering and Classification

As stated above, classification is the process of finding a set of models that describe and distinguish data classes and concepts, for being able to use the model to predict the class whose label is unknown. Clustering is different from classification as it builds the classes (which are not known in advance) based upon similarity between object features. Figure 2 outlines a general framework of an integration of clustering and classification process, implementing upon an exact training dataset. It should be mentioned, that integration of clustering and classification technique is useful even when the dataset contains missing values.

Generally, a comparative study of data mining classification technique and an integration of clustering and classification technique helps in identifying large datasets. The presented experiment shows that integration of clustering and classification techniques results in more accurate results than simple classification technique to classify datasets whose attributes and classes are given to us. It can also be useful in developing

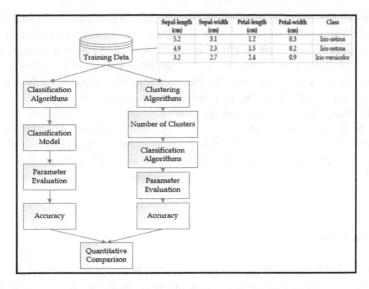

	Sepal-length (cm)	Sepal-width (cm)	Petal-length (cm)	Petal-width (cm)	Class
	5.2	3.1	1.2	0.3	Iris-setosa
	4.9	2.3	1.5	0.2	Iris-setosa
	3.2	2.7	2.4	0.9	Iris-versicolor

Fig. 2. Integration of clustering and classification

Table 3. Main differences between clustering and classification

Characteristic	Classification	Clustering
Visualized Example		
Short Description	We have a training set containing data that have been previously categorized. Based on this training set, the algorithms find the category that the new data points belong to.	We do not know the characteristics of similarity of data in advance. Using statistical concepts, we split the datasets into sub-datasets such that the sub-datasets have "similar" data.
Type of Data	Labeled data	Unlabeled data
Type of Learning	Since a training dataset exists, we describe this technique as supervised learning.	Since training dataset is not used, we describe this technique as unsupervised learning.
Example	We use a training dataset that categorized customers that have churned. Now based on this training dataset, we can classify whether a customer will churn or not.	We use a dataset of customers and split them into sub-datasets of customers with "similar" characteristics. Now based on these sub-datasets, we can identify a specific segment of customers that will buy a product.

rules when the dataset is containing missing values. Therefore, as clustering is an unsupervised learning technique, it builds the classes by forming a number of clusters to which instances belongs to, and then by applying classification technique to these clusters, we get decision rules which are very useful in classifying unknown datasets. We can then assign some class names to the clusters to which instance belongs to. This integrated technique of clustering and classification gives a promising classification results with utmost accuracy rate and robustness.

However, despite the fact that the clustering and classification techniques can be harmonically integrated, there are always many differences between these two (2) techniques, which make it possible to separate their real individual goals. Table 3 contains the main differences as for the characteristics between these two (2) basic types of data mining:

4 Conclusions

Nowadays, by knowing the importance of the data, in this paper our first goal was to focus on the procedures that have to do with the analysis and the processing of the various kinds of data. As for this sector, there were given sufficient and understandable definitions and clarifications, targeting on the separation between the different data mining techniques, with which we are daily come across. Generally, we are able to see that the use of data occupies more and more the technology sector. Every day, the data that is transferred among the companies, the organizations, the firms, and between individuals, is multiplied in a very large scale, creating difficulties as for the separation between important and non-important information. As it becomes comprehensible, the data mining field was created in order to sort, classify and exploit this data, so as to extract useful information or prepare data in such a way, for gaining the maximum usage out of it.

For that reason, we dealt with the field of data mining, trying to analyze with the most efficient and thorough way the techniques, the methodologies and the ways with which data mining is accomplished. It should be mentioned, that we gave significant importance to this part, as our main goal was to understand the existing data mining algorithms, meanings, as well as techniques and methods. However, we concentrated more on the fields of clustering and classification, as they constitute the data mining categories that are most important and most commonly used.

Moreover, it must be mentioned that data mining techniques based on statistics and machine learning, can significantly boost the ability to analyze data. Despite the potential effectiveness of data mining to significantly enhance data analysis, this technology is destined to be a niche technology unless an effort is made to integrate it with traditional database systems. This is because data analysis needs to be consolidated at the warehouse for data integrity and management concerns. Therefore, one of the key challenges is to enable integration of data mining technology seamlessly within the framework of traditional database systems.

Acknowledgements. The CrowdHEALTH project has received funding from the European Union's Horizon 2020 research and innovation programme under grant agreement No. 727560.

Athanasios Kiourtis would also like to acknowledge the financial support from the "Foundation for Education and European Culture (IPEP)".

References

1. Fayyad, U., Piatetsky-Shapiro, G., Smyth, P.: From data mining to knowledge discovery in databases. AI Mag. **17**(3), 37 (1996)
2. Chen, M.S., Han, J., Yu, P.S.: Data mining: an overview from a database perspective. IEEE Trans. Knowl. Data Eng. **8**(6), 866–883 (1996)
3. Agarwa, P., Alam, M.A., Biswas, R.: An efficient fuzzy data clustering algorithm for relational databases. Int. J. Eng. Sci. Technol. **1**(3), 8281–8288 (2011)
4. Kumar, V., Rathee, N.: Knowledge discovery from database using an integration of clustering and classification. Int. J. Adv. Comput. Sci. Appli. **2**(3), 29–33 (2011)
5. Data Mining. http://databases.about.com/cs/datamining/g/dmining.htm
6. Danso, S.O.: An exploration of classification prediction techniques in data mining: the insurance domain. Master Degree thesis, Bournemouth University (2006)
7. The Primary Tasks of Data Mining. http://www2.cs.uregina.ca/~dbd/cs831/notes/kdd/2_tasks.html
8. Witten, I.H., Frank, E., Hall, M.A., Pal, C.J.: Data Mining: Practical Machine Learning Tools and Techniques. Morgan Kaufmann, Burlington (2016)
9. Massive Data Mining (MDM) on Data Stream using Classification Algorithms. http://www.academia.edu/28451198/MASSIVE_DATA_MINING_MDM_ON_DATA_STREAMS_USING_CLASSIFICATION_ALGORITHMS
10. Cluster Analysis. http://en.wikipedia.org/wiki/Cluster_analysis
11. Regression. http://docs.oracle.com/cd/B28359_01/datamine.111/b28129/regress.htm
12. Chandola, V., Kumar, V.: Summarization-compressing data into an informative representation. In: Fifth IEEE International Conference on Data Mining, p. 8. IEEE (2005)
13. What kind of Data can be Mined. http://www.sqldatamining.com/index.php/data-mining-techniques/what-kind-of-data-can-be-mined
14. Tan, P.N.: Introduction to Data Mining. Pearson Education, India (2006)
15. Data Science Basics: What Types of Patterns can be Mined from Data? http://www.kdnuggets.com/2016/12/data-science-basics-types-patterns-mined-data.html
16. A Tutorial on Clustering Algorithms. http://home.deib.polimi.it/matteucc/Clustering/tutorial_html
17. Alfred, R., Kazakov, D.: Aggregating multiple instances in relational database using semi-supervised genetic algorithm-based clustering technique. In: ADBIS Research Communications (2007)
18. Clustering. http://databases.about.com/od/datamining/g/clustering.htm
19. Fung, G.: A Comprehensive Overview of Basic Clustering Algorithms (2001)
20. Data Clustering Algorithms. https://sites.google.com/site/dataclusteringalgorithms/
21. Cluster Analysis. http://en.wikipedia.org/wiki/Cluster_analysis
22. Andritsos, P.: Data Clustering Techniques Qualifying Oral Examination Paper, Department of Computer Science, University of Toronto (2002)
23. Omran, M.G., Engelbrecht, A.P., Salman, A.: An overview of clustering methods. Intell. Data Anal. **11**(6), 583–605 (2007)
24. Moore, A.W., Zuev, D.: Internet traffic classification using bayesian analysis techniques. In: ACM SIGMETRICS Performance Evaluation Review, vol. 33, no. 1, pp. 50–60. ACM (2005)

25. Han, J., Cai, Y., Cercone, N.: Concept-based data classification in relational databases. In: 1991 AAAI Workshop Knowledge Discovery in Databases, pp. 77–94 (1991)
26. Classification. http://databases.about.com/od/datamining/g/classification.htm
27. Witten, I.H., Frank, E., Hall, M.A., Pal, C.J.: Data Mining: Practical Machine Learning Tools and Techniques. Morgan Kaufmann, Burlington (2016)
28. Methods for Classification. http://sundog.stsci.edu/rick/SCMA/node2.html
29. Data Mining - Evaluation of Classifiers. http://www.cs.put.poznan.pl/jstefanowski/sed/DM-4-evaluatingclassifiersnew.pdf
30. Types of database management system and their evolution. https://www.analyticsvidhya.com/blog/2014/11/types-databases-evolution/
31. Different Types of Databases. http://www.my-project-management-expert.com/different-types-of-databases.html
32. Types of Database Management Systems. http://www.brighthub.com/internet/web-development/articles/110654.aspx
33. NoSQL Databases: An Overview. https://www.thoughtworks.com/insights/blog/nosql-databases-overview
34. Methods for Classification. http://sundog.stsci.edu/rick/SCMA/node2.html
35. Classification Methods. http://www.d.umn.edu/~padhy005/Chapter5.html
36. Michie, D., Spiegelhalter, D.J., Taylor, C.C.: Machine learning, neural and statistical classification (1994)
37. Mahajan, A., Ganpati, A.: Performance evaluation of rule based classification algorithms. Int. J. Adv. Res. Comput. Eng. Technol. (IJARCET) 3(10), 3546–3550 (2014)
38. Zhang, C., et al.: An up-to-date comparison of state-of-the-art classification algorithms. Expert Syst. Appl. 82, 128–150 (2017)

A Framework for Cloud Selection Based on Business Process Profile

Mouna Rekik[1,3(✉)], Khouloud Boukadi[1,3],
and Hanene Ben-Abdallah[2,3]

[1] Sfax University, Sfax, Tunisia
rekik.mona@yahoo.fr
[2] King Abdulaziz University, Jeddah, Kingdom of Saudi Arabia
[3] Mir@cl Laboratory, Sfax, Tunisia

Abstract. The lack of a system that assists in the business process out-sourcing to the cloud hinders the widespread adoption of this emerging computing paradigm. To the best of our knowledge, there is no system that has tackled the business process outsourcing to the cloud issues such as selecting the activities to outsource and the cloud services to support their execution based on objective and subjective assessments of cloud services. In this paper, we propose a system tackling these issues by using previous cloud users experience for the subjective assessment. The objective assessment is based on simulations using the well-known CloudSim toolkit. The aggregation of the objective and subjective assessment allows for more reliable cloud service selection. Furthermore, we propose an optimal deployment of business process activities by proposing a novel penalty based genetic algorithm while considering business process profile.

Keywords: BPO · Penalty-based GA · Cloud selection · Objective assessment · Subjective assessment

1 Introduction

Generally, the paradigm of Business Process Outsourcing (BPO) to the cloud is not a new trend. It is a common and a well-known business practice insuring among others, the gaining of the business added-value. In order to minimize BPO risks while gaining as much as possible from the cloud computing, enterprises tend to keep the execution of some business process activities in-premise and outsource the remain ones to the cloud. Obviously, it is a fastidious task to identify the business activities that should be outsourced. The selection of these activities requires dedicated method that insures minimizing risks, costs and enhancing business process performance. Furthermore, the proliferation and the diversity of cloud services hinder the selection and the discovery of suitable cloud services that fit potential cloud users requirements. This issue urges researchers to conduct deep studies in order to assist potential cloud users to make the most suitable decision related to the cloud service to adopt. A judicious cloud service selection requires the elaboration of credible and reliable techniques that include, among others, a prior evaluation of the services available on the cloud landscape. This evaluation implies two-fold approaches namely, an objective and a subjective

© Springer International Publishing AG 2017
M. Themistocleous and V. Morabito (Eds.): EMCIS 2017, LNBIP 299, pp. 97–110, 2017.
DOI: 10.1007/978-3-319-65930-5_8

assessment. An objective assessment is based on the evaluation of the QoS (Quality of Service) values such as the service availability and response time through monitoring tools, benchmark, or simulations [1–3]. On the other hand, the subjective assessment is based on the users' previous experience related to cloud adoption. A subjective assessment is generally elaborated through users rating provided via dedicated interfaces. The resulted ratings constitute the experimentation database on which potential users can rely for a subjective evaluation of the experimented services [4]. In our research case, the experimentation database corresponds to previous outsourcing to the cloud experience of business processes. More specifically, enterprise experts specify whether the previous outsourcing experience related to a variety of adopted cloud services is satisfying. Relying on this database to select cloud services to adopt is not a trivial task and requires performing specific analysis regarding the huge available amount of data. Indeed, outsourced business processes to the cloud have generally various profiles. We mean by profile the outsourced business process type, its information technology requirements and even the business process owner's preferences, etc. Thus, identifying the cloud users' business processes having similar profile with the potential user one is undoubtedly an efficient means that insures a more judicious selection of the cloud service.

The paper contribution is two fold:

- We propose a novel framework to assist potential cloud users to select the most suitable cloud services as well as the most appropriate business process activities to outsource. This framework is based on the aggregation of objective and subjective assessments of cloud services.
- We propose a penalty based genetic algorithm to identify business process activities to outsource and the cloud services to adopt based on three outsourcing factors, namely the cost, the response time and the risk. Penalty is attributed to each cloud service having an unsatisfied opinions from previous users experience. Indeed, the proposed framework encompasses a dedicated database related to previous business process outsourcing experiences. More specifically, we gather these experiences from three enterprises experts having outsourced their processes to the cloud and ask them to annotate for each experience whether the outsourcing results are satisfying or not satisfying.

The remainder of this paper is organized as follows: Sect. 2 presents the related work. Section 3 describes in detail the proposed framework whose evaluation is discussed in Sect. 4. Finally, Sect. 5 summarizes the work status and highlights its extension.

2   Related Work

Several cloud service selection methods, frameworks and approaches are proposed with a variety of selection aims and techniques. Besides, the outsourced application type (business process or a classic application) is considered among the prominent focus of different researches. Indeed, the application type influences without a doubt, the selection of the cloud service as each type has its specific requirements and profile.

In this section, we categorize the cloud selection approaches into two groups according to the studied application type: approaches dealing with the cloud selection related to classic application and approaches related to business processes outsourcing to the cloud.

2.1 Cloud Selection Approaches for Classic Applications

An extensive amount of researches covers methods and approaches are proposed to assist potential cloud users in the fastidious task of selecting the suitable cloud services based on benchmark tests, monitoring tools, simulations or/and user's feedback [6, 7]. Based on the available cloud selection approaches, we highlight three basic categories, namely, approaches based on objective, subjective or hybrid performance assessment.

Objective Assessment. The objective assessment of cloud services is based on a quantitative evaluation of services through testing and monitoring tools or benchmark tests. In this context, Patiniotakis et al. proposed a recommender entitled PuLSaR (Preference-based cLoud Service Recommender) that assists potential cloud users in the fastidious task of cloud service selection based on multi-criteria decision making methods [8]. The paper tackles linguistically the expressed preferences and cloud service properties that lack a clear value and lead to some vagueness that can only be captured using the Fuzzy Set Theory [9]. SMICloud framework [6] uses the Analytic Hierarchy Approach (AHP) to rank cloud services based on functional and non-functional properties. These properties correspond to SMI attributes proposed by the Cloud Service Measurement Index Consortium (CSMIC [6]).

Subjective Assessment. A variety of techniques and methods are proposed for a subjective assessment of cloud performance. This assessment considers the user's feedback and opinion to evaluate services they invoke. To proceed for a subjective assessment of cloud services, the authors in [4] proposed a framework that ensures the cloud performance evaluation through monitoring techniques, based on the user's feedback. More specifically, the framework gathers the users feedback related especially to IaaS services regarding different cloud service performance properties. Overall, some research works consider the cloud as similar to web services and thus, they use some rating-based reputation systems [10, 11] for cloud service selection purposes.

Hybrid Assessment. The purpose of hybrid assessment for evaluating cloud services based on both objective and subjective assessments, is to consider simultaneously the qualitative and the quantitative cloud service properties for a more trustable cloud services selection. In this regard, [12] proposed a ranking approach for cloud services based on NSGA SR which belongs to multi-objective optimization methods. The proposed approach ranks cloud services based on the user's feedback and on the quantitative assessment of the QoS attributes. In the same context, Fan et al. [13] proposed a multi-dimensional trust-aware cloud service selection mechanism based on evidential reasoning (ER) approach. The mechanism integrates both perception and reputation based on trust value.

Overall, we notice that few studies tackled the objective and subjective assessment of cloud services for selection purposes. Furthermore, there is no approach that consider the business process profile similarities between the Business Process Candidate for Outsourcing (BPCO) and the previous outsourced to the cloud business processes. Indeed, an appropriate cloud service selection that fits as much as possible the business process depends on several aspects of the previous outsourced business process and even of its owner (enterprise in our case). These aspects include the business process type, its information technology requirements, the commitment period, etc.

2.2 Business Process Outsourcing to the Cloud

Outsourcing business process to the cloud is still a challenging issue due to the multiple cloud risks. The best way to benefit from cloud computing, while avoiding potential risks, is to combine the cloud-based and the traditional business process management. This is achieved in the aim of keeping sensitive data within the enterprise boundaries and outsourcing the compute intensive activities that require powerful resources to the cloud [14]. Towards this end, researchers started examining how to transform the original business processes. In this context, [15, 16] proposed an automated transformation support necessary to split business processes according to data and activity distributions (in cloud/in premise) defined by the users. Similarly, [17] proposed an approach to decompose the business processes between the cloud and in-premise sides while preserving the data constraints. The proposed approach uses as outsourcing factors, the monetary, the privacy and the execution cost. In these approaches, the target cloud configuration is already defined. Overall, there is no method that assists the enterprise experts with the BPO decision by selecting the most suited business process activities to outsource as well as the cloud services to adopt based on a subjective and an objective cloud service assessment.

3   Our Proposed Framework

We propose in this paper a novel framework that assists enterprises/potential users aiming at outsourcing their business process to the cloud. This assistance is envisaged through a new framework proposing a set of new contributions. In contrast to the major existing cloud selection approaches, our framework considers both subjective and objective assessments for cloud service selection. In addition to the objective assessment elaborated via simulations used by CloudSim [18], we propose a more accurate subjective assessment through previous users experience related to the business process outsourcing to the cloud. Indeed, authors in [24] proposed a comprehensive system assisting enterprises in the business process outsourcing to the cloud. The system identifies business process to outsource and cloud services to support their execution. Based on the system recommendation, enterprises outsource the identified business processes to the selected cloud services. We extend the proposed system so it supports the experts opinions/satisfaction related to the outsourcing experience for every outsourced business process once the outsourcing to the cloud is achieved. Furthermore, our system allows to identify values related to each outsourced business process profile property. Accordingly, the experts annotate their satisfaction in a dedicated database that we name "Previous outsourcing experience". The possible annotation attributed by experts is satisfying/not satisfying. More specifically, we gather the cloud users experiences from three enterprises' experts having outsourced their processes to the cloud and ask them to annotate for each experience whether the outsourcing results are satisfying or not satisfying. The total number of gathered experiences is 300 and they are related basically to the same outsourcing period, namely from "01/01/2016" until "01/12/2016". Overall, the outsourced business processes cover sales management, personnel training sessions as well as oil drill creation using 3D techniques.

Once a new BPCO is inserted within the system, the latter fetches the outsourced business processes whose profiles are similar to the candidate business process one. Accordingly, the subjective assessment is based on similar business processes for a more accurate cloud service evaluation. Overall, the outsourced business processes profile has a variety of properties as well as a variety of outsourcing concerns. For instance, suppose a not satisfying outsourcing experience related to an outsourced business process managing customer sales. Suppose that this process is characterized by a very low workload intensity and whose cloud data-centers supporting its execution are located in America. This business process profile is not similar to a BPCO one having a high workload intensity and that the cloud data-center location is preferred to be in Asia. Consequently, relying on the user experience related to the customer sales business process is not convenient considering the BPCO profile.

To tackle this issue and to propose an accurate and a reliable subjective assessment, we propose a set of business process profile properties that should be considered. Afterwards, we calculate the profiles similarity to match similar BPCO profile with the already outsourced business processes. This step enables filter inappropriate users subjective assessment.

Once the inappropriate business process profiles are filtered, the framework applies the penalty-based genetic algorithm [19] to select the business process activities to outsource as well as the most appropriate cloud services to support the execution of the outsourced activities. The aggregation of the objective and subjective assessments is elaborated by simulating different activities' distribution (in cloud/in premise) using CloudSim simulator. The aim is to identify the distribution that minimizes the overall business process cost, risk and response time. Afterwards, we apply penalties to the cloud services having unsatisfied experience deduced from the "Previous outsourcing experience" database.

3.1 Profiles Similarity

As mentioned above, our aim is to match BPCO profile with those of already out-sourced business processes stored in the "Previous outsourcing experience" database. Other profiles in the database are not considered for the subjective assessment. We represent the profile as a vector $C = (d_1, \ldots, d_n)$ corresponding to a set of profile values along different profile dimensions [21]. To compute the similarity between two

profiles (the BPCO profile and each user's one), the distance between them is calculated following steps detailed below:

1. Compute the vector $D1, 6 = (\alpha_1, \ldots, \alpha_6)$ values representing the distance between
   the different profile dimensions (business process type, its security requirements, its
   workload type, the commitment period, and the enterprise familiarity with cloud
   services). The highest the value is, the less similar both profiles are. For instance,
   suppose that d_1 and d_2 correspond respectively to the business process type and the
   Commitment period profile dimensions. The values of α_1 and α_2 are computed by
   counting the edges separating the BPCO profile value of both dimensions and the
   same dimensions related to the outsourced business process profiles. Figure 1
   represents an extract of the business process profile illustrating the business process
   type and the commitment period profile dimensions. Suppose that the BPCO profile
   is $C_1 = $ (Purchase Request, days) and the profile C_2 of one user from the database is
   $C_2 = $ (Line provisioning, hours). Accordingly, $D_{1,2} = (\alpha_1, \alpha_2) = (4, 2)$.

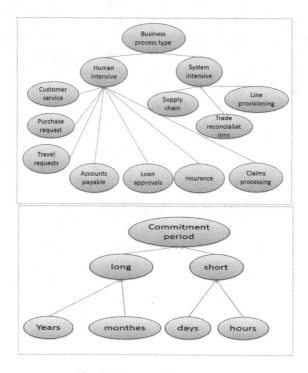

Fig. 1. Extract of the BP profile

2. Normalize the distance to obtain the percentage of the maximal distance related to
   each profile dimension. This normalization is ensured by calculating the maximum
   distance in each profile dimension (from the root element to the deepest leaf). For
   instance, the maximum distance max_2 in the commitment period is 3. Thus, the
   normalization of the distance within d_2 is $\frac{\alpha_2}{max\alpha_2}$.

3. Finally, compute the similarity value between profiles is achieved through Eq. 1.

$$sim(C_1, C_2) = \frac{1}{1 + D(C_1, C_2)}$$
(1)

where the distance D(C1,C2) is presented in Eq. 2.

$$D(C_1, C_2) \sqrt{(\frac{\alpha_1}{max\alpha_1})^2 + (\frac{\alpha_2}{max\alpha_2})^2}$$
(2)

It is straightforward to mention that we limit the above examples to only two profile dimensions for simplicity sake. However, we consider in our work the six profile dimensions to compute the profiles similarity.

3.2   Optimal Deployment of Business Process Activities in the Cloud

Our aim is to find out the best business process activities to outsource as well as the most appropriate cloud services to adopt. This problem is an NP-hard one, as the number of potential solutions increases, if the number of business process activities and cloud services grows. More specifically, we focus on selecting the optimal deployment of business process activities in the cloud regarding the outsourcing factors (reducing cost and response time and minimizing the outsourcing risks).

Statement of the Problem

- A business process is considered as a set of inter-related activities and control structure (e.g. sequence, loop, parallel, and choice). The activities and the control structure are considered as business process elements E_i where i is the element's identifier.
- If w^j is the outsourcing factor weight regarding the enterprise preferences, then $\sum_{j=1}^{3} w_j = 1$ where j is the number of outsourcing factors (risk, cost and response time).
- C is the cloud service supporting the execution of the business process element E_i.
- After filtering not similar business process profiles from the database, we compute the percentage P_{nofc} of the unsatisfied business process experiences.
- The total number n_{out} of outsourced business process elements is variable and cannot be determined. Indeed, a value of n_{out} able to insure the desired optimality in unknown
- A specific cloud service C may support the execution of different number ne of business process activities.

Penalty-Based Genetic Algorithm. The studied optimization problem can be resolved through evolutionary algorithms, such as genetic algorithms, neuro-evolution, genetic programming, etc. [22]. The herein presented approach applies the penalty based genetic algorithm. Indeed, genetic algorithms are considered as interesting algorithms tackling optimization problems based on the principle of survival in nature.

- Individual or genome encoding: is a prominent step in genetic algorithm. In our research case, we present an individual as a vector arranged as $<E_i, C, \ldots, E_j, C>$ where i and j correspond to the business process element identifier and C is the cloud service. Each cloud service C is expected to support the execution of the business process element E that it follows in the individual encoding.
- Genetic operators: in this paper, we use the single-point to elaborate the crossover operator. Overall, the crossover operator is controlled by a specific probability P_{cro}. The mutation is applied according to a predefined probability value $Pmut$. The mutation changes whether a business process element or a cloud service with others which are randomly selected.
- The fitness function: ensures the individual evaluation as it measures the correspondence level of each potential solution with the end-user preferences. The present work proposes several objective functions that constitute the building blocks of the fitness function. As mentioned above, the aim is to minimize the business process cost, response time and risks.

Some individuals are considered as infeasible as they are constituted of business process elements whose execution is supported by cloud services having unsatisfied users' opinion. Discarding these individuals is not an appropriate procedure as they potentially have some information necessary to evolve toward optimality. Thus, we propose a penalty function to change the fitness function in order to generate individuals comprising cloud services with satisfactory quality deduced from filtered business process profiles. The integration of penalties constitutes the cloud service subjective assessment. Equation 3 presents the fitness function related to individual I.

$$\text{Fitness(I)} = \begin{cases} F(I) & \text{if } I \text{ is feasible} \\ F(I) - P(I) & \text{otherwise} \end{cases} \tag{3}$$

F(I) is the objective function we define in Eq. 4.

$$MinimizeF(I) = F_c(I)w_c + F_{rt}(I)w_{rt} + F_{risk}(I)w_{risk} \tag{4}$$

where $F_c(I)$ is the cost objective function. Generally, the cost related to renting cloud services depends on various aspects, namely the resource allocation type as well as the resource usage rates. The resource allocation in our research case corresponds to the reserved one that supposes that the enterprise already knows the cloud service commitment period. Hence, the cloud service is rented in advance requiring one-time reservation fees besides the hourly rate usage. Equation 5 defines the objective function $F_c(I)$.

$$F_c(I) = p_a^d + f_{us} \tag{5}$$

where p_a^d is the in-advance reservation fees based on a specific cloud provider cost a, the corresponding cloud service type p as well as the adoption length time period d. The function f_{us} is the hourly usage cost fees presented in Eq. 6.

$$f_{us} = \sum_{i=1}^{m} k \times size(dej) \times D(i,j) \times |u_i - c_j| + \sum_{i=1}^{m} f \times t_c(a_i) \tag{6}$$

where $\sum_{i=1}^{m} k \times size(dej) \times D(i,j) \times |u_i - c_j|$ corresponds to the transfer of data dej fees between the business process elements supported by the cloud services and those on the enterprise side. If the business process elements and the related transferred data are on the same side, $|u_i - c_j| = 0$ and k corresponds to the transfer of data cost proposed by a specific cloud provider. $F_{risk}(I)$ is the risk objective function. Overall, outsourcing to the cloud may potentially lead to undesirable outcome especially when the outsourcing enterprise is not aware about the cloud risks. These risks are related basically to security concerns regarding the outsourced business process requirements. Hence, we analyze the cloud risks which consist essentially of: data breaches, data loss, account hijacking, insecure interfaces, and denial of services. Each risk influences some of the CIANA objectives (Confidentiality (C); Integrity (I); Availability (A); Non-repudiation (NR); Authenticity (AU)):

- Data Breaches (DB) influence Confidentiality;
- Data Loss (DL) influences Availability and Non-Repudiation;
- Account Hijacking (AH) influences Confidentiality, Integrity, Availability, Non-Repudiation and Authenticity;
- Insecure Interfaces (II) influence Confidentiality, Integrity and Authenticity;
- Denial of Service (DS) influences Availability.

The Eq. 7 present the formula used to compute the influence of each cloud risk (cr) on each business process activity a_j regarding its security requirements (CIANA objectives (CO)).

$$Inf(cr, a_j) = \sum_{i=1}^{5} (CO \times eval) \tag{7}$$

where $eval$ is equal to 1 if the activity a_j requires the objective (CO) and 0 otherwise. The aim of formula 8 [23] is to minimize the outsourcing of the activities the most influenced by cloud risks as well as the adoption of cloud services lacking security mechanisms required to mitigate these risks.

$$F_{risk}(I) = \sum_{k=1}^{n} \sum_{j=1}^{m} Inf(cr_j, ak) + (covmax - cov(cr_j)) \tag{8}$$

where $\sum_{k=1}^{n} \sum_{j=1}^{m} Inf(cr_j, a_k)$ corresponds to the total risks of outsourcing a specific set of business process activities and $(covmax - cov(cr_j))$ computes the difference in value of the overall existing security mechanisms and those provided by the cloud cr_j.

The objective function F_{rt} defined in (9) computes the response time of outsourcing a specific set of business process activities to cloud services.

$$F_{rt}(I) = t_c(A) + \sum_{i=1}^{m} \sum_{j=1}^{l} \frac{size(d_i, d_j) \times |u_i - c_j|}{b} \tag{9}$$

where $t_c(A)$ refers to the execution time of a set of activities A supported by cloud services. Computing the execution time is elaborated while considering the different business process control structures (sequence, loop, parallel, and choice). $\sum_{i=1}^{m} \sum_{j=1}^{l} \frac{size(d_i, d_j) \times |u_i - c_j|}{b}$ is the transfer time of date exchange between the cloud services and in-premise. $Size(d_i, d_j)$ is the size of the transferred data and b is the bandwidth size provided by a cloud service. The proposed equation of the Penalty(I) is presented in (10).

$$Penalty(I) = \sum_{i=1}^{3} w_i \sum_{j=1}^{m} P_{nofcj} \times \frac{ne}{nout} \tag{10}$$

The values of outsourcing factor weights w_i for the fitness function as well as for the penalty one are equal and attributed basically by the enterprise experts. The more the individual I is composed of cloud services C_j whose percentage of unsatisfied opinion is high, the more he is harshly penalized. Furthermore, the percentage of outsourced business process elements $\frac{ne}{nout}$ influences the penalty value. Indeed, the penalty value increases if the percentage of unsatisfied opinion related to a specific cloud service is considerable as well as the percentage of $\frac{ne}{nout}$ whose execution is supported by the same cloud service.

4　Experimental Evaluation

The evaluation of our system is based on a real business process related to an oil and gas enterprise. The average response time $Thrt$ of the business process instances, when executed in the enterprise, is 40 min and the average cost Thc is \$18.81/day. These values are related to one week observation (from 03/02/2016 to 10/02/2016) characterized by a high workload intensity. The profile of the studied business process is illustrated in Table 1.

4.1　Experiments 1

In this experiment, we validate our proposed profile similarity approach proposed in Sect. 2. The number of the "Previous outsourcing experience" database instances is 300. We interviewed 10 experts belonging to business research field. We asked these experts to select among the extracted opinions those related to business process whose profile is similar to the business process profile presented in Table 1. Figure 2 illustrates the F-score values when comparing the similar profiles identified based on our

Table 1. Business process profile

Workload intensity	High	
CIANA requirements of each activity	A1	(C), (I), (A), (NR)
	A2	(AU)
	A3	(C), (A)
	A4	(C), (NR)
	A5	(C), (I), (A), (AU)
	A6	(C), (NR)
	A7	(C)
	A8	(C), (NR)
	A9	(I)
	A10	(I)
Business process	Line provisioning	
Commitment period	Days	
Preferred datacenters location	US east	
Enterprise familiarity with cloud services	No	

system with those selected by experts for 10, 100, 200 and 300 outsourcing experience. According to Fig. 2, the F-score values are considerably high as they vary between 75.32 and 94.9. Hence, the results demonstrate that with small and a considerable number of opinions, our system can properly identify similar business process profiles.

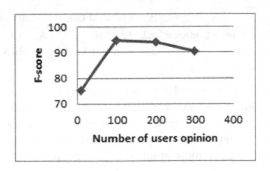

Fig. 2. F-score values

4.2 Experiments 2

Table 2 illustrates the parameter values used to experiment our penalty-based genetic algorithm. In this experiment, we asked the experts to weigh each outsourcing factor regarding their preferences (see Table 3).

Considering this interview results, each outsourcing factor weight is attributed the following values: 0.5 for the cost and the risk, the response time weight value is 0.25.

To evaluate whether the Penalty-based GA solution is of a satisfying quality, we compare the business process response time and cost values obtained from the

Table 2. Parameters used for our proposed penalty based genetic algorithm

Parameters	Description
Population size	50
Selection technique	Tournament selection
Termination condition	Number of generation = 100
Crossover probability	0.05
Mutation probability	0.015

Table 3. Outsourcing factor weights

Rank	Factor	Percentage of experts who select the corresponding outsourcing factor
1	Cost	100%
2	Risk	60%
3	Response time	60%

simulations using the CloudSim toolkit to the cost Thc and the response time $Thrt$ values. Furthermore, we compare the system results using the GA with the greedy algorithm (see Table 4).

Table 4. Comparison between greedy and penalty-based GA results

	Penalty-based GA	Greedy algorithm
Activities selected for outsourcing	A2, A3, A10	A1, A8, A9, A10
Response time cost	23.5 min $11.81/day	41 min $12.81/day

Table 4 shows that our Penalty-based genetic algorithm performs better than the greedy algorithm regarding the response time and the cost values. Furthermore, the identified response time and cost values are considerably less than the enterprise's ones. Thus, our system can be considered as a pertinent and a reliable means to assist in the fastidious task of selecting the most prominent business process activities and cloud services.

5 Conclusion

Outsourcing business processes to the cloud is seen to be an emergent and a prominent paradigm regarding the harsh economic competition. However, deciding about the most prominent and appropriate business process activities to outsource as well as the cloud service to adopt is not a straightforward task and require deep analysis and evaluation. Indeed, in the cloud landscape, there are many numbers of cloud services. Furthermore, each business process has its own profile which empowers the need to use suitable and appropriate approaches to make decisions. In this paper, we have presented a framework assisting potential cloud users to select the suitable cloud services able to support, in an optimal way, the execution of the outsourced business process activities. The proposed framework is based on both objective and subjective assessments of cloud services. We have conducted experiments in many different settings. The experimental results demonstrate that our system can be a reliable tool to assist potential cloud users to outsource to the cloud purposes.

References

1. Sheikh Mahbub, H., Sebastian, R., Max, M.: Towards a trust management system for cloud computing. In: Proceedings of the 2011 IEEE 10th International Conference on Trust, Security and Privacy in Computing and Communications, TRUSTCOM 2011, Washington, DC, USA, pp. 933–939. IEEE Computer Society (2011)
2. Chen, G., Bai, X., Huang, X., Li, M., Zhou, L.: Evaluating services on the cloud using ontology QoS model. In: Proceedings of 2011 IEEE 6th International Symposium on Service Oriented System (SOSE), pp. 312–317, December 2011
3. ur Rehman, Z., Hussain, O.K., Hussain, F.K.: Multi-criteria IaaS service selection based on QoS history. In: 2013 IEEE 27th International Conference on Advanced Information Networking and Applications (AINA), pp. 1129–1135, March 2013
4. ur Rehman, Z., Hussain, O.K., Parvin, S., Hussain, F.K.: A framework for user feedback based cloud service monitoring. In: 2012 Sixth International Conference on Complex, Intelligent, and Software Intensive Systems, pp. 257–262, July 2012
5. Bo, P., Lillian, L.: Opinion mining and sentiment analysis. Found. Trends Inf. Retr. 2(1–2), 1–135 (2008)
6. Garg, S.K., Versteeg, S., Buyya, R.: Smicloud: a framework for comparing and ranking cloud services. In: 2011 Fourth IEEE International Conference on Utility and Cloud Computing, pp. 210–218, December 2011
7. Zhang, M., Ranjan, R., Menzel, M., Nepal, S., Strazdins, P., Wang, L.: A cloud infrastructure service recommendation system for optimizing real-time QoS provisioning constraints. arXiv preprint arXiv:1504.01828 (2015)
8. Triantaphyllou, E.: Multi-criteria Decision Making Methods: A Comparative Study, vol. 44. Springer Science & Business Media, Heidelberg (2013). doi:10.1007/978-1-4757-3157-6
9. Zadeh, L.: Fuzzy sets. Inf. Control 8(3), 338–353 (1965)
10. Srivastava, A., Sorenson, P.G.: Service selection based on customer rating of quality of service attributes. In: 2010 IEEE International Conference on Web Services, pp. 1–8, July 2010
11. Li, L., Wang, Y.: A trust vector approach to service-oriented applications. In: IEEE International Conference on Web Services, pp. 270–277, September 2008
12. Arezoo, J., Leyli, M.: Cloud service ranking as a multi objective optimization problem. J. Supercomput. 72(5), 1897–1926 (2016)
13. Fan, W., Yang, S., Perros, H., Pei, J.: A multi-dimensional trust-aware cloud service selection mechanism based on evidential reasoning approach. Int. J. Autom. Comput. 12(2), 208–219 (2015)
14. Han, Y.-B., Sun, J.-Y., Wang, G.-L., Li, H.-F.: A cloud-based BPM architecture with user-end distribution of non-compute-intensive activities and sensitive data. J. Comput. Sci. Technol. 25(6), 1157–1167 (2010)

15. Duipmans, E.F., Pires, L.F., da Silva Santos, L.O.B.: Towards a BPM cloud architecture with data and activity distribution. In: 2012 IEEE 16th International Enterprise Distributed Object Computing Conference Workshops, pp. 165–171, September 2012

16. Duipmans, E.F., Pires, L., da Silva Santos, L.O.B.: A transformation-based approach to business process management in the cloud. J. Grid Comput. **12**(2), 191–219 (2014)

17. Povoa, L.V., de Souza, W.L., Pires, L.F., do Prado, A.F.: An approach to the decomposition of business processes for execution in the cloud. In: 2014 IEEE/ACS 11th International Conference on Computer Systems and Applications (AICCSA), pp. 470–477, November 2014

18. Calheiros, R.N., Ranjan, R., Beloglazov, A., De Rose, C.A., Buyya, R.: CloudSim: a toolkit for modeling and simulation of cloud computing environments and evaluation of resource provisioning algorithms. Softw. Pract. Exp. **41**(1), 23–50 (2011)

19. Hu, Y.-B., Wang, Y.-P., Guo, F.-Y.: A new penalty based genetic algorithm for constrained optimization problems. In: 2005 International Conference on Machine Learning and Cybernetics, vol. 5, pp. 3025–3029, August 2005

20. Gillen, A.L., Bozman, J.S.: White paper: business-critical workloads: Supporting business-critical computing with an integrated server platform. Technical report, Microsoft Corporation, April 2010

21. Liwei, L., Nikolay, M., Dong-Ling, X.: Context similarity metric for multidimensional service recommendation. Int. J. Electron. Commer. **18**(1), 73–104 (2013)

22. Kalyanmoy, D.: Multi-objective Optimization Using Evolutionary Algorithms, vol. 16. Wiley, Hoboken (2001)

23. Goettelmann, E., Dahman, K., Gateau, B., Dubois, E., Godart, C.: A security risk assessment model for business process deployment in the cloud. In: 2014 IEEE International Conference on Services Computing (SCC), pp. 307–314. IEEE (2014)

24. Rekik, M., Boukadi, K., Ben-Abdallah, H.: A comprehensive framework for business process outsourcing to the cloud. In: 2016 IEEE International Conference on Services Computing (SCC), pp .179–186. IEEE (2016)

Wildfire Prevention in the Era of Big Data

Nikos Athanasis[1(✉)], Marinos Themistocleous[1],
and Kostas Kalabokidis[2]

[1] Department of Digital Systems, University of Piraeus, Piraeus, Greece
athanasis@geo.aegean.gr, mthemist@unipi.gr
[2] Department of Geography, University of the Aegean, Mytilene, Greece
kalabokidis@aegean.gr

Abstract. Big Data analysis emerges as an innovative technology capable of providing solutions to complex problems of global concern, including prevention from natural disasters. In this study, we illustrate the contribution of Big Data technology for wildfire prevention. We describe the development of a web application that combines wildfire behavior simulations with Big Data analysis for alerting and notifying end users in case of a fire emergency. The article aims to contribute in highlighting the role of the Big Data for wildfire prevention, with a view, under certain conditions, to reduce the human, environmental and socio-economic losses that are caused.

Keywords: Big Data analysis · Wildfire prevention · Wildfire behavior modeling

1 Introduction

The widespread use of the worldwide web, combined with a drastic reduction of the manufacturing cost of devices for the storage and management of digital data, has led to the creation of an increasingly larger "digital galaxy" of information. The total amount of information in the world is estimated to have grown from 2.6 optimally compressed exabytes in 1986 to 15.8 in 1993, over 54.5 in 2000, and to 295 optimally compressed exabytes in 2007 [1], while 2.5 Quintillion bytes of data are created every day [2]. The number of internet-connected devices today is estimated approximately at 15 billion and is projected to the amount to 50 to 100 billion by the year 2020, forming an explosively growing "Internet of Things" [3].

Despite the digital explosion, complex problems of global concern such as forest fires continue to have serious environmental and socioeconomic effects, including occasionally the loss of human lives [4]. The influence of climate change trends and anthropogenic causes has radically increased the number of wildfires worldwide over the last decades [5]. Approximately 50,000 fires per year have occurred during the past three decades in the Euro-Mediterranean (Portugal, Spain, France, Italy, and Greece) region that results in about a half a million hectares of burnt area each year [6].

Many wildfire decision support systems have been developed and have been used in a number of different wildfire regions, including Spain [7], France [8], Italy [9], Turkey [10], Greece [4, 11]. The European Forest Fire Information System (EFFIS) [12]

© Springer International Publishing AG 2017
M. Themistocleous and V. Morabito (Eds.): EMCIS 2017, LNBIP 299, pp. 111–118, 2017.
DOI: 10.1007/978-3-319-65930-5_9

operates for current situation assessment, especially for post-wildfire burned area estimation. Functionalities such as fire danger forecast, fire behavior prediction, active fire detection, access to historical fire data and fire damage assessment are provided by fully integrated solutions such as the US (Wildland Fire Decision Support System, WFDSS) [13], the Canadian Wildland Fire Information System (CWFIS) [14] and the Greek wildfire prevention and management system AEGIS [15].

Nevertheless, the volume of digital data generated every day is so large and is growing at such a pace that its management with conventional storage and analysis methods is becoming increasingly difficult [16]. Furthermore, the geospatial calculations are often based on very large geographical input data sets to conduct the necessary spatiotemporal calculations which require a significant amount of processing time. Not only an enormous volume of geographic data is required and produced, but also such data change with high frequency. In many cases, the data processing and spatial analysis over the acquired data set require expensive investments in terms of hardware, software and personnel training among other overhead costs [11].

2 The Big Data Era

Big Data technology emerges as a technology capable of successfully addressing contemporary digital challenges. Big Data is high-volume, high-velocity, and/or high variety information assets that require new forms of processing to enable enhanced decision making, insight discovery and process optimization [17]. Even though the Big Data ecosystem integrates many platforms and software components, it is mainly based on distributed storage and processing of very large data sets on computer clusters. According to [18], there are three dimensions that characterize the challenges and opportunities of the large-scale data: (i) Volume (data volume); (ii) Velocity (data rate), and (iii) Variety (data variety). The Volume dimension refers to the amount of data generated and analyzed daily. The Velocity dimension relates to the speed at which the data are generated and to how often they are updated. Finally, the Variety dimension refers to the different types of data available. These definitions also determine the challenges of large-scale data: There is a huge volume of data, of different kinds and characteristics, which must be analyzed as soon as possible.

In the new digital reality, the amount of geospatial data generated by machines as a part of the Internet of Things will be even greater than that generated by people. Geospatial data collection is shifting from a data sparse to a data rich paradigm [19]. A huge number of mobile devices disclose information based on the current location of the user, contributing to the expansion of Volunteer Geographic Information (VGI) [20]. Whereas some years back geospatial data capture was based on technically demanding, accurate, expensive and complicated devices, we are now facing a situation where geospatial data acquisition is a commodity implemented in everyday devices such as smartphones used by many people. These devices are capable of acquiring environmental geospatial information at an unprecedented level with respect to greatly improved geometric accuracy, temporal resolution, and thematic granularity. Furthermore, Remote Automatic Weather Stations equipped with sensors collect meteorological conditions data in real time, continuous video streams from various study areas are being processed, while photographs, texts, and instant messages are regularly made public on social media such as Facebook and Twitter along with the exact spatial footprint of the content.

With the rise of social media we are also seeing vast amounts of data (e.g., Twitter feeds) that can be geo-tagged and assist in disaster management and emergency relief. In [21] Twitter streams have been evaluated to detect large-scale flooding events in Germany. In [22] Big Data technology is used for natural disaster monitoring and alerting system based on the social networking site, Twitter. In [23] the DiasterMapper tool is introduced, which is a CyberGIS framework that is able to synthesize data such as social media and socioeconomic data from many sources, to track disaster events, to produce maps, and to perform spatial and statistical analysis for disaster management. It uses a scalable distributed environment and a machine learning library to store, process and analyze massive social media data.

In this paper, we describe our effort to leverage Big Data technology for alerting and notifying mobile end users in case of a fire spread. In contrast to the aforementioned work based on social media information analysis, this work is based on spatial analysis of user location. The integration of Big Data analysis, Web-based GIS and wildfire modeling technologies into a web application creates state-of-the-art fire management services. To the best of our knowledge, our approach is the first attempt that blends Big Data technology with wildfire behavior modeling. The proposed methodology can form the basis of a general framework for confronting several natural disasters through Big Data analysis.

3   Methodology

Our approach relies on the development and utilization of a state-of-the-art platform, capable of alerting mobile end-users for upcoming fire spreads. The challenge is to constantly capture the exact location of the end-users and warn them in case they enter a "dangerous" area, i.e. when they approach an area simulated by a fire behavior simulation. The large number of end-users and the rapid change of their location presuppose the exploitation of scalable, efficient, timely and reliable state-of-the-art methods.

The proposed methodology combines the efficient tracking of the location of end-users with results of fire behavior simulations based on real-time fire ignitions. Fire behavior simulations are handled by the wildfire prevention and management system AEGIS, accessible to authorized end users through a web-based Graphical User Interface [15] or an innovative smartphone app, the AEGIS App [24]. At the core of the AEGIS platform, the Minimum Travel Time (MTT) algorithm runs as a powerful fire behavior prediction system [25]. The MTT algorithm can be used to compute potential short-term fire behavior characteristics (rate of spread, fireline intensity, time of arrival, flow paths, etc.). The MTT fire spread algorithm and the associated crown fire prediction models, as implemented through the FlamMap code libraries is by far the most widely used and tested fire simulation in the world [15].

Apart from the fire behavior modeling part, the Apache Hadoop[1] and Apache Spark[2] framework cluster computing frameworks are utilized, together with GIS (Geographical Information System) analysis. The Hadoop open-source work framework provides tools for organizing, managing and transforming large-scale data. On top of Hadoop runs the Hadoop Distributed File System (HDFS) which is a distributed file system designed to run on commodity hardware. Furthermore, the Spark Streaming module is utilized that enables scalable, high-throughput, fault-tolerant stream processing of live data streams. For GIS analysis and real-time handling of geospatial data, the ArcGIS GeoEvent Server[3] is utilized which acts as a GIS-based stream processing engine.

Figure 1 visualizes the methodology followed for utilizing Big Data for fire behavior modeling. Fire behavior simulations are conducted by end-users (i.e. firefighting personnel, emergency crews and other authorities) utilizing the AEGIS web app or the AEGIS app (Fig. 1-a). The results of the fire behavior simulations correspond to the simulated fire spread, fireline intensity and burned area of the simulation. These results are stored in a geodatabase (Fig. 1-b). At the same time, locations of end-users are ingested into the Hadoop cluster and analyzed by using the Hadoop cluster and the Spark streaming module (Fig. 1-c). Based on the area calculated for the arrival time of the fire spread, the GeoEvent module automatically creates areas of high

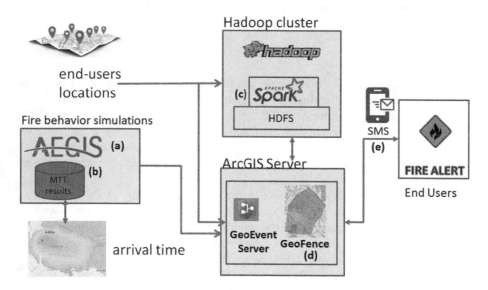

Fig. 1. Architecture of the proposed methodology

---

[1] http://hadoop.apache.org/.

[2] http://spark.apache.org/.

[3] http://server.arcgis.com/en/geoevent/.

risk called GeoFences (Fig. 1-d), tracks all footprints inside the areas and alerts the involved stakeholders if needed by SMS (Fig. 1-e).

4   Results and Conclusions

The Big Data technology is already exploited in areas such as Health, constructions, electronic commerce as well as in research issues of natural sciences, such as Biology and Physics [26]. Big Data analysis can improve the quality of life in modern mega-cities and contribute to the design of the "smart cities" of the future [27]. Considering that a very large part of the information handled in the modern digital world concerns geographical data, the contribution of the Big Data to the analysis of geographical data is a critical field of research.

The present work aims at exploiting Big Data's technology in forest fire simulation and in effectively alerting users while moving near high-risk areas. The application may potentially not only support civil protection and fire control services in the organization of innovative wildfire management plans but also contribute to the immediate and massive alert of end-users who are at risk during a fire outbreak.

A first prototype of the proposed approach will be tested by the local fire authorities in Lesvos Island, Greece throughout the 2017 wildfire season. The prototype does not allow the automatic registration of the end user's devices for tracking their location. Thus, currently, only ad-hoc tracking is supported. In a next step, we will develop a mobile application that will allow users to register their device in order to allow the continuous tracking of their location. Future work also includes wider application to a number of different geographic areas in Greece and expanding the scale of the individual testing areas.

Figure 2 shows the information flow for this prototype. The exact location of the user is tracked and analyzed in a continuous way. At the time the user enters the area specified by the fire behavior analysis, an alert is sent directly to his device in order to take the necessary actions.

The effectiveness of alerting end-users in real-time about wildfire ignitions simulations via a web-based application can be perceived as a significant technological advance for operational fire suppression activities. However, before moving from a research prototype to a more functional version of the application, privacy protection issues that may arise during the analysis of the information available must be highlighted, and ways to ensure the privacy of the information handled must be suggested.

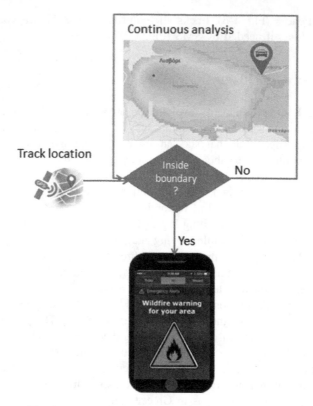

Fig. 2. Information flow for the example application

References

1. van der Zee, E., Scholten, H.: Spatial dimensions of big data: application of geographical concepts and spatial technology to the internet of things. In: Bessis, N., Dobre, C. (eds.) Big Data and Internet of Things: A Roadmap for Smart Environments. SCI, vol. 546, pp. 137–168. Springer, Cham (2014). doi:10.1007/978-3-319-05029-4_6
2. Chasparis, H., Eldawy, A.: Experimental evaluation of selectivity estimation on big spatial data. In: Proceedings of the Fourth International ACM Workshop on Managing and Mining Enriched Geo-Spatial Data, p. 8. ACM (2017)
3. Evans, D.: The internet of things: how the next evolution of the internet is changing everything. White Paper by Cisco Internet Business Solutions Group (IBSG) (2012)
4. Kalabokidis, K., Athanasis, N., Gagliardi, F., Karayiannis, F., Palaiologou, P., Parastatidis, S., Vasilakos, C.: Virtual Fire: a web-based GIS platform for forest fire control. Ecol. Inform. **16**, 62–69 (2013)
5. Pausas, J.G., et al.: Are wildfires a disaster in the Mediterranean basin? – a review. Int. J. Wildland Fire **17**(6), 713–723 (2008)
6. Kalabokidis, K., Athanasis, N., Vasilakos, C., Palaiologou, P.: Porting of a wildfire risk and fire spread application into a cloud computing environment. Int. J. Geogr. Inf. Sci. **28**(3), 541–552 (2014)

7. Alonso-Betanzos, A., et al.: An intelligent system for forest fire risk prediction and fire fighting management in Galicia. Expert Syst. Appl. **25**, 545–554 (2003)
8. Figueras Jové, J., Fonseca i Casas, P., Guasch Petit, A., Casanovas, J.: FireFight: a decision support system for forest fire containment. In: Teodorescu, H.-N., Kirschenbaum, A., Cojocaru, S., Bruderlein, C. (eds.) Improving Disaster Resilience and Mitigation - IT Means and Tools. NSPSSCES, pp. 293–305. Springer, Dordrecht (2014). doi:10.1007/978-94-017-9136-6_19
9. Losso, A., Corgnati, L., Bertoldo, S., Allegretti, M., Notarpietro, R., Perona, G.: SIRIO: an integrated forest fire monitoring, detection and decision support system–performance and results of the installation in Sanremo (Italy). In: Perona, G., Brebbia, C.A. (eds.) Modelling, Monitoring and Management of Forest Fires III, WIT Transactions on Ecology and The Environment, Southhampton, UK, pp. 79–90 (2012)
10. Gumusay, M.U., Sahin, K.: Visualization of forest fires interactively on the internet. Sci. Res. Essays **4**, 1163–1174 (2009)
11. Kalabokidis, K., Xanthopoulos, G., Moore, P., Caballero, D., Kallos, G., Llorens, J., Roussou, O., Vasilakos, C.: Decision support system for forest fire protection in the Euro-Mediterranean region. Eur. J. Forest Res. **131**, 597–608 (2012)
12. San-Miguel-Ayanz, J., Barbosa, P., Schmuck, G., Liberta, G., Schulte, E.: Towards a coherent forest fire information system in Europe: the European Forest Fire Information System (EFFIS). In: Viegas, D. (ed.) IV International Conference on Forest Fire Research, Luso, Coimbra, Portugal, 18–23 November 2002, pp. 5–16. Millpress Science Publishers (2002)
13. Noonan-Wright, E.K., Opperman, T.S., Finney, M.A., Zimmerman, G.T., Seli, R.C., Elenz, L.M., Calkin, D.E., Fiedler, J.R.: Developing the US wildland fire decision support system. J. Combust. **14**, 168473 (2011). doi:10.1155/2011/168473
14. Lee, B.S., Alexander, M.E., Hawkes, B.C., Lynham, T.J., Stocks, B.J., Englefield, P.: Information systems in support of wildland fire management decision making in Canada. Comput. Electron. Agr. **37**, 185–198 (2002)
15. Kalabokidis, K., Ager, A., Finney, M., Athanasis, N., Palaiologou, P., Vasilakos, C.: AEGIS: a wildfire prevention and management information system. Nat. Hazards Earth Syst. Sci. **16**(3), 643–661 (2016)
16. Villars, R.L., Olofson, C.W., Eastwood, M.: Big data: what it is and why you should care. White Paper, IDC, vol. 14 (2011)
17. Laney, D.: The importance of 'big data': a definition (2008)
18. Laney, D.: 3D data management: controlling data volume, velocity and variety. META Group Res. Note **6**, 70 (2001)
19. Li, S., Dragicevic, S., Castro, F.A., Sester, M., Winter, S., Coltekin, A., Pettit, C., Jiang, B., Haworth, J., Stein, A., Cheng, T.: Geospatial big data handling theory and methods: a review and research challenges. ISPRS J. Photogram. Remote Sens. **115**, 119–133 (2016)
20. Gao, S., Li, L., Li, W., Janowicz, K., Zhang, Y.: Constructing gazetteers from volunteered big geo-data based on hadoop. Comput. Environ. Urban Syst. **61**, 172–186 (2017)
21. Fuchs, G., Andrienko, N., Andrienko, G., Bothe, S., Stange, H.: Tracing the German centennial flood in the stream of tweets: first lessons learned. In: Proceedings of the Second ACM SIGSPATIAL International Workshop on Crowdsourced and Volunteered Geographic Information, pp. 31–38 (2013)
22. Dhamodaran, S., Sachin, K.R., Kumar, R.: Big data implementation of natural disaster monitoring and alerting system in real time social network using hadoop technology. Indian J. Sci. Technol. **8**(22), 1 (2015)

23. Huang, Q., Cervone, G., Jing, D., Chang, C.: DisasterMapper: a CyberGIS framework for disaster management using social media data. In: Proceedings of the 4th International ACM SIGSPATIAL Workshop on Analytics for Big Geospatial Data, pp. 1–6. ACM (2015)
24. Athanasis, N., Karagiannis, F., Palaiologou, P., Vasilakos, C., Kalabokidis, K.: AEGIS App: wildfire information management for windows phone devices. Procedia Comput. Sci. **56**, 544–549 (2016)
25. Finney, M.A.: Fire growth using minimum travel time methods. Can. J. For. Res. **32**, 1420–1424 (2002)
26. Chen, X., Lu, W., Liao, S.: A framework of developing a big data platform for construction waste management: a Hong Kong study. In: Wu, Y., Zheng, S., Luo, J., Wang, W., Mo, Z., Shan, L. (eds.) Proceedings of the 20th International Symposium on Advancement of Construction Management and Real Estate, pp. 1069–1076. Springer, Singapore (2017). doi:10.1007/978-981-10-0855-9_94
27. Moreno, M.V., Terroso-Sáenz, F., González-Vidal, A., Valdés-Vela, M., Skarmeta, A.F., Zamora, M.A., Chang, V.: Applicability of big data techniques to smart cities deployments. IEEE Trans. Ind. Inf. **13**(2), 800–809 (2017)

Big Data from a Business Perspective

Lasse Berntzen[1(✉)] and Milena Krumova[2]

[1] University College of Southeast Norway, Kongsberg, Norway
lasse.berntzen@usn.no
[2] Technical University of Sofia, Sofia, Bulgaria
mkrumova@tu-sofia.bg

Abstract. Big data provides new business opportunities. Companies can use big data to gain competitive advantage, and to make disruptive innovations. But also, data acquisition, transformation, integration, analysis and visualization can bring business opportunities. This paper looks at big data from a business perspective. Porter's value chain model is applied to big data, and is used as a framework to identify business opportunities. Some common revenue models from electronic commerce are presented and their applicability is discussed.

Keywords: Big data · Value chain · Revenue model

1 Introduction

Laney [1] defined big data as data having high volume, high velocity and/or high variety. High volume refers to large amounts of data demanding both specialized storage and processing. High velocity refers to streams of real-time data, e.g. from sensor networks or large-scale transaction systems. Finally, high variety is about dealing with data from different sources having different formats.

According to Marr [2], the real value of big data is not in the large volumes of data itself, but in the ability to analyse vast and complex data sets beyond anything we could ever do before. Due to recent advances in data analysis methods and cloud computing, the threshold for using big data has diminished.

Big data can be structured or unstructured. Structured data is simple to process, since the format is fixed. Transaction records, sensor readings and GPS (Global Positioning System) locations are examples of structured data. Unstructured data has no fixed format, and requires other means of processing. Unstructured data includes text, but also images and video content. Often unstructured data has some meta-data attached that provide some information about the unstructured data, e.g. tags describing the content, date and location of an image or a video.

Open data is data released under some license that gives users free access to the data. Open data is often shared through specialized thematic or organizational repositories. Open data made available through such repositories is predominantly structured data and it can be in a different format – human or machine readable. The existing technologies for data mining can be a good source for gaining business value based on big and open data.

© Springer International Publishing AG 2017
M. Themistocleous and V. Morabito (Eds.): EMCIS 2017, LNBIP 299, pp. 119–127, 2017.
DOI: 10.1007/978-3-319-65930-5_10

The most important source for open data is the world-wide-web itself. Not only does the world-wide-web provide access to huge amounts of documents, images and videos, but also acts as a platform for social media services. Recent research by the Open Data Institute [3] shows that companies use open data from government and non-government sources. Non-government sources include data generated by businesses such as opencorporates [4], not-for-profit associations such as p-lei.org [5], and many community projects such as DBpedia [6], DMOZ (Directory Mozilla) [7][1], GeoNames [8] and wikimapia [9]. This reflects the existence of a sharing economy, with companies frequently using data created by each other.

Our interest in big data value chain and revenue models is based on a real case [10], a project that developed a mobile platform for air-quality monitoring. One thing is to produce a set of mobile units collecting sensory information, another thing is to find a sustainable business model for the project. Should we sell the mobile units themselves, the data collected by the units, the results from using data analytics on the data, or the final visualizations of the data? Who would pay for such data, and what are the alternatives to sell the data?

2 The Big Data Value Chain

Porter [11] introduced the concept of the value chain as a general framework for thinking strategically about the activities in any business and assessing their relative cost and role in differentiation. The value chain is based on a set of primary activities to handle the transformation of inputs into finished products or services, as well as marketing and sales, and service. The support activities are firm infrastructure, human resources management, technology development and procurement. A visualization of the value chain is shown in Fig. 1.

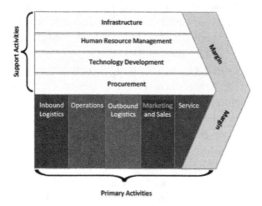

Fig. 1. Porter's value chain model

---

[1] The DMOZ website was closed in March 2017, but the editors set up a static mirror, and the data sets are still available.

Miller and Mork [12] applied Porter's value chain to big data by making a specific value chain for data. Their framework focused on the primary activities and specified three main activities: Data discovery, data integration and data exploitation. The data discovery phase includes the following activities:

- Collect and annotate: Create an inventory of data sources and the meta data that describe them.
- Prepare: Enable access to sources and set up access-control rules.
- Organize: Identify syntax, structure and semantics for each data source.
  The data integration phase has one activity:
- Integrate: Establish a common data representation of the data, Maintain data provenance.
  The data exploitation phase has three activities:
- Analyse: Analyse integrated data.
- Visualize: Present analytic results to decision makers as an interactive application that supports exploration and refinement.
- Make decisions: Determine what actions (if any) to take based on the interpreted results.

Curry [13] also applied Porter's value chain to big data. He used a similar approach by focusing on primary activities, and used the following activities: Data acquisition, data analysis, data curation, data storage and data usage.

Miller and Mork [12], and Curry [13] identified important activities of the data value chain, but their value chains do not consider the whole complexity of big data. Our framework addresses some missing aspects, and includes Porter's support activities.

2.1 Primary Activities

The primary activities consist of inbound logistics, operations, outbound logistics, marketing and sales, and service.

Inbound Logistics
The inbound logistics are about acquiring raw data. The raw data may come from many different sources including open data portals, and may be structured or unstructured. The data may also be delivered in real time. The inbound logistics will store the data (and metadata) for further processing and analysis, provide access, and set up access-control rules.

In some cases, it will be beneficial to do some pre-processing of the data close to their origins. This is particularly relevant for data produced in the Internet-of-Things. Cisco has named this pre-processing "fog computing" [14].

Operations
Operations are the activities required to transform raw data into a product useful for the end-user. According to Miller and Mork [12] it is necessary to establish a common data representation of the data. The transformation or analysis is done by software, using different approaches ranging from statistical processing to machine learning and neural networks.

Outbound Logistics

The outbound logistics handle the results of the transformation and analysis. The results need to be stored, visualized and disseminated. The results can then be used for decision-making. Miller and Mork [12] placed decision-making as part of operations. In our opinion, decision-making is done by the customers, and is therefore an outcome of the whole value chain.

Marketing and Sales

The results may be sold in the marketplace. The results may alternatively be released as open data (business open data). It is still possible to get profit through other revenue models. Revenue models are discussed later. Marketing is necessary to make potential users aware of the results. Often results are published on portals or broker platforms. If data or results are sold, there will be activities related to sales, contract negotiations, order handling, invoicing and accounting.

Service

Service typically includes data updates, training, and customer support. Customer support may be online help pages and guides, but also telephone support or real time chat functionality.

2.2   Secondary Activities

The secondary activities include procurement, human resource management, technological development, and firm infrastructure, Secondary activities may be candidates for outsourcing. Outsourcing is when a business buys services from another company instead of providing the services itself. Outsourcing may be beneficial for economic reasons, but also to obtain competence and robustness. A service provider can deliver specialized services that may be difficult to provide in-house. Outsourcing creates new business opportunities for companies that want to build specialized competence within a specific niche, and offer this competence to other businesses.

Procurement

Procurement of raw data may be an important activity requiring special skills on intellectual property rights and contract law. Such support activities are often outsourced due to the need for special competence. A smaller company may not be able to retain specialists in these fields. In other cases, it is necessary to make contractual agreements with data providers. A contract can regulate the use of data and payment conditions, but also set service level requirements, e.g. about availability/uptime and data quality.

Human Resource Management

Human resource management is also a candidate for outsourcing, just because this competence is getting more specialized due to regulatory environment, but also because companies may want some distance to human relations management functions in times of downscaling.

Technological Development

For big data, technological developments may be critical for competitive advantage. To outsource technology developments may be hazardous if the business relies on being ahead of its competitors. However, to develop own software is expensive. Modules can be bought from software developers and integrated with in-house developed software.

Firm Infrastructure

Infrastructure is buildings, computers, networks etc. Infrastructure is important, but may also be expensive. Infrastructure is a typical candidate for outsourcing. For big data companies outsourcing will often be about storage, network and computing power. Cloud computing can be used to ensure high availability and scalability.

We have so far looked at Porter's value chain model to identify primary and support activities. We have also discussed candidates for outsourcing among support activities. The next step is to look at business opportunities in the field of big data and some possible revenue models.

3 Big Data Business Opportunities

Revenue opportunities are closely connected to the value chain described above. Each part of the value chain can be separate units that can either be done in-house or outsourced to external partners.

According to Deloitte [15, 16], an open data value chain model distinguishes between five actors:

- Suppliers - Organisations that publish their data via an open interface, letting others use and reuse them.
- Aggregators - Organisations that collect aggregated open data and sometimes other proprietary data, typically on a particular theme, and find correlations, identify efficiencies, or visualise complex relationships.
- Developers - Organisations and software entrepreneurs that design, build, and sell web-based, tablet, or smartphone applications for individual consumption.
- Enrichers – Organisations - typically larger, established businesses - that use open data to enhance their existing products and services through better insights.
- Enablers - Organisations that facilitate the supply or use of open data, such as the AppsForX competition initiatives.

We now identify some specific activities related to the big data value chain, that may bring business opportunities. A modified visualization of the value chain is shown in Fig. 2.

Raw Data Production

Raw data can be produced from different kinds of sensors. In this context, a sensor is a device that produces data. Sensors can measure physical characteristics like temperature, weight, humidity, blood pressure etc., but can also be cameras or sound recorders. Sensor platforms can be developed and sold. The sensor data itself can be sold or alternatively be published as open data.

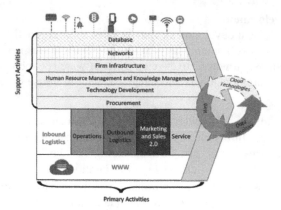

Fig. 2. Data driven business value chain model

Mining Data

Raw data can also be mined from the Internet. According to WorldWideWebSize [17] the indexed web now contains at least 4.54 billion pages. One valuable part of such data is social media, where citizens publish information relevant to their own situation. Data may be obtained through web site crawling or using API's (Application Programming Interfaces) for social media platforms.

Store Data

The raw data needs to be stored before transformation. The business can keep their own storage, but storage may also be rented from service providers, and stored in the cloud.

Analyse Data

Data analytics can be applied to produce the business value. Analytics can be used for diagnoses, predictions and trend discovery based on historical and current data.

Sell Results

Results can be sold directly to customers or through brokers/portals. An effective approach can be the use of CRM (Customer Relationship Management) 2.0 and gaining values through the web 2.0 channels, social media strategies, use social media marketing to listen, analyse, publish, and engage new and current business leads across networks.

Building Collaborative Networks of Businesses

Networks are pooling their business wisdom, sharing their experiences and learning from each other as never before [18]. This is an opportunity for business value co-creation, collaboration 2.0, etc.

Infrastructure

Big data creates business opportunities for companies providing network infrastructure and cloud computing (both processing and storage).

Human Resource Management (and Knowledge Management)

Since smaller companies often do not have specialized competence on human relations management, business opportunities exist for specialists in this field. The organizational knowledge base can contain the knowledge of the individual members of the organization applicable to the creation of value chain [19].

Technological Development

Software to handle big data may be developed in-house of bought from external providers. The development of software is a potential niche. Big data creates business opportunities for software developers to make software for a wide range of applications: Data acquisition, data transformation, data analysis and visualization.

Procurement

Procurement creates business opportunities for companies with specific competence in intellectual property rights and contract law.

4   Revenue Models

A revenue model explains how a business will make profit. The most obvious revenue model is to sell products or services. Transformation of raw data into new data sets or visualizations creates value for the customer. One example is to use data about consumer behaviour to decide on special offers or placement of products in stores. Another example is travel data to decide where to establish new airline connections or to change schedules for public transport. Such data can help the customer to compete and to generate more profit. Data or results can also be sold to a sponsor (e.g. a government organization) who then puts the data or results into the public domain.

But sales may not be the only revenue model. Laudon and Traver [20] propose some alternative models for use in e-commerce: Advertising, subscription, transaction fees and affiliates. These models may also be relevant for big data. A modified visualization of the value chain including the revenue models is shown in Fig. 3.

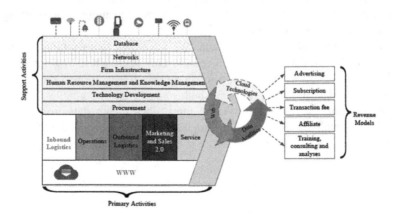

Fig. 3. Data driven business value chain and revenue model

Advertising

The results may be available for free, where customers will be exposed to adverts before getting access to the results. Several services use the advertising revenue model, like YouTube, Facebook and Google.

Subscription

A subscription model implies that customers subscribe to get access to results. This model is well suited for data that are updated.

Transaction Fee

A broker is someone that connects suppliers with consumers. By providing the necessary infrastructure to make this connection, it is possible to charge a transaction fee for establishing the connection. A broker can use a web site to connect suppliers and consumers of big data.

Affiliate

An affiliate receives a referral fee or percentage of sales by referring customers to another business. If a customer is downloading free data, and get an offer of buying a book relevant to the context, the affiliate can get some revenues. The publishing house O'Reilly has a good example of an affiliate program.

Training, Consulting and Analyses

In addition to the revenue models proposed by Laudon and Traver, it is also possible to collect revenues from training and consulting on big data.

5   Conclusions and Future Work

Most research on big data has focused on technological challenges and applications, not on business perspectives. Our motivation came from a specific project, where a sustainable business model was needed.

We used Porter's value chain model [11] as a framework to identify business opportunities related to big data. The analysis of the value chain identified many different opportunities ranging from producing units for data collection to visualizing results. The analysis also revealed several areas fit for outsourcing. Outsourcing may create niches for specialized businesses.

Big data is often published as open data. We have therefore examined alternative revenue models for big data and big data results. One approach is to get a sponsor to publish the data or results as open data. Another approach is to allow for advertising on the site where the data or results are published.

We aim to continue researching viable approaches to big data from a business perspective. This will be done through empirical studies combined with exploring alternative approaches to business models, e.g. business canvas modelling.

References

1. Laney, D.: 3D data management: controlling data, volume, velocity, and variety. Technical report. META Group (2001)
2. Marr, B.: Big Data – Using Smart Big Data Analytics and Metrics to Make Better Decisions and Improve Performance. Wiley, Hoboken (2015)
3. Open Data Institute: Research: Open data means business. https://theodi.org/open-data-means-business. Accessed 09 June 2017
4. Opencorporates: The Open Database of the Corporate World. http://opencorporates.com. Accessed 09 June 2017
5. P-lei.org.: http://p-lei.org. Accessed 09 June 2017
6. DBpedia: http://wiki.dbpedia.org. Accessed 09 June 2017
7. DMOZ (Directory Mozilla) static mirror web site. http://dmoztools.org. Accessed 09 June 2017
8. GeoNames. http://www.geonames.org. Accessed 09 June 2017
9. Wikimapia. http://www.wikimapia.org. Accessed 09 June 2017
10. Florea, A., Berntzen, L., Rohde Johannessen, M., Stoica, D., Naicu, I.S., Cazan, V: Low cost mobile embedded system for air quality monitoring - air quality real-time monitoring in order to preserve citizens' health. In: The Sixth International Conference on Smart Cities, Systems, Devices and Technologies, (SMART 2017), 25–29 June 2017, Venice, Italy. IARIA (2017)
11. Porter, M.: Competitive Advantage, Creating and Sustaining Superior Performance. The Free Press, Imphal (1985)
12. Miller, H.G., Mork, P: From data to decisions: a value chain for big data. IT Professional, January/February 2013. IEEE Computer Society (2013)
13. Curry, E.: The big data value chain: definitions, concepts, and theoretical approaches. In: Cavanillas, J.M., Curry, E., Wahlster, W. (eds.) New Horizons for a Data-Driven Economy, pp. 29–37. Springer, Cham (2016). doi:10.1007/978-3-319-21569-3_3
14. Cisco: Fog Computing and the Internet of Things: Extend the Cloud to Where the Things Are (White Paper). www.cisco.com/c/dam/en_us/solutions/trends/iot/docs/computing-overview.pdf. Accessed 09 June 2017
15. Delloite: Open Growth – Stimulating demand for open data in the UK. https://www2.deloitte.com/content/dam/Deloitte/uk/Documents/deloitte-analytics/open-growth.pdf. Accessed 09 June 2017
16. DXC.Technology: Open government data – learn why it will change your digital life forever. https://businessvalueexchange.com/wp-content/uploads/sites/14/2015/09/Open-Government-Data.pdf. Accessed 09 June 2017
17. WorldWideWebSize.com.: http://worldwidewebsize.com. Accessed 09 June 2017
18. Greer, J: Business models of the future: emerging value creation. http://www.accaglobal.com/content/dam/ACCA_Global/Technical/Future/pi-emerging-business-models-FINAL-26-01-2017.pdf. Accessed 09 June 2017
19. Wissenmanagement Forum: An Illustrated Guide to Knowledge Management. http://wm-forum.org/files/2014/01/An_Illustrated_Guide_to_Knowledge_Management.pdf. Accessed 09 June 2017
20. Laudon, K.C., Traver, C.G.: E-Commerce 2016: Business, Technology, Society, vol. 12. Pearson, London (2016)

Predicting Customer Churn: Customer Behavior Forecasting for Subscription-Based Organizations

Leonidas Katelaris$^{(\boxtimes)}$ and Marinos Themistocleous

University of Piraeus, Piraeus, Greece
{lkatelaris,mthemist}@unipi.gr

Abstract. Churn is the opposite of growth. Losing customers has serious impact on company's overall performance. More specifically, means lost in sales and revenue, but also negative sentiment and potential negative impact to organization's image for the competition. The increased importance of managing churn in subscription-based organizations, lead various efforts by subscription-based organizations to face the problem. Both, academic researchers and business practitioners, focusing on techniques around customer behavior forecasting. During the last years, various technologies have been used to forecast customer behavior in subscription-based organizations. To investigate further this area this paper aims to report on the research issues around customer churn and investigate previous customer churn prediction approaches in order to propose a new conceptual model for customer behavior forecasting.

Keywords: Churn prediction · Customer churn · Behavior forecasting · Prediction

1 Introduction

Customer churn is a critical and challenging issue affecting subscription-based organizations, in today's rapidly growing and competitive market. Customer churn is defined as the situation when a customer stops collaboration with a subscription-based company or service [1–3]. The reality is that, getting new customers is an obvious win [4], but due to the associated high cost of acquiring new customers [5], many subscription-based organizations are working on how to retain existing customers. Customer churn affects organizations' growth with impact on their overall performance, decrease in sales and revenue loses, potential cause for negative impact to organization's image in the competitive market.

Several prediction algorithms have been proposed to address the issue of churn, such as regression [6], neural networks [3,7], random forest [8] and support vector machines (SVM) [9,10] which differ in terms of statistical technique, number of variables in the model, time allocation between tasks etc. Although, the challenge of selecting the appropriate prediction method for each case still remains a research problem.

© Springer International Publishing AG 2017
M. Themistocleous and V. Morabito (Eds.): EMCIS 2017, LNBIP 299, pp. 128–135, 2017.
DOI: 10.1007/978-3-319-65930-5_11

This study contributes to the existing literature by investigating the issues around churn prediction in subscription-based organizations. More over, report and analysis of churn causes and financial parameters around customer attrition is made. According to that, an analysis of the characteristics for some of the most known prediction methods (regression analysis, neural networks, random forest, support vector machines) is made, to report positives and negatives for each technique.

Following is the *Results and Discussion* section, where we discuss the results of prediction approaches and make hypothesis to be investigate in next steps of our research and they will conclude to our conceptual model for customer behavior forecasting.

2   Customer Churn

2.1   What is it?

Churn is a major problem affecting subscription-based organizations, which occurs when a customer stops collaboration with a company for another competitor in the market [1–3]. The customer shifting between providers is described as churn rate and is one of the most challenging and critical problems faced by subscription-based companies during the last years [4,5,11]. Churn rate can be described as a Key Performance Indicator (KPI), which gives the number of customers' lost during a specified time period divided by the average total number of customers over that specific period of time [12].

2.2   Why Churn Happens?

As mention above (Introduction), churn is inextricably linked with company's prosperity and growth. Losing customers, lead subscription-based organizations to multiple negative consequences. Such negative consequences include the lower numbers in sales and negative impact on organization's image.

The reasons why customers leave a company varies, but those reasons who can be considered as most significant for churn are few. In an attempt to name those causes of customer churn, literature reports some, but three are the causes which can be characterized as the primary causes of churn [13–16].

- **Poor customer service experience:** Different studies [17–19] highlight the need for a good customer service in the company and how that affect in customer satisfaction rate. A poor customer service experience creates unsatisfied customers which will probably abandon the company. Customer service experience is connected with company's knowledge about their customers' needs and expectations. Prioritizing customer support in a way of delivering high quality products/services to customers is for sure a correct measure against churn.

- **Unsuccessful customer on-boarding:** On-boarding is very important for organization health and growth, because is where a company can make a big positive impact. In a customer life-cycle for company two are the most important milestones. The first is when a customer comes to your business and sings-up for your product/service. The second milestone is when a customer achieve his first success with your product/service.
- **Weak customer relationship building:** Customers need to hear that they are collaborating with a professional company with goals and ready to help them face a problem with their product/service. If customers stop hearing the company or receiving news about releases of products/services from the company, the company itself wont be able to get the right feedback from them. The consequence of not hearing your customers is the churn from that organization.

While the above causes are leading churn, the normative literature reports three types of churners [20]:

- **Active churner (voluntary):** This category of churners, describes the customers who decide to leave a provider for the next competitor in the market. Multiple reasons for this including low service experience, high cost, quality of service, etc.
- **Passive churner (non-voluntary):** This category describes churns that occur, due to provider's action to discontinue the contract itself.
- **Rotational churner (incidental):** This last category of churners describes, the customers who leave a company, without the knowledge of both parties (company or customer), where we have both parties (company or customer), may suddenly discontinue collaboration without prior notification. Reasons for this vary and may include financial problems, change of geographical location of customer where there is no service from current provider.

2.3 Customer Attrition: Investigating Customer Churn

For the most part, subscription-based organizations which are driven by promotional actions like telecommunication service providers [1,2,4,11], banks [10,21,22], online gaming industry [23,24] etc., are seen more vulnerable to churn. Despite the different strategies and tactics used by them to attract new customers, they are all often impelled by insights from historical data they have collected in their CRM systems. Although, the critical analysis of the literature around the area of churn prediction, will lead to assumption that, historical customer data does not always lead to accurate churn predictions.

Most of the subscription-based organizations, which analyze data in order to define their customer behavior are focusing on *Active churners* and secondly on *Passive churners* category. The *Active churners* category is the most important for subscription-based organizations followed by *Passive churners*, because include those subscribers who churn for reasons that may be avoided if a trustworthy forecasting process is used in the organization. The third churner's category is difficult to predict, because in that way a possible churn will be for unknown reasons or very difficult to predict.

Customer retention has quantifiable impact on revenue, while acquiring customers is critical to business growth [5], it make even more sense on new companies in the area of subscription-based services. Even more, increase in retention has more positive effects in a company. Customer retention specifically lows churn rates increases revenue and customer lifetime value (CLV) [25]. Many studies including Earl Sass er of the Harvard Business School alongside with Bain & Company [26] and Keiningham's [27] agree that a 5% increase in customer retention has impact in revenue of the company between 25 and 95%. Also studies [28] show that selling service/product to existing customers is 5x more likely to repurchase and 4x more likely to refer their friends, while selling to new customers is between 5–20%. According to that, existing customers are 50% more likely to try new products and spent even 31% more compared to newcomers.

2.4   Churn Prediction Approaches

Churn prediction had clearly been observed by subscription-based organizations in recent decade, as managing churn rate accurately will have major impact on organization's growth and prosper [5]. To accomplish this, various predictive modeling techniques had been proposed. Different churn prediction techniques have been used by subscription-based organizations to forecast their customer behavior, one of the most common is customer-centric approach [11], which is based on data stored in their CRM systems. The management and analysis of those data could give very useful insights for possible churners [29]. Different algorithms used to analyze historic data of customers and predict churn for the organizations. All those proposed algorithms depend on different methods and differ in their statistical approaches, amount of variables, in the prediction model etc. Below an overview of the most known churn prediction methods is given.

Random Forest: Random forest method, has been introduced by Breiman [8] and present a new and powerfully statistical classifier, which found to perform very well compared to many other classifiers. Random forest introduced a solution to decision trees' problem where small change in the data in most of the times results in very different series of splits, which in turn lowers accuracy when validating the training data [30]. Random forest uses a randomly selected subset of m predictors to grow trees on bootstrap sample of the training data. Followed, the large number of generated trees, each tree votes for the most popular class (m which present the number of the predictors is much smaller than the total number of variables in the model). Aggregating all votes from different trees, conclude to class label prediction. This technique is easy to implement because of the small number of parameters need to be set (m:predictors, total number of trees to be generated).

Regression: Regression technique is considered to be an effective technique for customer behavior forecasting [6,31]. Regression method uses IBM's SPSS

(Statistical Package for the Social Sciences) to calculate the standard error rate for the variables to be used in the regression model. Following the error rate calculation, only the variables with the most significance will be included in the model. While predicting non-churners linear regression model overcomes other predicting methods like decision trees and neural networks [32], fails to predict more difficult churns customers.

Neural Networks: Neural Networks are common in finding complexity of no-linear functions. Neural Networks have been taken into serious consideration by researchers and practitioners because they provide a prediction alongside with its possibility. In neural networks the idea is that each variable is associated with a weight. Then a combination of weighted variables process is running to develop the prediction model. The main disadvantage of neural networks in prediction process is that they need a large volume of data set and are time consuming while trying to weight the predictor variables. However, the achieved accuracy of prediction model in most of the times, overcomes in performance other prediction techniques like logic regression [32,33].

Support Vector Machines: Support vector machine is a very popular discrimination classifier [10,34]. Using support vector machine in a set of variables, each variable separated to each of two categories. Then the support vector machine training algorithm develops a model where it assigns new variables to one or the other category. The output of support vector machine prediction model is points in space divided by a clear gap. Support vector machine as prediction method is equivalent for data-set in which data are linearly separable, but the truth is that in real world data is often not linearly separable. In addition to performing linear classification,support vector machines use kernel function to perform non-linear classification end enhance feasibility of linear separation.

3   Results and Discussion

As we are driving into subscription economy, where customers looking new ways to engage with companies, customer behavior forecasting techniques become even more popular. In the previous section (*Churn Prediction Approaches*), a presentation and analysis of main facts for four of the most popular prediction techniques is made. Some of the prediction methods are performing better in noisy data (random forest) [8], while are not performing so well on data-sets where data are extremely unbalanced. Others (neural networks), overcome in accuracy of prediction but they need reasonable volume of data and time to make predictions.

Big and well-known players of the subscription-based domain like Netfix, Spotify, Amazon etc., are the real-life examples to subscription economy model shifting. For those companies churn is translated to millions loss of revenue and even more expenses in customer acquisition campaigns. All previously mentioned prediction methods present positives and negatives and differ in various terms, but all of them are performing their predictions on data where come from the CRM of the organizations and can be described as historical data. In this point we can make our hypothesis about a new conceptual model for predicting customer behavior.

3.1 Conceptualization of Comprehensive Prediction Model

At this point, we are not ready to choose the right prediction algorithm for our model, but we can describe the main components of the conceptual model of our comprehensive prediction model which will be using some machine learning techniques and it will be presented in the next steps of our research. The main conceptualization of our comprehensive prediction model is that, it will be able to define updated customer behavior patterns. All those patterns will be included in the prediction data-set development and enhance accuracy of churn prediction for more difficult customer churns.

Reports in literature agree that customer satisfaction is the key for decreasing churn [13,16]. Also is known that, all old customers come with their history trail. The difficult process is to create the trail for newcomers. We assume that the comprehensive prediction model it has to give more importance to the development of the data-set, which will train our machine learning algorithm.

Challenges in developing as much more accurate customer behavior pattern, include the creation of old customer profile from our historical data and combination of "ad-hoc" KPIs and other KPIs. We make a distinction between KPIs, in our prediction model. As ad-hoc KPIs we will describe a set of KPIs to be included in prediction model under some circumstances which will be well correlated to customer behavior pattern and they will not be included in every prediction. Another challenge is the development of the newcomers trail, which will be critical for our model success. During this challenging process we will try to define some of the "ad-hoc" KPIs and all the aspects to be included in the sign-up process. The way we will develop the sign-up process will give us the newcomers trail and the final customer behavior pattern to be taken into consideration our the comprehensive prediction model and proceed to do churn predictions.

4 Conclusion and Future Work

In this paper, report on the research issues around customer churn and issues around churn prediction in subscription-based organizations is made. Furthermore, churn causes and financial parameters around churn for the organizations are reported and analysed. According to that, an analysis of the characteristics for some of the most known prediction methods (regression analysis, neural networks, random forest, support vector machines) is made, to report positives and negatives for each prediction technique. Literature review gave us the confident for the next steps of our research which include the development of the comprehensive prediction model. In order, for the comprehensive prediction model to be able to define customer behavior patterns for the prediction model, we may define the KPIs for customer profiling.

Acknowledgment. The publication of this paper has been partly supported by the University of Piraeus Research Centre.

References

1. Huang, B., Tahar Kechadi, M., Buckley, B.: Customer churn prediction in telecommunications. Expert Syst. Appl. **39**(1), 1414–1425 (2012)
2. Yan, L., Wolniewicz, R.H., Dodier, R.: Predicting customer behavior in telecommunications. IEEE Intell. Syst. **19**(2), 50–58 (2004)
3. Hung, S.-Y., Yen, D.C., Wang, H.-Y.: Applying data mining to telecom churn management. Expert Syst. Appl. **31**(3), 515–524 (2006)
4. Amin, A., et al.: Comparing oversampling techniques to handle the class imbalance problem: a customer churn prediction case study. IEEE Access **4**, 7940–7957 (2016)
5. Hadden, J., et al.: Computer assisted customer churn management: state-of-the-art and future trends. Comput. Oper. Res. **34**(10), 2902–2917 (2007)
6. Noh, H., Kwak, M., Han, I.: Improving the prediction performance of customer behavior through multiple imputation. Intell. Data Anal. **8**(6), 563–577 (2004)
7. Tsai, C.-F., Lu, Y.-H.: Customer churn prediction by hybrid neural networks. Expert Syst. Appl. **36**(10), 12547–12553 (2009)
8. Breiman, L.: Random forests. Mach. learn. **451**, 5–32 (2001)
9. Huang, K., et al.: Sparse learning for support vector classification. Pattern Recognit. Lett. **31**(13), 1944–1951 (2010)
10. Farquad, M.A.H., Ravi, V., Bapi Raju, S.: Churn prediction using comprehensible support vector machine: an analytical CRM application. Appl. Soft Comput. **19**, 31–40 (2014)
11. Amin, A., Shehzad, S., Khan, C., Ali, I., Anwar, S.: Churn prediction in telecommunication industry using rough set approach. In: Camacho, D., Kim, S.-W., Trawiński, B. (eds.) New Trends in Computational Collective Intelligence. SCI, vol. 572, pp. 83–95. Springer, Cham (2015). doi:10.1007/978-3-319-10774-5_8
12. Techtarget.com: Churn rate, February 2017. http://searchcrm.techtarget.com/definition/churn-rate
13. Kumar, V., Reinartz, W.: Creating enduring customer value. J. Mark. **80**(6), 36–68 (2016)
14. Melgarejo Galvan, A.R., Clavo Navarro, K.R.: Big data architecture for predicting churn risk in mobile phone companies. In: Lossio-Ventura, J.A., Alatrista-Salas, H. (eds.) SIMBig 2015-2016. CCIS, vol. 656, pp. 120–132. Springer, Cham (2017). doi:10.1007/978-3-319-55209-5_10
15. Mwegerano, A.M., et al.: Managing customer issues through a support channel network (2014)
16. Keaveney, S.M.: Customer switching behavior in service industries an exploratory study. J. Mark. **59**, 71–82 (1995)
17. Stevenson, W.J., Hojati, M.: Operations Management, vol. 8. McGraw-Hill/Irwin, Boston (2007)

18. Sain, S., Wilde, S.: Customer Knowledge Management. Springer International Publishing, Cham (2014)
19. McColl-Kennedy, J.R.: Customer satisfaction, assessment, intentions and outcome behaviors of dyadic service encounters: a conceyfual model. In: Ford, J.B., Honeycutt Jr., E.D. (eds.) Proceedings of the 1998 Academy of Marketing Science (AMS) Annual Conference. DMSPAMS, pp. 48–54. Springer, Cham (2015). doi:10. 1007/978-3-319-13084-2_10
20. Lazarov, V., Capota. M.: Churn prediction. Bus. Anal. Course. TUM Comput. Sci. (2007)
21. Bloemer, J., De Ruyter, K., Peeters, P.: Investigating drivers of bank loyalty: the complex relationship between image, service quality and satisfaction. Int. J. Bank Mark. 16(7), 276–286 (1998)
22. McDonald, L.M., Rundle-Thiele, S.: Corporate social responsibility and bank customer satisfaction: a research agenda. Int. J. Bank. Mark. 26, 170–182 (2008)
23. Coussement, K., De Bock, K.W.: Customer churn prediction in the online gambling industry the beneficial effect of ensemble learning. J. Bus. Res. 66(9), 1629–1636 (2013)
24. Suznjevic, M., Stupar, I., Matijasevic, M.: MMORPG player behavior model based on player action categories. In: Proceedings of the 10th Annual Workshop on Network and Systems Support for Games, p. 6. IEEE Press (2011)
25. Abe, M.: Deriving customer lifetime value from RFM Measures: insights into customer retention and acquisition (2015)
26. Reichheld, F.F., Schefter, F.: E-loyalty: your secret weapon on the web. Harv. Bus. Rev. 78(4), 105–113 (2000)
27. Keiningham, T.L., et al.: The value of different customer satisfaction and loyalty metrics in predicting customer retention, recommendation, and share-of-wallet. Manag. Serv. Qual. Int. J. 17(4), 361–384 (2007)
28. Invesp: https://www.invespcro.com/blog/customer-acquisition-retention
29. Burez, J., Van den Poel, D.: CRM at a pay-TV company: using analytical models to reduce customer attrition by targeted marketing for subscription services. Expert Syst. Appl. 32(2), 277–288 (2007)
30. Friedman, J., Hastie, T., Tibshirani, R.: The Elements of Statistical Learning. Springer Series in Statistics, vol. 1. Springer, New York (2001)
31. Lawson, C., Montgomery, D.C.: Logistic regression analysis of customer satisfaction data. Qualit. Reliab. Eng. Int. 22(8), 971–984 (2006)
32. Hadden, J., et al.: Churn prediction: does technology matter. Int. J. Intell. Technol. 1(2), 104–110 (2006)
33. Au, W.-H., Chan, K.C.C., Yao, X.: A novel evolutionary data mining algorithm with applications to churn prediction. IEEE Trans. Evol. Comput. 7(6), 532–545 (2003)
34. Zhao, Y., Li, B., Li, X., Liu, W., Ren, S.: Customer churn prediction using improved one-class support vector machine. In: Li, X., Wang, S., Dong, Z.Y. (eds.) ADMA 2005. LNCS, vol. 3584, pp. 300–306. Springer, Heidelberg (2005). doi:10.1007/ 11527503_36

Digital Services, Social Media and Digital Collaboration

MOOCS' Potential for Democratizing Education: An Analysis from the Perspective of Access to Technology

Valéria F. Moura[1(✉)], Cesar A. Souza[1], José D. Oliveira Neto[2], and Adriana B.N. Viana[1]

[1] University of São Paulo, São Paulo, Brazil
{valeria.feitosa.vv,calesou,backx}@usp.br
[2] University of São Paulo, Ribeirão Preto, Brazil
dutra@usp.br

Abstract. MOOCs can be considered the fifth generation of distance education and can potentially provide access to quality education to students with disadvantaged background in developing countries. However, ensuring that MOOCs provide access to education to these students involves ensuring access to the necessary ICT infrastructure. The question is then whether the access to technology is truly democratic, especially in those countries. The aim of this study is then to assess the potential for democratization of quality education provided by MOOCs, considering the perspective of access to technology. To reach this objective, data obtained from 4.784 students from 27 universities in 9 Southern Hemisphere countries were analyzed. Results show that the potential for democratization of education provided by the MOOCs is limited by access barriers to technology, so that to fully realize their potential, governmental actions are necessary to democratize access to technology and promote the reduction of the digital skill gap.

Keywords: Massive Open Online Courses (MOOCs) · Higher education · Digital divide · ICT access policies

1 Introduction

The 21st century has been marked by a great volume of technologies and changes that create a great demand for quality in higher education. Distance education and online learning are largely mentioned as options for meeting this demand, especially Massive Open Online Courses (MOOCs), presented as a revolution for e-learning [1].

MOOCs can be considered as the fifth generation of distance education and have the potential to reach and assist students with disadvantaged background and in developing countries, providing access to good education for those who would not have it any other way [2]. However, ensuring that MOOCs provide access to satisfactory education for these students involves ensuring access to the necessary IT infrastructure. Without recognizing these fundamental economic and technological disparities, it is not possible to reach the population that most needs access to education [3]. Therefore, we may identify MOOCs as a promise to democratize education by using technology. But, on the other hand, the question is whether the access to technology is truly democratic. The

© Springer International Publishing AG 2017
M. Themistocleous and V. Morabito (Eds.): EMCIS 2017, LNBIP 299, pp. 139–153, 2017.
DOI: 10.1007/978-3-319-65930-5_12

objective of this study is to assess the potential for democratization of quality education provided by MOOCs. In order to reach the proposed objective, this article analyzes data obtained from 4.784 university students from 27 higher education institutions in 9 countries of the Southern Hemisphere (Brazil, Chile, Colombia, Ghana, Kenya, South Africa, India, Indonesia and Malaysia) through the Research on Open Educational Resources for Development (ROER4D) project, which aims to improve understanding on the use and impact of Open Educational Resources (OER) to improve educational policy, practice and research in developing countries. Thus, this initiative seeks to provide evidence-based research from several countries in South America, Sub-Saharan Africa and South/Southeast Asia [4].

The article is structured as follows: initially we present a contextualization of MOOCS, then we show the benefits and barriers related to MOOCs, mainly regarding technological barriers, later we present the methodology followed by the results of the study, and we end this article with the final considerations.

2   MOOCs Definition and Contextualization

The first Massive Open Online Course (MOOC) was conducted in 2008 by George Siemens and Stephen Downes [5]. MOOCs are characterized by the absence of formal requirements for enrollment and free participation, by the content being delivered entirely online in an asynchronous manner, by not requiring a link with universities and the lack of penalty for evasion [6]. By definition, MOOCs are offered in virtual environments, with online registration and use of videos, blogs, etc. [7], and may or may not be linked to universities. The "M" refers to "Massive", which means that thousands of people can simultaneously take courses [8]. This is the characteristic that differentiates MOOCs from other e-learning experiences, since it relates to the capacity and size of the network to generate new knowledge, thus reflecting participatory learning, respecting the diversity of the large number of participants [9]. The first "O" refers to "open", which means that participation in courses is not restricted by geographical location, age or financial resources. This aspect may include open technology, open software, open content, open evaluation process, open registration, and open educational resources [5]. The second "O" means "online," meaning that MOOCs are exclusively Internet-based courses. Finally, "C" refers to "Courses".

In general, there are two formats of MOOCs [10]: cMOOCs and xMOOCs, which have been widely adopted in the literature. The cMOOCs have a connectivist nature and are aligned with the principle of open education [11], they focus on the creation and generation of knowledge, emphasize connected and collaborative learning [12] and share the notion of free worldwide participation in a non-credit course [13]. On the other hand, xMOOCs follow a more traditional approach to learning through video conferences, short questionnaires and peer evaluations [13], focusing on knowledge duplication [12] and being more structured [8]. Despite this division, we may observe that MOOCs tend to be closer to one of the ends of this spectrum, but incorporate elements of both models [8].

Massive nature is the aspect that most differentiates MOOCs from other online learning experiences [14]. Scalability combined with openness enables MOOCs to increase accessibility to higher education [8, 12], democratizing education [15] and expanding the dissemination of knowledge [12] and the chances of creating enriching connections with people throughout the world. In addition, they provide a large amount of data, allowing a better understanding of students' behavior to improve their courses [16].

MOOCs are an innovation for higher education as they offer a fundamentally innovative way of learning [8, 12], encouraging universities to rethink the curriculum development process for more open and flexible educational models [14], enabling lifelong autonomous learning. In this context, MOOCs bring the need to redefine concepts to describe and understand user behavior while reaching a new audience and improving the quality of education [17].

3   Access to Technology: A Challenge to Be Overcome to Enhance the Potential of MOOCs

Despite the many benefits of MOOCs, we may observe that the phenomenon has experienced a fast cycle of hype and disappointment. Meisenhelder [3], for example, has a very critical position in relation to these courses, stating that they do not meet two basic principles to be truly transformative: they do not expand access to low-income students who are disproportionately excluded from the higher education system and they do not provide education with a greater emphasis on the learning needs of the individual. In fact, the main criticisms identified in the literature relate to high dropout rates [14, 18], and the limited impact on the access to education by disadvantaged groups [8, 18], factors that conspicuously reduce the potential of democratization of higher education proposed by MOOCs.

In developing countries, ensuring access to quality education through MOOCs for the population necessarily involves ensuring the access to the needed hardware and IT infrastructure. It is not possible to reach the population that most needs access to education without recognizing these economic and technological disparities [3].

Literature on distance learning has indicated the importance of adapting technology for high quality learning, suggesting that the technological tool should not be reduced to just putting courses online, and so it should contemplate a number of features that allow access to a large number of users, technologies focused on users and their needs, wireless technologies, Web 2.0 and video, tools that support synchronous and asynchronous communication, and tools for access to information and social networking [2]. Fortunately, most of these technologies and learning management tools are being integrated into higher education. MOOC platforms such as Coursera, Udacity, and EdX are making online environments more interactive, integrating videos, gamification systems and online tests [2].

However, these are not the only factors to be considered. We must also consider that there are several barriers related to Internet and hardware access and digital skills [19, 20], which reduce the potential for using MOOCs. In developing countries, despite the availability of technologies in urban areas, rural and poor populations continue to be deprived of this infrastructure and the literature has consistently considered the connection to broadband Internet as the main difficulty in using open educational resources in these regions [21, 22]. Even if the infrastructure is somehow present, the operational cost may mean that the less privileged groups do not have access to the Internet, either because of the difficulty to pay for high-speed Internet or to acquire the necessary equipment for access [23]. In fact, Internet access is not universalized and Internet penetration is 66.7% in South America, 45.2% in Asia and only 27.7% in Africa, [24] a situation that minimizes the potential of MOOCs to reach the population that most needs access to quality education.

When evaluated considering hardware access, MOOCs also have some barriers, since most participants do not have differences in their profiles only, but also in access devices, requiring the courses to be developed with technologies that adapt to different forms of access [12]. In addition, Amemado [2] argues that technology needs to be user-dominated to be fully utilized and that technology-savvy people are most likely to benefit from participating in MOOCs. None of these factors suggests that MOOCs cannot be widely disseminated beyond the limits of current university practice, but they entail serious additional costs.

4   Methodology

This study is characterized as quantitative and descriptive, and the data used here come from a survey that composes a research project developed for the Research on Open Educational Resources for Development (ROER4D) initiative, led by the University of Cape Town in South Africa and sponsored by the Canadian International Development Research Center (IDRC). The ROER4D project aims to improve the understanding of the use and impact of OERs to improve educational policy, practice and research in developing countries in order to ensure that educational policy development initiatives proposed by philanthropic foundations and governments are effectively achieving the expected results, that is, enabling accessible education in the Southern Hemisphere through the development of affordable, socially acceptable and high quality educational resources [4].

The ROER4D initiative seeks to provide evidence-based research from several countries in South America, Sub-Saharan Africa and South/Southeast Asia. Thus, all data from this survey come from 9 countries in these regions: Brazil, Chile, Colombia, Ghana, Kenya, South Africa, India, Indonesia and Malaysia. Next, we present information related to the selection of regions for data collection, definition of sampling, and elaboration of the instrument for data collection, survey and analysis.

The survey methodology was chosen for the ROER4D project as the most adequate way to collect, compare and explore hypotheses related to quantitative indicators related to OER and MOOCs use in the selected countries. Higher education professors and students were chosen as the target population, as this public is most acquainted in using and have easier access to the researched technologies.

4.1   Unit of Analysis and Sampling

The unit of analysis is composed of university students from 27 higher education institutions in the countries surveyed. The distribution of institutions by country is shown in Table 1. The selection of individuals was performed using the method of random

Table 1. Participant universities by country

Country	University
Brazil	Claretiano
	Universidade de São Paulo
Chile	Instituto de Estudios Bancarios Guillermo Subercas
	Universidad de Tarapacá
Colombia	Universidad Santo Tomás
	Universidad Nacional de Colombia
Ghana	Catholic Institute of Business and Technology
	Kwame Nkurumah University of Science and Tech.
	The University of Ghana
	University of Cape Coast
India	Gauhati University
	University of Delhi
Indonesia	Universitas Terbuka
	University of Mercu Buana
	University of Nasional
	University of Pancasila
Kenya	Great Lakes University
	Jomokenyatta University of Agriculture and Technology
	Maseno University
	Tangaza University College
Malaysia	Disted college
	KDU college
	University of Malaysia
	Wawasan Open University
South Africa	The University of Cape Town
	Unisa
	University of Pretoria

sampling, avoiding bias in the selection, as this method enables equal participation opportunity for all individuals [25]. To identify the necessary sample, 30 courses were selected in the semester of the collection through a random selection for each country. Next, teachers of these courses were approached in order to identify 10 teachers with courses with more than 30 students who were interested in participating in the research. Finally, 20 students from each class were randomly selected to answer the survey.

4.2   Data Collection Instrument

To elaborate the questionnaire for the survey, a comprehensive review of literature was carried out, as well as discussions with specialists in OER. Definitions of the constructs were centered on the factors that influence the adoption of OER and resulted in the identification of 71 variables. To evaluate content and face validity, 34 OER researchers (of whom 76% had six or more years of research and educational experience) were invited to evaluate the impact of each of the variables identified in the adoption of open educational resources (defined as "use" and/or "creation" of OERs) by responding to a questionnaire with 62 questions. Following this process, the questionnaire was simplified to a set of 25 questions.

To identify potential problems with data collection, a pilot questionnaire was applied to a sample of 63 English-speaking students and professors, 8 Portuguese-speaking students and professors, and 3 Spanish-speaking students and professors from all the institutions that composed the sample of the ROER4D research project. The pilot questionnaire allowed us to identify the need to include the description of the OER concept in the questionnaire, and several questions were revised to use only the term "educational resource" in order to eliminate hypothetical responses, since many respondents would have their first contact with the term "open educational resource" during the research. To test the updated version of the questionnaire, a second pilot test (in English and Spanish) was carried out with 34 teachers and 28 students from the sample of higher education institutions to be researched, generating small revisions and resulting in the final version.

4.3   Data Collection and Analysis

Data collection was carried out through a SurveyMonkey link sent to the email of the individuals selected for the survey. In cases where the students did not have access to the Internet, printed forms were delivered to respondents. The 25 questions were distributed in 3 dimensions: individual characteristics, which included, for example, questions such as age, gender, access to information technology and English proficiency; use of educational resources, which addresses matters that identify which educational resources are used, elaborated, modified and shared; and Open educational resources, which addresses matters that assess the specific perception of who uses OER. Considering the scope of this paper, there were 13 variables, as shown in Table 2.

Table 2. Research variables

Code	Variable	Type
X_1	Gender	Nominal
X_2	Country	Nominal
X_3	Digital skills	Ordinal
X_4	Internet access location	Nominal
X_5	Internet access device	Nominal
X_{6Cost}	Satisfaction with Internet cost	Ordinal
X_{6Speed}	Satisfaction with Internet speed	Ordinal
$X_{6Stability}$	Satisfaction with Internet stability	Ordinal
Y	Creation, use, modification and/or sharing of a MOOC	Nominal
Y_1	Creation of a MOOC (Y/N)	Binary
Y_2	Use of a MOOC (Y/N)	Binary
Y_3	Modification of a MOOC (Y/N)	Binary
Y_4	Sharing of a MOOC (Y/N)	Binary

Data were analyzed using SPSS software (v. 18), using descriptive statistics technique and analysis of significance indexes of the results through Chi-square test, using 5% significance level, an adequate technique for the types of variables as shown in Fig. 2 [25]. For the analyzes, variables Y, Y1, Y2, Y3, and Y4 were considered dependent and variables X1, X2, X3, X4 and X5 were considered explanatory variables.

5 Findings and Discussion

From the selected sample, a return of 4,784 respondents was obtained, distributed among the nine countries surveyed, as presented in Table 3.

Table 3. Total participants who have already created, used, modified and/or shared a MOOC

Country	Participants		Created, used, modified and/or shared a MOOC	Used a MOOC	Created a MOOC	Modified a MOOC	Shared a MOOC
	Qty	%					
South Africa	621	13.0%	39.5%	28.3%	4.8%	6.9%	3.1%
Brazil	287	6.0%	21.3%	12.2%	1.7%	2.8%	5.2%
Chile	293	6.1%	29.7%	17.1%	2.4%	5.5%	9.2%
Colombia	170	3.6%	47.1%	43.5%	0,6%	3,5%	12,9%
Ghana	817	17.1%	45.3%	32.3%	3.8%	7.8%	3.3%
India	437	9.1%	58.1%	24.3%	5.5%	19.9%	9.6%
Indonesia	645	13.5%	83.1%	27.3%	5.1%	43.7%	7.9%
Malaysia	716	15.0%	55.0%	20.0%	4.3%	25.6%	6.8%
Kenya	798	16.7%	38.5%	28.8%	8.9%	3.3%	2.0%
Total	4784	100.0%	48.8%	26.2%	4.9%	14.9%	5.6%

Nearly half of the respondents (48.8%) stated that they have already created, used, modified or shared a MOOC. When we evaluate each of the actions carried out with MOOC, it is possible to identify that most respondents used them (26.2%) and modified them (14.9%). It is also possible to observe that the Asian countries (India, Indonesia and Malaysia) are the ones that are doing the most with MOOCs. This result can be explained by the fact that teachers are at the forefront of the use of OER in Asian higher education and that Indonesia is making significant progress in the production and distribution of OER under the leadership of the University of Terbuka [26] which produced a series of materials in English and Indonesian and distributed them extensively to all students, including other universities in the country with the support of the UNESCO program on OER [27].

The countries in South America (Brazil, Chile and Colombia) are the ones that least carried out actions with MOOCs, being important to note that Colombia presents different results. This can be explained by the fact that, although Brazil has several OER initiatives and that Chile has a policy of free access, only Colombia has a document with national and institutional guidelines to promote and strengthen the production and management of OER, in addition to producing a document describing the open digital educational resources in Colombia [28]. The analysis of this information demonstrates the importance of establishing institutional and governmental policies to increase the use of OER, especially MOOCs.

Regarding to gender, women comprised 45.7% of the sample, men 54.1% and 0.2% declared themselves as others genders. Among men, 52.4% reported having used, created, modified and/or shared MOOCs, while among women this number dropped to 44.5%. However, when evaluating the results by country, it is possible to verify that the difference is statistically significant only in Kenya and S. Africa, both African countries, as can be seen in Table 4.

Table 4. Use, creation, modification and/or sharing of a MOOC by gender.

Country	Gender	Created, used, modified and/or shared a MOOC			Used a MOOC	Created a MOOC	Modified a MOOC	Shared a MOOC
		%	χ^2	p				
Brazil	Female	16.5%	2.791	0.095	9.9%	1.7%	0.8%	4.1%
	Male	24.7%			13.9%	1.8%	4.2%	6.0%
Chile	Female	29.3%	0.033	0.855	19.3%	2.1%	5.0%	5.7%
	Male	30.3%			15.1%	2.6%	5.9%	12.5%
Colombia	Female	40.8%	1.889	0.169	40.8%	0.0%	1.4%	11.3%
	Male	51.5%			45.5%	1.0%	5.1%	14.1%
Ghana	Female	44.3%	0.194	0.660	31.9%	3.3%	7.5%	3.6%
	Male	45.9%			32.5%	4.1%	8.0%	3.1%
India	Female	56.9%	0.290	0.590	20.9%	5.8%	19.6%	11.1%
	Male	59.4%			27.8%	5.2%	20.3%	8.0%
Indonesia	Female	82.6%	0.103	0.748	25.7%	4.6%	45.4%	8.3%
	Male	83.6%			28.2%	5.4%	43.0%	7.7%
Kenya	Female	32.4%	10.103	0.001	21.5%	8.1%	4.2%	1.7%
	Male	43.4%			34.8%	9.5%	2.5%	2.3%
Malaysia	Female	51.6%	4.344	0.114	20.5%	2.6%	23.9%	6.1%
	Male	58.7%			19.8%	6.1%	27.1%	7.6%
South Africa	Female	34.6%	8.536	0.003	22.8%	4.9%	6.0%	3.3%
	Male	46.3%			36.1%	4.7%	8.2%	2.7%

Respondents self-assessed their digital skills and 14.3% reported being on an advanced level, 52.4% intermediate level and 33.3% basic level. The distribution of digital skills by country is shown in Table 5. Apart from India, where most respondents stated that they possess a basic level of digital skills, most respondents from other countries stated that they had intermediate levels.

Table 5. Level of digital skills by country.

Country	Proficiency level		
	Basic	Intermediate	Advanced
South Africa	11.0%	54.6%	34.5%
Brazil	46.3%	47.0%	6.6%
Chile	34.8%	53.9%	11.3%
Colombia	17.6%	55.9%	26.5%
Ghana	31.8%	60.0%	8.2%
India	58.6%	34.3%	7.1%
Indonesia	21.1%	57.4%	21.6%
Malaysia	44.6%	46.4%	9.1%
Kenya	36.1%	55.0%	8.9%
Total	33.3%	52.4%	14.3%

As seen on Table 6, the level of digital skills influences the use and sharing of MOOCs, since the advanced level is the one that most performs this action and the results of the Chi-square test demonstrate a statistically significant difference. This result evidences the arguments presented by Amemado [2] that it is necessary to reduce the digital skills gap to ensure access to MOOCs.

Table 6. Level of digital skills by type of actions performed with a MOOC

Action performed with a MOOC	Level of digital skills			χ^2	p
	Advanced	Basic	Intermediate		
Used	32.9%	21.4%	27.5%	37.255	.000
Created	6.4%	4.3%	4.8%	4.631	.099
Modified	14.9%	15.0%	14.9%	.008	.996
Shared	7.9%	4.6%	5.6%	9.540	.008

Although the Internet access index is only 49.6% in the world [24], 99.4% of the respondents stated that they had access to the Internet (we must take into consideration that respondents represent a share of the population already engaged in higher education). Even when this result is evaluated by country, as shown in Table 7, we observe that there are no significant differences, even considering the differences in Internet penetration in the surveyed regions, which is 66.7% in South America, 45.2% in Asia and only 27.7% in Africa [24].

However, when assessing the access locations, we can identify that Ghana and Kenya have the lowest indexes of respondents who have Internet access at home. According to Gulati [21], in order to be successful in the use of educational resources, it is necessary to have Internet access at home, so we evaluated if there are differences in the use, creation, modification and sharing of MOOCs among those who have and do not have access to the Internet at home. However, the results demonstrate that only in Indonesia and South Africa, having Internet access at home significantly increases the variation, use, modification and sharing of MOOCs, as presented in Table 8.

Table 7. Internet access location

Country	Access to Internet	Access location				
		House of friends and families	Home	Coffee shops	Libraries	At work or educational institution
Brazil	100.0%	36.9%	96.5%	3.8%	9.4%	70.0%
Chile	99.7%	45.7%	90.4%	9.9%	23.9%	72.4%
Colombia	100.0%	45.9%	94.7%	13.5%	54.7%	84.1%
Ghana	99.4%	22.3%	43.2%	40.1%	35.7%	78.8%
India	99.1%	13.3%	68.4%	26.1%	22.9%	55.4%
Indonesia	99.2%	17.4%	80.8%	37.7%	31.6%	73.2%
Kenya	99.0%	39.8%	28.9%	34.2%	62.7%	85.7%
Malaysia	99.6%	32.5%	90.8%	17.2%	23.3%	74.6%
S. Africa	99.7%	28.3%	74.2%	16.4%	17.1%	85.3%

Table 8. Creation, use, modification and/or sharing of MOOCs by Internet access at home

Country	% Created, used, modified and/or shared a MOOC		χ^2	p
	With access at home	Without access at home		
Brazil	98.4%	96.0%	0.784	0.376
Chile	89.7%	90.8%	0.089	0.765
Colombia	95.0%	94.4%	0.026	0.872
Ghana	44.3%	42.3%	0.344	0.557
India	55.9%	85.8%	43.972	0.000
Indonesia	82.8%	70.6%	8.673	0.003
Kenya	31.9%	27.1%	2.146	0.143
Malaysia	91.1%	90.4%	0.117	0.732
South Africa	68.6%	77.9%	6.786	0.009

When questioned about satisfaction with Internet connection cost, speed and stability, a large percentage of respondents showed dissatisfaction, as shown in Table 9. This result shows that the countries surveyed need to invest in these requirements because it is not enough that the population has access to the Internet; the access must

Table 9. Level of satisfaction with Internet cost, speed and stability

Level of satisfaction	Cost	Speed	Stability
Very dissatisfied	10.8%	11.1%	11.2%
Dissatisfied	23.8%	26.7%	26.8%
Neither satisfied nor dissatisfied	14.3%	12.0%	16.9%
Satisfied	38.9%	39.2%	35.5%
Very satisfied	10.1%	10.4%	8.7%
Not applicable	2.1%	0.5%	1.0%

also be satisfactory and affordable, especially for using MOOCs, that often use a great amount of data (which can increase the cost for users) and that need Internet speed and stability so that users can follow the course.

Figures 1, 2 and 3 show the percentages of those who use MOOCs according to the levels of satisfaction with the Internet cost, speed and stability. Analyzing Figs. 1, 2 and 3, we can observe that respondents who are dissatisfied or very dissatisfied with the requirements of Internet cost, speed and stability use MOOCs less than others. Chi-square test results showed statistically significant differences (χ^2 cost = 26.224 and p = 0.000; χ^2 speed = 19.927 and p = 0.001; χ^2 stability = 28.160 and p = 0.000).

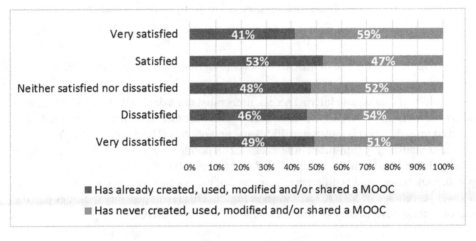

Fig. 1. Chart of use of MOOCs by level of satisfaction with Internet cost

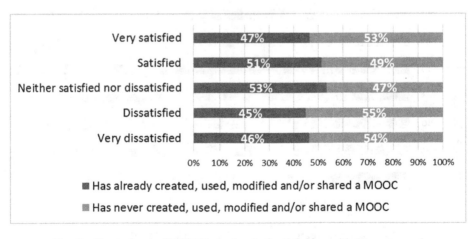

Fig. 2. Chart of use of MOOCs by level of satisfaction with Internet speed

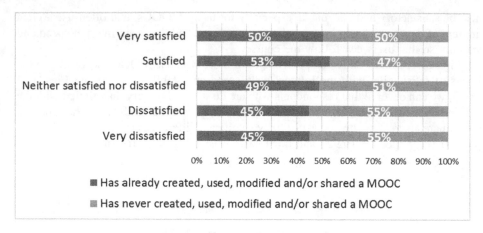

Fig. 3. Chart of use of MOOCs by level of satisfaction with Internet stability

In addition to adequate Internet access, users must use a device that provides access to all course tools for MOOCs to be successful. 74.3% of respondents use notebooks, 47.8% use desktops, 30.7% use tablets and 76.5% use cell phones. The results of the Chi-square test show statistically significant differences for respondents who access the Internet with notebooks and desktops (χ^2 notebook = 7.694 and p = 0.006; χ^2 desktop = 84.491 and p = 0.000), however, this difference indicates that the population using these devices has a lower percentage of MOOCs as shown in Fig. 4. This result possibly indicates that the use of mobile devices may be more relevant to the use of MOOCs.

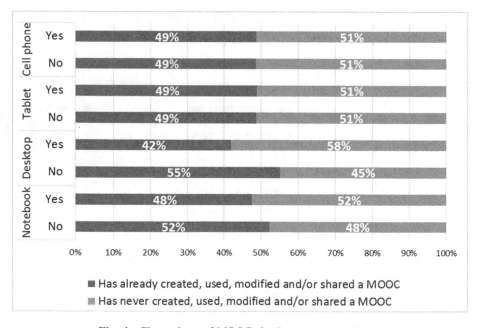

Fig. 4. Chart of use of MOOCs by Internet access device

6 Conclusions

Distance education, especially MOOCs, is considered as a potential solution for the democratization of access to quality education. However, when assessing this potential from the perspective of access to technology, we identified that 33.3% of the South America population, 54.5% of Asia population and 72.3% of Africa population cannot benefit from these courses, as they do not yet have Internet access. This study aimed to assess the potential of the democratization of quality education provided by MOOCs, considering the perspective of access to technology, by analyzing the data of university students in 9 different countries.

Significantly different results were presented among the countries surveyed, showing that countries in Asia are the ones that most use, create, modify and share MOOCs, and that countries in South America are the ones that least perform these actions. In all countries, it was possible to observe that the skills to use digital technologies, as well as the quality of access, influence the level of use of MOOCs.

Thus, for MOOCs to reach their full potential, government actions that democratize access to technology, taking into account the Internet cost, speed and stability, as well as actions aimed at reducing the digital skills gap, are determinant factors for the use of MOOCs as presented in the results of this study. Actions are also necessary to equalize access to educational resources between men and women in Kenya and South Africa, as the results show statistically significant differences in the use of MOOCs between men and women in those countries.

Considering that the use of mobile devices is high among the respondents of the research, it is important that the courses are developed with technologies that adapt to the different forms of access of the users, as presented by Teixeira et al. [12].

It is also relevant to highlight that the results presented in this study refer to a sample that has access to higher education and that, therefore, the results for the population that do not have access to this level of education can be even more meaningful, demonstrating that a great part of the potential population to be benefited by the phenomenon of MOOCs still remains disadvantaged due to the technological barrier.

References

1. Rosini, A.M., Palmisano, A., Roque, O.: MOOCS: where the learning process and use of it technology resources are heading toward. In: Proceedings of the 11th International Conference on Innovation and Management, pp. 1022–1028 (2014)
2. Amemado, D.: Integrating technologies in higher education: the issue of recommended educational features still making headline news. Open Learn. 29(1), 15–30 (2014)
3. Meisenhelder, S.: MOOC mania. NEA Higher Education, Fall 2013, pp. 7–26 (2013). http://www.nea.org/assets/docs/HE/TA2013Meisenhelder.pdf
4. Research on Open Educational Resources for Development (ROER4D). About ROER4D (2017). http://roer4d.org/about-us. Accessed 10 May 2017
5. Kennedy, J.: Characteristics of massive open online courses (MOOCs): a research review, 2009–2012. J. Interact. Online Learn. 13(1), 1–16 (2014)

6. Alemán de la Garza, L.Y., Sancho-Vinuesa, T., Gómez Zermeño, M.G.: Atypical: analysis of a massive open online course (MOOC) with a relatively high rate of program completers. Global Educ. Rev. **2**(3), 68–81 (2015)
7. Billington, P.J., Fronmueller, M.P.: MOOCs and the future of higher education. J. High. Educ. Theory Pract. **13**(3/4), 36 (2013)
8. Clair, R.S., Winer, L., Finkelstein, A., Wald, S., Finkelstein, A., Fuentes-Steeves, A.: Big hat and no cattle? The implications of MOOCs for the adult learning landscape. Can. J. Study Adult Educ. **27**(3), 65–82 (2015)
9. Knox, J.: Digital culture clash: "massive" education in the e-learning and digital cultures MOOC. Distance Educ. **35**(2), 164–177 (2014)
10. Rodriguez, O.: Two distinct course formats in the delivery of MOOCs. Turk. Online J. Distance Educ. **14**(2), 66–80 (2013)
11. Yeager, C., Hurley-Dasgupta, B., Bliss, C.A.: MOOCS and global learning: an authentic alternative. J. Asynchron. Learn. Netw. **17**(2), 133–147 (2013)
12. Teixeira, A., Mota, J., García-Cabot, A., García-Lopéz, E., De-Marcos, L.: A new competence-based approach for personalizing MOOCs in a mobile collaborative and networked environment. Rev. Iberoam. Educ. Distancia **19**(1), 143–160 (2016)
13. Zheng, S., Wisniewski, P., Rosson, M.B., Carroll, J.M.: Ask the instructors: motivations and challenges of teaching massive open online courses. In: 19th ACM Conference on Computer-Supported Cooperative Work and Social Computing, pp. 206–221 (2016)
14. Sánchez-Vera, M.-M., León-Urrutia, M., Davis, H.: Challenges in the creation, development and implementation of MOOCs: web science course at the University of Southampton. Comunicar **22**(44), 37–44 (2015)
15. Schmid, L., Manturuk, K., Simpkins, I., Goldwasser, M., Whitfield, K.E.: Fulfilling the promise: do MOOCs reach the educationally underserved? Educ. Media Int. **52**(5), 1–13 (2015)
16. Griffiths, R., Chingos, M., Mulhern, C., Spies, R.: Adopting MOOCS on campus: a collaborative effort to test MOOCS on campuses of the university system of Maryland. J. Asynchron. Learn. Netw. **19**(2), 1–15 (2015)
17. Nkuyubwatsi, B.: Evaluation of massive open online courses (MOOCs) from the learner's perspective. In: European Conference on E-Learning, ECEL, pp. 340–346 (2013)
18. Banerjee, A.V., Duflo, E.: (Dis)Organization and success in an economics MOOC. Am. Econ. Rev. **104**(5), 514–518 (2014)
19. Warusavitarana, P.A., Dona, K.L., Piyathilake, H.C.P., Epitawela, D.D., Edirisnghe, M.U.: MOOC: a higher education game changer in developing countries. In: Rhetoric and Reality: Critical Perspectives on Educational Technology, ASCILITE 2014, Dunedin, NZ, pp. 359–366 (2014)
20. Harb, I.: Higher education and MOOCs in India and the Global South. Change: Mag. High. Learn. **47**(43), 42–49 (2015)
21. Gulati, S.: Technology-enhanced learning in developing nations: a review. Int. Rev. Res. Open Distrib. Learn. **9**(1), 1–10 (2008)
22. Johnstone, S.M.: Open educational resources serve the world. Educ. Q. **28**(3), 15–18 (2005)
23. Wright, C.R., Dhanarajan, G., Reju, S.A.: Recurring issues encountered by distance educators in developing and emerging nations. Int. Rev. Res. Open Distance Learn. **10**(1), 1–25 (2009)
24. Internet World Stats: Internet Usage Statistics (2017). http://www.internetworldstats.com/stats.htm. Accessed 10 May 2017
25. Cooper, D.R., Schindler, P.S.: Business Research Methods, 11th edn. McGraw-Hill/Irwin, New York (2011)

26. Dhanarajan, G., Porter, D.: Open educational resources : an Asian perspective. Common-wealth of Learning and OER Asia (2013)
27. UNESCO: Promoting Open Educational Resources Across Indonesia (2015). http://www.unesco.org/new/en/communication-and-information/resources/news-and-in-focus-articles/all-news/news/promoting_open_educational_resources_across_indonesia/. Accessed 10 May 2017
28. Hoosen, S.: Survey on government's open educational resources (OER) policies. Prepared for the World OER Congress, by Sarah Hoosen of Neil Butcher and Associates for the Commonwealth of Learning and UNESCO, Vancouver, BC, Canada (2012). http://www.unesco.org/fileadmin/MULTIMEDIA/HQ/CI/CI/pdf/themes/Survey_On_Government_OER_Policies.pdf. Accessed 10 May 2017

Relationship Between Self-disclosure and Cyberbullying on SNSs

Jooyeon Won[1] and DongBack Seo[2(✉)]

[1] Global Studies on Management Information Science,
Chungbuk National University, Cheongju, Republic of Korea
wjy9243@cbnu.ac.kr
[2] Department of Management Information Systems,
Chungbuk National University, Cheongju, Republic of Korea
dseo@cbnu.ac.kr

Abstract. Since many people have used Social Networking Sites (SNSs), benefits and problems of using SNSs appear. One problem is traditional bullying has extended everywhere SNSs can reach, called cyberbullying. Researchers have investigated definitions and types of cyberbullying, but there is a paucity of cyberbullying related to self-disclosure. Self-disclosure is one of main activities users take to connect and interact with many people on SNSs. Through an empirical analysis, this paper investigates three things; the relationship between the frequency of self-disclosure and the victimization experience of cyberbullying, the factors of self-disclosure, and cyberbullying types based on seriousness. Results show that self-disclosure is influenced by self-control and self-esteem. Active self-disclosure affects the victimization experience of cyberbullying. However, passive self-disclosure doesn't affect the victimization experience of cyberbullying. Moreover, less serious cyberbullying victimization experience affects more serious cyberbullying victimization experience. Furthermore, impersonation, the most serious cyberbullying, is affected by all different types of cyberbullying.

Keywords: SNS · Self-disclosure · Cyberbullying · Self-control · Self-esteem

1 Introduction

Social Networking Sites (SNSs) have been developed and diffused worldwide since SixDegrees.com launched in 1997 [15]. For example, Facebook, a leader in SNS, surpassed 1 billion monthly active users [56], supporting both maintenance of existing social ties and the formation of new connections [16], and nearly 2.1 billion people have their SNS accounts [9]. This phenomenon has expedited with the rapid adoptions of smart devices such as smartphones and tablets, facilitating SNSs by making it affordable and wireless, as more than 3.5 billion users have accessed the internet through mobile, and 1.6 billion users among them are activated [9].

Although a person can open a SNS account with fake personal information (e.g., birthday), (s)he discloses him/herself to others in a certain level in order to join a network and maintain those relationships. On the other hand, there are people who

© Springer International Publishing AG 2017
M. Themistocleous and V. Morabito (Eds.): EMCIS 2017, LNBIP 299, pp. 154–172, 2017.
DOI: 10.1007/978-3-319-65930-5_13

disclose themselves by providing real personal information. Thus, self-disclosure is necessary when using a SNS, even though the depth varies depending on users [15].

By doing to, self-disclosure has studied in various ways. From a benefit perspective, self-disclosure is related to social attraction to build and maintain relationship on SNSs [14, 50, 52]. However, there is a paucity of research about self-disclosure on SNS from a cost perspective such as cyberbullying and identity-theft, even though these negative consequences are acknowledged [17, 54]. In short, self-disclosure has been studied lopsidedly. As self-disclosure shows more and more side effects, this research focuses on self-disclosure on SNSs and its negative aspects.

While there are various negative effects caused by self-disclosure, one of the noticeable problem is cyberbullying, an aggression that harmful or fatal threat through Internet or other technologies [66]. Moreover, lots of youngsters have witnesses cyberbullying on SNS in UK and involved in cyberbullying in US [24, 30]. Few researches infer that self-disclosure influences cyberbullying [17, 54], but there is no study about this relation in depth. As the investigation into cyberbullying is getting more important, this study formulates the premise that there is a certain relation between self-disclosure and cyberbullying.

This study proposes the relation between self-disclosure and cyberbullying based on Social Exchange Theory (SET). Referring to this theory, all interaction among people is regarded as a commodity exchange [6], and self-disclosure has been studied as the exchange with two main streams; one hand, most studies have focused on the benefit side, such as relationship maintenance, on the other hand, a few studies have reported the privacy intrusion from the cost side. However, this research suggests the possibility of cyberbullying as the cost induced by self-disclosure, beyond the aspects of privacy intrusion.

This paper pinpoints three main purposes; first, this study aims at proving the factors of self-disclosure, second, the influence of self-disclosure on cyberbullying will be discussed and the last, this research focuses to investigate the stage of cyberbullying. Above all, the issue why cyberbullying should be examined systematically is zeroed on by understanding the impact and the difference between the stage, classified into low cyberbullying versus high cyberbullying depending on seriousness.

In particular, this study is conducted from the victim's point, because if victim suffered through mobile devices mentally and physically, even though perpetrator abused victim without awareness of the peril, this harassment should be defined as cyberbullying and the result of this harassment is technically up to victim.

This research suggests one way to avoid cyberbullying, expecting the possibility before the occurrence, and alerts SNS users to the potential danger caused by self-disclosure. Moreover, users can regulate the self-disclosure to prevent cyberbullying.

This paper is organized into 5 parts. In the next chapter, theoretical background is provided with hypotheses, and then methodology and results are followed. The last part sums up the conclusion.

2 Theoretical Background

2.1 Social Exchange Theory

Social Exchange Theory (SET) proposed that various individual's social behaviors are useful interaction as exchanges [26]. Referring to this theory, people behave after measuring tangible or intangible benefits and cost through exchange [34, 47, 59]. While SET focused at the individual level by Homans [26], Blau extends SET to groups, as well as between groups [6], and both scholars regarded "communication" as one of the foundation between people.

To date, researchers have studied self-disclosure based on SET. Posey et al. studied self-disclosure in social media, probing that perceived online community trust (benefits) and privacy risk beliefs (cost) affected self-disclosure [47]. Lowry et al. studied privacy concerns as cost through self-disclosure via instant message [36]. As can be seen, privacy concern has been addressed mainly as a cost of self-disclosure, however, is it all costs caused by self-disclosure?

Interestingly, few studies have implied cyberbullying as the potential downside of self-disclosure. Cyberbullying and online harassment are risks of online communication [60], moreover, personal information which is uploaded online can be used for identity theft, stalking cyberbullying, or sexual harassment [64]. In addition, the number of following friends and highly frequency of posting affect cyberbullying victimization [14]. So far, cyberbullying has been surmised the potential dark side of self-disclosure. From this aspect, this research regards cyberbullying as the cost caused by self-disclosure and induces this cost with empirical data, expecting the danger of potential cyberbullying when the user of SNS discloses information about self.

2.2 Self-disclosure in SNSs

Self-disclosure refers to the act of revealing personal information one shares with others, including any information exchange about the self, such as personal states, or dispositions [7]. Member information that is released through self-disclosure enables social interaction among site members maintained by facilitating learning about each other [28]. Moreover, self-disclosure allows social networking sites to propagate their business models for greater prosperity by creating site members' own contents such as photos, videos, and status [12].

To have the benefits, site members are willing to disclose themselves, for example, posting, one of the way to disclose information in SNSs mainly, updated about 55 million on Facebook per a day [8]. Furthermore, 4.5 billion likes are made on Facebook every day, produced 3,125,000 new likes per each minute [25]. As can be seen, self-disclosure isn't a serious behavior in SNSs, rather than this behavior is quite spontaneous.

Although there are various ways to disclose member information by user oneself, no study examines the behavior of self-disclosure, distinguishing the types in depth. This paper suggests the classification of self-disclosure into two parts based on user behavior; active self-disclosure and passive self-disclosure. active self-disclosure refers to post text, pictures, and videos in user one's SNS wall or other user' s wall. Conversely, passive self-disclosure is the act such as clicking "Like" replying, and sharing contents. What the difference between active and passive self-disclosure is that the former one delivers user's own text, pictures, and videos on one's way, but the other is acted under the contents posted by other users.

Interestingly, self-disclosure has been investigated by providing positive functions like the development of relationships [33], the psychological well-being [27], social attraction [14], and the maintenance of connection on SNSs, however, previous studies also suggest side effects caused by self-disclosure, such as intellectual property [23], privacy issue [12, 36, 47, 63], and cyberbullying [14, 17, 23, 54, 60, 64]. Especially, self-disclosure has been supposed that affects cyberbullying, making people vulnerable without any precise verification. Regarding to this aspect, this paper set up premises that active and passive self-disclosure will affect cyberbullying.

In addition, even though people who use same SNSs, one enjoys self-disclosure frequently or not depending on who (s)he is, connoting that there are specific factors to influence self-disclosure. On this condition, this paper establishes self-control, self-esteem, and the frequency of profile setting as the factors of self-disclosure. These factors reinforce self-disclosure which is one of the social exchange based on SET.

2.3   Cyberbullying

Definition of Cyberbullying. The development of technologies including mobile devices allows school problem to extend the range from school to bedroom including cyberbullying [33]. As cyberbullying considers a new harassment caused by electronic devices, it has defined variously [33, 42]. For example, cyberbullying refers an aggression that harmful or fatal threat spreads through Internet, other technologies based on ICT [66]. Moreover, cyberbullying is an intentional and repetitive harassment including sexual text, photo, etc. [24], a deliberate and hostile behavior intended to harm people using the Internet by leveraging the imbalance of power between bullies and victims [36], or an intentional harm to involve the use of ICT such as e-mail, cellphone, etc. by an individual or group [5]. While various definitions exist alike, this paper adopt the definition by Von Marées and Petermann which is an intentional, repeated, and aggressive act or behavior carried out by a group or individual employing information and communication technology (ICT) as an instrument [62].

As cyberbullying is defined similarly, most scholars have agreed that cyberbullying has certain characteristics such as intention, repetitiveness, and imbalance of power [41, 61]. Specifically, a perpetrator harasses a victim intentionally and repetitively online when the perpetrator is more powerful than the victim, creating harmful contents by having the ability to higher levels of media literacy or a higher social status within a virtual community (e.g. designation) [24, 32, 41, 44, 61]. Moreover, some scholars include anonymity and public in the nature of cyberbullying in characteristics of cyberbullying [44].

Types of Cyberbullying. There are some kinds of the classification of cyberbullying. one hand, cyberbullying is classified into 7 types; flaming; harassment; cyberstalking; denigration; masquerade; outing and trickery; and exclusion [66]. However, Nocentini et al. reclassified above 7 types into four forms [44]. First, written-verbal behavior is the type of bullying which uses written or verbal forms through phone call, text message, SNS, etc. Second, visual behavior is to send and post photos or videos through cellphone or Internet, and third, exclusion is a specific intentional action to exclude someone from one group online. Last, impersonation is stealing or thieving somebody's name or account for changing, revealing, or deleting something [41, 43, 44]. This study adopts this category.

Although previous study presents 4 types of cyberbullying equally, each type is different depending on seriousness. Written-verbal and visual behaviors are low cyberbullying because these types could occur only for fun without clear intention. Such experiences were classified as written-verbal and visual behavior of cyberbullying, for example, unpleasant, unwanted or embarrassed messages or pictures online. The extent of damage through these types of cyberbullying is confined to emotional or psychologic injury, especially if a perpetrator bullies a victim once, then the damage might not be too serious.

However, exclusion and impersonation are classified into high cyberbullying. A victim who was excluded or impersonated online could noticed that (s)he was cyberbullied when a perpetrator excludes or impersonates the victim repetitively with clear intention as much as a victim notice. Thus, in this research, each cyberbullying has different seriousness and are influenced by self-disclosure.

> H1a: The frequency of active self-disclosure influences the victimization experience of written-verbal behavior strongly
>
> H1b: The frequency of active self-disclosure influences the victimization experience of visual behavior strongly
>
> H2a: The frequency of passive self-disclosure influences the victimization experience of written-verbal behavior strongly
>
> H2b: The frequency of passive self-disclosure influences the victimization experience of visual behavior strongly

Exclusion is measured through ignored experience online, which cannot notice if when victim suffered it once in his/her experience. Hence it is possible to predict a victim who written-verbal or visual cyberbullied excluded than a victim, and the metallic or psychologic injury might be more serious than low types of cyberbullying. Moreover, previous study showed student who is ignored in school has more excluded online, and it means that exclusion online and offline has significant a relationship. Thus, exclusion is more serious cyberbullying than written-verbal and visual behaviors of cyberbullying, and low cyberbullying could impact on high cyberbullying, as hypothesis 3 and 4 shows.

> H3a: The victimization experience of written-verbal behavior influences the victimization experience of exclusion strongly
>
> H3b: The victimization experience of written-verbal behavior influences the victimization experience of impersonation strongly

H4a: The victimization experience of visual behavior influences the victimization experience of exclusion strongly

H4b: The victimization experience of visual behavior influences the victimization experience of impersonation strongly

Moreover, impersonation is the most serious level because if a perpetrator accesses a victim's account to change personal information or post something, a victim could be damaged although (s)he doesn't know what it occurred. In addition, this damage includes psychologic, physical, and monetary harm. Regarding this aspect, this research assumed that each type of cyberbullying would affect impersonation.

H5: The victimization experience of exclusion influences the victimization experience of impersonation strongly.

2.4   Self-control

Self-control refers the ability to control one's behavior or emotion as one wants by oneself [35]. Self-control encourages people to behave ordinarily, unusually, or calmly, involving frustrating urges, desires [3, 42]. Furthermore, self-control operates a variety of mechanisms, then the primary self-controlled act facilitates subsequent acts [42].

As self-disclosure is a behavior by human being, self-control could influence the willing to disclose oneself, however, self-control has been considered in relation to self-disclosure rarely. One paper examined the relation between self-control and self-disclosure [29] with insufficient data sample, rejecting the assumption that people with high level of self-control have valid score to control over talking about themselves. On the other hand, another research proves that the high level of self-control involves interpersonal success [57] developed through self-disclosure. Consequently, this study set up the H6.

H6a: The low level of self-control influences the frequency of active self-disclosure in SNSs strongly

H6b: The low level of self-control influences the frequency of passive self-disclosure in SNSs strongly.

2.5   Self-esteem

Self-esteem is defined as a person's general self-evaluation of his/her value [65]. Self-esteem has been studied with interesting outcomes, such as satisfaction in relationships, work, and health [46]. People who have the high level of self-esteem tend to encourage themselves to exaggerate their successes and good traits [4].

Self-esteem is also regarded as a cause of self-disclosure [2, 10, 55]. Barker suggests that the most important motivation to use SNSs is keeping relationship among users, moreover people who have high level of self-esteem motivate more than others who are not [2]. Meanwhile, if men have the higher level of self-esteem, they disclose more, however, there is no correlation between self-disclosure and self-esteem for women [55] Furthermore, self-disclosure improves health and relationships, hence, people who have low self-esteem tend to avoid self-disclosure [10]. Based on the previous studies, H7 is established.

H7a: The high level of self-esteem influences the frequency of active self-disclosure in SNSs strongly
H7b: The high level of self-esteem influences the frequency of passive self-disclosure in SNSs strongly.

2.6 Profile Setting

To become a member of certain SNS such as Facebook and take a pleasure from these sites, each user creates one's own profile. For example, Facebook forces user to enter (nick)name, profile picture, relationship status (e.g. married, engaged, etc.), and so on. Profile setting on SNSs plays the role of connecting individuals from online to offline, and maintaining social context from reality to virtual as people incline to hold on relationships [15].

According to previous study, the more use of Internet and SNSs, the higher possibility to have profile online [51], moreover, one who desires to make or maintain relationships, the most important motivation to use SNSs [2], tends to invest more time in setting profile than those who are not [38]. In other words, SNS users with profile access SNSs more and disclose themselves more through various behaviors (e.g. posting, sharing, clicking the "like", etc.) in developing relationships. Thus, this paper posits the last hypothesis.

H8a: The frequency of changing profile setting influences the frequency of active self-disclosure strongly
H8b: The frequency of changing profile setting influences the frequency of passive self-disclosure strongly

All hypotheses is depicted in Fig. 1.

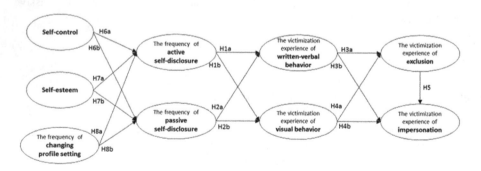

Fig. 1. The summary of hypotheses

3 Methodology

3.1 Participants

Young people tend to be less cautious about online activity such as posting or sharing private information through SNSs than whose are older [23]. Thus, a survey used to this paper was conducted among young people who are familiar with SNSs.

3.2 Measurement Instrument

To conduct the survey, all constructs consist of 3 to 5 items, and each question consists of 7 Likert scales from '(1) Great disagree' to '(7) Great agree' or from '(1) Never' to '(7) Usually' depending on constructs.

Survey questions are adopted through literature review and reformulated in context. Constructs, questions, and references are included in Table 1.

Table 1. Construct, question and reference

Construct	Question	Average	Standard deviation	Reference
Self-control*	I sometimes do risky things for the fun	2.831	1.648	Ha [21]
	I sometimes do things on my mood without thinking the result	3.840	1.640	
	I sometimes like to test myself	3.240	1.629	
	I sometimes do whatever brings me pleasure no matter what is the purpose	2.957	1.583	
Self-esteem	I sometimes think I am valuable as much as others	5.265	1.576	Rosenberg [49]
	I sometimes think I have a good personality	5.126	1.519	
	I am able to do things as well as others	5.206	1.604	
	Overall, I am satisfied with myself	4.935	1.631	
The frequency of changing profile setting	When I use SNSs, I '(1) Never' to '(7) Usually' update ...			Carpenter [11]
	... status (e.g. engagement, marriage, ...) on SNSs	2.385	1.590	
	... educational background (e.g. school, major, grade, ...) on SNSs	3.240	1.782	

(continued)

Table 1. (*continued*)

Construct	Question	Average	Standard deviation	Reference
	... my profile picture on SNSs	4.240	1.801	
The frequency of active self-disclosure	When I use SNSs, I '(1) Never' to '(7) Usually' post ...			Ross et al. [50]
	... text, photos, or videos on other people's wall	3.166	1.582	
	... text on my SNS wall	3.262	1.637	
	... photos or videos on my SNS wall	3.471	1.695	
The frequency of passive self-disclosure	When I use SNSs, I '(1) Never' to '(7) Usually' ...			
	... comment on other people's text, photos or videos	3.957	1.762	
	... click "like" for other people's text, photos or videos	4.569	1.939	
	... re-comment on other people's comment	4.486	1.913	
	... share text, photos or videos from other people's wall to my wall	3.431	1.907	
The victimization experience of written-verbal behavior	When I use SNSs, I have ...			Kwan and Skoric [31]
	... received insulting comments or messages	2.231	1.347	
	... seen comments or messages that damaged my reputation	2.394	1.49	
	... been mistaken because of comments or messages about me	2.329	1.461	
	... seen comments or messages that made me ridiculous	2.292	1.396	
	... felt betrayed by my friends who posted or commented about me on SNS that I want to keep a secret from others	2.142	1.286	
The victimization experience of visual behavior	When I use SNSs, ...			Den Hamer et al. [13]
	... I have been tagged or received with an unpleasant photos or videos on SNS	3.486	1.872	

(*continued*)

Table 1. (*continued*)

Construct	Question	Average	Standard deviation	Reference
	... my photo or video that I want to keep a secret has been uploaded on SNS	3.182	1.883	
	... a photo or video that I want to hide has been released on SNS	2.889	1.864	
	... a photo or video that made me ridiculous has been posted on SNS	3.018	1.894	
	... a photo or video that I don't want to open to the public has been posted on SNS	3.160	1.953	
The victimization experience of exclusion	When I use SNSs, I have been ...			Prinstein et al. [48]
	... no good responses even I posted or commented on SNS	3.578	1.742	
	... left out of a posting or comment on SNS even other users know that I want to join	2.172	1.238	
	... ignored by my "friends" on SNS	2.015	1.174	
The victimization experience of impersonation	When I use SNSs, someone has ...			Den Hamer et al. [13]; Kwan and Skoric [31]
	... posted or commented on SNS under my account	1.982	1.352	
	... updated my personal information under my account	2.009	1.407	
	... contacted my "friends" under my account	2.077	1.484	

*Self-control was coded reversely. **Introduction of each question

3.3 Data

The survey was conducted in May 2016 in South Korea. The survey invitation message was distributed to Korean students through the instant messenger application, and invitees distributed the message in the same way. For this reason, it was not known that how many people received the invitation message. By doing so, 358 individuals participated in the survey, and 33 data were removed due to invalid response, therefore, 325 data were analyzed. Among those, 192 data were gathered through Google docs, and 133 data were delivered at a public university.

The gender of 325 respondents consisted of 103 males (31.7%), and 222 females (69.2%). 51 freshmen (15.7%), 113 sophomores (34.8%), 71 juniors (21.8%), and 90 seniors (27.7%) were participated. Most of the respondents answered Kakaotalk (65.6%) as favorite SNS, and Facebook (51.1%), Instagram (26.2%) followed. 249 respondents (76.7%) have used SNSs more than 1 h a day, moreover, 119 respondents (36.6%) did more than 2 h. Detailed demographic information is presented in Table 2.

Table 2. Demographic data

Variable	Category	Respondents	Percentage (%)
Gender	Male	103	31.7
	Female	222	69.2
Grade	Freshman	51	15.7
	Sophomore	113	34.8
	Junior	71	21.8
	Senior	90	27.7
Favorite SNS (multiple responses)	Facebook	166	51.1
	Twitter	24	7.4
	Kakaostory	1	0.3
	Kakaotalk	213	65.5
	Instagram	85	26.2
	Others (e.g. community sites, etc.)	4	1.2
	None	4	1.2
Average time of SNSs usage per day	~0.5 h	24	7.4
	0.5~1 h	52	16
	1~1.5 h	78	24
	1.5~2 h	52	16
	2 h~	119	36.6

4 Result

Partial Least Squares 2.0 and SPSS were used because this research is exploratory research by using empirical survey [20, 22].

4.1 Data Analysis

The reliability and validity of the scales and measurement items were verified. The composite reliability and Cronbach's alpha indicate the reliability of the scales. The composite reliability must exceed 0.7 to be reliable [1]. Cronbach's alpha was evaluated for construct reliability, and each value is above the greatly acceptable 0.7 threshold [19, 45], and 0.6 (changing profile setting) is also acceptable [22].

The Average Variance Extracted (AVE) is calculated for convergent validity, and all AVE values are above the recommended minimum, 0.50 [19]. Moreover, t-value is qualified to test convergent validity [20].

Meanwhile, GoF is calculated by the square root of multiplication of the average of R-square (cf. Fig. 2) and average of AVE, and if it is greater than 0.1, the model is qualified [37, 58] (Table 3).

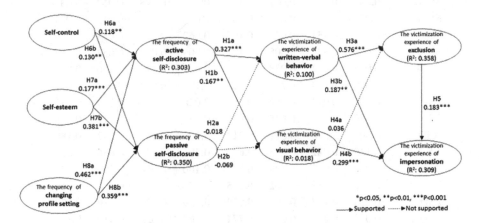

Fig. 2. The result of hypotheses testing

Table 3. Reliability and validity

	AVE	Composite reliability	Cronbach's alpha	GoF
Self-control	0.548	0.823	0.795	0.407
Self-esteem	0.798	0.940	0.916	
*Changing profile setting	0.550	0.781	0.610	
*Active self-disclosure	0.704	0.877	0.788	
*Passive self-disclosure	0.699	0.903	0.855	
**Written-verbal behavior	0.667	0.909	0.875	
**Visual behavior	0.667	0.909	0.872	
**Exclusion	0.679	0.863	0.760	
**Impersonation	0.735	0.893	0.820	

*The frequency of... **The victimization experience of...

Loadings of all items were calculated for convergent validity and discriminant validity. If each item is loaded onto each construct with the highest value, it is considered valid [12]. As can be seen in Table 4, each item is loaded with greatest value onto right construct, although some items have value lower than 0.6, except for the frequency of active and passive self-disclosure. For the reason, in Table 5, these two constructs were tested again after other items were eliminated.

Table 4. Validity-all constructs

Construct	Factor loadings							
	SC	SE	PS	SD	WV	V	E	I
SC_1	**0.829**	0.026	−0.068	−0.059	0.197	0.017	0.020	0.039
SC_2	**0.717**	0.006	0.041	0.231	−0.073	0.184	0.090	0.140
SC_3	**0.739**	0.111	0.132	−0.114	0.122	−0.027	0.050	0.064
SC_4	**0.802**	−0.123	−0.047	0.126	0.123	0.098	0.012	−0.033
SE_1	−0.011	**0.841**	0.058	0.248	−0.038	−0.019	−0.043	−0.084
SE_2	−0.042	**0.851**	0.047	0.197	−0.062	−0.003	0.044	−0.089
SE_3	0.048	**0.882**	−0.005	0.197	−0.009	−0.092	−0.052	−0.114
SE_4	0.037	**0.862**	0.038	0.142	0.052	−0.178	−0.096	−0.005
PS_1	−0.010	0.044	**0.744**	−0.01	0.192	−0.054	0.083	0.017
PS_2	−0.001	0.073	**0.556**	0.388	0.028	0.151	0.259	−0.031
PS_3	0.082	0.050	**0.635**	0.377	−0.173	0.213	−0.111	0.205
ASD_1	−0.003	0.076	0.071	**0.717**	0.205	0.045	0.017	0.180
ASD_2	0.119	0.082	0.315	**0.634**	0.201	−0.157	−0.036	0.200
ASD_3	0.018	0.009	0.429	**0.63**	0.160	−0.064	−0.105	0.248
PSD_1	−0.016	0.159	−0.002	**0.855**	0.078	−0.001	0.001	0.010
PSD_2	−0.025	0.257	0.004	**0.820**	−0.019	0.028	0.116	−0.138
PSD_3	0.013	0.295	−0.064	**0.774**	0.069	0.026	0.110	−0.154
PSD_4	0.191	0.151	0.179	**0.601**	−0.034	0.129	0.258	−0.048
WV_1	0.124	−0.118	0.150	0.090	**0.758**	0.235	0.042	0.049
WV_2	0.054	0.009	0.183	0.051	**0.713**	0.266	0.051	0.232
WV_3	0.128	0.073	−0.004	0.206	**0.690**	0.242	0.16	0.292
WV_4	0.138	−0.037	−0.028	0.183	**0.711**	0.379	0.232	0.031
WV_5	0.075	0.029	−0.036	0.035	**0.655**	0.244	0.194	0.171
V_1	0.143	0.065	−0.053	0.219	0.29	**0.518**	−0.074	0.293
V_2	0.096	0.071	−0.074	0.121	0.295	**0.753**	0.031	0.063
V_3	−0.042	−0.074	0.090	−0.034	0.224	**0.825**	0.080	0.216
V_4	0.095	−0.221	0.119	−0.094	0.271	**0.732**	0.058	0.168
V_5	0.075	−0.177	0.057	−0.069	0.188	**0.832**	0.111	0.128
E_1	0.039	−0.158	0.272	0.083	0.057	0.367	**0.614**	0.214
E_2	0.084	−0.014	0.012	0.192	0.407	0.008	**0.737**	0.157
E_3	0.118	−0.084	−0.007	0.072	0.508	0.007	**0.658**	0.162
I_1	0.038	−0.136	0.090	0.042	0.189	0.182	0.139	**0.765**
I_2	0.089	−0.097	0.046	−0.003	0.196	0.220	0.104	**0.788**
I_3	0.056	−0.083	0.041	0.022	0.161	0.184	0.093	**0.772**

SC: Self-control, SE: Self-esteem, *PS: Changing profile setting, *ASD: Active self-disclosure, *PSD: Passive self-disclosure, **WV: Written-verbal behavior, **V: Visual behavior, **E: Exclusion, **I: Impersonation/*The frequency of… **The victimization experience of…

Table 5. Validity-self-disclosure

	Factor loading	
	ASD	PSD
ASD_1	**0.614**	0.488
ASD_2	**0.728**	0.336
ASD_3	**0.849**	0.221
PSD_1	0.412	**0.750**
PSD_2	0.235	**0.864**
PSD_3	0.184	**0.869**
PSD_4	0.248	**0.657**

ASD: The frequency of
active self-disclosure
PSD: The frequency of
passive self-disclosure

Discriminant validity is supported if the square root of the AVE of each latent construct (with bold) is greater than its inter-construct correlation in the related columns and rows [18]. As Table 6 presented, the least square root of AVE, 0.740 on self-control, is higher than other inter-construct correlated values. Thus, all items and data have a discriminant validity.

Table 6. Discriminant validity

	SC	SE	PS	ASD	PSD	WV	V	E	I
SC	**0.740**								
SE	0.009	**0.893**							
PS	0.167	0.157	**0.742**						
ASD	0.197	0.251	0.509	**0.839**					
PSD	0.193	0.439	0.440	0.655	**0.836**				
WV	0.260	−0.038	0.198	0.315	0.196	**0.817**			
V	0.238	−0.175	0.200	0.122	0.041	0.607	**0.816**		
E	0.231	−0.119	0.217	0.282	0.192	0.598	0.478	**0.824**	
I	0.196	−0.217	0.216	0.244	−0.025	0.386	0.483	0.410	**0.857**

SC: Self-control, SE: Self-esteem, *PS: Changing profile setting, *ASD: Active self-disclosure, *PSD: Passive self-disclosure, **WV: Written-verbal behavior, **V: Visual behavior, **E: Exclusion, **I: Impersonation/*The frequency of... **The victimization experience of...

4.2   Hypotheses Testing

The frequency of active self-disclosure affects the victimization experience of written-verbal behavior (H1a) and visual behavior (H1b), however, the relation between the frequency of passive self-disclosure and both victimization experience of written-verbal behavior (H2a) and visual behavior (H2b) are insignificant.

Interestingly, the victimization experience of cyberbullying also affects itself. The victimization experience of written-verbal behavior not only affects the victimization experience of exclusion (H3a) but also impersonation (H3b). In addition, the victimization experience of visual behavior doesn't influence on the victimization experience of exclusion (H4a), but impersonation (H4b). Moreover, the victimization experience of exclusion also affects the victimization experience of impersonation (H5).

Low level of self-control (H6), high level of self-esteem (H7), and the frequency of changing profile setting (H8) affect the frequency of active self-disclosure and passive self-disclosure. The result of hypotheses testing is in Fig. 2.

5   Conclusion

SNSs have been spread worldwide, and users are enjoying these SNSs [11]. To become a member of SNSs and take a pleasure, users disclose personal information by themselves [34]. Self-disclosure has good influence mentally and physically [10], but some studies stressed self-disclosure can induce cyberbullying [17, 54]. This paper found the relation between self-disclosure and cyberbullying through an empirical analysis [17].

Cyberbullying is a new bullying through mobile equipment that time and place are not necessary when they are offline [41]. That is, the danger of cyberbullying could be serious more and more because perpetrator bully victim whenever and wherever online.

Interesting conclusion between self-disclosure and cyberbullying is resulted in. The frequency of active self-disclosure will affect the victimization experience of written-verbal behavior (H1a) and visual behavior (H1b), whereas the frequency of passive self-disclosure won't affect the victimization experience of written-verbal behavior (H5a), but (H5b) was negatively meaningful access, although this hypothesis was rejected. That is, passive self-disclosure has lower possibility to suffer cyberbullying than active self-disclosure.

Besides, it is interesting to analyze the result of the relation to the victimization experience of cyberbullying one another which has been dealt with yet. According to the result, the people who victimized by written-verbal behavior would have the possibility of exclusion (H3a) or impersonation (H3b), but the people who victimized by visual behavior would have possibility of impersonation (H4b), not exclusion (H4a). It is interpreted that written-verbal behavior regarded as harassment because the questions of written-verbal behavior include intentional action.

Another result is all victimization experience of cyberbullying affects the victimization experience of impersonation (H3b, H4b, and H5). Through this result, it is possible to predict that victim could have the experience of impersonation which is more serious problem, if victim has the experience of written-verbal or visual behavior and even exclusion in the future. This interpretation has involved that all cyberbullying victimization experience can be expanded. Especially, impersonation should be regarded as crime because this problem occurs when someone access or hack other user's account for revealing, stealing, even deleting.

Moreover, before this study was designed, H6 is that the people who have low level of self-control disclose themselves more than the high, but the result showed reverse relation; the high self-controlled people tend to disclose themselves frequently. As self-control is the ability to control feeling, emotion, and even acts as he or she wants by oneself [35], if there is no relation between self-control and self-disclosure in SNSs, it could explain that the self-disclosure in SNSs is an action through his or her will rather than unconscious.

Meanwhile, the relation between self-esteem and the frequency of active (H7a) and passive (H7b) self-disclosure proved; the higher level of self-esteem, the higher the frequency of active and passive self-disclosure, as previous studies proved. It means, the people who have high self-esteem disclose themselves online. Indeed, these people have posted texts, photos, etc. (active self-disclosure) and clicked "like", and shared something, etc. (passive self-disclosure) more than the lower.

The frequency of changing profile setting is proved as a factor of self-disclosure in SNSs. Although there is not enough study about the relation, both path co-efficient of the frequency of changing profile setting and the frequency of active (H8a) and passive (H8b) self-disclosure are higher than 0.5 and 0.3, showing a significant relation.

Even though there are many interesting conclusions, there are some limitations. First, factor loading of active and passive self-disclosure wasn't perfect, so factor loading is needed twice. Because there was no previous research about types of self-disclosure, category was divided arbitrarily, but each item showed interesting and different result. Second, the conclusion is difficult to prevail on the all generation and people because people who received the invitation message through instant message application were in researcher's school or their colleague. Thus, it is necessary to add more population in the future research.

References

1. Bagozzi, R.P., Yi, Y.: On the evaluation of structural equation models. J. Acad. Mark. Sci. **1** (1), 74–94 (1988)
2. Barker, V.: Older adolescents' motivations for social network site use: the influence of gender, group identity, and collective self-esteem. Cyberpsychol. Behav. **12**(2), 209–213 (2009)
3. Barkley, R.A.: ADHD and the Nature of Self-control. Guilford Press, New York (1997)
4. Baumeister, R.F., Campbell, J.D., Krueger, J.I., Vohs, K.D.: Does high self-esteem cause better performance, interpersonal success, happiness, or healthier lifestyles? Psychol. Sci. Public Interest **4**(1), 1–44 (2003)
5. Belsey, B.: Cyberbullying.ca (2004)
6. Blau, P.M.: Exchange and Power in Social Life. Wiley, New York (1964)
7. Boyd, D.: Why youth ♥ social network sites: the role of networked publics in teenage social life, youth, identity, and digital media. In: Buckingham, D. (ed.) The John D. and Catherine T. MacArthur Foundation Series on Digital Media and Learning, pp. 119–142. The MIT Press, Cambridge (2008)
8. Branckaute, F.: Facebook statistics: the numbers game continues. The Blog Herald (2016). http://www.blogherald.com/2010/08/11/facebook-statistics-the-numbers-game-continues

9. Bullas, J.: 33 social media facts and statistics: you should know in 2015. Jeffbullas.com (2015).   http://www.jeffbullas.com/2015/04/08/33-social-media-facts-and-statistics-you-should-know-in-2015

10. Cameron, J.J., Holmes, J.G., Vorauer, J.D.: When self-disclosure goes awry: negative consequences of revealing personal failures for lower self-esteem individuals. J. Exp. Soc. Psychol. **45**(1), 217–222 (2009)

11. Carpenter, C.J.: Narcissism on Facebook: self-promotional and anti-social behavior. Pers. Individ. Differ. **52**(4), 482–486 (2012)

12. Chen, R.: Living a private life in public social networks: an exploration of member self-disclosure. Decis. Support Syst. **55**(3), 661–668 (2013)

13. Den Hamer, A., Konijn, E.A., Keijer, M.G.: Cyberbullying behaviors and adolescents' use of media with antisocial content: a cyclic process model. Cyberpsychol. Behav. Soc. Netw. **17**(2), 74–81 (2014)

14. Elizabeth., C., Magdalena, I., Kevin, W., Cory, C., Nicole, P.: Will you be my friend? Computer-mediated relational development on Facebook.com. In: International Communication Association, San Francisco (2007)

15. Ellison, N.B., Boyd, M.D.: Social network sites: definition, history, and scholarship. J. Comput.-Mediat. Commun. **13**(1), 210–230 (2008)

16. Ellison, N.B., Steinfield, C., Lampe, C.: The benefits of Facebook "friends:" social capital and college students' use of online social network sites. J. Comput.-Mediat. Commun. **12**(4), 1143–1168 (2007)

17. Erder-Baker, Ö.: Cyberbullying and its correlation to traditional bullying, gender and frequent and risky usage of internet mediated communication tools. New Media Soc. **12**(1), 109–125 (2010)

18. Fornell, C., Larcker, D.F.: Evaluating structural equation models with unobservable variables and measurement error. J. Mark. Res. **18**, 39–50 (1981)

19. Fornell, C., Bookstein, F.L.: Two structural equation models: LISREL and PLS applied to consumer exit-voice theory. J. Mark. Res. **19**(4), 440–452 (1982)

20. Gefen, D., Straub, D.: A practical guide to factorial validity using PLS-graph: tutorial and annotated example. Commun. Assoc. Inf. Syst. **16**(1), 91–109 (2005)

21. Ha, C.: Validation of Korean version of the self-control scale (in Korea). Ph.D. thesis, Dankook University (2007)

22. Hair, J.F., Black, W.C., Babin, B.J., Anderson, R.E.: Multivariate Data Analysis, 6th edn. Pearson Educational Inc., Hoboken (2006)

23. Henderson, M., de Zwart, M., Lindsay, D., Phillips, M.: Legal risks for students using social networking sites. Aust. Educ. Comput. **25**(1), 3–7 (2010)

24. Hinduja, S., Patchin, J.W.: 2015 Cyberbullying Data. Cyberbullying Research Center (2015). http://cyberbullying.org/2015-data

25. Ho, K.: 41 Up-to-Date Facebook Facts and Stat, Wishpond (2015). http://blog.wishpond.com/post/115675435109/40-up-to-date-facebook-facts-and-stats

26. Homans, G.G.: Social behavior as exchange. Am. J. Sociol. **63**(6), 597–606 (1958)

27. Irwin, A., Taylor, D.A.: Social Penetration: The Development of Interpersonal Relationships. Holt Rinehart & Winston, Oxford (1973)

28. Jia, Y., Zhao, Y., Lin, Y.: Effects of system characteristics on users' self-disclosure in social networking sites. In: 2010 Seventh International Conference, Information Technology: New Generations (ITNG) (2010)

29. Johnson, J.A.: The "self-disclosure" and "self-presentation" views of item response dynamics and personality scale validity. J. Pers. Soc. Psychol. **40**(4), 761–769 (1981)

30. Kang, C.: Nine of 10 teenagers have witnessed bullying on social networks, study finds, Washingtonpost (2011). https://www.washingtonpost.com/business/economy/nine-of-10-teenagers-have-witnessed-bullying-on-social-networks-study-finds/2011/11/08/gIQAPqUq3M_story.html

31. Kwan, G.C.E., Skoric, M.M.: Facebook bullying: an extension of battles in school. Comput. Hum. Behav. **29**(1), 16–25 (2013)

32. Langos, C.: Cyberbullying: the challenge to define. Cyberpsychol. Behav. Soc. Netw. **15**(6), 285–289 (2012)

33. Li, Q.: Cyberbullying in schools: a research of gender differences. School Psychol. Int. **27** (2), 157–170 (2006)

34. Liu, Z., Min, Q., Zhai, Q., Smyth, R.: Self-disclosure in Chinese micro-blogging: a social exchange theory perspective. Inf. Manag. **53**(1), 53–63 (2013)

35. Logue, A.W.: Self-Control: Waiting Until Tomorrow for What You Want Today. Prentice-Hall, Englewood Cliffs (1995)

36. Lowry, P.B., Cao, J., Everard, A.: Privacy concerns versus desire for interpersonal awareness in driving the use of self-disclosure technologies: the case of instant messaging in two cultures. J. Manag. Inf. Syst. **27**(4), 163–200 (2011)

37. Martin, W., Odekerken-Schröder, G., Van Oppen, C.: Using PLS path modeling for assessing hierarchical construct models: guidelines and empirical illustration. MIS Q. **33**(1), 177–195 (2009)

38. McAndrew, F.T., Jeong, H.S.: Who does what on Facebook? Age, sex, and relationship status as predictors of Facebook use. Comput. Hum. Behav. **28**(6), 2359–2365 (2012)

39. Menesini, E., Nocentini, A., Palladino, B.E., Frisén, A., Berne, S., Ortega-Ruiz, R., Calmaestra, J., Scheithauer, H., Schultze-Krumbholz, A., Luik, P., Naraskov, K., Blaya, C., Berthaud, J., Smith, P.K.: Cyberbullying definition among adolescents: a comparison across six European countries. Cyberpsychol. Behav. Soc. Netw. **15**(8), 455–463 (2012)

40. Mishna, F., Khoury-Kassabri, M., Gadalla, T., Daciuk, J.: Risk factors for involvement in cyber bullying: victims, bullies and bully-victims. Child Youth Serv. Rev. **34**(1), 63–70 (2012)

41. Mishna, F., Saini, M., Solomon, S.: Ongoing and online: children and youth's perceptions of cyber bullying. Child Youth Serv. Rev. **31**(12), 1222–1228 (2009)

42. Muraven, M., Baumeister, R.F.: Self-regulation and depletion of limited resources: does self-control resemble a muscle? Psychol. Bull. **126**(2), 247–259 (2000)

43. Naruskov, K., Luik, P., Nocentini, A., Menesin, E.: Estonian students' perception and definition of cyberbullying. TRAMES **16**(4), 323–343 (2012)

44. Nocentini, A., Calmaestra, J., Schultze-Krumbholz, A., Scheithauer, H., Ortega, R., Menesini, E.: Cyberbullying: labels, behaviours and definition in three European countries. Aust. J. Guid. Couns. **20**(2), 129–142 (2010)

45. Nunnally, J.C.: Psychometric Theory. McGraw-Hill Inc., New York (1978)

46. Orth, U., Robins, R.W.: The development of self-esteem. Curr. Dir. Psychol. Sci. **23**(5), 381–387 (2014)

47. Posey, C., Lowry, P.B., Roberts, T.L., Ellis, T.S.: Proposing the online community self-disclosure model: the case of working professionals in France and the UK who use online communities. Eur. J. Inf. Syst. **19**(2), 181–195 (2010)

48. Prinstein, M.J., Boergers, J., Vernberg, E.M.: Overt and relational aggression in adolescents: social-psychological adjustment of aggressors and victims. J. Clin. Child Adolesc. Psychol. **30**(4), 479–491 (2001)

49. Rosenberg, M.: Society and the Adolescent Self-image. Princeton University Press, Princeton (1965)

50. Ross, C., Orr, E.S., Sisic, M., Arseneault, J.M., Simmering, M.G., Orr, R.R.: Personality and motivations associated with Facebook use. Comput. Hum. Behav. **25**(2), 578–586 (2009)
51. Sengupta, A., Chaudhuri, A.: Are social networking sites a source of online harassment for 60 teens? Evidence from survey data. Child Youth Serv. Rev. **33**(2), 284–290 (2011)
52. Sheldon, P.: I'll poke you. You'll poke me! Self-disclosure, social attraction, predictability and trust as important predictors of Facebook relationships. J. Psychosoc. Res. Cybersp. **3**(2), 5–15 (2009)
53. Jourard, S.M.: The Transparent Self. D. Van Norstand Company Inc., Princeton (1964)
54. Smith, P.K., Mahdavi, J., Carvalho, M., Fisher, S., Russell, S., Tippett, N.: Cyberbullying: its nature and impact in secondary school pupils. J. Child Psychol. Psychiatry **49**(4), 376–385 (2008)
55. Sprecher, S., Hendrick, S.S.: Self-disclosure in intimate: associations with individual and relationship characteristics over time. J. Soc. Clin. Psychol. **23**(6), 857–877 (2004)
56. Statista: Number of daily active Facebook users worldwide as of 2nd quarter 2016 (2016). http://www.statista.com/statistics/346167/facebook-global-dau
57. Tangney, J.P., Baumeister, R.F., Boone, A.L.: High self-control predicts good adjustment, less pathology, better grades, and interpersonal success. J. Pers. **72**(2), 271–324 (2004)
58. Tenenhaus, M., Vinzi, V.E., Chatelin, Y.M., Lauro, C.: PLS path modeling. Comput. Stat. Data Anal. **48**(1), 159–205 (2005)
59. Thibaut, W.J., Kelley, H.H.: The Social Psychology of Groups. Wiley, New York (1959)
60. Valkenburg, P.M., Peter, J.: Online communication among adolescents: an integrated model of its attraction, opportunities, and risks. J. Adolesc. Health **48**(2), 121–127 (2011)
61. Vandebosch, H., Van Cleemput, K.: Defining cyberbullying: a qualitative research into the perceptions of youngsters. Cyberpsychol. Behav. **11**(4), 499–503 (2008)
62. Von Marées, N., Petermann, F.: Cyberbullying: an increasing challenge for schools. School Psychol. Int. **33**(5), 467–476 (2012)
63. Wakefield, R.: The influence of user affect in online information disclosure. J. Strateg. Inf. Syst. **22**(2), 157–174 (2013)
64. Walrave, M., Vanwesenbeeck, I., Heirman, W.: Connecting and protecting? Comparing predictors of self-disclosure and privacy settings use between adolescents and adults. J. Psychosoc. Res. Cybersp. **6**(1), Article 3 (2012)
65. Weiten, W.: Psychology Themes and Variations. Wadsworth/Thomson Learning, Belmont (2004)
66. Willard, N.: Cyberbullying and cyberthreats. In: National Conference, Washington 4 (2005)

Software Product Line to Express Variability in E-Learning Process

Sameh Azouzi[1(✉)], Sonia Ayachi Ghannouchi[2], and Zaki Brahmi[1]

[1] ISITCom Hammam Sousse/Laboratory RIADI-GDL,
ENSI, Mannouba, Tunisia
`azouzi_sameh@yahoo.fr`
[2] ISG Sousse/Laboratory RIADI-GDL, ENSI, Mannouba, Tunisia

Abstract. As a consequence of the massive adoption of internet, many platforms such as Moodle, WebCT and Claroline aim to ease and improve the
teaching/learning process by means of taking advantage of internet technologies.
However, available systems do not satisfy all the needs of different institutions/
teachers, which push them to develop their own systems. Our contribution is the
proposition of a general model for collaborative learning processes. The proposed process is modeled with the BPFM (Business Process Feature Model)
notation which is a combination of BPMN (Business Process Model Notation)
and FM (Feature Model). In fact, BPMN offers almost no means to model
process variability. It becomes then necessary to find an efficient solution that
allows the fast development of systems and overcomes the afore-mentioned
issues. We strongly believe that adopting a software product line (SPL) ap
proach in e-Learning domain can bring important benefits. Knowing that
Business Process Management (BPM) is a potential domain in which Software
Product Line (SPL) can be successfully applied, we propose in this paper to use
the Business Process Feature Model (BPFM) notation that combines in a new
notation concepts coming both from feature modeling and from BP modeling to
create a reusable and reconfigurable e-learning process.

Keywords: E-Learning · Variability · Reconfiguration · Collaboration ·
Software Product Lines · Business process lines · FM · BPMN · BPFM

1 Introduction

Today, internet offers a wide range of new opportunities for development of education.
Taking advantage from the benefit of using the Internet, educational institutions use
some kind of e-learning platform (LMS). These platforms such as Moodle, claroline,
WebCT and Sakai aim to ease and improve the teaching-learning process. These
platforms offer different resources and tools that enable students to develop both
autonomous and collaborative learning. But it must be said that the needs of
universities/learners are rapidly changing and diversified. It becomes thus so hard to
available systems to satisfy all the needs of different institutions, which push them to
develop their own systems. Many recent studies [1–3] show that LMSs do not satisfy
all the needs of teachers and students which push them to use social networks, Web2.0,

© Springer International Publishing AG 2017
M. Themistocleous and V. Morabito (Eds.): EMCIS 2017, LNBIP 299, pp. 173–185, 2017.
DOI: 10.1007/978-3-319-65930-5_14

cloud based services and mobile applications in order to complement the lack of LMSs, and suggest that students need learning environments which are better adapted to their needs. In order to overcome these issues, we suggest a general e-learning process. It is a reconfigurable, reusable and collaborative learning process. This will allow the use of a new e-learning system based on BPM for Virtual Learning Environments. Its aim is to provide the functionality of management of documents which can be found in the respective zip packages on our website.

Learning process through the effective modelling of education pedagogies in the form of learning process workflows using an intuitive graphical flow diagram user-interface. One of the challenges in the adoption of the BPM concept is that, since there are differences between a learning process and a business process, it is not clear whether the concept of process in BPM is compatible with the learning process. The aim is to investigate the relationships and, through implementation, to verify if it is possible to apply BPM concept to online learning process management through using a pedagogical modelling perspective.

So, according to Helic et al. [5], collaborative learning processes are such learning processes where learning tasks are based on real-life tasks or authentic situations and typically require and motivate the co-operation or collaboration (co-construction and exchange of knowledge) of learners in a group. Since there is a strong similarity between a learning process and a business process in both user aspects as well as technical aspects, we believe that applying BPM to model and manage learning processes can be efficient, easy to develop and maintain, as well as powerful enough to support a wide range of common learning situations in e-learning systems [4]. But, the problem while using BPMN, is that process variability can only be modeled using one process model or separate and distinct process models. BPMN offers almost no means to model process variability [16]. Then, how resolve the problem of reconfiguration and variability? In SPL field, variability management is addressed. In this context, we suggest the use of Software Product Line (SPL) approach for the development of e-Learning applications. E-Learning applications could be implemented in a variety of settings and respond to the different needs of universities/learners. On the other hand, it was observed the importance of the practice of reutilization of business process models, making use of specific techniques such as the Business Process Line (BPL) that was originated from concepts of Software Product Line (SPL). As seen in SPL, in order to aid the creation, instantiation and evolution of BPLs, the usage of computational tools is also very important in the context of BPL, due mainly to the complexity and dynamism from businesses [14].

To do that, we propose to use the Business Process Feature Model (BPFM) notation that combines in a new notation concepts coming both from feature modeling and from BP modeling. The notation permits to represent activities, their partial execution order, and involved data objects if it is necessary [15]. A BPFM model collects all the possible BP variants, and via a configuration step it is possible to derive the most suitable one for the specific organization. Thereby, faced with the importance of modeling the variability of e-learning processes to manage the needs of various universities/teachers, this work, proposes to use the BPFM notation with the main objective of offering a general, reconfigurable, reusable and collaborative learning process. Besides that, an evaluation was done for the e-learning process according to two types of characteristics: functionalities (help tools, communication tools, teaching tools), quality of the system (ease of use, technology selection, adaptability and modularity), by experts in the field of e-learning.

This paper is structured as follows: In Sect. 2 we give a brief introduction to some basic concepts and describe in Sect. 3 the SPL approach. Section 4 reports relevant related works. Section 5 gives an overview of the BPFM approach, and Sect. 6 shows the use case. Finally, Sect. 7 reports conclusions and opportunities for further research.

2 Preliminaries

This section provides background on those concepts that will be used throughout this paper.

2.1 BPM (Business Process Management)

BPM is an acronym for Business Process Management. It is a key concept where business processes are put at the center of the reflection of the managers to find a culture based on the continuous improvement of the performances of the companies and where the technologies give new possibilities to a better management of the process. BPM is seen as an approach that aims to achieve a better overall view of all the company's business processes and their interactions in order to be able to optimize and automate them by using specific applications. Dominique Annet states that «By modeling processes, we aim to have an overall view of all the processes and their interactions in order to optimize the general functioning and to automate all that can be» [6].

When we talk about BPM life cycle, the four stages of this cycle are as follows [7]:

- (Re) design and analysis of the business process: the life cycle begins with the creation of business processes either "from scratch" or by reconfiguring an existing model.
- Configuration of the information system: the business process is implemented by configuring the information system of the organization.
- Execution and supervision of the business process: in this phase, the process is executed while controlling and supervising its execution.
- Diagnosis: This phase consists of learning knowledge from the business processes in execution and using them as input for possible improvements in business processes.

2.2 BPM and Learning Process

Business Process Management (BPM) includes methods, techniques, and software to design, enact, control, and analyze operational processes [6]. BPM is considered as an approach which aims to achieve a better overall view of business processes and their interaction in order to optimize them and automate the maximum of tasks taking advantages of devoted applications.

Helic [4] discusses the possibilities of using BPM technology for the management of collaborative learning processes. Otherwise, in e-learning scenarios, activities can be perceived as processes or workflows. A learning process represents a series of tasks or activities executed to achieve the individual or group learning goals.

Usually, in BPM a business process is defined using a graphical notation such as Petri net, YAWL (Yet Another Workflow Language), UML (Unified Modeling Language) and BPMN (Business Process Management Notation). BPMN is a comprehensible graphical notation for defining a learning process including learning activities, stakeholders, their roles and their interactions.

So, processes in e-learning context are similar to those in a business context also related to the BPM context [8].

Within this frame, several works supported the idea of modelling learning processes (collaborative) and confirm the fact that these processes can be presented by means of BPMN. According to Arnaud, the activities of learning introduce a serious need of follow-up and evaluation, which has many objectives including the fact that learners exploit at best the environment and succeed in reaching the goals of the learning activity that they realize (in terms of acquisition of knowledge, skills, etc.). Nonetheless, the domain of BP has methods and tools susceptible to serve this type of follow-up [9]. The authors in [10] suggested the BPM approach for modelling a learning process and integrated the SOA technologies to guarantee the aspect of collaboration and interaction in the learning process. Da Costa in [11] proposed a track for the development of learning devices based on scenarios by using the abstract frames of BPM. Adesina in [12] proposes an environment of virtual learning process named VLPE (Virtual Learning Process Environment) based on BPM. The idea is to propose an application "stand-alone" allowing modelling learning courses, which are afterward implemented via a transfer towards the BPEL execution language. Schneider in [13] modelled activities of learning with the notation BPMN.

We can conclude by considering that all of these research works concerning the systems of e-learning in the cloud and the modelling of the existing processes of e-learning suffer from some limits namely:

- The absence of a complete initiative or approach for the construction of an agile and reconfigurable process of learning adaptable to change and being able to evolve in order to meet the needs of learners;
- Lack of collaboration in learning processes and limited reuse possibilities from the processes deployed by other universities/teachers;
- The absence of an automated mechanism of communication, which informs people at the good moments when an intervention is required during the execution of the learning process.

Our objective is to adopt BPM and FM in one notation BPFM for the construction of a general model of the process of e-learning as a collaborative, reconfigurable and reusable model. In this paper, we are interested in modeling collaborative learning process and we focus on the use of synchronous web2.0 tools such as virtual classroom, webinar, video conferencing, with the BPMN notation and thus focusing on the first phase of the BPM cycle (Design). In fact, we focus on asynchronous and also synchronous collaborative learning processes because on the one hand, we observe that most LMSs lack synchronous collaboration. On the other hand, a BPM approach seems very useful for better management and orchestration of learning processes.

As a consequence from the mapping procedure, SPL knowledge not yet used by BPM researchers could be explored. We identified research challenges in BPM with respect to variability management and explored how this SPL knowledge may be useful to treat open issues in BPM. Given that SPL is becoming a mature discipline and a large body of knowledge has been accumulated, we assume that several approaches in SPL can be appropriate to solve similar variability problems in BPM.

3 Software Product Line and Feature Model

Software Product Lines (SPL) systematizes the reuse across a set of similar products that a software company produces. The main goal of SPL is obtaining a reduction of the overall development costs and times for the products derived from the product line. In SPL, a product is composed of a set of common features and a set of variable features. Common features appear in all products and variable features appear under demand of consumer's products. Observing a certain product of an SPL, although it is described as a set of fixed features, some features can be in use in a certain moment and some not. This is called runtime variability. *Feature Models* (FM) are one of the most used artifacts for modeling variability, that is, specifying which features are common and which are variable. A FM represents all possible products in an SPL in terms of features. There are several notations of FM, such as FODA [5], or J. Bosch [4]. A FM establishes a parental relationship between each feature, as shown in Fig. 1, that can be: (i) *Mandatory*: if a child feature node is defined as mandatory, it must be included in every product that contains the parent; (ii) *Optional*: if a child feature node is defined as optional, it can be included or not when its father feature appears in a product; (iii) *Alternative*: if the relationship between a set of children nodes and their father is defined as alternative, only one of the children features could be included in every father feature products; and (iv) *Or*: if the relationship between a set of children nodes and their father is defined as or, one or more of them could be included in every father feature products. In addition to the parental relations between features, a FM can also

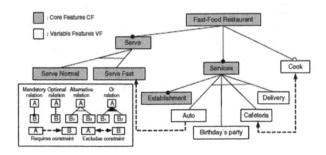

Fig. 1. Example of feature model

contain cross-tree constraints between couples of features. These are: (i) *Requires*: If a feature A requires a feature B, the inclusion of A in a product implies the inclusion of B in such product; and (ii) *Excludes*: if a feature A excludes a feature B, both features can not be part of the same product. An example of feature model is given in Fig. 1.

4   Software Product Line and Feature Model

Modeling variable BP is the ability to represent in a single model many alternative BPs sharing the same goal [15]. In order to describe variable BPs several approaches have been proposed. In [17], Groner et al. propose the adoption of BPMN elements to model the BPMT and the adoption of the model FM to model the variabilities of a BPL. In that study, each variability is represented in the BPMT in an execution flow of the business process.

In this case, it is not possible to explicitly distinguish the commonalities and the variabilities from the BPL in the BPMT, since the BPMN doesn't have specific elements for that purpose. In another work, Schnieders and Puhlmann [19] propose an extension for the BPMN, referred as vrBPMN, whose the objective is to explicitly represent the variabilities on business process models.

However, since this extension has many stereotypes, its use implies the need for the business domain engineer to know another notation, in addition to the FM and the original BPMN. Other notations to represent variabilities in business process were proposed in the literature, such as C-EPC [16], Configurable integrated EPC (C-iEPC) [3] or C-YAWL [18], however they are not based on the BPMN.

5   Business Process Feature Model

5.1   Overview

The BPFM approach is organized in four main steps (Fig. 2) and it results to be particularly suitable in situations in which an abstract definition of a process needs to be successively refined to consider specific aspects of the deployment context, such as the specific characteristics of the organization supporting the process itself. This is a quite

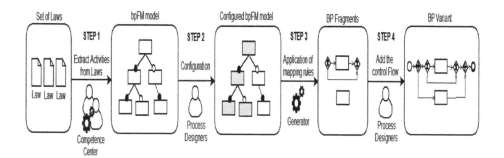

Fig. 2. Steps of the BPFM approach [4, 9]

common situation for processes supporting PA services to citizens. In such a case objectives and activities constituting the process are general and independent from the specific characteristics of the offices delivering the service itself. Nevertheless, the precise definition of the process, in terms of roles and ordering of the activities, depends from the deployment of related aspects such as for instance the organizational model. The input of the proposed approach are the laws regulating the provisioning of a service, while the final output will be a BP variant that can be deployed according to the characteristics of the service under analysis, and the organizational model of the Public Administration which delivers the service to the citizen. In particular, the approach is organized in 4 successive steps:

- The first step aims at defining a general model that can successively constitute the basis for the definition of a process variant for the specific deployment context. The model will be codified using the BPFM notation presented in the next section. This step includes knowledge acquisition through the study of legal and regulatory frameworks governing the delivery of the PA service under study. This step should be carried out only once for each service delivered by the PAs. The activity performed by a focus group or a competence center, will permit to derive a model that will include only the activities that have to be carried out, the relations among them, and the data structure they possibly get in input or produce in output (as said this information are codified in a BPFM model as illustrated in Fig. 2.
- The second step foresees the refinement of the previously defined model taking into account the specific needs of the service that have to be delivered and of its deployment context. Similarly to what it is done in feature modeling, this step foresees the definition of a *configuration* on the BPFM model which will permit to define a specific variant from the BP family.
- The third step takes in input activities and data objects resulting from the configuration defined in the previous step. Through the application of mapping rules we define, it is then possible to automatically derive BP fragments representing portions of the behavior that has to be completed to reach the goal of the service to be delivered.
- The last step concerns the derivation of the fully specified BP variant starting from the generated BP fragments. At this stage process designers add control flow relationships among the generated BP fragments, also taking into account the specific characteristics of the PA organization that need to deliver the service to citizens. It is worth mentioning that the same activity could be associated to different roles in different BP variants, as a result of possible different organizational models for different PA offices.

5.2   BPFM Notation

A BPFM [3, 15] model is constituted by a set of activities organized in a tree, where the root identifies the BP family that is described. Going up and down in the tree, BPFM introduces different levels of detail in the BP specification. In particular, each internal (non-leaf) activity denotes a sub-process, whereas the leaves represent atomic activities (tasks). In order to better specify operational details, the BPFM enables the user to specify atomic activities using the same meaning and symbols used by BPMN 2.0 for the task type. Connections between activities at different levels of the tree are called constraints (such as in standard FM). Constraints may be used to represent variability in two dimensions, (i) if each child activity must be inserted in each BP variant (it means the activity is selected in the configuration phase) and, in the case the activity is inserted, (ii) if it must or can be executed at run-time. This is how both the static and dynamic (execution time) inclusion of activities can be specified. Constraints can be binary or multiple depending on the number of child activities connected to a parent activity, they also specify a partial execution order of the activities. BPFM binary constraints are reported in the following:

- A *Mandatory Constraint* requires that the connected child activity must be inserted in each BP variant, and it also has to be included in each execution path (Fig. 3a).

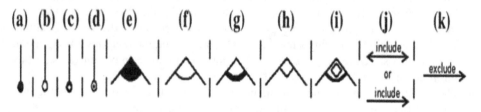

Fig. 3. BPFM constraints [4]

- An *Optional Constraint* specifies that the connected child activity can be inserted (or not) in each BP variant, and when included it is not necessary to include it in each execution path (Fig. 3c).
- A *Domain Constraint* requires that the connected child activity must be inserted in each BP variant but it is not necessary to include it in each execution path (Fig. 3b).
- A *Special Case Constraint* specifies that the connected child activity can be inserted (or not) in each BP variant. Nevertheless, if selected, it has to be included in each execution path (Fig. 3d).

Below is given the list of multiple BPFM constraints:

- An *Inclusive Constraint* requires that at least one of the connected child activities must be inserted in each BP variant, and at least one of them has to be included in each execution path (Fig. 3e).
- An *Optional Constraint* requires that exactly one of the connected child activities has to be inserted in each BP variant, and it is not necessary to include it in each execution path (Fig. 3f).
- A *Selection Constraint* specifies that exactly one of the connected child activities has to be inserted in each BP variant. Moreover it has to be included in each execution path (Fig. 3g).

- A *XOR Constraint* requires that all connected child activities are inserted in each BP variant, and exactly one of them has to be included in each execution path (Fig. 3h).
- A *XOR Selection Constraint* requires that at least one of the connected child activities is inserted in each BP variant, and exactly one of them has to be included in each execution path (Fig. 3i).

Finally, as in "traditional" Feature Model, it is possible to specify *Include* and *Exclude* relationships between activities (Fig. 3j, k). More details are given in [15].

6 Deriving BP Fragment and BP Variants for the E-Learning Process: Running Example

In this section, we show the development process of our e-Learning product line. To endorse our approach for variability modeling of learning process, we will consider here new models of learning process based on learning activities and interactions between learners themselves and between learners and teachers. To select the learning scenarios, we were based on Lebrun [17] and Monnard [18] models (Fig. 4). We will consider the scenario focusing on the variability perspective. We will discuss here only a small portion of our learning process. The description of the BPFM model resulting from the application of the notation to the learning process is introduced below. The resulting BPFM model has four levels. It includes 4 sub-processes. Figure 4 reports the BPFM model that we consider here for illustrative purposes. The first level activities are scripting, Inform, Interact, produce and formative evaluation.

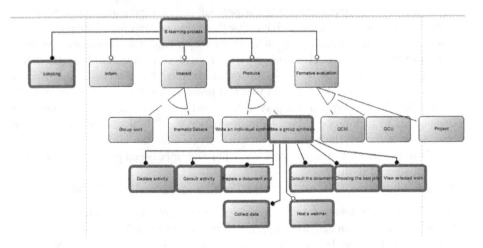

Fig. 4. A BPFM e-learning process model created via the BPFM tool

In detail, the sub-process "scripting" refers to the need to define the objectives of the proposed course. It is an activity connected to the root via a Mandatory Constraint since it has to be inserted in each BP variant and it has to be always executed.

The sub-processes: inform, interact, produce and formative evaluation are not mandatory; they are connected to the root via an optional constraint.

In sub-process "inform", besides the activities that the teacher offers to the students, he/she will provide them with access to a corpus of documents and web links related to the themes covered in his/her course. To prevent his/her students from losing themselves in a "shapeless mass" of files and links, the teacher organizes these documents by grouping them by theme and/or by activities. Access to these files and links is given at the right moments (weekly organization) and/or in the right places (thematic organization) of the course. In addition, he/she chooses for these files and links the formats best adapted to the needs of the course and the tools with which the students will access their contents.

The sub-process "interact" allows students to interact in a course; one possibility is to organize a thematic debate on a problematic issue. The aim is to make students adopt a scientific perspective during the debate. The objective is that the students better understand different points of view by identifying the arguments, with their greater or lesser scientific validity on the debated problematic issue, and thus improve their ability to argue. It is composed of two sub-activities: group work and thematic debate.

The sub-process "product" consists in teaching students how to produce a synthesis from several documents; a very different exercise from a summary. This type of activity is carried out individually or in groups to improve collaboration and interaction between learners using web2.0 collaborative tools. It is composed of two sub-activities: Write an individual synthesis and Write a group synthesis.

Finally, in the sub-process "formative evaluation", the students of the course make projects, and take QCMs or QCUs to validate the knowledge acquired during the course session.

We focus through our work on the aspect of collaboration and communication between the various actors of an activity. For that purpose, and for every process, we propose sub-processes and activities of learning which run in a collaborative context with the use of tools of collaborative work such as docs.google.com, Google drive, discussion forum and webinars. As an illustration of our idea, we choose the process "Produce" which is itself detailed into two sub-processes, which are: "Write an individual synthesis" and "Write a group synthesis" and we try to show the collaboration between actors in such process (Fig. 4). In this paper, we will more precisely focus on the modeling of the «Write a group synthesis» process for the high level of collaboration that it involves between learner and teacher and/or between peers to perform and accomplish some activities.

So, the sub-process "Write a group synthesis" is composed of eight sub-activities:

- Declare activity: refers to the need to define the description of the activity proposed by the teacher; it is an activity connected to the parent via a Mandatory Constraint.
- Consult activity: The learner consults the details of the activity proposed by the teacher; this activity is also connected to the parent via a Mandatory Constraint.
- The activities collect data, prepare a document and share with a teacher: The participating learners download the activity, divide into subgroups and start the work. Then they prepare a document and share it with the teacher; these activities are also connected to the parent via a Mandatory Constraint.

- The teacher consults the various works sent by the sub-groups and chooses the best work to make an oral presentation, then he/she launches a webinar.
- Host a webinar is an optional activity, then it is also connected to the parent via an Optional Constraint.

As soon as the BPFM is provided, several establishments can take advantage of it to define different configurations in order to derive variants which are better shaped to the universities/teacher's needs. We will discuss below the application of the proposed approach on a simple random example. Considering the Configuration Step 2 starting from the given BPFM, the BP designer, that was, in our case, the teacher has to define a novel configuration and he/she selected the activities that are needed to define the corresponding BP variant according to his/her needs.

The configuration includes all the mandatory activities as well as some optional activities taking into account the progress of learners in the course and the teacher's pedagogical learning strategy.

Then, considering the fragment derivation (Step 3), the selected configuration is automatically mapped into a set of BP fragments. The first level activities "scripting" and "produce" are mapped as sub-processes using composed activity mapping rules.

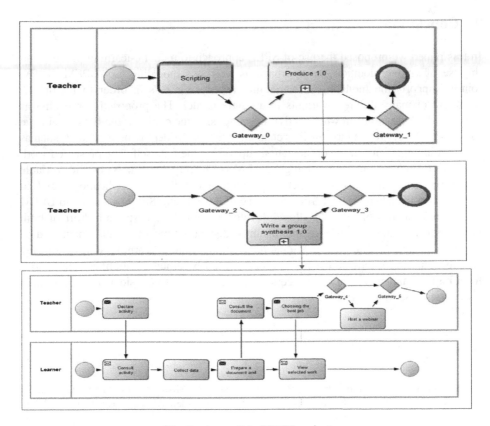

Fig. 5. A possible BPMN variant

According to the mapping rules, the sub-process "scripting" is mandatory, while the "produce" sub-process is optional but in our case, it was chosen. The sub-process "produce" is composed of two sub-activities: write an individual synthesis and write a group synthesis. In our case, we choose the sub-activity "write a group synthesis" and we choose all activities (mandatory and optional). Since we work in a collaborative context, we choose a configuration that contains the activity "host a webinar".

Finally, we derive the variant Step 4. At this stage, the BP designer needs to further refine the process introducing missing sequence flows, parallel execution constraints and events among the generated BP fragments, finally resulting in a fully specified BP variant. This step enables the user to introduce more details taking into account the possibility to allocate activities to different participants and roles. A possible variant generated to support reporting activities from the BPFM model and configuration is provided (Fig. 5). The derived process has two phases. Phase one is carried out by the teacher, while the second phase is carried out by two actors: teacher and learner. To summarize, as expected, the application of the approach allowed to reason on different aspects of the process at different times. It is moreover possible in the future to reuse previous modeling decisions to build a more specific-learning process model for a given university/teacher context.

7 Conclusion

In this paper, we proposed the use of SPLE approach to model e-learning process and the use of a novel notation and approach as well as a modeling tool BPFM. BPFM jointly supports the modeling of variability aspects for BPs in order to permit the inclusion of many different variants in a single model. The proposed approach was conceived for situations in which activities composing the configurable BP model have to be successively refined to consider characteristics of the deployment context, such as different arrangements of the organization supporting the BP itself. The presented work aims to overcome the shortcomings of LMSs, mainly by satisfying the variable requirements of universities/teachers, providing more flexible applications, and to benefit from the advantages of SPL engineering. Also, Using SPL approach and BPFM approach in such a broad field allows developers to reduce costs and effort of both production and maintenance, to significantly decrease time of development and to improve quality. As future work, we intend to improve our e-Learning product line and we are going to investigate the possibilities offered by cloud computing technologies for a better management of our process. This will lead as to consider it as a Business Process as a Service (BPaaS).

References

1. Stantchev, V., Colomo-Palacios, R., Soto-Acosta, P., Misra, S.: Learning management systems and cloud file hosting services: a study on students' acceptance. Comput. Hum. Behav. **31**, 612–619 (2014)

2. Conde, M.A., García, F., Rodríguez-Conde, M.J., Alier, M., García-Holgado, A.: Perceived openness of learning management systems by students and teachers in education and technology courses. Comput. Hum. Behav. **31**, 517–526 (2014)
3. Du, Z., Fu, X., Zhao, C., Liu, Q., Liu, T.: Interactive and collaborative e-learning platform with integrated social software and learning management system. In: Lu, W., Cai, G., Liu, W., Xing, W. (eds.) Proceedings of the 2012 International Conference on Information Technology and Software Engineering. Lecture Notes in Electrical Engineering, vol. 212, pp. 11–18. Springer, Heidelberg (2013). doi:10.1007/978-3-642-34531-9_2
4. Helic, D.: Technology-supported management of collaborative learning process. Int. J. Learn. Change **1**(3), 285–298 (2006). doi:10.1504/IJLC.2006.010971
5. Helic, D., Hrastnik, J., Maurer, H.: An analysis of application of business process management technology in e-learning systems. In: Proceedings of World Conference on E-Learning in Corporate, Government, Healthcare, and Higher Education, pp. 2937–2942 (2005)
6. van der Aalst, W.M.P., Hofstede, A.H.M., Weske, M.: Business process management: a survey. In: van der Aalst, W.M.P., Weske, M. (eds.) BPM 2003. LNCS, vol. 2678, pp. 1–12. Springer, Heidelberg (2003). doi:10.1007/3-540-44895-0_1
7. Gillot, J.N.: La gestion des processus métiers. Lulu.com (éditeur), vol. 372 (2007). ISBN 2952826609, 9782952826600
8. Issa, T., Isaias, A.M., Kommers, P.: Multicultural Awareness and Technology in Higher Education: Global Perspectives. Advances in Higher Education and Professional Development (AHEPD) Book Series. IGI Global, Hershey (2014)
9. Gavidia, A.R., Urbán, M.A.S., Barriocanal, E.G., Plazuelos, G.M.: Beyond contents and activities: specifing processes in learning technology. In: Current Developments in Technology-Assisted Education (2006)
10. Fang, C.F., Sing, L.C.: Collaborative learning using service-oriented architecture: a framework design. Knowl.-Based Syst. **22**(4), 271–274 (2009)
11. Da Costa, J.: BPMN 2.0 pour la modélisation et l'implémentation de dispositions pédagogiques orientées processus. Mémoire présenté pour l'obtention du master MLALT (2011)
12. Adesina, A.: Virtual Learning Process Environment (VLPE): A BPM Based Learning Process Management Architecture (2013)
13. Schneider, D.K.: Les approches scénarisation et la modélisation du workflow pédagogique (2011)
14. Terenciani, M.F., Landre, G.B., Paiva, D.M.B., Cagnin, M.I.: A plug-in for eclipse towards supporting business process lines documentation. In: 2015 IEEE/ACS 12th International Conference on Computer Systems and Applications (AICCSA), pp. 1–8 (2015)
15. Cognini, R., Corradini, F., Polini, A., Re, B.: Business process feature model: an approach to deal with variability of business processes. In: Karagiannis, D., Mayr, H., Mylopoulos, J. (eds.) Domain-Specific Conceptual Modeling, pp. 171–194. Springer, Cham (2016). doi:10.1007/978-3-319-39417-6_8
16. Vervuurt, M.: Modeling business process variability (2007)
17. Pohl, K., Böckle, G., van Der Linden, F.J.: Software Product Line Engineering: Foundations, Principles and Techniques. Springer Science & Business Media, Heidelberg (2005)
18. Reichert, M., Weber, B.: Enabling Flexibility in Process-Aware information Systems: Challenges, Methods, Technologies. Springer Science & Business Media, Heidelberg (2012)
19. Schnieders, A., Puhlmann, F.: Variability mechanisms in e-business process families. In: Abramowicz, W., Mayr, H.C. (eds.) BIS, vol. 85, pp. 583–601. GI (2006)
20. Gottschalk, F., Van Der Aalst, W.M., Jansen-Vullers, M.H., La Rosa, M.: Configurable workflow models. In. J. Coop. Inf. Syst. **17**(02), 177–221 (2008)

IT-Enabled Digital Service Design Principles - Lessons Learned from Digital Cities

Petr Štěpánek$^{(\boxtimes)}$, Mouzhi Ge, and Leonard Walletzký

Faculty of Informatics, Masaryk University,
Botanická 68a, 602 00 Brno, Czech Republic
stepanek.petr@mail.muni.cz, mouzhi.ge@muni.cz, qwalletz@fi.muni.cz

Abstract. With the rapid expansion of emerging digital technologies, digital service creation and delivery demand new and more structured ways to design, develop and manage the service sustainability. Although, there are various views and strategic aspects from different stakeholders for designing the digital services, the clear answer specifying how to design the digital service within IT architectures or how to re-use design processes learned from service design used in previous Digital City projects is still unknown. In order to derive the digital service design principles, we study the IT architectures that are related to digital services and revisit one typical Digital City - Barcelona. Based on the lessons learned from the Digital City project, we propose a set of design principles that can guide researchers and practitioners to design the digital services in a Digital City.

Keywords: Digital services · Digital city · IT-enabled services · Lessons learned · Service design principles

1 Introduction

Nowadays, urbanization has created extensive challenges of broadly meeting the demand of modern societies to ensure the quality of life, sustainability, and economic growth. Proposals to address these challenges with technology are usually associated with the term Smart City. In addition to technological advances, many emerging service industries create greater competitive advantages to be cost-effective and innovative [6]. Among others, transportation, and environment, the major elements of an urban planning include the implementation of information systems and providing a smart Information Technology (IT) environment. While Smart City is more focused on activities, smart citizens and their sustainability, the concept of Digital City is much more tightly connected to the IT and the goal of a Digital City is to use IT to provide a complete virtualization of the city and to make the city digital services ubiquitous [11]. In this paper, we consider the scope of the digital service within the IT-enabled smart services in the city.

© Springer International Publishing AG 2017
M. Themistocleous and V. Morabito (Eds.): EMCIS 2017, LNBIP 299, pp. 186–196, 2017.
DOI: 10.1007/978-3-319-65930-5_15

Digital services have been considered as one of the important components in a Digital City since the digital service is capable of connecting service providers, users, infrastructures, and communities in a common ecosystem to support the value co-creation [20]. Adapted from [22], Digital City can be explained in terms of a complex of services exchanged by a network of actors interconnected in order to share knowledge, resources, competences, and capabilities to perform better solution. The domain of digital services represents an emerging field to investigate how an understanding of networked architectures can offer a better satisfaction of stakeholders' needs.

The importance of IT has demonstrated that the capacity of Digital Cities programs to create value for stakeholders is directly related to a new information technology, processes, services and associated business and information architectures. To organize these components, many researchers and practitioners have described Enterprise Architecture (EA) concepts or frameworks such as TOGAF, focusing on increasing IT efficiency while continuing business innovation [14]. Organizations can use enterprise architecture frameworks to manage system complexity and align business and IT resources. Employed within a traditional enterprise context, EA assists in providing an integrated environment supporting the alignment of business and IT [21,23]. However, applications of EA concepts to digital government and public services are relatively rare. Especially, when we consider and assimilate diverse interests and objectives from different stakeholders, then the typical design processes like the TOGAF Architecture Development Method may require changes and adaptations to design digital services. Moreover, beyond IT architectures, there is still a lack of principles of how to design the digital services.

In this paper, we propose a set of design principles to create, develop and manage digital services. In order to derive the principles, we investigate and review the IT architectures in Smart and Digital City, and revisit the existing Digital City projects in Barcelona. This paper further develops the concept of digital service design, which can be used to integrate different service views and strategic aspects in Digital Cities.

The remainder of the paper is organized as follows, Sect. 2 reviews the IT architectures that integrate digital services for Digital City. Following the IT architectures, Sect. 3 revisits the Digital City projects and their digital services. Based on the review of IT architectures and previous digital services, Sect. 4 derives a set of principles that can guide users when designing and managing the digital services. Finally, Sect. 5 concludes the paper and outlines the future research in digital service design.

2   IT Architectures for Digital Service

The architectures of the digital services design can be generally viewed from following two perspectives. One perspective considers cognitive and creative capability and human skills in the Digital City, for example, knowledge networks, creative environment, social connections etc. [17]. The other perspective concerns the physical communication methods, digital channels and IT technologies for the digital service delivery [10]. We take the IT perspective and focus on the IT-enabled digital service design. We have thus reviewed different IT architectures concerning digital service development.

2.1   e-Trikala

The Trikala logical model is an n-tier e-Government model that was developed for governmental services in the Digital City of Trikala in Greece [16]. This model consists of 4 layers (Users layer, Service layer, Infrastructure layer and Information layer) and was further developed into a 5 layer model by adding a Business layer, renaming a Users layer to Stakeholders layer. The Stakeholders layer is used twice in a model, each time from different perspectives. Once as stakeholders providing the data and then as stakeholders using the end services. The evolved model then consists of Stakeholders layer (servants), Infrastructure layer, Information layer, Business layer, Service layer, and Stakeholders layer (end users).

The first layer is composed of companies, agencies and other communities connected to the MAN network. They provide the data as well as they are the end users of the city services (the last layer). The Infrastructure layer then uses MAN, metro Wi-Fi, information systems, phone center and public access points to transmit data into an Information level, where these public and private data are stored. The Service layer finally provides services as web portals, engines, web services or geospatial services that serve information from the Information layer for the end users to access it [5]. Business layer slightly steps out of the model. It doesn't work with data itself, it contains policies and operating rules, together with Enterprise Architecture of the digital city. In other words, Business layer defines how all the digital city systems will be designed, installed and interconnected.

2.2   CARS

Abdelwahab et al. [4] proposed a Cloud-Assisted Remote Sensing(CARS) model together with an Internet of Everything to create Everything as a Service [12] and Sensing as a Service [19]. This model consists of 4 layers, which are: Fog layer, Stratus layer, Alto-cumulus layer and Cirrus layer. Here follows a description of all layers with their specific functions.

In the Fog layer, there are all physical objects, machines and any other devices that can be connected to the internet. The layer has three main functions: (1) to provide heterogeneous networking and communication infrastructure to connect billions of objects; (2) to provide a unique identification of all objects through IPv6; (3) to provide a data aggregation points to serve as sensing clusters. The Stratus layer consists of thousands of clouds, whose main resources are sensors and SANs (Sensing and Actuating Networks). The Stratus layer provides an abstraction of physical objects from a Fog layer to Alto-cumulus layer. Specifically, main purposes of the layer are to: (1) abstract and virtualize physical SANs through virtual network embedding techniques; (2) handle and manage virtual SAN migration and portability across different clouds; (3) manage and ensure operations and functionalities of virtual SAN instances; (4) enable and manage (physical or virtual) SAN configurations to ensure network connectivity and coverage; (5) control the layers operations and functionalities to ensure that customers SLAs communicated from higher layers are met. The alto-cumulus layer serves as a mediator between the Cirrus layer (requests) and the Stratus layer (possibilities) in order to meet restrictions of both layers. There are four main functions of the Alto-cumulus layer: (1) serving as a point of liaison between Cirrus and Stratus layers by translating a policy and regulation requirements expressed by the Cirrus layer into domain-specific requirements understood by Stratus; (2) enabling business and payment transactions between Cirrus and Stratus layers by providing two-way brokerage services; (3) enabling and facilitating SLA negotiations between Cirrus and Stratus, and monitoring and ensuring that these SLAs are met; (4) coordinating and facilitating inter-cloud interactions, data exchange, task migration, and resource sharing across different Stratus clouds. The last layer is the Cirrus layer, which is an interface of the CARS architecture to the users. Its functions are to: (1) act as the customers' entry point to CARS systems by allowing them to specify their required services via SLAs and to select their desired service models; (2) allow CARS customers to set up their sensing task requirements and do whatever their chosen service model allows them to do (e.g. software deployment); (3) negotiate SLAs with customers and communicating them to Alto-cumulus layer; (4) provide on-line applications for remote data analysis to be used by customers to visualize the data in real time.

To sum CARS architecture up, it transfers data from sensors and actuators in Fog layer into cloud systems in the Stratus layer, that are coordinated and controlled by the Alto-cumulus layer to provide services to the end users in a Cirrus layer. The CARS architecture was developed to enable the use of different devices, systems, and standards in order to minimize a negative influence of distinctions.

2.3 SSC

A Sustainable Smart City (SSC) is an extension of a Smart City term emphasizing a sustainable objective of a Smart City. International Telecommunication Union proposed [15] an IT architecture for sustainable Smart City. This model has four layers: Sensing layer, Network layer, Data and Support layer, and Application layer.

The Sensing layer is composed of two parts. The first is the Terminal nodes (sensors, transducers, actuators, cameras, RFID readers, barcode symbols, GPS trackers, etc.) that monitor and control physical infrastructure of the city. The second one is a Capillary network (SCADA, sensor network, HART, WPAN, video surveillance, RFID, GPS related network, etc.) connecting terminals to a Network layer, enabling a data transfer. The Network layer purpose is to deliver data from the lower level to the upper layer and vice versa. In the Data and Support layer, data is stored in city data centers or the clouds waiting to be processed. The Application level then offers various services for a city management. Across all the four layers, there is a vertical layer to provide the operation, administration, maintenance and provisioning, and security function.

Calderoni [9] has further validated this IT architecture for a Sustainable Smart City.

2.4 SWIFT

Smart WSN-based Infrastructural Framework for smart Transaction (SWIFT) was developed due to the heterogeneity of devices connected to the city infrastructure [18]. This model provides a platform for a synergistic interaction of various devices and objects with an ability of simple scaling. It has three layers: S-WSN (Smart Wireless Sensor Network), SWIPE (Smart Wireless-based Pervasive Edifice) and SDCE (Smart Decision and Control Enabler).

The S-WSN layer is to sense surroundings data and aggregate them prior sending to the higher layer. S-WSN layer contains two types of devices. Sensing devices are sensor nodes (SN), that are usually cheap, low-end devices with limited computational capacity and energy power. These nodes observe its close surroundings in regular intervals and send observed data to smart cluster heads (SCH). A reason of existence SCH is that WSN are usually spatially large and its capability to communicate is limited to small distances. Therefore, SCHs are distributed in the whole WSN so as to split the network to clusters, and aggregate sensed data from these clusters to respective SCHs. SCHs are more powerful devices that are placed near a power source. They process incoming data and enhance it by adding contextual data prior sending to the SWIPE layer. The SWIPE layer is the core of the SWIFT architecture. Here all the devices interact to process, share or feed information. It contains several smart fusion nodes (SFN) that are conveniently located in the city to be powered and to collect information from all neighboring SCH as well as other smart devices. SFNs process received data according to contextual information and take appropriate decisions based on the fused data evaluation. Finally, processed data is stored in the SDCE cloud. The top SDCE cloud layer represents an interface for the citizens, stores and processes the data and provides a solution for context-based services requested by citizens. This layer offers SenaaS to its users based on EaaS concept.

The four IT architectures above are focused on different aspects of the digital services. e-Trikala mainly targets on how to design the service to satisfy the needs from stakeholders. SSC positions the digital services in the top application layer and it is more focused on digital service delivery. CARS focuses on the IT infrastructure such as system model and architecture, which serves as the base for developing digital services. SWIFT is developed to coordinate heterogeneous digital devices, which can be transited, inter-operated, integrated or optimized. Along with the reviewed four architectures and their focused aspects in digital service, we have summarized the results in Table 1.

Table 1. IT architectures for digital service

	Service design	Delivery standards and principles	System models and architecture	Transition and operation
IT architectures	e-Trikala [16]	SSC [15]	CARS [4]	SWIFT [18]
Aim	Standards	Governance planning	System modeling	Maintain integrate optimize
Role/area	Strategy	Governance	Architect	Operation
Outcomes	Service portfolio	Service governance	System architecture and artifacts	Procedures
Value	Standardization	Decision guidelines risk reduction	Deployment efficiency	Reduced costs resources

3 Revisit Digital City Project

There has been a variety of digital city projects in the real world. A European Digital Cities assessment model has been developed to profile and benchmark European middle-sized and large cities [13]. The recent city profiling can be found [3], which contains 77 middle-sized and 90 large cities in the assessment of 2015. Another assessment was made by the Intelligent Community Forum [2]. The evaluation is based on the annual surveys answered by more than 100 cities and regions on five continents, in which 21 Smart Communities are chosen to provide more detailed data and then 7 the most successful communities are studied more by independent research companies and by detailed personal inspections to evaluate the Intelligent community of each year.

In this paper, we have selected one of the most discussed digital cities in recent years, which is Barcelona. We will revisit their provided digital services and summarize the lessons learned from the digital city project.

3.1 Barcelona: The Digital City

Barcelona, the capital city of Spain was able to overcome an economic crisis in the 1980s and developed into one of the fastest evolving cities in Europe [8]. It started with building a city infrastructure followed by managing of housing, energies, transportation and environmental issues and more complex tasks.

There are several reasons for the transformation Barcelona city into a European smart city leader. We will point out here 3 most important ones. The first one is a use of IT as a support of city processes for them to become more efficient, effective, transparent and accessible. The second reason is an interconnection of a city council, companies, educational institutions and city inhabitants, which is called a knowledge society. This interconnection resolves in mutual understanding of different needs of respective groups, which is essential for enabling a control over changing a state of support and cooperation, or competition, depending on the goal. The third reason is processing, storing and mainly sharing of information (Barcelona provides Open Data from a public sector). These are the main 3 pillars modern Barcelona is built on.

IT technologies enable automation and optimization of a huge number of different tasks. For instance, monitoring public as well as private transportation, sensing free parking places, optimization of energy consumption, waste management, sensing an air pollution, energy efficient lighting and so on. Some of these services are operated across the city, other are implemented only in smaller parts.

There exist the 22@ projects that contain several city districts in Barcelona. The main district is called 22@Barcelona and its purpose is to develop a model of knowledge, sustainable, innovative city ensuring a high quality of life. This place serves as a meeting point for city council, companies, universities and city inhabitants trying to work together on the common goal on one hand and compete among each other to support innovations. Other 22@ projects, called Living Labs, serve as test beds for companies to test pre-commerce services. It means, in these city parts, companies use their technologies and services in a real scenario, but just on a small scale to test it. Inhabitants living in 22@ districts have access to the new technologies and services that can use.

Despite the 22@ projects, all the city is covered by citywide Wi-Fi allowing citizens a ubiquitous connection to the network. Citizens have therefore access to all the city information and Open Data everywhere in the city. Together with the connection to the network, city council operates websites fostering a transparency and making easier for the citizens as well as companies to access required data.

To ensure the functioning of such a huge system as is in Barcelona city, there needs to be a reasonable number of sensors generating the data that is transferred using a city infrastructure to respective places, where are processed and stored. After this, processed data can be evaluated and combined in different ways to offer an intended service. For managing such a network of networks, Sentilo platform [1] was developed specifically for the city of Barcelona.

Until now, we were describing strictly a top-down approach of development and management of services in the city, which was the first step of the city development. During the implementation of different technologies and approaches, some bottom-up phenomena started to appear [7]. The 22@ districts unintentionally influenced inhabitants and separated them from other districts. After some time, a reaction (it is called a negative feedback in complex systems) appeared from communities in the city and acted against the separation. The same effect was observed when speaking about infrastructures. Except for the city infrastructure that was created using a top-down approach, people started to use a shared, decentralized network as a cheap way to connect to Wi-Fi. The last two bottom-up phenomena were shared co-working places either to support some community and share the inspiration or to share technology and tools and generating and sharing of Open Data by the citizens that require them.

4   Digital Service Design Principles

Based on the reviews from the IT architectures and lessons learned from the selected digital cities, we have derived following principles that can be used in a digital service design.

1. **To develop digital services, new IT technologies should be used as a key supporting pillar in analyzing, understanding, and managing all the variables that are involved in the processes of satisfying stakeholders' needs**
   IT technologies cannot be simply considered a tool to support the connection among different digital service entities. It is important to use (new) IT technologies to analyze the smart services using the conceptual framework for example related to the Complex Adaptive System, in which the entities in the smart services are interconnected (e.g. one entity actions change the context for other entities).

2. **A robust infrastructure needs to be developed to support digital services in the city**
   To transfer the data from sensors to data stores, analytical tools, specific services and then back to stakeholders, to support interconnection of services and create a ubiquitous internet connection, a Digital City needs to create robust, powerful and scalable infrastructure throughout the city.

3. **The complex of used IT technologies and the data should be managed by a specialized software**
   There is a reasonable number of IT devices, tools and data used within the city that need to be managed and controlled. For this purpose, cities should use some platform which can enable it (e.g. Sentilo [1] in Barcelona).

4. **Interconnections between city council, companies, educational institutions and city inhabitants need to be identified and modeled to design the digital service chain**
   Digital city connects service providers, users, infrastructures, and communities in a common ecosystem to support the value co-creation. As a key component in a Digital City, a digital service should be designed as a complex of services exchanged by a network of actors interconnected in order to share knowledge, resources, competences, and capabilities to perform a better solution.

5. **Information sharing such as open data should be used to integrate the digital services**
   Service separation has been one of the main disadvantages in the city. For example, in different services of the city council, photo taking is usually one of the important steps for issuing the documents. Because of the service and

resource separation. Citizens may have to take the same photo for different documents. Thus, information sharing can facilitate and re-use the resources among different digital services. Furthermore, open data offer more opportunities for IT companies to design and develop digital services. IT companies are able to access more open data such as traffic, health, weather etc. to develop ubiquitous applications.

6. **Top-down and bottom-up approaches need to be combined when developing digital services**
   The city and IT companies usually have a close collaboration. The city can be a service provider to provide policy or infrastructure support to IT companies, at the same time, the city can be also a service user e.g. using the IT system service from IT companies. Same as the city, IT companies can also possess two roles, a service provider, and a service user. Therefore, when the city is the service provider, the digital service design can use the top-down approach to facilitate the service deployment. Also, when the city becomes a service user, IT companies can fulfill the city needs from the bottom-up. The combination of the top-down and the bottom-up approaches are needed when designing and developing digital services.

7. **Digital services need real-world test beds to verify and improve them in the pre-commerce environment before they are deployed in a large-scale**
   Based on the lessons learned from 22@ districts in Barcelona, it can be seen that it is important to run a pilot study or living lab to test the digital services. When creating a new digital service in the city, several aspects such as citizen's adaptation, service efficacy, sustainability and especially the effects of the digital services on other existing facilities or services need to be understood. Therefore, before the large-scale deployment, the digital services should be tested within a controlled environment. In this controlled environment such a living lab, the digital service can be continuously improved and further designed.

These derived principles above can be used to guide the researchers and practitioners to design services in the Digital City. Most of the principles are particularly focused on the IT perspective, which aim to leverage the IT technologies in digital cities and facilitate the IT-enabled digital service design.

5   Conclusion

In this paper, we have reviewed a variety of IT architectures that can be used to develop digital services in Digital Cities, which are e-Trikala, CARS, SSC and SWIFT. The four IT architectures are focused on different aspects of the digital service design with different outcomes and roles. We have summarized and positioned them for designing the digital services to create an overview of their divergence. The overview can help a city council to decide which architecture a city wants to use. Furthermore, we have revisited one typical Digital City project in Barcelona, in which the digital services and service deployments are described in detail.

Based on the review of the related IT architectures for digital services and one Digital City case study, we have proposed a set of design principles to create, develop and manage digital services. These principles can be used in other digital cities to improve or facilitate the digital service design procedure. We suppose that taking into account proposed principles, it can help the city to prevent major situations during implementation of digital services. As future work, we plan to conduct more case studies for other smart cities and develop an architecture for the digital service design. We will also further validate the proposed principles in real-world scenarios such as conducting field study or action research in the certain Digital City.

References

1. Sentilo. http://www.sentilo.io/wordpress/. Accessed 03 June 2017
2. Intelligent Community Forum (2017). http://www.intelligentcommunity.org/. Accessed 03 June 2017
3. Smart City Profiling, Benchmarking and Ranking in Europe (2017). http://www.smart-cities.eu/. Accessed 03 June 2017
4. Abdelwahab, S., Hamdaoui, B., Guizani, M., Rayes, A.: Enabling smart cloud services through remote sensing. IEEE Internet Things J. **1**(3), 276–288 (2014)
5. Anthopoulos, L., Fitsilis, P.: From online to ubiquitous cities: the technical transformation of virtual communities. In: Sideridis, A.B., Patrikakis, C.Z. (eds.) e-Democracy 2009. LNICSSITE, vol. 26, pp. 360–372. Springer, Heidelberg (2010). doi:10.1007/978-3-642-11631-5_33
6. Anttiroiko, A.-V., Valkama, P., Bailey, S.: Smart cities in the new service economy: building platforms for smart services. AI Soc. 1–12 (2014)
7. Attour, A., Thierry, B.-H., Capdevila, I., Zarlenga, M.I.: Smart city or smart citizens? The Barcelona case. J. Strategy Manage. **8**(3), 266–282 (2015)
8. Bakıcı, T., Almirall, E., Wareham, J.: A smart city initiative: the case of Barcelona. J. Knowl. Econ. **4**(2), 135–148 (2013)
9. Calderoni, L.: Distributed smart city services for urban ecosystems. Ph.D. thesis, Alma, Giugno (2015)
10. Chourabi, H., Nam, T., Walker, S., Gil-Garcia, J.R., Mellouli, S., Nahon, K., Pardo, T.A., Scholl, H.J.: Understanding smart cities: an integrative framework. In: 2012 45th Hawaii International Conference on System Sciences, pp. 2289–2297, January 2012
11. Cocchia, A.: Smart and digital city: a systematic literature review. In: Dameri, R.P., Rosenthal-Sabroux, C. (eds.) Smart City. PI, pp. 13–43. Springer, Cham (2014). doi:10.1007/978-3-319-06160-3_2
12. Duan, Y., Fu, G., Zhou, N., Sun, X., Narendra, N. C., Hu, B.: Everything as a service (XaaS) on the cloud: origins, current and future trends. In: 2015 IEEE 8th International Conference on Cloud Computing, pp. 621–628, June 2015
13. Giffinger, R., Fertner, C., Kramar, H., Kalasek, R., Pichler-Milanović, N., Meijers, E.: Smart cities: ranking of European medium-sized cities. Vienna University of Technology, Vienna (2007)

14. Helfert, M., Ge, M.: Developing an enterprise architecture framework and services for smart cities. In: Proceedings of the 25th Interdisciplinary Information Management Talks, Trauner Verlag (2017)
15. I.T.U. (ITU): Setting the framework for an ICT architecture of a smart sustainable city (2015)
16. Tsoukalas, I.A., Anthopoulos, L.G.: The implementation model of a digital city. The case study of the digital city of Trikala, Greece. J. E-Gov. **2**(2), 91–109 (2006)
17. Nam, T., Pardo, T.A.: Conceptualizing smart city with dimensions of technology, people, and institutions. In: Proceedings of the 12th Annual International Digital Government Research Conference: Digital Government Innovation in Challenging Times, dg.o 2011, pp. 282–291. ACM, New York (2011)
18. Nandury, S.V., Begum, B.A.: Smart WSN-based ubiquitous architecture for smart cities. In: 2015 International Conference on Advances in Computing, Communications and Informatics (ICACCI), pp. 2366–2373. IEEE (2015)
19. Perera, C., Zaslavsky, A., Christen, P., Georgakopoulos, D.: Sensing as a service model for smart cities supported by internet of things. Trans. Emerg. Telecommun. Technol. **25**(1), 81–93 (2014)
20. Lusch, R.F., Vargo, S.L., Tanniru, M.: Service, value networks and learning. J. Acad. Mark. Sci. **38**(1), 19–31 (2010)
21. Clark, T., Barn, B.S., Oussena, S.: A method for enterprise architecture alignment. In: Proper, E., Gaaloul, K., Harmsen, F., Wrycza, S. (eds.) PRET 2012. LNBIP, vol. 120, pp. 48–76. Springer, Heidelberg (2012). doi:10.1007/978-3-642-31134-5_3
22. Vargo, S.: Toward a transcending conceptualization of relationship: a service-dominant logic perspective. J. Bus. Ind. Mark. **34**(5/6), 373–379 (2009)
23. Šaša, A., Krisper, M.: Enterprise architecture patterns for business process support analysis. J. Syst. Softw. **84**(9), 1480–1506 (2011)

Enterprise Social Network Success: Evidences from a Multinational Corporation

Bruno Faria[1](✉) (iD) and Rui Dinis Sousa[2] (iD)

[1] University of Minho, 4804-533 Guimarães, Portugal
a69174@alunos.uminho.pt
[2] Information Systems Department, ALGORITMI Research Centre,
University of Minho, 4804-533 Guimarães, Portugal
rds@dsi.uminho.pt

Abstract. In a globalized world, where companies operate across different locations and work becomes increasingly complex, collaboration in a diversity of ways is required among employees to perform tasks more effectively. Following a case study methodology that involved six interviews across three different country locations, this research addresses the phenomenon of Enterprise Social Networks (ESN) in a multinational corporation with a focus on the assessment of ESN success.

The findings show that the company, while trying to assess the success of Yammer, the freemium social networking service at use, has mainly relied on analytics tools to measure usage through indicators such as the total number of users. However, the extent to which ESN is used does not provide a complete picture of ESN success. Business value from that ESN usage is another dimension to be considered to assess success. Therefore, the study of specific ESN usage scenarios that are perceived to have a trackable impact on business results can be used to assess ESN business value on top of ESN usage to fully understand ESN success.

Keywords: Enterprise Social Networks · Benefits · Success criteria · Success assessment

1 Introduction

As the name suggests, Enterprise Social Networks are social networks that were tailored to meet specific requirements of the organizational context, to engage and connect employees, to boost collaboration, communication, information exchange, and to create a community feeling among users [1, 2]. Different authors use different denominations when they address ESN topics, e.g., Social Networking Technologies [3] or Enterprise Social Networking Systems [4]. An evidence of a young research field in development [5]. Usually, these platforms support several social media functionalities such as status updates, microblogging, groups and communities, instant messaging, or content management. ESN also provide personal profiles, the possibility to like and comment content and to follow or unfollow different users [6]. By increasing interaction between employees and encouraging collaboration and communication,

M. Themistocleous and V. Morabito (Eds.): EMCIS 2017, LNBIP 299, pp. 197–203, 2017.
DOI: 10.1007/978-3-319-65930-5_16

knowledge management infrastructures are being incorporated with social features to capture tacit, social and individual knowledge [6, 8]. These technologies became very important in large and distributed companies to support knowledge sharing among individuals, teams and units spread by different geographical locations and time zones [8]. Employees choose to meet new people instead of just reaching out to people they already know, sharing work and non-work-related content, and using the platform to spread messages to larger audiences [9]. However, there is a small understanding of how ESN can be used in organizational simple work practices [2]. Organizations have difficulties to assess the potentialities of Enterprise Social Software (ESS) in general, in which ESN are included. Nevertheless, the lack of knowledge about these technologies makes organizations to implement them without clearly defining the strategy and the expected business outcomes [10]. Therefore, this work, following a case study methodology, intends to provide some insights on ESN, pursuing an answer to the following research question "How to assess the success of Enterprise Social Networks?", in the context of a multinational corporation.

2   Enterprise Social Networks Success

Delone and McLean IS success model [11] is one of the first research initiatives trying to provide a holistic overview on the topic of Information Systems success measurement. An extensive literature review has highlighted a set of metrics used over the years to assess IS success in categories as system quality, information quality, use, user satisfaction, individual impact, and organizational impact. However, Richter et al. [12] argue that existing theories and models fail by being too theoretical as they don't provide examples on how success can be measured in a practical way.

Large multinational organizations are increasingly dependent on successful knowledge sharing among individuals, teams, and units because of their high degree of geographical dispersion throughout locations and time zones, what has led to the adoption of ESN [8]. Even though there are several models explaining the adoption of technology, "acceptance is not equivalent to success, but rather a necessary precondition to success" [12]. Success seems to be a much broader concept then just having users using the tool.

As more ESN start to be used in organizations, it becomes important to understand how the success of these platforms can be measured. If on one side decision makers have to justify their Information Technology (IT) investments and the owner of the system wants to improve the usage of the platform, on the other side, the added value of the ESN use must also be understood by users [12]. Herzog et al. [13] support the same idea as they refer many IT executives are pressured to clearly show the benefits of using such technologies recurring to understandable methods and indicators. However, the impact of using those technologies is not easy to prove.

Steinhueser et al. [14] propose a set of indicators and barriers to measure ESS expenditures, assets, use and organizational performance impacts. A McKinsey study has presented a matrix highlighting the increase in the value added for different types of Enterprise 2.0 technologies for different levels of penetration of the technologies in the organization. The results come from a survey conducted between 2007 and 2015 in

1500 companies. However, the study doesn't provide any guidelines about how these returns can be assessed for specific technologies in specific organizations [15].

A set of measures for the ESS success and group them in two categories: usage and business value [13]. To tackle the inexistence of an "integrated and easy to apply approach" to measure ESN success, a very similar approach that identifies measures for seven specific ESN usage scenarios [12]. The resulting Success Measurement Framework is also structured in two dimensions: usage and business value. The authors distinguish between the measurement of the usage of the platform, that usually can be assessed using analytics platforms, and the measurement of the organisational impact caused by ESN usage. Business value can be measured in the form of business cases analysis or return on investment.

3 Research Methodology

3.1 Organizational and Technological Context

This study was conducted in a company with around 24 000 employees across more than 120 countries. The company has introduced Yammer (http://yammer.com/), an Enterprise Social Network provided by Microsoft, in 2014. After a series of initiatives to introduce the platform and engage users, the network keeps growing and has already 22 000 users (May 2017). The company has also access to Tryane Yammer Analytics (https://tryane.com/en/yammer_analytics.html), a web-based analytics tool that provides insights into Yammer activity at global and group level.

3.2 Interviews

As the final goal of the study was to understand how to assess success, the interviews were developed around this topic in a semi-structured way. Four interviews were conducted in person and two interviews were conducted using Skype for Business. All the interviews were recorded and, then, transcribed using oTranscribe (http://otranscribe.com/), a freemium web application suited for the task, and, then, analyzed using QDA Miner Lite (http://provalisresearch.com/products/qualitative-data-analysis-software/freeware/), a freemium qualitative text analysis software. Some emails were also exchanged to clarify some aspects from the interviews.

3.3 Interviewees

Six semi-structured interviews were conducted to five IT employees and to one product manager from Mexico. These five IT employees are the ones that have been responsible for tracking ESN success in the company along the years. INT-1 and INT-2, were the project leads of the initiatives undertaken to implement and launch Yammer. One interviewee, INT-3, was part of both teams during that process. INT-4 is the current owner of the tool and INT-5 is the head of the IT sub-department responsible for the collaboration technologies. INT-6, Head of Product Management in Mexico, is the only non-IT interviewee. He developed a competition on Yammer that is the current best example of how Yammer has impacted business results in this corporation.

4 Assessing ESN Success

4.1 Analytics Tools

Analytics tools are a good resource to evaluate the health of a network. According to INT-1, the level of engagement - percentage of users accessing the platform -, the number of new users and the number of active groups, on daily and weekly basis, were some of the indicators used to measure the adoption of the platform, and, therefore, the health of the network. *"If you have fifty or over engagement, you are doing really well with your users. We looked anywhere that 30-50% range of engagement as being very good. So, we definitely did analytics and looked at them weekly and provided the status report for our management."* (INT-1). For INT-2, on a high level, the main success criteria was to enable the exchange of knowledge and, then, to achieve a sustainable usage: *"we achieved it, we are not great, but we are in a good track, this is what we achieved, so more than 40% of our usage rate"*. INT-5 also believes the main target was to *"Get users on and get them engaged"*, because the biggest success criteria is to keep the engagement and the excitement with the platform high. *"Social networking is only working with participation (...) from a huge community"* (INT-5).

Based on the assumption users *"wouldn't be active if it[Yammer] was not something that helps them"* (INT-3), the increasing of engagement means new users found the network beneficial for them. However, it is a challenge to retrieve meaning out of the analytics and clearly understand what are the specific reasons triggering increasing usage. *"We can increase now 1% [of engagement], but what does that 1% mean? And then, maybe, we don't know why that increase is happening. [If] We cannot measure that, how can we replicate it?"* (INT-4).

Because social networks live from the excitement and the contributions of their user bases, understanding the reasons behind usage is important because they could be advertised and make other users to join and to start contributing for the network. *"1% increase is because, suddenly, Japan started using it a lot. Maybe, in the global scale, it's not that much. But if we replicate that in other regions, then, that 1% means 10%, and that would be significant."* (INT-4).

Analytics tools like Tryane Yammer Analytics provide a set of indicators as the number of comments, likes, most active discussions, more active groups, or more discussed topics. That information is helpful by providing insights into the level of collaboration, communication and information exchange inside the tool.

4.2 Limitations of Analytics Tools

There is a set of inherent benefits to the usage of social networks that can't be clearly measured recurring to analytics tools. By using Yammer, *"people feel integrated, but what exactly does that mean? A remote user, in a remote location, that does not talk too much with his team? Or does that apply even on the region level or in a global team?"(INT-4)*. It seems important to understand these aspects in a deeper level. *"I think that if you don't measure something, then there is no way to improve. And that's the challenge usually we have. That's the challenge also with those benefits.".*

On the other side, if we want to relate Yammer to productivity gains, the task can get even harder. INT-2 expresses the difficulty in measuring the productivity gains because there are different aspects that can influence productivity. INT-3 follows the same idea: *"the effect of collaboration cannot be directly putted to euros science".* INT-4 adds: *"The benefits are there, but if we want to put a number, a money value, to that, that might be a little bit more tricky."* To execute this kind of analysis, the solution seems to resort on high level studies that have focused on the topic. McKinsey's study "Taking the measure of the networked enterprise" [15] relates the percentage of penetration of different Enterprise 2.0 technologies to companies results. *"I came back to the situation of 20/30 million yearly that should bring for [the company]. (...) We had, in that year, also a huge productivity gain, when we started to do it properly. But I would not be so bold to say all of that is related to that [Yammer], but some portion of it, definitely, is based on Yammer. It would be wrong to say there is a real correlation".*

4.3   Resorting to Specific Usage Scenarios

One case from the Mexican Sales Organization is a good example on how benefits from using Yammer can be assessed. Globally, one of the organizational targets for the sales agents is the performance of product demonstrations, however, in the Mexican organization, there wasn't a strong culture of product demos. On average, a sales agent performed 0,3 demonstrations per day. Even though, in a survey answered by sales agents and sales managers, 40.5% said that more than half of the demonstrations resulted in one products sale.

Because product demonstrations seem to be related to sales, - *"If you don't do demos, you lose money."* (INT-6) -, there was the idea to create a competition among sales agents to promote product demonstrations in an exciting way. The competition was developed from February 2016 till December 2016 and Yammer was the driving platform used. In a brief way, each sales agent should upload demo photos or videos in the Yammer group and, then, he or she would be awarded points depending on the product being demonstrated. Strategic products were worth more points. The number of likes received in a specific demo would also be converted in points. Sales agents had the possibility to challenge others to perform demos. If the challenge wasn't accepted, points would be deducted from the challenged sales agent's overall score.

In the end of the competition, prizes were awarded to the agents in the top of the ranking. The results showed an increase on the rate "number of demos per sales agent, per day" from 0,3 to 1,40. The top 10 sales agents also achieved a sales result 3% above the plan. Furthermore, the competition was important for the sales guys that worked in remote areas to feel integrated in the team and to make a change on sales agents' mind: making product demonstration wasn't an imposition, but something natural to do.

The Yammer group used during the competition was week after week among the five most active groups in the network, what can be correlated with the following statement: *"I do believe by reviewing the Yammer analytics for groups and individuals that you should be able to directly correlate increased collaboration with relationships to new innovations, improved processes, decreased implementation times, successful product launches, increased sales and even happier employees"* (INT-1).

Initiatives like this one provide a good example on how the success of ESN can be assessed in a more understandable way. In this case, the business was positively impacted due to a higher number of product demonstrations and good sales results.

Even though it's not possible to directly link the platform with the achieved business results, Yammer was the platform used to engage users in the competition.

5   Discussion and Conclusion

The company mainly relies on the information provided by Tryane Yammer Analytics to assess the success of the network. Even though there is an overall concern to assess the business impact from using such technology, the reality is that the main success criteria is based on the total number of users and the percentage of active users. These indicators are specifically identified as measures of the usage of the platform.

However, there is an understanding that analytics platforms are limited when there is the need to assess a new set of benefits that can't be measured using analytics tools. Log analysis, among other analytic methods, can be used to measure the usage of a platform [12]. However, when business value needs to be assessed, authors seem to resort among other, in a set of qualitative methods as users interviews [12]. This evidence expresses the need to select the approach depending on the intended assessment. These ideas were referred by INT-3 when "asking the users" is presented as a valid and non-quantitative alternative to assess the success of the network. Indeed, a quantitative assessment of ESN business impact seems difficult, mainly when organizational global results are desired. Measuring reduced times to find information or reduced amounts of emails seem to be more attainable indicators [12].

The existence of successful Yammer initiatives that have a trackable (direct or indirect) impact in business results provide an evidence of ESN success. The demonstrations competition in Mexico illustrates how Yammer social features can be useful to accomplish organizational targets.

In summary, the evidence from data collection in this multinational corporation suggests that the assessment of ESN success has relied mainly on information from usage. A clear concern to assess the impact of the ESN on business value appears later. Innovations are only worth it if they are used in work processes, thus, their benefits have to be clearly explained to end-users so they can understand how to take advantage of such tools for their own work [16]. Therefore, companies should start assessing the ESN success from the very first beginning to understand its impact on work routines and business results, addressing both usage and business value. A clear understanding of these aspects would not only be a clear proof of concept for ESN adoption, as it would also be important for users to understand why ESN is used and ultimately how the organization can benefit from it.

Acknowledgements. This work has been supported by COMPETE: POCI-01-0145-FEDER-007043 and FCT – Fundação para a Ciência e Tecnologia within the Project Scope: UID/CEC/00319/2013.

References

1. Bell, G.: Enterprise 2.0: bringing social media inside your organization: an interview with Monika Wencek, Senior Customer Success Manager at Yammer. Hum. Resour. Manag. Int. Dig. **20**(6), 47–49 (2012)
2. Riemer, K., Tavakoli, A.: The role of groups as local context in large enterprise social networks: a case study of Yammer at Deloitte Australia. In: Business Information Systems Working Paper Series at the University of Sydney (2013)
3. Ortbach, K., Recker, J.: Do good things and talk about them: a theory of academics usage of enterprise social networks for impression management tactics. In: Proceedings of the 35th International Conference on Information Systems, pp. 1–13 (2014)
4. Qi, C., Chau, P.Y.K.: An empirical study of the effect of enterprise social media usage on organizational learning. In: Proceedings of the Pacific Asia Conference on Information Systems (2016)
5. Wehner, B., Ritter, C., Leist, S.: Enterprise social networks: a literature review and research agenda. Comput. Netw. **114**, 125–142 (2016)
6. Leonardi, P.M., Huysman, M., Steinfield, C.: Enterprise social media: definition, history, and prospects for the study of social technologies in organizations. J. Comput. Commun. **19**(1), 1–19 (2013)
7. Anderson, S., Mohan, K.: Social networking in knowledge management. IT Prof. **13**(4), 24–28 (2011)
8. Ellison, N.B., Gibbs, J.L., Weber, M.S.: The use of enterprise social network sites for knowledge sharing in distributed organizations: the role of organizational affordances. Am. Behav. Sci. **59**(1), 103–123 (2015)
9. Dimicco, J., Millen, D.R., Geyer, W., Dugan, C., Brownholtz, B., Muller, M.: Motivations for social networking at work. In: ACM Conference on Computer Supported Cooperative Work 2008, pp. 711–720, April 2016
10. Drakos, N., Mann, J., Rozwell, C.: Magic quadrant for social software in the workplace. Gartner **207256**, 2–27 (2010)
11. DeLone, W.H., McLean, E.R.: Information systems success: the quest for the dependent variable. Inf. Syst. Res. **3**(1), 60–95 (1992)
12. Richter, A., Heidemann, J., Klier, M., Behrendt, S.: Success measurement of enterprise social networks. Wirtschaftsinformatik **20**, 1–15 (2013)
13. Herzog, C., Richter, A., Steinhüser, M., Hoppe, U., Koch, M.: Methods and metrics for measuring the success of enterprise social software – what we can learn from practice and vice versa. In: ECIS 2013 Completed Research, pp. 1–12 (2013)
14. Steinhueser, M., Herzog, C., Richter, A.: A process perspective on the evaluation of enterprise social software. In: 2nd European Conference on Social Media, pp. 429–436, July 2015
15. Bughin, J.: Taking the measure of the networked enterprise. McKinsey Q. **10**, 1–4 (2015)
16. Frambach, R.T., Schillewaert, N.: Organizational innovation adoption: a multi-level framework of determinants and opportunities for future research. J. Bus. Res. **55**(2), 163–176 (2002)

Go Vendla Go! Creating a Digital Coach for the Young Elderly

Christer Carlsson[1], Anna Sell[1], Camilla Walden[1], Pirkko Walden[1(✉)],
Siw Lundqvist[2], and Leif Marcusson[2]

[1] Institute for Advanced Management Systems Research,
Åbo Akademi University, Turku, Finland
{christer.carlsson,anna.sell,camilla.walden,
pirkko.walden}@abo.fi
[2] Linneus University, Kalmar, Sweden
{siw.lundqvist,leif.marcusson}@lnu.se

Abstract. The proportion of ageing citizens is high and increasing in most EU countries and there is a growing political pressure to make sure that the costs for the elderly care programs do not grow out of bounds. The focus of the ageing population programs is at the 75+ age group. The younger age group – the "young elderly" that is the focus in our research – does not get much attention. Recently, we have noticed a growing insight that preventive programs could be helpful as healthier young elderly will help produce healthier seniors (the 75 + age group) which over time will have significant effects on the costs for health and social care for the ageing population. Our study is a synergistic combination of two timely research areas: digitalisation and the ageing population. The focus combination of digital wellness services and the young elderly is unique. We build on a research program that is operating since January 2014. Our current research in progress aims at finding out how a digital coach – a personal trainer called Vendla – can be worked out for digital wellness services.

Keywords: Digital coaching · Young elderly · Physical wellness · Wearables

1 Introduction

1.1 Young Elderly

The proportion of ageing citizens is high in most EU countries and decision makers expect it to grow rapidly to reach about 20% of the population in the next decade. There is growing political pressure to find means to cope with the predicted (very) significant increase of the costs for the care programs for the ageing population. Not surprisingly, this is now tending towards finding trade-offs between the costs and the substance of the care programs (in plain terms, how to find ways to reduce the programs without too heavy political costs). The focus of these programs is at the 75+ age group as statistics and experience shows that the need for coordinated efforts is largest in this age group. The younger age group – the "young elderly" that we singled out – does not get much attention, which most probably could be a costly mistake. The "young elderly" is the 60–75 age group, which we distinguish from "the seniors". The

© Springer International Publishing AG 2017
M. Themistocleous and V. Morabito (Eds.): EMCIS 2017, LNBIP 299, pp. 204–209, 2017.
DOI: 10.1007/978-3-319-65930-5_17

young elderly represent 18–23% of the population in most EU countries; it is a large segment of the population and will be about 97 million EU citizens by 2020 [8]. The young elderly have been ignored by authorities because they are rather healthy, rather active and rather well connected socially, i.e. to the extent that the authorities do not feel any political pressure to allocate resources to them. It is also rather strange that the young elderly have not been identified as a market for digital services. The digital service developers, providers and distributors have their sights on the teenagers and young adults in the belief that they represent the growing digital service market. It appears that this "common wisdom" is widely accepted and the market actors are ready to compete fiercely for market shares. Then there is the digital wellness service market for young elderly with 97 million potential customers and – so far – with no or marginal competition. We believe that the young elderly represent a very interesting area for research and development work on digital services.

The young elderly age group have some risk to suffer functional impairment because of advancing age. This makes them motivated to spend time and resources to hold off functional impairment as long as possible, even to invest time and resources in order to improve on selected wellness functions. This was well put by one of our previous focus group members – "*it is more fun to get older if you are in good shape*".

Digital wellness services should be designed and built to help the young elderly to improve and maintain their functional capacity over time, and thereby their independence; these are aspects of life that today's citizens value and esteem.

Digital wellness services will be proactive work among the young elderly. We propose that services that will reduce the risk for serious and chronic illness that correlates with age will contribute to significant saving on the costs for health and social care in the senior population. The large numbers of ageing citizens linearly translate to large annual cost reductions, typically hundreds of millions (euro) in Finland and close to a billion (euro) in Sweden.

We build on a research program in Finland (WellS for Wellness for Senior Citizens) that is operating since January 2014 and has collected a number of research results [cf. 1, 2, 7, 10]. We are collaborating with a research group in Sweden to extend this research program and to compare the types of wellness services that will be adopted and used in Finland and Sweden. The design of a digital coach called Vendla is the first step in this joint research program.

2 Wellness

Wellness does not have an unequivocal definition, partly because it is a subjective experience. Wellness can for example be defined in the following way; to be in sufficiently good shape of mind and body to be successful with all requirements of everyday routines in the present context. The concept context is relative to what the young elderly want to achieve – some want to be able to achieve the same as in their work life; others may have redefined what they want to achieve after retirement; then there are limitations induced by illness, retirement pay, social factors, etc.

Donatella et al. [3 p. 7] define wellness as "An active process of becoming aware and making choices to create a healthier lifestyle in all of life's dimensions". Wellness is a construct with several dimensions; it is a holistic approach with the aim to improve a person's wellbeing. Wellness is in many studies divided into four dimensions: (i) cognitive wellness, (ii) physical wellness, (iii) social wellness, and (iv) mental wellness. Besides the four dimensions, some studies also include spiritual, occupational, environmental, cultural and climate dimensions [cf. 7].

Because wellness is a subjective experience the conceptualizing of the dimensions vary, as well as the context in which the research has been done. We consider wellness as a balance between the different dimensions that together form the core of wellness. In the current research project we focus mainly on the physical dimension of wellness when working out the digital coach Vendla.

3   Digital Wellness Services

Digital wellness services build on digital fusion of data from sensor systems, apps and software built and implemented for activity bracelets and smart mobile phones [1, 2]. Digital fusion forms information and knowledge both on smart phones and as cloud services (that support the services on smart phones); data, information and knowledge form the digital wellness services [1, 2]. The service designs build on co-creation processes in which the user needs, the user cognitive abilities and the user contexts decide the substance of the services [4].

The vision we have is that digital wellness services will attract users who gradually find out that the services can be meaningful parts of their daily routines. In this way, wellness routines – that are built and supported by digital wellness services – become part of everyday life. The adoption of wellness routines will progress over time and will follow processes that build on key results in the behavioural sciences and in sociology.

Differently from the adoption of wellness routines, the adoption of digital services is a faster moving process that offers interesting research. The basic conceptual frameworks for the acceptance and use of digital services are the TAM, UTAUT and UTAUT2 of which there are many variations. Studies of the user acceptance of digital wellness services build on surveys with large groups of young elderly. Our first results show that the UTAUT2 [9] framework is quite useful but requires user specific adaptations and modifications [1].

Results from work on digital wellness services show that generic data collected through smartphone applications (steps per day, walked distance, estimated burn of calories, etc.) is not interesting for more than on average 6 months before it becomes tedious, unsustainable routine [1]. We found more active use and better adoption rates when activity bracelets were added to the platform as (i) the data included more features (sleep patterns, heart rate, programmed exercise, etc.) and (ii) regular update of the software added more and improved features at 3–5 month intervals [1]. In Information Systems research, we have been advocating for quite some time that mobile services will work better if they are worked out with the actual users (not for some ill-defined potential users). As digital wellness services will be close to the daily life of the users it is a good proposition that the wellness services should be designed *with* the users.

4 Vendla – A Digital Coach

A digital coach is at our present stage of research an intelligent support function that is integrated with a system of digital wellness services to process data collected from activity bracelets to information and knowledge for the users [5, 6]. We found out that the digital coach will need algorithms for digital fusion (collecting and processing data from multiple devices and sources to information, then producing meaningful information through information fusion and then processing data and information through ontologies to produce knowledge and insight for guidance and support) and we also found out that automatic methods for digital fusion still are somewhat underdeveloped.

4.1 Objectives and Research Question

When developing the digital coach we have in mind to start with the physical wellness dimensions with the option to later include social wellness.

Our study is a synergistic combination of two timely research areas: digitalisation and the ageing population. The focus combination of (i) digital wellness services and (ii) the young elderly is unique, which allows us to work out unique contributions. Even a quick browsing of available studies gives serious and weighty arguments for activating the young elderly to start building wellness routines as interventions in their daily routines. We will contribute to and support the building of these routines with digital wellness services.

We found out in previous research that users rather quickly loose interest in the digital services. A finding related to currently available technology is, that wearable fitness technology in the present design of bracelets, does not meet the needs of the young elderly. For users to sustain interest in the wearables, the data needs to be value adding. It needs to go beyond step counts and sleep patterns, which after a while become predictable and therefore uninteresting.

Possibilities to improve the data can be found in predictive analytics, explanatory analytics and visualizations [as defined by INFORMS]. Predictive analytics could be used to develop suggestions for wellness-improving actions in everyday life, based on previous data, and visualizations could combine data from different sources to receive a better overview of an individual's wellness. Simple examples are insight on when sleep disruptions are more likely ("you always sleep poorly when you have forgotten your daily exercise"). The suggestions for successful scheduling of wellness activities based on previous activities and counsel on what to do in different problem situations ("you have 10 min; you can do this chair yoga program" or "I'm sorry you cannot sleep; try these simple breathing exercises"). Thus, we suggest that a personal trainer in the form of a digital coach would be the solution for the young elderly to sustain interest in the digital wellness service.

We intend to answer the following research question: How can a digital coach – a personal trainer called Vendla - be worked out for digital wellness services?

Our aim is,

1. to develop Vendla - a digital coach - for wellness services for the young elderly through co-creation with groups of young elderly volunteers, and
2. to study the adoption of Vendla among the young elderly and to find out if (and how) the help from Vendla forms wellness interventions in daily routines.

5   Research Agenda and Conclusions

In our research, we will concentrate on developing Vendla to work with smartphones and activity bracelets. Digital coaching of wellness services will support the young elderly to choose and build daily routines of wellness services and to continue to use them for years to come.

We will look for partners within groups of SMEs in both Finland and Sweden with the aim to find 12–15 companies that can form an embryo for a "wellness"-ecosystem. We make use of an agile service systems development method with the companies in order to develop functioning combinations of wellness services and coach-functions, which the companies are able to develop into digital services and applications.

Based on our experience we have it that developing digital services should take place together with the users; we talk about a co-creation process. Our aim is to put together groups of young elderly volunteers who will work on developing wellness-services (physical in the first phase). The groups will include 20–30 participants in both Finland and Sweden, and the number of participants in the two countries will be together around 120. This will be a longitudinal study that will run for a minimum of 36 months. Besides the companies and the young elderly we will conduct in-depth studies with personal trainers in the two countries as well as include personal trainers to work in close contact with us in order to follow and improve the digital coaches' performance.

The requirements on the digital coach which we see as important are the following: (i) Vendla needs to be aware of the physical wellness routines of the user; (ii) Vendla needs to know the relevant context information; (iii) Vendla needs to guide and/or assist the user about what to do in order to maintain/improve his/her physical health, (iv) and Vendla needs to be part of the user's everyday routines, i.e. Vendla needs to have learning capabilities in order to stay in tune with the user's physical wellness needs [cf. 5, 6].

References

1. Carlsson, C., Walden, P.: Digital wellness services for "young elderly"- a missed opportunity for mobile services. J. Theor. Appl. Electron. Commer. Res. **11**(3), 20–34 (2016)
2. Carlsson, C., Walden, P.: Digital wellness for young elderly: research methodology and technology adaptation. In: Proceedings of the 28th Bled eConference, eWellBeing, Bled 2015, pp. 239–250 (2015)

3. Donatelle, R., Snow, C., Wilcox, A.: Wellness: Choices for Health and Fitness, 2nd edn. Wadsworth Publishing Company, Belmont (1999)
4. Gronroos, C.: Service logic revisited: who creates value? and who co-creates? Eur. Bus. Rev. **20**(4), 298–314 (2008)
5. Klaassen, R., op den Akker, H.J.A., Lavrysen, T., van Wissen, S.: User preferences for multi-device context-aware feedback in a digital coaching system. J. Multimodal User Interfaces **7**(3), 247–267 (2013). doi:10.1007/s12193-013-0125-0
6. Kulyk, O., op den Akker, R., Klaassen, R., van Gemert-Pijnen, L.: Let us get real! an integrated approach for virtual coaching and real-time activity monitoring in lifestyle change support systems. The Sixth International Conference on eHealth, Telemedicine and Social Medicine, eTELEMED 2014, pp. 211–216 (2014)
7. Sell, A., Walden, C., Walden, P.: My wellness as a mobile app. identifying wellness types among the young elderly. In: Proceedings of the 50th Hawaii International Conference on System Sciences, pp. 1473–1483. IEEE (2017)
8. United Nations Department of Economic and Social Affairs: Population ageing and sustainable development, no. 4, 2014. http://www.un.org/en/development/desa/population/-publications/pdf/popfacts/Popfacts
9. Venkatesh, V., Thong, J.Y., Xu, X.: Consumer acceptance and use of information technology: extending the unified theory of acceptance and use of technology. MIS Q. **36**(1), 273–315 (2012)
10. Walden, C., Sell, A.: Wearables and wellness for the young elderly - transforming everyday lives? In: Proceedings of the 30th Bled eConference, Digital Transformation – from Connecting Things to Transforming Our Lives, Bled 2017. ISBN 978-961-043-1 (2017)

Social Networks and Web Pages
Used by Regional Municipalities
in the Czech Republic

Libuse Svobodova[(⊠)]

Department of Economics, Faculty of Informatics and Management,
University of Hradec Kralove, Hradec Kralove, Czech Republic
libuse.svobodova@uhk.cz

Abstract. The article focuses on the usage of social networks in Czech regional cities. The first part is devoted to introduction and to the marketing of towns and villages, communication with citizens and its modern forms, which take place in the online environment via social networks. The main part of the article is devoted to the survey and monitoring of selected social networks used by the regional cities in the Czech Republic. The aim of the article is to evaluate the current state of usage of social networks in the regional cities. Article will focus on use of social networks, link on web sites to the one social network and Facebook in the connection with the regional cities. Conclusion summarizes the findings. In the processing of the article were used primary and secondary sources.

Keywords: Communication · Interconnection · Internet · Municipality · Social network

1 Introduction

Modern times bring increasing demands for innovation and trends in the world of social networks, which are already widespread among people around the world. A large number of people cannot imagine living without social networks. Social networks have become a means of maintaining social relationships, gathering information, branding, and creating advertising. Social networks are today a phenomenon and become an important communication channel in all countries of the world. Social networks connect our everyday activities, and more and more users and supporters are constantly growing, with the young generation in particular spending more and more time on them. Social networks are so popular because we do not want anything if the price for internet access is neglected. How it will look like when we had to pay for every message or shared photo? We would certainly not write with our ten friends daily and not share every minute that we eat or eat. Social networking is one of the fastest growing methods of internet marketing and communication. A few years ago we could hardly imagine such an extension of one platform among users as is nowadays enhanced by, for example, Facebook or Twitter. The article will focus on the use of social networks in the Czech regional municipalities.

© Springer International Publishing AG 2017
M. Themistocleous and V. Morabito (Eds.): EMCIS 2017, LNBIP 299, pp. 210–218, 2017.
DOI: 10.1007/978-3-319-65930-5_18

The organization of the paper is as follows: firstly introduction is provided followed by described research methodology. Then a theoretical background with focus on definitions of key terms like social media and social networks. The key part brings results from the questionnaire investigation and observation which was run at the first quarter of 2017 from the Faculty of Informatics and Management, University of Hradec Králové. Examined areas relate to the use of social software applications and web pages for communication purposes between regional municipalities and citizens. Conclusion summarizes the findings.

2 Methodology and Goal

The goal of this study is to explore the use of the internet and social networks by all Czech regional municipalities. The possibilities of future development of municipalities on social networks will be revealed on the basis of the research of the current situation of the municipalities of the Czech Republic. To accomplish this goal, it will be necessary to explore secondary resources for online communication of municipalities and towns and to reveal the use of social networks on the territory of the municipalities of the Czech Republic through quantitative research. The following goals and scientific questions will be set.

- Goal 1: To map the current situation of social networks in the regional municipalities in the Czech Republic.
- Goal 2: To determine the use of web site and links on social networks by Czech regional cities.
- Goal 3: To identify types of social networks that are used by Czech regional municipalities.
- Goal 4: To identify the use of Facebook by selected municipalities.
- Scientific question 1: Each regional city use at least one social network.
- Scientific question 2: Minimally half of regional cities have a link to the one social network.
- Scientific question 3: Each regional city use at least two social networks.
- Scientific question 4: Facebook is the most widespread in the regional cities.

Primary and secondary sources were used in the processing of the article. Primary sources were obtained within the survey, which was conducted by teacher and student at the Faculty of Informatics and Management at the University of Hradec Králové. The investigation was done in March 2017. A student was searching in social networks for communities which were created to meet the needs of municipalities. Student focused also on presented information about municipalities on these networks. Student was assigned to all regional, statutory or district municipalities. In total there were 72 of those municipalities. In the article will be presented results only about regional municipalities.

The cities went from Sect. 3, Sect. 1 of Act 128/2000 Coll. on Municipalities, where it is stated that the municipality, which has at least 3,000 inhabitants is a city.

As for secondary sources, they comprised websites of selected surveys and also official statistics from the Czech Statistical Office and Eurostat, technical literature, information gathered from professional journals, discussions or participation at professional seminars or conferences. Then it was necessary to select, categorize and update available relevant information from the collected published material.

3 Literature Review

Next part will be focused on the legal statues of the municipality in the Czech Republic, official statistics of use of Internet by citizens and other literature review of use Internet and social networks by municipalities abroad.

3.1 Legal Status of the Municipality in the Czech Republic

The legal status of the municipality in the Czech Republic is indicated in Act No. 128/2000 coll. law on Municipalities [1]. According to current legislation, the city is characterized as a territorial community of citizens who have the right to self-government, where they form a territorial unit that is defined by the boundary of the municipality. Municipalities are also legally listed as towns, cities and statutory towns. The capital of Prague is the only exception, where the principle of special legal regulation applies. Prague has the status as the capital of the Czech Republic, the region and the municipality and the status of the city districts of Prague are regulated.

The Czech Republic is divided into 14 regions.

3.2 Social Media and Municipalities

Marketing communication is very important for community marketing. Marketing communication of towns and municipalities informs about the activities and products of the city. The city office and its tools work on the target groups so that urban activities, products and information are perceived positively by the public. The most important thing is to know the current needs and wishes of the population and to try to anticipate changes in the needs, preferences and expectations of the entities in the area of the consuming of the products provided by the municipalities. Marketing concepts in municipalities and cities can arise problems. Cities are dynamic, frequent and unpredictable for changes. Furthermore, they are not always completely homogeneous when some budgetary decisions are made at government level, such as motorways, repair of historical monuments, investments and others [10].

Web sites are still the most widely used information tool for municipalities and cities at present time. Regions, cities, and the vast majority of municipalities make full use of web sites and consider them an inseparable part of the structure of their information systems. Websites allow direct information transfer, image building, relationships and advertisement serving. Municipalities have an obligation to use the Internet to the extent stipulated by law. However, the Internet now offers a wealth of opportunities to municipalities that they can use to support their activities. Towns and municipalities' websites should meet the basic three functions:

- Informative (static communication)
- Interactive (dynamic communication)
- Presentation [22].

Today there is a great boom among social networks, which are very widespread and popular and have a significant role in marketing. For this reason, social networks offer the opportunity to reach a large circle of potential citizens, visitors and customers at a very low financial cost, with the advantage of this period being a huge boom. Social marketing brings ease of content publishing, targeting accuracy, measurement of user responses, speed, fast change and rapid dissemination among users, especially when multimedia content is breakthrough. In today's hurried time, it is good on social networks to prefer images, videos, charts, and various multimedia content to classic text submissions that people no longer have time to do.

Majumdar [13] focused on the recent advancements in communications technology that have made possible for many local government agencies, like regional transportation planning organizations, to use social media tools like Facebook, Twitter, Flickr, YouTube, and others to provide information to the public, educate them, and seek their inputs and ideas for meaningful decision making in transportation projects. Jiang et al. [11] focused on the Using the E-leadership theory as the conceptual Framework. Their study examined strategic communicators' perceptions of the impact of social media use on their work, leadership behaviors, and work-life conflict. From their study (N = 458) the use of YouTube in professionals' work, social media use in media relations, employee communications, and cause-related marketing/social marketing were significantly, positively associated with participants' perceptions of the enhancing impact of social media use. The use of Facebook and YouTube in strategic communication, the use of social media in environmental scanning, as well as the positive and negative impact of social media use all significantly and positively predicted communication professionals' leadership behaviors. Also public affairs/governmental relations professionals who were frequent users of social media for their work reported a high level of strain-based work-life conflict.

Study of Park, Lee et al. [14] investigated the utilization of Facebook by local Korean governments for the purposes of tourism development. The results indicate that most local Korean governments actively manage online tools such as Facebook, to communicate with the public and offer a wide range of information to promote tourism. Taking into consideration the growing popularity of social media in North American countries, study [4] aims to perform a comparative analysis of the use of Facebook as a communication strategy for encouraging citizen engagement among local governments in The United States, Canada and Mexico. With regards to the three dimensions used in all regions to measure online citizen engagement, in general terms, the "popularity" and "virality" dimensions are the most common, while the "commitment" dimension is still underutilized. DePaula, Dincelli [5] focused on Facebook. They have done an empirical analysis of local government social media communication. They have focused on the models of e-government interactivity and public relations. Sandoval-Almazan focused with next authors [16] on the concept of Smart cities and social media. Ure [18] delves into management models, communicative proposal and the strategies of the government to engage with citizens on social networks, more than half a decade of incorporating the resources of Web 2.0 in the official communication from the governmental institutions, through the case study two local urban authorities were compared, Buenos Aires (Argentina) and Bologna (Italy). It analyzes the contents of the publications of the institutional accounts of both cities on Twitter, Facebook and YouTube for a period of six weeks between the months of October and December 2014, elaborated and classified in categories in order to show the recognition level and empowerment of users as citizens. On the development of medialitization of local government from 1989 to 2010 focused in Sweden Djerf-Pierre and Pierre [6]. Bellström et al. [2] focused on Facebook usage in a local government in Sweden. Data were collected between May 2015 and July 2015. It was focused on a content analysis of page owner posts and user posts. Picornell et al. [15] exploring the potential of phone call data to characterize the relationship between social network and travel behavior. In the article we will focuses on the next selected social networks

- Facebook
- Twitter
- YouTube
- Instagram.

The topic was solved in the Czech Republic in the previous year by Svobodová, Dittrichová [17].

4 Results

The following chapter will focus on comparison of the statistics obtained from the regional cities of the Czech Republic. A detailed survey of the individual social networks was examined. Facebook, YouTube, Instagram and Twitter were conducted.

4.1 Use of Social Networks in the Regional Municipalities in the Czech Republic

It is possible to state that all of the surveyed Czech regional towns have official websites and have links to at least one social network on Facebook, see Tables 1 and 2. The tables below gives a comprehensive overview of social networks types that are used by regional cities and also what social networks are linked to the websites of the Czech regional cities.

The most widespread social network used by the Czech regional cities is the social network Facebook, which is used by all regional cities, and also all regional cities have a direct link to their social network sites on their website. YouTube is the second most commonly used social network site. 85% of Czech regional cities use YouTube, with a total of 57% linking to YouTube on their websites. The survey found that in the Czech Republic, the YouTube social network does not use two regional cities, namely Liberec and České Budějovice. It is also important to link this social network to the regional cities' websites. YouTube, which is being used by the Czech cities, links eight of its ten cities to its website. In particular, the regional city of Pardubice and Jihlava use

Table 1. Use of social networks by regional cities in the Czech Republic (Source: own elaboration)

Municipality	Number of citizens	Facebook	YouTube	Instagram	Twitter
Prague	1 267 449	Yes	Yes	Yes	Yes
Brno	377 028	Yes	Yes	Yes	Yes
Ostrava	292 681	Yes	Yes	No	Yes
Pilsen	169 858	Yes	Yes	Yes	No
Liberec	103 288	Yes	No	Yes	No
Olomouc	100 154	Yes	Yes	Yes	Yes
České Budějovice	93 513	Yes	No	No	No
Ústí nad Labem	93 248	Yes	Yes	No	No
Hradec Králové	92 891	Yes	Yes	Yes	Yes
Pardubice	89 638	Yes	Yes	No	Yes
Zlín	75 171	Yes	Yes	Yes	Yes
Jihlava	50 714	Yes	Yes	No	No
Karlovy Vary	49 326	Yes	Yes	Yes	Yes

Table 2. Interconnection of social networks and web pages by regional cities in the Czech Republic (Source: own elaboration)

Municipality	Facebook	YouTube	Instagram	Twitter	Others
Prague	Yes	Yes	No	Yes	Yes
Brno	Yes	Yes	Yes	Yes	No
Ostrava	Yes	No	No	Yes	No
Pilsen	Yes	Yes	Yes	No	No
Liberec	Yes	No	Yes	No	Yes
Olomouc	Yes	Yes	Yes	No	Yes
České Budějovice	Yes	No	No	No	No
Ústí nad Labem	Yes	Yes	No	No	Yes
Hradec Králové	Yes	Yes	No	Yes	No
Pardubice	Yes	No	No	Yes	No
Zlín	Yes	Yes	No	Yes	No
Jihlava	Yes	No	No	No	No
Karlovy Vary	Yes	Yes	Yes	No	No

YouTube but do not have links to this social network on their websites. Instagram is the third most widely used social network for the regional cities of the Czech Republic. Instagram uses 57% of the regional cities of the Czech Republic, where only 36% of the regional cities have links to this social network on their website. Regarding the interconnection of Instagram with the websites of the Czech regional cities using the Instagram social network are eight when the cities of Ostrava, České Budějovice, Ústí nad Labem, Pardubice and Jihlava do not use this social network. Only 5 cities have direct links on their websites.

Twitter is the least used social network by the regional cities of the Czech Republic. Twitter uses a total of 64% of the Czech regional cities, with only 42% referring to Twitter on their website. In the table we can see that Twitter's social network in the Czech Republic uses eight regional cities, where the regional cities of Pilsen, Liberec, České Budějovice, Ústí nad Labem and Jihlava do not use this social network. Linking to the social network Twitter has six of the eight regional cities of the Czech Republic using this social network.

It was also investigated with what other social networks are the regional cities of the Czech Republic connected to their web sites. There are four regional cities in the Czech Republic, including Prague, Liberec, Olomouc and Ústí nad Labem. The regional city of Prague refers to its social network "Linkedin and Snapchat", Olomouc refers to the social network "SoundCloud", Liberec and Ústí nad Labem refer to their websites on the social network "Google+". From the table above, it is possible to see that there are only six regional cities in the Czech Republic that use all social networks under examination, such as Prague, Brno, Olomouc, Hradec Králové, Zlín and Karlovy Vary. The interconnection of websites with social networks is mainly used by the regional cities of the Czech Republic.

5   Conclusion, Discussion and Recommendations

Just as the popularity of social networks is growing, their use by the public administration increases as well with the development of information technologies. Czech citizens and businesses can communicate with the government and local authorities through e-mail, they can browse their websites, where required forms can be found, downloaded or directly filled in and sent electronically.

The current situation of the use of social networks by all regional municipalities was based on the quantitative research and monitoring of data. Types of social networks used by the municipalities and whether these social networks are interconnected with the municipalities' websites was monitored.

The main purpose of this article was to elaborate an analysis of the use of individual selected social networks by the regional cities of the Czech Republic. Selected topic is not often solved by the Czech researches. The monitoring of the use of social networks by the Czech municipalities was not currently mapped and presented to the auditorium. In the processing of the article it was necessary to focus on the detailed collection of data gained from the visit and observation of individual social networks.

Scientific question 1: Each regional city use at least one social network.

Conducted research from a selected sample of municipalities in the Czech Republic revealed that municipalities use at least one social network to a large extent.

Scientific question 2: Minimally half of regional cities have a link to the one social network. The municipalities refer to their websites to individual social networks. It was found that all regional cities have interconnection to Facebook. As for the connection to the other social networks examined, very different results were found here. At least cities refer to the Instagram.

Scientific question 3: Each regional city use at least two social networks.

Conducted research from a selected sample of municipalities in the Czech Republic do not revealed that municipalities use at least two social networks. České Budějovice and Jihlava use only Facebook.

Scientific question 4: Facebook is the most widespread in the regional cities.

The survey of regional municipalities revealed that Facebook is the only one social network that is used by all regional towns.

The analysis of websites of individual selected municipalities revealed that municipalities also use other than selected social networks. There is even interconnection of these social networks with the websites of some surveyed municipalities. In addition to compared social networks are used also Linkedin, Google+ , Snapchat, Flicker, or SoundCloud.

Based on the findings from research and analysis of social networks in the Czech municipalities, it would be good for municipalities and towns to use more interconnections not only for websites but also for interconnections between individual social networks that are important in today's modern times.

Acknowledgement. This paper is supported by specific project No. 2103 "Investment evaluation within concept Industry 4.0" at Faculty of Informatics and Management, University of Hradec Kralove, Czech Republic. Thanks to help student Marta Martinová.

References

1. Act No 128/2000 Coll., on Municipalities
2. Bellström, P., Magnusson, M., Pettersson, J.S., Thorén, C.: Facebook usage in a local government: a content analysis of page owner posts and user posts. Transf. Gov.: People, Process Policy **10**(4), 548–567 (2016)
3. Czech Republic map. http://czechrepublicmap.facts.co/czechrepublicmapof/czechrepublic map.php
4. Del Mar Galvez-Rodríhuez, M., Haro-de-Rosario, A., Del Carmen Caba-Pérez, M.: A comparative view of citizen engagement in social media of local governments from North American Countries. In: Handbook of Research on Citizen Engagement and Public Participation in the Era of New Media, pp. 139–156 (2016)
5. DePaula, N., Dincelli, E.: An empirical analysis of local government social media communication: models of e-government interactivity and public relations. In: ACM International Conference Proceeding Series, 08–10 June 2016, pp. 348–356 (2016)
6. Djerf-Pierre, M., Pierre, J.: Mediatised local government: social media activity and media strategies among local government officials 1989–2010. Policy Politics **44**(1), 59–77 (2016)
7. Internet World Stats: Internet Stats and Facebook Usage in Europe March 2017 Statistics. http://www.internetworldstats.com/stats4.htm#europe
8. Eurostat: Individuals regularly using the internet. http://ec.europa.eu/eurostat/tgm/table.do?tab=table&init=1&plugin=1&pcode=tin00091&language=en
9. Eurostat: Individuals using the internet for participating in social networks. http://ec.europa.eu/eurostat/tgm/table.do?tab=table&init=1&language=en&pcode=tin00127&plugin=1
10. Janečková, L., Vaštíková, M.: Marketing měst a obcí. Praha: Grada, 1999. Města a obce. ISBN 8071697508

11. Jiang, H., Luo, Y., Kulemeka, O.: Strategic social media use in public relations: professionals' perceived social media impact, leadership behaviors, and work-life conflict. Int. J. Strat. Commun. **11**(1), 18–41 (2017)
12. Kadeckova, V.: Usage of social networks by Czech and Slovak cities. Diploma thesis. University of Hradec Kralove (2017)
13. Majumdar, S.R.: The case of public involvement in transportation planning using social media. Case Stud. Transp. Policy **5**(1), 121–133 (2017)
14. Park, J.H., Lee, C., Yoo, C., Nam, Y.: An analysis of the utilization of Facebook by local Korean governments for tourism development and the network of smart tourism ecosystem. Int. J. Inf. Manag. **36**(6), 1320–1327 (2016)
15. Picornell, M., Ruiz, T., Lenormand, M., Ramasco, J.J., Dubernet, T., Frías-Martínez, E.: Exploring the potential of phone call data to characterize the relationship between social network and travel behavior. Transportation **42**(4), 647–668 (2015)
16. Sandoval-Almazan, R., Cruz, D.V., Armas, J.C.N.: Social media in smart cities: an exploratory research in Mexican municipalities. In: Proceedings of the Annual Hawaii International Conference on System Sciences, March 2015, pp. 2366–2374 (2015)
17. Svobodová, L., Dittrichová, J.: Používání sociálních médií v municipalitách České republiky. In: XIX. mezinárodní kolokvium o regionálních vědách.: Sborník příspěvků. Brno: Masarykova univerzita, pp. 824–831 (2016). doi:10.5817/CZ.MUNI.P210-8273-2016-106, http://is.muni.cz/do/econ/soubory/katedry/kres/4884317/Sbornik2016.pdf#page=824
18. Ure, M.: The communication of public administration in social networks: the case of the cities of Buenos Aires and Bologna. Palabra Clave **19**(1), 240–270 (2016)

e-Government

Impact of eGovernment on Citizen Satisfaction: A Case of Federal Government Agencies in Pakistan

Muhammad Akmal Javaid[1]
and Muhammad Irfanullah Arfeen[1,2(✉)] (iD)

[1] QASMS, Quaid-i-Azam University, Islamabad, Pakistan
[2] United Nations University-Operating Unit on Policy Driven
Electronic Governance (UNU-EGOV), Guimarães, Portugal
arfeen@unu.edu

Abstract. The present study is an attempt undertaken to examine the relationship between eGovernment services and citizens' satisfaction in the context of federal agencies in Pakistan. Almost all the developing nations are still struggling to whether initiate these services or fully benefit from the already initiated e-services. This study proposed government to citizen satisfaction model considers the role of website content, trust, security or privacy, e-readiness, and quality of services while measuring the citizen satisfaction. Five hypotheses have been developed. A total of 500 questionnaires have been distributed of which only 302 received back with a response rate of 60.4%. On the collected data, quantitative and qualitative tests have been applied. Besides this it has been concluded form the analysis that all hypotheses were supported based on the contemporary research findings. The findings and recommendations can be successfully utilized for the betterment of the e-services and public service delivery tools.

Keywords: Agencies · eGovernment · Citizens · Satisfaction · E-services · Pakistan

1 Introduction

eGovernment is known as an essential element in the modernization of any government. This increased level of efficiency and effectiveness will eventually result in increased citizen satisfaction [1]. Government of Pakistan established Electronic Government Directorate (EGD) in 2002 which was transformed into National Information Technology Board, hereafter NITB, in 2014 decided to merge Pakistan computer bureau (PCB) and EDG into one whole [2]. The aim of the directorate was to achieve much needed transparency, fairness, efficiency, effectiveness and accountability. However, the fruitfulness of the eGovernment services is strongly dependent on the willingness of the customers to adopt, and their ability to avail from the eGovernment services [3]. eGovernment evaluation is referred to as a process of observing and measuring the ability of an eGovernment system to achieve its predetermined objectives [4]. It is the aim of the study to analyze the satisfaction from the

© Springer International Publishing AG 2017
M. Themistocleous and V. Morabito (Eds.): EMCIS 2017, LNBIP 299, pp. 221–237, 2017.
DOI: 10.1007/978-3-319-65930-5_19

citizen's point of view [5]. Furthermore, the study will conclude the efforts undertaken by the federal government agencies (FGA) in Pakistan and their impact on the direct beneficiaries or the people. The focus would be on the Government to Citizen (G2C) model of eGovernance.

Information and Communications Technologies (ICT) have influenced the human life, be it trade, services, manufacturing, government, education, research, entertainment, culture, defense, etc. [6]. A readiness index developed by United Nations determines the relative ranking of the eGovernance readiness index. Pakistan has many success factors in his pockets such as NADRA and DGIP has successfully implemented the eGovernance mechanism [5]. However, Provinces department have also started the application of ICT in governance. Punjab has successfully implemented the Land Record Management System (LRMS). Beside this the government has also allocated weight to the future initiatives including online tax filing, arm licensing, and identity registration [7].

For this study, the research questions are narrated below:

1. What role eGovernment services play in the public service delivery?
2. What impact e-services leave on Service Quality and Service Delivery?

2 Literature Review

eGovernment ensure its citizens certain advantages such as fairness and transparency in the processes of the governance, efficiency in the delivery of the government services, simplification of the complex procedures, improvement in office management, and friendly attitude of the public personnel serving in the public offices [8]. But yet the efforts of adoption of eGovernment have challenges especially in the developing countries [9, 10] such as lack of infrastructure, awareness among the citizens, poor human resource capacity, lack of technical skills, ineffective governing regulations and the expansive technology for the delivery of the services through the eGovernment portals.

In developing countries, one of the most important reasons for the low-level of adoption of eGovernment services is that the needs and requirements of citizens are ignored. In the South Asian region, most portals and government web sites remained dormant in 2010 [9]. In 2010 and 2012, UN eGovernment world surveys ranked Pakistan 146th and 156th, respectively. However, as a whole the South Asian region regressed in the 2012 survey and remains far below the world average. The rankings of South Asian countries in eGovernment survey [11] are as shown in Table 1.

2.1 eGovernment and Pakistan

According to Pakistan Telecommunication Authority, PTA (2014) and National Database and Registration Authority (NADRA) some of the major projects under the umbrella of eGovernment are, Sahulat, Cellular Village Connectivity, National Rabta Portal, NADRA kiosk and E-Pakistan Vision 2020 at national and provincial levels.

Table 1. Pakistan eGovernment ranking in South Asia

World eGovernment ranking					
Country/year	2005	2008	2010	2012	2014
Afghanistan	168	167	168	184	173
Bangladesh	162	142	134	150	148
Bhutan	130	134	152	152	143
India	87	113	119	125	118
Iran	97	108	103	105	100
Maldives	77	95	92	95	94
Nepal	126	150	153	164	165
Pakistan	136	131	146	156	158
Sri Lanka	94	101	111	115	74

Source: [9, 11]

PTA has also reported the telecom indicator, as cellular connectivity is an important tool in the eGovernment cycle for the delivery of the service, with a total density of 70.87% in April, 2016 and 70.33% in March of the same year. Whereas the tele-density stood at the 69.05% in April, 2016 compared to 68.51% in March of the same year [7]. Local Line mainly hosted by the Pakistan Telecommunications Limited (PTCL) stood at 3,141,700 connections with a downfall from the 3,172,344 in the last year. This downfall is mainly because of the increased ease of access and features introduced by the cellular companies (PTA 2016). A severe blow was brought to the landline connection by the introduction of the 3G and 4G services by the government which has abolished the monopoly of internet connectivity by landline operators. However, the total telecom investment in the economy stood at $1001.0 billion (PTA 2016).

One important factor of eGovernment is the design of websites. Those websites are preferable which are easy to access and easy to use for the customers. Web-based services are provided to the public on government Web sites [5]. Mostly government websites are citizen-centric. These are prepared on the needs and demands of the public. The citizen-centric features include search capability, links to other governmental bodies and personalized interfaces [7, 12]. Based on the above discussion, the following hypotheses have been proposed:

H1: eGovernment website design will have a positive influence on citizen's satisfaction.

H2: Perceived Trust in eGovernment services will have a positive influence on citizen's satisfaction.

H3: Available ICT infrastructure and its awareness (E-readiness) have a positive impact on citizen's satisfaction.

H4: Quality of eGovernment Service has a significant positive effect on citizen's Satisfaction.

H5: Security of the eGovernment services has a positive influence on Citizen's Satisfaction from eGovernment services.

A successful initiative by the government is dependent on the support of the citizen. Many developing countries eGovernance adoption has remained only on paper than in the field [13]. Although, the reformer of the public administration has been keen to the adoption of the eGovernance mechanism as they ensure effectiveness, efficiency and responsiveness in the administration, a goal of the services delivery [14–16].

2.2   Theoretical Framework

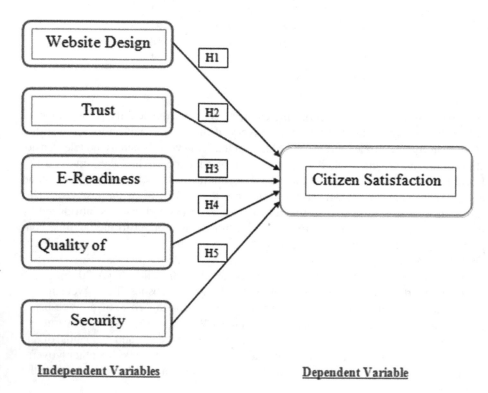

Fig. 1. Proposed theoretical framework – Government to Citizen (G2C)

3   Research Methodology

The study undertakes the evaluation of the e government in two dimensions, i.e. the government perspective and citizen's perspective. The framework refers to the service availability and its quality to enhance and in cases ensure the quality service delivery to the citizens as a whole [17]. It (Framework; shown in Fig. 1) has five variables i.e. Service Quality, Trust, E-readiness, Website Design, Security or Privacy concern. Each of the variables has its relevance to the other variable/variables. The relevance can directly be assessed and in cases can be assessed in underlying way for example the relevance between the service quality and e-readiness is somewhat associated but cannot be measured in quantitative terms because of its complexity [18]. The demand side includes those variables which are related to needs of the citizens and are demanded by the citizens in the form of the eGovernment services [7]. The frame work (shown in Fig. 1) and its other three variables, Trust, Security, and website are related. Although the existing researches have exerted enormous weight behind these variable but none were able to fully undertake and analyses the area of influence of these stated variables [19].

The study is aiming to find out the impact which e-services exert on the citizen satisfaction. The study undertakes both the qualitative and quantitative approaches and relies on the triangulation to conclude the results. In testing the hypotheses, which are developed in line with the extensive study of the literature, the study describes the positivism approach [5, 20–22]. The researcher's own ideas cannot influence the study as the main focus has been on the objectivity of the research. The findings and results are backed by the data interpretations which further rely on the validity and reliability. In order to make sure the reliability and validity of the data the long standing rich test has been applied to validate the stated fact [23]. The results of these tests are in line with the long standing historical values which serve as a recognized source of acceptance and rejection. In addition to this the research ethics have been duly followed in all the stages of the research whether it is data collection, analysis and interpretation [17]. As a triangulation study the data for qualitative and quantitative analysis has been collected through relevant techniques.

The data has been collected from methods through the use of internet, social network sites and hard form [24]. Qualitative data has been collected by use of interview techniques. For sampling and the sampling size the purposive method of the sampling has been adopted for the study. As the e-government is still developing in Pakistan and limited class from the population has known how about it so, somewhat educated class has been targeted for the sampling.

The study has been conducted through a sample size of some 300 respondents who were asked to fill a questionnaire and besides this some e-government practitioners were also interviewed for the purpose of qualitative data collection [5, 25–27]. Keeping in view the convenience and other limitations of the study the researcher has decided to conduct eight interviews from the e-government practitioner who are mainly civil servants of the government of Pakistan. Qualitative data had been conducted by conducting in-depth interviews with the government officials from different ministries, departments and autonomous agencies which have recently applied e-government technologies in the services delivery [17].

For data collection a questionnaire was adopted from the literature and it was refined further by tasking the experts considerations into account besides this it is also important that how the questionnaire is disseminated and administered to the respondents [15]. For that reasons the same questionnaire was uploaded on the Google Docs (a web feature of Google incorporation) and its link has been mailed to the potential respondents [28]. After the initial mail a reminder to those respondents who have not yet filled the questionnaire has been mailed after a gap of two weeks [29]. A similar reminder was followed after another gap of two more weeks.

4 Data Analysis and Results

Demographic data was obtained against eight variables during the process of obtaining data from the employees through questionnaire [17]. The demographic data was obtained to have a better and systematic understanding about the research sample. All the demographic variables are discussed as follows;

The first variable used for obtaining demographic data was gender. 68.2% of citizens are male while 31.8 are female in the study. The frequency shows that the questionnaires filled by male citizens were 206 in count while 96 women were available for questionnaires (see Table 2).

Table 2. Gender profile of the respondents

Gender profile		Frequency	Percent	Valid percent	Cumulative percent
Valid	Female	96	31.8	31.8	31.8
	Male	206	68.2	68.2	100.0
	Total	302	100.0	100.0	

The third variable used to collect demographic data was education. The data was collected from the citizens and 55.3% employees have bachelor's degree while 44.7% had master's degree (see Table 3). The distribution of the sample participants was in this manner that 167 members of the research had bachelor's degree while the other 135 had a master's degree.

Table 3. Education status of respondents

Education		Frequency	Percent	Valid percent	Cumulative percent
Valid	Bachelors	167	55.3	55.3	55.3
	Masters	135	44.7	44.7	100.0
	Total	302	100.0	100.0	

The internet usage experience variable was designed on the likert scale for a better data understanding and concise results. The experience scale was divided into 4 points i.e. 1–3 years, 4–6 years, 7–10 years and 11 years or above. In this study 17 participants lie in the age range of 1–3 years that makes a percentage of 5.6 (see Table 4). Then 102 citizens has the experience between 4–6 years making it only 33.8%. 26 citizens are of experience range between 7–10 years making it 8.6% of the whole study and 157 citizens have an experience of 11 years and above making it a 52% of the entire research.

The scale of Pakistan eGovernment portal awareness was a yes/no scale. In this study all the participants are well aware with the Pakistan eGovernment system and services (see Table 5). They may not be using it but they all are well aware of such services.

Table 4. Internet usage experience of respondents

Internet usage experience					
		Frequency	Percent	Valid percent	Cumulative percent
Valid	1–3	17	5.6	5.6	5.6
	4–6	102	33.8	33.8	39.4
	7–10	26	8.6	8.6	48.0
	11 and above	157	52.0	52.0	100.0
	Total	302	100.0	100.0	

Table 5. E-portal awareness among respondents

Pakistan eGovernment portal awareness					
		Frequency	Percent	Valid percent	Cumulative percent
Valid	Yes	302	100.0	100.0	100.0

Measurement of Inter-correlation Among Variables

The next step was to calculate inter-correlations among the variables. In order to do so, a correlation matrix was generated using SPSS software. Through the statistical methods applied over data collected the variables were studied for correlation. The correlation coefficient measure the strength of linear relationship between two variables. The correlation coefficient ranges between +1 and −1. The closer the correlation is to +1 and −1, the closer to perfect linear relationship. Detail of relationship among variables is showing under in Table 6.

Table 6. Inter-correlation among variables

Correlation						
	CS	WD	ER	SEC	QOS	TST
CS	1					
WD	.491**	1				
ER	.973**	.469**	1			
SEC	.177**	.320**	.180**	1		
QOS	.403**	.821**	.403**	.268**	1	
TST	.307**	−.014	.280**	.186**	−.057	1

Website design and citizen satisfaction has a strong positive relationship as the Pearson correlation value stood at 0.491**. Similarly, e-readiness also reflects a strong positive relation with dependent variable citizen satisfaction as the value of "r" is 0.973**. However, security has a weak positive relationship with the dependent variable as the correlation value is 0.177**. On the other hand, quality of service and trust also have a positive relationship with citizen satisfaction as the correlation coefficient value for both of the independent variables relationship is 0.403** and 0.307** respectively. All these relationships between independent and dependent variable are significant as the p value is 0.000 for each relationship, which is less than 0.05.

Regression Analysis

Linear regression technique was used to analyze the significance of variables and their relationships. The factor R^2 shows that how much impact or effect does one variable have on the other variable (see Table 7).

Table 7. Regression analysis

Regression Variable	R square	Adj R square	b	t	F	Sig.
Combine impact of IVs on CS	.953	.951			1171.802	.000
WD			.093	3.971		.000
ER			.937	61.264		.000
SEC			−0.18	−1.299		.195
QOS			−.043	−1.930		.055
TST			.047	3.366		.001

Combine Impact of Independent Variables on Citizens' Satisfaction

If the combine impact of website design, e-readiness, trust, security, and quality of service on citizen satisfaction is checked through regression it can be observed that all the variables explain 95% of citizen satisfaction as the value of adjusted R square is 0.951 (see Table 8). This impact is highly significant as the value of F is 1171.802 and p value 0.000. The beta value explains in the impact of each variable in the presence of all the independent variables. Security and quality of service have a negative relationship with citizen satisfaction as the beta value of −0.18 and −0.043 respectively. While trust has a weak positive relationship with citizen satisfaction as the beta value is 0.047 but the relationship of security and trust with citizen satisfaction is insignificant as the p value is 0.195 and 0.055 respectively which is greater than 0.05.

Table 8. Combine impact of independent variables on citizens' satisfaction

Individual IV regression with CS						
Variable	R square	Adj R square	b	T	F	Sig.
WD-CS	.241	.238	.491	9.755	95.169	.000
ER-CS	.948	.947	.973	73.680	5428.670	.000
SEC-CS	.031	.028	.177	3.123	9.751	.002
QOS-CS	.163	.160	.403	7.633	58.260	.000
TST-CS	.094	.091	.307	5.589	31.239	.000

Qualitative Analysis

In this study the researcher has heavily relied on the interview transcripts which were developed by the researcher after a thorough and detailed discussion with the eGovernment practitioners. Anwer et al. [7] has analyzed various factor to answer these questions and concluded that these initiatives have a positive impact on the citizen satisfaction, however, the response to these findings and their intensity also vary differently [3]. These views are also endorsed by some other researchers [18, 19, 24, 30]. Moreover, the eGovernment adoption and its impact on citizen satisfaction are illustrated by the answer of one of the interviewees in the following words when he was asked to comment on the eGovernment and its impact on citizen satisfaction.

> *"In my opinion, the Government has been trying to automate the existing system by introducing e-services in different departments, ministries and attach departments. But unfortunately, the governmental will is very poor which consequently leads to a poor pace of the eGovernment services introduction. Imagine this we have been working on the eGovernment since 2000 when NADRA was established, it was followed by establishment of EGD in 2002 and later on the same was molded into NITB in 2014. However, ironically there are only a few examples of eGovernment projects out of which many are still in their development phases. The offered services also lack many basic requirements such as their website is faulty or often remains hanged or busy, the lack of provision of updated contents is another such problem, similarly, the government has failed to ensure the safety and security of these services. All these failures on the government end are leading to the poor service delivery and poor users' satisfaction."*

In addition to this the quality of the services is also another key area which is being heavily commented by the respondents, in the words of one interviewee the quality of services is also lacking some basic grounds, he commented on this in the following manner; *"The quality of these services still needs improvement. The improvement is required in front office and back office channels the quality of these services is a big challenge for the government. The current quality challenges are incapacity of the officials, the hardware and software incapability and poor organization of the systems to provide these services."*

Further the same respondent commented; *"In public sector financial management the compatibility issue is so convoluted that the e-practitioners are failed to develop an integration model. Each department and institution has its own procedure to follow. State bank has its own system, accounting offices has its own software and version, FBR is using oracle based system and same goes for E&D which has a different system. They all lack synchronization. Basically the government has created Islands of eGovernments or patch works of the government. It is the need of hour that government must launch a compatibility mechanism. So that, all the institutions can work in harmony. So that, the tax and financial administration can bring the desired fruit."*

Similarly when they were asked about the trust and usability of these services in e-services and its efficacy the respondents replied in the following words; *"Yes people are capable of using these services. But the issue is with interface development which is not in line with their requirements. For example the question of language doesn't mean that someone is illiterate rather it's the responsibility of the government to provide them suitable channel for the development of the services. On the other hand, the Para-lingual website of the government provides only two options which are either Urdu or English but what about the provincial languages like Balochi or Sindhi's who are eloquent in speaking English but yet they are unable to benefit from these services. The solution of this problem could be better quality of the graphical or interactive user interface which encourage user to use services with an increased pace of service quality."*

In these themes one thing appears to be common that every variable is in the right direction but it needs improvement and reinforcement to strengthen its impact and also to provide the people more efficient service. This view is endorsed by many other researchers as well [24, 30, 31], however, their view slightly differ from the variable point of view.

The respondents were also asked about the e-readiness which reflects the governmental ability and the readiness of the resources to offer the needed services through the eGovernment domain the respondents expressed their views in the following manner; *"Presently, the government has ensuring citizens security by enacting the preventive legislation. The recent enactment of cyber crime bill is one of the key elements of this strategy. In the present age of technology where the e-information of citizens is at great risk of being hacked, government is still performing poorly to take some bold steps. However, few measures have been taken to ensure the security of e-information. In the previous year, few lunatic students who identified themselves with heroic names were able to hack the website of the ministry of health. Such happenings can only curtail by means of strict surveillance of the social media and by enacting necessary legislations."*

The biggest hurdle in readiness endeavor is the attitude of the employees. The traditional bureaucratic organization and the attitude of the office bearer is basically hindering the overall development of the eGovernment to its full [32]. Despite of the political government will and the increased financial allocation the eGovernment projects are in doldrums. This is mainly because of the poor or no interest on part of the bureaucracy as opined by the one respondent that the government is failed to change the attitude of its employee.

Security/Privacy of E-Services and Its Impact on CS

The security and privacy concerns are among those variables which were significantly talked about by the interviewees. The word tree (see Fig. 2) reflects this phenomenon by reflecting the most repeated words in word frequency which are government, security, services and information; suggests that the security puts a great challenge to the offered e-services.

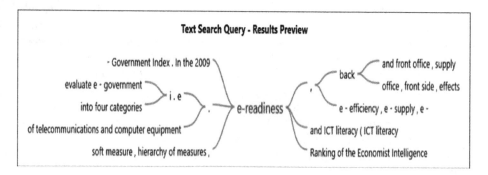

Fig. 2. Word tree - e-readiness

Similarly a respondent concluded the same question in the following manner;

"...it refers to protection of the information systems, assets and the control of access to the information itself. It is a vital component in the trust relationship between citizens and government. Security issues may present the largest obstacle to the development of eGovernment services. Thus, security policies and standards that meet citizen expectations are an important step toward addressing these concerns."

E-Readiness and Its Impact on CS

The e-readiness is one of the decisive factors in the study of the eGovernment and its impact on the citizens' satisfaction. The following readiness talk by the respondent is on such example;

".....United Nations issued eGovernment Readiness Ranking using measures of human capacity, infrastructure and access to information and knowledge. Pakistan ranked 131st i.e. pretty much at the bottom. In the following years, regional countries like Bangladesh and Iran improved their rankings, While Pakistan went down to 146th place in the 2010 eGovernment Development Rankings of the UN and slipped further down to 156th place in the 2012 UN eGovernment Index. In the 2009 E-Readiness Ranking of the Economist Intelligence Unit, Pakistan ranked 66th out of 70 countries assessed. But the present trends are showing some what good trends such as the same report published by UN in 2014 shows a better view where Pakistan among the second categories countries. Similarly, the improvement in other indicators i.e. ease of doing business is an obvious outcome of the better governance."

Testing of Hypothesis

H1: The e-government website design has a positive influence on citizen satisfaction in Pakistan.

H1 hypothesis states positive relationship between website design and citizen satisfaction. Correlation matrix shows a significant strong relationship between the two variables of $-.491^{**}$. Moreover the relationship is highly significant with p-value < 0.01. The hypothesis is accepted on the basis of both correlation matrix and regression.

H2: Available ICT infrastructure and its awareness (E-Readiness) have a positive impact on citizen's satisfaction.

H2 hypothesis states positive relationship between E-readiness and citizen satisfaction. Correlation matrix shows positive correlation of $.973^{**}$ between the two variables. Moreover the relationship is highly significant wit p-value < 0.01. The hypothesis is supported by the correlation matrix and regression analysis so it is accepted.

H3: Security of the e-government services has a positive influence on Citizen's Satisfaction from E-government services.

According to H3 hypothesis security has a positive relationship with citizen satisfaction of employees. Correlation matrix shows weak positive correlation of $.177^{**}$ between the two variables. The relationship is positive, moreover the relationship is significant with p-value < 0.01. The hypothesis has been supported by correlation matrix and regression analysis.

H4: Quality of service has a positive effect on citizen satisfaction in Pakistan.

According to H4 hypothesis quality of service is positively related to citizen satisfaction. Correlation matrix shows positive correlation of .403[**] between the two variables. The relationship is positive, moreover the relationship is significant with p-value < 0.01. The hypothesis has been supported by correlation matrix and regression analysis.

H5: Perceived Trust in e-government services exert a positive influence on citizen's satisfaction in Pakistan.

According to H5 hypothesis trust is positively related to citizen satisfaction. Correlation matrix shows positive correlation of .307[**] between the two variables. The relationship is positive, moreover the relationship is significant with p-value < 0.01. The hypothesis has been supported by correlation matrix and regression analysis.

All hypotheses are accepted in the light of the quantitative and qualitative analysis which not only conclude that perceived Trust in eGovernment services exert a positive influence on citizen's satisfaction in Pakistan, but also helped in the identification of the other themes which are helpful for future work, as they can act as preset guidance for the new study.

Discussion

Majority of the citizens whether literate or illiterate lack the English language ability which is the prime language of almost all the e-services websites. This feature helps the citizens to effectively use the services as they are able to see the contents in their native language. The citizens of Pakistan are willing to use these services in their local languages and with a lower literacy rate and poor standards of education in the country it becomes essential aspect which needs government considerations [3]. Despite of the fact, that government has established EGD fifteen years ago, yet eGovernment is in its developing phase in Pakistan.

The Ministry of IT is trying to learn from the pilot projects and it has great results in some areas [33]. The ministry is focusing to automate the process to make the quality of service better and effective. In addition to this the employees of the government lack ICT skills to effectively deliver the services to the citizens. They prefer to avoid the computer related task. This behavior of the employees need modifications by means of building their capacity to use a system effectively and for developing public value [34]. The information quality is also hugely impact by this incapacity of the employees as they usually failed to understand the impact of their task. Consequently, the website remains inefficient un-updated and lacks the needed contents which waits in the drawer of a particular employee instead of being uploaded on the website. Despite of initiation of the use of the ICT services in the public service delivery many government employees rely on the traditional paper based work which happens to be way more easy for them as they are well trained in it, but the same traditional method proves to be inefficient and costly when compared with the modern methods of service delivery such as eGovernment services [31]. The employees as well as their government must realize that it is the need of the hour to follow the slogan of the reinventing government which was chanted in lieu of reforming the traditional method of government business.

The availability of the resources to use services at home and work also impacts the citizen's level of satisfaction. These resources include the internet facility, computers or other devices such as tablets, mobiles, and handheld computers to access the websites, eventually use the government services [31]. Additionally, the awareness about the services and their use also affect the satisfaction of an individual. For that reason the objective of the government is to educate the people and make them able to use these services and benefit from that. MoITT is actively pursuing for the fulfillment of this objective, it has prepared many training modules for the capacity building of public sector employees.

In a society which is suffering from multiple ills such as illiteracy, instability, terrorism, extremism, and poor infrastructure the adoption and application of the e-services becomes too tough. In addition to this the prevailing energy crisis further deteriorates the situation as the whole eGovernment infrastructure is dependent of the availability of the electricity. With a average load shedding of some six to eight hours in urban areas, not to mention of rural areas where load shedding further plunges to the peak, the application and use of e-services is nothing more than an utopian thought.

Recommendations and Conclusion

In the light of the current research work some suggestions are made to improve the current state of eGovernment and their usability. These suggestions can work as a guiding principle for the policy makers, practitioner, public sector managers, think tanks, researchers and academics. The key recommendations are as follows:

The government of Pakistan needs to realize that there are different levels of e-service delivery, so that the citizens can be categorized accordingly and genuine effort can be made to help citizen avail their needed set of e-services. For example the for financial matters, e-filing and online submission of tax returns can be designed more effectively by targeting those customer who are currently under the domain of FBR instead of general public.

The citizen classification can be made according to certain demographic indicators such as their education level, background, and age and internet usage experience. Besides this, another variable for the categorization can be identified by analyzing the e-service usage level. As the eGovernment is still new to Pakistan, therefore, the citizens are at the initial stage of using these e-services. Thus, it becomes essential that for a greater usability rate the citizens should be made aware of the benefits of the e-services. Similarly, at the transaction level the government needs to build citizens trust in internet and then in government so that the citizen may use the services in a secure and trustworthy environment.

eGovernment services bring desirable fruits for all of the stakeholders only when they are used at a large scale for specified services delivery [18, 19]. In order to effectively introduce these service citizens must be taken into confidence and should be familiarized with these services first for the effective implementation and afterwards increased level of citizen satisfaction.

It is also essential that in order to boost the citizen satisfaction the government must educate its citizen about the presence of online service. This will eventually result in increased adoption rate and eventually to increased citizen satisfaction [35]. Without creating the necessary awareness among citizen the higher level of satisfaction cannot be achieved as they are not tilted towards the usability and efficacy of the newly enacted e-services. In other words they must also be ensured about the benefits of these e-services in terms of time and cost.

Another significant factor which can affect the citizen satisfaction is the availability of the infrastructure that enables the citizen to use these services. It is also important to note that in many developing countries including Pakistan the citizen lack necessary ICT infrastructure in terms of internet connectivity and devices. This makes everything stagnant as despite of the fact that the citizens are aware and willing to use e-services yet they lack the capacity to use these services. The statistics of the UN survey state that only 11% of the total population of Pakistan is internet users [11]. The recent sale of third generation and fourth generation of internet connectivity brought the revolution in the internet usage but yet the government needs to ensure the availability of e-services in remote and under developed areas. For that reasons government must establish internet hubs at town and Tehsil levels where citizen can access these services with necessary assistance.

This study also reveals some concerns of the citizens about trust and security in government and of their private information. The citizens concerned with the information and transaction stage are concerned about the trustworthiness of the e-services providers. Further the government must improve the security conditions of the e-services so that the increased usability can be ensured. This increased usability will further result in the increased cost and time benefits. Similarly it is the need of the hour that government of Pakistan must bring efficiency and effectiveness in its service delivery and also enhance the quality of good governance. To materialize this government of Pakistan needs to overcome the prevailing hindrances in the way of adoption of the e-service. These hindrances include lack of internet facility, incapacity of the citizens, poor ICT infrastructure, lack of political and bureaucratic will, red tape, and rigid system which is resistant to change to name a few. Government of Pakistan must also enlighten the minds of its citizens about the perceived benefits of internet and also help them to have increased interest in the service delivery.

According to the findings the citizens are reluctant to use these services because they feel insecure. The government must enact the necessary legislation to ensure a vibrant security mechanism which is in line with the needs of the current cyber environment.

The research reveals that according to the citizen perceptions, the information available on the government website is not updated and links accessible on the government web site are broken [19]. Therefore, citizens do not trust in the information available on the government web site, which ultimately leads to low usage of available eGovernment services. The GoP should give surety to the citizens about the accessibility and trustworthiness of information provided on the government website to make sure that their trust in the government websites, website contents and updated information lead to higher level of citizen satisfaction from eGovernment services [18].

Research Limitations

Although, the current research work is designed to evaluate the current level of eGovernment services and their impact on the citizen satisfaction yet the model can be improved by including further variables in it. These include, service availability, digital divide, and social divide. If included these three variables can help in proposing a successful government to citizen evaluation model which will further help in the doing a research work which will bring more actual, sophisticated and unbiased results.

Future Dimensions

This study aims to evaluate the current level of e-services and their impact on the citizen satisfaction from their perspective. The constructed model helps in measuring the level of satisfaction from the services. The model measures the citizen satisfaction from five different possible dimensions which are in line with the societal dynamics of Pakistan. A single criterion cannot give the overall picture of the citizen satisfaction so an integrated mechanism is devised to measure the overall dynamics. The study concludes that the overall performance of the available services is satisfactory; however, the government must ensure the provision of the services to the remote and underdeveloped areas. For that reasons the country needs to enhance its effort to transform its traditional method to the modernize one. It is also important that the existing traditional bureaucratic system is also causing a hindrance in this transformation as they lack the necessary skill and will for transformation for their own vested interest. The services quality as suggested by the SERVQUAL model contributes more to the citizen satisfaction than any other model. These factors as a whole explain that the government of Pakistan has to enhance its EGDI ranking. Currently, Pakistan stood at seventh position in the list of nine South Asian countries leaving only Nepal and Afghanistan behind.

Acknowledgment. This paper is a result of the project "SmartEGOV: Harnessing EGOV for Smart Governance (Foundations, methods, Tools)/NORTE-01-0145-FEDER-000037", supported by Norte Portugal Regional Operational Programme (NORTE 2020), under the POR-TUGAL 2020 Partnership Agreement, through the European Regional Development Fund (EFDR).

References

1. Nograšek, J., Vintar, M.: eGovernment and organisational transformation of government: black box revisited? Gov. Inf. Q. **31**(1), 108–118 (2014)
2. National Information Technology Board, NITB: Nitb.gov.pk (2014). http://www.nitb.gov.pk/. Accessed 29 July 2016
3. Malik, B., Mastoi, A., Gul, N., Gul, H.: Evaluating citizen e-Satisfaction from eGovernment services: a case of Pakistan. Eur. Sci. J. **12**(5) (2016). http://dx.doi.org/10.19044/esj.2016.v12n5p346
4. Syamsuddin, I.: Evaluation of eGovernment initiatives in developing countries: an ITPOSMO approach. Int. Res. J. Appl. Basic Sci. **2**(12), 439–446 (2011)
5. Alawneh, A., Al-Refai, H., Batiha, K.: Measuring user satisfaction from eGovernment services: lessons from Jordan. Gov. Inf. Q. **30**(3), 277–288 (2013). http://dx.doi.org/10.1016/j.giq.2013.03.001
6. Arfeen, M.I., Kamal, M.M.: eGovernance implementation model: case study of the federal government agencies of Pakistan. In: European, Mediterranean & Middle Eastern Conference on Information Systems 2013 (EMCIS 2013), 17–18 October 2013, Windsor, United Kingdom (2013)

7. Anwar, M., Esichaikul, V., Rehman, M., Anjum, M.: eGovernment services evaluation from citizen satisfaction perspective. Transform. Gov.: People Process Policy **10**(1), 139–167 (2016). http://dx.doi.org/10.1108/TG-03-2015-0017
8. Laudon, K., Laudon, J.: Management Information Systems, 11th edn. Prentice-Hall/University of Manchester, Upper Saddle River/Manchester (2009)
9. UN eGovernment Survey: eGovernment for People. United Nation, New York (2012)
10. Gupta, M., Jana, D.: eGovernment evaluation: a framework and case study. Gov. Inf. Q. **20** (4), 365–387 (2003)
11. UN eGovernment Survey: eGovernment Survey 2014. United Nation, New York (2014)
12. Segovia, R.H., Jennex, M.E., Beatty, J.: Paralingual web design and trust in eGovernment. Int. J. Electron. Gov. Res. (IJEGR) **5**(1), 36–49 (2009)
13. Kumar, V., Mukerji, B., Butt, I., Persaud, A.: Factors for successful eGovernment adoption: a conceptual framework. Electron. J. eGov. **5**(1), 63–76 (2007)
14. Cook, A., Sheikh, A.: Descriptive statistics (part 2): interpreting study results. Prim. Care Respir. J. **8**(1), 16–17 (2000)
15. Gibson, W., Brown, A.: Working with Qualitative Data. SAGE, London (2009)
16. Sivia, D., Skilling, J.: Data Analysis. Oxford University Press, Oxford (2006)
17. Wray, N., Markovic, M., Manderson, L.: "Researcher saturation": the impact of data triangulation and intensive-research practices on the researcher and qualitative research process. Qual. Health Res. **17**(10), 1392–1402 (2007)
18. Rehman, M., Esichaikul, V., Kamal, M.: Factors influencing eGovernment in Pakistan. Transform. Gov.: People Process Policy **6**(3), 258–282 (2012)
19. Ovais Ahmad, M., Markkula, J., Oivo, M.: Factors affecting eGovernment adoption in Pakistan: a citizen's perspective. Transform. Gov.: People Process Policy **7**(2), 225–239 (2013)
20. Alghamdi, I., Goodwin, R., Rampersad, G.: eGovernment readiness assessment for government organizations in developing countries. Comput. Inf. Sci. **4**(3) (2011). http://dx.doi.org/10.5539/cis.v4n3p3
21. Atkinson, P., Delamont, S.: SAGE Qualitative Research Methods. SAGE, London (2010)
22. Banerjee, S., Katare, J.: A quality assessment index for evaluation of district eGovernance websites. Int. J. Electron. Gov. **8**(2), 140 (2016). http://dx.doi.org/10.1504/ijeg.2016.078122
23. Creswell, J., Plano Clark, V.: Designing and Conducting Mixed Methods Research. SAGE Publications, Thousand Oaks (2007)
24. Rehman, M., Kamal, M., Esichaikul, V.: Adoption of eGovernment services in Pakistan: a comparative study between online and offline users. Inf. Syst. Manag. **33**(3), 248–267 (2016)
25. Verdegem, P., Verleye, G.: User-centered eGovernment in practice: a comprehensive model for measuring user satisfaction. Gov. Inf. Q. **26**(3), 487–497 (2009). http://dx.doi.org/10.1016/j.giq.2009.03.005
26. Zhou, Y.: Voluntary adopters versus forced adopters: integrating the diffusion of innovation theory and the technology acceptance model to study intra-organizational adoption. New Media Soc. **10**(3), 475–496 (2008)
27. Trainor, A., Graue, E.: Qualitative Methods in the Social and Behavioral Sciences. Routledge (Taylor and Francis), New York (2013)
28. Tsohou, A., Lee, H., Irani, Z., Weerakkody, V., Osman, I., Anouze, A., Medeni, T.: Proposing a reference process model for the citizen-centric evaluation of e-government services. Transform. Gov.: People Process Policy **7**(2), 240–255 (2013)
29. Yin, R.: Case Study Research. Sage Publications, Thousand Oaks (2003)
30. Weerakkody, V., Irani, Z., Lee, H., Hindi, N., Osman, I.: A review of the factors affecting user satisfaction in electronic government services. Int. J. Electron. Gov. Res. **10**(4), 21–56 (2014)

31. Haider, Z., Shuwen, C., Burdey, M.: eGovernment project obstacles in Pakistan. Int. J. Comput. Theory Eng. **8**(5), 362–371 (2016)
32. Alshehri, M., Drew, S.: A comprehensive analysis of eGovernment services adoption in Saudi Arabia: obstacles and challenges. Int. J. Adv. Comput. Sci. Appl. **3**(2) (2012). http://dx.doi.org/10.14569/ijacsa.2012.030201
33. MoITT: Moitt.gov.pk (2008). http://www.moitt.gov.pk/. Accessed 20 Aug 2016
34. Shuler, J., Jaeger, P., Bertot, J.: Editorial: eGovernment without government. Gov. Inf. Q. **31** (1), 1–3 (2014)
35. Varavithya, W., Esichaikul, V.: Dealing with discretionary decision making in eGovernment: an open government approach. Electron. Gov. Int. J. **3**(4), 356 (2006). doi:10.1504/eg. 2006.010799

Public Participation and Regulatory Public Policies: An Assessment from the Perspective of Douglasian Cultural Theory

Flavio Saab[1](✉), Gustavo Cunha Garcia[2], Jonathan Soares Pereira[1], and Paulo Henrique de Souza Bermejo[1]

[1] University of Brasilia (UnB), Brasília, Brazil
flaviosaab@gmail.com, jonathansprj@gmail.com,
paulobermejo@unb.br
[2] Pontifical Catholic University of Rio de Janeiro (PUC-RJ),
Rio de Janeiro, Brazil
gcgfla@gmail.com

Abstract. The objective of this paper is to evaluate how public consultation encourages participation from different points of view and if there is a real transfer of power to the society in policy-making. Public participation is a means of inclusion that allows decision-makers to know the real needs of society, and it promotes the elaboration of public policies, leading to greater acceptance. Public participation must be encouraged and the opinions of various groups in the society must be considered. This paper adopted the perspective of Douglasian Cultural Theory to investigate the contribution of public consultations in the formulation of regulatory policies for the food sector in Brazil. The results of this research indicate that public participation remains restricted to some groups in society and that the public authority should dedicate more efforts to offer mechanisms that foster more plural and diversified public participation.

Keywords: Public consultation · Management of ideas · Electronic government · Public administration · Cultural theory · Douglasian Cultural Theory · Public participation · Social participation

1 Introduction

Public participation or social participation is one of the cornerstones of democracy and has been used since the beginning of the 1960s as a means of redistributing government power to society [1]. Among its main objectives are to involve citizens in decision-making and to better understand the real needs and priorities of society. In addition, it is an important instrument of social inclusion in increasing the quality of public policies [2, 23], since it gives decision-makers access to data and information that enables the mapping and measurement of policy impacts before they occur in practice [3, 4].

However, different public participation processes and instruments have particularities that may cast doubt, in some cases, on their democratic character [5]. There are processes by which participation is a mere bureaucratic procedure and there are processes where there is a real transfer of power from government to society, which

© Springer International Publishing AG 2017
M. Themistocleous and V. Morabito (Eds.): EMCIS 2017, LNBIP 299, pp. 238–249, 2017.
DOI: 10.1007/978-3-319-65930-5_20

contributes to the outcome of government intervention [1]. When the public participation process is inadequate, the public policy outcome can be affected by distortions, such as government capture, corruption, or government capture by economic agents [6].

Lodge et al. [7] employed cultural theory to develop arguments to evaluate the process of meat inspection in Germany. This approach can contribute to the analysis of public participation in the formulation of public policies. Linsley et al. [8] relied on these arguments to assess whether certain public consultation processes in the United Kingdom were designed to promote change or merely to reinforce the status quo and existing practices. In another study, Perri 6 and Swedlow [9] analyzed the potential contributions of cultural theory to public administration studies and concluded that the conceptualization and testing of cultural theory can bring enormous gains to public administration research.

Considering the possibility of contributing to the knowledge of public administration and the need to study the democratic character of government and public responsiveness, the present research was aimed at determining, from the Douglasian Cultural Theory point of view, the extent to which the process of public consultation encourages participation in, and empowers society through, the formulation of regulatory policies.

This remainder of this article is divided into four parts. In the second section, relevant concepts related to Douglasian Cultural Theory, which contributed to the development of the research at hand, are discussed. In the third section, the main methodological aspects used in conducting the research are presented. In the fourth section, the results obtained and the discussion resulting from the application of Douglasian Cultural Theory to the analysis of public consultations are presented. Lastly, the main results of the research and its contributions to the theory and practice of public administration are discussed.

2   Theoretical Framework

2.1   Cultural Theory

Douglas et al. [10] worked on a sociological theory regarding different forms of religions, points of view, and ideologies, in an attempt to relate this set of "beliefs" to different societies. This theory divides a society into four "voices" living in a retrenchment process and disputing among themselves [7]. The development of the theory involved the creation of a diagram (Fig. 1) that serves as a classification scheme for social relations in two dimensions, which are relatively independent [11]. The dimension shown vertically in the diagram is called the *grid* and refers to the behavior of individuals in a society. The dimension represented horizontally is called the *group* and refers to the involvement of individuals with established social groups [10]. Initially, the theory was developed with the nomenclature *grid-group analysis*, but once it was accepted as a theory, it began to receive other names, such as *cultural theory* [7], *cultural theory grid-group* [12], *Douglasian cultural theory* [8], and *theory of cultural bias* [9].

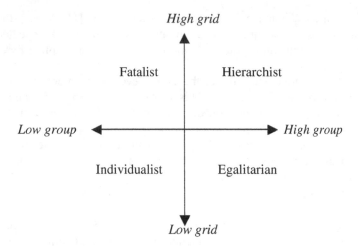

Fig. 1. Four forms of social groups (Source: Douglas et al. [10], p. 101)

Douglasian Cultural Theory, the term adopted in this article, places culture as the central aspect in the explanation of social life [12], and it is based on the idea that the ways in which individuals view the world and organize themselves are related issues [13]. Each individual adheres to one or more lifestyles relating to a combination of cultural biases—that is, beliefs and values—and to patterns of social relations [13].

In accordance with Douglasian Cultural Theory, the *grid* relates to the obligations of the individual to others—that is, the degree of social control that society has on the individual [11]. The *group* is related to belonging to a social group [14] and measures the extent to which an individual's behavior is influenced by the social groups to which the individual belongs [13]. The definition of values for the *grid* and *group* dimensions results in four types of "voices," represented by each quadrant shown in the diagram below: fatalist, hierarchist, individualist, and egalitarian [10].

To Douglas et al. [10], individuals in the *individualist* quadrant believe in the functioning of the market and see corruption, selfishness, and aggression as elements of human social life, in contrast to the joyful and simple elements of nature [11]. Individuals are free to transact in a market system, where recognition will come to those who are deemed deserving in a competitive environment [12]. Those individuals will reject coercion in favor of competition and autonomy [13].

The individuals in the *egalitarian* quadrant see nature as fragile and man as essentially good, until corrupted by institutions, such as the market [10]. This quadrant combines the presence of strong groups with low regulation of behavior. These groups are maintained by close relations between members, with small differences between them. For egalitarians, justice is represented by equal results, and the failures are the responsibility of the system [12, 13].

According to Douglas et al. [10], individuals in the *fatalist* quadrant see no logical sense or beauty in nature and see man as inconsistent and unfair. In this approach, justice will not be achieved and there is no way to promote improvements to society. The failure is not the fault of the individuals or the system but the result of fate [13].

Individuals in the *hierarchist* quadrant see the world as controllable, nature as stable, since it is subject to limits, and man as adaptable [10]. Further, rules are necessary to strengthen the collective over the individual and, hence, the division of labor, differentiation of roles, and hierarchy are typical of this group. Justice occurs when there is equality before the law and failure is a result of noncompliance with the rules [12].

The classification proposed via the perspective of cultural theory allows for analyzing actions and behaviors associated with the four "voices" in a continuous process in which each "voice" can attempt to persuade the other with arguments, and these interactions will influence the patterns of social relations. Therefore, arguments based on cultural theory have been developed and used by researchers in the field of public administration, as will be presented in the next section.

2.2   Related Studies

Lodge et al. [7] developed proposals expecting to making observable and testable implications of cultural theory in the evaluation of the meat inspection process in Germany. Their central objective was to test the application of theory, as opposed to explaining governmental failures in the activity of inspecting meat. To evaluate the propositions, the authors classified the main arguments identified in newspapers, whose subject matter was related to the quality of meat in Germany, as the "voices" indicated by cultural theory.

Subsequently, Linsley et al. [8] applied Douglasian Cultural Theory to examine the processes of public consultation in policy- making. They analyzed the participation of society in three public consultations undertaken by the Financial Reporting Council (FRC), the body responsible for regulation, finance, and auditing in the United Kingdom. Based on Douglasian Cultural Theory, they developed four propositions derived from those developed by Lodge et al. [8], interpreted the content of the responses of participants, and analyzed the process of participation in public consultations. Linsley et al. [8] suggested the production of new research on policies and public consultations, based on Douglasian Cultural Theory, to recommend ways to ensure the encouragement of full participation in public consultations. They argued that this would enable the four "voices" to respond to the consultations and be heard, thus avoiding regulatory self-capture, understood by the authors as a situation in which the point of view of the regulatory authority is perpetuated.

Thus, it appears that related studies based on Douglasian Cultural Theory have indicated that the use of this theory to analyze social relations can offer important gains for public administration research, especially to help clarify the process of public participation in the formulation of public policies. From the perspective that public policies can be understood as the result of relations and confrontations between groups of individuals in society and that the Douglasian Cultural Theory classifies these groups into "voices", its use becomes an alternative for the understanding of policy-making.

3 Methodological Aspects

3.1 Classification of Research

This paper discusses explanatory research using Douglasian Cultural Theory to explain the relationship between the phenomenon of public participation and public policies. In addition, it examines two public consultations from *Brazilian Health Surveillance Agency*—ANVISA, a government institution that regulates the production of goods and services that can pose health risks—to contribute to the general understanding of this instrument of public participation. Furthermore, the research has a documentary and qualitative characteristic, since the data analyzed were from official documents that present the manifestations of the government and the participants representing the society in public consultations [15, 16].

3.2 Method

Public consultation on-line is a mechanism based on open innovation, used by public institutions, with the objective of increasing democratic potential and strengthening social participation [24]. The procedures of the research involved the analysis of all public contributions in two public consultations conducted by ANVISA and the evaluation of the responses made by the institution in official documents on the subject. Specifically, the research investigated public consultations 255 and 256, conducted in September 2016, dealing with the regulation of the production and nutrition labeling of industrialized food. These public consultations were selected due to their theme (which is not technical) and the fact that ANVISA has already made demonstration building journal about the events received.

ANVISA conducted the two public consultations in a virtual environment, using electronic forms, to collect contributions from society as a whole. The process lasted for 30 days. The forms allowed each participant to suggest changes to draft regulations previously submitted by the agency. According to ANVISA, PC 255 had 93 participants, of which 57 (61.29%) identified themselves as individuals and 36 (38.71%) as legal entity. PC 256 had 50 participants, 25 (50%) of whom identified themselves as individuals and the other 25 (50%) as legal entity [17, 18].

For this research, with the aid of an electronic spreadsheet, the researchers analyzed data from 143 participants and excluded 8 duplicate cases. The comments of the remaining 135 participants were analyzed and each participant was classified as fatalist, individualist, hierarchist, or egalitarian, according to the typology of Douglasian Cultural Theory. The process of classification was conducted separately by three researchers. When the classification were identical, there was definition by consensus. When there was no consensus, the researchers discussed the arguments used by each for joint establishment of the classification.

After classifying each participant, organizing the data, and consolidating the information, the results were evaluated through an analysis (confirmation or refutation) of the propositions developed by Linsley et al. [8], listed in Table 1.

Table 1. Propositions from Douglasian Cultural Theory

Proposition 1 (P1): The participation of "voices" in public consultations	(A) The four "voices" will manifest in the public consultation process
	(B) As the hierarchy has a greater preference for regulation, hierarchists will have greater participation in the process
	(C) Given the fatalists' isolation from society, they will participate less in the public consultation process
Proposition 2 (P2): The arguments presented by each of the "voices"	The contributions made by each of the four "voices" will have common characteristics in the public consultations
Proposition 3 (P3): The "voice" of the regulatory body influences the outcome of the public consultation	The point of view of the regulatory body will influence the outcome of the public consultations
Proposition 4 (P4): There will be participants with hybrid arguments as a means of ensuring partial support for their contribution	Some participants will present hybrid arguments to gain at least some support for their contribution

4   Results and Discussion

Using the four propositions obtained in the study of Linsley et al. [8], the results of each proposition will be analyzed and discussed in light of the analysis of the data obtained from PC 255 and PC 256, carried out by ANVISA.

4.1   Proposition 1: Participation of "Voices" in Public Consultations

According to Linsley et al. [8], Douglasian Cultural Theory indicates that all "voices" should participate in the public consultation process in different measures. The aim of Proposition (P1) was to analyze the participation of the hierarchist, egalitarian, fatalist, and individualist "voices" in the process.

Of the 135 participants, 70 represented the hierarchist, 59 represented the egalitarian, and 6 represented the individualist (Table 2). There was no participant with a fatalist dominant discourse. The absence of participants representing the fatalist and the low number of participants representing the individualist led to the rejection of part (a) of P1. Even though the theoretical propositions recognize that participation among various groups should not be equal, the absence of fatalists and the low involvement of individualists could mean that the public consultation process did not encourage participation and did not motivate a plurality of views in the construction of the policy.

The rejection of part (a) of P1 confirms the results obtained by Linsley et al. [8], but differs from the results of Lodge et al. [7]. Linsley et al. [8] identified low participation of egalitarians (4.4%) and fatalists (0.7%), which led the authors to conclude that the proposition was partially rejected, owing to the under-representation of these two groups. Otherwise, Lodge et al.'s [8] results confirmed part (a) of P1, causing them to conclude that the four "voices" were involved in the process.

Table 2. Dominant and subordinate voices (elaborated by the authors)

Dominant and subordinate points of view	PC 255		PC 256		Total	
	Qty	%	Qty	%	Qty	%
Dominant hierarchist	42	47.70%	26	55.30%	68	50.37%
Dominant hierarchist with fatalist (hybrid)	2	2.30%	0	0.00%	2	1.48%
Dominant individualist	4	4.50%	0	0.00%	4	2.96%
Dominant individualist with egalitarian (hybrid)	2	2.30%	0	0.00%	2	1.48%
Dominant egalitarian	2	2.30%	1	2.10%	3	2.22%
Dominant egalitarian with hierarchist (hybrid)	36	40.90%	20	42.60%	56	41.48%
Total	**88**	**100.00%**	**47**	**100.00%**	**135**	**100.00%**

From the analysis of the results shown in Table 2, it is also possible to detect high participation of hierarchists, who accounted for 51.85% of all participants, in both public consultations. This result confirms part (b) of P1, since it indicates the prevalence of hierarchist participants by their "preference for better regulation and greater oversight" [8]. This result confirms the results obtained by Linsley et al. [8], which identified the participation of hierarchists in the order of 39%, and those obtained by Lodge [7], which identified the participation of this group in the order of 56.6% and 59.1%, respectively, in the two surveys conducted.

Part (c) of P1, which shows low fatalist participation in the public consultation process, is confirmed by the results; no fatalists participated in the two public consultations. Linsley et al. [8] identified only one fatalist participant, which also led them to confirm this proposition. However, in Lodge et al.'s [7] study, fatalists had participation rates of 13.3% and 8.1%, which were higher than the 10% and 6.6% for egalitarians, implying the rejection of the proposition.

4.2 Proposition 2: Arguments Presented by Each of the "Voices"

According to Linsley et al. [8], the common arguments of each of the points of view in different public consultation processes must be identified. The expectation of the second proposition (P2) was that the hierarchist, egalitarian, fatalist, and individualist arguments would be preserved in different manifestations. An important issue here was the presence of a considerable number of participants who participated in both consultations.

Table 3 illustrates that the main arguments of the hierarchy were raised in both public consultations. They argued that the regulation should reduce information asymmetry, expressed concern about the vulnerability of consumers, and advocated expanding the rules to other situations. Table 4 indicates that egalitarians also preserved their main arguments that the rules can create problems for a section of society, can generate high financial costs, and should be relaxed. However, egalitarians argued only in PC 255 that the current rules are sufficient to handle the problem.

Table 3. Sumary of hierarchists themes (elaborated by the authors)

Theme	PC 255	PC 256
Advocated regulations that reduce information asymmetry	Yes	Yes
Advocated expanding the rules to other situations	Yes	Yes
Expressed concern about the vulnerability of consumers	Yes	Yes

Table 4. Sumary of egalitarians themes (elaborated by the authors)

Theme	PC 255	PC 256
The rules can create problems for a section of society	Yes	Yes
The rules can generate high financial costs	Yes	Yes
The current rules are sufficient	Yes	No
The rules need to be flexible	Yes	Yes

Although the analysis of the preservation of the fatalists' and individualists' arguments could not be carried out, as these "voices" did not participate in the two public consultations, it is possible to recognize the confirmation of P2. Hence, the arguments put forward by the participants classified under the same "voices," in general, were preserved in the two public consultations.

The results of the research of Linsley et al. [8] and Lodge et al. [7] also indicate the confirmation of P2. Linsley et al. [8] observed that the arguments were preserved in three public consultations and concluded that "there is a sufficient degree of common characteristics to confirm the second proposition." Yet Lodge et al. [7] recognized that the fact that the main arguments presented in the first consultation were reinforced in the second consultation also confirmed P2.

4.3   Proposition 3: The "Voice" of the Regulatory Body Influences the Outcome of the Public Consultation

The third proposition (P3) was proposed by Linsley et al. [8] for identifying signs of self-capture, which consists of the perpetuation of the point of view of the regulator, under the influence of the institution itself, in decision-making, due to the patterns and values cultivated by it.

The reports of contributions made by ANVISA show that the regulatory agency presented a predominantly hierarchist "voice", with a propensity to regulate behavior through the establishment of rules and limits. Generally speaking, while ANVISA did not accept contributions suggesting deleting or relaxing the rules, it welcomed arguments promoting greater clarity in the proposed rules. The reports prepared by ANVISA show that the accepted contributions presented a predominantly hierarchical characteristic. In PC 255, ANVISA received 219 valid contributions; 152 (69.4%) were not accepted, 29 (13.2%) were accepted, and 38 (17.4%) were accepted in part. In PC 256, ANVISA received 196 valid contributions; 86 (44%) were not accepted, 66 (34%) were accepted, and 44 (22%) were accepted in part [19, 20] (Table 5).

Table 5. Response of ANVISA to contributions (elaborated by the authors)

Position of ANVISA in public consultations	PC 255	PC 256
Accepted contributions	13.20%	34%
Contributions accepted partially	17.40%	22%
Contributions not accepted	69.40%	44%

The analysis shows that ANVISA preserved its point of view, which favored the establishment of rules and limits for the conduct of society. It is important to recognize that, to some extent, the agency received suggestions from society members to change the proposed rules, but the confirmation of P3 was due to the hierarchical "voice" of ANVISA that prevailed in the public consultation process. It is important to highlight that, in Linsley et al. [8], the same proposition was also confirmed.

4.4 Proposition 4: Some Participants Present Hybrid Arguments as a Means of Gaining at Least Partial Support for Their Contribution

The fourth proposition (P4) was intended to assess whether individuals who represent a "voice" were willing to compromise their view of the world by offering compensations to other "voices" in order to have at least part of their contribution accepted.

The data in Table 2 indicate that 60 (44.4%) of the participants in the two public consultations presented hybrid "voices," and the egalitarian "voice" with hierarchist characteristics was dominant, accounting for 56 participants (41.5%). The arguments displayed that egalitarians defended their group, but used some hierarchist characteristics, supporting the adoption of rules and limits for the agents. These characteristics resulted in the confirmation of P4 and may be linked to the fact that egalitarians were part of a hierarchical discourse in an effort to argue for the protection of their group, albeit by means of rules and limits.

Linsley et al. [8] also confirmed P4 owing to a "significant number of respondents who indicated a hybrid characteristic", in the order of 38.9%. However, P4 was not confirmed by Lodge et al. [7], who found that only 11.67% of the arguments presented a hybrid character.

5   Conclusion

In the two public consultations analyzed in this paper, three of four "voices" that represent the points of view of society were identified, which implied rejection of part (a) of P1. Egalitarian and hierarchist were the groups that had significant representation in both public consultations, whereas the individualist group had low representation and the fatalist group remained silent. This discrepancy in representation can generate a suboptimal result in the process of public participation, since a public policy based on a few points of view may not meet the aspirations of society and will not have the support needed for its implementation.

The absence or low participation of fatalists is insufficient to constitute an unexpected result, since this is an isolated group whose members believe that their contributions do not affect public policy. Furthermore, the low representation of individualists should be further investigated. The timing of the public consultations may have been a factor in the disincentive for individualists and fatalists to participate. By the time the regulatory agency held these public consultations, at an advanced stage of the policy's preparation, the discussion was limited to the question of having more or fewer rules, with little margin for profound changes, such as the absence of rules [21]. It is possible to hypothesize that the later the process of participation occurs, the lower the propensity will be of a review of any position. By "late", one can understand how that process in which there is public consultation is carried out when there is already a previous positioning of the governmental body.

Through the analysis of P2, it was possible to observe that the "voices" were consistent and maintained the same arguments in both consultations. However, this finding implies that each "voice" ended up offering arguments and solutions that were restricted to their vision of the world. Governments often face "wicked problems" that require "clumsy solutions" that can only be achieved by the combination of contributions between different "points of view" [22].

The evaluation of P3 identified that despite the broad process of public participation, there was a maintenance of the point of view of a self-regulatory agency in the formulation of regulatory policies. Despite offering society members a channel of participation and having, in some measure, received some public contributions, ANVISA maintained the dominance of its "voice". According to Linsley et al. [8], when the government influences the decision-making process with its own standards and values, this is an example of self-capture.

In analyzing P4, there was a significant number of participants with a hybrid vision. The majority represented egalitarians with a secondary hierarchist vision. It is possible that the intent of egalitarians was to pass through the hierarchy in an attempt to achieve some of their objectives, considering that the advanced stage of the regulatory process and the hierarchical profile of ANVISA could lead to the denial of claims differing dramatically from the rules and limits already imposed by regulation.

The results of this paper indicate practical and theoretical contributions. The use of the Douglasian Cultural Theory in public participation has shown to be applicable and with high potential for understanding the process of policy-making. From the practical point of view, public participation through public consultation still requires more effort from public managers to offer mechanisms that foster more plural and diversified public participation. In addition, public managers should seek ways to ensure participation and to incorporate the contributions of all "voices" into public policies.

Some limitations should be highlighted in the present paper, such as the subjectivity inherent in Douglasian Cultural Theory for classification of "voices", the small number of contributions and the advanced stage of public consultation could be important issues for future research.

Furthermore, studies examining public consultations or other participatory tools, as participative budget, with more participants in other economic sectors and that preferably contemplate public participation at the beginning of the policy-making process could be useful in validating the use of Douglasian Cultural Theory to evaluate the participation of the various "voices". Other factors that may influence the engagement of the different "voices" can also be studied, such as the design of the public consultation, the timing of the public consultation, the dissemination strategy, the amount of information available, and the communication strategy used in the public consultation. In addition, studies examining the influence of one or more of these factors on the number and diversity of contributions would be particularly useful in indicating a possible path to broaden public participation and, consequently, to adopt more democratic public consultation processes.

Acknowledgments. The authors would like to thank the National Council for Scientific and Technological Development (CNPq- Brasil) - Process 402789/2015-6 for the financial support in conducting this study. Besides that, the authors are grateful to EMCIS's reviewers for their valuable contributions.

References

1. Arnstein, S.R.: A ladder of citizen participation. J. Am. Inst. Plan. **35**(4), 216–224 (1969). https://doi.org/10.1080/01944366908977225
2. Enserink, B., Koppenjan, J.: Public participation in China: sustainable urbanization and governance. Manag. Environ. Qual. Int. J. **18**(4), 459–474 (2007). https://doi.org/10.1108/14777830710753848
3. Fischer, F.: Citizen participation and the democratization of policy expertise: from theoretical inquiry to practical cases. Policy Sci. **26**(3), 165–187 (1993)
4. Reed, M.S.: Stakeholder participation for environmental management: a literature review. Biol. Cons. **141**(10), 2417–2431(2008). https://doi.org/10.1016/j.biocon.2008.07.014
5. Laird, F.N.: Participatory analysis, democracy, and technological decision making. Sci. Technol. Human Values **18**(3), 341–361 (1993). https://doi.org/10.1177/016224399301800305
6. Perez, M.A.: A administração pública democrática. Editora Forum, Belo Horizonte (2004)
7. Lodge, M., Wegrich, K., Mcelroy, G.: Dodgy kebabs everywhere? Variety of worldviews and regulatory change. Publ. Adm. **88**(1), 247–266 (2010). https://doi.org/10.1111/j.1467-9299.2010.01811.x
8. Linsley, P., McMurray, R., Shrives, P.: Consultation in the policy process: Douglasian cultural theory and the development of accounting in the face of crisis. Publ. Adm. **94**(4), 988–1004 (2016). https://doi.org/10.1111/padm.12212
9. Perri 6, Swedlow, B.: An institutional theory of cultural biases, public administration and public policy. Publ. Adm. **94**(4), 867–880 (2016). https://doi.org/10.1111/padm.12296
10. Douglas, M., Thompson, M., Verweij, M.: Is time running out? The case of global warming. Daedalus Time **1322391**(2), 98–107 (2003)
11. Spickard, J.V.: A guide to Mary Douglas's three versions of grid/group theory. Sociol. Anal. **50**(2), 151–170 (1989). https://doi.org/10.2307/3710986
12. Mamadouh, V.: Grid-group cultural theory: an introduction. GeoJournal **47**(3), 395–409 (1999). https://doi.org/10.1023/A:1007024008646
13. Bale, T.: Towards a "cultural theory" of parliamentary party groups. J. Legis. Stud. **3**(4), 25–43 (1997). https://doi.org/10.1080/13572339708420527
14. Caulkins, D.D.: Is Mary Douglas's grid/group analysis useful for cross-cultural research? Cross Cult. Res. **33**(1), 108–128 (1999). https://doi.org/10.1177/106939719903300107

15. Gil, A.C.: Métodos e técnicas de pesquisa social. 6thedn. Editora Atlas SA, São Paulo (2008)
16. Günther, H.: Qualitative research versus quantitative research: is that really the question? Psic: Teor e Pesq **22**(2), 201–210 (2006). https://www.scopus.com/inward/record.uri?eid=2-s2.0-37849185433&partnerID=40&md5=e136a24ef26d7f27a6601083dbbe9b13
17. ANVISA, Agência Nacional de Vigilância Sanitária: Relatório de análise da participação social n. 40/2016 (2016a). http://portal.anvisa.gov.br/documents/10181/2955920/Relatório+de+Análise+da+Participação+Social+%28RAPS%29+-+CP+255_2016.pdf/b2736244-3782-41cf-99bb-f230f7292528?version=1.0
18. ANVISA, Agência Nacional de Vigilância Sanitária: Relatório de análise da participação social n. 41/2016 (2016b). http://portal.anvisa.gov.br/documents/10181/2955920/Relatório+de+Análise+da+Participação+Social+%28RAPS%29+-+CP+256_2016.pdf/41097e1e-8aeb-4f81-8599-2bba18a57ec6?version=1.0
19. ANVISA, Agência Nacional de Vigilância Sanitária: Relatório de análise de contribuições em consulta pública, CP n. 255/2016 (2016c). http://portal.anvisa.gov.br/documents/10181/2955920/Relatório+de+Análise+das+Contribuições+%28CP+255%29.pdf/4afbaff7-ce19-45b4-abef-ed78529fccd7?version=1.0
20. ANVISA, Agência Nacional de Vigilância Sanitária: Relatório de análise de contribuições em consulta pública, CP n. 256/2016 (2016d). http://portal.anvisa.gov.br/documents/10181/2955920/Relatório+de+Análise+das+Contribuições+%28CP+255%29.pdf/4afbaff7-ce19-45b4-abef-ed78529fccd7?version=1.0
21. Macintosh, A.: Characterizing e-participation in policy-making. In: Proceedings of 37th Annual Hawaii International Conference on System Sciences, no. C, pp. 1–10. Computer Society Press (2004)
22. Rayner, S.: Uncomfortable knowledge: the social construction of ignorance in science and environmental policy discourses. Econ. Soc. **41**(1), 107–125 (2012). https://doi.org/10.1080/03085147.2011.637335
23. Santos, A.C., Zambalde, A.L., Veroneze, R.B., Botelho, G.A., Souza Bermejo, P.H.: Open innovation and social participation: a case study in public security in Brazil. In: Kő, A., Francesconi, E. (eds.) EGOVIS 2015. LNCS, vol. 9265, pp. 163–176. Springer, Cham (2015). doi:10.1007/978-3-319-22389-6_12
24. Martins, T.C.M., de Souza Bermejo, P.H.: Desafio de ideias para o governo aberto: o caso da Polícia Militar de Minas Gerais - Brasil. Cadernos Gestão Pública e Cidadania **21**, 303–324 (2016). http://dx.doi.org/10.12660/cgpc.v21n70.59470

Is e-Government Serving Companies
or Vice-Versa?

Anton Manfreda[✉]

Faculty of Economics, University of Ljubljana, Ljubljana, Slovenia
anton.manfreda@ef.uni-lj.si

Abstract. Information technology and systems always had an important role in organizations and presented a competitive advantage if properly used. However, in the last few years the technological progress and new business models seems to be escalating. Companies are challenged with changes in consumer behavior, emergence of new organizations with digital business models. Although public administration is treated as an area of having no competition, and it would be anticipated that these challenges are less relevant, the external environment is also forcing it into changes. These changes are seen through the increasing number of digitalized services, emphasizing the importance of bringing e-Government towards the people etc. However, it is often overlooked that digitization is not only about the technology itself, but it should include the business process management perspective as well. The purpose of this research is thus to present important concepts related to e-Government and demonstrate its efficiency through the micro company perspective.

Keywords: e-Government · Business process management · Digitalization · Case study

1 Introduction

E-business has significantly changed in the last decades. Its original role was to make business processes more efficient, and with time, it has extended to all spheres of business and personal lives. Although, information technology and systems always played an important role in organizations and presented a competitive advantage, the technological progress and new business models being developed in the last few years seem to particularly reshaping the complete industry. Changes in consumer behavior, emergence of new so called digital organizations are challenging existing companies and forcing them into changes.

Though the digitization initiative started in the 1980s it has become important over the last few years and will become even more significant in the next decades [1]. However, many organizations are having problems familiarizing with and adjusting to this new era and the same applies to public administration.

Although public administration within single country has no competition and therefore contemporary challenges may be less relevant, the environment is also forcing it into changes. Citizens are namely expecting increasing number of digitalized services and bringing e-Government services towards them.

© Springer International Publishing AG 2017
M. Themistocleous and V. Morabito (Eds.): EMCIS 2017, LNBIP 299, pp. 250–260, 2017.
DOI: 10.1007/978-3-319-65930-5_21

The paper will thus present an overview on some of the most important issues related to e-Government and a case renovating e-Government in a selected country. Given that the digitalization may be wrongly considered as merely applying a new technology into an organization without considering concept of business process management, the paper will present the digitalization of e-Government from the perspective of a micro company.

The purpose of this research is to examine the efficiency of e-Government through the micro company perspective. The main objective is thus to demonstrate the extent to which digitization in public administration is following the objectives such as reducing costs, shortening business cycles and improving the quality of public services judged by the micro company.

This paper thus start with the reviews of the latest issues and challenges that are influencing the development of e-Government, namely digitalization, cloud computing and big data. In the second part the project of renovating e-Government in a selected country is presented together with the case study of a micro company that is significantly involved in using several e-Government services. Finally, some concluding remarks are outlined together with future research possibilities.

2   Literature Overview

2.1   Digitalization and Business Process Management

Organizations are nowadays facing with the challenges that are driven by the new technologies, innovations or the advent of new online based companies. Contemporary technological trends, such as cloud computing, social media, internet of things and data analytics together with the complexity of coordinating all these aspects are bringing several new challenges. One of the main challenges of the existing companies is how to transform their business models into the digital ones.

Digitalization is becoming the most significant technological trend faced globally [2]. It is affecting individuals, organizations, communities and entire nations. Different countries, industries and sectors are varying in the level to which they are making use of new technologies and digitalization. However, digitalization offers enormous opportunities for all, yet it is also bringing new issues particularly related to the human labor.

The digital transformation era is sometimes called even as a new revolution. It is claimed that digital transformation will transform business processes, the customer experience and the entire business model and consequently improve competitive success. It has been claimed decades ago that business models from the industrial age are not suitable to deal with the upcoming challenges of the information age [3]. Nevertheless, innovative business models were always claimed to be the key reason behind the success of several corporations [4].

However, digitalization may be considered as merely applying a new technology into an organization without considering whether existing business processes are efficient enough. Different researchers already showed that organizations can improve their performance by accepting process view of their business [5–7] where possible approaches to increase process view proposed were BPM, lean manufacturing, six sigma, BPR, total quality management etc. [8]. Yet, with the digital transformation era, these concepts were put slightly behind.

There are several issues related with the digital transformation. Despite the fact that digital transformation is sometimes claimed to be just another buzzword, it is evident that the business world is abundant of important changes.

2.2   Cloud Computing

One of the main challenges in the last decade is related to the cloud computing and how to use effectively use it. Allowing access from anywhere and anytime, cloud computing successfully upgrade the initial promise of World Wide Web and it offers a new way for individuals and organizations to communicate and work by using internet services.

Cloud computing is now perceived as a global trend and has in the last decade obtain attention from both practitioner and academic communities. Although the development of cloud computing has not reached the maturity level, there is still a lack of research on it [9]. However, in the last years, the field of research moved from understanding what cloud computing is to examining its proper and efficient usage [9]. In the initial years, researchers focused more on the definition of cloud computing and namely studied how cloud computing could be applied to existing solutions. Therefore, most of the articles were dealing with its conceptualization. In the last years, the research has moved to the technical dimensions of cloud computing including architectural designs. Even though cloud computing has achieved excessive progress in the last years; it is expected to continue with its growth [10].

Cloud computing offers additional business models namely Infrastructure as a Service, Platform as a Service, and Software as a Service. Due to the emergence of these new models organizations are examining how to adapt their business structure to a particular cloud computing model [11].

The positive effects of cloud computing have also encourage to start a National computer cloud (NCC) project in selected country. The project was presented in August 2014 and it started in December 2015 while the first system, namely custom's system at the Financial Administration of the selected country was launched at the end of January 2016. NCC is intended for mainly for state institutions that are using it as an appropriate service model to fulfil their objectives, while the Ministry of Public Administration owns the infrastructure.

However, besides all positive implications of cloud computing, there are also several important issues associated with cloud computing like reliability, security, privacy, legal matters, data ownership and long-term sustainability [12].

2.3   Big Data

Big data is a concept that originated from the need of large organizations like Yahoo, Google, and Facebook to analyze large amounts of data [13]. It presents a concept that is related to the increased volume of data that are difficult to store, process and analyze using the traditional database technologies. Although, the term big data is quite new in the field of information technology, several researchers and practitioners have already used it in previous literature. It was for instance referred to a large volume of scientific data for visualization [14]. Currently, several different definitions of big data exist. It was also defined as the amount of data that is hardly efficiently store, manage and process [15]. Recently, different explanation from 3V like Volume, Variety, and Velocity to 4V like Volume, Velocity, Variety and Veracity have been offered to define big data [16, 17]. The 4V definition of big data is commonly recognized since it denotes both the meaning and necessity of big data as well.

Big data has also a large research potential and therefore it is getting substantial devotion from academia and practitioner communities. The latter is becoming even more important in the digital world where the amount of data that is generated has escalated in the last few years. Consequently, this expansion of data is offering many new challenges [18]. Walmart alone handles more than 1 million transactions every hour, and it is estimated that this value results in more than 2.5 Petabytes of data each hour.

Big data analytics helps social media and several private or public agencies to discover the behavioral patterns of people even the hidden ones [19]. Furthermore, big data analytics also enables decision makers to take a valuable decision by improved understanding of customers and products. Nonetheless, data analytics also supports acquiring the knowledge about market trends. In addition, the benefit of data analytics is also to timely identify potential risks and opportunities for an organization.

Big data and big data analytics presents enormous potential for different applications. One of the main sources for producing huge amounts of data are namely IoT, multimedia and social media. Also cloud computing and big data analytics are related since big data analytics enables users to process queries across multiple datasets and receive results in a timely manner, while cloud computing provides the underlying data processing platforms [17].

Additionally, it has been forecasted that there will be an enormous increase in demand for big data skills in the near future. It is been even expected that the increase in demand for these skills will grow by 160% in the United Kingdom alone [18].

3 Case Study

3.1 Methodology

The main research question is related to the balancing governmental digital services with the organizational ability to fulfil governmental requests. Further, the intention is to analyze reasonableness of different services and tasks provided and required by the state. Although several services are obligatory, it does not affect the research question regarding the proper administrative burden in organizations by the state.

A longitudinal case study was selected as an appropriate methodological approach. A detailed case study allows a better understanding of the research question. Since case studies dealing with digital e-governmental services are quite rare and missing, this qualitative approach was chosen to offer additional insights into the topic that is becoming more and more important.

A waste management company WasteRec (the name is fictional although the analysis is real) in a Central European country was chosen for the purpose of this research for several reasons. WasteRec is a micro organization involved in highly regulated industry. As a micro organization it has to accomplish almost the same amount of reporting tasks specified by the state; however, doing it by less employees. The glitches on both sides, governmental and organizational, are therefore even more evident in micro organizations. In large corporations either problems or system anomalies may be concealed by the higher amount of employees doing administrative staff. In micro organization the focus is on its core business; and therefore, any additional (although minor) administrative work loaded by the state has more evident influence and consequences.

The second reason for choosing WasteRec is its core business, namely waste management. The industry is highly regulated due to the environmental impact. Digitalization has important consequences even in an industry where labor work is still dominant, mainly due to the prescribed administrative activities by the public institutions.

For this case study usual techniques were applied [20, 21], namely examining internal documentation and performing interviews with decision makers. Dealing with micro organization was helpful since decision maker was highly involved in all working processes and was therefore well informed about strengths and weaknesses in WasteRec and industry related problems as well.

The case analysis was conducted between January 2017 and March 2017; however observing a company for other purposes was done even before.

3.2   Bright Side Perspective: Renovating e-Government Services

Before presenting the WasteRec view, it is important to examine the e-Government services in the country where WasteRec operates. The country received several prestigious international awards for its e-services including recognition of excellence in public administration The United Nations Public Service Award. The award was granted a decade ago for developing government-to-business (G2B) services in the category "Improving service delivery in the public sector".

Yet, quite passive role in the next years in the area of e-Government development placed the country in the lower middle of e-Government Development Index in Europe (UN EGDI). The latter is evident from the Fig. 1 since there was no changes on the e-Government portal in a decade from 2006 to 2016, while business environment, technology and user requirements and habits considerably changed.

Therefore, the project of renovating the e-Government portal took place as well. The renovated portal of e-Government in selected country was launched in November 2015 (beta version), while the final version took place in 2016. The effort for renovating e-Government services in last few years resulted in the improved ranking, namely achieving 21st place among 193 countries [22].

Despite the fact, that the new portal seems to be well accepted among users according to the number of registered users in the first year and submitted e-applications as it is evident from the Table 1, there is still more than 95% of the population in the selecting country not using any of the services. However, from the

Fig. 1. Development of e-Government portal in the selected country in the last 15 years

Table 1. Comparison of user activity on the renovated and old e-Government portal in the selected country.

	New portal (first year)	Old portal
Registered users	30.000	110.000 (in 10 years)
Submitted e-applications	30.267	20.000 (per year)
Posts on bulletin board	25.000	N/A
Announcing public events and meetings	15.000	N/A
Questions, inquiries	4.600	8.000 (per year)

Public administration perspective halving the number of questions or inquiries from different users is aligned with the promise of e-business to reduce the number of time spent on different activities.

Currently, most visited e-services on the new portal refer to obtaining a certificate of criminal record, application for registration of temporary residence, application for the basic rights of public funds (child allowance, a kindergarten, a state scholarship), subsidy for reduced kindergarten payment, while most used informative pages are related to obtaining identity card or a passport and registering a residence.

Although there are several popular applications and services at the portal, an important rethink about the citizens needs is missing. At the moment the portal enables 321 activities, 491 documents in pdf and doc format and 110 e-applications. Due to the cultural and language differences, some of them are translated into three languages.

This high number of services and documents reflects the fact that renovating public administration and the portal is often done merely as renovating a webpage, while the processes behind remained unchanged. Yet, all this numbers refer to the so called inner part of the e-government portal, while the whole public sector consists of several additional public agencies as it is evident from the next section.

3.3   User Side Perspective: Using the Services

Bearing in mind the companies and their responsibility for reporting to different state agencies for either merely the statistical purposes or other obligatory reporting reveals additional problem of increasing the number of reports that have to be submitted to several agencies. Since e-documents and applications are at least in the theory easy to manage, several agencies are yearly including new reports. It seems that with the advent of e-applications and e-documents their complexity is not reducing, causing additional obligation for companies.

WasteRec was established in 1990 starting its business in a paper age. In the first years, the majority of business was oriented mostly in obtaining new customers and suppliers. Administrative tasks were focused on accountancy merely. However after the year 2000 the regulation on the waste management field become more strict regarding obtaining permissions for doing business. Additional reporting also took place. However, the majority of reporting was still in the paper format limited to few pages. After 2010 several new regulations, laws and acts came into force. One of the first was obligatory reporting of payroll expenses several other business related payments and through the prescribed governmental portal presented in Fig. 2 below.

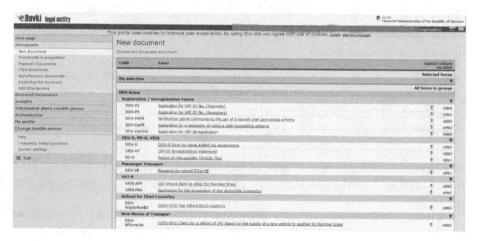

Fig. 2. Available documents in the electronic tax operation portal.

At the same time new documents were introduces by the Environment Agency, firstly in the paper form. Every single shipment has to be recorded and details like place of origin, destination place, date of shipment, mean of transport, remark and recycling method had to be included. After a year, the Agency prepared online form, as it is evident from the Fig. 3. After the form was prepared, several additional fields were included. According to the WasteRec manager, fulfilling one form take approximately 3–5 min if all the details are already prepared. According to him, the major problem lays in duplicating the entry details. Another problem related to this system is its complexity in terms of material codes set by the Agency and constantly adding new

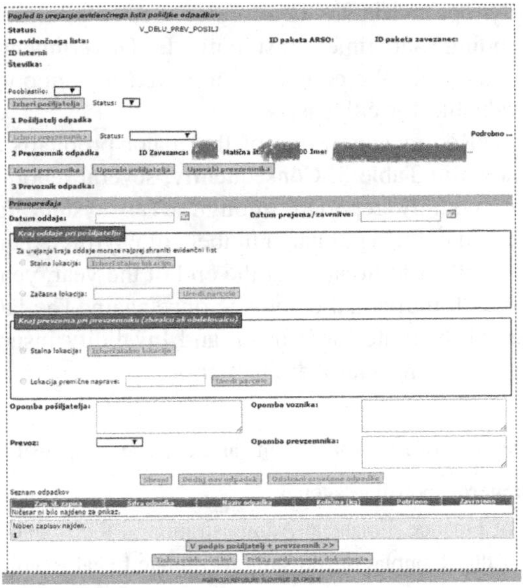

Pogled in urejanje evidenčnega lista pošiljke odpadkov
Status: V_DELU_PREV_POSILJ
ID evidenčnega lista: ID paketa ARSO: ID paketa zavezanec:
ID interni:
Številka:
Pooblastilo:
Izberi pošiljatelja Status:
1 Pošiljatelj odpadka
Izberi prevzemnika Status: Podrobno ...
2 Prevzemnik odpadka ID Zavezanca: Matična št: Ime:
Izberi prevoznika Uporabi pošiljatelja Uporabi prevzemnika
3 Prevoznik odpadka:
Primopredaja
Datum oddaje: Datum prejema/zavrnitve:
Kraj oddaje pri pošiljatelju
Za urejanje kraja oddaje morate najprej shraniti evidenčni list
Stalna lokacija:
Začasna lokacija:
Kraj prevzema pri prevzemniku (zbiralcu ali obdelovalcu)
Stalna lokacija:
Lokacija premične naprave:
Opomba pošiljatelja: Opomba voznika:
Prevoz: Opomba prevzemnika:
Seznam odpadkov
Ničesar ni bilo najdeno za prikaz.
Noben zapisov najden.
1
V podpis pošiljatelj + prevzemnik >>

Fig. 3. An example of empty form (first page) in IS-Waste system.

requirements making it impossible to automate fulfilling these forms for micro company. The latter is not a problem for large corporation having enough resources for powerful ERP software. Although, even larger companies are complaining against the system and unreal expectations and demands set by the Agency.

In the last years, the number of required activities and submitted documents online is rising although not constantly as it is evident from Table 2. It is quite clear that having more than 200 merely e-documents available (without counting paper reports) and monitoring all these documents with tracking possible changes presents an important part of daily activities not related to core business. In the table, different systems are presented with the number of offered e-documents or applications together with the number of documents that were submitted by WasteRec. In the last years the number is quite constant; however it has escalated in the last 10 years. Separate

Table 2. Number of e-documents available by different public systems and submitted by the company WasteRec.

Public system	Available (2017)	Submitted (2005)	Submitted (2014)	Submitted (2015)	Submitted (2016)
eTax	64	12	357 + 1	378 +1	213 + 1
APLRS	10	1	5 + 1	12 + 1	12 + 1
M4	1	1	1	1	1
ZZZ	14	3	6	1	1
IS-Waste	5	0	548 + 3	690 + 3	687 + 3
e-Government	110	0	0	0	0
Total	**204**	**17**	**922**	**1087**	**919**

numbers present yearly reports and are therefore not treated as ordinary documents, but rather as quite demanding, and time consuming. E- Government portal – although renovated – is not used within the company; however, it is important to add that its services are mostly valuable for citizens.

According to WasteRec manager, one of the major problems is reporting to different Agencies as shown in Table 3. Consequently, several report are duplicated. All payroll expenses are monthly reported through eTax system for each individual employee. At the same time, the aggregate numbers of payroll expenses are reported to the APLRS system as well. Additionally, at the end of the year, yearly data is reported to the M4 system, for each individual employee once again. The latter system is set to change in 2018 when the Institute for Pension and Invalidity Insurance will automatically obtain the data from Financial Administration.

Table 3. Overview of relevant systems and agencies responsible for them.

System	Administrator/responsible agency
eTax	Ministry of Finance, Financial Administration of the Republic
APLRS	Agency of the Republic of Slovenia for Public Legal Records and Related Services
M4	Institute for Pension and Invalidity Insurance
ZZZ	Health Insurance Institute
IS-Waste	Ministry of the Environment and Spatial Planning, Environment Agency
e-Gov	Ministry of Public Administration

4   Discussion

The case of WasteRec company presents an important issue in public e-service development on G2B segment. As stated by WasteRec manager the number of required reports and other administrative task is rising from year to year. 10 years ago, when most of the reporting was done in the paper format, *"all the Agencies were complaining about their shortage of employees since they had to go through all these paper forms entering the data into their own systems. Consequently, several additional employees were employed to manager all the reports"*. However, with the advent of e-services and automatic process reporting on a state level, there were several surplus employees. In order to maintain their employment additional reports were invented which resulted in additional workload passed to organization.

WasteRec manager admits that company internal information system saved a lot of time in preparing all this documents and reports; however the unstable legislation changing to often has an important impact on IS development. Although, the whole IS development is outsourced, it is still time consuming and expensive. In 2014 the Tax legislation affecting WasteRec changes twice. *"We have completely changed our IS to align it with the legislation requirements which were valid only for two months. After that, legislation was reverted into original state, since there were too many problems with it; causing us to revert to old system once again. Quite huge investment was in place only for two months"*. Yet, the amount of reporting is not the only problem.

According to WasteRec manager all these duplicated reporting clearly shows that e-Government, particularly in the area of G2B is not working as it should. This observation is aligned with the study performed a decade ago in US stating that e-government is not providing enough high-quality services to citizens, particularly if compared to e-business services [23].

It seems that several Agencies are not aware of the work process in others or there is a lack of political consensus regarding renovating the whole processes in the public administration. *"It is quite annoying to report the same data to different Agencies, especially since we don't have an employee working only with the reports. And if you have luck, you are selected at the end of the year by the Statistical Office of the Republic asking you to report everything once again."*

The paper thus presented an overview on some of the most important issues related to e-Government, namely digitalization, cloud computing and big data. The concepts related to e-Government that were important a decade ago significantly differ from the topical concepts. Digital transformation is becoming a new trend today and it applies to e-Government as well. Therefore, the paper presented a case renovating e-Government in a selected country. Since the digitalization is often done merely as applying new technology into an organization without considering the internal processes or complete business model, the paper presented the digitalization of e-Government from the perspective of a micro company.

From the described case, it is evident, that digitalization in public administration is often focused on renovation the visual part of public portals, while bureaucratic procedures and processes often remains unchained.

The paper has also some limitations. Firstly, it refers to one single central European country and one organization. Future research should include also larger corporations and their perspective on the development of e-Government services. Moreover, other countries should also be included in order to compare cultural differences and perceptions regarding e-Government. Lastly, as stated by WasteRec manager *"we know a Laffer curve is representing the relationship between rates of taxation and the government revenue, and at a certain point there is a negative effect"*, the future research should also focus on identifying the optimal level between governmental requirements and organizational ability to willingly fulfil these requirements.

5   Conclusion

The paper focused on presenting important contemporary issues that are related to e-Government, namely digitalization, cloud computing and the big data. Particularly digitalization is seen as an urge today; and therefore the paper presents the case of digitalizing e-Government services in a selected country. However, given that digitalization often refers to applying merely a new technology into existing organizations without considering the possibility of optimizing and renovating business processes, the paper also presented the perspective of a micro company as a user of these digitalized services. The comparison of offered e-services, required e-services and used e-services revealed several problems and issues that are related to digitalizing existing e-governmental services and exposed the need for coordinating government actions.

References

1. Gerth, A.B., Peppard, J.: The dynamics of CIO derailment: how CIOs come undone and how to avoid it. Bus. Horiz. **59**, 61–70 (2016)
2. Leviäkangas, P.: Digitalisation of Finland's transport sector. Technol. Soc. **47**, 1–15 (2016)
3. Venkatraman, N., Henderson, J.C.: Real strategies for virtual organizing. MIT Sloan Manag. Rev. **40**, 33 (1998)
4. Afuah, A.: Business models: a strategic management approach (2004)
5. Davenport, T.H.: Process Innovation: Reengineering Work Through Information Technology. Harvard Business Press, Boston (1993)
6. Sidorova, A., Isik, O.: Business process research: a cross-disciplinary review. Bus. Process Manag. J. **16**, 566–597 (2010)
7. Harmon, P.: Business Process Change: A Manager's Guide to Improving, Redesigning, and Automating Processes. Morgan Kaufmann Publishers, Amsterdam (2003)
8. Ramirez, R., Melville, N., Lawler, E.: Information technology infrastructure, organizational process redesign, and business value: an empirical analysis. Decis. Support Syst. **49**, 417 (2010)
9. Bayramusta, M., Nasir, V.A.: A fad or future of IT?: a comprehensive literature review on the cloud computing research. Int. J. Inf. Manag. **36**, 635–644 (2016)
10. Wang, H.H.: Study of data security based on cloud computing. Adv. Mater. Res. **756–759**, 1097–1100 (2013)
11. Sultan, N.: Cloud computing: a democratizing force? Int. J. Inf. Manag. **33**, 810–815 (2013)
12. Singh, A.K., Mishra, R., Ahmad, F., Sagar, R.K., Chaudhary, A.K.: A review of cloud computing open architecture and its security issues. Int. J. Sci. Technol. Res. **1**, 65–67 (2012)
13. Garlasu, D., Sandulescu, V., Halcu, I., Neculoiu, G., Grigoriu, O., Marinescu, M., Marinescu, V.: A big data implementation based on grid computing. In: 2013 11th RoEduNet International Conference (RoEduNet), pp. 1–4. IEEE (2013)
14. Cox, M., Ellsworth, D.: Managing big data for scientific visualization. In: ACM SIGGRAPH, pp. 146–162 (1997)
15. Manyika, J., Chui, M., Brown, B., Bughin, J., Dobbs, R., Roxburgh, C., Byers, A.H.: Big data: the next frontier for innovation, competition, and productivity (2011)
16. Gandomi, A., Haider, M.: Beyond the hype: big data concepts, methods, and analytics. Int. J. Inf. Manag. **35**, 137–144 (2015)
17. Hashem, I.A.T., Yaqoob, I., Anuar, N.B., Mokhtar, S., Gani, A., Khan, S.U.: The rise of "big data" on cloud computing: review and open research issues. Inf. Syst. **47**, 98–115 (2015)
18. Yaqoob, I., Hashem, I.A.T., Gani, A., Mokhtar, S., Ahmed, E., Anuar, N.B., Vasilakos, A. V.: Big data: from beginning to future. Int. J. Inf. Manag. **36**, 1231–1247 (2016)
19. Raghupathi, W., Raghupathi, V.: Big data analytics in healthcare: promise and potential. Health Inf. Sci. Syst. **2**, 1 (2014)
20. Yin, R.: Case Study Research: Design and Methods, 3rd edn. SAGE Publications, Thousand Oaks (2003)
21. Eisenhardt, K.M.: Building theories from case study research. Acad. Manag. Rev. **14**, 532–550 (1989)
22. UN: E-Government Survey 2016. United Nations, New York (2016)
23. Morgeson, F.V., Mithas, S.: Does e-Government measure up to e-Business? Comparing end user perceptions of US Federal Government and e-Business web sites. Publ. Adm. Rev. **69**, 740–752 (2009)

Public Transparency in the Brazilian Context: An Integrative Review

Pedro de Barros Leal Pinheiro Marino[✉],
Grazielle Isabele Cristina Silva Sucupira,
Wender Rodrigues de Siqueira,
and Paulo Henrique de Souza Bermejo

Universidade de Brasília, Brasília, Brazil
pedroblpmarino@gmail.com,
graziellesucupira@gmail.com,
wendersiqueira@gmail.com, paulobermejounb@gmail.com

Abstract. The present work reviews literature on public transparency in Brazil published from 1990 to 2017. The objectives were to identify different forms of transparency, identify the outcomes of public transparency, and which outcomes were achieved. The research involved five electronic databases and did not establish search limitations related to journal status or research design. The results show growing academic interest in the subject; the predominance of empirical research on municipal and executive entities; focus on budget transparency and evaluation of electronic websites; and limited research on the outcomes of public transparency. Theoretical researches about Brazil evaluate transparency's outcomes better than researches from other countries, suggesting that different contexts lead to different interpretations of its impact. Although public transparency is part of the Brazilian research agenda, empirical research must improve its methodological approach, address other forms of transparency, and investigate the outcomes of public transparency.

Keywords: Open government · Public transparency · E-government · Systematic literature review

1 Introduction

Public transparency can have a range of effects, such as promoting citizens' trust in their government, reducing public corruption, and improving financial performance [1–4]. The topic gained popularity through a normative prism, especially in regard to government abuse, democracy strengthening, and the emergence of new technologies, becoming a unanimous concept [5]. However, despite its popularity, research on transparency has failed to produce tools for measuring, evaluating, and comparing public transparency practices as well as investigating the determinants of transparency initiatives' success [6].

The public transparency debate can be divided into two related streams focusing on different aspects of the government [7]: (1) politics, especially how and when transparency contributes to the amount of democracy in a government, and (2) administration, especially how and when transparency contributes to executive competence.

© Springer International Publishing AG 2017
M. Themistocleous and V. Morabito (Eds.): EMCIS 2017, LNBIP 299, pp. 261–274, 2017.
DOI: 10.1007/978-3-319-65930-5_22

Although the issue has received much attention in recent years, there is no conclusive definition of public transparency [8]. In this work, public transparency is broadly defined as the extent to which external actors are afforded access to information about how public organizations function [9].

Due the variety of contexts (political, administrative, institutional, cultural, and demographic) in which public transparency occurs, evaluating it could be problematic [7]. For example, although some studies found a positive relationship between transparency and less corruption [2, 10, 11], others indicate that this occurs only when certain conditions are met, such as sufficient education and free press [12]. In countries with endemic corruption, transparency can erode institutional confidence and lead to public resignation due to the belief that corrupt practices are impossible to contain [13].

The importance of context in evaluations of transparency is reinforced by Cucciniello et al. [9], who point out that further studies of transparency in Africa and Latin America must be conducted in order to understand how the results of such studies vary in different contexts. The relevance of such studies is reinforced by the low level of corruption control [14] and high perception of corruption [15] in those regions.

Although few Latin American countries were able to approve federal laws guaranteeing access to public information in the past, similar initiatives have had varying levels of success in recent years [16]. For example, in Brazil, this subject has received attention from academics, experts, the media, and the government in recent years [17]. The right to access public information was guaranteed by the Brazilian Federal Constitution of 1988 as well as the Fiscal Responsibility Law, Transparency Law, and Information Access Law [18, 19]. According to da Costa Bairral et al. [20], despite Brazilian researchers' increased interest in this topic, little is known about the level of public disclosure of information in Brazil.

To better understand Brazilian public transparency, literature published from 1990 to 2017 is reviewed in this study. This work is guided by three research questions: What forms of public transparency are identified in the literature? What outcomes has the literature attributed to public transparency? What outcomes does public transparency tend to be (un)successful in achieving?

The study is divided into five sections. In the next section, the methodology of this study is explained. Then, the results of the literature review are presented. Next, the results and conclusions are discussed, followed by the references.

2   Research Method

This paper reviews the literature on public transparency in Brazil between 1990 and 2017. This period was chosen as it aligns with Cucciniello et al.'s [9] study, which examined studies published from 1990 to 2015 and identified a gap in public transparency studies in Latin America. This review is necessary to determine whether this subject was discussed previously and, if so, how it was approached. An integrative review method was used because it allows researchers to analyze knowledge obtained in previous research [21].

Five electronic databases were used: Scielo, Spell, Redalyc, Web of Science, and Science Direct. The search terms included "government transparency," "public transparency," "public disclosure," "public information transparency," "public information disclosure," "transparency and government," "public sector transparency," "administrative transparency," "transparent government" (and its Portuguese translation), and "Brazil." These terms were searched across all search indexes. There were no limitations concerning publication status or research design. We searched for works published from 1990 to 2017. The database searches and analysis were carried out in May 2017.

2.1   Review Method and Coding

Following the procedures indicated in the previous section, we found 2,689 articles. In order to identify articles with direct relevance to the topic under investigation, the integrative review process described by Botelho et al. [21] was used. First, the search was conducted in two Brazilian databases (Spell and Scielo) and three worldwide databases (Redalyc, Science Direct, and Web of Science) using the descriptors indicated above. Second, repeated articles at each base were eliminated. Third, the titles, abstracts, and keywords of studies were read to verify whether the articles dealt with public transparency in Brazil, and studies that did not fulfill that criterion were excluded from the sample. In case of doubt, the full text of the article was read. Articles were excluded mostly because transparency was not the main subject, instead using the search terms in a broad sense (i.e., "governments must have transparency"), dealing with countries other than Brazil, or only mentioning the search terms in the titles of their references. The number of articles found in each database after exclusion is indicated in Table 1. After consolidating the five databases, 129 articles were identified as relevant. However, there were 31 repetitions, resulting in a total of 98 articles dealing with transparency in the Brazilian context that underwent analysis.

Table 1. Database articles

Database	Articles	Articles without repetition	Articles after screening
Redalyc	304	252	46
Spell	662	310	58
Scielo	444	287	17
Science direct	1265	972	3
Web of science	14	12	5
Total	2689	1833	129

Studies were coded according to the following criteria: author(s), publication year, title, journal, level of government, geographical context, sphere of government, field of occupation, research method, research objectives, form of transparency addressed, and effects of transparency. Table 2 outlines the criteria in greater detail.

The authors classified all studies according to these criteria. This procedure was used to ensure reliability in terms of how each study was coded (according to the

Table 2. Extraction form category

Category	Extraction
Author, publication year, title, journal	Article's bibliographic data
Level of government	Municipal, state, or national
Geographical context	Countries in which the study is located
Sphere of government	Executive, legislative, or judicial
Research method	Theoretical or empirical (qualitative, quantitative, experimental)
Research objectives	Exploratory, descriptive, or explanatory
Form of transparency	According to Cucciniello et al. [9], transparency can be administrative, political or budget-related, and it can be involved in decision-making, policy-making, and policy outcomes
Outcomes of transparency	According to Cucciniello et al. [9], transparency can have effects on both citizens and the government: Effects on citizens: legitimacy, participation, trust in government, satisfaction Effects on government: accountability, decreased corruption, improved performance, improved decision-making process, improved financial management, increased collaboration between governments

research design used, the form of transparency addressed, and the transparency outcomes identified). If there were doubts concerning an article's codification, all authors reviewed it and came to a consensus.

3   Results of the Integrative Review

The 98 articles identified in the databases were published in 61 journals. Of these journals, 43 only published one article, 10 published two, and three published three. The remaining articles are distributed among five journals: seven articles in *Revista de Administração Pública* (ISSN 0034-7612), five articles in *Revista Catarinense de Ciência Contábil* (ISSN 1808-3781), five articles in *Revista Contemporânea de Contabilidade* (ISSN 2175-8069), five articles in *Revista do Serviço Público* (ISSN 2357-8017), and four articles in *Enfoque: Reflexão Contábil* (ISSN 1984-882X). The distribution of publications according to Qualis—the classification of journals defined by Coordination for the Improvement of Higher Education Personnel (CAPES) related to the Brazilian Ministry of Education—was as follows: A2 periodicals published 19 articles on public transparency in Brazil; B1 published 32; B2 published 20; B3 published 16; B4 published 6; and journals that could not be classified published 5 articles. The Qualis classification is divided in eight groups (by order of quality), they are: A1, A2, B1, B2, B3, B4, B5, and C. The publications were published in various types of journals, including those focused on government administration, accounting sciences, and the public sector (law, social sciences, or public service).

The production on the theme was carried out by 221 researchers. Table 3 presents the 27 authors who produced more than one work. The number in parentheses represents the number of articles for which they were the first author. Only 9 of the 27 authors published more than one article for which they were the first author, corresponding to 23% (23) of the identified articles. These researchers are associated with nine different Brazilian institutions: UFSC (Orion Augusto Platt Neto), UDESC (Fabiano Maury Raupp), UFV (Robson Zuccolotto), IFRN (Fábia Jaiany Viana de Souza), São Paulo City Hall (Otávio Prado), UFRJ (Cláudia Ferreira Cruz), UFF (José Maria Jardim), UnB (José Matias Pereira), and CGU-RJ (Temístocles Murilo de Oliveira Júnior).

Table 3. Authors associated with the most publications

Authors	Number of articles
Orion Augusto Platt Neto (4)	7
Flávio da Cruz (0)	5
Fabiano Maury Raupp (3); Marco Antônio Carvalho Teixeira (0); Robson Zuccolotto (3)	4
Aneide Oliveira Araújo (0); Fábia Jaiany Viana de Souza (3); Lino Martins da Silva (0); Maurício Corrêa da Silva (0); Maurício Vasconcellos Leão Lyrio (1); Otávio Prado (2)	3
Claudia da Silva Jordão (0); Cláudia Ferreira Cruz (2); Ernesto Fernando Rodrigues Vicente (1); Georgete Medleg Rodrigues (0); Joel de Lima Pereira Castro Junior (0); José Antônio Gomes de Pinho (0); José Dionísio Gomes da Silva (0); José Maria Jardim (2); José Matias Pereira (2); Leonardo Ensslin (0); Marie Anne Macadar (1); Mario Vinícius Claussen Spinelli (0); Rogério João Lunkes (0); Sandra Rolim Ensslin (0); Temístocles Murilo de Oliveira Júnior (2); Valmor Slomski (1)	2

Note: The number in parentheses after each author's name represents the number of publications for which they were the first author.

Most of the articles were published in Portuguese in Brazilian journals. One was published in Spanish and five were published in English in journals from other countries. However, the authors of those works were Brazilian.

Figure 1 shows the distribution of publications on public transparency in different time periods between 1990 and 2017. From 1990 to 2004, only two articles were published, both of which used a theoretical approach to discuss the role of transparency in the public sector [22, 23]. In the following five years (2005–2009), scientific investigation into the subject increased substantially, with 20 articles identified. The apex of production was between 2010 and 2016, with 21 articles published in 2016 alone. In 2017, no works on public transparency in Brazil were published until the beginning of this research.

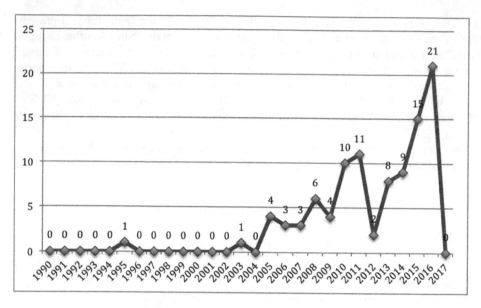

Fig. 1. Public transparency publications between 1990 and 2017

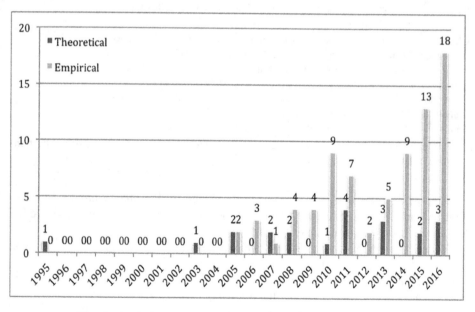

Fig. 2. Analytical methods used in transparency research by year

Most articles were classified as empirical (77), with significantly fewer classified as theoretical (21) (Fig. 2). Although initially research focused on the conceptual aspects of transparency, more recent studies have focused on empirical research; the number of empirical studies has constantly grown, while the number of theoretical studies has remained stagnant.

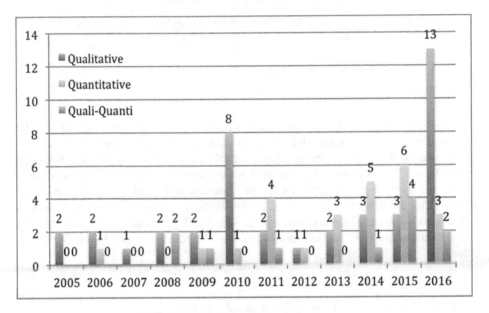

Fig. 3. Types of empirical studies by year

Figure 3 presents the methods used in empirical research, demonstrating that most studies are qualitative (41), followed by quantitative (25) and qualitative-quantitative (11). Regarding their objectives, most studies are descriptive (51), and fewer are exploratory (35) and explanatory (4) (Table 4).

Table 4. Research objectives

Objectives	Number of articles
Exploratory	24
Exploratory, descriptive	11
Descriptive	38
Descriptive, explanatory	2
Explanatory	2
Total	77

Regarding the level of government (Table 5), 45 studies focused on public transparency at the municipal level, 22 focused on the national level, 17 focused on the state level, and 14 studies either addressed public transparency in a general sense or addressed transparency at more than one level of government.

Table 5. Level of government studied

Level of government	Number of articles
Municipal	45
State	17
National	22
Nonspecific	14

Regarding the geographic area of study (Table 6), most studies were not specific about the region of Brazil on which they focused, either because they could be classified as national surveys or because they were related to broad concepts or entities that are not limited by geography. The studies that were specific (even if they included more than one region) were concentrated in the southern and southeastern regions. Few works focused on the northeastern and center-western regions of Brazil, and no published works focused on the northern region. Specifically, the studies were concentrated in the Brazilian states of Santa Catarina (14), Rio Grande do Sul (9), Minas Gerais (7), Paraná (7), and São Paulo (6).

Table 6. Geographical area of study

State	Number of articles
Distrito Federal	2
Espírito Santo	5
Goiás	1
Minas Gerais	7
Pernambuco	2
Paraná	7
Rio de Janeiro	5
Rio Grande do Norte	3
Rio Grande do Sul	9
Santa Catarina	14
São Paulo	6
Nonspecific	55

Note: Because some studies investigated transparency in more than one state, the sum of the right column is larger than the total number of articles in this literature review.

Table 7 presents the articles that specifically analyze a certain sphere of government (executive, legislative, or judicial). Most papers focused on the executive branch of government (65) or were not specific (27). Few papers focused on the legislative (4) and judicial branches (2).

Table 7. Sphere of government

Sphere of Power	Number of articles
Executive	65
Legislative	4
Judicial	2
Nonspecific	27

3.1 Forms of Transparency Identified in the Literature

Due to the diverse types of information that can promote transparency in the public sector, Cucciniello et al.'s [9] framework was used to systematize the results. This framework identifies three forms of transparency according to the object of transparency. Budget transparency refers to the disclosure of information about the government's financial situation and how public resources are used. Administrative transparency relates to the disclosure of information related to the administrative or bureaucratic activities of public organizations. Political transparency relates to information about elected representatives. Transparency can be classified according to the public sector activities to which they refer, such as the decision-making process, formulation of policy content, and policy outcomes.

Table 8 shows that 66 articles can be classified according to the proposed framework, and the others (32) were not specific about the form of transparency. Most studies focused on the object of transparency (59 articles), especially budget transparency (43 studies) and administrative transparency (15 articles). Only one study focused on political transparency, and seven focused on the transparency of public sector activities. In general, the articles focused on verifying the level of information dissemination on the institutions' websites, analyzing compliance with legal obligations and recommendations, and identifying the degree of transparency using various measurement models.

3.2 Effects Attributed to Public Transparency

This section presents the effects of public transparency identified in previous research. Many empirical studies focused on entities' level of transparency but did not address the relationship between transparency and its expected effects. Most of the studies examining the effects of transparency are theoretical and identify the positive relationship between transparency and social control by citizens (7), citizen participation (6), reduction of government corruption (5), government performance (2) and institutions' legitimacy in society (1).

Table 8. Forms of transparency identified in the literature

Forms of transparency	Number of articles
Type of transparency	
Administrative	15
Budget	43
Political	1
Transparency of activities in the public sector	
Decision-making process	1
Formulation of policy content	2
Policy outcomes	4
Unspecified/general	32

Three empirical papers analyzed the relationship between transparency and its expected effects, with one finding a positive relation with social development [24], one finding a positive relation with accountability [25], and one finding a correlation with irregularities in public administration [26]. In the latter study, no sophisticated statistical technique beside correlation was used to identify a causal relationship between the variables (Table 9).

Table 9. Effects attributed to public transparency

Effects on...	Theoretical	Empirical	Effect
...citizens/society	14	1	
Legitimacy	1		+
Participation	6		+
Social development		1	+
Social control	7		+
...government	7	2	
Accountability		1	+
Performance	2		+
Irregularity		1	Mixed
Less corruption	5		+

4 Discussion

This work emphasizes the increase in research on public transparency in Brazil. The research has been published in several administration, accounting sciences, and public sector journals, with Qualis classifications ranging from A2 to B4. Several authors focused on public transparency, but few published multiple studies about the subject.

Most articles included in this study were empirical in nature. However, several theoretical articles were published during the period under investigation, which indicates consistent uncertainty about what transparency is and what it can do [9]. As for the methods, qualitative studies with exploratory and/or descriptive objectives are most common, in line with the results of other similar literature reviews [27, 28]. When compared with worldwide publications on public transparency, it is apparent that publications about public transparency in Brazil are significantly less experimental [9]. Quantitative research does not use robust statistical techniques, instead measuring just the level of transparency. Few studies use transparency as an independent variable.

Most studies focused on the southern and southeastern regions of Brazil, and none focused on the north, which is a research gap. Regarding the sphere of government analyzed, most research focused on executive transparency. Therefore, future research should investigate transparency in the legislative and judicial branches of government.

Although transparency can be considered as a broad concept, many studies focus on information disclosure, fiscal responsibility, and laws granting the public access to information. A small number of papers analyzed transparency related to education—specifically, university management [29–32].

Most studies analyzed budget transparency, similar to international research [9]. This can be attributed to the ease with which accounting and financial data can be accessed and handled. Few studies investigated the relationship between public transparency and its outcomes. As transparency was used as a dependent variable and most papers exclusively investigate the level of public transparency, few empirical studies attempted to establish a relationship between transparency and its outcomes on society and public administration. Most research that shows these relationships was theoretical in nature. New outcomes of transparency not mentioned by Cucciniello et al. [9] were identified, including social development, social control, and irregularities in public administration.

Comparison of the results of international [9] and Brazilian research concerning the outcomes of public transparency revealed that international theoretical studies found mixed effects of public transparency on citizens, while Brazilian studies found only positives effects. The same was true for effects on the government, except for irregularity, which has a mixed effect. These results show that, in the Brazilian context, transparency outcomes tend to be better evaluated by theoretical research than international research.

5 Conclusions

This research analyzed research on public transparency in Brazil published between 1990 and 2017 to identify the forms of transparency mentioned in the literature, the outcomes attributed to transparency, and which outcomes tend to be achieved. In general, studies assessed institutions' disclosure of information on their websites, analyzed adherence to legal obligations or recommendations, verified the level of public transparency, and did not address the relationship between transparency and its outcomes.

Although the right to access public information in Brazil was guaranteed by the 1988 Federal Constitution [18, 19], this work reveals that little academic attention was paid to public transparency in the first few years after that law was passed. From 2005 onwards, there was a substantial increase in research related to public transparency, reaching its apex between 2010 and 2016, with 21 articles published in 2016 alone. In line with the findings of Medeiros et al. [17], scholars continue to be interested in public transparency in Brazil.

Although at the beginning of the period under investigation, articles primarily focused on the conceptual aspects of transparency, today they are focused on empirical analysis. The present review points out that the literature on public transparency in Brazil is concentrated on municipal government entities and public transparency models based on websites and fiscal reports.

Many Brazilian empirical studies do not address the relationship between transparency and its outcomes, contrary to international literature [9]. These relationships were addressed primarily in theoretical studies, which emphasized social control by citizens, citizen participation, reduction in government corruption, government performance, and institutions' legitimacy in society. In addition, theoretical researches about the Brazilian context have a better evaluation of transparency´s outcomes than international researches, suggesting that different contexts lead to different interpretations of its impact.

Further empirical studies focusing on Brazil must improve its methods, analyze other forms of transparency, and address the relationship between transparency and its outcomes concerning the government and society.

This research is limited because the research terms used may not have fully covered the scope of public transparency in Brazil. Future research must analyze citizens' perception of the degree of transparency of public entities and investigate reasons for compliance (or lack of compliance) with legislation guaranteeing freedom of information since most previous studies have focused on the level of transparency on the institutions' websites. In addition, future studies should fill the gap in the research concerning public transparency in the northern region of Brazil. Also, it is necessary to analyze e-government as an instrument for improving public transparency and promoting democratization and interaction between citizens and the government in the context of representative democracy.

Acknowledgments. The authors thank the National Council for Scientific and Technological Development (*CNPq, Brazil*) – Process 402789/2015-6 for their support of this research.

References

1. Benito, B., Bastida, F.: Budget transparency, fiscal performance, and political turnout: an international approach. Public Adm. Rev. **69**(3), 403–417 (2009)
2. Bertot, J.C., Jaeger, P.T., Grimes, J.M.: Using ICTs to create a culture of transparency: e-government and social media as openness and anti-corruption tools for societies. Gov. Inf. Q. **27**(3), 264–271 (2010)
3. Welch, E.W., Hinnant, C.C., Moon, M.J.: Linking citizen satisfaction with e-government and trust in government. J. Public Adm. Res. Theory **15**(3), 371–391 (2005)
4. Worthy, B.: More open but not more trusted? The effect of the Freedom of Information Act 2000 on the United Kingdom central government. Int. J. Policy Adm. Inst. **23**(4), 561–582 (2010)
5. Mabillard, V., Zumofen, R.: The complex relationship between transparency and accountability: a synthesis and contribution to existing frameworks. Public Policy Adm. **32**(2), 1–19 (2016)

6. Cruz, N.F., Tavares, A.F., Marques, R.C., Jorge, S., Sousa, L.: Measuring local government transparency. Public Manage. Rev. **18**(6), 1–28 (2015)
7. Meijer, A., Hart, P., Worthy, B.: Assessing government transparency: an interpretive framework. Adm. Soc. 1–26 (2015)
8. Raupp, F.M., Pinho, J.A.G.: Review of passive transparency in Brazilian city councils. Rev. Adm. **51**(3), 288–298 (2016)
9. Cucciniello, M., Porumbescu, G.A., Grimmelikhuijsen, S.: 25 years of transparency research: evidence and future directions. Public Adm. Rev. **77**(1), 32–44 (2017)
10. Cruz, N.F., Marques, R.C.: Scorecards for sustainable local governments. Cities **39**, 165–170 (2014)
11. Cuillier, D., Piotrowski, S.J.: Internet information-seeking and its relation to support for access to government records. Gov. Inf. Q. **26**(3), 441–449 (2009)
12. Lindstedt, C., Naurin, D.: Transparency is not enough: making transparency effective in reducing corruption. Int. Polit. Sci. Rev. **31**(3), 301–322 (2010)
13. Bauhr, M., Grimes, M.: Indignation or resignation: the implications of transparency for societal accountability. Governance **27**(2), 291–320 (2014)
14. World Bank: Worldwide Governance Indicators (WGI) project. Interactive Data Access (2017)
15. Transparency International: Corruption Perceptions Index 2016 (2017)
16. Pinto, J.G.: Transparency policy initiatives in Latin America: understanding policy outcomes from an institutional perspective. Commun. Law Policy **14**(1), 41–71 (2009)
17. Medeiros, S.A., Magalhães, R., Pereira, J.R.: Lei de Acesso à Informação: Em busca da transparência e do combate à corrupção. Informação Informação **19**(1), 55–75 (2014)
18. Abdala, P.R.Z., Torres, C.M.S.O.: A Transparência como Espetáculo: Uma análise dos portais de transparência de estados brasileiros. Adm. Pública e Gestão Soc. **8**(3), 147 158 (2016)
19. Silva, D.J.M., Segatto, J.A.C., Silva, M.A.: Disclosure no serviço público: análise da aplicabilidade da lei de transparência em municípios Mineiros. Rev. Catarinense da Ciência Contábil **15**(44), 24–36 (2016)
20. Bairral, M.A.C., Silva, A.H.C., Alves, F.J.S.: Transparência no setor público: uma análise dos relatórios de gestão anuais de entidades públicas federais no ano de. Rev. Adm. Pública **49**(3), 643–675 (2015)
21. Botelho, L.L.R., Cunha, C.C.D.A., Macedo, M.: O método da revisão integrativa nos estudos organizacionais. Gestão e Soc. **5**(11), 1–16 (2011)
22. Jardim, J.M.: A face oculta do leviatã: gestão da informação e transparência administrativa. Rev. do Serviço Público **119**(1), 137–152 (1995)
23. Pereira, J.M.: Reforma do estado e controle da corrupção no Brasil. Rev. Adm. Mack. **4**(1), 39–58 (2003)
24. Abreu, W.M., Gomes, R.C., Alfinito, S.: Transparência fiscal explica desenvolvimento social nos estados Brasileiros? Soc. Contab. e Gestão **10**, 54–69 (2015)
25. Filho, J.R.F., Naves, G.G.: A contribuição do Sistema Integrado de Administração Financeira do Governo Federal (SIAFI) para a promoção da accountability horizontal: a percepção dos usuários. Braz. Bus. Rev. **11**(61), 1–28 (2014)
26. Casalecchi, A.R.C., Oliveira, E.M.: As auditorias da CGU e a transparência licitatória dos municípios paulistas. Cad. Gestão Pública e Cid. **15**(56), 49–62 (2010)
27. Roza, M.C., Machado, D.G., Quintana, A.C.: Análise bibliométrica da produção científica sobre contabilidade pública no Encontro de Administração Pública e Governança (ENAPG) e na Revista de Administração Pública (RAP), no período 2004-2009. Contexto **11**(20), 59–72 (2011)

28. Ribeiro, H.C.M.: Doze anos de estudo da Revista de Administração Pública à luz da Bibliometria e da Rede Social. Rev. Ciências Adm. **20**(1), 137–167 (2014)
29. Bezerra, R.O., Borges, L.J., Valmorbida, S.M.I.: Análise das prestações de contas na internet da Universidade do Estado de Santa Catarina. Rev. Gestão Univ. na América Lat. **5**(1), 66–82 (2012)
30. Platt Neto, O.A., Cruz, F.: Proposta de um Painel de Informações Sintéticas sobre as Universidades: Aplicação do 'Raio-X' na Universidade Federal de Santa Catarina. Rev. Contemp. Contab. **1**(12), 109–126 (2009)
31. Platt Neto, O.A., Cruz, F., Vieira, A.L.: A evolução das práticas de uso da internet para divulgação das contas públicas na Universidade Federal de Santa Catarina. Gestão Univ. na América Lat. **3**(1), 1–14 (2010)
32. Platt Neto, O.A., Cruz, F., Vieira, A.L.: Transparência das contas públicas: um enfoque no uso da internet como instrumento de publicidade na UFSC. Rev. Contemp. Contab. **3**(5), 135–146 (2006)

e-Government and e-Inclusion.
The Role of the Public Libraries

Mariusz Luterek[✉]

Faculty of Journalism, Information and Book Studies,
University of Warsaw, Warsaw, Poland
m.luterek@uw.edu.pl

Abstract. The aim of this paper is to present and discuss opinions of the librarians on the role of public libraries as intermediaries in access to public information and services in Poland. The assumption is that Polish public libraries, following the example of United States and United Kingdom, could fulfill the role of intermediaries in access to e-government, empowering citizens and promoting e-inclusion. The data collection was executed through an online survey, which was subjected to libraries in three voivodeships.

Keywords: Public information · Public services · e-Government · Public libraries · Poland

1 Introduction

Transformational, digital, or electronic, government (e-government) is the use of information and communication technologies in providing information and services by public authorities. The concept itself was named in the mid-1990s, as the result of introducing government to the World Wide Web, although computers were used in public sector much earlier [2]. Since then the understanding of what e-government is has changed significantly. At first, it was "a tool for information and service provision to citizens", then a way to "enhances the capacity of public administration through the use of ICTs to increase the supply of public value", and "continuous innovation in the delivery of services, public participation and governance through the transformation of external and internal relationships by the use of information technology", to finally be understood as "use and application of information technologies in public administration to streamline and integrate workflows and processes, to effectively manage data and information, enhance public service delivery, as well as expand communication channels for engagement and empowerment of people" [3]. That means, that at the beginning the implementation of e-government was supply driven, then it switched to citizen centric approach, to finally involving different groups of stakeholders in planning and improving e-government solutions [4].

Surprisingly, public authorities often forget that in some social groups access to ICTs is limited to the point of digital divide, which refers not only to technology, but also ICT literacy (using digital technology, communication tools, and/or networks appropriately to access, manage, integrate, evaluate, and create information).

© Springer International Publishing AG 2017
M. Themistocleous and V. Morabito (Eds.): EMCIS 2017, LNBIP 299, pp. 275–283, 2017.
DOI: 10.1007/978-3-319-65930-5_23

That's why e-inclusion is necessary in order for people to be able to use and benefit from e-Government applications [5]. Nowadays, as many authors discuss, e-government and e-inclusion should be seen as two parallel processes of government intervention to support a socially inclusive development strategy [6].

Much works related to the problem of e-inclusion in the case of e-government have been published. That includes problems like: e-inclusion gaps [8], digital literacy [9], or e-exclusion of certain groups [10, 11], among others.

Quite naturally public libraries emerged as intermediaries in access to public information and services in many countries, especially in United States and United Kingdom. That's why, looking for inspirations on how to address the problem of e-inclusion in case of e-government, one could take a look at American the system of Federal Depository Libraries (FDLs), established in 1962. It's goal was, and still is, to facilitate access to public records, however, since early 1990-ties, main weight of their activities has shifted from paper to electronic documents [1]. In a survey from 2011 a new digital landscape was described: 82.4% of public libraries reported that providing access to government websites and services is important; 91.8% of them were helping people understand and use government websites; 96.6% - helping people in applying for e-government services; and 70.7% - helping people in completing e-government forms [7]. However, as research done by Information Policy & Access Center shows, there are some common obstacles, which libraries meet in trying to provide access to public information and services: too few workstations to meet patron demand; workstation time limits that do not allow enough time for patrons to complete their e-government forms, seek government information, among others [12].

In Poland, since 2001, when Freedom of Information act was established, we could observe a very irregular growth of e-government. The rise of regional e-services portals and their nation-wide equivalent (ePUAP), the re-do of PESEL registry and, as a result, introduction of new eID, the enhancement of online presence of public offices – all led to bringing Polish public sector to information society. It became important to ensure that people at risk of digital exclusion have appropriate opportunities to enter into relations with the public sector, for example by taking on the mediating functions by the public libraries.

2   The Aim and Research Method

The aim of this paper is to present results of a survey on opinions of the librarians on the role of public libraries as intermediaries in access to public information and services. Data collection was executed through a questionnaire, which was subjected to libraries in three voivodeships: Lesser Poland and Świętokrzyskie (these are the provinces of respectively the biggest and the smallest development of Internet access in households) [13], as well as Mazovian Voivodeship (chosen because of its internal differentiation). All municipal and rural-municipal public libraries in these provinces received invitations to participate in the study. Libraries which agreed to participate in the study were asked to complete a questionnaire, available in both paper and electronic form (the suggestion was for the questionnaire to be filled by the most knowledgeable employee). The main objective of the survey was to investigate the possibility of performing by a given library the role of mediator in access to information and public services.

The survey was divided into 6 sections: (1) Imprint (10 questions concerning the library name, its location, and the area, the number of users and workstations); (2) Familiarity with the e-government (one matrix question); (3) Library and mediation in access to public services (in total 7 questions, including one open); (4) Library and mediation to public information (5 questions, including one open); (5) Library as an institution cooperating with other public institutions (7 questions, including 3 open); (6) Conclusion and closing remarks (1 open question). In addition, each question was accompanied by the field for adding an open comment. Specific questions within each section are discussed in respective sections of this paper.

It is important to clarify, that this paper is based on much broader project, which was financed by the National Science Center Poland, through grant based on the Decision No. DEC-2011/03/D/HS2/01124. Its goal was to analyze all sides of e-government-library-citizens triangle, and relations between them.

3   Findings

The study involved a total of 112 libraries from three voivodships (Lesser Poland, Mazovian and Świętokrzyskie), where positive answer to the invitation to participate in the survey was given by respectively 41, 53 and 18 libraries. It should be noted that the rate of return places these three regions somewhat differently: the highest percentage of libraries participating in the study was reported in Lesser Poland voivodeship (21%), followed by Świętokrzyskie (18%) and Mazovian (16%). Interestingly, the survey was filled more willingly by the libraries located in small towns - 50 of them came from the place with the population up to 5 thousand, the next 38 - from 5 to 20 thousand. This gives a bit more than 78.5%.

3.1   Library Staff and Their Knowledge of e-Government

One of the key aspects of the mediation by the public libraries in access to information and public services is to hire staff with appropriate knowledge and skills in this area. In response to the question on how the representatives of libraries assess their knowledge on e-government the largest part of them (47) said that it is poor. When we take into account the 18 answers "I know from hearsay" and 1 indicating "I do not know what it is", it turns out that respondents from almost 59% of libraries declare lack of adequate preparation substantive to perform such a function. It should also be noted, that this question, as well as another in this part of the questionnaire lead the respondents to make kind of self-evaluation, which as shown by the experience of all studies of this type usually leads to overestimating the percentage of declarations that the respondents see as more positive. For a better understanding of the phenomenon, we applied here two solutions of verification.

Firstly, in the above question the term "e-administration" was used, instead of "e-government". It seems that in the Polish language usage, especially outside big cities, "e-government" (no equivalent in Polish) could be an unfamiliar term. Moreover, the understanding of the term "e-administration" (*e-administracja*) with a minimum knowledge of new technologies can be deduced on the basis of other words that begin with the prefix "e-"(e.g. e-mail).

Secondly, a series of control questions were used, which allowed to verify self-general esteem declared by the respondents. The first question in this group was the question on the familiarity of the e-PUAP platform. It is a solution that is fundamental for the functioning of the system of electronic public services in Poland. Available from 2010, the e-PUAP is (or - eventually should be) an integrated access point to all the information and public services in Poland. Meanwhile, 42 libraries declared good knowledge of e-administration, but only 28 equally good knowledge of the e-PUAP. At the same time, as many as in 19 cases the answer "I do not know what it is" was given. It seems that the declared level of librarians' knowledge concerning e-administration exceeds reality.

Another form of integrated access to public e-services are regional portals, created since 2002 under the joint strategy "Gates of Poland" (*Wrota Polski*). In comparison to the e-PUAP platform, knowledge on these portals should be much more widespread. And indeed, in relation to the declarations obtained in respect of the e-PUAP platform, in the structure of responses to regional portals we can see a slight shift to declarations indicating somewhat greater level of knowledge. Interestingly, one library more than in the previous two questions, declared here a very good knowledge of the issue. This may indicate that in this library there is an employee who for some reason is particularly associated with the portal.

Finally, the last important issue in the context of access to public services which we asked for, is familiarity with the so-called trusted profile. Information about it can be found at the starting site of the e-PUAP platform. As it can be read, it is a free mobile electronic signature, which should be used while using electronic services. After creating an account on the platform it is necessary to submit a request for profile confirmation. Then, you have to go to the nearest public office with the identity document, or in some cases - login through bank account. In practice without a trusted profile you cannot effectively use the e-PUAP platform. So it is an informative verification question, which shows the actual knowledge of the respondents. And so, as long as respondents from 28 libraries declared good knowledge of the e-PUAP, for the trusted profile there were only respondents from 20 libraries. The majority (respondents from 41 libraries) declared poor knowledge. There may be several reasons for such results. Firstly, the respondents could considered "familiarity" as familiarity with the applied computer solutions. Secondly, declarations of the knowledge in the field of the e-PUAP platform could be overstated.

The Freedom of Information act is the main piece of legislation defining the rules concerning access to information which are being the responsibility of the public sector institutions, including public libraries. The declarations of knowledge of librarians in the case of this issue were much higher than in the case of public services. Interestingly, 10 respondents indicated that "I know from hearsay" and one – "I do not know what it is". Taking into account that the act dates back to 2001 and it regulates the important area of accessing information, it was unexpected.

One of the latest trends in e-administration is a so-called open administration. In Poland, its result is a portal DanePubliczne.gov.pl created on the basis of the amended Freedom of Information Act, which established the rules of reusing public information. It takes the form of the Central Repository of Public Data, created to share data sets from the institutions of public administration. Its aim is to give means for citizens and business to create new solutions, applications and services. This is equivalent to sites such as the British data.gov.uk or the US www.data.gov. Re-use is indeed not a key issue from the point of view of the information activities of libraries, which explains why the obtained data show the limited interest in this subject among respondents, and above all - no need to update their knowledge in this area.

Another important issues related to e-administration is e-democracy. It covers issues such as e-consultations, e-regulation, or e-voting. They allow the citizens to e-participate in making various kinds of decisions at the central, regional and local levels. Initiatives of this type ensure the transparency of public institutions and give citizens the feeling that their opinion matters. According to the answers of the respondents as many as 33 of them show that they only "hearsay" this issue and 16 declared their ignorance on the subject. However, it seems that such structure of responses is partly due to the use of the term e-democracy. During study visits in selected libraries it turned out that they often use e-questionnaires, which is the simplest form of e-consultation.

3.2 Library and Mediation in Access to Public Services

Strategies for the development of e-government in the European Union refer to a particular set of public services for citizens and business whose level of advancement is used to compare the progress of digitization of public administration since the *eEurope* strategy was published. In the case of services to citizens they are: settlement with the tax office; services offered by employment offices; obtaining social benefits and pensions; obtaining personal documents; vehicle registration; obtaining a building permit, access to library catalogues; reporting to the police; changing the place of residence, services offered by the registry offices and submission of documents to higher education institutions (HEIs). During our study the respondents were asked whether in any of these services they offer assistance in their implementation. The most commonly declared service, as many as 99 times, was access to the catalogues of other libraries. However, taking into account the specific characteristics of library work, it may be surprising that 13 libraries do not offer assistance in this case.

The second indication was the submission of documents for the HEIs - 68 cases, the third - the services offered by employment offices (47 times), then the settlement with the tax office (35) and signing up for a visit to the doctor. In other cases aid is offered sporadically, which can be easily explained by the lower level of their computerization (or even unavailability online). The structure of the replies to this question can also indicate that librarians are more willing to provide assistance in the case of services which probably they had the opportunity to use before (which would explain a surprisingly large number of indications for services related to the submission of documents to the HEIs). It is also worth noting that 11 libraries are not prepared to provide any assistance for those services.

The surveyed libraries in a much lesser extent declare their readiness to assist in access to public services for entrepreneurs - as many as 76 of them responded "none of the above". Less than one-fifth of the respondents pointed to "submitting data to the statistical office" and "settlement with the tax office". This seems to be completely understandable - in Poland public libraries have little relationships with the business. Therefore, even these few indications should arouse positive surprise.

In the next section of the questionnaire the respondents were asked to indicate the level of interest in the implementation of individual support while realizing services to users of the libraries. It turned out that help is most rarely sought in the case of services such as: building permits (95.5% of responses - not at all), reporting to the police and changing the place of residence (94.6% each); services offered by the registry offices (93.8%), vehicle registration (92.9%) and obtaining social benefits and pensions (90.2%). It is worth noting, however, that these are the services for which the process of computerization was not launched at a time when the study was performed, or it was only possible to obtain forms. At the same time, however, in the case of some services, particularly those that are highly computerized or are particularly important in people's daily lives, the users' interest in assistance provided by librarians was observed. In particular it regards the service of settlement with the tax office - for more than 40% of libraries such inquiry appeared several times a year (the specifics of the service should be noticed- it is carried out only between January and April). This is of course one of the few public services in Poland, which allows the realization fully online in the whole country, without visiting the office. More often (35.7% - several times a year, 21.4% - several times a month, 1.8% - several times a week) the support in dealing with labor offices is sought.

In the case of services for entrepreneurs the indications were much less optimistic. In the vast majority of them this type of question did not appear at all (indications "not at all" on a level above 90%). The only exception occurred to be a service "transmission of statistical data" - 21 libraries received such requests several times a year, while one - several times a month. As it was already written, much less interest in the services for this group of customers is understandable due to the nature of Polish public libraries.

The last question in this part of the questionnaire concerned the main difficulties in mediating of the public libraries in access to information and public services. Indications can be divided into two groups: internal barriers and external barriers. In the first one the most frequently indicated barriers were too small number of employees (84 libraries), insufficient knowledge of workers (62), or financial constraints (59) and housing constraints (46). Interestingly, most rarely indicated answer was insufficient number of computers with Internet access for users, which proves the success of computerization programs implemented for years in Polish libraries. This can be concluded that real barriers may be in (a) the mentality of librarians, (b) external factors. The latter include determinants of demand - a lack of interest on the part of users (77 responses) and relations with the authorities - the perception of libraries as only the lending institution of books (58).

3.3   Library and Mediation in Access to Public Information

The fifth part of the questionnaire was devoted to mediation by libraries in access to public information, which can be understood as local information or as presented in the Freedom of Information Act, e.g. in the Public Information Bulletin. As it was written, some indications in this area appeared in the comments of respondents to questions in other parts of the questionnaire. The first question in this section was open and said: "If the library offers help in finding public information online (information provided by public institutions), what information is most commonly of interest to users?" Because it was not an obligatory question, it was completed by only 84 respondents (75%), although in fact it looks a little worse: some entries were "not applicable", "hard to say" or even "I do not know". In the responses most frequently appeared: information on job vacancies and insurance.

In response to a question about whether the library offers help in understanding/ interpretation of public information found in the Internet, only 42 times "yes" answer was given. It seems that this is an issue that will require further, in-depth research. On the one hand, such a large number of "no" answers can be explained by the lack of willingness of librarians to take on the responsibility for any consequences resulting from errors of interpretation. On the other hand, one should try to understand what librarians understand by the interpretation. Is finding information on the Internet with the help of library staff (and including the definition of the found answers as relevant) not a form of interpretation?

When the respondents were asked about whether they conduct courses/trainings/ workshops facilitating the use of online public services we also achieved a surprisingly small number of positive responses - only 20. Comments added to other questions repeatedly referred to the courses for seniors, computer courses, opening an e-mail account etc. This means that the libraries feel much more confident in the area of spreading computer competences, than in the case of information competences.

3.4   Library as an Institution Cooperating with Other Public Institutions

The last group of questions contained in the questionnaire concerned the issues related to the forms of cooperation of public libraries with other public institutions. According to the declarations of the respondents, the most common cooperation covers relations with educational institutions - 90 indications. As the second group they indicated the district offices, municipalities etc. (58), but in this case it is associated with treating by librarians formal and legal relationships with the competent authority as the cooperation. Next non-government organizations (50 times), employment offices (35), district offices (25) and tax offices were indicated. In particular, cooperation with units of employment offices and tax offices here has a positive connotation- the comments from respondents indicate that they are mostly various types of meetings/shifts of representatives of these offices in the library. Among others emerged: centers of culture, municipal social welfare centers, volunteer fire brigade, the house of social assistance. Interestingly, the most common initiators of such cooperation are the libraries themselves. For the question which public institutions offered in the past cooperation, the most often response was "none of the above." The comments also make it clear that the librarians usually come with their ideas outside and look for partners for cooperation. Concerning the future cooperation libraries would like to work with employment offices (38 responses), educational institutions (35), district offices (24), and non-government organizations (21).

Finally, on the question whether in the library there are meetings of users with representatives of other public institutions organized, only 39 respondents answered positively. Most often they were representatives of employment offices, municipal offices, tax offices, police and municipal police.

4 Conclusions

The conducted study was exploratory and shows, that so far the level of interest of Polish public libraries arranging access to information and public services is very low. However, libraries sometimes do not even know that they serve such activities, and are even a role model of good practice - the case of City Library in Oświęcim, comes to mind. As part of its services was, among others, not only Local Information website, but also radio station, where local news are broadcasted.

In most cases was observed an approach which can be summarized as follows: users do not need it, in the library there is not enough place, no enough employees and there are too serious financial shortages. When we correlate this with visits to libraries, e.g. in London (struggling with exactly the same problems) and libraries, e.g. in French Montpellier (modern, rich, rich in human resources and premises), it turns out that the main factor for engaging libraries in innovative solutions is seeing only the traditional role of the library by all stakeholders: users, librarians and government officials. Only a few try to break common stereotypes in this regard.

Polish Libraries are working on promoting e-inclusion in relation to digital skills and providing access to technology, however are still lacking in helping patrons to access public information and services online. Although they provide explanation to this situation with too small number of employees, insufficient knowledge of workers or financial and housing constraints, in comparison they do not differ that much from American and British Libraries – which allows to assume, that it is more the case of perspective and willingness.

References

1. Keeping America Informed: The U.S. Government Printing Office, Washington D.C. (2012)
2. Scholl, H.J.: e-Government: information, technology and transformation. In: Scholl, H. J. (ed.) Electronic Government: A Study Domain Past its Infancy, pp. 11–32. M.E. Sharpe, Armonk (2010)
3. United Nations e-Government Survey 2016: E-Government in Support of Sustainable Development. Department of Economic and Social Affairs UN, New York (2016)

4. Singh, S., Castelnovo, W.: e-Government: a time for critical reflection and more? In: Singh, S., Castelnovo, W. (eds.) Leading Issues in e-Government Research, vol. 2, pp. vii–xv. Academic Conferences and Publishing International Limited, Reading (2015)

5. Oyekunle, R., Akanbi-Ademolake, H.B.: An Overview of e-Government Technological Divide in Developing Countries (2016). https://www.researchgate.net/publication/3049547 26_An_overview_of_e-government_technological_divide_in_developing_countries

6. Sahraoui, S.: e-Inclusion as a further stage of e-government? Transform. Gov. People Process Policy 1(1), 44–58 (2007)

7. Bertot, J.C., McDermott, A., Lincoln, R., Real, B., Peterson, K.: 2011–2012 Public Library Funding and Technology Access Survey: Survey Findings and Report. Information Policy & Access Center, University of Maryland College Park, College Park (2012)

8. Becker, J., Niehaves, B., Bergener, P., Räckers, M.: Digital divide in Egovernment: the Einclusion gap model. In: Wimmer, M.A., Scholl, H.J., Ferro, E. (eds.) EGOV 2008. LNCS, vol. 5184, pp. 231–242. Springer, Heidelberg (2008). doi:10.1007/978-3-540-85204-9_20

9. Ferro, E., Gil-Garcia, J.R., Helbig, N.: The digital divide metaphor: understanding paths to IT literacy. In: Wimmer, M.A., Scholl, J., Grönlund, Å. (eds.) EGOV 2007. LNCS, vol. 4656, pp. 265–280. Springer, Heidelberg (2007). doi:10.1007/978-3-540-74444-3_23

10. Niehaves, B., Becker, J.: The age-divide in e-government — data, interpretations, theory fragments. In: Oya, M., Uda, R., Yasunobu, C. (eds.) I3E 2008. ITIFIP, vol. 286, pp. 279–287. Springer, Boston, MA (2008). doi:10.1007/978-0-387-85691-9_24

11. Niehaves, B., Gorbacheva, E., Plattfaut, R.: The digital divide vs. the e-government divide: do socio-demographic variables (still) impact e-government use among onliners? In: Gil-Garcia, J.R., (ed.): e-Government Success Factors and Measures: Theories, Concepts, and Methodologies, pp. 52–65. IGI Global (2013)

12. Library e-Government and Employment Services and Challenges. Information Policy and Access Center. http://www.ala.org/research/sites/ala.org.research/files/content/initiatives/plftas/2011_2012/egovsvcs-ipac.pdf

13. Czapiński, J., Panek, T. (eds.): Social Diagnosis 2011. Objective and Subjective Quality of Life in Poland. Contemporary Economics Quarterly of University of Finance and Management in Warsaw, vol. 5, no. 3 (2011)

Positive Emotion and Social Capital Affect Knowledge Sharing: Case of the Public Sector of the Kingdom of Bahrain

Isa Yousif Alshaikh[ID], Anjum Razzaque[✉][ID],
and Mahmood Saeed Mustafa Alalawi[✉][ID]

College of Business and Finance of Ahlia University,
Manama, Kingdom of Bahrain
isa.alshaikh.88@gmail.com, anjum.razzaque@gmail.com,
mahmood.alawi@gmail.com

Abstract. Many scholars attempted to describe one or more of the following three: Positive emotion (PE), social capital (SC) and knowledge sharing (KS). However, only a few researches have tried to discover a relationship that links the three domains. Numerous studies have asserted that PE in SC environments, particularly in virtual reality group' formation would stimulate a KS behavior. Whereas, another domain is tackled in this study; the holistic view of these three variables. Scholars have long believed that the existence of a KS culture facilitated by SC and PE. Thus, career progression. This empirical research; conducted on a number of selected public sector organizations in the Kingdom of Bahrain has proved the existence of a positive relationship between the Three tested domains. There was no clear evidence that this relationship existed in the virtual level community. As a result linear regression analysis was applied to assess the relationship between the three domains of this study. The findings led to a modified conceptual model viable for practical implications.

Keywords: Knowledge sharing · Positive emotions · Social capital

1 Introduction

Past research on social capital (SC), knowledge sharing (KS) and positive emotions (PE) was reviewed from existing yet current literature and was re-assessed for its holistic relation in the Kingdom of Bahrain (KoB), i.e. to understand how KS can be influenced by SC and PE. This is due to the lack of organizational direction or strategy to support KS. The expatriate workforce that tends to form the majority of the expertise in professional jobs in the region is a challenge. As the threat of expatriate turnover can lead to the loss of knowledge, many of them believe that their knowledge is their power and do not share it. This negatively influences KS. This study will also inspect if KS is valued throughout organizations, and if decision making, productivity, and innovation are driven by KS quality. This research's findings could be used to boost current, local researches and to trigger the initiative for further and more comprehensive research in future. In the KoB, it is commonly belief that knowledge mainly exists at a formal

© Springer International Publishing AG 2017
M. Themistocleous and V. Morabito (Eds.): EMCIS 2017, LNBIP 299, pp. 284–293, 2017.
DOI: 10.1007/978-3-319-65930-5_24

organizational and team levels, driven by individual initiatives. The existence of SC in VCs is somewhat limited. There are several that can be investigated where locals actively participate in the international VC in KS. It is also understood that many of the barriers to KS is based on the concept; knowledge power syndrome, which could result in knowledge retention by individuals. The second issue is somewhat related to the integration of knowledge management strategies. Hence, sharing initiatives within the organizations' goals and strategic approaches is vague or non-existent. The final issue is related to the lack of VC and the internal desire to share knowledge, as it is believed that many individuals were brought-up to acquire and use knowledge rather than share them with others. With this being the introduction of this research study; the next section (Sect. 2) is the critique of a literature review. Section 2.1 in Sect. 2 discusses PE, while Sect. 2.2 discusses SC and Sect. 2.3 KS. Section 3 discusses how the measurement was designed and developed, (Fig. 1). Section 4 elaborates on the research methodology of this study. Section 5 discusses the path scholars took to perform the data analysis and hence discuss the empirical findings. Section 6 concludes a summary of the entire article and demonstrated the practical implications conceivable for the future of the KoB and other business sectors.

Fig. 1. Research model

2 Literature Review

This study is mainly concerned with the three bespoke domains and how various literature described their main definitions, based on how they are related and on what basis each domain is implemented. Past literature reviewed indicated the relationship between those domains but none showed the relationship between all the domains as described in the conceptual model of this section.

2.1 Positive Emotion

As reported by [21] that an emotion has been defined as the mental state that could be an effective as well as a cognitive nature, in expressing a physical way, and leads to actions and behaviors that were an expression of, or way to coping with, this mental state [4]. It is important to distinguish emotions from attitudes, as attitudes (evaluative judgments of an object) can be a mental state as well [4]. As [3] concluded that employees who reported more frequent levels of PE tend to be more socially integrated in their organization, which is more likely to lead to higher engagement than those who reported less positive emotions [39, 40, 42]. These psychological resources generated by all employees experiencing PE could lead to employee attitudes such as emotional engagement [11, 17]. This employee engagement level would not only affect individual employees but could also impact other team members' motivation and emotions levels, which in turn would positively be influenced by organizational change. [5, 6]. It was argued by [17] that while the absence of PE could limit thought-action repertoires to instinctual human behavior (e.g., fight or flight) that may lead to more short-term thinking and undesirable organizational outcomes. In conclusion, PE are a pre-requisite for encouraging the sharing of knowledge behavior [33] so that SC of resources can be shared amongst the KS parties.

2.2   Social Capital

As stated by [28] SC can be described as the human element, such as learning, trust, and innovation, which is created and further enhanced during interpersonal interaction in SC concept [11]. SC is "the stock of active connections among people: the trust, mutual understanding and shared values and behaviors that bind the members of human networks and communities and makes cooperative action possible" [12, 22]. The importance of networks plays a role in active connections among people is obvious within organizations or companies [11]. As stated by [28] the basis that forms intellectual capital can be viewed as being created through two generic processes: combination and exchange [28]. Combination is combining elements of knowledge that were previously unconnected and finding a new way to combine them using some previous association [28]. While, exchange gives different entities the opportunity to combine, reframe, and formulate different experiences and unique perspectives [28]. An additional element in the development of Intellectual Capital is the dynamic nature of knowledge, content, creation, and learning that occur best when people are reacting to someone else's thinking; thus networks and other communities of learning are part of knowledge program architecture [32]. Organizations can consider nurturing opportunities for exchange for intellectual capital [28]. Industrial engineers and others have worked over the past century to fine tune the workplace [28]. Also [28] looked at the office layout including configuration and arrangement and office décor. She reported that all dimensions has significance that she illustrated through corporation [25]. Efforts made to improve organizational performance through innovation in the physical environment happen across industries. It was observed that in many organizations connections, and thus SC get created when companies simply help people be in the same place at the same time. As [45] stated that SC is "the sum of the actual and potential resources embedded within, available through, and derived from the network of relationships possessed by an individual or social unit" [30]. The SC through social or virtual media provides a strong interpersonal relationship that could influence the level and quality of shared knowledge [44]: the base of SC: trust, identifications and commitments for cooperative norms [45]. As [30] observed, SC is structure of network and resources to examine the individuals' pro-social behavior in organizations, such as KS, knowledge contribution [18, 23], boundary-spanning or brokerage [43], and mobility [43]. As reported by [37], a community, group of people performing mutual obligation. Virtual communities (VC) are developed on interests and values [31]. SC: a theory is based on the structural dimension, relational dimension and cognitive dimension where structural dimension demonstrates is overall relations, relational dimension reflects participations via trust, identification and norms of reciprocity and cognitive dimension reflects participants' ability to communicate via common language and vision [34, 45] reported that affective commitment is an emotional significance: individual members attach to the membership of work teams; due to identification [2, 10] obligated to engage in future actions [33, 34, 44]. Social identification could nurtures one's motivation to sharing knowledge. [30]. People do not attempt to contribute knowledge unless they recognize themselves as being part of the team and perceive their contribution to be conducive to their welfare or career [2, 10, 15]. Individual affection from identification leads to a positive trust in a group: therefore elevates team KS [6]. Team cooperative norms represented corporate shared values among team members of: willingness to value diversity, openness of critics, and expectation of reciprocity [33, 34, 44]. The team cooperative norms affect individual sharing behavior in two ways: firstly motivating individuals to adhere to team expectation; and diminishing the potential competition resulting from KS [45]. Team norms guide individual behavior. Individuals perceive to the pressures to conform to team norms of their teams. [14].

2.3   Knowledge Sharing

It was reported by [16] that knowledge as "authenticated information" [1], i.e. knowledge as processed information in individuals' minds: personalized information [1]. KS is one of Knowledge Management's processes, including: knowledge creation, generation, transfer, usage and acquisition, [1, 13, 26, 38]: "*knowledge life cycle within an organization*" [7], i.e. knowledge flows [41]. KS is individual learning through combining and exchange of knowledge with a growing need for a system for KS [41]. Some organizations achieve KS through a culture of organizational communication [8]. According to [27] KS maximizes organizations to generate solutions for competitive advantage [35]. In this context, KS was defined as a social interaction culture, involving the exchange of employee knowledge, experiences, and skills through the whole organization [27]. KS is a set of understandings\providing employees' access to relevant information and then building and using knowledge networks [20]. Moreover, KS occurs at the individual and organizational levels [27]. For individuals, KS is chatting to colleagues to aid in problem solving. Organizational KS is capturing, organizing, reusing, and transferring experiential knowledge residing in an organization [27]. KS is essential for enhancing innovation performance: hence, reducing redundant learning [9, 38]. Firms successfully promote a KS culture by directly incorporating knowledge in strategies and changing employee attitudes towards a KS culture [11, 27]. Virtual communities (VC) without a supply of relevant knowledge are unable to satisfy members: so cannot nurture [24, 41]. Knowledge is shared as community related information, ideas, suggestions and expertise among members [45]. Participation are viewed as active - providing information, i.e., contributing by producing an outcome valued by others, or passive (receiving information only [36]. KS, implies active involvement from both givers and receivers, it promotes interaction and learning among the members who are capable of making sense of the provided information [19].

3   Measurement Model

This study has three integrated domains by its conceptual model as explained ahead. The PE theory was based on the principle that individual with more frequent levels of PE tends to be more socially integrated in their organization, which is more likely leading to higher engagement than those who reported less PE [21]. (2) The SC theory is the human element, such that learning, trust, and innovation, which is created and further enhanced during interpersonal interaction in SC [11]. The main elements of this theory is based on structural capital, relational capital, and cognitive capital concepts. The structural capital pronounces the environment or media such as the VC or places where SC activities takes place. The relational capital plays a significant role in the relationship established between peers or between the organizations and their customers or partners. It explains the emotional significance that individual members attach to the membership in their work teams. (3) The KS domain could result from a more dynamic activity such as the SC interaction and coactivity and driven by individual with more PE. Social identification could motivate KS; in contrast, distinct and contradictory identities within communities may set up barriers to KS. The possible lack of perceived shared cognition by most individuals could lead to reluctance to share knowledge with their counterparts. Hence, individual members are more likely to share their knowledge with one another when they demonstrate shared cognition of the work.

Figure 1 (model) is theatrical founded from existing SC theory in relation PE that can lead to KS and hence advance career progressions. The model showed that there are two hypothesis. The first hypothesis related PE and SC while the second relates SC and KS. The empirical result show that there is a correlation between the three domains, i.e. variables, which is based on the analytical results from the regression analysis. From the obtained results, it can be concluded that the SC domain had an impact on KS and PE, proving that both hypothesis are accepted. This study is based on two hypotheses hat are also depicted in Figure one. These two hypotheses are:

- H1. Participants' positive emotion have positive significance on social capital.
- H2. Participants' Social Capital has positive significance on Knowledge Sharing.

4   Research Method

This is a literature driven research that initiated with a thorough literature review to identify a gap in research. The gap was reflected into the appropriate hypotheses and henceforth this is a deductive approach, which was conducted in the KoB's Ministry of Interior (MoI). Aim of research was to prove that the theories are related with SC. The adapted survey was distribution, based on non-probability sample where convenience sample technique was applied when selecting a group of MoI volunteering staff from various departments. A pilot study was samples on 30 voluntary participants and was selected from MoI and listed approximately 1 month period to validate the survey's content. The results from the pilot study expressed that no items needed to be removed except few clarification grammar wise. The amended survey was then distributed to 300 participants, active in KS communities or members in committees or teams. Active in public safety and law enforcement of the KoB and its headquarters are in The Diwan Fort (Manama Fort) in Manama. Study tested two hypotheses: method was the best of its kind; portrayed in its research design, based on a top-down research method, i.e. quantitative method [26]. Survey content was adopted from previously research, previously validated and tested. 400 surveys were distributed at General Directorate of Nationality of MoI. Sample size of 300 and another 100, for contingency sake, was calculated to generalize the population of the MoI. Estimated total population of employees at the General Directorate of Nationality, Passport and Resident of the MoI are about 1200 employees. After the pilot study the main data collection led to data analysis for the pilot study reported that the survey's items were reliable and valid to resume survey distribution without changes.

5 Data Analysis and Discussion

This research aimed: to validate the proposed model composed of PE, SC and KS depicted in Fig. 1. SPSS statistical tool helped analyze data with results reported. Analytical analysis expressed reliability and validity tests. Pearson correlation were also performed.

Table 1. Frequencies of the independent variables (gender, age, education)

Variables		Gender	Age	Education
N (Number of)	Valid responses	300	300	300
	Missed responses	0	0	0
Skewness	Valid responses	.326	.288	1.765
Sd. error of skewness	Missed responses	.141	.141	.141

Table 1 above illustrates results of 300 respondents, i.e. 100% response rate. There is only 1 value missing in the education question. Highest skewness is found in education and the lowest in age. Standard error of skewness is similar among all three independent variables, which means they are equally the same. The skewness figures determine the level of normal distribution, which is ideally 1 for 100% normal distrusted data. Results obtained for S divided over K are both above 1.96 and they fall in the positively skewed distribution pattern. For example, Z distribution for gender is [S] .326/[K] .141 = 2.31, which is greater than 1.96 and hence, rejects the null hypothesis of none skewed pattern. The overall results obtained are still within the acceptable range to do F, T and regression analysis test even if it is slightly positively skewed. Age frequency distribution showed that 48% fall between age 35 and 44 and 31% between 25 and 34: i.e. 80% of the respondents' age: typical age distribution in the KoB in many organizations. The small sample size for age ranges from 55–74 and 65–74 could represent a problem when performing tests such as t test. This can be resolved by combining the range 55–74 with 45–54 age range. Frequency distribution for education showed that around 79.60% of the participants are at a bachelor's level. 19.40% are the master's level. 1% are at a doctoral' level. Reliability and domain statistics tests that were conducted showed that the Cronbach alpha for the 3 domains variables is greater than .5, i.e. data is reliable and valid. The domain statistics showed the means and standard deviations of the 3 domain variables. The results displayed that all domain scores where between scale 3 to 4.

Table 2. Pearson correlation for dependent variables

Domains		Positive emotion	Social capital	Knowledge sharing
Positive emotion	Pearson correlation	1	.540[**]	.305[**]
	Sig. (2-tailed)		.000	.000
	N	300	300	300
Social capital	Pearson correlation	.540[**]	1	.538[**]
	Sig. (2-tailed)	.000		.000
	N	300	300	300
Knowledge sharing	Pearson correlation	.305[**]	.538[**]	1
	Sig. (2-tailed)	.000	.000	
	N	300	300	300

**Correlation is significant at the 0.01 level (2-tailed).

The Pearson correlation (Table 2 above) shows the results obtained from the SPSS output. This is a univariate analysis mainly used to determine the correlation levels between dependent variables such as the three domains listed above (PE, SC, and KS). The results pointed out a clear positive relationship between the 3 listed domains and 2-tailed significant less than 0.01. The results rejected the null hypothesis and proved that there is a positive relation between PE, SC and KS variables. The total score from all the questions for each domains have been aggregated prior to running the analysis. This was done first on Excel sheet and later imported into SPSS data sheet. Multiple linear regression tested the relations. The first regression analysis was done for the PE domain as the dependent variable to the 2 independent variables. It was concluded that PE have a positive impact on all independent variables tested. The highest impact or relationship was between the PE and SC, which confirms the related hypotheses. The second regression analysis was done for the SC domain as the dependent variable. The results obtained showed a different relationship between both SC and PE, and then SC and KS from the proposed conceptual model. The third and last regression analysis was used for the KS domain as the dependent variable. It was concluded that SC had the highest effect on KS, which proves the related hypothesis that SC drives KS.

6   Conclusion and Implications

This study is based on the case of MoI of the KoB and is limited to a public sector with a small selected sampling size, even though this generalizes over the total population of MoI staff. The SC, as per the literature review, was mainly based on the existing VC of practices. However, the selected samples in this study were not members of any social communities related to KS and they work at organizational and individual levels through formal teams. This study was based on the SC theory, PE and KS with the main objective being to test the relationship between these three domains identified as per the conceptual model in Fig. 1, tested statistically using linear regression and then to prove the assumed hypotheses. The research outcome accepted both of the hypothesis as the relationship identified between SC l, KS and PE. It is recommended that further research is required on a larger sample size and more organizations such as private sectors to have better validity and reliability of the results and findings. Also, policy makers in the KoB can utilize the findings of this study for improving their KS processes. Furthermore, this model can also be tested in other domains and sectors.

References

1. Alavi, M., Leidner, D.E.: Review: knowledge management and knowledge management systems: conceptual foundations and research issues. MIS Q. **25**(1), 107–136 (2001)
2. Alavi, M., Kayworth, T., Leidner, D.E.: An empirical examination of the influence of organizational culture on knowledge management initiatives. Paper presented at the Presentation for KBE Distinguished Speaker Series, Queen's School of Business (2003)
3. Avey, J.B., Wernsing, T.S., Luthans, F.: Can positive employees help positive organizational change? Impact of psychological capital and emotions on relevant attitudes and behaviors. J. Appl. Behav. Sci. **44**, 48–70 (2008)
4. Bagozzi, R P , Gopinath, M , Nyer, P U : The role of emotions in marketing. J. Acad. Mark. Sci. **27**(2), 184–206 (1999)
5. Bakker, A.B., van Emmerik, H., Euwema, M.C.: Crossover of burnout and engagement in work teams. Work Occup. **33**, 464–489 (2006)
6. Bergami, M., Bagozzi, R.P.: Self-categorization, affective commitment and group self-esteem as distinct aspects of social identity in the organization. Br. J. Soc. Psychol. **39**(4), 555–577 (2000)
7. Birkinshaw, J., Sheehan, T.: Managing the knowledge life cycle. MIT Sloan Manag. Rev. **44** (1), 75–83 (2002)
8. Buckman, R.H.: Knowledge sharing at Buckman labs. J. Bus. Strategy **19**(1), 11–15 (1998)
9. Calantone, R.J., Cavusgil, S.T., Zhao, Y.: Learning orientation, firm innovation capability, and firm performance. Ind. Mark. Manag. **31**(6), 515–524 (2002)
10. Chen, I.Y.L.: The factors influencing members' continuance intentions in professional virtual communities– a longitudinal study. J. Inf. Sci. **33**(4), 451–467 (2007)
11. Connelly, C.E., Kelloway, K.: Predictors of employees' perceptions of knowledge sharing cultures. Leadersh. Organ. Dev. J. **24**(5/6), 294–301 (2003)
12. Cohen, D., Prusak, L.: In Good Company: How Social Capital Makes Organizations Work. Harvard Business School Press, Boston (2001)
13. Davenport, T., Prusak, L.: Working Knowledge: How Organizations Manage What They Know. Harvard Business School Press, Boston (1998)
14. Dragoni, L.: Understanding the emergence of state goal orientation in organizational work groups: the role of leadership and multilevel climate perceptions. J. Appl. Psychol. **90**(6), 1084–1095 (2005)
15. Ellemers, N., Kortekaas, P., Ouwerkerk, J.W.: Self-categorisation, commitment to the group and group self-esteem as related but distinct aspects of social identity. Eur. J. Soc. Psychol. **29**(2–3), 371–389 (1999)

16. Ford, D.P.: Knowledge Sharing: Seeking to Understand Intensions and Actual Sharing. Queen's University Kingston, Ontario, Canada (2004)
17. Fredrickson, B.L.: Positive emotions and upward spirals in organizations. In: Cameron, K.S., Dutton, J.E., Quinn, R.E. (eds.) Positive Organizational Scholarship, pp. 241–261. Berrett-Koehler, San Francisco (2003)
18. Gibson, C., Vermeulen, F.: A healthy divide: subgroups as a stimulus for team learning behavior. Adm. Sci. Q. 48(2), 202–239 (2003)
19. Hendriks, P.: Why share knowledge? The influence of ICT on the motivation for knowledge sharing. Knowl. Process Manag. 6(2), 91–100 (1999)
20. Hogel, M., Parboteeah, K.P., Munson, C.L.: Team-level antecedents of individuals' knowledge networks. Decis. Sci. 34(4), 741–770 (2003)
21. Van Den Hooff, B., Schouten, A., Simonovski, S.: What one feels and what one knows: the influence of emotions on attitudes intentions towards knowledge sharing. J. Knowl. Manag. 16(1), 148–151 (2012)
22. Hsu, M.-H., Ju, T.L., Yen, C.-H., Chang, C.-M.: Knowledge sharing behavior in virtual communities: the relationship between trust, self-efficacy, and outcome expectations. Int. J. Hum. Comput. Stud. 65, 153–169 (2007)
23. Kankanhalli, A., Tan, B.C.Y., Wei, K.K.: Contributing knowledge to electronic knowledge repositories: an empirical investigation. MIS Q. 29(1), 113–143 (2005)
24. Kosonen, M.: Knowledge sharing in virtual communities – a review of the empirical research. Int. J. Web Based Communities 5(2), 144–163 (2009)
25. Lieber, R.B.: Cool offices. Fortune 200–210 (1996)
26. Liebowitz, J., Megbolugbe, I.: A set of frameworks to aid the project manager in conceptualizing and implementing knowledge management initiatives. Int. J. Project Manag. 21(3), 189–198 (2003)
27. Lin, H.: Knowledge sharing and firm innovation capability: an empirical study. Int. J. Manpow. 28(3/4), 315–332 (2007)
28. McGrath, R., Sparks, W.: Knowledge, social capital and organizational learning. Int. J. Knowl. Cult. Change Manag. 5(9), 125–129 (2005–2006)
29. Ministry of Interior (2016). http://www.interior.gov.bh/dcfault_cn.aspx, http://www.interior. gov.bh/default_en.aspx. Accessed 13 May 2017
30. Nahapiet, J., Ghoshal, S.: Social capital, intellectual capital, and the organizational advantage. Acad. Manag. Rev. 23(2), 242–266 (1998)
31. Nguyen, L., Torlina, L., Peszynski, K., Corbitt, B.: Power relations in virtual communities: an ethnographic study. Electron. Commer. Res. 6(1), 21–37 (2006)
32. Pasternak, B.A., Viscio, A.J.: The Centerless Cooperation. Simon & Schuster, New York (1998)
33. Razzaque, A., AlAlawi, M.: Role of positive emotions for knowledge sharing in virtual communities like knowledge management tools. In: 2nd International Conference on Emerging Trends in Multidisciplinary Research (ETMR 2016), Bangkok, Thailand. KS Global Research, Bangkok (2016)
34. Razzaque, A., Eldabi, T., Jalal-Karim, A.: Physician virtual community and medical decision-making: mediating role of knowledge sharing. J. Enterp. Inf. Manag. 26(1), 500–515 (2013)
35. Reid, F.: Creating a knowledge sharing culture among diverse business units. Employ. Relat. Today 30(3), 43–49 (2003)
36. Ridings, C., Gefen, D., Arinze, B.: Some antecedents and effects of trust in virtual communities. J. Strateg. Inf. Syst. 11, 271–295 (2002)

37. Rothaermel, F.T., Sugiyama, S.: Virtual internet communities and commercial success: individual and community-level theory in the atypical case of Time-Zone.com. J. Manag. **27** (3), 297–312 (2001)
38. Scarbrough, H.: Knowledge management, HRM and innovation process. Int. J. Manpow. **24** (5), 501–516 (2003)
39. Staw, B.M., Barsade, S.G.: Affect and managerial performance: a test of the sadder-but wiser vs. happier-and-smarter hypotheses. Adm. Sci. Q. **38**, 304–331 (1993)
40. Staw, B.M., Sutton, R.I., Pelled, L.H.: Employee positive emotion and favorable outcomes at the workplace. Organ. Sci. **5**, 51–71 (1994)
41. Wasko, M.M., Faraj, S.: It is what one does: why people participate and help others in electronic communities of practice. J. Strateg. Inf. Syst. **9**(2), 155–173 (2000)
42. Wright, T.A., Staw, B.M.: Affect and favorable work outcomes: two longitudinal tests of the happy-productive worker thesis. J. Organ. Behav. **20**, 1–23 (1999)
43. Xiao, Z.X., Tsui, A.S.: When brokers may not work: the cultural contingency of social capital in Chinese high-tech firms. Adm. Sci. Q. **52**(1), 1–31 (2007)
44. Yu, A., Hao, J., Dong, X., Khalifa, M.: Revisiting the effect of social capital on knowledge sharing in work teams: a multilevel approach. In: International Conference on Information Systems (ICIS) 2010 Proceedings, AISeL (2010)
45. Yu, T.-K., Lu, L.-C., Liu, T.-F.: Exploring factors that influence knowledge sharing behavior via weblogs. Comput. Hum. Behav. **26**(1), 32–41 (2010)

Innovative and Strategic Solutions to Foster the Public Server Valorization Policy

Denise Lucena Sousa Balbino[1,2(✉)]

[1] Public Management, University of Brasília, Brasília, Brazil
denise.unb.df@gmail.com
[2] Public Policies and Governmental Management of the Government
of the Federal District, Brasília, Brazil

Abstract. This paper discusses which factors are favorable or inhibitive to the implementation of innovation as a motivator of labor motivation, through a study of the perceptions from people internal and external to the public organization, with the purpose of promoting the valorization of the public server. The model considered will be based on the open innovation concept. After analyzing the results obtained with the application of a questionnaire, the State Department of Planning, Budget and Management of the Federal District (SEPLAG) received suggestions for actions to motivate the manager to implement the innovative profile presented in the research results and to broaden the theme in the public service.

Keywords: Innovation · Motivation · Public server's valorization

1 Introduction

Innovation in the Brazilian public sector lacks studies and scientific research, technologies for its diffusion and financial resources. However, in view of the challenges pointed out, the present study is justified by the need for solutions that add value to the actions of the public server before the citizen.

The article considered the union of the public server motivation, with the innovation implementation and the intention of producing information for the decision making of the public managers. The relationship in question was given after the analysis of the results obtained with the application of the questionnaire, from the perspective of the motivational theories mentioned in sequence in a holistic and concomitant way.

The attempt to know the perceptions, satisfactions, expectations and opinions of several questioned actors is closely linked to the context studied. It has the aim of improving the motivation and the valorization of the public server, through new ideas, using research of perception of open innovation, having in view of the Brazilian scenario, rooted in obstacles, in the face of the political and budgetary crisis faced.

Currently, it is perceived that the rigidity and the formal principles of public administration, the organizational structure, the current political system, the destructive

---

D.L.S. Balbino—State Department of Planning, Budget and Management of the Federal District.

M. Themistocleous and V. Morabito (Eds.): EMCIS 2017, LNBIP 299, pp. 294–304, 2017.
DOI: 10.1007/978-3-319-65930-5_25

critics made by citizens, the lack of motivation of the public servers, the budget constraint and the absence of new technologies in the service are barriers to the diffusion of innovations. For this, the present research will analyze individual and correlated constructs capable of demystifying the false perception that has been culturally rooted over the years, which will support an approximation between the citizen and the server and puncture factors inhibiting innovation and the valorization of public servers.

The motivation of the public servant depends on the valorization that he receives from the company and the citizen, however, for this actions to take place, it takes a change in the leaders profile. Because, mostly part they are not innovative.

It is also worth mentioning that, currently, the SEPLAG has a female Secretary of State, however, in other organs there is a need to expand the female occupation in strategic or leading position in the government, as, according to data annexed from the System of Human Resources Management until may/2017, only 24.83% of the political nature leadership were occupied by women and the other positions of special nature are mostly occupied by the male gender. That makes it difficult to implement the innovation according to the results presented in the research.

This study will provide the identification of current perceptions of innovative ideas that make it possible to construct the public server's valorization policy, as well as other citizens interested in coparticipating in the expansion of the use of innovative solutions in the public field, which is why it is important to emphasize the receptivity and the paths of innovation.

2 Theoretical Reference

The concept of innovation (Kühl and da Cunha 2013) occurs when there is the implementation of something new or significantly improved, either product – or service –, process, marketing method or organizational method.

The definition of innovation given by Schumpeter (1997) is considered by Nuchera et al. (2002, p. 55) as the most classical of definitions. The OECD (2005, p. 56), much quoted today, follows the line of Schumpeter, defining innovation, briefly, as the implementation of something new or significantly improved, both a product (or service), as a process, a marketing method or organizational method.

The study in question analyzed in greater depth the comparison of the perception of groups of public servants, but inserting other stakeholders to promote co-participation, since it was intended to investigate the characteristics of the phenomenon in the place where it occurs (Vergara 1997), but also aimed at Support the phenomenon of open innovation (Seminar and Inova 2008). This perspective, called by Chesbrough (2003) as a model of open innovation, maintains that its process must be more collaborative, seeking to access knowledge from several external actors, the said stakeholders.

Software companies were studied (Pinheiro and Tigre 2015) regarding the support of innovation in services. The learning process still depends heavily on "learning by doing" and "learning by using". They also highlighted which organizational changes are required for the use of more advanced Information and Communication Technology tools, as follows.

The analysis reveals that the organizational rigidity and the behavior of aversion to the use of new technologies are significant barriers to the diffusion of innovations.

Crowdsourcing is a value-creation strategy whereby companies directly capture the contribution of their consumers to the creation of new products and solutions, increasing the capacity for innovation. However, entrepreneurs were not yet prepared for the implementation of crowdsourcing-based strategies (Murakam 2015).

The theme of crowdsourcing as an effective means of empowering communities with the potential to engage individuals in innovation, self-organization remains limited (Smith et al. 2016).

The results of a doctoral thesis (Dias 2014) that mapped the conditions under which journalism awards in Brazil are organized, resulting in the conclusion that there is a valorization of the guidelines and devaluation of the professional and the profession.

The article entitled "Well-being, happiness and structural crisis of neoliberalism: an interdisciplinary analysis through the lens of emotions" (Pilkington 2016) concluded that the scientific value of emotions is rarely recognized and that sociology of emotions is rapidly growing.

Pilkington noted that there is a phenomenon of manipulation of emotions happier or sadder than usual depending on the news feed of social networks.

Still according to Pilkington's (2016) survey, individuals are more sensitive to losses than gains in terms of economic growth, there is a strong asymmetry in positive economic growth and recessions. The use of management control systems and operational management techniques, focusing on diagnostic control systems (Nisiyama 2011), used to monitor and reward the achievement of specific objectives as essential management tools to transform desired strategies into strategies, Such as the process of standardization and formalities in internal procedures that are perceived negatively to the innovation process (Harari 1997), neutralizing the favorable aspects in the process of introducing new products.

In a China case study (Elfstrom 2017), considered home to the world's largest working class, exploiting governments vulnerable to workers demonstrations, demonstrating expanded public policies after the protests pressure, it was concluded that they provide openings for innovation in the workplace if they are in the management creative leaders.

Finally, the study of the meaning of voluntary work (Borchardt 2016) in a Lutheran church: identification with Lutheran ethics; seriousness and image of the institution; unit; conditions and willingness to participate; encouragement of others; relation to the profession; pleasant relationship and recognition.

3 Methodology

Vergara (1997) named the taxonomy used in the classification of the methodology of this work, as much as the ends as means. As for the purposes, the methodology used is descriptive and as to the means, it is a bibliographical, documentary, telematized and field research. The study in question is of a qualitative approach since, in synthesis, the data obtained refer to the semistructured questionnaires applied.

It is worth mentioning that for the use of correlations weighted in this investigation, it was intended to observe the degree of statistical significance, using the SPSS Statistics software, which performed the analysis of variance (ANOVA) of the answers as defined by Professor Bardin (2009) as a set of analysis techniques applied to the description of the content being studied.

Data collection from this study was done through Google Drive, which houses the Google Forms application, which was used to create the online form. The collection phase began with the elaboration of the questionnaire, sent to the participants by social media. The questionnaire was sent in February of 2017 and the total of 224 responses were sent until March of the same year.

The inclusion criteria was different for three groups stratified by per se characteristics: the first group included the public servants of the union, totaling 32 participants, the second considered only the public servants of the Federal District, a total of 112 participants and, for the last of other citizens who do not hold public positions, resulting in 80 participants, so that the elements obtained could be impartial and focused on open innovation.

The questionnaire referred to (Appendix) had a semi-structured script set up to guide the researcher, containing 19 closed questions and 1 open, elaborated with the purpose of obtaining the relevant data regarding the general and specific objectives of the research.

3.1 Characterization of Motivation Adopted in Research

Motivational theories seek to identify the sources of pleasure that workers find in their work environment, and they may be in the individual, in the work environment or both (Paschoal and Tamayo 2005), so the questions were theoretical constructs to validate the results obtained.

It was also considered the concept of motivation as a complex psychological process, that results from an interaction between the individual and the environment that surrounds him (Latham and Pinder 2005).

The classical motivational theories of Maslow (Hesketh and Costa 1980) and Herzberg (Pocinho and Gouveia Fragoeiro 2012) were considered in the present study.

In the consolidated democracies, an operation of responsibility not only, not so much, "vertically", in relation to a number of public powers, except, retrospectively, at the time of the elections, but "horizontally" relatively autonomous – that is, other institutions – that have the ability to question, and eventually punish "improper" ways of the position occupant in question to fulfill its responsibilities. Make text above and possibly draw a distinction between two dimensions. A first, vertical accountability, concerns electoral control in relation to voters and elected representatives. The second, horizontal accountability, is benchmarked to the control exercised by various powers of the State and other institutions with capacity for supervision and punishment.

The evaluation of the server will take into account its motivation and the opinion of several stakeholders on the subject, and a horizontal analysis of the perception of the control of these actors is required. That transparency can be classified into two dimensions, each according to the O'Donnell (1991) stated: "In consolidated democracies, accountability operates not only "vertically" in relation to those who elected the occupier of a public office (except, retrospectively, at the time of the elections), but "horizontally" in relation to a network of powers Relatively autonomous (ie other institutions) who have the ability to question, and possibly punish, "improper" ways for the occupier of the post in question to fulfill his/her responsibilities. From the above text it is possible to draw the distinction between the two dimensions. The first, vertical accountability, concerns electoral control in relation to voters and elected representatives. The second, horizontal accountability, refers to the control exercised by the various powers of the State and other institutions with capacity for supervision and punishment."

Thus, the result of this research can serve as an indicator of democratic quality, pointing to the perception of the employees from different spheres and among the citizens.

The participation of the citizen in the suggestion of innovative ideas to motivate the public servant, with public policies of approach of the groups mentioned goes beyond the democratic inclusion, but a factor of change in the culture of the public innovation, because it shows a direct relation of cause and effect.

In this sense, the questions present in the questionnaire of this research were theoretical constructs to validate the results obtained, influenced by the motivational theories cited in the present work.

4   Results and Discussion

The questionnaire applied in this research is presented in Appendix and can be analyzed in several ways. However, throughout this survey, the focus of innovation was based on perceptions of participation, satisfaction, environment, incentive and motivation of the public server among the study participants, to define favorable and inhibiting factors.

The variable referring to question 1 demonstrates that most of those interviewed know SEPLAG's competencies.

Nonetheless, when this variable is stratified by the characteristics *per si*, it points out that the level of transparency about the competences of SEPLAG has significance only among the district public servers.

Still in the same sense, the sequential question identified that among the public servers of the Federal District, there are more servers that do not know their assignments when compared to servers from other spheres.

Question 4 deals with the classification of genders, and it can be pointed out that there is no relevance between the female and male optics regarding the dissatisfaction with the practices and policies of public server remuneration, despite the dissatisfied majority being female.

The variable "schooling", when stratified to the classification "sex", shows that the female gender has a higher degree of qualification than the male.

Question 5 correlated the level of schooling with satisfaction mentioned in question number 7. Thus it is perceived that the higher the level of schooling, the lower is the satisfaction regarding the remuneration practices and policies of the public server. For of the 34 forms answered by people with the average level, two responded to be totally satisfied. However, out of 66 responses from seniors, only three responded to be fully satisfied.

Of the total of 101 questionnaires filled out by people with a postgraduate degree, only five answered to be totally satisfied, and there were no responses of "Completely Satisfied" to the masters and doctoral degree.

In summary, we can observe the perception of total satisfaction with the remuneration practices and policies of public servers inversely proportional to the level of schooling of the participants.

In this sense, by studying the significance of the age group, we can see that the more age, the more qualification and the lower satisfaction with the remuneration.

Still talking about question 6, regarding the age group of the participants, when correlated with question 19, which refers to unconventional work environments and/or tools to stimulate innovation in the public service, results in the absence of the age group of 26 to 35 years old who answered "no".

It is also collected that the age group of 26 to 35 years is more prone to environments and/or non-favorable tools to stimulate innovation in the public service. When the same variable is stratified to gender, it is noticed that the male gender is more resistant to innovation in the public service in environments and/or unconventional tools.

In the same sense, analyzing the results under the aegis of men, it is observed that the majority of male public servers participating in the research also do not believe in innovation without financial resources.

Analyzing the occupants of strategic or managerial positions, most feel that innovation in the district public service does not depend on financial resources.

The same question 7, which deals with the age group of the participating public, when stratified with the characteristic of the occupation of strategic position or of leadership in the government (question 18), noticed the perception that the age has growth directly proportional to the occupation in positions in the government.

It is deduced that at present the strategic positions have been occupied mostly by those over 35 years of age, with less propensity for innovation.

Question 8 brings the popular participation in district public policies, which in its overwhelming majority responded by agreeing to increase participation in the debate.

Thus, there is a consensus on increasing popular participation in district public policies.

Among those who answered "no" to increasing popular participation in district public policies, all are above the age of 35 and have the following common answers: they believe that the Federal District does not have quality service (question 10); believe that there is no environment conducive to creativity, experimentation and implementation of new ideas that could generate a differential for the public service of the Federal District (question 11); everyone thinks that the bureaucracy is a hindrance (question 14); everyone thinks that the innovation depends on political decision (question 15); and everyone believes that it is necessary to improve server motivation (question 17) (Fig. 1).

Question 11 showed that there is a perception of the environment favorable to creativity, experimentation and the implementation of new ideas that can generate a differential for the public service of the Federal District to the majority of the participants.

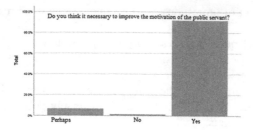

Fig. 1. Motivation of the public servers

It is important to emphasize that when the answers are extracted only among public servers of the Federal District, most believe that there is no favorable environment for innovation.

The terminology "bureaucracy", when used in a derogatory way, as synonymous with the lack of efficiency and slowness of the public machine, constitutes an obstacle to the implementation of innovation, even for the interviewees who have already held a strategic position or head of government.

In question 15, which inquired about the main factor to have an organizational innovation, most of the answers, including occupiers of strategic position or leadership, meant the "political decision" factor.

Question 9 sought to identify what causes the most dissatisfaction in the public career, leaving an open field for other responses. Of the 144 responses of the civil servers, 57 indicated "the lack of recognition of the importance of their work by the organ itself", of the 80 responses of non-servers, only 6 answered "the lack of recognition of the importance of their work by the organ itself", with the majority responding to "the absence of a merit-based quality-of-work evaluation system".

It is noted that the public server seeks recognition of the importance of his work by the organ itself and the citizen meritocracy, that is, results.

Regarding the responses without the stratification of the public servers group, that is, considering all the answers provided, it can be observed that wages do not have relevance in the variable "public career dissatisfaction".

Question 10, resulted in the perception of the lack of public service quality by 85.3% of respondents.

Question 12 presented the existence of an incentive for the district public server to have a continuing education and had as an implication in the majority of the answers the inexistence of the incentives mentioned, including when stratifying the opinion of those who were already occupying a strategic position or head of government.

Thus, it can be affirmed that there is no statistical relevance in the difference of the answers between those who have held strategic positions and in the government and the other interviewees, that is, the majority believes that there is no incentive in the comment.

Concerning Question 16, the result obtained from the majority was the perception of the indispensable participation of all members of a public organization, because everyone has responsibility for the results to be achieved.

It is observed that both the participation increasement of the members in a body and the increase of the popular participation constitute the majority of the answers.

For the question about the need for improvement in the motivation of the public server, the most significant of the respondents answered yes, if it becomes necessary.

It is important to emphasize that the opinions of servers and non-servers when compared are unisonic, that is, both believe that it is necessary to improve the motivation of the public server.

With due prominence in this paper, question 20 – which holds the variable "greatest need for innovation in the public sphere" – had the significance of most responses as "all alternatives", which were classified as: "In administrative procedures"; "In the economy of public resources"; "In organizational culture"; and "Legislation". Statistically highlighting the need for innovation, especially in organizational culture.

Finally, the results point to a favorable perception of the implementation of innovation in all the areas presented, thus consolidating the wide acceptance of public innovation by the various participating groups.

5   Conclusions and Recommendations

The purpose of this study was to identify the perception of valorization of the public server with the implementation of innovation in order to produce information for decision making.

From this research, it is possible to infer that the level of transparency regarding the competencies of the State Department of Planning, Budget and Management of the Federal District in the groups of participants is greater among public servers than among citizens.

In this sense, it can be concluded from the research that there is room for expansion of the democratic quality indicator as studied by O'Donnell, especially in the other spheres of the district and among citizens.

However, the present study demonstrated that the salary issue in gender perception is not relevant, as proposed by the "Teto de Vidro" literature phenomenon.

Thus, it is possible to point out that among the servers of the Federal District, which SEPLAG is part of, there are more servers that do not know their assignments when compared to servers from other spheres. Based on this perception, it is possible to infer that the level of transparency with respect to the responsibilities of the State Department of Planning, Budget and Management of the Federal District in the groups of respondents is greater among public servers than among citizens.

In this sense, it can be concluded that there is room for expansion of the democratic quality indicator.

Table 1 summarizes the conclusion of the research presenting the factors that favor or not the innovation in the public service.

As follows, to encourage innovative actions to motivate SEPLAG`s public servants it is suggested to implement actions to qualify low-level servers and to opportunize the qualification of male servers occupying strategic positions in the innovation theme, include at the core of innovation the participation of women's groups from 26 to 35 years, create regulation that provide for the filling of at least 50% of positions of

Table 1. Factors that favor versus factors that do not favor innovation

Factors favoring innovation	Factors that do not favor innovation
Most women have a higher degree of professional qualification	The more dissatisfaction with the lower remuneration is the degree of schooling
The age group of 26 to 35 years are more confident in environment and/or unconventional tools	The older the lower the satisfaction with the remuneration
The female gender is more environmentally friendly and/or unconventional tools	The male gender is prone to the environment and/or conventional tools
To have more occupants of strategic positions or of leadership with the age group of 26 to 35 years	Occupants of strategic or managerial positions have more than 35 years
The public server fosters popular and server participation	District servers do not have a favorable perception of new ideas
Reduction of bureaucracy	Absence of political decision
Motivate the server with the recognition of its work by the organ	Absence of quality in the public service
Improve the participation of all servers in the management of ideas	Lack of incentive to continuing education

strategic position occupied by women, use media resources to bring the citizen closer to the public servant, regulating Decrees dealing with de-bureaucracy, to implement in innovation awards managed by SEPLAG recognition for innovative ideas and create opportunities for all employees to participate in managerial ideas.

It is concluded that the results achieved point to some shortcomings that may allow SEPLAG and Pubic Service opportunities for future studies.

Thus, it is not possible to rule out the existence of other variables and even other constructs other than those indicated in the accepted results.

Appendix

Public server satisfaction survey
1. Do you know what the Secretary of Planning, Budget and Management of the Federal District is doing?*
2. Are you a public server?*
3. Are you a district public server?*
4. What is your gender?
5. What is you educational level?*
6. What is your age group?*
7. How satisfied are the public server's remuneration practices and policies?*
8. Do you agree to increase popular participation in district public policies?*

(*continued*)

(continued)

Public server satisfaction survey
9. What causes most dissatisfaction in the public career?*
10. Do you think the Federal District has a quality public service?*
11. Do you think that there is a favorable environment for creativity, experimentation and implementation of new ideas that can generate a differential for the public service of the Federal District?*
12. Do you think there is an incentive for the district public server to have a continuing education?*
13. Do you think innovation in the district public service depends on financial resources?*
14. Do you think that bureaucracy hinders public service innovation?*
15. Do you think that an organizational innovation depends primarily on:*
16. Do you think that the participation of all the members of a public organization is indispensable, since everyone has responsibility for the results to be achieved?*
17. Do you think it necessary to improve the motivation of the public server?*
18. Have you ever held a strategic or leading position in government?*
19. Unconventional environments and/or tools can stimulate innovation in the public service?*
20. Choose the option of greater need for innovation in the public sphere:*

References

Bardin, L.: Análise de Conteúdo. Edições 70, LDA, Lisboa (2009)

Borchardt, P.: Meanings of volunteer work : a study with members of a lutheran institution. **17** (5), 61–84 (2016). https://doi.org/10.1590/1678-69712016/administracao.v17n5p61-84

Chesbrough, H.: Open Innovation: The New Imperative for Creating and Profiting from Technology. Harvard Business School Publishing, Boston (2003)

Dias, R.: Meritocracia na midiocracia: reflexões sobre Prêmios em Jornalismo na cultura profissional jornalística. Revista Famecos **21**(2), 595–621 (2014)

Elfstrom, I.M.: A Dissertation Presented to the Faculty of the Graduate School of Cornell University in Partial Fulfillment of the Requirements for the Degree of Doctor of Philosophy, January 2017

Harari, O.: Ten reasons TQM doesn't work. Manag. Rev. **86**(1), 38–44 (1997)

Hesketh, J.L., Costa, M.T.P.M.: Construção de um instrumento para medida de satisfação no trabalho. Revista de Administração de Empresas **20**(3), 59–68 (1980). https://doi.org/10.1590/S0034-75901980000300005

Kühl, M.R., da Cunha, J.C.: Obstacles to implementation of innovations in Brazil: how different companies perceive their importance. Braz. Bus. Rev. **10**(2), 1–24 (2013)

Latham, G.P., Pinder, C.C.: Work motivation theory and research at the dawn of the twenty-first century. Annu. Rev. Psychol. **56**(1), 485–516 (2005). https://doi.org/10.1146/annurev.psych.55.090902.142105

Murakam, L.C.: O Crowdsourcing Como Fator De Competitividade: Uma Investigação Em Pequenas Empresas Do Setor Da Moda, pp. 138–155 (2015)

Nisiyama, E.K.: Uso dos Sistemas de Controle Gerencial, Técnicas de Gestão e o Desempenho de Empresas do Setor de Autopeças, no. 11, pp. 57–83 (2011)

Nuchera, A.H., Serrano, G.L., Morote, J.P.: La gestión de la innovación y la tecnologia en las organizaciones. Pirámide, Madrid (2002)

OCDE – ORGANIZAÇÃO PARA COOPERAÇÃO E DESENVOLVIMENTO ECONÔMICO: Manual de Oslo: proposta de diretrizes para coleta e interpretação de dados sobre inovação tecnológica (2005). Disponível em: http://download.finep.gov.br/imprensa/manual_de_oslo.pdf. Accessed 16 Nov 2016

Paschoal, T., Tamayo, A.: Impacto dos valores laborais e da interferência família: trabalho no estresse ocupacional. Psicologia: Teoria E Pesquisa **21**(2), 173–180 (2005). https://doi.org/10.1590/S0102-37722005000200007

Pilkington, M.: Well-being, happiness and the structural crisis of neoliberalism: an interdisciplinary analysis through the lenses of emotions. Mind Soc. **15**(2), 265–280 (2016). https://doi.org/10.1007/s11299-015-0181-0

Pinheiro, A.O.M., Tigre, P.B.: Proposta de investigação sobre o uso de softwares no suporte à inovação em serviços. Rae **55**(5), 578–592 (2015)

Pocinho, M., Gouveia Fragoeiro, J.: Satisfação dos Docentes do Ensino Superior. Acta Colombiana de Psicología **15**(1), 87–97 (2012)

Schumpeter, J.A.: Teoria do desenvolvimento econômico: uma investigação sobre lucros, capital, crédito, juros e o ciclo econômico. Nova Cultural, São Paulo (1997). (Coleção:Os Economistas)

Seminar, O.I., Inova, A.: Open Innovation and Open Business Models. World Trade (2008)

Smith, K.L., Ramos, I., Desouza, K.C.: Economic resilience and crowdsourcing platforms. J. Inf. Syst. Technol. Manag. **12**(3), 595–626 (2016). https://doi.org/10.4301/S1807-17752015000300006

Vergara, S.C.: Projetos e relatórios de pesquisa em administração. Atlas, São Paulo (1997)

Healthcare Information Systems

Digital Transformation in the Pharmaceutical Compounds Supply Chain: Design of a Service Ecosystem with E-Labeling

Alexandra Ângelo[1]([⊠]), João Barata[1,2], Paulo Rupino da Cunha[2], and Vasco Almeida[1]

[1] ISMT, Miguel Torga Institute, Coimbra, Portugal
{alexandraangelo,jbarata,vascoalmeida}@ismt.pt
[2] Department of Informatics Engineering, CISUC,
University of Coimbra, Coimbra, Portugal
rupino@dei.uc.pt

Abstract. We propose the design of a digital service ecosystem for the pharmaceutical compounds supply chain. Our method of inquiry is the canonical action research and a retail pharmacy provides the setting. A comprehensive review of existing literature about compounding of medicines is provided and six digital services identified: Supply management, Product traceability, Quality management, Order management, Digital assistant, and Product experience. The new services are supported by dynamic QR code identification and mHealth technologies. Preliminary results suggest that the digital ecosystems offer an opportunity to implement electronic labels (e-labels) in pharmacies, improve medicine information quality, and restore a broken link between medicine customer and medicine producer. Our findings can assist service design and service innovation in pharmaceutical supply chains. Moreover, they can support retail pharmacies in dealing with the increase of medicine compounding, address regulatory pressure for e-labeling, and to take advantage of their proximity to local communities.

Keywords: Digital transformation · Digital ecosystem · E-labeling · QR code · Pharmaceutical compounding · Service innovation

1 Introduction

Pharmacies are key elements in healthcare systems, establishing links between the governmental health policies, the pharmaceutical industry, hospitals, doctors, and customers [1]. Moreover, retail (or community) pharmacies offer one of the most immediate contacts for healthcare information, especially relevant for the elders and population in rural areas. Modern retail pharmacies sell medical devices, perform clinical analyses, and include a plethora of healthcare services to local populations, for example, fitness, nutrition, and healthcare guidance. Yet, the compounding and manufacturing of medicines to serve the population is still now, as it was in its creation, a central element of the pharmacies mission [2, 3].

© Springer International Publishing AG 2017
M. Themistocleous and V. Morabito (Eds.): EMCIS 2017, LNBIP 299, pp. 307–323, 2017.
DOI: 10.1007/978-3-319-65930-5_26

Pharmacy information systems, sometimes named as patient medication records, have functionalities that include *"medicine labelling, patient medication records, decision support for drug interactions and other warnings, stock control, ward inventory management, order processing and functions to support pharmacy manufacturing processes in hospitals"* [4]. Several authors studied this specific type of healthcare information systems, for example [5], with the introduction of barcode scanners in pharmacies, [6] assessed the user satisfaction in a pharmacy group, while [7] found limitations in pharmacy decision support software, including pharmacists awareness of the system functionalities to support decisions. However, most studies address the use of information technology (IT) by the pharmacy staff, not considering the customer as an active participant in the pharmacy IS. Furthermore, a large number of studies consider the context of hospital pharmacies [8–10], not evaluating the case of retail pharmacies and the information flows between them and the customers, that are so crucial in the context of retail pharmacies.

More recently [11] propose a disease management platform to support online pharmaceutical care services for chronic patients. The authors identified positive results for pharmacists and patients in their online interaction, offering important guidelines to the development of new online services for pharmacies. The provision and use of medicine involves multiple information exchanges between (1) the prescriber and the pharmacy, (2) the pharmacy and its suppliers, and (3) the pharmacy and its customer. Therefore, medicine information is critical for human safety during the entire lifecycle of selection, production, and medicine use. In the case of pharmaceutical compounds – produced for the need of a specific patient, pharmacies are responsible for the product information and labeling. Moreover, they are also responsible for monitoring the usage of the medication.

The creation of digital ecosystems is a possible solution to know more about the customers and solve their *"life event needs"* [12]. The digital age offers new forms to innovate organizational services including the client interface, intra/inter organizational service delivery, and supporting technology [13], for example, the increasing use of mobile devices that opened new avenues for research in mobile health (mHealth). To achieve this goal, some authors proposed methods for service design evolving through business analysis and service strategy definition, requirements of service delivery system, modeling, and validation [11, 14, 15].

Medicine labels provide essential information for the customer. Electronic labeling (e-labeling) refers to the digital access to medicine information, replacing or extending the common paper leaflets and stickers. Yet, we could not find studies on how to (1) create digital ecosystems in retail pharmacies exploring the potential of e-labeling, (2) that promote IT-enabled service innovation centered in the medicine supply chain. What emerges from the use and adaptation of pharmacy IT and the (formal and informal) pharmaceutical processes by all of its users [16], is a challenge for healthcare IS, which motivated us to pose the following research questions:

1. *Which digital services could be created to empower the user of pharmaceutical compounds and increase prevention, useful for the entire supply chain?*
2. *How to create a digital service ecosystem for pharmaceutical compounds in retail pharmacies that explore the potential of e-labeling and mHealth?*

The remaining of this paper is structured as follows. Next, we present background knowledge, namely, in pharmaceutical compounding and e-labeling. Afterwards, we introduce our research approach that is the canonical action research [17] conducted in a retail pharmacy. We proceed with the study presentation and conclude our paper stating the study limitations and future steps.

2 Background

2.1 Pharmaceutical Compounding

Compounding of medicines is an essential part of quality healthcare and the basis of pharmacy since its creation [3]. According to the U.S. Food and Drug Administration (FDA), "[compounding] *is generally a practice in which a licensed pharmacist, a licensed physician, or, in the case of an outsourcing facility, a person under the supervision of a licensed pharmacist, combines, mixes, or alters ingredients of a drug to create a medication tailored to the needs of an individual patient*" [2]. There are many reasons to tailor medications, for example, when (1) a specific drug is not available in the market, (2) the end user requires changes in the composition, for example, due to specific components' allergies, (3) it is necessary to change the form of the drug, for example, a child that can't swallow a pill and needs liquid preparation, (4) and when the product is allowed to be prepared by pharmacists.

The importance of this topic is increasing all over the world, as presented in a recent Forbes article [18]. The new therapies and the implications for human safety are two essential reasons. Moreover, policy reforms were accelerated after the recent deaths that resulted from drug contamination and insufficient quality procedures [18]. Consequently, new regulation for pharmaceutical compounding is being produced worldwide.

National and international regulations must be considered. For example, in Portugal where our research takes place, Ordinance 594/2004, issued in July 2nd [19], establishes best practices for pharmaceutical compounding about personal, installations and equipments, documentation, raw materials, packaging, compounding procedure, quality control, and labeling. Additionally, deliberation 1985/2015, issued in November 2nd, defines substances that cannot be compounded by pharmacies, ordinance 769/2004 issued July 1st establishes the selling price; dispatch 18694/2010 [December 16th] approves the compounded drugs that are supported by the national Government. Examples of international regulations include the European Pharmacopoeia [20], applicable in 37 European countries and used in over 100 countries around the globe.

There are several implications for IS research in pharmaceutical compounding. For example: traceability of product/raw materials; quality procedures to prepare the product; product information; and adoption of IT in healthcare. According to [21] "*technological developments such as electronic prescribing and the availability of electronic decision support systems can effectively implement compliance with labelled conditions of use and safety precautions in the prescription process. It will be one of the major challenges to make labelling easily available and suitable for use in such systems*". In the emerging era of digital transformations, product labeling is a key aspect to ensure human safety and process auditability, especially in the case of e-labeling that is accessible with mobile devices.

2.2   Electronic Labeling of Pharmaceutical Compounding

Pharmaceutical products must comply with multiple guidelines and regulations for labeling, for example, (1) the Company Core Data Sheet [22], (2) the EU SmPC [23], (3) the DailyMed – official provider of FDA label information, and (4) the Structured Product Labeling (SPL) – defines the content of human prescription drug labeling in an XML format [24]. Patient information sheet describe possible adverse effects of treatment, and Consumer Medication Information (CMI), namely, in the form of leaflets provided to the customer. Retail pharmacies are essential sources of CMI, however, as identified by [25] "*the content, format, reading level, and excessive length of CMI are disconcerting*". Moreover, these authors state that "*private sector initiatives to provide useful CMI have failed. Research is needed on effective information selection and presentation in terms of effects on comprehension, retention, and appropriate patient actions to derive optimal drug benefit*".

Electronic labeling is a priority in different parts of the globe [26]. One of the reasons is to create measures for anti-counterfeiting, for example, using QR codes [27]. Almost twenty years ago, [28] stated that "*technology available today could allow every physician and pharmacist to easily obtain the most recent prescribing information electronically via the Internet. It is suggested that the package insert, that is, the printed professional labeling which accompanies the actual product, is an obsolete method for information dissemination*". However, the traditional stickers are the most common solution nowadays in pharmaceutical compounds, with a clear increasing pressure to upgrade to digital labels.

At a national level, Portuguese law 594/2004 defines rules for labeling raw materials (internal use) and finished products (customer use). Raw materials packaging must identify the material, supplier, fabrication details, conservation conditions, warnings, and expiration date. Finished products must include the patient name, formula (magistral formulas that are specific to the patient), batch/lot number, expiration date, conservation conditions, special instructions (e.g. external use), mode of administration, dosage, pharmacy, and name of the technician that made the preparation [19]. Due to the importance of pharmaceutical compounding for healthcare, the legal authorization to do it must include a proposal of label text.

The Portuguese initiatives follow European guidelines provided by HMA/CMDh - Co-ordination Group [29]. CMDh agreed that product information and risk minimization material could be provided via the use of QR codes but "*cannot replace the inclusion of the statutory information (e.g. printed package leaflet)*" [29]. According to the article 62 of Directive 2001/83/CE, it is already possible to use symbols and pictograms in outer packaging and drugs leaflets. However, there are issues related with multilingual support (with a single QR code), the physical location of the digital information (that must remain active and protected), and possible discrepancies between the printed label and continuous updates that digital information allows.

Additionally, the CMDh guidelines are not specific to pharmaceutical compounds – for example, they require the inclusion of the URL along with the QR code, something that proves more difficult in small packages (e.g. single dose) and it suggests that QR codes *"could be included in the outer carton and/or the package leaflet"*, that are usual in industrially produced medications, but not in pharmaceutical compounds [29]. Nevertheless, CMDh states that it will monitor the decision and maintain the discussion after gaining more experience in electronic labeling of medications.

On the one hand, electronic prescription is increasing all over Europe using QR codes. Some countries have already established guidelines for the use of QR codes in medications, for example, Spain [30]. On the other hand, this form of prescriptions is in the scope of doctor-pharmacy relation, not including the patient. The literature provides evidences of European digital efforts in prescription of medications but there are also risks for data protection and barriers to cross-national prescriptions, namely, the costs for pharmacies [31].

QR codes and 2D barcodes in packaging can be used (1) to access web pages with information about the product, (2) to identify batch number and expiration date, with specific functionalities for the visually handicapped, (3) for manufacturing and stock control, and (4) for safety purposes regarding falsification of drugs [29]. mHealth and e-labeling already captured the attention of IS researchers. For example, [32] developed an app that provides verbal instructions to assist elderly patients medication, especially important when taking multiple drugs. The expansion of cloud platforms opens possibilities for new solutions that include QR codes to reduce medication errors by the elderly [33]. The mobile system proposed by [33] has five main functions (1) to configure personal data, (2) to decode QR codes attached to drug-bags, (3) a pill-dispensing assistant, (4) a medication reminder, and (5) medication recording. These authors concluded that the major challenges are user privacy and information security. It becomes clear that governments and the pharmaceutical industry consider electronic labeling a priority [34] but the studies made so far do not explore the potential of integrating information along the entire medicine supply chain, involving medicine producers, customers, prescribers, and healthcare partners.

The ongoing process of digitalization in pharmaceutical compounds presents challenges. We highlight the need of standardization [23, 26], the need to understand how people use the technology, including "app stickiness" [35], consistency of bioequivalent medication information [36], the privacy and information security [31, 33], regulatory issues [29], and data quality. The accessibility and increase of information presents serious risks of over warning: *"while a high number of labeled ADEs* [adverse drug events] *is not necessarily indicative of drug's true toxicity, the presence of such excess data still may induce information overload and reduce physician comprehension of important safety warnings"* [37].

In spite of the growing necessity, studies about pharmaceutical e-labeling are scarce, for example, Google Scholar returns 9 results with the keywords ("pharmaceutical compounds" OR "pharmacy compounds" OR "pharmaceutical compounding" OR "medicine compounds" OR "medicine compounding") AND ("mobile health" OR "e-labeling" OR "e-labeling" OR "electronic label"). Additional research is needed to address the above mentioned open issues in the supply chain of retail pharmacies.

3   Research Approach

Since we aimed at developing information systems in a specific context of pharmacies and study social changes in this process, action research was the selected research approach. Amongst its different forms, canonical action research (CAR) as described by [17] is one of the most widely used and documented [38]. Similarly to other forms of action research, CAR aims at solving a relevant organizational problem while contributing to scientific knowledge [39]. Action research is well suited for complex social situations that require an understanding as a whole, as typically found in IS [40], well suited for context-centric research [41]. Canonical action research requires the identification of the client-system infrastructure and five main steps [17, 38, 42] represented in Fig. 1.

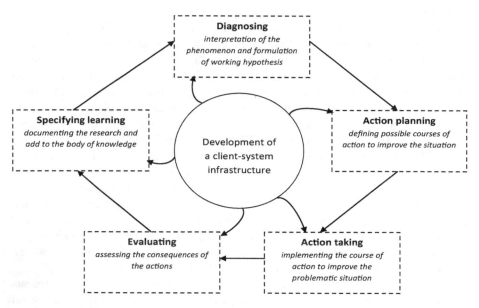

Fig. 1. Canonical Action Research Cycle (adapted from [17])

Theory has an important role during the entire action research cycle to guide researchers' and practitioners' actions, and constantly challenging the results. In our cycle we started with an in-depth review of existing literature, considering regulations and scientific articles. We gathered inspiration in the theory of digital ecosystems and service design for our action plan [12, 13, 15]. Moreover, we have followed the principles specifically proposed by [38] to ensure rigor and validity in our action research cycle: *Principle of the Researcher–Client Agreement; Principle of the Cyclical Process Model; Principle of Theory; Principle of Change through Action; and Principle of Learning through Reflection.*

4   Research Setting and Action Planning

Our client is a retail pharmacy located in the centre of Portugal. Currently, the company has eleven employees and a gross income above 2.3 M€ per year. Since its foundation, the pharmacy has a strong commitment to research, advanced training, and community service. More recently, the pharmacy implemented an online commerce platform and started to invest in new digital services. Examples of the services include pilot projects with Internet-of-Things to monitor health parameters, and the creation and maintenance of a healthcare database that relates symptoms and possible solutions. The pharmacy does not have in-house development capabilities, but has created strong partnerships with the IT industry and universities in the region.

The owner of the pharmacy contacted us to assist in their digital transformation in the pharmaceutical compounds that are produced to end users and to other pharmacies. The labeling process was completely manual and the pharmacy wanted to improve (1) label use for product traceability and (2) the quality of information to the customer. The idea was submitted to a national co-funded project (Portugal 2020) aimed at increasing research and development in small and medium sized companies.

The CAR diagnosis phase was conducted with the pharmacy owner and technical manager. Data collection included interviews, document analysis, and observation. Simultaneously, two researchers conducted a literature review to identify applicable regulations and cases in the literature, as described in Sect. 2.

The strategy of the pharmacy is to differentiate their business with digital services. Additionally, they plan to generate additional revenue by selling digital services to other pharmacies in Portugal and abroad, justifying their recent investing in a cloud infrastructure. A summary of the pharmacy goals in the recently started co-funded project follows:

1. Gain a deeper understanding of e-labeling regulations and mHealth application to retail pharmacies;
2. Identify potential digital services, key functionalities, and applicability to the specific case of pharmaceutical compounds;
3. Develop a model of digital service ecosystem and a pilot case;
4. Improve traceability of pharmaceutical compounds providing complete and rigorous information to the patient, consequently, improving safety;
5. Improve online access to medicine information, using the potential of cloud and mobile health;
6. Evaluate the potential of the model of the digital ecosystem to change the business services.

According to the pharmacy technical manager *"I do not see e-labeling as a substitution to stickers, rather a way of providing better/different information to the consumer, differentiate our products, and improve our internal processes"*. He added that *"Portuguese pharmacies do not have a digital infrastructure to implement e-labeling and the national institutions did not yet create conditions to lead e-labeling of pharmaceutical compounds (...) our pharmacy is ahead of the competition and we have the possibility to create new digital services at a supranational scale (...). The increasing market of healthcare tourism are opportunities for our pharmacy (e.g. digital service that translate medicine information)"*. However, *"e-labeling of compounds is completely new to our pharmacy and to other pharmacies; we need to know how we can innovate and design these services"*.

We outlined an action plan agreed on by researchers and practitioners to improve the pharmacy situation according to the co-funded project proposal, approved by the end of 2016. The pharmacy already had an enterprise system to manage all operational data such as stocks, sales, or customers orders. Our purpose at this stage is to complement the operational backbone already in place with an integrated digital service ecosystem [43].

5   Proposing a Digital Service Ecosystem for Pharmaceutical Compounding

Based on our literature review and field diagnosis, we created a model for the digital service ecosystem presented in Fig. 2.

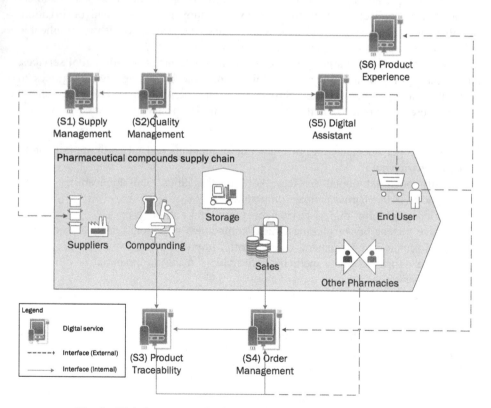

Fig. 2. Digital ecosystem landscape for pharmaceutical compounding

Figure 2 includes a simplified flowchart of pharmaceutical compounds supply chain (in the middle), starting with the purchase of raw materials from suppliers, compounding, storage, and sales, which can be initiated by end user request, other pharmacies request, or own decision, for example, for cosmetics sold in the pharmacy. There are services exclusively accessible by pharmacy technicians and others exposed to external stakeholders of the retail pharmacy (dotted lines). The actors involved in the digital service ecosystem for pharmacy compounds are:

- *Supplier*: Receives order for medicine (raw materials) and information about product experience and quality management procedures (Service S1);
- *Pharmacy Technician (Compounding, Storage, and Sales)*: Main responsible for production (S3) and quality (S2). The retail pharmacy must provide the medicine compound information to customers (via S5) and suppliers (via S2) in the supply chain;
- *Other Pharmacies*: Pharmaceutical compounds can be sold to other pharmacies that, in turn, resell to the end user. These actors require complete product information, place orders, and view quality procedures;
- *End user*: Who uses the medicine requires pharmaceutical support (digital assistant) and may need to contact with the pharmacy (e.g. orders – S4 and complaints/inquiries – S6).

After identifying the main actors of the digital ecosystem we made a joint-reflection by researchers and practitioners to identify (1) the main functionalities of each digital service, and (2) their possible interactions (e.g. product experience – S6 send data to quality management – S2). The functionalities model is outlined in Fig. 3.

The key functionalities of each digital service portrayed in Fig. 3 are:

- *Supply management (E – externally accessible)*: Allows interaction between the pharmacy and its suppliers. Access filtered information about product experience (e.g. customer complaints, adverse reactions) and is accessible to medicine suppliers. It allows B2B pharmacy-supplier cooperation;
- *Product traceability (I – Internally accessible)*: Generate compounds data, QR codes, and the formulas for each patient;
- *Quality management (I)*: Ensures complete product information and characteristics. It requires double validation (two pharmacists) of data – inspired in common aeronautical quality procedures;
- *Order management (E)*: Allows creating orders by other pharmacies, follow the production stage, and provide product information for resellers;
- *Digital assistant (E)*: Includes access to the e-labels via mobile device, specific information for using the medicine in different media (e.g. sound support for patients with vision problems); Alert messages;
- *Product experience (E)*: Supports product use records and inquiries (e.g. questions about reactions with other medicine).

The specification of our digital services is compliant to regulations. For example, the fields in the patient information sheet are configurable (there are differences in medicine regulations across countries, even inside the European Union) and allows multiple language selection, which is important for international supply chains and

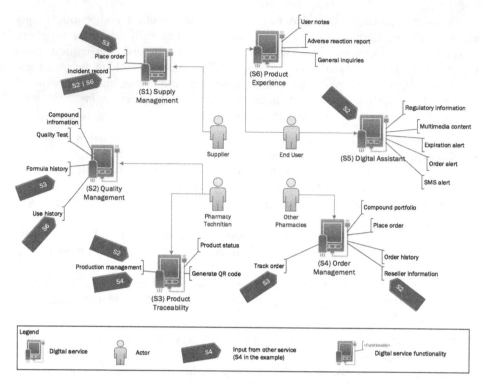

Fig. 3. Digital services functionalities and main interactions

healthcare tourists. Moreover, there is a clear separation between internal and external services to enforce data confidentiality, communication security, and service auditability that may occur in the future by governmental agencies.

We used dynamic QR codes in the compound packing. By dynamic we mean that the web page opened by the (same) QR code changes its content according to the lifecycle phase of the medicine compound. For example, while the product is in production, the digital label provides technical information, specifications of raw materials, available quantities, and compound formula. When it is shipped to other pharmacies, it exposes functionalities for the reseller. After the sale to end user, the QR code activates the digital assistant and the product experience services.

The development of the digital platform (e.g. interface design, coding, testing...) is sourced to an IT team and is out of the scope of our paper. At this stage, we only have a prototype and we have not yet contacted with suppliers or other pharmacies in the supply chain. Nevertheless, we obtained interesting information from pharmacy staff and three customers selected by them. According to the pharmacy manager, the main benefits of this model is the product traceability, and customer focus: *"pharmacy compounds are one of the less digitalized processes in pharmacy practice but the potential is significant – we need to achieve the same quality level of industrial medicine production"*. One of the interviewed customers suggested that this platform should be prepared to receive data from biometric sensors, namely for quality management (S2) and product experience (S6). She told us that *"even if only the prescriber can adjust the formula, the evolution of specific parameters could provide guidance to doctors and medicine producers (...)"*. A customer suffering from multiple allergies told us that product experience could help in the identification of possible allergic reactions when combining different drugs simultaneously, providing a valuable data source for the suppliers and to adjust formulas to each individual. The evaluation and discussion of our model for digital service ecosystem is presented in the next section.

6 Discussion

To address our first research question we proposed six integrated digital services to promote interaction within medicine compound supply chain: Supply management, Product traceability, Quality management, Order management, Digital assistant, and Product experience. With respect to the second research question, we defined key functionalities that use e-labels with QR code. Figure 4 describes the changes in the destination URL of the QR code according to the lifecycle phase of the medicine.

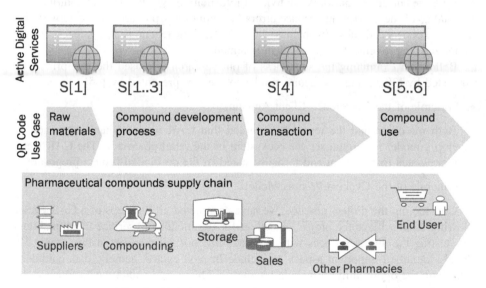

Fig. 4. Digital services activation by the QR code

The QR code points to a web page that dynamically offers digital services according to the lifecycle stage: material supply (S[1]), production (S[1..3]), sale (S[4]), and use (S[5..6]). Although the pharmaceutical compound QR code and e-label is generated at the beginning of the compound development (for each specific formula) we also considered QR codes for the components to be mixed, with the purpose of improving traceability and access by the suppliers (components QR code link to a page with supplier information). The order service (S4) is activated when the compound is sent to another pharmacy or, alternatively, the digital assistant/product experience is activated if the destination is the end user. The order management service also allows recording delivery to the end user. Consequently, all actors of the supply chain know, in real time, the current stage of the medicine compound. A QR code is generated for each order to ensure differentiation of the batch; however, the QR code has a parameter that identifies the base formula, allowing tracing all deliveries of the same product composition.

We identified digital services to increase prevention (e.g. alerts of product expiration date) and empower the user to provide product feedback. Then, we gathered preliminary insights from pharmacy technicians and interviewed three regular customers selected by the pharmacy managers to evaluate potential benefits (e.g. interaction between customer and retail pharmacy) and threats (e.g. excessive information, confidentiality, website failure) of the digital service ecosystem. The typical pharmacy-customer (end user or other pharmacies) relations do not have the characteristics of the doctor-patient relations, namely, the interaction during the entire treatment and the comprehensive knowledge of the health situation of the patient. Further, in large retail pharmacies, it is possible that the customer never buys the compound from the same technician. The creation of a digital ecosystem in retail pharmacies can improve the current scenario. An individual information system [44] can contribute to the quality of the service and reduce errors in compounds preparation. Moreover, the patient experience details can help to identify adverse reactions and suggest, when authorized, improvements to the patient treatment.

Bellow, in examining the soundness of our research, we used the five principles suggested by [38], using a structure similar to the one presented by [45].

- Principle of the Researcher–Client Agreement

Both the client and the researchers agreed that CAR was a suitable approach to develop a model for digital service ecosystem in the retail pharmacy. The CAR steps were explained to the client and formally stated in the co-funded project proposal.

- Principle of the Cyclical Process Model

Considering the project timeline, we have completed all the steps of a CAR cycle, allowing us to achieve the project objectives for the model and service functionalities. Yet, testing the digital services with real data was not yet possible at this stage. We already identified important aspects to include in next cycles, namely, data confidentiality and the opportunity to include Internet-of-Things and social media functionalities in the digital assistant.

- Principle of Theory

A literature review was the starting point for our research. The client constantly participated in theory discussion, particularly in legal frameworks. mHealth is a popular research area in healthcare information systems and recent studies address the topic of online pharmacy services and user empowerment [11]. The potential interest of QR codes have been suggested for pharmacies [34], however, we could not find any study addressing e-labeling of pharmaceutical compounds that are produced for each individual patient. The outcomes of this research were evaluated by researchers and practitioners according to the lens of the literature review previously conducted.

- Principle of Change through Action

The pharmacy is changing the labeling process and developing new digital services supported by mobile technologies. For the sake of simplicity and restrictions in paper size we excluded details of the IT platform outsourced to implement the model and currently under development. However, we describe its digital services.

- Principle of Learning through Reflection

The model results from a joint-reflection between researchers and practitioners. We gathered positive feedback and deepened the analysis of our findings in the current discussion section:

1. *From the point of view of the user.* The model includes digital services and rich information that was not available in the traditional medicine leaflets. Moreover, that information can have multimedia characteristics to improve access to people with disabilities. The digital assistant provides specific SMS alerts, for example, perishable medicine expiring. It is also possible to include estimation of stock rupture for all the compounds under use, simplifying the orders and ensuring the availability of the required product [11]. The main concerns raised for the product experience service (S6) were about data confidentiality (e.g. personal data accessible to medicine suppliers) and the need to adapt product experience to different pathologies. For example, diabetes requires recording insulin levels, while asthmatics may need to record the frequency of the episodes of asthma. Our recommendation is to create a back-office service that allows configuring the parameters for each user. The most relevant risks pointed to the digital assistant are data quality and "overwarning".

2. *From the point of view of the pharmacy.* The digital services of each medicine compound are accessible via the same link, which means that a single QR code can be used for the entire compound lifecycle. Moreover, the mHealth platform can assist (1) pharmacy staff dealing with user requests (individual end users or other pharmacies), (2) managing raw materials of compounds, (3) tracing all pharmaceutical compounds – a difficult task with traditional barcodes that limit their functionality to the pharmacy staff, (4) getting real time information from product use via product experience service, and (5) creating a "digital memory" (contextual data of each customer) of the medicine compounded by the pharmacy. The "digital memory" can be useful in the future to evaluate the effectiveness of pharmaceutical compounds and adjust the formula to each patient.

Our model addresses the three priorities of the European Green Paper for mHealth, namely, (1) prevention, (2) user empowerment, and (3) quality of life [46]. Moreover, it can be used by mHealth providers to improve their apps for pharmaceutical compounds. Nevertheless, as recommended by EU regulation 207/2012 on electronic instructions for medical devices [47], documented risk assessment is vital to digital transformation in pharmaceutical contexts. Risk assessment should cover the digital skills of the end user, impact caused by temporary unavailability of the website, backups, and performance of the system. The infrastructure of the digital ecosystem must comply with requirements of protection against intrusion, reducing downtime and display errors, allowing content to be accessible with freely available software [47]. Although anecdotal at this stage of model specification, pharmaceutical staff and a small group of customers with chronicle diseases found our proposal promising to help pharmaceutical compounding go digital.

7 Conclusion

We presented a model for a digital service ecosystem supported by e-labeling of pharmacy compounds with dynamic QR codes. There is a lack of IS studies in retail pharmacies. Our research addresses this gap by proposing new e-labels that are accessible to multiple stakeholders of the medicine supply chain: medicine suppliers, pharmacy technicians, other pharmacies, and end users.

We encountered difficulties in this project, namely, the regulatory complexity of medicine labels that vary with each country and the potential increase of work to pharmacy technicians to create product information. One way to address these difficulties is to make e-labeling useful in different parts of compound production and not only for end users of the medicine. The label can extend the typical leaflet use (informative) to include interaction between the customer (individual/company)-pharmacy, and pharmacy-supplier.

To our knowledge, this is the first study of mHealth that address the specific case of pharmaceutical compounds supply chain in retail pharmacies. Nevertheless, there are limitations that must be stated. First, we restricted the context of our research to retail pharmacies and it would be interesting to extend our study to other settings such as hospitals (e.g. nutrients compounds). Second, this is a single case in a leading retail pharmacy; therefore, future studies should include other types of pharmacies, in distinct competitive environments (e.g. rural pharmacies with higher number of elder patients). Third, we gather initial insights from pharmacy staff and customers but we couldn't get feedback from official authorities, for example, regarding the certification of the proposed system due to the lack of regulations in this emergent field. Moreover, we interviewed a restricted number of participants to validate our model; it is necessary to proceed with a larger sample and study the service use by different profile of users (e.g. age, mobile experience, type of healthcare needs). Forth, qualitative studies are complex and the positive results must be carefully evaluated due to the potential risk of the Hawthorn effect, suggesting that the observed participants behavior could be "*related only to the special social situation and social treatment they received*" [48]. We already planed a second CAR cycle to address these opportunities testing the digital services with real data. Future research will also include an extension of the digital assistant/product experience services for healthcare professionals, namely, doctors, nurses, and nursing homes.

Acknowledgements. This work has been partially funded by European Regional Development Fund (ERDF), Centro 2020 Regional Operational Programme, Portugal 2020.

References

1. Taylor, D., Mrazek, M., Mossialos, E.: Regulating pharmaceutical distribution and retail pharmacy in Europe. In: Regulating Pharmaceuticals in Europe: Striving for Efficiency, Equity, and Quality, pp. 196–212 (2004)
2. FDA: Human Drug Compounding. https://www.fda.gov/Drugs/GuidanceCompliance RegulatoryInformation/PharmacyCompounding/default.htm. Accessed 16 May 2017
3. Allen, L.V.: Contemporary pharmaceutical compounding. Ann. Pharmacother. **37**, 1526–1528 (2003)
4. Goundrey-Smith, S.: Pharmacy Management Systems. In: Goundrey-Smith, S. (ed.) Information Technology in Pharmacy, pp. 151–173. Springer, London (2013). doi:10.1007/978-1-4471-2780-2_6
5. Nanji, K.C., Cina, J., Patel, N., Churchill, W., Gandhi, T.K., Poon, E.G.: Overcoming barriers to the implementation of a pharmacy bar code scanning system for medication dispensing: a case study. J. Am. Med. Inform. Assoc. **16**, 645–650 (2009)
6. Batenburg, R., Van Den Broek, E.: Pharmacy information systems: the experience and user satisfaction within a chain of Dutch pharmacies. Int. J. Electron. Healthc. **4**, 119–131 (2008)
7. Hines, L.E., Saverno, K.R., Warholak, T.L., Taylor, A., Grizzle, A.J., Murphy, J.E., Malone, D.C.: Pharmacists' awareness of clinical decision support in pharmacy information systems: an exploratory evaluation. Res. Soc. Adm. Pharm. **7**, 359–368 (2011)
8. El Mahalli, A., El-Khafif, S.H., Yamani, W.: Assessment of pharmacy information system performance in three hospitals in eastern province. Saudi Arabia. Perspect. Heal. Inf. Manag. **13**, 1b (2016)
9. Khlie, K., Abouabdellah, A.: A study on the performance of the pharmacy information system within the Moroccan hospital sector. In: 3rd International Conference on Logistics Operations Management (GOL). pp. 1–7. IEEE (2016)
10. Mahoney, C.D., Berard-Collins, C.M., Coleman, R., Amaral, J.F., Cotter, C.M.: Effects of an integrated clinical information system on medication safety in a multi-hospital setting. Am. J. Heal. Pharm. **64**, 1969–1977 (2007)
11. Lapao, L.V., da Silva, M.M., Gregorio, J.: Implementing an online pharmaceutical service using design science research. BMC Med. Inform. Decis. Mak. **17**, 1–14 (2017)
12. Weill, P., Woerner, S.L.: Thriving in an increasingly digital ecosystem. MIT Sloan Manag. Rev. **56**, 27–34 (2015)
13. Barrett, M., Davidson, E., Prabhu, J., Vargo, S.L.: Service innovation in the digital age: key contributions and future directions. MIS Q. **39**, 135–154 (2015)
14. Goldstein, S., Johnston, R., Duffy, J., Rao, J.: The service concept: the missing link in service design research? J. Oper. Manag. **20**, 121–134 (2002)
15. Immonen, A., Ovaska, E., Kalaoja, J., Pakkala, D.: A service requirements engineering method for a digital service ecosystem. Serv. Oriented Comput. Appl. **10**, 151–172 (2016)
16. Paul, R.J.: Challenges to information systems: time to change. Eur. J. Inf. Syst. **16**, 193–195 (2007)
17. Susman, G.I., Evered, R.D.: An assessment of the scientific merits of action research. Adm. Sci. Q. **23**, 582–603 (1978)
18. Gulfo, J.: Pharmaceutical Compounding: The FDA Is Not The Problem. https://www.forbes.com/sites/realspin/2016/08/29/pharmaceutical-compounding-the-fda-is-not-the-problem/#1ed3715978df. Accessed 16 May 2017
19. Portuguese Ministry of Health: Ordinance 594/2004, 2nd July - Portuguese Law. Diário da República, 1.ª série-B. 129, 3441–3445 (2004)

20. Council of Europe: European Pharmacopoeia (Ph. Eur.) 9th Edition. https://www.edqm.eu/en/european-pharmacopoeia-9th-edition. Accessed 16 May 2017

21. Fontaine, A.L.: Current requirements and emerging trends for labelling as a tool for communicating pharmacovigilance findings. Drug Saf. **27**, 579–589 (2004)

22. Nahler, G.: Company core data sheet CCDS. In: Nahler, G. (ed.) Dictionary of Pharmaceutical Medicine, p. 33. Springer, Vienna (2009)

23. European Commision: A Guideline on Summary of Product Characteristics (SmPC) (2009)

24. HL7: Structured Product Labeling (SPL) Implementation Guide with Validation Procedures (2016)

25. Winterstein, A.G., Linden, S., Lee, A.E., Fernandez, E.M., Kimberlin, C.L.: Evaluation of consumer medication information dispensed in retail pharmacies. Arch. Intern. Med. **170**, 1317–1324 (2010)

26. Songara, R.K., Sharma, G.N., Gupta, V.K., Gupta, P.: Need for harmonization of labeling of medical devices: a review. J. Adv. Pharm. Technol. Res. **1**, 127–144 (2010)

27. Fei, J., Liu, R.: Drug-laden 3D biodegradable label using QR code for anti-counterfeiting of drugs. Mater. Sci. Eng., C **63**, 657–662 (2016)

28. Martin, I.G.: Electronic labeling: a paperless future? Drug Inf. J. **32**, 917–919 (1998)

29. CMDh: CMDh position paper on the use of QRD codes to provide information about the medicinal product 59, 1–5 (2015)

30. AEMPS: Utilización de códigos quick response (QR) para proporcionar información sobre los medicamentos. Agencia española Medicam, 1–4 November 2015

31. Kierkegaard, P.: E-prescription across Europe. Health Technol. (Berl) **3**, 205–219 (2013)

32. Mira, J.J., Guilabert, M., Carrillo, I., Fernández, C., Vicente, M.A., Orozco-Beltrán, D., Gil-Guillen, V.F.: Use of QR and EAN-13 codes by older patients taking multiple medications for a safer use of medication. Int. J. Med. Inform. **84**, 406–412 (2015)

33. Tseng, M.H., Wu, H.C.: A cloud medication safety support system using QR code and Web services for elderly outpatients. Technol. Heal. Care. **22**, 99–113 (2014)

34. Lockwood, W.: QR codes: a digital avenue for PMI. Comput. Pharm. **33**, 12–15 (2013)

35. Furner, C.P., Racherla, P., Babb, J.S.: What we know and do not know about mobile app usage and stickiness: a research agenda. Int. J. E-Serv. Mob. Appl. **7**, 48–69 (2015)

36. Duke, J., Friedlin, J., Li, X.: Consistency in the safety labeling of bioequivalent medications. Pharmacoepidemiol. Drug Saf. **22**, 294–301 (2013)

37. Duke, J., Friedlin, J., Ryan, P.: A quantitative analysis of adverse events and "Overwarning" in drug labeling. Arch. Intern. Med. **171**, 944–946 (2011)

38. Davison, R., Martinsons, M.G., Kock, N.: Principles of canonical action research. Inf. Syst. J. **14**, 65–86 (2004)

39. Chiasson, M., Germonprez, M., Mathiassen, L.: Pluralist action research: a review of the information systems literature. Inf. Syst. J. **19**, 31–54 (2009)

40. Baskerville, R.: information systems with action research: Investigating information systems with action research. Commun. Assoc. Inf. Syst. **2**, 1–32 (1999)

41. Davison, R.M., Martinsons, M.G.: Context is king! Considering particularism in research design and reporting. J. Inf. Technol. **31**, 241–249 (2016)

42. Lindgren, R., Henfridsson, O., Schultze, U.: Design principles for competence management systems: a synthesis of an action research study. MIS Q. **28**, 435–472 (2004)

43. Andersen, P., Ross, J.W.: Transforming the LEGO group for the digital economy. In: ICIS 2016 Proceedings, Dublin, Ireland (2016)

44. Baskerville, R.: Design Theorizing Individual Information Systems. Proceedings of the Pacific Asia Conference on Information Systems (PACIS). pp. 1–12 (2011)

45. Malaurent, J., Avison, D.: Reconciling global and local needs: a canonical action research project to deal with workarounds. Inf. Syst. J. **26**, 227–257 (2016)

46. European Commission: Green Paper on mobile Health ("mHealth") (2014)
47. European Commission: Regulation (EU) 207/2012 on electronic instructions for use of medical devices (2012)
48. French, J.R.P.: Field experiments: changing group productivity. In: Miller, J.G. (ed.) Experiments in Social Process: A Symposium on Social Psychology, pp. 81–96. McGraw-Hill (1950)

Acquiring the Ontological Representation of Healthcare Data Through Metamodeling Techniques

Athanasios Kiourtis$^{(\boxtimes)}$, Argyro Mavrogiorgou,
Dimosthenis Kyriazis, and Marinos Themistocleous

Department of Digital Systems, University of Piraeus, Piraeus, Greece
{kiourtis,margy,dimos,mthemist}@unipi.gr

Abstract. The recent trends in healthcare ICT technologies promise high quality of care, whereas the variety and diversity of healthcare data increase the challenges of information exchange. With healthcare systems facing poor communication and information exchange difficulties, interoperability has become the holy grail of Health IT. While several techniques and researches are conducted to face this challenge, a global solution is still missing, as only particular scenarios are being currently addressed. This paper emphasizes on achieving semantic interoperability across multiple electronic health records (EHRs), through ontologies and Model-driven Engineering techniques. A multi-stepped approach is proposed that transforms EHR datasets into syntactic models and metamodels, in order to be finally transformed into their semantic structure through a mechanism that translates them into a common ontological representation.

Keywords: Interoperability · Ontologies · Model-driven Engineering · Electronic Health Records · Medical standards

1 Introduction

In recent years, the explosion of available ICT services has led to the creation of a variety of sensors and applications supporting personalized care. Currently, our lives are becoming increasingly connected to our devices, other people and a variety of things. Smart machines get smarter, and a new IT reality must evolve with technology architectures and platforms to support the advancement of a digitally connected world. Gartner [1] forecasts that 8.4 billion connected devices will be in use worldwide in 2017, up 31% from 2016, and will reach 20.4 billion by 2020, while total spending on endpoints and services will reach almost $2 trillion in 2017. Most of these devices are designed to monitor patients' daily activities, to detect sleep patterns and emotions [2], while measuring vital signs [3]. Many applications and systems [4] have been developed based on wearable data to detect and predict patient health anomalies, to study the human behavior, and to manage therapy.

Currently, the data provided is heterogeneous and operate independently, whilst the value emerging from their exploitation is limited. Most of the traditional data mining

© Springer International Publishing AG 2017
M. Themistocleous and V. Morabito (Eds.): EMCIS 2017, LNBIP 299, pp. 324–336, 2017.
DOI: 10.1007/978-3-319-65930-5_27

algorithms are not able to deal with both the scale and the heterogeneity of wearable data in healthcare [5], thus it is getting increasingly common for preliminary indications of diseases to be missed. According to [6] medical errors such as delays in diagnosis, preventable surgical complications and medication overdoses are a leading cause of death and injury in the United States. An estimated 80% of the most serious medical errors can be linked to poor communication between clinicians.

Furthermore, today's Electronic Health Records (EHRs) are not offering the desired value to the citizens' health [7], despite offering the potential to transform the health care system from a mostly paper-based industry to one that utilizes clinical and other pieces of information to assist providers in delivering higher quality of care to their patients. Shortly, electronic medical records seem to be the current trend in health care, with findings of many physicians, allied health professionals, pharmacists and hospitals using some form of electronic recording of patient data. Capturing information such as patient demographics, progress notes, problems, medications, vital signs, past medical history, immunizations, laboratory data, and radiology reports, while linking it with other data in EHRs, would be of benefit for learning about outcomes of prevention strategies, diseases, and efficiency of patient pathway management. EHR systems can include many potential capabilities, with three particular functionalities holding great promise in improving the quality of care and reducing costs at the healthcare system level: clinical decision support (CDS) tools, computerized physician order entry (CPOE) systems, and health information exchange (HIE).

One of the biggest obstacles to communication occurs when there are multiple ways of describing a single concept. Thus, it is clear that the problem is the difficulty of data exchange between systems, and data incompatibility. Interoperability is the only sustainable way to help partners to collaborate harmoniously and deliver quality healthcare [8]. Interoperability has become the holy grail of health IT, however there is little consensus on how to get there. A recent study [9] estimated that savings of approximately $78 billion could be achieved annually if data exchange standards were utilized across the healthcare sector. Though interoperability is a technological issue, it's not just a technological issue. As it is seen in other industries, interoperability requires all parties to adopt certain governance and trust principles, and to create business agreements and highly detailed guides for implementing standards. The unique confidentiality issues surrounding health data also require the involvement of lawmakers and regulators. Tackling these issues requires multi-stakeholder coordinated action, and that action will only occur if strong incentives promote it. It becomes clear that a standard vocabulary is needed, for recognizing the complexity of the processes that surround healthcare when aiming towards interoperability of healthcare information.

Interoperability has yet to become a reality, despite such overwhelming support for the free flow of patient data between caregivers. Many challenges - technical, financial and procedural - remain as healthcare moves toward interoperability. One of the biggest hurdles is getting the technology to the point where it will allow the different EHR systems to talk to one another. Hence, we propose an approach for transforming the collected healthcare data into a generic ontological scheme, enabling easier and more interoperable data exchange. In short, a syntactic and semantic model is represented for different EHR datasets of medical standards, where these datasets will be expressed into ontological rules, after being translated into syntactic models. The proposed approach will facilitate the aggregation of the heterogeneous data coming from multiple EHRs to make better-informed health related decisions and support data sharing and integration for better decision making.

This paper is organized as follows. Section 2 describes the related work regarding interoperability, and EHRs, while comparing them with our proposal. Section 3 describes the proposed approach for getting the ontological representation of different EHRs, while Sect. 4 is addressing the challenges of the future internet, analyzing our conclusions and future plans.

2 Related Work

2.1 Interoperability Levels and Barriers

Interoperability is the ability of two parties, either human or machine, to exchange data or information [10]. Interoperability among components of large-scale, distributed systems is the ability to exchange services and data with one another, based on agreements between requesters and providers [11].

In general, at the very top of an "interoperability scale", there exist three (3) levels of health information technology, each one subdivided as foundational, structural, and semantic [12]. Extensive sharing and exchange of information requires that at least two levels of interoperability be reached:

- *Foundational interoperability* allows data exchange from one information technology system to be received by another and does not require the ability for the receiving information technology system to interpret the data.
- *Structural interoperability* is an intermediate level that defines the structure or format of data exchange (i.e., the message format standards) where there is uniform movement of healthcare data from one system to another such that the clinical or operational purpose and meaning of the data is preserved and unaltered.
- *Semantic interoperability* refers to the ability for information shared by systems to be understood at the level of formally defined domain concepts so that the information is computer processable by the receiving systems.

Generally, an established common vocabulary and a unified representation are required elements for syntactic interoperability. However, semantic interoperability is by nature harder to achieve, as it is not usually enough to use the same words, but also to interpret the words with the right meaning. Semantic interoperability takes advantage of both the structuring of the data exchange and the codification of the data including vocabulary so that the receiving information technology systems can interpret the data [13]. This level of interoperability supports the electronic exchange of patient summary information among caregivers and other authorized parties via potentially disparate electronic health record (EHR) systems and other systems to improve quality, safety, efficiency, and efficacy of healthcare delivery. Semantic interoperability is essential for automatic computer processing which will enable the implementation of advanced clinical applications such as Electronic Health Records (EHRs), laboratory systems, and intelligent decision support systems.

In the literature [14], three (3) different categories of barriers can be met when trying to achieve interoperability:

- *Conceptual barriers* refer to the syntactic and semantic differences of information to be exchanged, concerning the modelling and the the level of the programming.
- *Technological barriers* refer to the incompatibility of information technologies, concerning the standards to present, store, exchange, process and communicate the data.
- *Organizational barriers* refer to the definition of responsibility and authority, concerning the incompatibility of organization structures.

2.2 Electronic Health Records (EHRs)

Health records provide a way for exchanging and storing patient-specific information [15, 16] and facilitating information analytics [17–19]. Hence, an EHR is not just a placeholder for data, but also provides additional functionality. EHRs focus on the total health of the patient - going beyond standard clinical data collected in the provider's office and inclusive of a broader view on a patient's care. EHRs are designed to reach out beyond the health organization that originally collects and compiles the information. They are built to share information with other health care providers, such as laboratories and specialists, so they contain information from all the clinicians involved in the patient's care. The National Alliance for Health Information Technology stated that EHR data "can be created, managed, and consulted by authorized clinicians and staff across more than one healthcare organization". The use of EHRs enables transfer of patient data across any kind of border as information moves with the patient - to the specialist, the hospital, the nursing home, the next state or even across the country, in order to be shared and used more effectively for quality assurance, disease surveillance, public health monitoring and research [20, 21].

Generally, each clinical statement is an observation, thus it becomes possible for two statements about the same event to disagree with each other. Such disagreements can often be resolved if the context or provenance of each statement is recorded. Consolidating all information available on individual patients in their single EHR has been a subject of extensive discussions [22], in order to explore ways to use EHRs data to monitor health. EHRs are designed to be accessed by all people involved in the patients' care - including the patients themselves. Active problems, drug dosages, side effects, allergies, laboratory tests, observations, treatments, therapies, medical alerts and drug interactions become available electronically to doctors, enabling that information to be incorporated into electronic health-care systems [23, 24]. In recent years, several EHR medical standards that enable structured clinical content for exchange purposes have been developed, such as DICOM SR [25], HL7 [25–27], openEHR [26, 27], ISO EN 13606 [25–27], GEHR [26], EuroRec [28] and epSOS [29].

The ISO EN 13606 reference model for EHR communication sets out a useful hierarchical structure for clinical information in the context of exchanging clinical information between parties [30]:

- *Composition*: The composition is a set of information committed to one EHR by a healthcare provider relating to a specific clinical encounter. Each composition shares common metadata such as the author, subject (patient), date/time, location, notes, laboratory test reports, or clinical assessments.
- *Folder*: Compositions may be grouped together into folders and subfolders. Folders may be used as containers for various purposes, grouping together the records by episode, care team, clinical specialty, condition, or period.
- *Entry*: Each composition comprises a number of entries, also known as clinical statements. An entry is the information recorded in the EHR because of a single clinical action, observation, interpretation, or intention.
- *Section*: Entries may be grouped together in sections. A section is a grouping of related data within a composition usually under a heading such as Presenting History, Allergies, Medication, and Plans.
- *Element*: The leaf node of the EHR hierarchy is an element, which is a single data value, such as systolic blood pressure, a drug name, or body weight.
- *Cluster*: Related elements may sometimes be grouped into clusters, such as systolic and diastolic blood pressures that may be grouped into a cluster that represents one item in an entry.

2.3 Healthcare Interoperability Mechanisms and Techniques

In the last years, many projects are dealing with interoperability of EHR information systems, proposing solutions based on specific medical standards and technologies in order to satisfy the needs of a particular scenario.

A group of approaches uses XML technologies for achieving interoperability, such as HL7 CDA where representation models are represented in XML and their mappings are defined by using XSLT rules. In [31] the interoperability among different health care systems is riched by annotating the Web Service messages through archetypes defined in OWL. The same researchers presented an approach [32] based on archetypes, ontologies and semantic techniques for the interoperability between HL7 CDA and ISO 13606 systems. Archetypes are represented in OWL, but the problem is approached by using HL7 RIM as basic and common information model. The authors in [33] show how the HL7 Virtual Medical Record (vMR) standard can be used to design and implement a Data Integrator (DI) component. The latter collects patient information from heterogeneous sources and stores it into a personal health record, from which it can then retrieve data, hypothesising that the HL7 vMR standard can properly capture the semantics needed to drive evidence-based clinical decision support systems. The work of [34] must be mentioned, where the authors present a solution based on the Enterprise Service Bus (ESB). The latter is translated into the healthcare domain using a messaging and modelling standard, which upholds the ideals of HL7 V3, combined with a SNOMED CT. Given the importance of the semantic interoperability, the main clinical standardization organizations, ISO, HL7 and open-EHR, are making an effort for harmonizing their specifications. Such approach is the ISO 13606 named "Reference archetypes and term lists" that provides an informative guide in order to represent clinical information codified according to HL7 and open-EHR by using ISO 13606 structures. In addition to this, Detailed Clinical Models (DCM) [35] have been used for defining clinical information independently of a specific clinical standard but with the aim of offering the possibility of being transformed into other medical standards.

Despite the development of medical standards, interoperability and data integration are still open issues. [36] is an example of research efforts that deal with integrating medical standards in order to ensure the interoperability between the HL7 and the IEEE 1451 standards when monitoring the patient data. The authors in [37] proposed an ontology to describe the patient's vital signs and to enable semantic interoperability when monitoring patient data. In the same direction, [38] proposed an ontology driven interactive healthcare with wearable sensors, to acquire context information at real time using ontological methods by integrating external data in order to prevent disease. Moreover, in [39] a framework is presented for the semi-automated integration and linkage of heterogeneous data sets by generating OWL representations of these datasets that can be used for querying and reasoning. What is more, [40] suggests a Cloud Health Information Systems Technology Architecture (CHISTAR) that is able to achieve different types of interoperability with a generic design methodology that uses a reference model that defines usual purpose set of data structures and an archetype model that defines the clinical data logic. The work of [41] should be noted, where the NIST Big Data model was enriched with semantics in order to smartly understand the collected data, and generate valuable information by correlating scattered medical data deriving from multiple wearable devices. In [42, 43], following the concepts of Domain Specific and Domain Agnostic languages, according to how much something could be generalized, we have developed an approach that focuses on how is it possible to gather data from a specific source, to understand how does the latter perform to its data lifecycle and how does this dataset uses the extracted knowledge combined with unlabelled data.

In summary, each particular solution helps to provide access to the patient clinical information for specific clinical organizations. One of the disadvantages of these approaches is that they do not propose frameworks and methods easy to apply to different medical standards. Consequently, in this paper transformation techniques for the interoperable use of data in different services, locations and contexts will be implemented, combined with an envisioned ontology-based approach for achieving semantic interoperability. In short, in our approach, metamodels and ontologies will be used to provide a formal syntactic and semantic model, for representing EHR datasets of multiple medical standards. Thus, a translation mechanism that will transform heterogeneous medical data of different nature and format into a common ontological representation will be built, using the Web Ontology Language (OWL) [44].

3 Proposed Approach

Our approach proposes a comprehensive generic syntactic and semantic mechanism to cope with heterogeneous data deriving from different EHRs, by obtaining the onto-logical representation of EHRs' data.

As depicted in Fig. 1, the proposed architecture aims at harmonizing the healthcare data of multiple medical standards in order to be processed and transformed into a generic ontological scheme, enabling easier and more interoperable data exchange. In more details, the developed mechanism provides a formal semantic model for repre-senting the EHR datasets of medical standards, using the Protégé framework [45] to define OWL ontologies. With such a mechanism, the different EHR datasets are expressed into ontological rules, including detailed and Domain Specific definitions of clinical concepts. More specifically, the transformation process consists of the fol-lowing steps:

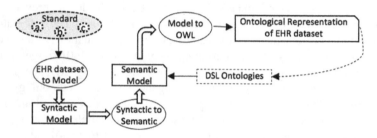

Fig. 1. Obtaining the ontological representation of EHRs' data

1. Each EHR dataset of each different standard is expressed as a syntactic model, through an EHR parser.
2. The syntactic model is transformed into a semantic model by using a set of rules that have been predefined for the "syntactic to semantic" model mapping.
3. The semantic model along with the predefined Domain Specific (DSL) ontologies, is transformed into the ontological representation of its standard, through a "model to OWL" ontologies transformation.

This process is repeated for all the EHR datasets of the different medical standards, resulting into each EHR's ontological representation.

Our solution combines a series of technologies, namely, ontologies and meta-models. The input to our transformation process is an EHR dataset and the output is the ontological representation of the EHR dataset, respectively. The architecture of this solution is depicted in Fig. 2. Generally, two different layers can be distinguished, the *Metamodel Layer* and the *Ontology Layer*.

- The *Metamodel Layer* contains the models and the metamodels corresponding to the EHR datasets. In this layer, the transformation mappings are formalized.
- The *Ontology Layer* comprises a series of ontologies that model EHR-related knowledge for the different medical standards. In this layer, the transformation of the EHR datasets into ontologies is performed.

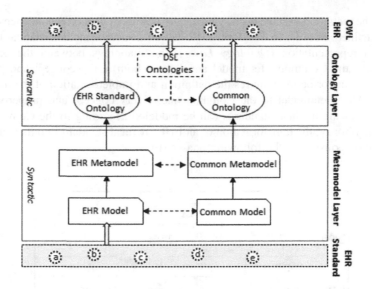

Fig. 2. Architecture of the proposed approach

In more details, as depicted in Fig. 2, the architecture of the proposed approach consists of the following:

EHR Standard
The EHR Standard component is assumed that it contains multiple datasets of different medical standards, which will be translated into OWL Ontologies through their transformation from the *Metamodel* and the *Ontology Layer* accordingly.

Metamodel Layer
The *Metamodel Layer* is based on the idea of Model-driven Engineering (MDE) about using models at different abstraction levels for developing systems. Generally, such approaches facilitate the development of formal and maintainable solutions, constituting an optimal technological infrastructure for achieving our goal. In MDE, model transformation is the process of converting a model M1 conforming to metamodel MM1, into a model M2 conforming to metamodel MM2. In recent years, several model transformations languages have been defined, such as RubyTL [46], a rule based hybrid transformation language for defining transformation rules in both declarative and imperative ways. In this work, model transformations expressed in RubyTL will be used for the transformation of the models, while the latter will be represented into OWL ontologies using the Protégé Framework, along with the model to text language MOFScript [47]. More particularly, in our case, we assume that a model is an instance of a particular metamodel, which is why the ontologies developed for the various EHR medical standards will be expressed out of the metamodels. For this purpose, the Ontology Definition Metamodel [48] standard, which defines the semantics of the transformation of models to OWL ontologies, will be used.

Once the metamodels have been syntactically obtained, the correspondences among them are defined. In order to transform EHR data of a specific medical standard into its ontological representation (*Ontology Layer*), the mappings between the particular standard and the Common Metamodel are defined both at the concept and property levels. This transformation requires the definition and implementation of the mappings from the EHR Metamodel to the Common one. Every concept and property of the metamodels of the medical standards can be modeled according to the Common representation. Shortly, the transformation of an EHR medical standard into its metamodel representation consists of the following phases (Fig. 3):

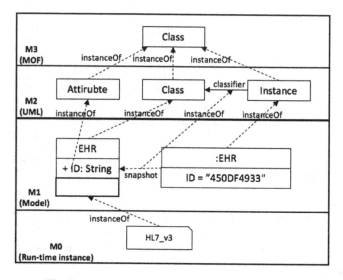

Fig. 3. Internal architecture of the metamodel layer

1. The EHR standard (M0) input is transformed into its Model representation (M1).
2. The Model representation (M1) is transformed into a generalized way following the UML model representation (M2), where the different instances of the model representation are categorized into different classes.
3. All the classified instances of the UML model are being categorized into a generic Class, getting their final Metamodel representation using the MOF standard (M3), in order to model the abstract syntax of the UML model.

With regards to the Common model, it is transformed into the Common Metamodel following the same mechanism. Both Metamodels are given in their Syntactic expression to the *Ontology Layer*, to be transformed into their Semantic ontological representation.

Ontology Layer
The *Ontology Layer* provides the formal semantics of our domain, and is composed of a series of OWL ontologies developed for the multiple EHR medical standards. In more

details, the *Ontology Layer* is built by identifying the common and disjoint knowledge defined in the ontologies of the multiple EHR medical standards, so it could be considered as a global EHR ontology. In the *Ontology Layer*, the detection of the equivalent concepts and data types is supported by the ontology integration methodology developed in [49]. More specifically, three different types of ontologies can be seen:

- The *EHR Standard Ontology* refers to the ontologies that are constructed through the known and easily identifiable categories of the healthcare data that each medical standard represents. For instance, concepts such as ACTIVITY and ISM_-TRANSITION are defined only in OpenEHR. Thus, they were added to the EHR Standard Ontology as a single concept.
- The *Common Ontology* refers to the ontologies that are constructed through the common parts and structures of the different medical standards, which are identified during the parsing and the transformation process. For instance, concepts such as FOLDER, COMPOSITION, SECTION, or ELEMENT are common to several medical standards. Thus, they were added to the Common Ontology as a single concept.
- The *DSL Ontology* refers to the ontologies that refer to a specific healthcare domain, and can be used/reused to identify possible healthcare data that cannot be a priory categorized or identified.

In our case, the metamodels gathered by the *Metamodel Layer* are being transformed into their ontological representation using the Protégó Framework, in combination with ontologies, information and metadata belonging to the DSL Ontologies.

OWL EHR

In the OWL EHR component, the datasets of the different medical standards have been transformed into their ontological representation, while they provide metadata information in order to form the DSL Ontologies.

As a result of the described process, the ontological representation of different medical standards has been formed, and can be integrated into different healthcare systems, so as heterogeneous EHR datasets can be transformed into understandable, interoperable and shareable knowledge that can be used for a better communication between healthcare professionals. It must be noted that the constructed ontological representations are then stored into an Ontology–Based Data Repository (GraphDB [50]), that can be easily retrieved and used.

4   Conclusions

Healthcare-related systems make interoperability among different platforms, records, sources, and devices, a challenging task. The challenges to achieving semantic interoperability transcend the technical field, as there are cultural, social, policy and economic barriers to data sharing. In this paper, we proposed a generic semantic data architecture that copes with the EHR data heterogeneity by suggesting a mechanism confronting the challenge of heterogeneous data that is spread across different health care providers and systems. A translation mechanism that translates heterogeneous data of different nature and format into a common ontological representation was presented, using metamodels and OWL ontologies for providing a formal syntactic and semantic model, and representing EHR datasets of multiple medical standards. The architecture used in this work distinguishes the *Ontology Layer* that comprises a series of ontologies that model EHR-related knowledge for the different medical standards, and the *Metamodel Layer* that contains the metamodels corresponding to the EHRs datasets. The whole transformation process is located in the *Ontology Layer* by means of applying model-to-model and model-to-text transformations rules.

The presented implementation is an innovative approach of data interoperability between multiple medical standards, as recent researches are designed to give a solution to specific problems. It must be noted that EHR technology has shown to be effective in transforming the quality, safety, and efficiency of care in health care organizations that have implemented it successfully. However, successful implementation is not easy. Integration of EHR technology into clinical workflow, the adoption strategies used when implementing EHR technology, and technology upgrades and continuous quality improvement are all issues when seeking to implement and use EHRs to store and manage clinical information.

The main objective remains to share the same semantics of data to ensure a smart interpretation for better decisions. Currently, we are working on the evaluation of different mapping languages from both the Semantic Web and Model-driven Engineering areas, since having a formal representation of these mappings will be useful for extending our approach to additional medical standards. Furthermore, we are also evaluating the developed mechanism, by testing it with multiple EHR datasets, of various medical standards. Our future work will focus on the definition of more elaborated scenarios and on the implementation of advanced algorithms by collaborating with medical experts. We aim to develop a generic and extensible architecture, capable of dealing in the future with additional medical standards, addressing security and privacy constraints.

Acknowledgements. The CrowdHEALTH project has received funding from the European Union's Horizon 2020 research and innovation programme under grant agreement No. 727560. Athanasios Kiourtis would also like to acknowledge the financial support from the "Foundation for Education and European Culture (IPEP)".

References

1. Gartner Press Release. http://www.gartner.com/newsroom/id/3598917
2. Kwon, J., Kim, D.H., Park, W., Kim, L.: A wearable device for emotional recognition using facial expression and physiological response. In: 2016 IEEE 38th Annual International Conference of the Engineering in Medicine and Biology Society (EMBC), pp. 5765–5768. IEEE Press (2016)
3. Dionisi, A., Marioli, D., Sardini, E., Serpelloni, M.: Autonomous wearable system for vital signs measurement with energy-harvesting module. IEEE Trans. Instrum. Measur. **65**(6), 1423–1434 (2016)

4. Koshti, M., Ganorkar, S., Chiari, L.: IoT based health monitoring system by using raspberry pi and ECG signaly. Int. J. Innov. Res. Sci. Eng. Technol. **5**(5), 8977–8985 (2016)
5. Kalra, M., Lal, N.: Data mining of heterogeneous data with research challenges. In: Colossal Data Analysis and Networking (CDAN), pp. 1–6. IEEE (2016)
6. Medical errors drop with improved communication during hospital shift changes. https://source.wustl.edu/2014/11/medical-errors-drop-with-improved-communication-during-hospital-shift-changes/
7. European Citizens' Digital Health Literacy. http://ec.europa.eu/public_opinion/flash/fl_404_en.pdf
8. Domingo, M.C.: Managing healthcare through social networks. IEEE Comput. **43**(7), 20–25 (2010)
9. Mead, C.N.: Data interchange standards in healthcare IT-computable semantic interoperability: now possible but still difficult, do we really need a better mousetrap? J. Healthc. Inf. Manag. **20**(1), 71–78 (2006)
10. New European Interoperability Framework. https://ec.europa.eu/isa2/sites/isa/files/eif_brochure_final.pdf
11. Beig, L., Eghbali, N., Ghavamifar, A., Kharrat, M.: A multi-dimensional framework for inter-organizational collaboration in a knowledge network. Arch. Des Sci. **66**(1) (2013)
12. Venkatasubramanian, K.K., Nabar, S., Gupta, S., Poovendran, R.: User-driven healthcare: concepts, methodologies, tools, and applications. In: IGI Global (2013)
13. Heatlhcare Interoperability. https://healthcareit.me/2015/04/07/the-four-definitions-of-interoperability/
14. Amor, R.: Interoperability for Enterprise Software and Applications (2010)
15. An Introduction to Electronic Health Records. The McGraw-Hill Companies (2011)
16. Jensen, P.B., Jensen, L.J., Brunak, S.: Mining electronic health records towards better research applications and clinical care. Nat. Rev. Genet. **13**(6), 395–405 (2012)
17. Electronic Health Records. http://www.healthworldnet.com/link-directory/stay-healthy/electronic-health-records.html
18. Potential of electronic personal health records. http://www.ncbi.nlm.nih.gov/pmc/articles/PMC1949437/
19. Kush, R.D., Helton, E., Rockhold, F.W., Hardison, C.D.: Electronic health records, medical research, and the tower of babel. New Engl. J. Med. **358**(16), 1738–1740 (2008)
20. Rynning, E.: Public trust and privacy in shared electronic health records. Eur. J. Health Law **14**(2), 105–112 (2007)
21. Kalra, D.: Electronic health record standards. Schattauer GMBH-Verlag, Stuttgart (2006)
22. Jaspers, M., Knaup, P., Schmidt, D.: The computerized patient record: where do we stand? Methods Inf. Med. **45**(1), 29–39 (2006)
23. Friedman, D.J., Parrish, R.G., Ross, D.A.: Electronic health records and US public health: current realities and future promise. Am. J. Publ. Health **103**(9), 1560–1567 (2013)
24. Eichelberg, M., Aden, T., Riesmeier, J., Dogac, A., Laleci, G.B.: Electronic health record standards–a brief overview. In: Proceedings of the 4th IEEE International Conference on Information and Communications Technology (2006)
25. Begoyan, A.: An overview of interoperability standards for electronic health records. USA: society for design and process science (2007)
26. Kalra, D.: Electronic health record standards, pp. 136–144. Schattauer GMBH-Verlag (2006)
27. Schloeffel, P., et al.: The relationship between CEN 13606, HL7, and openEHR. HIC 2006 and HINZ 2006: Proceedings, p. 24 (2006)
28. EuroRec. http://www.eurorec.org/
29. epSOS. http://www.epsos.eu

30. Benson, T., Grieve, G.: Principles of health interoperability: SNOMED CT, HL7 and FHIR. Springer, Heidelberg (2016)
31. Dogac, A., et al.: Artemis: deploying semantically enriched Web services in the healthcare domain. Inform. Syst. **31**(4), 321–339 (2006)
32. Schulz, S., Udo, H.: Part-whole representation and reasoning in formal biomedical ontologies. AI Med. **34**(3), 179–200 (2005)
33. Marcos, C., et al.: Solving the interoperability challenge of a distributed complex patient guidance system: a data integrator based on HL7's virtual medical record standard. J. Amer. Med. Inform. Assoc. **22**(3), 587–599 (2015)
34. Ryan, A., Eklund, P.: A framework for semantic interoperability in healthcare: a service oriented architecture based on health informatics standards. Stud. Health Tech. Inform. **136**, 759 (2008)
35. Goossen, W., Goossen-Baremans, A., Van Der Zel, M.: Detailed clinical models: a review. Healthc. Inform. **16**(4), 201–214 (2010)
36. Kim, W., Lim, S., Ahn, J., Nah, J., Kim, N.: Integration of IEEE 1451 and HL7 exchanging information for patients' sensor data. J. Med. Syst. **34**(6), 1033–1041 (2010)
37. Lasierra, N., et al.: A three stage ontology-driven solution to provide personalized care to chronic patients at home. J. Biomed. Inform. **46**(3), 516–529 (2013)
38. Kim, J.H., Lee, D.S., Chung, K.Y.: Context-aware based item recommendation for personalized service. In: Proceedings of International Conference on Information Science Applications, pp. 595–600. IEEE Computer Society (2011)
39. Bouamrane, M.M., Tao, C., Sarkar, I.N.: Managing interoperability and complexity in health systems. Methods Inf. Med. **54**(1), 1–4 (2005)
40. Bahga, A., Madisetti, V.K.: A cloud-based approach for interoperable electronic health records (EHRs). IEEE J. Biomed. Health Inform. **17**(5), 894–906 (2013)
41. Mezghani, E., Exposito, E., Drira, K., Da Silveira, M., Pruski, C.: A semantic big data platform for integrating heterogeneous wearable data in healthcare. J. Med. Syst. **39**(12), 185 (2015)
42. Kiourtis, A., Mavrogiorgou, A., Kyriazis, D., Maglogiannis, I., Themistocleous, M.: Towards data interoperability: turning domain specific knowledge to agnostic across the data lifecycle. In: 30th International Conference on Advanced Information Networking and Applications Workshops (WAINA), pp. 109–114. IEEE (2016)
43. Kiourtis, A., Mavrogiorgou, A., Kyriazis, D., Maglogiannis, I., Themistocleous, M.: Exploring the complete data path for data interoperability in cyber-physical systems. Int. J. High Perform. Comput. Netw. (IJHPCN) (2016, in press). Inderscience Publishers
44. OWL. https://www.w3.org/TR/owl-guide/
45. Protégé Framework. http://protege.stanford.edu/
46. Cuadrado, J.S., Molina, J.G., Tortosa, M.M.: RubyTL: a practical, extensible transformation language. In: Rensink, A., Warmer, J. (eds.) ECMDA-FA 2006. LNCS, vol. 4066, pp. 158–172. Springer, Heidelberg (2006). doi:10.1007/11787044_13
47. MOFScript. https://eclipse.org/gmt/mofscript/
48. Kendall, E., Dutra, M., Jacobs, J., Schwab, S.: The Ontology Definition Metamodel (ODM) and Motivation for a Semantic Meta-Object Facility (SMOF) (2006)
49. Fernández-Breis, J.T., Martiinez-Bejar, R.: A cooperative framework for integrating ontologies. Int. J. Hum.-Comput. Stud. **56**(6), 665–720 (2002)
50. GraphDB. http://graphdb.ontotext.com/

Digital Wellness Services and Sustainable Wellness Routines

Christer Carlsson[✉] and Pirkko Walden

Institute for Advanced Management Systems Research, Turku, Finland
{christer.carlsson,pirkko.walden}@abo.fi

Abstract. Our focus is on digital wellness services for the "young elderly" (the 60–75 years old) age group and we are working out ways to turn the adoption and use of the services to sustainable wellness routines. Wellness routines will help young elderly people to improve and maintain their independence and their functional capacity. The interest for digital services – mostly distributed as apps on smartphones – appears to wane after a few months, which is a problem, as wellness routines will produce health effects only if sustained for months and years. Digital coaching will help users to build good and effective, sustainable wellness routines. We identify potential early adopter groups and work out the functionality of digital coaching for wellness services.

Keywords: Digital coaching · Digital wellness services · Young elderly

1 Introduction

The proportion and socioeconomic impact of ageing citizens is already high in most EU countries (cf. [28]) and there is growing political pressure to find long-term strategies for the ageing population.

The young elderly [the 60–75 years old] age group represents 18–23% of the population in most EU countries; this is a large segment, recent statistical estimates show that it will be about 97 million EU citizens by 2020. In the Nordic countries there were close to 4 million citizens in the young elderly age group in 2016 (Table 1).

Table 1. Young elderly in Nordic counties [20]

Country	Women	Men	Total
Denmark	505 981	484 533	990 514
Faroe Island	4 022	4 178	8 200
Greenland	2 897	3 819	6 716
Finland	556 889	512 961	1 069 850
Åland	3 046	2 768	5 814
Iceland	25 319	25 412	50 731
Norway	422 212	418 641	840 853
Sweden	842 441	825 736	1 668 177
Nordic	*2 025 067*	*1 950 610*	*3 975 677*

© Springer International Publishing AG 2017
M. Themistocleous and V. Morabito (Eds.): EMCIS 2017, LNBIP 299, pp. 337–352, 2017.
DOI: 10.1007/978-3-319-65930-5_28

The costs for health and social care for this age group is estimated to be around 100–400 B€ on an annual basis (variations between countries are expected to be high). It should be possible to cut this cost by 10–15% annually with digital technology. Furthermore, given their income level and accumulated wealth, the young elderly represents very large and growing markets for digital wellness services, especially in higher per capita income EU countries.

The focus on the *young elderly* is a new concept and an interesting target market for digital services. The society needs to have a strategy and priorities for the young elderly, which are – and need to be – different from the strategy and priorities for the senior [75+] age group. A majority of the young elderly are reasonably healthy, active and socially interactive and do not require much intervention or support from the health and social care systems of the society. The interpretations of "relatively" differ between countries and cultures. In a research perspective, the young elderly provide fertile ground for collaborative research on innovative solutions for large user groups, for testing designs, benefits and impacts of digital wellness services.

This paper addresses the potential for a new, significant market for digital services – the young elderly. It identifies key features of early adopters among the young elderly; it shows that wellness services should be digital and will contribute to sustainable wellness routines with the support from digital coaches, and it shows that digital coaching is a new and important research area for digital wellness services.

The existence of opportunities to progress beyond the state-of-the-art in research and to work out service innovations with good business potential is interesting for industry/university collaboration efforts. We have collected first-hand knowledge and experience of the young elderly market in Finland through the *Bewell* project 2014–15, which was part of the Data to Intelligence (D2I) SHOK programme (Tekes 340/12) (cf. [4–7]). A first result of the *Bewell* showed that wellness services are needed and useful for individual young elderly; a second result was that wellness services need to be digital and accessible through smartphones as this is both an efficient and cost-effective way of distribution. A third result emphasized the need for an ecosystem of designers, developers, builders and maintainers of the digital services as there probably will be hundreds of thousands of users. This again, as a fourth result, requires an industry and university collaboration network. A fifth result, is that there is a potential for building a young elderly EU market, which may both grow very fast and be very large.

The ultimate aim for the digital wellness services is to help form and support interventions in the daily routines of young elderly so that the interventions form sustainable wellness routines. In turn, these routines will help to preserve physical wellness for the young elderly (there are actually more wellness dimensions, cf. Section 4) and help to build the basis for healthier elderly years.

The rest of this paper follows a storyline. In Sect. 2, we summarize some background facts and results from the *Bewell* research project. In Sect. 3, we give a brief overview of Information Systems research methodology to motivate our multi-method approach. In Sect. 4, we have a summary of literature on digital wellness and in Sect. 5, we show results from a study of young elderly on digital wellness services in the Åland Islands. In Sect. 6, we work out some key principles for digital coaching, and in Sect. 7 there is a summary with some conclusions.

2 Background

Our work with young elderly groups in *Bewell* (cf. [4]) has shown that it is quite easy to get enthusiastic support for wellness activities on smartphones in the short term. We have also learned that we need long-term impacts on daily routines to get long-term health effects from the digital wellness services. The long-term impact turned out to require digital wellness service infrastructures and ecosystems that are able to self-organize, tailor, grow, maintain and sustain the services in the years to come.

In the *Bewell*, we worked out a plan in four phases to form, support and sustain the interventions that would help shape wellness routines among large groups of young elderly:

- Develop and validate digital wellness services for young elderly suitable to be implemented, monitored, and maintained for (very) large groups of users
- Develop and validate a virtual coach for the digital wellness services
- Develop and validate adoption models for digital wellness services and coaching among young elderly
- Develop and validate sustainable solutions for ecosystems and business models to initiate exploitation/implementation/deployment programmes for the wellness services

We have gradually found out that a useful instrument would be a virtual wellness coach for the young elderly; we called this coach VAIDYA [Vaidya in Sanskrit means "physician"] and worked out the following generic features and characteristics for the young elderly context (cf. [4, 5]).

- Collects (i) individual data, (ii) group/cluster data, (iii) big user group data and develops information and knowledge on-line, in almost real-time
- Adaptive coaching on individual, group/cluster, general insight levels
- Coaching is built with different granularities (i) context dependent, (ii) individual wellness, goals, (iii) "good practice" as compiled from updates (discrete, continuous) of wellness activities from large and growing user groups
- Adaptive in short term to develop recommendations for (i) individuals, (ii) groups/clusters
- Adaptive to context, new needs and requirements, new unobtrusive technology (devices, software, user interfaces, feedback modalities)
- Adaptive to national language, cultural habits, changes in legislation
- Designed to be handled by ecosystem of SME:s

In general, working with digital coaches contributes some more features to the development and use of digital wellness services.

- Virtual coaching is continuous and is changing dynamically
- Virtual coaching needs to be maintained, supported, enhanced, changed, developed (sometimes radically)
- Unobtrusive technology: virtual coaches on smart phones + omnivore platform that will have numerous interface solutions for sensor data + data from wearables + data from technology innovations

- New intelligent ICT environments; intelligence, that builds on (i) soft computing [computational intelligence beyond machine intelligence]; (ii) approximate reasoning [to work out information and knowledge from imprecise data], and (iii) soft ontology [to work out knowledge from imprecise data, information]
- Co-creation of new digital wellness services – good theory, good models for carrying this out with 20–25 member user groups
- Action design research [ADR] as methodological framework for co-creation of digital services
- Ideal wellness profiles worked out as a basis for recommendations
- Maturity models for digital wellness services for individuals and organisations to benchmark progress and "good practice'

The VAIDYA coach aims to improve quality of life on an individual level by increasing wellness. It aims to improve the everyday routines of groups of young elderly with context and user group adaptive wellness routines. The digital wellness services - with the integrated VAIDYA coach - will intervene in the everyday routines of the young elderly and support the formation of wellness routines.

Intervention in the young elderly age group requires less effort, fewer resources and less cost than intervention in the senior age group, and intervention among young elderly will significantly reduce the need for intervention in the senior age group (in which intervention will require the use of health and social care systems at high and increasing cost).

The WHO defines wellness as "the complete mental, physical as well as social well-being of a person or groups of persons in achieving the best satisfying or fulfilling life and not merely the absence of disease or any form of infirmity (cf. [30]). We will avoid a conceptual debate (cf. [21, 26]) and use the following definition: *wellness – to be in sufficiently good shape of mind and body to be successful with all everyday requirements.*

The young elderly age group will have some risk to suffer functional impairment because of advancing age. This risk builds on common sense and spreads through common wisdom – most of the young elderly have first-hand experience of somebody in the family or among their friends, colleagues at work or business associates that suffer from functional impairment after having reached the same age group. This makes them motivated to spend time and resources to hold off functional impairment and to invest time and resources to improve on selected wellness functions.

Wellness is not dependent on strictly regulated health or social care data that carries confidentiality and privacy limitations. Nevertheless, wellness data and the results of using digital wellness services could be available as background information when working out health care issues in the large health care support systems.

The research goal – *to work out how digital wellness services can contribute to sustainable wellness routines for young elderly* – will be reached through an understanding of four different elements. These include, (i) the importance of wellness routines for the young elderly; (ii) who the early adopters of digital wellness services could be; (iii) how digital wellness services can be sustainable with the support of digital coaches, and (iv) how digital wellness services can contribute to sustainable wellness routines. The understanding builds on work with groups of young elderly (totalling about 100 semi-structured interviews), a survey of 101 young elderly respondents, cross-fertilization of several conceptual frameworks and outlines of roadmaps to sustainable wellness routines for young elderly (cf. [4–7] for details). We have set the context with background facts in Sects. 1 and 2; we also found a first understanding of (i) as stated by one of our young elderly co-workers – *it is nicer to get old if you are in good shape.*

3 Information Systems Research Methodology

We have found, in a decade of studies of the development and launching of mobile value services (cf. [2, 7, 8])), two common denominators of most services: (i) they have been designed by software specialists for some ill-defined group of potential users, and (ii) they have failed to gain acceptance in the market. Sometimes they failed even after heavy promotion and marketing, because the intended customers saw no need for the services (cf. [8]). In the following, we will work out some IS methodology contributions to the design and implementation of digital wellness services.

In Information Systems research, we have been advocating for quite some time that mobile services are more often useful when we design them in co-operation with actual users and not for some loosely defined potential users (cf. [31]). As the digital wellness services will be "close" to the daily life of the users, it is a good proposition to use the same approach as for mobile services.

The IS methodologies that are useful are the design science [DS] and action design research [ADR] that are useful approaches for interaction with the actual users (cf. [23]). Both these methodologies are multi-disciplinary as they draw upon behavioural action research, group dynamics, psychology and sociology and are highly compatible with modern software engineering methodologies.

We should add the notion of co-creation of services to the DS and ADR to build joint work of the young elderly users, software developers working on wellness service platforms (cf. [13]), and researchers working on wellness service designs and virtual coaches. Gronroos [13] first developed the co-creation methodology in marketing research which can be adapted to serve as an extension of the multidisciplinary IS methodology.

The adoption of wellness routines will progress over time and will probably follow some processes originally developed in behavioural sciences and in sociology (e.g. cf. [3]).

Differently from the adoption of wellness routines, the adoption of digital services is a faster moving process that is part of much published research in the last two decades. The basic conceptual frameworks for the acceptance and use of digital services are the TAM, UTAUT and UTAUT2 of which there now are 100 + variations (cf. [31]). The user surveys of wellness services that are part of the testing and validation phases normally build on the UTAUT framework but with user specific adaptations and modifications (cf. Sect. 4).

Work with Finnish young elderly user groups [several groups with about 20 participants; in total about 100 participants] showed that there are specific services that need to be worked out and implemented (cf. [4–6]). The experience with the user groups supports the notion that an omnivore platform (a platform that "eats every-thing", i.e. will support multiple devices, applications, services, links, user profiles, etc.) is a good IS basis for large-scale implementation of digital wellness services. Tests with such a platform, that works with the Android and iOS operating systems, proved the notion; the *Wellmo* platform is now operational and in commercial use. We also learned that that most (or all) of the back-end support needs to be built as cloud services. Fortunately enough, modern cloud services (cf. [14, 15]) offer the necessary functionality, the ability to scale up to much larger user groups and the flexibility to offer support for a large variety of digital services.

Results from our work on digital wellness services show that generic data collected through smartphone applications (steps per day, walked distance, estimated burn of calories, etc.) becomes tedious routine in about 3–6 months (cf. [4–6]). Then the digital wellness services will be forgotten and disused. We found more active use and better adoption rates when activity bracelets were added to the platform as (i) the data included more features (sleep patterns, heart rate, programmed exercise, etc.) and (ii) regular update of the software added more and improved features at 3–5 month intervals (cf. [4–6]). Further research results now show that personalized digital coaching could activate the users, improve the adoption rate of the services and con-tribute to a sustained use of the digital wellness services.

There is a need for analytics methodology to build algorithms for data fusion and information fusion, and then for soft ontology to build knowledge fusion; the fusion drives digital coaching (cf. Sect. 6). The fusion technologies should produce advice and instructions from sensor data generated by activity bracelets, smart watches and smart phones.

It is of course self-evident that the digital wellness services will not be viable unless the service developers and -providers can build viable business around them. A pos-sible approach is the STOF conceptual framework (cf. [3]) for business models in the digital service industry. The use of a Business SCRUM methodology (cf. [3]) will make it possible to activate networks of SMEs as service developers and –providers, for service maintenance and further development. Here it will suffice to make the point and refer to the Business SCRUM as the methodology; details on the building and maintenance of ecosystems will be worked out in forthcoming papers.

As we now have worked out the IS methodology to be used and outlined the technology basis for the digital wellness services we will move on to the research goals (ii)–(iv).

4 Digital Wellness

There is a reasonable logic for working out digital wellness services for the young elderly (cf. research goal (i)). Improved health in this age group will significantly raise the probability for continued improved health in the senior age group. Ill-health among large numbers of senior citizens turns out to be very expensive (in Finland the health care costs for the 65+ age group was 3.8 B€ in 2014; statistics show that a Finnish citizen on average spends 80% of their lifetime health care costs during the last 10 years of his/her life) (cf. [25]).

There has been lively debate over the years about the dimensions of wellness; we will cover some of the aspects even if we focus on physical and intellectual wellness; UCR [27] has compiled the following seven dimensions (here abbreviated):

- Social Wellness is the ability to relate to and connect with other people in our world.
- Emotional Wellness is the ability to understand ourselves and cope with the challenges life can bring.
- Spiritual Wellness is the ability to establish peace and harmony in our lives.
- Environmental Wellness is the ability to recognize our own responsibility for the quality of the air, the water and the land that surrounds us.
- Occupational Wellness is the ability to get personal fulfilment from our jobs or our chosen career fields while still maintaining balance in our lives.
- Intellectual Wellness is the ability to open our minds to new ideas and experiences for personal decisions, group interaction and community betterment.
- Physical Wellness is the ability to maintain a healthy quality of life that allows us to get through our daily activities without undue fatigue or physical stress.

Adams [1] and Els and de la Rey [11] show the need for holistic wellness models and they have tested this approach with a large empirical study. A holistic wellness approach suggests that there could be trade-offs and compromises among the wellness dimensions – time to improve on intellectual wellness could be taken out of exercise time to build physical wellness, or both could be compromised with to instead log time on social wellness. This invites analytical modelling for optimal compromises (cf. [21]) which will be the theme of a forthcoming study.

There are of course challenges to introduce digital wellness services; the first challenge is that common wisdom has it that young elderly do not have smartphones and do not use smartphone applications. Statistics now show that smartphones are becoming affordable general purpose instruments and will be even more so by the year 2020 (the mobile connection subscriptions are more than 100% of the population in most EU countries; the proportion of smart phones is closing on 70% in several EU countries). Recent statistics [17] shows that the number of smartphones in Finland was 6.029 million in 2016 (about 73% of the young elderly have a smartphone, cf. Sect. 5); on an average Finnish consumers had 20 apps installed on their smartphones, of which 4 were paid apps (less than the EU average).

A second challenge is the doubt that digital wellness services will be at all attractive to the young elderly. This follows on a belief we have found in the market for mobile value services (cf. [2, 3, 7, 8]). The common wisdom is that (i) elderly people will not learn how to use services on mobile phones; (ii) they have no real use for mobile services in their daily routines; (iii) advanced technology fits (mostly) young people – and (iv) if elderly people use the services they create the wrong brand image for the service developers. We have now been running a research and development project for digital wellness services with support from two associations for elderly with more than 100 000 members; our findings show that the mobile service market beliefs are misconceptions.

A third challenge is to make the digital wellness services sustainable and through them help form sustainable wellness routines. Wellness routines will be sustainable parts of daily routines if the time spent on/with them makes sense, i.e. they contribute enough to keep a young elderly in sufficiently good shape of mind and body to be successful with all everyday requirements. There may be some challenge to follow up on this contribution, as it will be partly objective and measurable but partly subjective. Digital wellness services will be sustainable if they form sustainable wellness routines at reasonable cost. This is in line with [19] where new services support mobile health care and telehealth. The contribution of an individual digital service will also be partly objective and partly subjective; technological innovations will often produce more advanced services at lower cost and existing services will be replaced, which may change wellness routines. The sustainability of digital services is addressed in [18], where the key driver is the shift of the value creating process from *value chains* to *value networks* where different actors simultaneously contribute to value co-creation processes, which we have in mind for the digital wellness services.

A <u>fourth</u> challenge is to work out (i) valid, theory-based results on how the design of digital wellness services will match multiple wellness criteria; then to work out (ii) empirically verifiable results on the intervention with digital wellness services; then the final step will be (iii) empirical verification on how digital wellness services will form sustainable wellness routines. This will require large-scale research projects over several years with variations of digital wellness services (and supporting wellness technology) and large young elderly user groups; getting funding for this research has proved to be a challenge.

We have collected data with a survey among young elderly in the Åland Islands to test some of the ideas and proposals we have on digital wellness services; a summary of the results is in Sect. 5. We have also carried out semi-structured interviews with about 100 young elderly in three regions in Finland to collect insight on the possible use of digital wellness services (cf. [24, 29]).

5   Acceptance of Digital Wellness Services

We cooperated with the association for elderly in the city of Mariehamn (in the Åland Islands) and asked them to invite their young elderly to participate in a survey in the fall 2015 – winter 2016; we collected 101 usable answers (26.6% response rate) (see [4, 5] for details). There are two challenges we needed to test (the first and second challenges, Sect. 4): (i) young elderly do not have smartphones and they do not use app-based services; (ii) digital wellness services may not/will not be attractive to young elderly.

We first collected descriptive data on the sample. The proportion males/females is 44.6/54.5%; 83.1% of the respondents belong to the young elderly, and a further 14.9% are a bit older. Of the sample 65.3% are married and 14.9% are widowed; 77.2% have a university or technical/commercial degree, 20.8% have a basic education.

In the sample, 75.2% are retired and 23.8% are working full- or part-time or with voluntary work. The most typical annual incomes before tax are <30 k€ (51.5%), 30–40 k€ (19.9%), 40–50 k€ (9.9%) and >50 k€ (17.8%). These profiles are typical for the Åland Islands and are representative for their group of young elderly; here we will use these profiles as a snapshot – we will add to the profiles in the following to get a fuller description of the young elderly.

A majority of the respondents (about 73%, but not all) use smartphones; we could confirm this by the result that 72.9% of the respondents use mobile apps for navigation, weather forecasting, Internet search, etc. (which require smartphones).

We added to the profiles by asking how useful, easy to use and valuable mobile apps are for them following the Davis [10] and Venkatesh et al. [31] structure of questions. For the about 70 respondents that use mobile applications we found that the adoption of mobile apps scored high on a 5-grade Likert scale on several items:

- mobile apps are useful in my daily life [4.32];
- I will continue to use mobile apps [4.19];
- mobile apps help me to carry out my tasks faster [4.08];
- using mobile apps helps me to carry out important tasks [3.94];
- I can use mobile apps without assistance [3.91];
- I have the necessary knowledge to use mobile apps [3.87];
- It is easy for me to learn to use mobile apps [3.79];
- I can use the mobile apps I need with the phone I have [3.75].

The results give us some insight to build on: (i) the young elderly use of smartphones is sufficient to launch digital wellness services; (ii) the young elderly are confident users of mobile apps, which is a prerequisite for getting the wellness services adopted. This settles the first challenge ("young elderly do not have smartphones") and as mobile apps are digital services the second challenge ("digital wellness services may not/will not be attractive to young elderly") finds an answer for the sample in the Åland Islands. The research continues, but so far, we can stick to the proposal that digital wellness services could be developed and offered on smartphones for the young elderly.

Then we moved on to get an understanding of what perceptions the young elderly have of two wellness dimensions, physical and intellectual wellness. A number of proposals scored high on a 6-grade forced scale (101 respondents):

- intellectual challenges are important for my wellbeing [4.91];
- I get sufficient intellectual stimulation from my everyday life [4.61];
- my physical health has been good compared to people around me [4.38];
- my resistance to illness is good [4.24];
- the amount of information I have to process in my daily life is suitable for me (not too much, not too little) [4.20];
- I expect my physical health to remain good [4.14];
- I expect my physical health to deteriorate with increasing age [3.94].

The results we got show that the young elderly have clear perceptions of both wellness dimensions; thus, it makes sense to develop digital wellness services.

Then we need to find out if there are any characteristics that could single out the most promising potential users as we want to get good, strong initial adoption of the wellness services. Our idea was that relations between socio-economic characteristics, attitudes toward the use of mobile applications and perceptions about wellness would help us to identify promising potential users, see details in [4, 5].

We first run a factor analysis with 19 statements on mobile applications and 11 statements on wellness. The results gave an indication for possible sum variables (cf. Table 2) which we tested for reliability by calculating Cronbach's alpha coefficients; we reached the target value ($\alpha > 0.7$) for the constructed sum variables (their names show the groupings we found).

Table 2. Reliability analysis for six sum variables, Cronbach's alpha

SUM variable	Cronbach's alpha
Mobile_apps_positive	0.913
Mobile_apps_experienced	0.926
Mobile_apps_social	0.729
Mobile_apps_value	0.820
Physical_wellness_positive	0.858
Intellectual_wellness_positive	0.791

We run a non-parametric Mann-Whitney U-test in order to explore possible differences between gender and age (combined to two categories: – 69 years; 70 –). We continued this with the highest level of education (–higher vocational school; technical/commercial degree + university), marital status (single; in a relationship) and current work status (working (full-time, part-time, volunteer work); retired). We combined this with annual income (– 30000 €; 30001 –) and the level of experience of using mobile applications (routine; experienced).

We found that there are significant differences in the positive attitudes to using mobile apps between the two age groups; the younger age group was more positive (cf. *mobile_ apps_ positive*). The more educated group was experienced in using mobile apps (cf. *mobile_ apps_ experienced*). The more educated group was socially active in using mobile apps (cf. *mobile_ apps_ social*). The group with higher income gave more value to mobile apps (cf. *mobile_ apps_ value*). Males with higher income had a positive perception of (their) physical wellness (cf. *physical_ wellness_ positive*). The group with a more active work status had a positive perception of their intellectual wellness (cf. *intellectual_ wellness_ positive*). We had of course to take some care with making conclusions from the collected data; the sample of 101 respondents is not a fully random sample and the survey was not fully reliable, as we had no possibility to check the circumstances under which the respondents answered the questions. Nevertheless, the results are interesting, as the young elderly represent a new field of study.

We can summarize the insight we gained from the material relative to our overall vision as follows: (i) to get young elderly interested in digital wellness services, (ii) to get them to adopt them, and (iii) to make the services part of their daily routines, we should start with young elderly, who are,

- Active in full time/part time/volunteer work and advanced users of mobile apps and <70 years
- Experienced users of mobile apps and more educated
- Males with good physical health and income >30 k€ per year
- More educated and find mobile apps good value for the price

It makes sense to get the process started with people who are positive to get to know the services and are willing to invest some time and money in trying out digital wellness services. It is easy to get across people with negative attitudes (cf. [4]) and we have some experience of how that will kill any service adoption even before it gets started.

There is another option in [32], where a synthesis of several studies of physical wellness routines of 2500 elderly Swedes (the age group 65–84; there are 2 million people in this group) over several years shows some interesting results. The findings are that 12% of the female and 14% of the male show no activity to improve their physical wellness. There are 69% of the female and 64% of the male, who show regular physical wellness activity at low or medium intensity (the minimum recommendation is 150 min at medium intensity per week). The remaining about 20% show regular physical wellness activity at medium or high intensity. Only regular physical activity at high intensity will have long-term health effects. There is a large group of (young) elderly, who are potential users of the digital wellness services - both high intensity exercisers and exercisers who need to improve the intensity of their regular routines.

6 Digital Coaching

In early work with the young elderly in the *Bewell* project, (cf. [6, 7]) we found that interest in using digital wellness services on smartphones started to wane after 2–3 months of continued use; when the services combine with activity bracelets the period of active use doubled to about 4–6 months. These are early indicators but have something to do with the fact that there is a one-time cost for the activity bracelets (the smartwatches cost around 350 €, the more advanced bracelets 120–250 €, the cheap bracelets 40–50 €). This follows the pattern, that if you have paid for a bracelet you keep up the wellness routines in order to get value for money you have spent. An Insight report [16] found that users abandon 95% of applications on smartphones within 1 month of download; Gartner [12] found that the abandonment rate of smartwatches is 29% per month; this corresponds to 30% per month for fitness trackers in a study with more than 9500 participants in Australia, U.K. and the U.S.

We noted in Sect. 1 that this creates a dilemma for building wellness routines. The wellness routines need to be sustained for at least 3–5 years in order to get health effects. For the young elderly the ideal is to keep up the wellness routines for 15–20 years, i.e. well into senior years. There is a need to build a variety of innovations, new features and advanced, intelligent functionality into the digital wellness services to make them interesting to use for years. This supports the proposal we had in Sects. 1 and 2 to create ecosystems of SME:s (with large corporations for cloud services and infrastructure) for continuous design, development, maintenance and adaptation of the digital wellness services to the changing needs of the users and the changing conditions of the society. In another words, we will need a digital wellness industry; this may not be possible in the short term even if digital services do not show long development cycles (cf. [9]).

A quicker way to deal with this dilemma is *coaching*, i.e. to make variety and innovative, advanced functionality part of the digital wellness services. The material we worked out in Sect. 5 already gave us some insight about the young elderly and this should be a starting point for giving the users individual advice and guidance (coaching) on how to improve their wellness routines. The coaching should be a function of how advanced the users are and what their goals and objectives are. It is of course a possibility to engage personal trainers/coaches but they mostly turn out to be too expensive.

There are some first solutions for coaching in the market for activity bracelets and we came across the following material on the data collected with the Polar A360 bracelet and then processed to become information on the supporting Polar Flow website (cf. Fig. 1). There is also some indication on how the information could be further processed to knowledge, but this is not offered on the bracelet (as we have in mind).

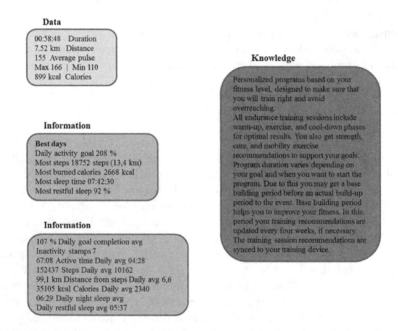

Fig. 1. Data, information and knowledge through an activity bracelet and supporting web site

The data collected (cf. Fig. 1) covers the duration and distance for an exercise (walking, jogging, and running), the average, max and min pulse during the exercise to show intensity and a calculated burn of calories. The processing of this data on the Polar Flow web-page gives information on daily and weekly wellness activities which is combined with sleep patterns, but the web-page offers no conclusions on the meaning of that information, neither any advice on how to use it. The knowledge part offers advice on personalised activity programs, but again not using any input from the data collected nor from the information produced.

The digital coaching methods, algorithms and software solutions should be answers to this problem – digital coaching should be synergistic, produce usable advice, and support users of digital wellness services; we will turn back to VADIYA to find out how we could do this. It appears that there are a few development steps needed to progress from the present state of the Polar Flow.

We worked out generic, functional requirements for the digital coach called VADIYA in Sect. 1; here we will summarize key features we found that got support in Sects. 2–4 and that we have explored further (cf. [4, 5]):

- The coach builds on digital fusion: fusion of heterogeneous data sets ((i) individual data, (ii) group/cluster data, (iii) big user group data) from 100 + applications and devices; algorithms to build information from data sets; fusion of sets of information with computational intelligence methods; ontology (fuzzy, soft) to build knowledge from sets of information; approximate reasoning and soft computing for knowledge fusion
- Coaching is built with data, information and knowledge of different granularities (i) context dependent; (ii) individual wellness, goals; (iii) "good practice"
- Coaching is adaptive to language, habits, legislation
- VAIDYA coaching is continuous and is built to change dynamically; should be maintained, supported, enhanced, changed, developed (sometimes radically)
- VAIDYA is unobtrusive technology: virtual coach on smart phone + omnivore platform that will have 100+ interface solutions for sensor data + data from wearables + new technology innovations
- VAIDYA builds on ideal wellness profiles that are worked out; for individuals, for groups/clusters of similar service users, for culture and traditions, etc.; deviations from the ideal profiles worked out with (fuzzy) MCDM methods to offer recommendations of how to make optimal steps towards the ideal profiles

Schmidt et al. [22] developed a digital personal coach to help fitness tracker users to personalize their training. The coaching builds on partially observable Markov processes and Markov decision processes that use data from the fitness trackers; when reported the coach was in first prototype development. The implementation and use of Markov processes may offer some challenges and the Markov processes do not match the functionality we listed for VAIDYA.

The digital fusion works on data, which will be multi-dimensional in each one of the two different wellness aspects. Some of the physical wellness data can be collected with sensors but other parts, such as data on intellectual wellness, will need a gamification interface and support (cf. [14]). Part of the data needs to be in real time; part of it can be daily and weekly summaries. We will need digital fusion to operate with heterogeneous data from different contexts and for different forms of wellness. All data and information from digital fusion should be worked out in an understandable and tailorable form for young elderly users; this is known as knowledge fusion.

The fusion methods and technology will add value to the digital wellness services:

- data fusion offers summary statistics from (several) apps and devices, goal attainment over days, weeks, months
- information fusion produces trends, deviations from trends, targets and target revisions; shows levels usually reached by similar service users
- knowledge fusion compiles activities to combine, add, enhance information for advice and support; tailors enhanced activity programs

In work with young elderly groups (cf. [6, 7]) discussion quite often focused on wellness as a sum of different activities and it was pointed out, "my best (individual) wellness is not achieved by maximizing my physical wellness, but by finding some smart combination of intellectual, physical and social wellness". Further study showed that wellness could be synergistic (cf. [1, 11]), i.e. that combinations of different forms of wellness could be "more than the sum of the individual wellness forms". As an example, the sport orienteering (as demonstrated to us) can produce synergistic wellness: (i) it offers cognitive and intellectual challenges in finding the optimal paths through the forest; (ii) it is one of the physically most demanding sports; (iii) the closely-knit "family of orienteers" – which is also an international family – promotes social wellness.

Similar versions of synergistic wellness activities can be found among the young elderly – concerts offer intellectual, emotional and social wellness; volunteer help and support of senior citizens offer physical, emotional and social wellness; singing in choirs offer emotional and social wellness, and so on.

The digital coach VADIYA should provide advice and support for synergistic wellness activities - and for hybrids of wellness activities that work out innovative and sometimes unique combinations that would offer new challenges for the young elderly users. There is of course the question of time to be spent on and support for trade-off between wellness activities, i.e. how to get the best wellness value for time allocated to various activities (which sounds too much like "productivity of working time" - a period the young elderly should have left behind).

7   Summary and Conclusions

We noted that the proportion of ageing citizens is high in most EU countries but that the young elderly has been ignored by the politicians worrying about the ageing population as they are "too active, and in too good shape" to request any budget-funded support from the society.

Thus, there are two cases of missed opportunities. There is a potential market for digital (mobile) services that represents 18–23% of the population in most EU countries; the young elderly will be 97 million by 2020 in the EU countries (a market that should get some business attention). Interventions that create sustainable wellness routines among the young elderly will reduce the probability for serious illness among the senior citizens. In order to work out the interventions we use wellness as the target concept – *to be in sufficiently good shape of mind and body to be successful with all everyday requirements*.

The research goal we stated – *to work out how digital wellness services can contribute to sustainable wellness routines for young elderly* – we reached by way of building an understanding of four different elements of digital wellness services for young elderly. The first element elaborates the point – *it is nicer to get old if you are in good shape*, which is an individual target; on the society level the arguments turn towards the fact that there will be serious challenges to carry the costs of the ageing population if people contract illnesses and live long lives. The second element identifies who the first potential users are for digital wellness services - those (physically, socially) active, well-educated and with higher income; those with regular wellness routines but carried out at low or medium intensity levels. The third element introduced digital coaching as a way to make digital wellness services innovative and adaptive to the users and a changing context of use. The fourth element showed that value co-creation forms and builds sustainable wellness routines, and that digital wellness services can be instrumental to form routines if they create user value at reasonable cost.

There were some interesting details in the study of young elderly in the Åland islands. It points to some first groups of supportive users of digital wellness services that may be generic and identifiable in other regions (and maybe countries). The first group is young elderly who are active in full time/part time/volunteer work, experienced users of mobile apps and are <70 years. The second group is young elderly who are experienced users of mobile apps and have a second level engineering or commercial degree. The third group is young elderly who are males with good physical health and an income >30 k€ per year. The fourth group is young elderly who have a second level engineering or commercial degree and find mobile apps good value for their cost. There is, of course, no problem if we can find potential users who belong to all four groups.

References

1. Adams, T.B.: The Power of Perceptions: Measuring Wellness in a Globally Acceptable, Philosophically Consistent Way. Wellness Management (2003). www.hedir.org
2. Bouwman, H., Carlsson, C., de Reuver, M., Hampe, F., Walden, P.: Mobile R&D prototypes – what is hampering market implementation. Int. J. Innov. Technol. Manag. **11**(1), 18 (2014)
3. Bouwman, H., Vos, H., Haaker, T.: Mobile Service Innovation and Business Models. Springer, Heidelberg (2008)
4. Carlsson, C., Walden, P.: Digital wellness services for the young elderly – a missed opportunity for mobile services. JTAECR **11**, 20–34 (2016)
5. Carlsson, C., Carlsson, J.: Interventions to form wellness routines among young elderly. In: Proceedings of the 29th eBled Conference, pp 406–418. Digital Wellness Track (2016)
6. Carlsson, C., Walden, P.: Digital wellness for young elderly: research methodology and technology adaptation. In: Proceedings of the 28th eBled Conference (2015)
7. Carlsson, C., Walden, P.: Performative IS research - science precision versus practice relevance. In: Proceedings of PACIS 2014, Section 9–7 #575, 12 pp (2014)
8. Carlsson, C., Walden, P.: From MCOM visions to mobile value services. In: Clarke, R., Puchar, A., Gricar, J. (eds.) The First 25 Years of the Bled eConference, pp 69–91. University of Maribor, Bled (2012)
9. Competing in the 2020: Winners and Losers in the Digital Economy. Harvard Business Review Analytic Services, March 2017
10. Davis, F.D.: Perceived usefulness perceived ease of use and user acceptance of information technology. MIS Q. **13**(3), 319–340 (1989)
11. Els, D.A., de la Rey, R.P.: Developing a holistic wellness model. SA J. Hum. Resour. Manag. **4**(2), 46–56 (2006)
12. Gartner report on wearables. http://www.gartner.com/newsroom/id/3537117

13. Gronroos, C.: Service logic revisited: who creates value? And who co-creates? Eur. Bus. Rev. **20**(4), 298–314 (2008)
14. Hamari, J., Koivisto, J., Sarsa, H.: Does gamification work? – A literature review of empirical studies on gamification. In: Proceedings of HICSS 2014, pp. 3025–3034. IEEE (2014)
15. Hashema, I.A.T., Yaqooba, I., Anuara, N.B., Mokhtara, S., Gania, A., Khanb, S.U.: The rise of "big data" on cloud computing: review and open research issues. Inf. Syst. **47**, 98–115 (2015)
16. http://www.nuance.com/ucmprod/groups/enterprise/@webnus/documents/collateral/nc_020 218.pdf
17. http://www-teleforum-ry.fi/Mobile-content-market-in-Finland-2012-2016-desk.top.pdf. Accessed 28 Feb 2017
18. Li, A.Q., Found, P.: Towards sustainability: PSS, digital technology and value co-creation. Procedia CIRP **64**(2017), 79–84 (2017)
19. Lin, F.-R., Hsieh, P.-S.: Analyzing the sustainability of a newly developed service: an activity theory perspective. Technovation **34**(2014), 113–125 (2014)
20. Numbers of young elderly in the Nordic countries. http://91.208.143.100/pxweb/norden/ pxweb/en/Nordic%20Statistics/Nordic%20Statistics__Population__Population%20projectio ns/POPU06.px/?rxid=fff-33fb5-841d-4c27-8e7f-8b85c23036af. Accessed 29 May 2017
21. Rachele, J.N., Washington, T.L., Cuddihy, T.F., Barwais, F.A., McPhail, S.M.: Valid and reliable assessment of wellness among adolescents: do you know what you are measuring? Int. J. Wellbeing **3**(2), 162–172 (2013)
22. Schmidt, B., Eichin, R., Benchea, S., Meurish, C.: Fitness Tracker or Digital Personal Coach: How to Personalize Training, UBICOMP/ISWV 2015, Osaka, Japan, pp. 1063–1067 (2015)
23. Sein, M.K., Henfridsson, O., Sandeep, P., Rossi, M., Lindgren, R.: Action design research. MIS Q. **35**(1), 37–56 (2011)
24. Sell, A., Walden, C., Walden, P.: My wellness as a mobile app. Identifying wellness types among the young elderly. In: Proceedings of the 50th Hawaii International Conference on System Sciences, pp 1473–1483. IEEE (2017)
25. Statistics Finland 2014 (2016) http://www.stat.fi
26. Student Health and Counselling Services, UC Davis (2015). https://shcs.ucdavis.edu/ wellness. Accessed 28 Mar 2015
27. UCR report on wellness dimensions. https://wellness.ucr.edu/seven_dimensions.html
28. United Nations Department of Economic and Social Affairs. Population ageing and sustainable development, No. 2014/4 (2014). http://www.un.org/en/development/desa/ population/publications/pdf/popfacts/-Popfacts_2014-4. Accessed 21 Mar 2015
29. Walden, C., Sell, A.: Wearables and wellness for the young elderly - transforming everyday lives? In: Proceedings of the 30th eBled Conference, Digital Wellness Track, Bled 2017, pp 637–650 (2017)
30. World Health Organization. Preamble to the Constitution (2014). www.who.int. Accessed 28 Mar 2015
31. Venkatesh, V., Thong, J.Y.L., Xu, X.: Consumer acceptance and use of information technology: extending the unified theory of acceptance and use of technology. MIS Q. **36**(1), 157–178 (2012)
32. Wallén, M.B., Ståhle, A., Hagströmer, M., Franzén, E. Roaldsen, K.S.: Motionsvanor och erfarenheter av motion hos äldre vuxna, Karolinska Institutet, Stockholm, March 2014

Ontology Based Possibilistic Reasoning for Breast Cancer Aided Diagnosis

Yosra Ben Salem[1,2(✉)], Rihab Idoudi[1,2], Karim Saheb Ettabaa[2],
Kamel Hamrouni[1], and Basel Solaiman[2]

[1] Ecole Nationale d'Ingénieurs de Tunis,
BP 37, Le Belvedere, 1002 Tunis, Tunisia
yosra.bensalem@gmail.com
[2] IMT Atlantique, 655 Avenue du Technopôle, 29200 Plouzané, France

Abstract. Medical diagnosis is a very complex task in the case where information suffer from various imperfections. That's why doctors rely on their knowledge and previous experiences to take the adequate decision. In this context, the case based reasoning (CBR) paradigm aims to resolve current problems basing on previous knowledge. Using ontologies to store and represent the background knowledge may notably enhance and improve the CBR semantic effectiveness. This paper proposes a possibilistic ontology based CBR approach in order to perform a possibilistic semantic retrieval algorithm that handles ambiguity and uncertainty problems. The approach is implemented and tested on the mammographic domain. The target ontology is instantiated with 113 real cases. The effectiveness of the proposed approach is illustrated with a set of experiments and case studies.

Keywords: Case Based Reasoning · Knowledge management · Possibilistic ontology · Breast cancer diagnosis · Case retrieval

1 Introduction

Breast Cancer is a confused and chronic disease that requires permanent control diagnosis for death risk prevention. In fact, it is considered as the most cause of death in women under 45 years aged [1]. According to the Breast Cancer Organization [1], about 12% of US Women have a breast cancer in her life. In 2017, it has been estimated that about 30% of new cancer cases diagnosed are breast cancers. Nevertheless, it is indicated that the death rate has been lessened since 1989 thanks to medical care advances and earlier detection. Considering the early investigation of small lesions in the first pathological stage can prevent significantly malignant complications. That's why developing Computer Aided Diagnosis (CAD) systems or Clinical Decision Support Systems (CDSS) can meaningfully help doctors to have better diagnosis accuracy.

Breast Cancer diagnosis is a complex task and a confused problem. It depends, in almost cases, on the doctor's background knowledge. Case based Reasoning (CBR) is one of the Artificial Intelligence (AI) paradigms [2] used for problem solving and decision making based on past experiences. It reproduces the human reasoning process

© Springer International Publishing AG 2017
M. Themistocleous and V. Morabito (Eds.): EMCIS 2017, LNBIP 299, pp. 353–366, 2017.
DOI: 10.1007/978-3-319-65930-5_29

to formulate a problem and to search an adequate solution. Several studies apply the CBR paradigm in breast cancer diagnosis. Zia et al. [3] have been interested in the case retrieval for a CBR applied to breast cancer data. Boroczky et al. [4] exploit CBR in CDSS for the similar cases retrieval applied to a breast MRI dataset. Darzi et al. [5] proposed an approach aiming at feature selection for breast cancer diagnosis based on a genetic algorithm and CBR. The CBR is used to evaluate a selected subset of features. Sharaf-elDeen et al. [6] proposed a method that combines CBR and RBR (Rule Based Reasoning) concepts applied to breast cancer diagnosis.

In CBR, a case is represented by the couple (problem, solution), such that for a given problem, a past solution is revised to solve it. This concept has been successfully exploited in medical domain [7]. In the mammographic context, a case is represented by the couple (description, diagnosis) where description represents clinical and radiological information.

Although any CBR system relies on a set of specific previous experiences, its reasoning power depends on the background general knowledge about the domain [8]. Ontologies represent a paradigm of knowledge representation and formalization. They can enhance the capabilities of CBR [9]. For this reason multiple researches are interested in combining the two paradigms. Garrido et al. [10] proposed an ontology based CBR approach applied in decision making problem. Samwald et al. [11] used ontology paradigm to model pharmacogenomics knowledge employed in an ontology based reasoning system. El-Sappagh et al. [12] used the SCT standard medical ontology as a domain knowledge specification CBR. This ontology is used to model diabetes cases [13] in a standard representation. Amine et al. [14] proposed an approach for breast cancer classification combining the CBR and ontology paradigms using clinical data.

The most critical steps in CBR are the domain cases representation and the cases retrieval. In this paper we focus on these two steps. In fact, Experts, usually, use their knowledge to describe a case with natural language. This description suffers, often, from imprecision and ambiguity. Furthermore, the imaging supports used to examine cases, are also affected by multiple imperfections (incompleteness, imprecision, uncertainty, etc....) due to the imaging conditions, sensor's resolution, patient position, etc.... As a result, these imperfections can induces the effectiveness and the accuracy of the CDSS when not treated in the adequate way. That's why, a CBR system should handle imperfect knowledge for effective results. Crisp ontologies are inadequate to model imperfections related to the domain knowledge.

To overcome this problem, Zadeh introduced the fuzzy sets theory [15] that models the imprecise information by defining a membership function which express information imprecision. The main limit of this theory is its incapacity to deal with the other imperfection types, especially uncertainty that influences almost problem solving process. For this reason, Zadeh extended his studies on fuzzy sets to introduce the possibility theory [16] with the main strength to handle various imperfection types.

Multiple studies have been interested in exploiting fuzzy sets theory in CBR. In [17] a CBR approach is proposed using fuzzy characterization. Ekong et al. [18] used a neuro-fuzzy-CBR approach for depression diagnosis. El-Sappagh et al. [19] used a fuzzy ontology based CBR for diabetes diagnosis. Otherwise, several studies used the possibility theory in CBR problem solving in multiple domains. Alsun et al. [20]

proposed a CBR approach based on possibility theory for medical diagnosis. In [21] authors used the possibility theory to make possibilistic rules used to similarity measure in CBR system.

In this paper, we propose an approach that take advantages of the three paradigms: CBR, Ontology and Possibility Theory for a Breast Cancer CDSS. Our contribution consists is the two crucial steps of the CBR: the domain case representation and the cases retrieval. The rest of the paper is structured as follows: In the following section we introduce the basic notions related to our study. In the third section, we explain our proposed approach. In the last section, and before concluding, we present the principle experimentation and evaluation results.

2   Basic Notions

In this section, we define concepts which are used in our study.

2.1   Case Based Reasoning

Case Base Reasoning is a methodology used in problem solving and some decision making. It is based on the exploitation of previous solutions adopted for similar cases. The fundamental idea of a CBR is "Similar problems have similar solutions" [21]. A case base is constructed from solutions adopted for previous cases. In a case base, a case is a conceptual knowledge representation of a situation. Formally a case is represented by a couple of terms as: $Ci = (Pi, Si)$, where Pi is a problem description and Si is the solution specification. Various definitions have been made for case base reasoning. Generally, it is defined as "a recent approach to problem solving and learning" [22]. In [23], Leake defines the CBR as "reasoning by remembering". Riesbeck and Schank [24] defined the CBR as solving new "problems by adapting solutions that were used to solve old problems". CBR is defined as a cyclic model called "the 4 Rs model" [22] as shown in Fig. 1. The First R stands for similar cases retrieving; the second represents the reuse of the solution of a similar case that can solve the studied problem; the third R refers to the potential revision of the solution reused; and finally the fourth R is the ending step which aims to retain a new solution for the new case studied. Then, a new solution for a new case is added to the case base in order to enrich it.

To retrieve potentially similar cases, a similarity measure is, necessary, performed. This measure depends, principally, on the data type used (numerical, textual, etc.…). Generally, the similarity measure is determined by the distance of each feature description defined in the two representations of the case studied and the case in the base.

2.2   Ontology Paradigm

Ontology is a recent concept used principally in semantic web application. It is defined as hierarchical specification [25] of a particular reality conceptualization. Specifically,

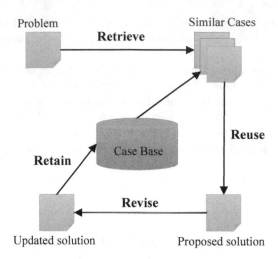

Fig. 1. The cyclic model of the CBR [22]

it is a formal conceptualization of a domain describing its knowledge in terms of concepts and relations between concepts [26]. That's to say that it is used to formalize knowledge, in a generic form. Accordingly, its main advantage is the sharing and reuse of the represented knowledge. The ontology paradigm is widely exploited in medical fields to capture knowledge and to conceptualize medical lexicons.

An Ontology O is specified by a four-tuple:

$$O = \{C, R, I, A\},$$

where:

- C: a set of concepts,
- R: a set of relations,
- I: a set of instances,
- A: a set of axioms

Concepts are characterized by labels, attributes and relations. Labels designate the names used to identify them. Attributes are features describing concepts in details. Relations specify the link between concepts. They are used to give a general structure of the ontology [27].

2.3   Possibility Theory

The possibility theory is a recent theory developed to deal with imperfect data. It was introduced by Zadeh [16] and developed by Dubois and Prad [28]. It represents an extension of the fuzzy sets introduced, yet, by Zadeh previously [15] to handle imprecision and ambiguity affecting information. Nevertheless, possibility theory was introduced to handle multiple forms of imperfections (imprecision, ambiguity, incompleteness, etc.…). It models the uncertainty of an ambiguous knowledge representation in order to overcome the classical theory limits. Classical (crisp) sets theory is defined by the Eq. (1) to express if an element is a member or not of a given set.

$$\Omega_A(x) = \begin{cases} 1 \ if \ x \in A \\ 0 \ if \ x \notin A \end{cases} \tag{1}$$

where, Ω represents a set of events, and A is a subset such that $A \subseteq \Omega$.

In fuzzy set, an element can be partially member of a set. That means that the element is characterized by a membership degree defined in [0,1]. In possibility theory, this membership function is called possibility measure which represents a coefficient attributed to a case in order to specify the degree of its possibility occurrence. According to this theory, if an event is sure, it is certainly possible; however, if it is possible, it is uncertain. Formally, the possibility theory is defined by the Eq. (2).

$$\pi : \Omega \to [0, 1]; x \to \pi(x) \tag{2}$$

Possibilistic reasoning has been applied in multiple domains to be close to the human reasoning in a decision making process when information are generally ambiguous and uncertain.

Usually, the possibilistic measure is coupled to a necessity measure (noted N) which indicates the degree of the certainty of a case characterized by a specific possibility distribution. N is defined also in [0,1]. It is defined by Dubois and Prad [29] by the Eq. (3).

$$N(A) = \Pi(\Omega) - \Pi(A^C) \tag{3}$$

where: A represents a subset of the univers (Ω); $\Pi(\Omega)$ represents the height of the possibility distribution; and A^c represents the complementary subset of A.

If N(C) equals to "0", that implies the total uncertainty of the case C; and if the N (C) equals to "1", that indicates that the certainty of the case is absolute.

3   Proposed Approach

Our proposed approach, aims to provide a CBR system based on possibilistic ontology in order to assist doctors in the breast cancer diagnosis. As shown in the Fig. 2, the approach is structured on four modules:

- Case preparation
- Case representation
- Case retrieval
- Case base Ontology enrichment

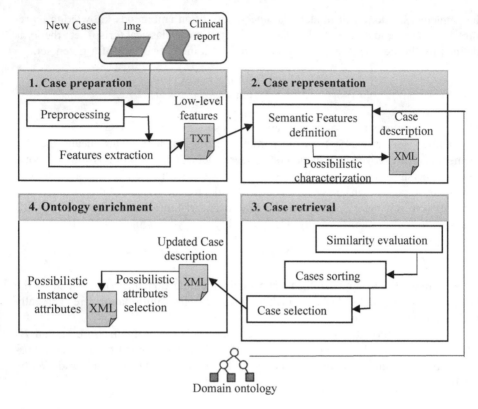

Fig. 2. The proposed approach

3.1 Case Preparation Module

This model aims to prepare data sources to be structured in local ontology representation. It consists in collecting the different features that describe the patient's case. They were extracted from two different data sources:

- The first one is a textual description which represent a clinical specification defined by the doctor in an interviewing process made with the patient to examine the pathological situation according personal information (age, history, risk factor, etc....) and several symptoms observations (pain, abnormal situation, etc....).
- The second one is a radiological specification determined by a medical imaging source used to represent the internal structure of an organ in order to examine any presence of abnormal entities.

This model is considered as a preprocessing step. Its main purpose is to extract relevant data from sources. Textual features are extracted from textual description and are, then, normalized to be formally represented by attributes and their corresponding values. For example: In the textual description the patient's age is specified as 52; this value is transformed in [50, 59]. Furthermore, the preprocessing step of the radiological images aims to extract visual features describing the pathological zone if it exists.

To this end, we exploit the methodology used in [30] to extract visual features from a mammographic. First, the image is filtered in order to eliminate the noise and enhance its quality. Then, it is segmented to get the pathological regions. Several computations are applied to extract information related to visual features (size, diameter, localization, etc....) of these regions. An example of this procedure is given by the Fig. 3.

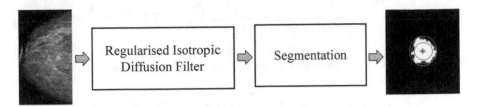

Fig. 3. The image preprocessing step

All extracted features represents the low-level features that are collected in a TXT file to be used in the next module.

3.2   Case Representation Module

In the mammographic context, a case is represented by a couple (description, diagnosis) where the description is the collection of the clinical observation and the radiological information. This description is specified from the TXT file constructed in the previous module by defining semantic features from both the textual and the visual features. This semantic specification describes the pathological situation of the patient, considering imperfections that can affect used information. To this end, firstly, semantic features are defined from the visual features extracted in the previous module using a rule base, constructed beforehand according to expert considerations. For example *"If the Circularity index is equal to 1, Then, The form is circle"* is a rule used to specify if a given from is circle. Then, to handle imperfections (imprecision, ambiguity, uncertainty), some attributes (contour, shape, size, distribution, localization) are specified as possibilistic attributes. Afterward, a couple of possibility and necessity measure are assigned to each attribute. These values determine the correctness degree of a semantic feature. For this purpose, a possibility distributions are used. This possibility distributions are defined in light of several criteria introduced by the expert concerning the parameters computed previously. Figure 4 depicts an example of a possibility distribution defined for the form of masses according the diameter parameter. To have the final possibility measure value for an attribute we select the min value of the possibility measures according to the different parameters. If there is a missed value for a certain attribute, its value take 0. Furthermore, for the necessity measure we used the function defined by Dubois and Prade [29] by the Eq. (3).

Semantic features are defined according a domain knowledge ontology. In breast cancer context, there is several ontologies (BCGO [31], Mammography Annotation Ontology [32], Mammo Ontology [33]) developed in order to formalize the domain knowledge. In our study we use the "Mammo" Ontology proposed, firstly, by Taylor

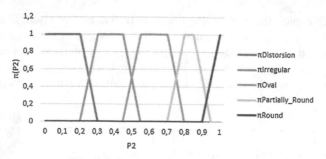

Fig. 4. Possibility distribution for the form of a mass according to its diameter

and Toujilov [33], and then, enriched by Idoudi [34]. The main advantage of this ontology is its concept richness and its high structure-level. Figure 5 shows the "Mammo" enriched Ontology structure. Semantic features are collected in an xml file according concepts defined in this reference ontology for the case description. This file contains an instance of the Ontology.

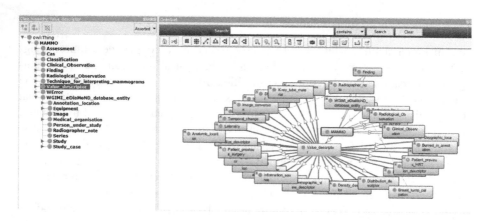

Fig. 5. The "Mammo" domain ontology structure

3.3   Case Retrieval Module

Considering the cases representation methodology, all cases have the same description structures. So, the comparison process can be performed by measuring similarities between the description instances.

The case retrieval process is operated according three steps: the similar cases retrieval; the sorting of retrieved cases considering weights assigned to attributes; and finally, the diagnosis selection step.

Similar Cases Retrieval. The query case is represented by (D_q, X) where D_q is the description of the case and X is the diagnosis required. Similar cases retrieval suppose that similar cases description involved similar diagnosis. Consequently, at a first time, the query case description is compared to each case description defined by ontology individuals. A syntactic similarity measure is used for the comparison process. It is used to select the potentially similar cases having similar value. It is computed by the Eq. (4)

$$Sim(D_q, D_k) = \sum_{i=1}^{n} \omega_i \, Sim(C_{qi}, C_{ki})$$ (4)

where ω_i represents the weights assigned to the different concepts specifying the description part of a case; and the $Sim(C_{qi}, C_{ki})$ term represents the similarity measure between two concepts, computed by the Eq. (5).

$$Sim(C_{qi}, C_{ki}) = \frac{\sum_{j=1}^{m} S_{sa}(A_{qij}, A_{kij})}{m}$$ (5)

where m is the number of the attributes characterizing a given concept; and the $S_{sa}(A_{qij}, A_{kij})$ represents the syntactic similarity measure between two corresponding attributes of the two concepts. It is considered as the possibility measure that express the correspondence degree of the two cases. This measure is defined for crisp attributes by the Eq. (6)

$$S_{sa} = \begin{cases} 1 \ if \ val(A_{qij}) = val(A_{kij}) \\ 0 \ if \ val(A_{qij}) \neq val(A_{kij}) \end{cases}$$ (6)

For possibilistic attributes the Eq. (7) proposed by Jenhani et al. [35]:

$$S_{IA}(\pi_i, \pi_j) = 1 - \frac{D_M(\pi_i, \pi_j) + Inc(\pi_i, \pi_j)}{2}$$ (7)

where: D_M represents a Minkowsk's Distance [36]. For example, the Manhattan's distance defined by Eq. (8):

$$L_1(\pi_i, \pi_j) = \frac{1}{N} \sum_{k=1}^{|N|} |\pi_i(x_k) - \pi_j(x_k)|$$ (8)

And $Inc(\pi_i, \pi_j)$ represents the Inconsistency measure between π_i and π_j. This measure considers the inconsistency of the intersection of the two possibility distributions.

At a second time, the query case description is confronted to the diagnosis of each retained case to get the most suitable diagnosis for the query case. To this end, we apply the principle of the possibility theory by estimating a possibility measure and a necessity measure for each potential diagnosis using possibility distributions and a rule base for corresponding some description features with diagnosis attributes, defined beforehand.

Diagnosis Sorting Step. Retained diagnosis are sorted according their possibility and necessity measures. Diagnosis having the highest values of the couple (possibility, necessity) are arranged, firstly, then, the possibility measure is considered to arrange the rest of diagnosis.

Diagnosis Selection Step. At the end of this process, we retain cases having a possibility measure grater than a threshold value equal to 0.7 (fixed by the domain expert) in order to provide the most relevant diagnosis for the expert. The latter, evaluate, then the potential diagnosis, choose one of them, revise it to be in concordance with the case description. Consequently, the xml file containing the ontology instance is updated with the diagnosis revised preparing it to be added to the reference ontology.

3.4 Ontology Enrichment Module

The aim of this module is to enrich the "Mammo" Ontology with the defined instance. As it is defined with both crisp and possibilistic attributes, a two step instantiation process is performed.

- First, a new instance is added, to the reference ontology with the crisp values. The different attributes defined in the instance file are mapped to the corresponding attributes defined in the ontology.
- Second, attributes with possibilistic values are mapped to an other xml file, which is assigned to the corresponding ontology instance.

4 Experimentations and Results

The proposed approach is tested on a data set containing 113 cases. This data set is decomposed on two subsets: the first is used to populate the "Mammo" Ontology in order to build a case base containing 56 cases; the second is used to evaluate the performance of the proposed methodology constituted by 57 cases.

Figure 6 depicts a use scenario scheme of our proposed approach. Table 1 represents an extract from the obtained result of the potential diagnosis retrieval process. Cases are sorted by their possibility measures. Two sorting results are presented to the domain expert: first representation shows the sorting results according to possibility measure of the case's description and the second shows sorting results according to the possibility measure of the case's diagnosis. Domain experts estimates that the relevant results must have a value greater than 0.7. As shown in the Fig. 6, the domain expert evaluate the presented results, select the most suitable diagnosis to the case situation. After that, the case representation is updated with the diagnosis specification. Finally the instance is mapped to the "Mammo" domain ontology in order to enrich it.

To evaluate the results performance we use the confusion matrix evaluation (cf. Table 2). It is a 2×2 matrix that contains the correct and incorrect decisions making according to some measures rate:

- True Positive (TP): both the CBR system and the doctor estimate a malignant case.
- False Positive (FP): the CBR system estimates a malignant case but the doctor estimates a benign case.
- True Negative (TN): both the CBR system and the doctor estimate a benign case.
- False Negative (FN): the CBR system estimates a benign case but the doctor estimates a malignant case.

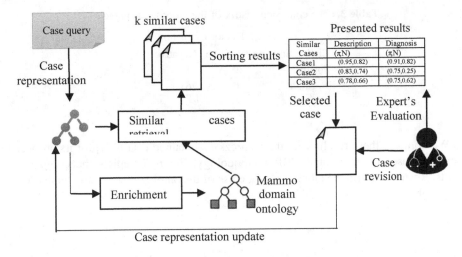

Fig. 6. Application scenario scheme of the proposed approach

Table 1. Extract from the similar cases retrieved

	Description	Diagnosis
Cases	(π, N)	(π, N)
Similar case 1	(0.95,0.82)	(0.91,0.82)
Similar case 2	(0.83,0.74)	(0.75,0.25)
Similar case 3	(0.78,0.66)	(0.75,0.62)
...	...	...

Table 2. The confusion matrix definition

	Doctor estimation	
CBR estimation	Positive	Negative
Positive	TP	FP
Negative	FN	TN

Table 3 depicts results obtained for the proposed approach. The different TP, FP, TN, and FN are used to compute the precision (P), specificity (S), the accuracy (A), and the recall (R) rates according to the equations below:

$$P = \frac{TP}{TP+FP} \quad S = \frac{TN}{TN+FP} \quad A = \frac{TP+TN}{TP+TN+FP+FN} \quad R = \frac{TP}{TP+FN}$$

Computing these rates, we obtain: P = 100%; S = 100%; R = 96.7%; and A = 98.2%. The different calculated rates prove the effectiveness of the proposed approach in breast cancer aided diagnosis using possibilitic based ontology CBR.

Table 3. The confusion matrix of the proposed approach

	Doctor estimation	
CBR estimation	Positive	Negative
Positive	30	1
Negative	0	26

We compare these results with the Boroczky's obtained results [4] for its proposed approach based on traditional CBR methodology. Figure 7 depicts the comparison results. Results confirm the efficiency of the use of the ontology paradigm coupled with the possibility theory in the CBR approaches.

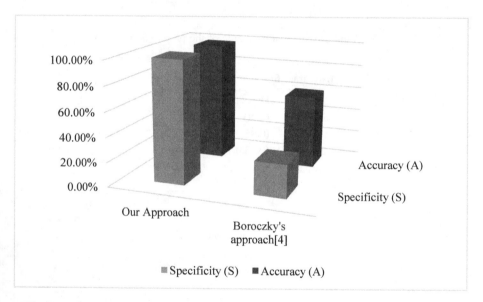

Fig. 7. Comparison results between the proposed approach and Boroczky's [4] approach

5   Conclusion

This paper presents a new approach for the breast cancer aided diagnosis. This proposed approach aims to use a CBR based on a possibilistic ontology. The basic novelties of this proposed approach are: (1) the case representation with a domain ontology instance characterized by possibilistic values; and (2) the cases retrieval from the ontology instances characterized also by possibilistic measures that express the degree of the adequacy of a certain diagnosis for a new query case. Our approach has achieved a performance of 98.2% in term of accuracy. The presented results show the effectiveness of the proposed approach in breast cancer diagnosis. As a future work, we propose to minimize as possible the expert intervention and so, automate the system. We will, principally, interest in augmenting the case base, such that, more the system dispose of various cases, more the retrieval results are suitable and effective to solve a given diagnosis problem.

References

1. Breast Cancer Organization, U.S. Breast Cancer Statistics, March 2017. http://www. breastcancer.org/symptoms/understand_bc/statistics
2. Begum, S., Ahmed, M., Funk, P., Xiong, N., Folke, M.: Case-based reasoning systems in the health sciences: a survey of recent trends and developments. IEEE Trans. Syst. Man Cybern. 7(1), 39–59 (2010)
3. Zia, S.S., Akhtar, P., Javid, T., Mughal, A., Mala, I.: Case retrieval phase of case-based reasoning technique for medical diagnosis. World Appl. Sci. J. 32(3), 451–458 (2014)
4. Boroczky, L., Simpson, M., Abe, H., Drysdale, J.: Observer study of a prototype clinical decision support system for breast cancer diagnosis using dynamic contrast-enhanced MRI. Am. J. Roentgenol. 200, 277–283 (2013)
5. Darzi, M., AsgharLiaei, A., Hosseini, M., Asghari, H.: Feature selection for breast cancer diagnosis: a case-based wrapper approach. Int. J. Med. Health Biomed. Bioeng. Pharmaceutical Eng. 5(5), 220–223 (2011)
6. Sharaf-elDeen, D.A., Moawad, I.F., Khalifa, M.E.: A breast cancer diagnosis system using hybrid casebased approach. Int. J. Comput. Appl. 72(23), 14–19 (2013)
7. Marlinga, C., Montanib, S., Bichindaritzc, I., Funkd, P.: Synergistic case-based reasoning in medical domains. Expert Syst. 41(2), 249–259 (2014)
8. Dendani, N., Khadir, M., Guessoum, S.: Use a domain ontology to develop knowledge intensive CBR systems for fault diagnosis. In: International Conference on Information Technology and e-Services (ICITeS), pp. 1–6 (2012)
9. Amailef, K., Lu, J.: Ontology-supported case-based reasoning approach for intelligent m-Government emergency response services. Decis. Support Syst. 55(1), 79–97 (2013)
10. Garrido, J.L., Hurtado, M.V., Noguera, M., Zurita, J.M.: Using a CBR approach based on ontologies for recommendation and reuse of knowledge sharing in decision making. In: Eighth International Conference on Hybrid Intelligent Systems, pp. 837–842 (2008)
11. Samwald, M., Antonio, J., Giménez, M., Boyce, R., Freimuth, R., Adlassnig, K.P.: Pharmacogenomic knowledge representation, reasoning and genome-based clinical decision support based on OWL 2 DL ontologies. BMC Med. Inf. Decis. Mak. 15, 12 (2015)
12. El-Sappagh, S., Elmogy, M., El-Masri, S., Riad, A.: A diabetes diagnostic domain ontology for CBR system from the conceptual model of SNOMED CT. In: The Second International Conference on Engineering and Technology, pp. 1–7 (2014)
13. El-Sappagh, S., Elmogy, M., Riad, A., Zaghloul, H., Badria, F.: A proposed SNOMED CT ontology-based encoding methodology for diabetes diagnosis case-base. In: The Ninth International Conference on Computer Engineering and Systems, pp. 184–191 (2014)
14. Amin, E., Abdrabou, M.L., Salem, A.M.: A breast cancer classifier based on a combination of case-based reasoning and ontology approach. In: International Multiconference on Computer Science and Information Technology, pp. 3–10 (2010)
15. Zadeh, L.: Fuzzy sets. Inf. Control 8(3), 338–353 (1965)
16. Zadeh, L.: Fuzzy sets as a basis for a theory of possibility. Fuzzy Sets Syst. 1(1), 3–28 (1978)

17. Abdul, M., Muhammad, A., Mustapha, N., Muhammad, S., Ahmad, N.: Database workload management through CBR and fuzzy based characterization. Appl. Soft Comput. **22**, 605–621 (2014)
18. Ekong, V., Inyang, U., Onibere, E.: Intelligent decision support system for depression diagnosis based on neuro-fuzzy-CBR hybrid. Modern Appl. Sci. **6**(7), 79–88 (2012)
19. Hullermeier, E., Dubois, D., Prade, H.: Model adaptation in possibilistic instance-based reasoning. IEEE Trans. Fuzzy Syst. **10**(3), 333–339 (2002)
20. Alsun, M.H., Lecornu, L., Solaiman, B., Le Guillou, C., Cauvin, J.M.: Medical diagnosis by possibilistic classification reasoning
21. El-Sappagha, S., Elmogy, M., Riadc, A.M.: A fuzzy-ontology oriented case-based reasoning framework for semantic diabetes diagnosis. Artif. Intell. **65**(3), 179–208 (2015)
22. Aamodt, A., Plaza, E.: Case-based reasoning: foundational issues. Methodol. Variations Syst. Approaches **7**(1), 39–59 (1994)
23. Leake, D.B.: Case-Based Reasoning: Experiences, Lessons, and Future Directions. MIT Press, Cambridge (1996)
24. Riesbeck, K., Schank, R.C.: Inside Case-Based Reasoning. Artificial Intelligence Series, 1st Edition (1989)
25. Manzoor, U., Balubaid, M.A., Zafar, B., Umar, H., Khan, M.S.: Semantic image retrieval: an ontology based approach. Int. J. Adv. Res. Artif. Intell. **1**(4), 1–8 (2015)
26. Gruber, T.: Towards principles for the design of ontologies used for knowledge sharing. Int. J. Hum. Comput. Stud. **43**(5–6), 907–928 (1995)
27. Gruber, T.R.: A translation approach to portable ontologies. Knowl. Acquisition J. **4**(5), 199–229 (1993)
28. Dubois, D., Prade, H.: Theory of Possibility an Approach to Computerized Processing of Uncertainty. Plenum Press, Berlin (1988)
29. Dubois, D., Prade, H.: An alternative approach to the handling of subnormal possibility distributions: a critical comment on a proposal by Yager. Fuzzy Sets Syst. **24**(1), 123–126 (1987)
30. Ben Salem, Y., Idoudi, R., Hamrouni, K., Soleiman, B., Bousetta, S.: Image based ontology learning. In: 11th International Conference on Intelligent Systems: Theories and Applications, pp. 1–5 (2016)
31. Branici, A.: Représentation et raisonnement formels pour le pronostic basé sur l'imagerie médicale microscropique. Application à la graduation du cancer du sein. Ph.D. thesis, Université de Franche-Comté (2010)
32. Bulu, H., Alpkocak, A., Balci, P.: Uncertainty modeling for ontology-based mammography annotation with intelligent bi-rads scoring. Comput. Biol. Med. **43**(4), 301–311 (2013)
33. Taylor, P., Toujilov, I.: Mammographic knowledge representation in description logic. In: Riaño, D., Teije, A., Miksch, S. (eds.) KR4HC 2011. LNCS, vol. 6924, pp. 158–169. Springer, Heidelberg (2012). doi:10.1007/978-3-642-27697-2_12
34. Idoudi, R., Ettabaa, K.S., Solaiman, B., Mnif, N.: Association rules based ontology enrichment. Int. J. Web Appl. **8**(1), 16–25 (2016)
35. Jenhani, I., Ben Amor, N., Elouedi, Z., Benferhat, S., Mellouli, K.: Information affinity: a new similarity measure for possibilistic uncertain information. In: Mellouli, K. (ed.) ECSQARU 2007. LNCS, vol. 4724, pp. 840–852. Springer, Heidelberg (2007). doi:10.1007/978-3-540-75256-1_73
36. Hertz, T.: Learning distance functions: algorithms and applications. Ph.D. thesis, pp. 9–14 (2006)

Information Systems Security and Information Privacy Protection

A New FDD-Based Method for Distributed Firewall Misconfigurations Resolution

Amina Saâdaoui$^{(\boxtimes)}$, Nihel Ben Youssef Ben Souayeh, and Adel Bouhoula

Digital Security Research Unit, Higher School of Communication (Sup'Com),
University of Carthage, Tunis, Tunisia
{amina.saadaoui,nihel.benyoussef,adel.bouhoula}@supcom.tn

Abstract. Firewall is one of the most commonly used techniques to protect a network, it limits or provides access to specific network segments based on a set of filtering rules that should be configured with respect to the global security policy. Nevertheless, the security policy (SP) changes frequently due to business or application needs, and this change often impact firewall configurations by generating new conflicts between different rules. Therefore, discovering and removing automatically the configuration errors is a serious and an unavoidable task. This problem has been addressed through a variety of approaches from firewalls rules analysis to firewall configuration verification, but existent solutions have, essentially, three drawbacks: First, most of them did not analyze all relations between all rules in such a way, some classes of configuration errors could be uncharted. Second, the distinction between syntactic intentional anomalies and effective misconfigurations is not generally highlighted. Third, although anomalies resolution is a tedious and error prone task, it is generally done manually by the network administrator. In this paper, we address this problem using a data structure (FDD: Firewall Decision Diagram). We propose a new approach to detect and correct automatically misconfigurations in a distributed environment. We demonstrate the applicability and scalability of our method by the use of a Satisfiability Solver. The first results we obtained are very promising.

Keywords: Firewall · Security Policy (SP) · Misconfiguration · Anomalies · FDD · Distributed environment · Inference system · SAT solver

1 Introduction

The complexity of networks is constantly increasing, as it is the size and complexity of firewall configurations. These Firewalls examine the traffic of network against an ordered list of filtered rules, generally, defined by network administrator according to the global security policy (SP). Over time, the exponential growth in network traffic, services and applications has led to a growth in rule-sets size and a growth in firewall complexity. Moreover, in multi-firewall environment, with hundreds of firewalls, it is difficult to verify detect and correct manually misconfigurations that can arise between different rules. Therefore,

© Springer International Publishing AG 2017
M. Themistocleous and V. Morabito (Eds.): EMCIS 2017, LNBIP 299, pp. 369–383, 2017.
DOI: 10.1007/978-3-319-65930-5_30

after each SP change, automated misconfigurations management to keep firewalls configured properly, without impacting network availability and IT productivity is essential for any enterprise.

In this paper, we consider the following problem: In a multi-firewall environment network, where each firewall can accumulate hundreds of changes over the years, how can we analyze detect and fix firewalls misconfigurations? As an example, consider the network topology shown in Fig. 1. We have three firewalls that delimit three subdomains (configurations of firewalls 1, 2 and 3 are shown in Fig. 2). The SP that should be implemented is described as follows: **SP_v1: Deny access from net1 to net2 except http access from machine M1 to subnet21; Allow all traffic from net3 to net2.** We can note here that for traffic from net_1 to net_2, passing through $FW2$ and $FW3$ respectively, the SP is properly implemented. However, if the SP is modified to: **SP_v2: Allow access from net1 to net2 except http access from machine M1 to subnet21; Deny all traffic from net3 to net2**, then this change requires a complete evaluation of firewalls configurations and we can note that rules of $FW1$, $FW2$ and $FW3$ are no longer implemented with respect to the SP. To deal with this problem, many solutions have been proposed but they have, essentially, the following drawbacks:

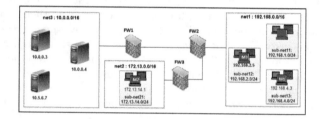

Fig. 1. Network topology

- In a multi-firewall environment, mostly we consider anomalies between only two firewalls in a given network path which cannot give a precise idea on real conflicts that can arise between different rules of different firewalls and obviously will not help to fix them. In a distributed environment these anomalies could exist between rules of different firewalls. For example, if we have traffic from net_3 to net_2, the path followed by this traffic contains $Fw1$, $Fw2$ and $Fw3$ so we should analyze relations between rules of these firewalls.
- Some Studies deal only with pairwise filtering rules. In such way, some other classes of configuration anomalies could be uncharted. For instance, if we consider the network path composed by firewalls $Fw1$ and $Fw2$, rule r_1 from $Fw1$ is overlapped with three other rules (r_2 and r_3 from Fw_1 and r_3 from $Fw2$). Therefore any correction technique should take under consideration all relations between all rules otherwise the correction process will not help to obtain the *required* action.

- Some Studies did not distinguish between intentional syntactic anomalies and real configuration errors. For example, we can note that the filtering rules of firewall $Fw3$ are conform to SP as filtering rules of firewalls are, generally, processed from the top down, and the first match wins. Here, when traffic from $sub-net11$ tries to access $sub-net21$, it will be blocked through r_{33}. Although no misconfiguration is identified, most related studies [3,7,8] present the conflict between rules r_{33} and r_{43} as a purely syntactic anomaly, since these two rules handle common packets with different actions.

In this paper, we propose a new approach to discover and fix misconfigurations in a multi-firewall environment. Our approach takes advantages of the sequential application of firewall rules modeling their relations in a firewall decision diagram (FDD). This data structure is presented by Gouda and Liu in [12,17]. Given the structure of FDD and firewall rules, we can discover precisely anomalies and by considering SP we can decide whether an anomaly is intentional or if it is a real configuration error. In this line of research, our earlier work deals with formal analysis of single firewall configuration. We proposed a new approach to correct misconfigurations in a single firewall by modifying some rules fields, modifying their order, removing some rules. The contributions of this work can be summarized as follows: (1) We use a formal method using inference systems and a SAT solver to deal with this problem. (2) We extract and decide if an anomaly is a real misconfiguration or an intended anomaly in distributed environment by using the FDD. (3) When the usual corrections of configuration errors are the insertion of new filtering rules, which is a practice not often optimal nor safe, we try to meet the challenge of first altering the distributed firewall rules in place (deleting, modifying fields, or even swap rules) to optimize configuration while preserving the behavior of firewalls intact (do not generate other errors). This paper is organized as follows: Sect. 2 presents a summary of related work. Section 3 overviews the formal representation of firewall configurations and security policies and details FDD structure. In Sect. 4, we articulate our approach to detect and correct firewall misconfigurations. In Sect. 5.1, we present the implementation and evaluation of our method. Finally, we present our conclusions and discuss our plans for future work.

2   Related Work

Intra and Inter-firewalls Anomalies Detection: Al-Shaer and Hamed [3] introduce a framework for discovering anomalies in centralized and distributed firewalls. They analyzed relations between rules using a state diagram that allows to identify anomalies and couple of rules involved in these anomalies or couple of firewalls (in case of inter-firewalls anomalies detection), this differs from our method that considers all rules and not only pairwise ones. Hu et al. [14] propose a new anomaly management framework (FAME) that facilitates the systematic detection and resolution of firewall policy anomalies by considering the analysis of relations between all rules in the firewall configuration. The proposed idea to

Table 1. Firewall Configuration-FW1

Order	Action	Srce_@	Dest_Port	Dest_@	Protocol
r_{11}	accept	10.0.0.3	80	172.13.14.1	*
r_{21}	accept	10.0.0.0/16	80	172.13.14.0/24	TCP
r_{31}	deny	10.0.0.3	*	172.13.14.0/24	TCP
r_{41}	accept	192.168.4.0/24	80	172.13.14.0/24	TCP

Table 2. Firewall Configuration-FW2

Order	Action	Srce_@	Dest_Port	Dest_@	Protocol
r_{12}	deny	192.168.2.0/24	80	172.13.14.0/24	TCP
r_{22}	deny	192.168.1.0/24	80	172.13.14.0/24	TCP
r_{32}	accept	10.0.0.3	80	172.13.0.0/16	TCP
r_{42}	accept	192.168.2.0/24	80	172.13.14.0/24	TCP

Table 3. Firewall Configuration-FW3

Order	Action	Srce_@	Dest_Port	Dest_@	Protocol
r_{13}	accept	192.168.2.0/24	80	172.13.0.0/16	TCP
r_{23}	deny	192.168.1.0/24	80	172.13.14.0/24	TCP
r_{33}	deny	192.168.1.0/24	80	172.13.14.0/24	*
r_{43}	accept	192.168.1.0/24	80	172.13.0.0/16	TCP
r_{53}	accept	192.168.4.0/24	80	172.13.0.0/16	TCP
r_{63}	accept	192.168.4.0/24	80	172.13.0.0/16	TCP

Fig. 2. Firewalls configurations

resolve anomalies is based on calculating a risk level that permits, in some cases, users to manually select the appropriate strategies for resolving the conflict. In such a way, the administrator can make wrong choices. So, the administrator decides manually if an anomaly is a misconfiguration. Unlike that, our method incorporates SP which allows deciding, automatically, whether an anomaly is intentional or a real configuration error. Authors in [7, 19] introduced a method of analyzing packets from the filtering rule list by using the concept of Relational Algebra and a 2D box model to show a simulation of packets by rectangular boxes and identify anomalies and relations between rules. Abbes et al. present in [1] a method for firewall anomalies discovering. They represent filtering rules in a tree data structure called FAT that allows identifying anomalies between rules two by two. In opposition, in our work we also represent relations between rules in a data structure, but additionally we identify anomalies by considering all rules. Authors in [1, 6, 14] proposed methods to manage a single firewall rules. This differs from our method that takes into account all firewalls in a given path in the network because even if each firewall in the network is well configured, anomalies could arise between rules of different Firewalls. Prior work on Inter-firewalls rules analysis [11, 13] focused on the analysis of relations between pairwise rules of two firewalls in a given network path. However, in reality it is common that a network path contains more than two firewalls and anomalies could happen between more than two rules in these firewalls. The precise indication of all firewalls and all rules involved in a misconfiguration will help to fix them easily without creating new misconfigurations. FIREMAN [21] is a static analysis toolkit to check anomalies in firewalls. The tool can handle large set of firewall rules since it uses an efficient BDD. This tool can only show there are anomalies between one filtering rule and preceding rules, and cannot identify all rules involved in the anomaly. We note that most of existing studies [1, 2, 9] focused on the anomaly discovery issue. However, they did not propose methods to resolve these anomalies. In [20] authors proposed a new approach for correcting anomalies within filtering rules. But the correction is not totally automatic it is assisted by the administrator to yield a required precision in reflecting the adopted SP.

Firewall Configuration Verification. Liu [16] proposes a firewall verification method. The method takes as input a firewall configuration and a given property, then outputs whether the firewall configuration satisfies the property. Matsumoto and Bouhoula [18] propose a SAT based approach for verifying firewall configurations with respect to the security policy requirements. This method checks the correctness of the firewall configuration whether it contains anomalies or not. FINSAT [4,5] incorporates ACL (Access Control List) conflict analysis procedure for detecting various types of ACL rule conflicts in the model using Boolean satisfiability (SAT) analysis. The conflicts are reported as "error(s)" in case of SAT result with satisfiable instances. Then, the Network administrator need to reconfigure by himself the ACL rules depending on the results. The objectives of our work are different. We aim first to discover all misconfigurations by considering the requirement of SP then to fix these misconfigurations automatically with respect to SP. So, our work involves two aspects: Rule analysis aspect and firewall verification aspect.

3   Preliminaries

Firewall Configuration. A single firewall configuration is a finite sequence of filtering rules of the form $FC = (r_i \Rightarrow A_i)_{0 < i < N+1}$. A filtering rule consists of a precondition r_i which is a region of the packet's space, usually, consisting of source address, destination address, protocol and destination port. Each right member A_i of a rule of FC is an action defining the behavior of the firewall on filtered packets: $A_i \in \{accept, deny\}$.

Security Policy. A security policy SP is presented as a finite unordered set of directives defining whether packets are accepted or denied. For example, a directive could be as follows: **A network net1, except the machine A, has the right to access to the FTP service provided by a server S located in the network net2.** We consider also two sets, SP_accept and SP_deny where SP_accept consists of packets accepted to pass through the set of directives SP and SP_deny is the subset of denied packets.

Firewall Decision Diagram (fdd) of a Single Firewall. The firewall decision diagram (fdd) as defined in [12,17] is an acyclic and directed graph that has the following properties: There is exactly one node in fdd that has no incoming edges. This node is called the root of fdd. The nodes in fdd that have no outgoing edges are called terminal nodes. fdd is the union of direct paths dp_i. So we have:

$$fdd = \bigcup_{i(i:1 \to m)} dp_i.$$
$$dp_i = dp_i.srce \land dp_i.protocol \land dp_i.dest \land dp_i.port \land dp_i.rules \land dp_i.action.$$

- $dp_i.src$ is the range of source address represents by the direct path dp_i.
- $dp_i.dst$ is the range of destination address represents by the direct path dp_i.
- $dp_i.port$ is the range of port number represents by the direct path dp_i.

– $dp_i.protocol$ is the range of protocols represented by the direct path dp_i.
– $dp_i.rules$ is the set of rules from the firewall configuration that match the domain of packets represented by this direct path, $dp_i.rules = \{r_{ki}\}_{(k:1 \rightarrow l)}$, where r_{1i} is the first rule in the firewall configuration applied on the domain of dp_i. The action of this direct path is the action applied by r_{1i}.
– $dp_i.action=$ the action of this direct path dp_i.

FDD of a Path in a Distributed Environment. A network path $path_i[src, dst]$ is composed by an ordered set of firewalls through which the traffic flow from the source src to the destination dst. $path_i = \{fc_j, n <= j <= m\}$. Let $Paths$ be the set of all possible paths in our network. $Paths = \{path_i, 1 <= i <= k\}$.

A FDD of a path $path_i$ is constructed using the collection of rules of different firewalls fc_j belonging to this path. Therefore, The FDD of the set $Paths$ of our network could be represented as follows: $FDD(Paths) = FDD = \bigcup_{\{0<i<N+1\}} fdd_i$, where each fdd_i is the FDD of the path $path_i$, so FDD is the union of fdd_i of each path in the network. The properties already defined for a direct path in a single firewall remains the same, only for sets $dp_i.rules$ and $dp_i.action$. In fact, we have to specify for each rule the firewall that belongs to it. Therefore, we define direct path $dp_j \in fdd_i$ as follows: $dp_j = dp_j.srce \land dp_j.dest \land dp_j.port \land dp_j.protocol \land dp_j.rules \land dp_j.action$ where $dp_j.rules = \{r_{h_j}^k\}$ here k is the index of the each firewall through which the traffic flow in the path $path_i$. The action of each direct path depends on the actions of each first rule handled by this direct path from each firewall in this path, so we have:

– $dp_j.action = accept$ if $\forall r_{1j}^k \in dp_j.rules$, $action(r_{1j}^k) = accept$.
– $dp_j.action = deny$ if $\exists! r_{1j}^k \in dp_j.rules$, $action(r_{1j}^k) = deny$.

Figures 3 and 4 show, respectively, fdds of two paths: $P1 = Path[net3, net2] = \{Fw1, Fw2, Fw3\}$ and $P2 = Path[net1, net2] = \{Fw2, Fw3\}$ for the network shown in Fig. 1. We consider the following functions:

– $dom(dp_i)$ is a function that maps each dp_i into the subset of packets $p \in P$ and represents the set of packets handled by dp_i. $dom(dp_i) = Packets\{dp_i.srce \land dp_i.protocol \land dp_i.dest \land dp_i.port\}$.
– $idx(r)$ this function returns the index of the rule r.

4   Our Approach for Resolving Misconfigurations

In our previous work, we presented an Inference system that allows discovering misconfigurations in distributed environment. In fact, in multi-firewalls an anomaly could happen between different firewalls in a network path if they apply different actions on the same traffic. Therefore, by using the data structure FDD already defined for each path, we can determine if a direct path in a given

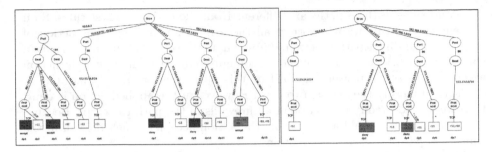

Fig. 3. FDD-P1 (FW1,FW2,FW3) (Color figure online)

Fig. 4. FDD-P2 (FW2,FW3) (Color figure online)

fdd_n contains an anomaly and if this anomaly is a real misconfiguration. So we define anomaly as follows: A direct path $dp_i \in FDD$ presents an anomaly iff $\exists r_{m_i}^k \in dp_i.rules$ where $act(r_{m_i}^k) \neq act(r_{m_i}^h)$ where $h \neq k$. So we have two types of misconfigurations: Total and partial misconfigurations.

- **TMC:** A direct path $dp_i \in fdd_n$ is totally misconfigured *iff* it presents an anomaly and **all** the packets mapped by this path apply a different action as applied in SP on these packets.
- **PMC:** A direct path $dp_i \in fdd_n$ is partially misconfigured *iff* it presents an anomaly and **some** packets mapped by this path apply a different action as applied in SP on these packets.

After discovering misconfigurations, we will try to fix them using one of five correction techniques detailed in the next five subsections. To facilitate this process, we define a new set FR_x, *Faulty rules*. For each direct path dp_x, we define the set of faulty rules correspondent to this direct path, we call it FR_x. The set of rules FR_x of each direct path depends on its action, so we have:

- $FR_x = \bigcup_i \{r_i, r_i = r_1{}_x^f \forall f\}$ if $dp_x.action = accept$ and $dp_x \in TMC$.
- $FR_x = \bigcup_i \{r_i, r_i = r_1{}_x^f \wedge action(r_i) = deny \forall f\}$ if $dp_x.action = deny$ and $dp_x \in TMC$.

These two sets define rules that we should modify in order to correct the direct path action. Therefore, for all fixed misconfigurations $FR_x = \emptyset$. We define a new set FR which represents the set of all faulty rules of all totally misconfigured direct paths. $FR = \bigcup_x FR_x$ where $dp_x \in TMC$.

Remove-Rule Inference System. Once all misconfigurations have been discovered, we start their correction process. First, we will try to correct Total misconfigurations by removing misconfigured rules using the inference system shown in Fig. 5. We should unsure that removing a given rule will not create other misconfigurations. We can remove a rule only if this rule exists in a decision path as a first rule, then this path is totally misconfigured and the action of second rules in these paths are different from the actions of first rules. So if we remove this rule we will correct all these misconfigurations. The action of misconfigured direct path determines if we should correct one rule or all rules in this direct path. For example, if the misconfigured direct path applies the action accept to the set of packets mapped by this path, then we should fix at least one rule to modify the decision to deny. However, if the applied action is deny, then we should modify the action of all firewalls in the path to accept. Because from a source we should always apply the action accept by each firewall in the concerned path to reach a destination.

The rules of the system shown in Fig. 5 apply to triple (FR, FR^*, FR_x) whose first component FR is the set of faulty rules of each dp_x in FDD as defined in the previous section, whose second component FR^* represents an updated version of FR after removing all rules after verifying if the third inference rule *remove* is applied or not. In fact, the second inference rule *Parse* allows to parse all faulty rules FR_x of each dp_x, then if the precondition of the third inference rule *remove* is applied we update the three components FR, FDD and TMC by using the function $\backslash_{remove}$ which allows to modify the three components as follows:

- FR: $\forall x$ we remove r_i from FR_x if $dp_x.action = deny$, else if $dp_x.action = accept$, $FR_x = \emptyset$ because in this case, we should correct at least one rule to obtain the action deny.
- TMC: if $FR_x = \emptyset$ then we remove dp_x from TMC, the direct path has the action required by SP.
- FDD: we remove r_i from each $dp_y.rules$ $\forall y$ and if $FR_y = \emptyset$ we modify the direct path action.

The condition to apply this inference rule is shown in Fig. 6. In general, we should verify before removing each rule if the misconfiguration will be fixed and we should be sure that we will not generate new misconfigurations. The inference rule *Follow* is applied when other inference rules could not be applied. The **Stop** rule is applied when we parse the faulty rules correspondent to totally misconfigured direct paths from the set TMC.

$$\textit{Init}\ \frac{}{FR, FR^*, \emptyset}$$

$$\textit{Parse}\ \frac{\{FR_x\} \cup FR, FR^*, \emptyset}{FR, FR^*, FR_x}$$

$$\textit{Remove}\ \frac{FR, FR^*, \{r_i\} \cup FR_x}{FR \backslash_{remove} \{r_i\}, FR^* \backslash_{remove} \{r_i\}, FR'_x}\ \text{if (remove-condition)}$$

$$\textit{Follow}\ \frac{FR, FR^*, \{r_i\} \cup FR_x}{FR, FR^*, FR_x}\ \text{if remove is not applied}$$

$$\textit{Stop}\ \frac{\emptyset, FR^*, \emptyset}{FR^*}$$

Fig. 5. Inference system for removing rules

$$\textit{if}\ \forall dp_y \in FDD, \textit{where}\{r_i\} \in dp_y.rules \begin{cases} r_1y^f \neq_f r_i \quad \vee \\ r_1y^f =_f r_i \wedge dp_y \in TMC \wedge (\exists r2y^f, act(r2y_f) \neq act(r_i)) \ \vee \\ dp_y.act = accept \wedge r_1y^f =_f r_i \wedge dp_y \notin TMC \wedge (\exists r2y^f, act(r2y_f) = act(r_i)) \end{cases}$$

$$\textit{where} \begin{cases} FR'_x = \emptyset & if(dp_x.act = accept) \\ FR'_x = FR_x \setminus \{r_i\} & if(dp_x.act = deny) \end{cases}$$

Fig. 6. Remove-condition

Modify-Action Inference System. After changing the action of a rule we should not generate new misconfigurations. So, we should verify first if **all** the direct paths in all $fdd_n \in FDD$ that have this rule as a first rule are totally misconfigured. If it is the case, we can change the action of the rule under consideration and using this one modification we will correct all misconfigured direct paths that have this rule as a first one.

The rules of the system shown in Fig. 7 apply to triple (FR, FR^*, FR_x) whose first component FR is the set of faulty rules of each dp_x in FDD, whose second component FR^* represents an updated version of FR after modifying actions of rules after verifying if the third inference rule $modify$ is applied or not. In fact, the second inference rule $Parse$ allows to parse all faulty rules FR_x of each dp_x, then if the precondition of the third inference rule $modify$ is applied we update the three components FR, FDD and TMC by using the function $\backslash_{modify}$ which allows to modify the three components as follows:

- FR: $\forall x$ we remove r_i from FR_x if $dp_x.action = deny$, else if $dp_x.action = accept$, $FR_x = \emptyset$.
- TMC: if $FR_x = \emptyset$ then we remove dp_x from TMC.
- FDD: we modify the action of r_i from each $dp_y.rules$ $\forall y$ and if $FR_y = \emptyset$ we modify the direct path action.

The condition to apply this inference rule is shown in Fig. 8. In general, we check if the rule under consideration verifies two properties in all paths then we can modify the action of the rule. These two properties are: r_i is not the first rule to be applied in direct paths that mapped this rule or it is the first rule and the direct path is totally misconfigured.

$$
\begin{array}{lll}
Init & \dfrac{}{FR, FR^*, \emptyset} & \\[2ex]
Parse & \dfrac{\{FR_x\} \cup FR, FR^*, \emptyset}{FR, FR^*, FR_x} & \\[2ex]
Modify & \dfrac{FR, FR^*, \{r_i\} \cup FR_x}{FR \backslash_{modify} \{r_i\}, FR^* \backslash_{modify} \{r_i\}, FR'_x} & \text{if (modify-condition)} \\[2ex]
Follow & \dfrac{FR, FR^*, \{r_i\} \cup FR_x}{FR, FR^*, FR_x} & \text{if modify is not applied} \\[2ex]
Stop & \dfrac{\emptyset, FR^*, \emptyset}{FR^*} &
\end{array}
$$

Fig. 7. Inference system for modifyin rules-actions

$$
\begin{array}{l}
if\ \forall dp_y \in FDD, where\{r_i\} \in dp_y.rules \begin{cases} r_1 y^f \neq_f r_i \quad \vee \\ r_1 y^f =_f r_i \wedge dp_y \in TMC \end{cases} \\
where \begin{cases} FR'_x = \emptyset & if(dp_x.act = accept) \\ FR'_x = FR_x \setminus \{r_i\} & if(dp_x.act = deny) \end{cases}
\end{array}
$$

Fig. 8. Modify-condition

Swap-Rules Inference System. Before swapping two rules, we need to test and to verify if this modification will generate new misconfigurations between one of the swapped rules and other rules. To address this challenge, we use the FDD as the core data structure. An FDD gives us a precise idea if the swap of the rules will correct the misconfigurations or not. In Fig. 9 we propose an inference system that presents necessary and sufficient steps to correct total misconfigurations by swapping two rules.

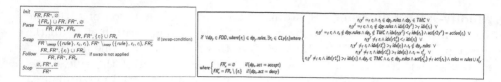

Fig. 9. Swap rules inference system

Fig. 10. Swap condition

The rules of the system shown in Fig. 9 apply to triple (FR, FR^*, FR_x) whose first component FR is the set of faulty rules of each dp_x in FDD, whose second component FR^* represents an updated version of FR after modifying actions of rules after verifying if the third inference rule *swap* is applied or not. In fact, the second inference rule *Parse* allows to parse all faulty rules FR_x of each dp_x, then if the precondition of the third inference rule *swap* is applied we update the three components FR, FDD and TMC by using the function $\backslash_{swap}((rules, rc, ri) :)$ which allows to modify the three components as follows:

- FR: $\forall x$ we remove r_i from FR_x if $dp_x.action = deny$, else if $dp_x.action = accept$, $FR_x = \emptyset$.
- TMC: if $FR_x = \emptyset$ then we remove dp_x from TMC.
- FDD: we swap two rules r_c and r_i and if $FR_x = \emptyset$ we modify the direct path action.

The condition to apply this inference rule is shown in Fig. 10. We should verify if we can swap rule r_i and rules from the set the set $CL_y(r_i)$ which is the candidate-rules list, rules from this list can be used to correct misconfigurations. In fact, for each dp_y, $CL_y(r_i)$ is composed by rules belonging to the same direct paths as r_i and having r_i as a first rule in these paths also they should have different action to this rule.

Field Modification Inference System. The rules of the inference system shown in Fig. 11 apply to three components (PMC, TMC, FDD). The first component is the set of partial misconfigurations discovered. The second component TMC is the set of total misconfigurations not fixed using methods depicted in previous subsections. The third component FDD is the set of fdd_n of all paths of the network. *Parse_PMC* is used to parse the set of partial misconfigurations (i.e., dp_i that have partially an action not exactly the same as defined by SP). This inference rule will divide dp_i into two sets, the first one is the set of paths that have the *correct* action as defined by SP ($dp_i \setminus SP_{dp_i.act}$), this set will replace the direct path dp_i in FDD, and the second one is the path dp_i' represents the subset of dp_i that should be fixed and will be added to the set of total misconfigurations, by this inference rule we transform the partial problem in dp_i into a *total* problem (misconfiguration) in dp_i' which will be added to the set TMC. The inference rule *Correct* deals with each direct path from TMC and according to the required action by SP we will add new rules in one or some firewalls in this path. The first case, when the required action is *deny* (i.e., $dp_i.act = accept$), then we add only one rule in the first firewall of this direct path, the rule $r_{1j}{}^{1'}$ should have the action deny and will be added to the set $dp_i.rules$. The second case, when the required action is *accept*, in this case we have to modify the action of each firewall that have the action deny to accept in this direct path. Therefore, we insert new rules $r_{1j}{}^{k'}$ in each firewall that applies the action *deny* on packets handled by the direct path dp_j.

$$
\begin{array}{ll}
Init & \dfrac{}{PMC, TMC, FDD} \\[2ex]
Parse\_PMC & \dfrac{\{dp_i\} \cup PMC, TMC, FDD}{PMC, TMC \cup dp'_i, update_{FDD}(i, dp_i \setminus SP_{|dp_i.act})} \; where \; dp'_i = dp_i \cap SP_{|dp_i.act} \\[2ex]
Correct & \dfrac{\varnothing, \{dp_j\} \cup TMC, FDD}{\varnothing, TMC, update_{FDD}(j, dp'_j)} \; where \; \forall k \begin{cases} dp'_j.rules = r'_1 j^1 \cup dp_j.rules \wedge action(r'_1 j^1) = deny & if(dp_j.act = accept) \\ dp'_j.rules = r'_1 j^k \cup dp_j.rules \wedge action(r'_1 j^k) = accept & if(dp_j.act = deny \wedge action(r1 j^k = deny)) \end{cases} \\[2ex]
Stop & \dfrac{\varnothing, \varnothing, FDD}{FDD}
\end{array}
$$

Fig. 11. Field modification inference system

5   Implementation and Evaluation

5.1   Case Study

We have chosen to apply our approach on the case study shown in Sect. 1. The SP is described as follows: Allow access from net1 to net2 except http access from machine M1 to subnet21; Deny all traffic from net3 to net2.

As defined in Sect. 3, *Path* is the set of all possible paths from a source to a destination by considering *SP*. In this case, we have: $P1 = Path[net3, net2] = \{Fw1, Fw2, Fw3\}$ and $P2 = Path[net1, net2] = \{Fw2, Fw3\}$. Figures 3 and 4 show, respectively, the FDD of two paths $P1$ and $P2$.

Discovering Distributed Firewalls Misconfigurations: We proceed to the discovering of misconfigurations using the inference system previously described in Sect. 4. We parse all paths of all FDDs, for each path we verify if we have anomaly or not and if this anomaly is an effective misconfiguration:

– In path $P1 = Path[net1, net2] = \{Fw1, Fw2, Fw3\}$: In this path we have four total misconfigurations (colored in red in Fig. 3), in direct paths dp_1, dp_3, dp_7 and dp_9. Also we have a partial misconfiguration in direct path dp_{12} (Colored in green), in fact, the traffic from machine M1 192.168.4.3 will be accepted by direct path dp_{12} even if we precisely indicated in SP that this traffic should be rejected, this misconfiguration is partial because other traffic from $net1$ will be allowed which is conform to SP. The SP is **partially** violated in this case.
– In path $P2 = Path[net3, net2] = \{Fw2, Fw3\}$: In this path we have two total misconfigurations (colored in red in Fig. 4), in direct paths dp_2 and dp_4.

Distributed Firewalls, Misconfigurations Resolution: After discovering process has been established, we will proceed in this section to the resolution of these misconfigurations automatically and in contrast with SP.

TMC in dp_1 in $P1$: According to the process of correction explained in Sect. 4, the set of faulty rules FR of this direct path contains rules r_{11}, r_{32}, the correct action is deny, therefore we can fix at least one of these rules to fix this total misconfiguration. So, we start by verifying if we can remove the rule r_{11}, it is not the case because r_{21} have the same action as r_{11}, so removing r_{11} will not fix the problem in this direct path and it is the same case for rule r_{32}. Then, we verify if we can modify the action of these rules, we note that rules r_{11} and r_{32} exist in other direct paths and these paths does not present any anomaly. Therefore, we try to apply the swap inference system, the set of candidate rules $CL = \{r_{31}\}$, according to the FDD swapping r_{11} and r_{31} will not only correct this misconfiguration but also the second misconfiguration in dp_3 in path $P1$ and will not generate new misconfigurations. Therefore, for these two misconfigurations we will use the swap-technique.

TMC in dp_7 in $P1$ and dp_2 in $P2$: r_{12} exists only in dp_7 from $P1$ and dp_2 from $P2$ and these two direct paths are totally misconfigured. The set of faulty rules of these direct paths contains rule r_{12} only, also the second rule in these direct paths is the rule r_{42} which have a different action from r_{12}, so by removing this rule (i.e., r_{12}) we will correct these two misconfigurations and we will not generate new misconfigurations.

TMC dp_9 in $P1$ and dp_4 in $P2$: The set of faulty rules of these two direct paths contains rules r_{22} and r_{23}. We should correct these two rules because they have the action deny. According to the process of correction explained in Sect. 4, we should start by verifying if we can remove these rules, it is the case for rule r_{22} but not for rule r_{23} because r_{33} have the action deny, so removing r_{23} will not fix the problem. So, we remove r_{22}. Then, for rule r_{23} we verify if we can modify the action of this rule, we note that r_{23} exists only in these misconfigured direct paths. So by changing the action of this rule (i.e., r_{23}) we will correct these misconfiguration and we will not generate new misconfigurations.

PMC in dp_{12} from Path $P1$: This misconfiguration is *partial* so we use the method "field modification" to fix this problem. The intersection between DP_{12} ad SP_{deny} can be represented as follow: $BSP = DP_{12} \cap SP_{deny} =$ branch represented by these values: $[@srce, port, @dest, protocol] = [192.168.4.3, 80, 172.13.14.0/24, TCP]$ Therefore, DP_{12} could be represented as follow: $DP_{12} = (DP_{12} \backslash BSP) \cup (DP_{12} \cap BSP)$. Then using our inference system shown in Fig. 11 we use first the inference rule $Parse$ to divide this direct path into two sub-FDDs where the first $(DP_{12} \backslash BSP)$ represents paths which are conform to SP and the second one $DP_{12} \cap BSP$ is the totally misconfigured path. Then to correct $DP_{12} \cap BSP$ we use $Correct$, this inference rule will add new rules with new action at each direct path that contains the total misconfiguration.

5.2 Tool Evaluation

Complexity: For n rules in FC, there can be a maximum of $2n - 1$ outgoing edges for a node. Therefore, the maximum number of paths in a constructed FDD is $(2n - 1)^d$, where d is the number of fields in each rule. After the construction of FDD, the process of misconfigurations discovering and removing, is done on direct paths elements $dp_i.rules$. Therefore, for our inference systems, the complexity is equivalent to the complexity of operations in an ordered list. Thus, in our case, the complexity of each inference system is equal to $O(n^d)$. Given that d is typically small (generally we have 4 or 5 fields) our inference systems have a reasonable response time in practice.

Implementation and Experimental Results: In order to better assess the effectiveness of our approach, we implemented the techniques and inference systems described earlier in a software tool, using a Boolean satisfiability (SAT) based approach. This approach reduces the verification problem into Boolean formula and checks its satisfiability. We have chosen also the Java developing language. On the other hand, the verification of the satisfiability of Boolean expressions is performed using Limboole [15]. To evaluate a practical value of our inference systems, we have implemented them based on the FDD approach and we tested our developed tool using the rule collections of the open-source rules available at emerging threats (ETOpen) rule sets [10]. Our tool demonstrates the scalability of proposed inference systems, we have also conducted a set of experiments to measure their performance, our tool has proved a stable performance showing acceptable processing time (the average processing time is some seconds) to the treatment of complex combination of thousands filtering rules. We have also conducted a set of experiments to measure the performance of our inference systems. The experiments were run on an Intel Dual core 1.6 GHz with 2 Gbyte of RAM. It supposes that we have IPv4 addresses with net-masks and port numbers of 16 bit unsigned integer with range support. Figure 12 summarizes our results. We consider time treatment factor that we review by varying the number of rules. In overall terms, we consider the average processing time, of the main procedures of FDD construction, misconfigurations detection and

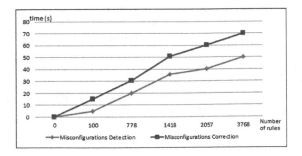

Fig. 12. Processing time

correction. In the end, our tool has proved a stable performance showing acceptable processing time to the treatment of complex combination of rules.

6   Conclusion

The prevalent use of firewalls in network security emphasizes the importance of efficient and optimal configuration. This paper describes two problems. The first, is firewall misconfiguration discovering. In fact, we propose a method to discover and distinct real configuration errors. The second is misconfigurations resolution by using a formal method and a data structure (FDD). Specifically, we presented a classification of misconfigurations (total or partial) and propose a set of inference systems that allow optimal and safe correction of these conflicts, without generating new misconfigurations, through the analysis of the rule relations basing on FDD structure. The efficacy and scalability of our approach has been demonstrated and the first results we obtained are very promising. While the current approach primarily focuses on the detection and correction of firewalls configuration errors. As a future work, we are working on extending our approach in order to handle other network security components misconfigurations like IDS.

References

1. Abbes, T., Bouhoula, A., Rusinowitch, M.: Detection of firewall configuration errors with updatable tree. Int. J. Inf. Secur. **15**, 1–17 (2015)
2. Adiseshu, H., Suri, S., Parulkar, G.M.: Detecting and resolving packet filter conflicts. In: Proceedings IEEE INFOCOM 2000, Tel Aviv, Israel, 26–30 March 2000, pp. 1203–1212 (2000)
3. Al-Shaer, E.S., Hamed, H.H.: Modeling and management of firewall policies. IEEE Trans. Netw. Serv. Manag. **1**(1), 2–10 (2004)
4. Bera, P., Ghosh, S.K., Dasgupta, P.: Policy based security analysis in enterprise networks: a formal approach. IEEE Trans. Netw. Serv. Manag. **7**(4), 231–243 (2010)
5. Bera, P., Ghosh, S.K., Dasgupta, P.: Integrated security analysis framework for an enterprise network - a formal approach. IET Inf. Secur. **4**(4), 283–300 (2010)
6. Bouhoula, A., Trabelsi, Z., Barka, E., Benelbahri, M.A.: Firewall filtering rules analysis for anomalies detection. IJSN **3**(3), 161–172 (2008)
7. Chomsiri, T., Pornavalai, C.: Firewall rules analysis. In: Security and Management, pp. 213–219 (2006)
8. Cuppens, F., Cuppens-Boulahia, N., Alfaro, J.G.: Detection and removal of firewall misconfiguration. In: CNIS IASTED, Phoenix, AZ, USA, November 2005
9. Eppstein, D., Muthukrishnan, S.: Internet packet filter management and rectangle geometry. CoRR, cs.CG/0010018 (2000)
10. Etopen ruleset (2016)
11. Alfaro, J.G., Cuppens, F., Cuppens-Boulahia, N.: Analysis of policy anomalies on distributed network security setups. In: Gollmann, D., Meier, J., Sabelfeld, A. (eds.) ESORICS 2006. LNCS, vol. 4189, pp. 496–511. Springer, Heidelberg (2006). doi:10.1007/11863908_30

12. Gouda, M.G., Liu, A.X.: Structured firewall design. Comput. Netw. J. (Elsevier) **51**(4), 1106–1120 (2007)
13. Hall, S., Ngoup, L., Villemaire, R., Cherkaoui, O.: Distributed firewall anomaly detection through LTL model checking. In: 2013 IFIP/IEEE International Symposium on Integrated Network Management (IM 2013), pp. 194–201, May 2013
14. Hu, H., Ahn, G.-J., Kulkarni, K.: Detecting and resolving firewall policy anomalies. IEEE Trans. Dependable Secur. Comput. **9**(3), 318–331 (2012)
15. Limboole sat solver (2016)
16. Liu, A.X.: Formal verification of firewall policies. In: ICC, pp. 1494–1498 (2008)
17. Liu, A.X., Gouda, M.G.: Diverse firewall design. IEEE Trans. Parallel Distrib. Syst. (TPDS) **19**(8), 1237–1251 (2008)
18. Matsumoto, S., Bouhoula, A.: Automatic verification of firewall configuration with respect to security policy requirements. In: CISIS, pp. 123–130 (2008)
19. Mukkapati, N., Bhargavi, Ch.V.: Detecting policy anomalies in firewalls by relational algebra, raining 2D-box model. IJCSNS Int. J. Comput. Sci. Netw. Secur. **13**(5), 94–99 (2013)
20. Yazidi, A., Bouhoula, A.: On assisted packet filter conflicts resolution: an iterative relaxed approach. In: 41st IEEE Conference on Local Computer Networks, LCN 2016, Dubai, United Arab Emirates, 7–10 November 2016, pp. 35–42 (2016)
21. Yuan, L., Mai, J., Su, Z., Chen, H., Chuah, C.-N., Mohapatra, P.: Fireman: a toolkit for firewall modeling and analysis. In: Proceedings of the 2006 IEEE Symposium on Security and Privacy, Washington, DC, USA. IEEE Computer Society (2006)

A Security Approach for Data Migration in Cloud Computing Based on Human Genetics

Hamza Hammami[✉], Hanen Brahmi, Imen Brahmi,
and Sadok Ben Yahia

Faculty of Sciences of Tunis, University of Tunis El Manar, LIPAH-LR11ES14,
2092 Tunis, Tunisia
hamza.hammami@aol.fr, hanen.brahmi@yahoo.fr

Abstract. Cloud computing technology is flexible, cost effective and reliable for the provision of IT services to businesses and individuals through means of internet. Clearly beneficial in terms of costs, this technology has gained immediate popularity. However, security concerns have slowed its expansion. It is possible that the full adoption of cloud computing is not appropriate in some cases, for security reasons related to confidentiality and data integrity. Cryptographic methods that could reduce these risks to acceptable levels, however, were developed. In this article, we introduce a method implementing encryption based on human genetics, more particularly on protein biosynthesis. The attractive coupling between the encryption of content and biosynthesis protects data against unauthorized access. The experiments show that our proposal provides a good balance between the integrity and confidentiality of data.

Keywords: Cloud computing · Cost effective · Security · Confidentiality · Integrity · Cryptographic · Human genetics · Protein biosynthesis

1 Introduction

IT departments are not always armed with all the expertise and skill required to establish and maintain an architecture of efficient and secure storage. Cloud computing is timely as an option to solve this problem [6, 11]. This simple concept offers the outsourcing of information and services of a company beyond its own limitations in warehouses of a large size. Experienced employees administer physical sites. The headquarters of the largest players in the cloud (Amazon, Google, Sales force, Microsoft, etc.) present their solutions and provide support for data storage. Note that the pooling of these computer resources causes a reduction in costs for a member company [2].

The widespread adoption of cloud computing does not exclude permanent and unpredictable dangers: The company loses control over his information, so many questions about the legality or safety remain. The risks associated with the adoption of this technology by companies are numerous [9]. Security remains the primary concern for migration towards cloud computing [2]. We focus on data. In fact, a multitude of threats from various backgrounds, specifically affluent target data [2]. These are deported to warehouses out of business or at home. Customers sometimes demonstrate

© Springer International Publishing AG 2017
M. Themistocleous and V. Morabito (Eds.): EMCIS 2017, LNBIP 299, pp. 384–396, 2017.
DOI: 10.1007/978-3-319-65930-5_31

against the reluctance of this migration, marked by the sudden disappearance of physical contact or alteration of data. If sensitive data fall into the wrong hands, the consequences can be disastrous. In addition, some system administrators have legitimate reasons to access data. This intimate contact with the information opens the door to malicious acts.

The literature reviews on the subject [3, 9, 10] consider that integrity and confidentiality are two vectors contributing to the availability of migrating data from an original host to a destination one in the cloud. On the one hand, integrity provides assurance that the data has not been altered in an unauthorized manner during the storage phase, treatment or migration [8]. Moreover, the confidentiality offers the guarantee that data cannot be disclosed to others, as service or unauthorized materials [3]. The preservation of confidentiality and data integrity seeks efficiency and adaptability as cryptography mechanisms.

The rest of the article is organized as follows. In Sect. 2, we review previous work. Our proposal is detailed in Sect. 3. The results of experiments showing the usefulness of the proposed approach are presented in Sect. 4. The last section concludes this paper and pins down several issues of future work.

2   Related Work

In this section, we focus on presenting some approaches to the literature to circumvent security problems of computer related data in the cloud. Specifically, we show the approaches that try to secure data migration.

For example, in [4] the approach was based on two steps: *(i)* dividing and storing data on a server chosen arbitrarily, and *(ii)* restoring data. Through these steps, data is ready to be transferred, stored and processed safely since they are encrypted. However, this approach generates a data redundancy problem.

Otherwise, [7] proposed a two-step approach based on: *(i)* encrypting the data using a symmetric key; and *(ii)* securing the key in the cloud. For this purpose, the authors employed a method to encrypt and decrypt the symmetric key generated by the utilizer from its identifier. This method was not sufficient for an open system like the cloud. Indeed, these groups of users shared and accessed the same data. However, the method was limited to a user.

Furthermore, [8] have suggested CLOUD SEC as a public key-based solution. The authors used two layers in order to separate key management and encryption techniques. This approach offered two major advantages: *(i)* a flexible deployment of a data-sharing scenario, and *(ii)* a solid security for outsourced data on servers in the cloud. The encryption mechanism and the decryption of data could be deployed intensively on processors, which resulted in wasting resources-something that cloud providers did not want.

Moreover, [9] put forward a method that provided users an effective means, allowing it to perform periodic audits of distance integrity. This method did not require a local data backup. The proof of ownership was based on three aspects: *(i)* level of security, *(ii)* public auditing, and *(iii)* performance. This method was limited by storage requirements and calculation of the client terminal and the size of outsourced data.

In [10] three encryption algorithms (AES, DES, RSA) were proposed to secure data during transmission in a network. The authors highlighted the execution time consumed at the time of encryption and decryption of data. [2, 3] suggested an architecture composed of three end users *(i)* a client, *(ii)* a cloud administrator, and *(iii)* a trusted third party acting as an intermediary between the client and the cloud services. To this end, they introduced the SCSM algorithm in order to ensure the security of data during migration in the network.

In [14] the authors put forward an efficient model to secure data sharing in the cloud. The proposed model consisted of a user, an authority, a hybrid cloud and an owner. The data was stored at a private cloud and shared in an encrypted way. The used encryption technology was keyword based. The keys were generated by authority and given to a user group for encryption and decryption. The model had some issues as if authority were fake, so data were insecure. Also, it was costly to use the model.

In [15] the authors proposed a performed randomness test on various eight encryption techniques, namely RC4, RC6, MARS, AES, DES, 3DES, Two-Fish and Blowfish.

The authors in [16] suggested a dynamic user revocation and a key refreshing scheme based on a cipher policy and an attribute based on the encryption technique. In this technique, the utilizer could be removed anytime without changing keys and also could refresh keys without re-encrypting data.

In [17] the authors put forward a system consisting of three entities: cloud broker, client and cloud storage. A broker would handle encryption, a hash key, decryption and local database management. According to the available cloud space, the client files were partitioned into segments. Hash values of segments were generated. When a client needs its file, it will send a request to the broker, who will then download the file, partition the file into segments and calculate the hash values. To check the hash values of data integrity, they are matched before uploading and after downloading. If this matches, data are untampered.

In [18] white papers of many organizations described three types of data security models in the cloud. The First model consisted of key generation, and the encryption on data was performed by the data owner itself. However, this model resulted in a high overhead for the data owner. The second model described the encryption performed by the data owner and the key generation by the cloud service provider. Unfortunately, the cloud service provider was fake, so the data would be insecure hands. The third model encryption and the key generation was controlled by the cloud service provider. If the cloud service provider was fake, then data would be endangered.

In Table 1, we present a comparative study of the approaches of cloud computing security mentioned above.

Table 1. Comparison between security approaches of cloud computing

Approaches	Confidentiality	Overhead	Authorization	Encryption	Dual verification	Cost effective
Approach proposed in [14]	Yes	No	No	Yes	No	Yes
Approach proposed in [15]	Yes	No	Yes	Yes	Yes	No

(continued)

Table 1. (*continued*)

Approaches	Confidentiality	Overhead	Authorization	Encryption	Dual verification	Cost effective
Approach proposed in [16]	Yes	May be	May be	May be	No	Yes
Approach proposed in [17]	Yes	No	No	Yes	No	No
Approach proposed in [18]	No	No	No	Yes	No	Yes
Our proposed approach	Yes	No	Yes	Yes	Yes	Yes

In this article, we look at different ways of tackling the data security problem during their migration from the original host to the destination host in the cloud. We seek to ensure that its data integrity always remain confidential. The main idea behind our approach comes from the findings of several previous works [1, 5]. They have exploited human genetics to discover and develop innovative solutions in order to solve technological dilemmas. For example, [1] proposed an edge detection algorithm for image processing. In addition, [5] implemented a behavior selection mechanism for mobile robots. Similarly, the main idea behind our approach is based on an immune system, precisely in protein biosynthesis.

3   Our Proposed Approach

We contemplate the use of a biological system, especially human genetics, to benefit from its advantageous discoveries in order to solve the security problem of the cloud environment. Indeed, human genetics is a rich source of inspiration for new ideas allowing solving various technological problems and inspiring algorithms so as to find efficient IT solutions for complex problems. This is why we are looking into human genetics and more specifically to protein biosynthesis. The latter is a collection of cells, molecules and nucleic acids that collaborate together to realize a protein recipe demanded by the cell. It takes place in the nucleus of a cell which will then emerge from this nucleus to go into the cytoplasm where it will be translated into proteins for the purpose of organizing and setting up cells as well as maintaining and adjusting its good functioning.

Our approach is based on the same principle as that of protein biosynthesis. We consider data migrating to the cloud as cells. The original host is assimilated to a kernel. The destination host receiving and running the data is similar to a cytoplasm. By analogy, we apply all the rules and steps of biosynthesis to the new defined entities.

3.1   Original Host

The original host is the owner of the data that are preparing to migrate. Three operations are carried out beforehand to guarantee the protection of the data against any analysis aimed at their disclosure during their migrations towards the cloud. These security operations arise in three phases in the following order: *(i)* DNA encoding, *(ii)* translation, and *(iii)* fragmentation. In what follows, we explain each phase in more details.

Phase of DNA Encoding: In this phase, we apply a coding method based on human genetics. This method consists in encoding the data into DNA nucleotides forming genetic information. The genetic code uses a DNA fragment consisting of a sequence of elementary links, called nucleotides. In order to implement this type of encoding we follow these steps:

- First, user data are converted into an ASCII code giving a binary string denoted by C_0.
- Then, the binary code C_0 chain is converted to base 4. Thus, we get a string {0, 1, 2, 3} denoted by C_1.
- Finally, C_1 is converted into a DNA chain designated by C_2. The latter is composed of a set of nucleotides. In biosynthesis, the transition to DNA encoding based on the nucleotides is done through four types of coding: A, C, G and T.

In biosynthesis, there are four ways of encoding: encoding type A, encoding type C, encoding type G, and encoding type T. Each character in base 4 is thus transformed into a nucleotide according to the selected encoding type (Fig. 1).

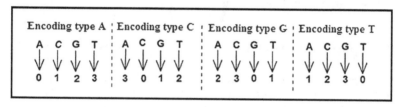

Fig. 1. Phase of DNA encoding

Figure 2 shows a complete example of this encoding step from the binary code to the DNA one. In this example, the T type encoding is used.

Phase of Translation: This phase is intended for a protein manufacturing plant by providing the recipe of the encoded proteins. The principle of the translation is described below. This is called the transform coding regions of DNA into an RNA chain. Therefore, the letters "A", "C", "G" and "T" are replaced by the letters "U", "G", "C" and "A". Translation is a process of decoding the information contained in an RNA to obtain corresponding amino acids. First, the RNA code is divided into groups of 3 nucleotides. Indeed, the amino acids are represented in an array of three dimensions where each acid is identified by three nucleotides: Code (1st nucleotide), code (2nd nucleotide), and Code (3rd nucleotide). In order to calculate the code for the amino acid, we use Eq. 1:

$$Amino\ acid = Code[(1^{st}nucleotide-1)*16] + [Code(2^{nd}nucleotide-1)*4] + [code(3^{rd}nucleotide)] \tag{1}$$

Table 2 presents the means of constructing chains of amino acids. Indeed, the letters tagging nucleotides are replaced by the codes in Eq. (1). Amino acids are stored in an array of 3 three dimensions. Hence, we multiply by 16 the first nucleotide to the projection on the lines. In addition, the second nucleotide is multiplied by 4 to the projection on the columns. Moreover, both previous results are added to the third nucleotide to fix the index of the exact amino acid. By obtaining this last, we can determine the corresponding amino acid based on Table 3.

Table 2. Code of each nucleotide

Nucleotide	Code
U	1
C	2
A	3
G	4

Example: Figure 2 illustrates an RNA code translation example of amino acids. The first group of three nucleotides is (GCU). Hence, Eq. (1) corresponds to an amino acid. We replace the nucleotides by their codes based on Table 2. We obtain the amino acid = $[(4-1) * 16] + [(2-1) * 4] + [1] = 53$. In Table 3 "53" corresponds to the index of the amino acid "A1". Hence, the group of nucleotides (GCU) should be replaced by "A1".

Phase of Fragmentation: This phase comprises two steps:

- **Fragmentation step:** We read the amino acids obtained in the translation step. At first, we cross the obtained amino acid sequence. Whenever an amino acid O1, O2 or O3 is detected, it is stopped. Thus, we consider the sequence read as a fragment. In biology, the amino acids O1, O2 and O3 are called fragment terminators, denoted stop *code*. Then we go to the rest of the sequence to read a new fragment. At the end of each fragment an amino acid is added, whose code is the playing order of this fragment. **Example:** In Fig. 2, we work on the A1C2L2R5G2T1O3G1 amino acids. The stop code is detected as O3. Hence, the first fragment is: A1C2L2R5G2T1O3. After that, the amino acid F1 of index 1 is added at the end of the first fragment, indicated as the first fragment.

- **Encoding Step:** For each obtained fragment we apply an encoding using a private key which is the unique identifier of the migrated data. Take for example a fragment composed of 8 amino acids A1C2L2R5G2T1O3F1 and the key 31407. The encoding consists in adding one by one the amino acids and the key numbers (for example the first acid is replaced by the amino acid code whose code is (A1) $+3 = 53 + 3 = 56$). At the end of this phase, the migrated fragments will be locked using a shared key generated between the original and remote hosts. This shared key is generated by a Diffie Hellman method [13]. This latter represents a specific method of securely exchanging cryptographic keys over a public channel. The detailed method is listed as follows:
  - Original host chooses prime P at random and finds a generator g.

Table 3. Table for encoding into amino acid

Index	Amino acid	Codon	Anticodon	Index	Amino acid	Codon	Anticodon
1	F1	TTT	UUU	33	I1	ATT	AUU
2	F2	TTC	UUC	34	I2	ATC	AUC
3	L1	TTC	UUA	35	I3	ATA	AUA
4	L2	TTC	UUG	36	M1	ATG	AUG
5	S1	TCT	UCU	37	T1	ACT	ACU
6	S2	TCC	UCC	38	T2	ACC	ACC
7	S3	TCA	UCA	39	T3	ACA	ACA
8	S4	TCG	UCG	40	T4	ACG	ACG
9	Y1	TAT	UAU	41	N1	AAT	AAU
10	Y2	TAC	UAC	42	N2	AAC	AAC
11	O1	TAA	UAA	43	K1	AAA	AAA
12	O2	TAG	UAG	44	K2	AAG	AAG
13	C1	TGT	UGU	45	S5	TGT	UGU
14	C2	TGC	UGC	46	S6	TGC	UGC
15	O3	TGA	UGA	47	R5	AGA	AGA
16	W1	TGG	UGG	48	R6	AGG	AGG
17	L3	CTT	CUU	49	V1	GTT	GUU
18	L4	CTC	CUC	50	V2	GTC	GUC
19	L5	CTA	CUA	51	V3	GTA	GUA
20	L6	CTG	CUG	52	V4	GTG	GUG
21	P1	CCT	CCU	53	A1	GCA	GCA
22	P2	CCC	CCC	54	A2	GCC	GCC
23	P3	CCA	CCA	55	A3	GCA	GCA
24	P4	CCG	CCG	56	A4	GCG	GCG
25	H1	CAT	CAU	57	D1	GAT	GAU
26	H2	CAC	CAC	58	D2	GAC	GAC
27	Q1	CAA	CAA	59	E1	GAA	GAA
28	Q2	CAG	CAG	60	E2	GAG	GAG
29	R1	CGT	CGU	61	G1	GGT	GGU
30	R2	CGC	CGC	62	G2	GGC	GGC
31	R3	CGA	CGA	63	G3	GGA	GGA
32	R4	CGG	CGG	64	G4	GGG	GGG

- Original host chooses $X \leftarrow_R \{0, 1 \ldots P - 2\}$ and sends P, g and $\hat{X} = g^X (mod\ P)$ to remote host.
- Remote host chooses $Y \leftarrow_R \{0, 1, \ldots, P - 2\}$ and sends $\hat{Y} = g^Y (mod\ P)$ to original host.
- Original host and remote host both compute $k = g^{XY}\ (mod\ P)$. original host does that by computing $\hat{Y}^X$ and remote host does this by computing $\hat{X}^Y$.
- They then use k as a key to exchange messages using a shared key encryption scheme.

Figure 2 shows a complete example of this stage of fragmentation of amino acids.

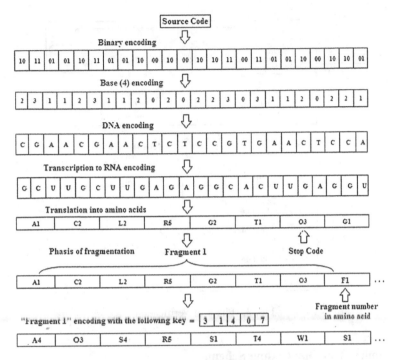

Fig. 2. Phase of original host

3.2 Remote Host

Once the encryption operation is completed, and upon reception in the destination host, the migrated fragments will undergo the next two operations: *(i)* matching, and *(ii)* verification.

Phase of Matching: In this phase, the fragments are first unlocked by the same key shared between original and remote hosts. Then, the encrypted fragments are hashed without having to decode them, using the Hash Message Authentication Code (HMAC) [12]. This latter is a specific type of message authentication code involving a cryptographic hash function in combination with a secret cryptographic key. The hashed code locked using the shared key of the migrated data is sent to the original host. The same hash is made to the initial code in the original host. Then both of the two hashes are compared.

Phase of Verification: To ensure this result, a procedure of verifying the legitimacy of the code starts. This phase consists in communicating with the remote host by sending the result of matching the considered legitimate after the hash calculation in the original host. This sent code is also encrypted using the unique identifier of the migrated data. According to the result of the matching, if the two codes are the same, proceed to the storage phasis of the migrated code. Otherwise, there will be a retransmission request code. Figure 3 shows the phase of verification.

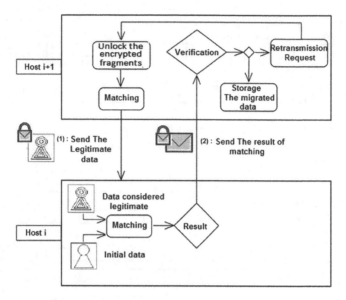

Fig. 3. Phase of verification

3.3 Complexity of Our Coding Scheme

In this section, we analyze the theoretical complexity of our coding scheme. Our approach complexity is analyzed as $O(n^2)$, which is the phasis of DNA encoding. In fact, amino acids encoding has a complexity $O(n^2)$. The phasises of translation, and fragmentation have a complexity $O(n)$. For these steps, in the worst case, the number of termination codons is greater than 64. The remote host steps have the same complexity as the original host. Then these steps' complexity is evaluated in $O(n^2)$. Thus, our coding scheme complexity is analyzed, in the worst case, as follows:

$T(n) = Max (O(n), O(n^2))$. We note that n represents the number of the amino acids in the migrated data.

4   Experimental Evaluation

Our experimental study is reposed in twofold: First, we study the performance of the GS-Cloud in terms of execution time. Second, we show the security aspects of our proposed scheme according to data confidentiality, data authentication, and data integrity.

4.1   Performance of GS-Cloud

In a first part, we study the execution time of our GS-Cloud approach for users with data sizes of 153 KB, 196 KB, 312 KB, and 868 KB.

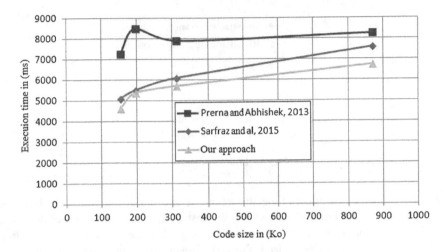

Fig. 4. Comparison of our approach with other approaches

First, we have chosen to compare our algorithm with two other algorithms (RSA [10], SCSM [2, 3]) fitting in the same trend and in the same data characteristics. Figure 4 depicts the execution time consumed by the GS-Cloud in order to develop the steps of: DNA encoding, translation, and data fragmentation considered for sizes, vs. the RSA and SCSM algorithms. We find that our algorithm is more efficient. The counting mechanism adopted by the GS-Cloud is more effective than the RSA and the SCSM. This can be explained by the genetic coding to win in terms of data size. However, the RSA and SCSM algorithms are handicapped by high traffic in the network because of massive communications between the three end users.

Then, we analyze the GS-Cloud in order to determine the source of the delay. Therefore, we perform two tests in the case of reception without attacks. In the first case, we send the data-encoded amino acids without fragmentation. In the second case, we send the same code but with fragmentation. Figure 5 clearly demonstrates that the delay is due mainly to the fragmentation of the code phase. This stage consumes the majority of execution time compared with other steps, *i.e.* the DNA encoding and translation.

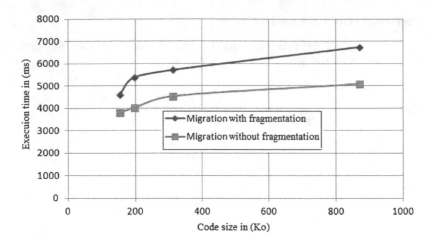

Fig. 5. Source of delay

Finally, we study the effect of the attacks on the execution time in Fig. 6. The case of an attack without migration follows the shortest time. We explain the result by the lack of waiting time for the communications messages between source and destination hosts after verification codes. Indeed, the codes received by the destination host are similar. In the case of migration in the presence of attacks, we see that we get ever shorter times.

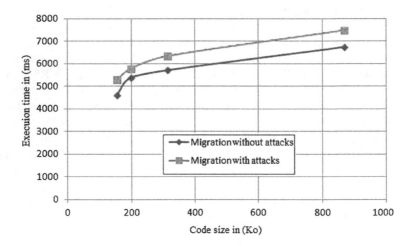

Fig. 6. Effect of attacks

4.2 Security Analysis

This section discusses how the GS-Cloud matches such security requirements separately from the security level and the efficiency of data integrity check. The proposed approach focuses on protecting the encryption of sensed data, and the goal is to securely deliver the concealed migrated data from original to remote hosts. Our approach aims to ensure data confidentiality, authentication and integrity.

Data Confidentiality: Migrated data can be exploited by a malicious adversary to violate the confidentiality of the sensitive data. For this purpose, all the migrated data are encrypted with a secret key and locked with a shared key that is shared by the original and remote hosts. In addition, data are encrypted in the form of amino acids. Hence, malicious adversary must first have the shared key to unlock and must second have the secret key to decode the migrated data. Even if the malicious adversary has the keys of the decoding, it cannot decode data unless it has all the IDs of fragments contained in the code. These are dispersed in the network. Therefore, it is difficult to have all of them for the attacker. Indeed, the malicious adversary needs to know all of *(i)* the IDs of the fragments; *(ii)* the decoding key fragments; and *(iii)* the shared key, in order to be able to constitute the total DNA encodes to decode the amino acids.

Data Authentication: Data authentication allows the remote host to verify that the data was really sent by the original host. In our approach, we verify data authentication by the encoding step in the phase of fragmentation. In the two-party communication case of original and remote hosts, data authentication can be achieved through a purely symmetric mechanism. The original and remote hosts share a secret key to compute a Message Authentication Code (MAC) of all migrated data. When a message with a correct MAC arrives, the remote host knows that it must have been sent by the original host.

Data Integrity: The exchanged data can be modified by a malicious adversary to violate the integrity of the sensitive data. In our approach, we palliate these threats via a matching phase. Thus, to ensure the integrity of all the exchanged data, the HMAC function is used to maintain the integrity of the migrated data. The original host can detect any changes performed by a malicious adversary including the verification information, by checking the MAC value using its shared key. If the migrated data are found to be modified, then they will be discarded, and an error indication "reception error" is displayed, so the storage cannot be done.

5   Conclusion

In this article, we have focused on data security approaches during their migration to the cloud to tackle the following challenges: *(i)* a time-cost performance, *(ii)* a low integrity due to the data corruption, and *(iii)* a meager privacy because of the disclosure of data. Thus, we introduced an approach called GS-Cloud based on protein biosynthesis. The experiments we conducted showed that our algorithm outperforms the algorithms forming part of the same trend of approaches for different sizes of tested data. In the case where the attacker modifies a single amino acid fragment of *(i)* the remainder of the code is damaged, and *(ii)* the migrated data must be retransmitted. Future work will include improving prospects of this approach by strengthening the autonomy of the data migration. We plan to guarantee data the ability to protect itself during migration. Therefore, the matching calculation is no longer required in order to ensure safety.

References

1. Ballet, P.: Intérêts mutuels des systèmes multi-agents et de l'immunologie: applications \`A l'immunologie et au traitement d'images. Ph.D. thesis, L'Université de Bretagne Occidentale (2000)
2. Brohi, S.N.: Secure cloud storage model to preserve confidentiality and integriy. Ph.D. thesis, University of Technology Malaysia, Advanced Informatics School (2015)
3. Brohi, S.N., Bamiah, M.A., Chuprat, S., Ab Manan, J.-L.: Design and implementation of a privacy preserved off-premises cloud storage. J. Comput. Sci. **10**(2), 210–223 (2014)
4. Chen, D., He, Y.: A Study on Secure Data Storage Strategy in Cloud Computing
5. Ishiguro, A., Watanabe, Y., Kondo, T., Uchikawa, Y.: Decentralized consensus-making mechanisms based on immune system-application to a behavior arbitration of an autonomous mobile robot. In: Proceedings of IEEE International Conference on Evolutionary Computation, pp. 82–87 (1996)
6. Jensen, M., Schwenk, J., Gruschka, N., Iacono, L.: On technical security issues in cloud computing. In: Proceedings of the IEEE International Conference on Cloud Computing, CLOUD 2009, Bangalore, India, pp. 109–116 (2009)
7. Kaaniche, N., Boudguiga, A., Laurent, M.: ID based cryptography for cloud data storage. In: Proceedings of the IEEE 6th International Conference on Cloud Computing, CLOUD 2013, Santa Clara Marriott, CA, USA, pp. 375–382 (2013)
8. Kaaniche, N., Laurent, M., El Barbori, M.: CloudaSec : a novel public-key based framework to handle data sharing security in clouds. In: Proceedings of the 11th International Conference on Security and Cryptography, SECRYPT 2014, Vienna, Austria, pp. 5–18 (2014a)
9. Kaaniche, N., El Moustaine, E., Laurent, M.: A novel zero-knowledge scheme for proof of data possession in cloud storage applications. In: Proceedings of the 14th IEEE/ACM International Symposium on Cluster, Cloud and Grid Computing, CCGrid 2014, Chicago, IL, USA, pp. 522–531 (2014b)
10. Mahajan, P., Sachdeva, A.: A study of encryption algorithms AES, DES and RSA for security. Global J. Comput. Sci. Technol. Netw. Web Secur. **13**(15), 14–22 (2013)
11. Wang, C., Wang, Q., Ren, K., Lou, W.: Ensuring data storage security in cloud computing. In: Proceedings of the 17th International Workshop on Quality of Service, IWQoS 2009, Charleston, South Carolina, pp. 1–9 (2009)
12. Anamika, S., Vishal, S.: Implementing data storage in cloud computing with HMAC encryption algorithm to improve data security. Int. J. Adv. Res. Comput. Sci. Softw. Eng. **5** (8) (2015)
13. Rameshwari, M., Pramod, K.: Cloud computing security improvement using Diffie Hellman and AES. Int. J. Comput. Appl. **118**(1), 0975–8887 (2015)
14. Li, J., Li, J., Liu, Z., Jia, C.: Enabling efficient and secure data sharing in cloud computing. Concurrency Comput.:Pract. Exper. (2013). Wiley
15. Mohamed, E.M., EI-Etriby, S.: Randomness testing of modem encryption techniques in cloud environment. In: 8th International Conference on Informatics and Systems (2012)
16. Xu, Z., Martin, K.M.: Dynamic user revocation and key refreshing for attribute-based encryption in cloud storage. In: International Conference on Trust, Security and Privacy in Computing and Communications. IEEE (2012)
17. Varalakshmi, P., Deventhiran, H.: Integrity Checking for Cloud Environment Using Encryption Algorithm. IEEE (2012)
18. Amazon Web Services.: Encrypting Data at Rest in AWS. https://aws.amazon.com/whitepapers

Using Homomorphic Encryption to Compute Privacy Preserving Data Mining in a Cloud Computing Environment

Hamza Hammami$^{(\boxtimes)}$, Hanen Brahmi, Imen Brahmi,
and Sadok Ben Yahia

Faculty of Sciences of Tunis, University of Tunis El Manar,
LIPAH-LR11ES14, 2092 Tunis, Tunisia
hamza.hammami@aol.fr, hanen.brahmi@yahoo.fr

Abstract. Cloud computing refers to an information technology infrastructure where data and software are stored and processed in a remote data center, accessible as a service through the Internet. Typical data centers within these fields are large, complex and often noisy. Further-more, privacy preserving data mining is an important challenge. It is required to protect the confidentiality of data sources during the extraction of frequent closed patterns. In fact, no site should be able to learn contents of a transaction at any other site. The work carried out in this paper deals with this problem. In this context, we suggest an approach that combines the extraction of frequent closed patterns in a distributed environment such as the cloud. We aim at maintaining the privacy of the sites during the data mining task in a cloud environment based on homomorphic encryption. The Simulation results and performance analysis show that our mechanism requires less communication and computation overheads. It can effectively preserve data privacy, check data integrity, and ensures high data transmission efficiency.

Keywords: Cloud computing · Privacy · Data mining · Confidentiality · Frequent closed patterns · Homomorphic encryption

1 Introduction

During the last decade, with the standardization of the Internet and the development of broadband networks, the computer world has popularized a new paradigm: *cloud computing*. Indeed, cloud computing brings a lot of benefits for businesses [1] such as: *(i)* rationalization and cost reduction, *(ii)* increased flexibility to the end user, *(iii)* usage billing, *(iv)* more efficient use of internet technology resource, and *(v)* data centers and high-performance storage bases. Thanks to these added-values, the recourse to the cloud is becoming more remarkable. In fact, billions of data are exchanged or stored in virtual spaces. This large volume of collected data is characterized by thousands of recording lines stored in a size of a few gigabytes. However, worsened with this huge volume of data, the privacy issues of data mining techniques have become very painful. In this context, preserving privacy is an important challenge. For example, consider a scenario in which two or more sites owning confidential databases wish to run a data mining algorithm on the union of their databases without revealing any unnecessary

M. Themistocleous and V. Morabito (Eds.): EMCIS 2017, LNBIP 299, pp. 397–413, 2017.
DOI: 10.1007/978-3-319-65930-5_32

information. In this scenario, it is required to protect privileged information. Consequently, it is also necessary to enable its use for research or for other purposes. In particular, although the sites realize that combining their data has some mutual benefit, none of them is willing to reveal their database to any other sites.

Hence, the challenge here is: How can we mine the data across distributed sources securely or without disclosing data to others?

This challenge has actually interested a lot of researchers whose primary purpose is to preserve the privacy of data sources during the extraction of frequent closed patterns from a distributed environment by suggesting new protection techniques and approaches.

To tackle this issue, we introduce a novel approach preserving privacy mining called Cloud-PPDM. In this respect, we take into account a main concern, namely maintaining privacy during closed frequent patterns mining in a distributed environment such as the cloud. To do this, we introduce a novel data privacy mining scheme based on homomorphic encryption. The scheme adopts a symmetric-key homomorphic encryption to protect data privacy and combine it with a homomorphic signature to check the integrity of data aggregation. In addition, during the decryption of aggregated data, the master of these sites is able to classify the encrypted and aggregated data based on encryption keys. Our experimental results reveal that the proposed approach is efficient on both runtime performances and security criteria.

The remainder of the paper is organized as follows. In Sect. 2, we describe the related work on privacy preserving data mining. In Sect. 3, we detail some notations that rely on cryptography. Section 4 describes our approach, which can extract frequent closed patterns in a cloud environment while preserving the constraints of privacy by using designed homomorphic encryption. Section 5 gives some tests to illustrate the performance of our approach. Finally in Sect. 6, we summarize our work and we sketch issues of future work.

2 Related Work

Data mining preserving privacy includes a variety of methods to extract useful knowledge from data, without divulging sensitive information on involved individuals. The challenge is to find effective models that meet these constraints. In the following, we survey some work allowing to deal with this problem. Four main categories of Privacy Preserving Data Mining (PPDM) methods have been identified [10–14]:

- Anonymization-based PPDM [15]: The anonymization technique implements generalization and suppression methods to generate an individual record indistinguishable within a group of records.
- Perturbation-based PPDM [16]: In this way, the statistical information computed from the perturbed data does not differ from the statistical information computed from the original data to a larger extent.
- Randomization-based PPDM [17]: The randomization technique implements data distortion techniques for adding little noise in the actual data.
- Cryptography-based PPDM [14]: Cryptographic algorithms are ideally meant for scenarios where multiple parties collaborate to: *(i)* compute results, *(ii)* share non sensitive mining results, *(iii)* and avoid disclosure of sensitive information.

Table 1 summarizes the main advantages as well as the limitations of the (PPDM) techniques.

Table 1. Advantages and limitations of PPDM techniques

Technique	Advantages	Limitations
Anonymization-based PPDM [15]	Hidden identity or sensitive data about record owners	Heavy loss of information
Perturbation-based PPDM [16]	Different independently preserved attributes	Original data values cannot be regenerated
Randomization-based PPDM [17]	Simple and useful for hiding information about individuals	This method does not deal with multiple attribute databases
Cryptography-based PPDM [14]	Better privacy comparing to randomized approach	Heavy calculations (in terms of computation time and memory consumption)

In the following, we only put the focus on the work based on cryptography. The cryptography-based PPDM technique usually guarantees a very high level of data privacy. In [18], the authors addressed the problem of secure mining of association rules over horizontally partitioned data, using cryptographic techniques to secure the shared information. Their solution was based on the assumption that each party would first encrypt its own patterns utilizing commutative encryption, then the already encrypted patterns of every other party. Later on, an initiating party would transmit its frequency count, plus a random value, to its neighbor. The latter would add its frequency count and pass it to other parties. Finally, a secure comparison would take place between the final and initiating parties to determine whether the final result was greater than the threshold plus the random value.

In addition, the authors in [19] dealt with the problem of association rule mining in vertically partitioned data. In other words, its aim was to determine the item frequency when transactions were split across different sites, without revealing the contents of individual transactions. The security of the protocol for computing the scalar product was analyzed.

Furthermore, the authors in [32] put forward an encryption scheme based on substitution cipher techniques in order to preserve the privacy of the transactional data used for outsourcing association rule mining. However, they considered that the association rules mining would be centralized on a single provider, which had to receive the different pattern frequency count and perform all the association rules mining tasks. In contrast, to avoid such overhead imposed on a single provider, the master miner in this scheme would mine the strong association rules on a global level by sending count queries to the data providers while avoiding to store any part of the data locally.

Moreover, the writers in [33] suggested a privacy-preserving model that merged the secure multiparty computation and differential privacy to preserve the privacy of the statistical operations (*i.e.*, count and aggregate count). However, it was not clear how this approach could be applied to handle association rules mining given that the division operations had to be performed between parties in a secure way in order to validate the minimum support and confidence.

Otherwise, the authors in [34] proposed to tackle the problem of outsourcing the association rule mining task within a corporate privacy-preserving framework by suggesting an encryption scheme based on substitution ciphers called, *RobFrugal*.

In addition, the authors in [20] focused on the use of encryption techniques to build a secure protocol (multi-party computation) to perform this task. The principle of this approach was to use a communication protocol between sites based on asymmetric cryptography arising protocols through solving the discrete logarithm problem. This protocol would ensure anonymity by commutative cryptography, and therefore would guarantee the preservation of the privacy of data owners. This method ensured secure communication while respecting the privacy of sites. However, it did not ensure the integrity of the exchanged data between the sites. In the case of a malicious site, false information may be generated and subsequently sent to the next site. The latter could not detect any modification, leaching the end, to a miscalculation.

Besides, the writers in [21] put forward an approach to transform original data using an encryption function associated with a signature. The key ensured and verified the authenticity of the message and its integrity. This approach was based on homomorphic encryption whose properties allowed performing various operations on encrypted data without knowing the plaintext data.

Added to that, the authors in [22] suggested a method based on Secure Multiparty Computation (SMC). The SMC was a set of cryptographic techniques that permitted the calculation of any function on a set of data distributed among multiple entities. Each entity had a portion of the data. Common calculation had to be done so that neither party could guess, in any manner, the data of other entities from the results and its own data. The limit of this method stood in the fact that communication complexity would exponentially increase as far as the number of distributed sites rose.

In [23], the authors proposed a method based on public key cryptosystems (asymmetric ciphers). A public-key (asymmetric key) based algorithm used two separate keys: a public key and a private one. The public key was utilized to encrypt the data, and only the private key could decrypt the data. A form of this type of encryption was called RSA [24]. It was widely used for secured websites that carry sensitive data such as username, passwords, and credit card numbers. A disadvantage of using public-key cryptography for encryption is speed. There have been other popular secret-key encryption methods, which are significantly faster than any currently available public-key encryption method.

In addition, the authors in [25] put forward an approach based on the Elliptic Curve Cryptography (ECC) [6] and the ElGamal cryptosystem [26]. These approaches would avoid multiple cipher operations on each site in order to ensure secure communication between different sites.

In this respect, there are various advantages and disadvantages of using cryptography techniques to ensure privacy preservation data mining [31]. These advantages and disadvantages are [31]:

– Advantages:
  • Robust
  • Sender and recipient authentication

- Anonymity
- Fairness
- Accountability
- Integrity in storage
- Disadvantages:
  - Taking a long time to figure out the code
  - Overall cryptography as a long process

Generally cryptographic techniques are ideally meant for such scenarios where multiple parties collaborate to compute results or share non sensitive mining results and thereby avoiding the disclosure of sensitive information. However, the major drawback for using cryptography techniques to ensure privacy during the mining task is the execution time. Owing to its usability and importance, preserving the privacy of data in a cloud computing environment still presents a thriving and compelling issue. In this respect, the main thrust of this paper is to propose a novel approach, called Cloud-PPDM, to ensure privacy preserving data mining. Our approach is based on cryptographic techniques in order to improve the performances in terms of execution time. Moreover, the Cloud-PPDM approach relies on mining closed itemsets within a cloud computing environment. The main idea behind our approach comes from the conclusion drawn from the data mining community that focuses on the lossless reduction of itemset mining over cloud computing data. In fact, the extraction of the latter requires less memory and running time. Table 2 summarizes the surveyed approaches dedicated to the cryptography based PPDM.

Table 2. Advantages and limitations of cryptography-based PPDM

Technique	Advantages	Limitations
Kantarcioglu and Clifton [18]	- Incorporating cryptographic techniques to minimize information shared, while adding little overhead to mining task	- Very successfull false information for malicious sites
Vaidya and Clifton [19]	- Efficient method for computing scalar product while preserving privacy of individual values	- Boolean association rule mining - Difficulty to compute scalar product while preserving privacy
Moez et al. [20]	- Anonymity by commutative cryptography - Increased security by asymmetric cryptography	- No integrity of exchanged data between sites - Easily transmitted false information in case of malicious site
Canard et al. [21]	- Anonymity approach for security of respondents identity and decreasing linking attack	- No sufficient protection against attribute disclosure by homogeneous attack and background knowledge attack
Chang et al. [22]	- Safety - Security - Trust-worthiness	- Exponential rising communication complexity with the number of sites

(continued)

Table 2. (*continued*)

Technique	Advantages	Limitations
Approaches proposed in [23, 24]	- For public-key cryptosystems, no need for exchanging keys, thus eliminating key distribution problem - No need for private keys to be transmitted or revealed to anyone - Ability to provide repudiated digital signatures	- High execution time public-key cryptosystems
Approaches proposed in [6, 25, 26]	- Preserving privacy, taking advantage of elliptic curve Cryptography and ElGamal cryptosystem	- Poor scalability in terms of dataset size and number of sites
Wong et al. [32]	- High security with low data transformation cost - Secure encryption scheme taking advantage of substitution cipher - Minimization of demands in resources	- One-to-n item mapping cannot be directly applied since it is effectively a one-to-one item mapping
Zhang et al. [33]	- Stronger privacy than current efficient secure multiparty computation approaches - Better accuracy than current differential privacy approaches while maintaining efficiency	- Weakness of direct use of differential privacy in privacy-preserving data mining against collision attack
Giannotti et al. [34]	- Adding weighted support in original item support transactions to reduce fake transaction table and storage overhead - Robustness against guessing attack and man-in-the-middle attack	- This approach is proposed only for information holders; however individual record owners should additionally have the rights and obligations to ensure their own particular private information

3   Cryptography Techniques

In this section, we provide the definition of some notations that rely on the cryptography and secure communication used in our work.

3.1   Homomorphic Encryption

A homomorphic encryption system provides the ability to perform various treatments on encrypted data without using the decryption operation [2]. Furthermore, homomorphic encryption schemes ensure secure aggregation. In fact, they allow data aggregation to be performed on encrypted data. In homomorphic encryption, certain aggregation functions such as the sum and the average can be applied on the encrypted data, reducing, significantly, the workload of the sites in the network. The data is encrypted and sent to the master site. The last site applies the aggregation function on the encrypted data. The master site receives the encrypted aggregated result and decrypts it. A homomorphic encryption scheme allows arithmetic operations on ciphertexts. These latter are the result of encryption performed on a plaintext using an algorithm, called a cipher. One example is a multiplicatively homomorphic scheme, where the decryption of the efficient manipulation of two ciphertexts yields the multiplication of the two corresponding plaintexts. Homomorphic encryption schemes are especially useful whenever some parties do not have the decryption key(s), while the other parties need to perform arithmetic operations on a set of ciphertexts. In the following, we present a description of the elliptical curve cryptography (ECC) and the signature scheme.

3.2 Elliptic Curve Cryptography

Elliptical Curve Cryptography (ECC) is a public key encryption technique based on elliptic curve theory that can be used to create faster, smaller and more efficient cryptographic keys [6]. The ECC generates keys through the properties of the elliptic curve equation instead of the traditional generation method as a product of very large prime numbers. This technology can be used in conjunction with public key encryption methods, such as the RSA [6] and the Diffie-Hellman [7]. According to some researchers, the ECC can yield a level of security with a 164-bit key, while other systems require a 1024-bit key to achieve the same security level [8]. Mainly, the ECC helps to establish equivalent security with lower computing power and battery resource usage. Consequently, it is becoming widely used for mobile applications.

3.3 Signature Scheme

A signature is a piece of information ensuring authenticity of messages between two parties without any shared secret information in advance [9]. The sender creates the signature by using their private key, while the receiver verifies a signature by using the sender's public key. An aggregate signature scheme is a method for combining n signatures from n different signers on n various messages into a single signature. Indeed, the latter will convince the verifier that the n signers have signed the n original messages. In the next section, we discuss our proposed approach that takes advantage of these cryptography techniques in order to preserve the privacy of data sources during the extraction of frequent closed patterns from a distributed environment such as cloud computing.

4 Cloud-PPDM Approach to Ensure Privacy Preserving Data Mining

In this section, we describe the problem statement. Then we present our Cloud-PPDM approach that is based on two components:

1. The first component uses our proposed Dist-CLOSE algorithm to extract frequent closed patterns with privacy preserving.

2. The second provides a security scheme associated with Dist-CLOSE, in order to ensure privacy concerns. In Algorithm 2, we show the details of this component.

4.1   Problem Statement

The need to ensure the confidentiality of data sources during the extraction of frequent closed patterns from a distributed environment such as the cloud is a hot research topic of data mining community. Each site in this environment has a private transaction database DB_i. The goal is to extract frequent closed itemsets in a distributed environment. In the meanwhile, no $site_i$ should be able to learn: contents of a transaction at any other $site_n$, what patterns are supported by any other site, or the specific value of support for any items at any other site, unless that information is revealed by the knowledge of one's own data and the final result. Furthermore, we are interested in using homomorphic encryption and aggregate signature scheme toolkits to construct a secure multi-party computation protocol to perform this task.

4.2   Background

Along this sub-section, we introduce basic definitions for closed pattern mining, on which our work relies.

Basic Definition 1. (Extraction context) An extraction context is a triplet $\mathcal{K} = (\mathcal{O}, \mathcal{I}, \mathcal{R})$, where $\mathcal{O}$ represents a finite set of objects, $\mathcal{I}$ is a finite set of items and, $\mathcal{R}$ is a binary (incidence) relation (i.e., $\mathcal{R} \subseteq \mathcal{O} \times \mathcal{I}$). Each couple $(o, i) \in \mathcal{R}$ expresses that the object $o \in \mathcal{O}$ contains the item $i \in \mathcal{I}$.

Definition 2. (Closure operator) Let $\mathcal{K} = (\mathcal{O}, \mathcal{I}, \mathcal{R})$ be a data mining context, $\mathcal{O}$ a set of transactions, $\mathcal{I}$ a set of items, and $\mathcal{R}$ a binary relation between transactions and items. For $O \subseteq \mathcal{O}$ and $I \subseteq \mathcal{I}$, we define:

$$f(O) = \{i \in I \,|\, \forall o \in O, \, (o, i) \in \mathcal{K}\}$$

$$g(I) = \{o \in O \,|\, \forall i \in I, \, (o, i) \in \mathcal{K}\}.$$

f(O) associates with *O* the items common to all transactions $o \in O$, and *g(I)* associates with *I* the transactions related to all items $i \in I$. The operators $\gamma = f \circ g$ and $\gamma' = g \circ f$ are the Galois closure operators.

The closure operator γ induces an equivalence relation on the power set of items portioning it into disjoint subsets called equivalence classes [3]. The largest element (*w. r.t.* the number of items) in each equivalence class is called a closed itemset and is defined as follows:

Definition 3. (Closed frequent itemset) An itemset $I \subseteq \mathcal{I}$ is said to be closed if and only if $\gamma(I) = I$ [4]. The support of I, denoted by *Supp*(I), is equal to the number of objects in $\mathcal{K}$ that contain I. I is said to be frequent if *Supp*(I) is greater than or equal to a user-specified minimum support threshold, denoted Minsup. The frequency of I in $\mathcal{K}$ is equal to $\frac{Supp(\mathcal{I})}{|\mathcal{O}|}$.

4.3   Global Architecture

The Cloud-PPDM allows extracting the frequent closed patterns in a cloud environment while preserving the constraints of privacy by using homomorphic encryption that we have designed. In this respect, the Cloud-PPDM follows the general principle presented in the algorithms that generate the frequent closed itemset such as the CLOSE algorithm [5]. Generally, the steps of the Cloud-PPDM are detailed as follows: Firstly, the initialization process of the communication protocol is invoked. Secondly, the master site, *i.e.* the site which launches the mining task, distributes the list of 1-itemset candidates. Therefore, the different sites run, concurrently, a local algorithm described in Fig. 1, which generates their *closure* and *support*. At this step, the communication protocol is lunched in order to communicate the results to the master site. Now, the master site has at a hand the set of local *closures* as well as local *supports* of the candidate items. The master site can now generate the global *support* by making the sum of local *supports*. The global *closure* is computed by making the intersection of local *closures*. In this way, the master site can generate the candidates of higher size. Then at this level, the master site repeats the above steps whenever it can generate candidates of higher size. Algorithm 1 shows the details of the our proposed approach. In Table 3, we present the definition of some notations used throughout Algorithm 1.

Algorithm 1: Dist-CLOSE: Distributed Extraction of Frequent Closed Itemsets with Privacy Preserving

Input: n: Number of sites; K: Extraction context;
Minsupp: Minimal threshold of support;
master: Boolean *flag* : Set to true if the current site is the master one, otherwise it is set to false;
Begin
    Initialize(n);
    If master **then**
        $FFC_1.generators \longleftarrow \{$ 1-*itemsets* $\}$;
    For $(k \longleftarrow 1; FFC_K.generators \neq \emptyset; k + +)$ **do**
        If master **then**
        Distribute(FFC_k, n);
        Receive(FFC_k);
        $FFC_k{}^L \longleftarrow$ *Gen-Local(FFC$_k$)*;
        Communication Protocol (FFC$_k{}^L$)
        If master **then**
        $FFC_k{}^G \longleftarrow$ *Collect(FFC$_k{}^L$)*
        $FF_k \longleftarrow$ *Gen-global(FFC$_k{}^G$)*
        $FFC_{k+1} \longleftarrow$ *Gen-Generator FF_k*;
    Result:$\cup_K FF_k$
End

The *Gen-Local* procedure receives a Frequent Closed Candidates (FFC_k) unit of candidate k-groups containing the k-*generator* candidates of the iteration k in argument. It computes the local *support* and *closure* of each *generator*. This procedure is run on all sites. The *Communication protocol* procedure receives the set of candidates with their *closures* and *supports*. Thus, the *communication protocol* is executed in order to transfer the results to the master site while ensuring privacy preserving. The *Gen-Global* procedure receives the set of FFC_k^L obtained by executing the protocol of communication, and generates the global *support* by making a sum of local *supports*

Table 3. Deftnition of some notations used throughout Algorithm 1

Notation	Deftnition
FF_K	Set of frequent closed itemset of k-size
FFC_K	Set of frequent closed itemset candidates of k-size
FFC_K^L	Set of local frequent closed itemset candidates k-size
FFC_K^G	Set of global frequent closed itemset candidates k-size

and the global *closure* by making the intersection between the local *closures* received previously. Then the master site can run the infrequent itemsets given *minsupp*. At this step, the master site executes the *Gen-Generator* procedure to generate the candidates of size $k + 1$ and it returns the set of this candidates. This process will be repeated until the *Gen-Generator* procedure generates an empty set. As a final step, the master site executes a procedure so as to generate a generic base of exact association rules.

4.4   Communication Protocol

The goal of our approach is to extract frequent closed itemsets while ensuring the maintenance of privacy between the various sites. The communication protocol consists of four procedures: (1) Setup, (2) Encrypt-Sign, (3) Aggregate, and (4) Verify. The **Setup** procedure is to prepare and install necessary secrets for the master and each site. When a site s_i decides to send sensed data to its site s_{i+1}, it performs the **Encrypt-Sign** and sends the result to the site. Once the site s_n receives all results from its sites, it activates the **Aggregate** to the received data, and then sends the final results (aggregated ciphertext and signature) to the master. The last procedure is **Verify**. First, the master site extracts the individual sensed data by decrypting the aggregated ciphertext. Afterwards, the master verifies the authenticity and integrity of the decrypted data based on the corresponding aggregated signature. The details of our approach are detailed as follows:

1. **Setup phase:** For each site s_i, the master generates (Sv_i, Sx_i) by KeyGen procedure 1 based on the approach proposed in [9], where $(Sv_i = v_i)$ and $Sx_i = x_i$ (MSpk, MSsk). These keys are generated by KeyGen procedure 2. The latter is based on the approach proposed in [27], where the Master Site public key (MSpk) = (n, g, k) and the Master Site secret key (MSsk) = (p, p_g). After that, the (MSpk) is loaded to s_i for all sites i.
2. **Encrypt-sign phase:** This procedure is triggered when a site s_i decides to send its sensed data to the site s_{i+1}. At the end, the site s_i sends the pair ciphertext and the signature (c_i, ∂_i) to site s_{i+1}.
3. **Aggregate phase:** The Aggregate procedure is launched after the site Aggregator s_n has gathered all ciphertext signature pairs.
4. **Verify phase:** When receiving all the ciphertexts and signatures (C', ∂') from the aggregator site s_n, the master can recover and verify each sensing data via the following steps: First, the master decrypts the aggregate result using its private key. Additionally, the master needs to reverse the mapping from the point on the elliptic curve to the aggregate result. To verify the signature, the master computes a point

on the curve using the received signature, the decrypted aggregate result, and the integer k. If the calculated x-coordinate of the point is the same as $r(x)$, then the signature is verified. The master makes sure that all data are generated by legitimate sites and included in the aggregate. In Algorithm 2, we show the details of the communication protocol.

Algorithm 2: Communication Protocol: privately collect messages from parties

---

1. **Setup Phase**
   KeyGen procedure 1 :
   For a user, pick random $x \leftarrow Z_p$, and compute $v = g^x$. The user's public key is $v \in \mathbf{G1}$, and secret key is
   $x \in Z_p$
   KeyGen procedure 2 :
   p and q are a large primes
   K, the bit length of prime p
   $n = p^2 q$, the modulus $g \in Z/nZ$ s.t. $p | \mathrm{ord}p_2(g)$
   $g_p = g \bmod p^2$
   Public-Key:(n,g,k), Secret Key: p, g_p
2. **Encrypt-Sign**
   Encoding : $m \in \{0, 1, ..., 2^{k-2}\}$, a message $r \in Z/nZ$, a random integer $c_i = g^{m+rn} \bmod n$, a ciphertext
   Signature : $\partial_i = x_i \times h_i$ where $h_i = x_i = H(\partial_i)$
3. **Aggregate Phase**
   Aggregated Ciphertext:
   $C' = \sum_{i=1}^{n} c_i$
   Aggregated Signature:
   $\partial' = \sum_{i=1}^{n} \partial_i$.
   Send the aggregated result (C', ∂') to the master
4. **Verify Phase**
   When receiving (C', ∂') from the aggregator site s_n, master can recover and verify each sensing data via the following steps:
   Decryption of C' :
   $M' = L(c^{p-1} \bmod p^2) L(g_p^{p-1} \bmod p^2)^{-1} \bmod p$
   Master obtains M' by decrypting C'.
   Master obtains m' from M' through the reverse function rmap():
   $m' = \mathrm{rmap}(M') = m_1 + m_2 + ... + m_n$.
   Master obtains each sensing data from m'.
   Master site verifies each ∂^i via checking whether the equation
   $e(g, \partial) = \prod_{i=1}^{k} e(v_i, h(m_i))$ holds or not.

---

5 Evaluation

In this section, we experiment the effectiveness and scalability of our proposed approach. In Subsect. 5.1, we present the security analysis and performance evaluation of our communication protocol. Subsection 5.2 describes the experimental environment and the characteristics of the datasets used to evaluate the performance in this work. Finally, in Sect. 5.3, we describe the experimental results and give analysis.

5.1   Security Analysis

In this section, we illustrate the performances of our approach in terms of integrity, freshness and confidentiality of the exchanged data between all sites. The exchanged data can be exploited by a malicious adversary to violate the confidentiality of the sensitive data. In our approach, we palliate these threats via the encryption phase. Also, to ensure the integrity of all exchanged data, each data message is sent only once from the original source. A signature is attached to each message. The signature is computed using the private key that is only known to the source such that the report cannot be forged when it is kept at other sites.

We use the Elliptic Curve to provide message and aggregate integrity in addition to data confidentiality. Each site is pre-loaded with the appropriate elliptic curve parameters, the master public key and a network wide random integer. The integer is used to generate a new key (k) at set intervals. This ensures that the signatures are additive and secure against attacks. At the start of each round, each site chooses a private key and computes the appropriate public key. Choosing a private key is straightforward and requires the site to pick an integer in the field of the elliptic curve. A new public and private key pair is necessary during each round of processing, because it will only take two signatures for a malicious site to determine another site private key. Clearly, if another message is signed with the same private key, then that signature will not be secure. We add another level of security by signing the message and then encrypting it before sending it to the next level. If a site signs the same message with the same key, then another site can determine the private key. The signature scheme is designed such that all signatures can be combined via a simple arithmetic operation. This makes the amount of work required from a master site very small and thus well suited for Privacy Preserving Data Mining (PPDM). The exchanged data are optimized to work with homomorphic encryption and aggregated signatures. The aggregator site waits for a certain amount of time, and when the aggregator has received data, they will add the ciphertexts, which are the digital signature and the public keys. At the end, the master receives only one exchanged data, which consists of one ciphertext corresponding to the sum of the readings of all sites. Besides, it receives one signature corresponding to the sum of data and the sum of the public keys of all sites. Then the master can decrypt the message and verify its integrity using the sum of signatures and the sum of public keys.

5.2   Test Environment and Datasets

In this paper, all simulation work is done in Java. Our simulation is run on the Amazon EC2 cloud computing platform. To show the performance of our proposed approach, we use High-CPU Medium Instances which have 1.7 GB of memory, 5 EC2 compute units (2 virtual cores with 2.5 EC2 compute units each), 320 GB of local instance storage, and 64-bit platforms. In addition, we select various types of datasets, dense and sparse, from the UCI KDD machine learning repository such as: Mushroom [28], Connect [29], C73D10K [30], and T40I10D100K [35] in our experimentation. Table 4 describes the dataset characteristics.

Table 4. UCI dataset characteristics: nature, number of objects, average size of objects, and number of items

Dataset	Mushroom	Connect	C73D10 K	T40I10D100K
Nature	Dense	Dense	Dense	Sparse
Number of objects	8124	67 557	10 000	100 000
Average size of objects	23	43	73	40
Number of items	127	129	2178	1000

5.3 Results and Analysis

To determine the efficiency of our approach, we measure the processing time consumption of the Cloud-PPDM with regard to the approach proposed in [20] fitting in the same trend and in the same data characteristics. We start with the approach proposed in [20] in order to determine the consumed time, for dense and sparse datasets with a various number of sites equal to three, four, and five.

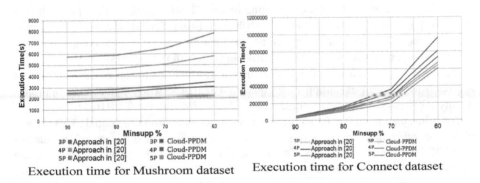

Execution time for Mushroom dataset   Execution time for Connect dataset

Fig. 1. Execution time of Cloud-PPDM vs approach proposed in [20]

In Figs. 1 and 2, the vertical axis represents the execution time of our Cloud-PPDM approach vs the approach proposed in [20], respectively on the Mushroom and Connect datasets, and the horizontal axis is exploited to present the variations in the execution time according to the number of sites P for various minsup. We note that P represents the number of sites. According to Fig. 1, we can analyze these results as follows: for example, for the Mushroom dataset with a minsup equal to 60% and with a number of sites equal to three sites, the Cloud-PPDM approach requires an execution time equal to 2,218 s to generate the result, while the other approach requires 3,494 s. Furthermore, for the Connect dataset with a minsup equal to 90% and with a number of sites equal to four sites, the execution time passed by the Cloud-PPDM approach to arrive at the result is 324,216 s, whereas the vs. approach passes 453,415 s. We can interpret also through Fig. 1, that the total processing runtime keeps increasing linearly as the number of minsup decreases. This is mainly due to the fact that the calculation time to generate the frequent closed itemsets will increase, when the value of the minsup decreases. In this case, the communication and distribution management time becomes

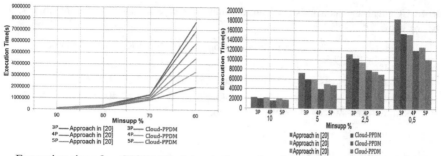

Execution time for C73D10K dataset   Execution time for T40I10D100K dataset

Fig. 2. Execution time of Cloud-PPDM vs approach proposed in [20] respectively on C73D10K and T40I10D100K datasets

negligible with respect to this calculation time. Subsequently, this will remarkably increase the execution time of the algorithm.

In Fig. 2, we show the execution time of the Cloud-PPDM vs the approach proposed in [20], respectively on the C73D10K and T40I10D100K datasets. This figure clearly demonstrates that our approach has the shortest execution time compared to the adversarial approach for each of C73D10K and T40I10D100K datasets. For example, in the case of three sites for the C73D10K dataset with a minsup equal to 80%, our approach requires an execution time equal to 197.614 s, whereas the other approach requires 327, 143 s. Moreover, in the case of five sites for the T40I10D100K dataset with a minsup equal to 0.5%, our approach needs an execution time equal to 100.068 s, while the other approach requires 127.583 s. Otherwise, in the case of the dataset T40I10D100K we can observe a reconciliation between the curves. In this case the execution time varies according to the number of sites. We can notice that if we increase the number of sites for the same threshold, the execution time will decrease.

The total communication cost of the Cloud-PPDM depends on the number of sites. The cost of each run based on the number of items n is as follows: s_1 sends the sensed data to its site s_{i+1}. This latter sends also the sensed data to the next site s_{i+2}. Once the site s_n receives all results from its sites, it will send the final results to the master. Then the communication cost of the Cloud-PPDM is O(n), where n represents the number of sites. Generally, the cost of maintaining privacy depends primarily on the number of sites, the size of the exchanged messages, the number of calls to the communication protocol, and the number of candidates in each iteration.

In this sub-section, we have presented an experimental study (Figs. 1 and 2) on the Cloud-PPDM approache and the one proposed in [20] for the extraction of frequent closed itemsets in a distributed environment while preserving the privacy of data owners. We have performed different tests on the datasets of different types and sizes, to evaluate the performance of our approach with respect to the approach proposed in [20]. According to these experiments, we conclude that the Cloud-PPDM approach mining has the shortest time to ensure privacy mining, compared to the approach proposed in [20].

6 Conclusion

Through this paper, we have introduced a new secure scheme associated with the Dist-CLOSE algorithm which takes advantage of homomorphic encryption. This scheme offers the advantage of carrying out the mining task while guaranteeing security and anonymity. In addition, this scheme protects the confidentiality of data sources during the extraction of frequent closed patterns from a distributed environment such as cloud computing, without revealing information that compromises the privacy of individual sources. In summary, we show that it is possible to achieve good individual security with a communication scheme.

Through extensive experiments carried out on benchmark datasets, we show the effectiveness of our proposed scheme on both runtime performances and security analysis.

Future work will include improving prospects of this approach by strengthening the autonomy of the exchanged data between sites. We plan to give the data the ability to protect itself during the exchange. Hence, the verification calculation by the master site is no longer required in order to ensure safety.

References

1. Muller, S.D., Holm, S.R., Sondergaard, J.: Benefits of cloud computing: literature review in a maturity model perspective. Commun. Assoc. Inf. Syst. **37** (2015). Article no. 42
2. Hayward, R., Chiang, C.C.: Parallelizing fully homomorphic encryption for a cloud environment. J. Appl. Res. Technol. **13**(2), 245–252 (2015). ISSN 1665-6423
3. Bastide, Y., Taouil, R., Pasquier, N., Stumme, G., Lakhal, L.: Mining frequent patterns with counting inference In: KDD Conference, pp. 66–75 (2000)
4. Zitouni, M., Akbarinia, R., Ben Yahia, S., Masseglia, F.: A prime number based approach for closed frequent itemset mining in big data. In: 26th International Conference on Database and Expert Systems Applications, DEXA 2015 Valencia, Spain (2015)
5. Ben Yahia, S., Mephu Nguifo, E.: Approches d'extraction de règles d'association basées sur la correspondance de Galois. Ingénierie des systèmes d'information **9**(3–4), 23–55 (2004)
6. Kumarn, D.S, Suneetha, C.H., Chandrasekhar, A.: Encryption of data using elliptic curve. Int. J. Distrib. Parallel Syst. (IJDPS) **3**(1) (2012)
7. Gajbhiye, S., Karmakar, S., Sharma, M.: Diffie Hellman key agreement with elliptic curve discrete logarithm problem. Int. J. Comput. Appl. **129**(12) (2015). (0975 8887)
8. Moumita, R., Nabamita, D., Jyoti, K.A.: Point generation and base point selection in ECC: an overview. Int. J. Adv. Res. Comput. Commun. Eng. **3**(5) (2014)
9. Boneh, D., Gentry, C., Lynn, B., Shacham, H.: Aggregate and verifiably encrypted signatures from bilinear maps. In: Biham, E. (ed.) EUROCRYPT 2003. LNCS, vol. 2656, pp. 416–432. Springer, Heidelberg (2003). doi:10.1007/3-540-39200-9_26
10. Vassilios, S.V., Elisa, B., Igor, N.F., Loredana, P.P., Yucel, S., Yannis, T.: State of the art in privacy preserving data mining. SIGMOD Rec. **33**, 50–57 (2004)
11. Wang, P.: Survey on privacy preserving data mining. Int. J. Digit. Content Technol. Appl. **4** (9) (2010)

12. Thakur, D., Gupta, H.: An exemplary study of privacy preserving association rule mining techniques. Int. J. Adv. Res. Comput. Sci. Softw. Eng. **3**(11) (2013). P.C.S.T., BHOPAL C.S Dept., India

13. Nithya, C.V., Jeyasree, A.: Privacy preserving using direct and indirect discrimination rule method. Int. J. Adv. Res. Comput. Sci. Softw. Eng. **3**(12) (2013). Vivekanandha College of Technology for Women Namakkal India

14. Lipmaa, H.: Cryptographic techniques in privacy preserving data mining, University College London, Estonian Tutorial (2007)

15. Hussien, A., Hamza, N., Hefny, H.: Attacks on anonymization-based privacy-preserving: a survey for data mining and data publishing. J. Inf. Secur. **4**(2), 101–112 (2013)

16. Li, Y., Chen, M., Li, Q., Zhang, W.: Enabling multilevel trust in privacy preserving data mining. IEEE Trans. Knowl. Data Eng. **24**(9), 1598–1612 (2012)

17. Li, X., Yan, Z., Zhang, P.: A review on privacy-preserving data mining. In: IEEE International Conference on Computer and Information Technology (CIT), pp. 769–774 (2014)

18. Kantarcioglu, M., Clifton, C.: Privacy preserving distributed mining of association rules on horizontally partitioned data. In: ACM SIGMOD Workshop on Research Issues in Data Mining and Knowledge Discovery, pp. 24–31 (2002)

19. Vaidya, J., Clifton, C.: Privacy preserving association rule mining in vertically partitioned data. In: 8th ACM SIGKDD International Conference on Knowledge Discovery and Data Mining, pp. 639–644. ACM Press (2002)

20. Moez, W., Poncelet, P., Ben Yahia, S.: A novel approach for privacy mining of generic basic association rules. In: ACM First International Workshop on Privacy and Anonymity for Very Large Datasets, Join with CIKM 2009, France, pp. 45–52 (2009)

21. Canard, S., Desmoulins, N., Devigne, J., Le Hello, D.: Anonymisation des données. Document de travail de l'objet de recherche: trust identity and privacy (2012)

22. Chang, X.-Y., Deng, D.-L., Yuan, X.-X., Hou, P.-Y., Huang, Y.-Y., Duan, L.-M.: Experimental realization of secure multi-party computation in an entanglement access network (2015)

23. Natarajan, R., Sugumar, R., Mahendran, M., Anbazhagan, K.: Design a cryptographic approach for privacy preserving data mining. Int. J. Innov. Res. Sci. Eng. Technol. **1**(1) (2012)

24. Saxena, S., Kapoor, B.: State of the art parallel approaches for RSA public key based cryptosystem. Int. J. Comput. Sci. Appl. (IJCSA) **5**(1) (2015)

25. Patel, S.J., Punjani, D., Jinwala, D.C.: An efficient approach for privacy preserving distributed clustering in semi-honest model using elliptic curve cryptography. Int. J. Netw. Secur. **17**(3), 328–339 (2015)

26. Jitarwal, Y., Mangal, P.K., Suman, S.K.: Enhancement of elgamal digital signature based on RSA & symmetric key. Int. J. Adv. Res. Comput. Sci. Softw. Eng. **5**(5) (2015)

27. Okamoto, T., Uchiyama, S.: A new public key cryptosystem as secure as factoring. In: Proceedings of the Annals International Conference on the Theory and Applications of Cryptographic Techniques (EUROCRYPT 1998), pp. 308–318 (1998)

28. ftp://ics.uci.edu/emerz/mldb.tar.Z

29. http://archive.ics.uci.edu/ml

30. ftp://fpt2.cc.ukans.edu/pub/ippbr/census/pumps

31. Rathore, B.S., Singh, A., Singh, D.: A survey of cryptographic and non-cryptographic techniques for privacy preservation. Int. J. Comput. Appl. **130**(13) (2015). (09758887)

32. Wong, W.K., Cheung, D.W., Hung, E., Kao, B., Mamoulis, N.: Security in outsourcing of association rule mining. In: Proceedings of the 33rd International Conference on Very Large Data Bases (VLDB), pp. 111–122 (2007)

33. Zhang, N., Li, M., Lou, W.: Distributed data mining with differential privacy. In: Proceedings of the IEEE International Conference on Communications (ICC), pp. 1–5 (2011)
34. Giannotti, F., Lakshmanan, L., Monreale, A., Pedreschi, D., Wang, H.: Privacy-preserving mining of association rules from outsourced transaction databases. IEEE Syst. J. 7(3), 385–395 (2013)
35. ftp://fpt2.cc.ukans.edu/pub/ippbr/census/pumps/pumbs90ks.zip

The Security of Internet of Things: Current State and Future Directions

Laila Dahabiyeh$^{(\boxtimes)}$

Department of Management Information Systems, School of Business,
University of Jordan, Amman, Jordan
Laila.dahabiyeh@ju.edu.jo

Abstract. Although the technical approach to securing the Internet of Things (IoT) witnesses considerable scholarly efforts, approaching the phenomenon from a social lens received scant attention. In this paper, I address this gap by perceiving the security of IoT as a social process. I view IoT security resulting from collective efforts among different actors. By examining such efforts in three cases; connected cars, wearable technologies, and smart home I shed light on the social process of IoT security and derive valuable insights that can enlighten future directions.

Keywords: Internet-of-Things · Connected cars · Smart home · Wearables · IS security

1 Introduction

The emergent phenomenon of the Internet of Things (IoT) promises huge transformations in business and society. 'Things' that have been viewed as standalone objects are now powered with computing capabilities that make them more powerful, dynamic and interactive. For example, ovens can now be controlled remotely using a smart phone to adjust temperature and change the cycle. Smart refrigerators keep track of their contents and automatically create a shopping list that can be transmitted to mobile phones. Wearable technologies such as Fitbits enable users to have more control over their health.

The IoT continues to grow. The number of connected devices is estimated to reach more than 50 billion in 2020 [1]. This growth is revolutionizing businesses with a wide belief that IoT will unlock new revenue opportunities [2]. Nonetheless, in the middle of these high promises, it is crucial to remember that the IoT is also accompanied with a new wave of security threats which without addressing, the value of this innovation can be seriously diminished [3, 4].

The security aspect of IoT cannot thus be ignored. While significant work has been done, and continues, on the technical aspect of IoT security [5] such as authentication schemes, network protocols, cryptographic mechanisms, it is noticed that its socio-organizational feature received scant attention. It is well-recognized that IS security entails both technical and socio-organizational solutions [6]. In this paper, I seek to take a first step towards addressing this gap. I view IoT security resulting from collective efforts among different actors. By examining such efforts (in the US) in three

© Springer International Publishing AG 2017
M. Themistocleous and V. Morabito (Eds.): EMCIS 2017, LNBIP 299, pp. 414–420, 2017.
DOI: 10.1007/978-3-319-65930-5_33

cases; connected cars, wearable technologies, and smart home I aim to shed light on the social process of IoT security and derive some insights that can enlighten future directions. Those three cases were selected because of the richness of the data available, and because they are seen among the top important IoT applications where they span across various industries.

2 Security in an Interconnected World

Technological innovations are pushing for more connectivity, and this has significant implications on how information systems can be secured against emergent security threats. Acknowledging that we live in an interconnected world propel us to shift our perception of security from a mere technological aspect to consider the network of actors (e.g. individuals, organizations and technologies) that work together to prevent security threats [7].

Higher connectivity is associated with a higher vulnerability to security threats because threats can originate from multiple sources [3, 8, 9] making security a challenging task. The security of IoT is even more complex because it does not only inherent security problems associated with the Internet but also because it brings its own security challenges [10]. For instance, flexibility and usability requirements of IoT necessitate light-weight and decentralized means of authentication that deviate from traditional security means [11, 12].

Facing this complexity, there is no doubt that individual efforts are not sufficient to achieve best security results and that collective effort is necessary [7]. Although IoT is often described as an ecosystem [3, 11, 12] little is known about the interactions between the ecosystem's different actors to achieve security. In what follows, I offer three cases on IoT to gain better understanding on *how* securing the IoT actually happen. This knowledge is valuable because it will enable the identification of potential bottlenecks in the process, and allow offering insights and recommendations for future directions.

2.1 The Case of Connected Cars

Technological innovations are sweeping the automotive industry in efforts to improve the driving experience, reduce fuel consumption and enhance safety. An emergent mode of transport is connected cars. Cars have become connected through various electronic systems such as infotainment and safety monitory tools. Connected cars promise a broad range of benefits from providing information about traffic jams and alternative routes to automatic emergency call upon accidents. However, with opportunities come challenges and connected cars have become the next target for security attacks.

Interests in automotive *cyber* security arose when in 2013 two security researchers, Charlie Miller and Christopher Valasek, demonstrated how they were able to hack a car, disable its braking system and take control over the steering system along with other things (e.g. turn the engine off, honk the horn) [13]. More recently, the same researchers wirelessly hacked a Jeep Cherokee through its Internet-connected entertainment system causing Fiat Chrysler to recall 1.4 million vehicles in July 2015 [14].

These demonstrations exposed the inapplicability of the current prevention measures (e.g. locks, alarm systems) in ensuring connected cars' security and triggered a need for developing new prevention measures that match the revolution in the transport industry. Security efforts went on different directions. The automobile industry formed an Information Sharing and Analysis Center to exchange information and effectively counter threats on a timely basis. The National Highway Traffic Safety Administration (NHTSA) solutions were more technical in nature and were focused on securing vehicle-to vehicle (V2V) communications. Their proposed solutions involved three technologies: symmetric encryption systems, group signature systems, and asymmetric public key infrastructure systems [15]. In their turn, legislators sent letters to 17 major automakers (e.g. General Motors, Ford, Toyota, Honda, Nissan, Volvo, Mercedes-Benz) and NHTSA asking for clarifications on the industry's security efforts. The responses to these letters revealed the different security directions being taken to secure the novel technology and the need to consolidate these efforts, clarify roles and responsibility, and build a national strategy [16]. Accordingly, several legislative hearings were conducted [17, 18, 19, 20] to examine the current state and propose solutions.

These hearings revealed the challenging task of securing connected cars (and the internet of things in general). Some legislators refused to support a bill because it had several weaknesses such as giving more responsibilities to NHTSA without allocating additional necessary funds for the agency to take on the extra work, failing to name an enforcing agency that would ensure car manufacturers compliance with security standards, and setting no minimum requirements for best practices or acknowledging the need for updating them in accordance with emerging threats and technologies. Such weaknesses were addressed by other bills. Moreover, the same bill improved vehicle safety at the expense of environmental safety. It granted automakers pollution credits in exchange of adding safety features. In addition, legislators were against the unilateral approach in writing the bill and argued had bipartisan approach been followed, many of the weaknesses would have been addressed leading to a stronger bill and faster process to achieve security [19].

Opponents further emphasized the bill's current security approach does very little in protecting the car and can, in fact, makes it more vulnerable. The bill prohibited all unauthorized access to vehicle data ignoring the fact that security researchers can hack the car for research purposes which contributes significantly in making it more secure. Negotiating these bills is still underway.

Besides focusing on a legal policy, other actors perceived the security of connected cars as a human resource problem and suggested developing automotive cybersecurity programs and degrees to develop the skills and talents needed in facing this new phenomenon. First attempts in achieving that can be seen in auto-specific hackathon such as CyberAuto Challenge. The event which is held annually brings together automotive engineers, government engineers, students and white hat hackers, and constitutes a learning environment where actors can gain and apply their knowledge and experiment on real cars to identify possible security threats and propose solutions that help in designing more secure cars. The event further provides an opportunity to expose current engineers to the cyber community and develop interest and awareness around auto-cyber security issues [21].

2.2   The Case of Wearable Technologies

As with the case of connected cars, the research community played a key role in raising attention to the lack of effective prevention measures for securing wearable technologies. In 2008, academic researchers demonstrated how they were able to attack a defibrillator and change its operations [22]. In 2011 researchers illustrated how attackers can intercept insulin pump signal and change the blood-sugar level read on the device alarming the person to adjust their insulin dosage which can be fatal over time [23].

These strong pieces of evidence of the possibility of penetrating wearable health devices and posing threats to human life triggered actions to develop better prevention measures for wearable devices. In a workshop organized by the Federal Trade Commission (FTC), various participants (academic, security analysts, technology experts, legal experts, service providers) aired their concerns about the security of wearable health devices and started negotiating how best to achieve security [24]. Technical solutions such as encrypting the data stored in the devices and while in transit, use of passwords, biometric and smartcard to limit unauthorized access, were proposed. Other actors focused on legislative solutions emphasizing that mobile health applications are not governed by Health Insurance Portability and Accountability Act (HIPAA) and wearables are not subject to security breach notification laws. Companies therefore have no legal obligation to make public disclosure of hacking incidents. Because actors vary in their interests, there was a clear tension between the need to ensure the usability of these devices and the need to secure them. Some actors did not favour technical measures that obliged the use of passwords to protect these devices and the data they contain. They argued that physicians did not favour them either because they saw them as an obstacle towards using the devices and therefore improving patients' health. Furthermore, others challenged the belief that publicly announcing breaches on wearable devices would make consumers more aware of the risks involved in this technology as they believed that consumers have become "alert fatigue" and accustomed to continuous hacking incidents. There was consensus however on that regulatory barriers and outdated laws were impeding not supporting the advancement of healthcare innovations.

Securing wearable technologies is in its early stages, actors are still finding their way on the best means to ensure security. Nonetheless, the FTC's release of security practices and recommendations to be taken by manufacturers [25] can be seen as an attempt to introduce some legal certainty that can protect the new innovation from legal liability because it ensures companies that they are following reliable and trusted guidelines suggested by a regulatory agency.

2.3   The Case of Smart Home

Attention to the security risks associated with smart home technology was drawn when in January 2012 hackers exploited a vulnerability in TRENDnet IP camera and spied into users' homes exposing the private lives of hundreds on the internet [26]. Following this practical evidence, efforts to secure the technology took place.

The FTC workshop on security and privacy in a connected world involved a panel dedicated to discussing home automation systems [24]. Critical issues were raised by participants about smart home security, and different solutions were offered. Some opted for technical solutions such as applying better security standards (such as ZigBee and Z-wave) in wireless home network. While others focused on the importance of organizational approaches as well technical ones. They saw the problem rising from the fact that most of the companies that offer smart home products were not expert in security. Accordingly, their suggestion was a change in organizational structure and hiring policies to recruit security experts in order to build more secure products. Another proposition was consumer-focused and suggested that educating consumers and creating awareness of the security risks associated with the technology can help in preventing attacks.

Reaching an agreement on how to approach the security of smart home products was difficult to achieve. Security experts believed that the proposition to focus on consumers to attain security through education and awareness programs would achieve very little. They supported their argument by drawing on computer security field where immense efforts had been undertaken to educate users, and still security problems remain. For those actors, the real problem lied in the products and the fact that manufacturers themselves do not understand the technology and its security implications. They argued that building strong security requires considerable computing power and storage capacity which significantly consume energy lowering the product's battery life. Vendors, who aim for value and convenience, therefore do not take security seriously and tend to rely on the security of the home network to prevent threats. In their view, securing smart home technology should start from the vendors who ought to be more responsible and develop expertise in security. The marketing manager at GE Appliances agreed that there is a trade-off between convenience and security. Nonetheless, he refused the perception of vendors as passive and considering security as an afterthought. He contended security by design approach is followed and many devices work only within acceptable parameter ranges making the products more secure [24].

3   Discussion and Conclusion

Different insights can be inferred about the security of IoT from the above three cases. First, the security of IoT is still in its infancy. Actors are still exploring the phenomenon, its potentials and consequences after which they will be able to identify the best security approach to follow.

Second, there is no one best solution to securing the IoT. Rather it is better to approach the phenomenon from a *systems of solutions* mentality where a collection of solutions, such as technical, organizational, legislative, are needed to face security threats.

Third, the cases demonstrated the important role of evidence in gaining attention to security. This is of significance since security was often stepped-aside because these technologies have not witnessed *real* hacking incidents yet. It was only when researchers practically demonstrated and hacked these technologies that actors started to take security seriously. Unfortunately, this confirms that IoT security (as with IS security in general) remains mainly reactive rather than proactive.

Fourth, the IoT is a multifaceted phenomenon and this led the security aspect to be *crowded out* by many other issues such as environmental safety, the privacy of collected data, and recall notices which all incited conflicts between actors and prolonged the security process. Consequently, for collective security efforts to be more efficient it has to be *focused*. Taking dedicated steps towards achieving security yield faster results than trying to incorporate security along with other matters.

Fifth, finger-pointing to throw the blame for insufficient security will only impede the security process and make the IoT more vulnerable. Securing the IoT mandates the involvement of different actors. Collective effort is necessary and without cooperation and coordination, security will only come to a halt. Actors therefore are expected to participate in different security initiatives. Organizations, for example, besides working on improving their internal security practices to develop securer products, need to participate in legislative hearings to provide their views on any proposed law. Accordingly, securing the IoT should not be confined to certain boundaries. Rather it should be approached in a more flexible way to allow the involvement of various stakeholders. This will increase the knowledge base and allow a better understanding of how the interconnectivity of these devices is influencing security which should result in better decisions.

Finally, security should not inhibit innovations in IoT, and actors should aim for a balance between security and convenience. As the phenomenon is new and still emerging, premature binding legislation can be seen constraining innovative work. A wait-and-see approach is thus preferable.

The security challenges for IoT remain. As the technology evolves new challenges are expected to emerge. While this research shed some light on the contested security process in IoT and offered valuable insights for future directions, many questions arise that constitute avenues for future research. For instance, how can we ensure the security of IoT products when many of them are developed offshore? How can we approach security without damaging innovation? And how incentive systems can be modified to incite manufacturers, who are currently driven by convenience and time to market, to follow security by design approach? In addition, this research looked at security efforts in only three applications of IoT. Examining security efforts in other important applications such as smart city and smart grid can offer new insights to the phenomenon.

References

1. Statista: Internet of things (IoT): number of connected devices worldwide from 2012 to 2020 (in billions) (2017). https://www.statista.com/statistics/471264/iot-number-of-connected-devices-worldwide/. Accessed 17 Jan 2017
2. Jacobsson, A., Boldt, M., Carlsson, B.: On the risk exposure of smart home automation systems. In: 2014 International Conference on Future Internet of Things and Cloud (FiCloud), pp. 183–190. IEEE (2014)

3. Mukhopadhyay, S.C., Suryadevara, N.K.: Internet of things: challenges and opportunities. In: Mukhopadhyay, S.C. (ed.) Internet of Things. SSMI, vol. 9, pp. 1–17. Springer, Cham (2014). doi:10.1007/978-3-319-04223-7_1
4. Dhillon, G., Backhouse, J.: Current directions in IS security research: towards socio-organizational perspectives. Inf. Syst. J. **11**(2), 127–153 (2001)
5. Dahabiyeh, L.: Networks of cybercrime prevention: a process study of the credit card. In: ECIS 2015 Research-in-Progress Papers, Paper 6 (2015)
6. Mookerjee, V., Mookerjee, R., Bensoussan, A., Yue, W.T.: When hackers talk: managing information security under variable attack rates and knowledge dissemination. Inf. Syst. Res. 22(3), pp. 606–623, Sep 2011
7. Jing, Q., Vasilakos, A.V., Wan, J., Lu, J., Qiu, D.: Security of the internet of things: perspectives and challenges. Wirel. Netw. **20**(8), 2481–2501 (2014)
8. Skarmeta, A., Hernández-Ramos, J.L., Bernabe, J.B.: A required security and privacy framework for smart objects. In: ITU Kaleidoscope: Trust in the Information Society (K-2015), pp. 1–7. IEEE (2015)
9. Zhang, Z.-K., Cho, M.C.Y., Shieh, S.: Emerging security threats and countermeasures in IoT. In: Proceedings of 10th ACM Symposium on Information, Computer and Communications Security, pp. 1–6. ACM (2015)
10. Tracking & Hacking: Security & Privacy Gaps Put American Drivers at Risk. A report written by the staff of Senator E.J. Markey (D-Massachussetts), February 2015
11. Internet of Things. Hearing before the Subcommittee on Courts, Intellectual Property, and the Internet of the Committee on the Judiciary. House of Representatives. 29 July 2015. (ed.) (2015)
12. The Internet of Cars. Joint Hearing before the U.S. House of Representatives, Committee on Oversight and Government Reform. Subcommittee on Information Technology and Subcommittee on Transportation and Public Assets. 18 November 2015 (ed.) (2015)
13. Examining Ways to Improve Vehicle and Roadway Safety. Hearing Before the Subcommittee on Commerce, Manufacturing, and Trade of the Committee on Energy and Commerce. 21 October 2015 (ed.) (2015)
14. Understanding the Role of Connected Devices in Recent Cyber Attacks. Joint Hearing before the U.S. House of Representatives, Committee on Energy and Commerce. Subcommittee on Communications and Technology, and the Subcommittee on Commerce, Manufacturing, and Trade. 16 November 2016 (ed.) (2016)
15. FTC: Internet of Things - Privacy and Security in a Connected World Workshop (2013). https://www.ftc.gov/news-events/events-calendar/2013/11/internet-things-privacy-security-connected-world

The Privacy Paradox in the Context of Online Health Data Disclosure by Users

Chrysanthi Kosyfaki[(⊠)], Nelina P. Angelova, Aggeliki Tsohou, and Emmanouil Magkos

Department of Informatics, Ionian University,
Tsirigoti Square 7, 49100 Corfu, Greece
{pl3kosy, pl3angk, atsohou, emagos}@ionio.gr

Abstract. The privacy paradox phenomenon corresponds to the inconsistency between the privacy concerns and the actual behavior of online users. The existence of the phenomenon has been studied in many fields, such as for social networks and especially for a variety of forums and online communities. This paper questions its existence in the context of sensitive data, and more specific in the health data area via a survey which took place in Greece. Given that health-related information is perceived as sensitive personal data it is often excluded from discussions and sharing. This research aims to unravel the paradox's existence regarding the disclosure of health data information when individuals visit related online forums and communities.

Keywords: User privacy · Privacy paradox · Health data · Data disclosure

1 Introduction

The past decade has been characterized by the explosion of the web and online social networks. The users' tendency to disclose personal information, together with their expressed concerns about their privacy's protection, raised the phenomenon of the privacy paradox. People commonly give away their belongings only when an expected gain comes from this act, and this gets bigger than the profit of keeping them. When this does not apply, we are dealing with the paradox phenomenon.

With the technological evolution, nowadays, there are plenty of online sites and communities that refer and discuss about any kind of information. Therefore, there are many sites in which users can search information, share problems and thoughts, and ask others for help. The pitfall is that in order to participate in this kind of conversations, one ought to talk openly about his/her issue, and disclose personal sensitive data of him/her. In Greece, it is common for people to visit sites and online communities where health issues are being discussed. Some examples of such sites include Mammyland[1] and Iatronet[2]. The members of these communities communicate through comments, ask questions about their problems and give specific information about their

---

[1] http://mammyland.com/.
[2] http://iatronet.gr/.

© Springer International Publishing AG 2017
M. Themistocleous and V. Morabito (Eds.): EMCIS 2017, LNBIP 299, pp. 421–428, 2017.
DOI: 10.1007/978-3-319-65930-5_34

health status and their personal details, thus disclosing much about their health condition and their identity.

Our Contribution. This paper aims to reveal if this kind of disclosure of medical data seen in forums and online communities, reflects a privacy paradox existence. In order to explore that, we developed and conducted a web survey conducted in three stages [11–13] with participants from Greece. To the best of our knowledge, this is the first time that this issue is being studied in the context of health data. The results of our study question the existence of the privacy paradox phenomenon and show an emotionally driven behavior from the side of the users. The paper is structured as follows: In Sect. 2, privacy and health data definitions are being given, while the privacy paradox phenomenon is also being discussed. In Sect. 3, the aims and the results of our study are presented, while in Sects. 4 and 5, the results are discussed and conclusions are drawn.

2   Privacy, Health Data and the Privacy Paradox's Phenomenon

Privacy, as Alan Westin contends, *"provides individuals and groups in society with a preservation of autonomy, a release from role-playing, a time for self-evaluation and for protected communication"* [1]. Private data is defined in the 1998 Act[3] as *"data which relate to a living individual who can be identified – (a) from those data, or (b) from those data and other information which is in the possession of, or is likely to come into the possession of, the data controller, and includes any expression of opinion about the individual and any indication of the intentions of the data controller or any other person in respect of the individual"*. Sensitive data are private data that are capable of revealing information about ethnic or racial origin, political opinion, religious or other similar beliefs, memberships, physical or mental health details, personal life and/or criminal or civil offences [7, 8]. Health data on the other hand, is defined as any information used to provide, manage, pay and/or report on the services used across the entire healthcare system [9, 10].

The *privacy paradox* suggests that while Internet users are concerned about privacy, their behaviors do not mirror those concerns [14]. Many researchers have studied the privacy paradox phenomenon and a single emerging development to explain this has yet to emerge. The first attempt to understand this phenomenon was in 2006 by Barnes [3]. In their study, they mainly explain the teenagers' behavior in online social networks and especially the information that they choose to share with their online friends. Furthermore there are several studies that explore the existence of the phenomenon in other fields of social sciences [16], like Norberg et al. [9], who eventually established the definition of privacy paradox in 2007 [6, 9]. Literature provides insights that the existence of the phenomenon relates to the users emotions. In 2014, Kehr et al. [5] tried to prove that the emotional state of users is responsible for the existence of the phenomenon. Moreover, another research by Kehr et al. [15] adopt the same approach.

---

[3] https://www.privacy.org.nz/the-privacy-act-and-codes/the-privacy-act/.

Our findings are in line with this approach and provide evidence that emotional factors are significant for the existence of the phenomenon in the context of health data.

While researchers study the privacy paradox phenomenon, it is known and generally accepted that individuals increasingly use information blogs, social networks and personalized applications for various reasons, such as curiosity and socialization. In the case of health data, people have organized online communities and forums to exchange opinions and knowledge, to discuss their health issues and support each other. While many researchers are trying to tackle with this situation and question the existence of the phenomenon in the field, two questions are naturally raised and are targeted by our survey [2, 4]:

- *Are people aware of the importance of their health- related data?* If one cannot understand the importance of this kind of own data, then he/she will not be able to efficiently value and, consequently, protect it. Do people who disclose their health-data acknowledge their importance? Do they know the processing and handling that their data may go through by those who own the copyrights of the data posted and shared through their websites? If they don't, then we cannot talk about a privacy paradox phenomenon. If they do, a second question needs to be answered.
- *How individuals assess that value of their health data?* If the value they assign reflects their online behavior, then we cannot talk about a privacy paradox either. The only case where a privacy paradox's existence exists, is if people's disclosing behavior does not align with the value that themselves assign to their data.

3   An Empirical Study

In an attempt to answer the questions made above, understand what forces people to participate in health-data disclosing discussions and compare the gain of this participation to people's value about their data in order to come to a conclusion about the phenomenon existence, we conducted a survey in Greece, which included the invitation of participants who are users of social networks such as Facebook as well as members of online communities. Our sample included 163 participants. Most of them (about 74.6%) belonged to the age-group of 18–40, which was our main target group.

When participants were asked if they know what kind of their data is categorized as sensitive, 76% claimed that they do. But, when they were asked to shortly name the categories they think that fall into the space of sensitive personal information, there were few answers in which health data were mentioned (see Table 1). These answers indicate that almost 25% of the participants didn't have any idea about what is sensitive data. Our results coincide with other studies, that also show unawareness and ignorance from the side of the users [4].

When asked how important the privacy of their health data information is to them, although many claimed that this kind of protection is indeed of primary importance, 1/10 did not seem to agree. In order to understand what makes people participate in online forums and look for information about their health, we asked how many of our participants had ever visited a web page or forums about health information. Among

Table 1. Answers to the question of which they believe that personal data are

Answers	Percentage
Properties and belongings	**63.00%**
Credit card PIN codes	**14.00%**
Medical data	**10.00%**
Religion beliefs	9.00%
Criminal records	4.00%

the participants 65% said that they have visited such sites at least once in their lifetime in order to search for information about health data issues. When asked about the reasons that led them to this kind of online behavior, their answers varied but they all had something in common: they included some kind of emotional factor (see Table 2).

Table 2. Reasons for visiting health forums or online communities

Answers	Percentage
Quick information	**57.00%**
Anxiety/fear	**20.00%**
Curiosity	**10.00%**
Conversations with homeopathic	9.00%
Valid diagnosis	4.00%

Those who have visited these web pages were additionally asked if they had ever participated in a discussion with other members of the online community or forum. The majority of the subjects (90.1%) declined that they have ever shared and posted information about their health issues. Furthermore, the participants were asked about the way that they posted or visited this kind of sites (i.e., using username, using a nickname or anonymously). The same question was made hypothetically to the participants that have never visited and posted in such communities. They were particularly being asked, which of the three options they would choose, if they were to comment and participate in online conversation: anonymously, using a nickname or by giving their true personal information (see Fig. 1). They were asked to justify the reasons behind their choice (see Table 3).

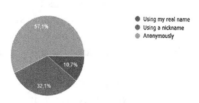

Fig. 1. Participants' behaviors with regards to personal identification (Source: our survey)

Table 3. Elaborated replies on choice selection (real name, nickname, anonymity) with regards to personal identification

Answers
I don't think that it is necessary to disclose my true personal details
I would prefer to participate anonymously in order to avoid an association of my comments with my identity
Conversations where real names are being used tend to be more serious
I don't regularly use my real personal information, I am trying to avoid it when possible
I'm ashamed
My identity does not change the diagnosis
I have nothing to hide
For my identity's protection

From these answers, it can be derived that the majority (57.1%) of the participants have concerns about the protection of their identity and privacy, and they would prefer anonymity in order to feel more protected.

70% of the participants that have posted, answered that they used a nickname, 20% used their real name and 10% were anonymous. But when we asked them if it was necessary to disclose personal details their answers were positive. The majority of people said that they gave their email to create an account to those pages and without this, the login was impossible. The participants were asked to describe the reasons that they preferred nicknames and they believed that it is a way to protect their privacy (Table 4). However, when asked if they feel safer by using a nickname, almost 85,7% said they do not (see Fig. 2). Moreover, in the answer about which of the data that they have given is real, almost everyone said that they have given their true personal information.

Table 4. Reasons why users prefer nicknames

Answers	Percentage
Privacy	**70.00%**
It is not necessary for someone to give his/her real name	**18.00%**
Possible recognition	**9.00%**
Unwilling to giving my real name	3.00%

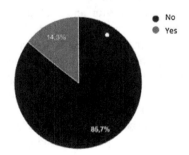

Fig. 2. Do you feel safer by using a nickname? (Source: our survey)

From the survey we understand that users are confused about which is the most appropriate way to post their thoughts (about 42.9% believe that it is safer for someone to use their real name while almost the same percentage has the opposite opinion).

Those who claimed that they have shared their problems asking for help, either from other members of the online community or from specialists (doctors), were asked what do they think that they gained in return from this kind of disclosure and sharing. Their answers were once again driven by their emotions (see Table 5), in line with what other research's results indicate [5].

Table 5. Replies regarding the gain that individual perceive to take from their visit

Answers	Percentage
Relief	**52.00%**
Useful opinion exchanging	**25.00%**
Courage	**12.00%**
Security	7.00%
Additional anxiety	4.00%

The participants were also asked to compare their perception about the privacy loss caused by their participation to the online sites with the gain that they think that this offered them in a return. With this question we actually called the participants to determine the existence or non-existence of the privacy paradox phenomenon in the field. If their gain makes the disclosure worthy, then we cannot talk about a privacy paradox, as people seem to disclose information about their health conditions for something fair in return. From the responses 65% of those who answered that they have commented and exchanged opinions in this kind of social forums feel that it was definitely worth the price.

4   Discussion

Many people who participated at our survey, claim that they value and care about their sensitive data, like the data exposing their health status, conditions and issues. Nonetheless, they give away such information in forums, communities, professionals' web pages and applications. Some of the questions that our study tried to address related to what leads users to this behavior and the participation in those activities. What do people gain in return so that they assume that it reflects the "label of importance" that they put upon their health data? Answering such questions, brings us one step closer to understanding the privacy's paradox phenomenon and question its existence. Here, we have two possible situations that would prove the nonexistence of a privacy paradox in the field: either the gain seems very high from the point of view of the users (i.e., high as the value the assign to the data), or their health data is not of much importance to them. As the beliefs of the users are a key to answering this question regarding the existence of the privacy-paradox phenomenon, our empirical study has shown the following scenarios:

- **Scenario 1:** We are dealing with many online users that do not consider their health data as sensitive, due to lack of awareness, ignorance or misinformation. Given that these users do not assign high value to their data they are not making great efforts to efficiently protect it. This leads us to the conclusion that we cannot talk about a privacy paradox phenomenon, as users don't understand the importance and value of their data and they are not putting them in danger knowingly.
- **Scenario 2:** For those who value their data and know the data's importance, we can see that their online behavior is emotionally driven. The most frequent visitors to forums were women (52%) who were worried about their health or their relatives' health. Therefore, we cannot talk about a privacy paradox, as emotional overload can affect the users' judgment and rational decision making and make them more prone to disclosure.
- **Scenario 3:** If a case does not fall in the above two categories, then we still cannot accept the phenomenon's existence. This is because in the last question made in the survey, the answers of the participants showed clearly that the benefit of sharing personal information is worth the disclosure of health information.

From the findings of our survey, thus, we can summarize and argue that:

The privacy paradox does not exist in the field of health data disclosure made by online users in communities and forums.

The only case in which we could assume that the privacy paradox seems to exist, is when users said that they had participated in conversations, after they had given away personal details during their registration. But we should take into consideration that they were emotionally driven during the time of the disclosure, something that takes us back to the Scenario 2. As we mentioned above we are not the first who study the phenomenon from the point of view of users' emotions and emotional state [5, 15]. However, to the best of our knowledge, it is the first time that a study addresses the existence or non-existence of the phenomenon in the health field and contributes to the general research upon users' behavior and the existence of the privacy paradox. We should note that our findings come from a relatively small sample of participants in Greece and they be considered as a starting point giving spark to other similar studies in different countries.

5   Conclusions

In this research we questioned and studied the existence of the privacy paradox when users disclose their health-related data in online communities. To the best of our knowledge, this is the first analysis of the privacy paradox in the health field. The outcome of our research is that the privacy paradox phenomenon does not exist in this context. First, in many cases the users do not seem to understand the value of this type of data and thus they disclose them due to lack of awareness. Second, during the time of the disclosure, they seem to be emotionally driven. When someone is emotionally charged, it is difficult for him/her to make the right decision and behave rationally. Of course there are some limitations like the kind of participants (mainly women) and the group age (18–40). Thus because this group uses more frequently the social networks, where we published our survey. As a future work, it would be interesting to analyze this phenomenon globally, to understand whether the privacy paradox phenomenon in this area exists or not, in countries besides Greece.

References

1. Westin, F.A.: Privacy and Freedom. Ig Publishing, New York (1968)
2. Acquisti, A., John, L.K., Loewenstein, G.: What Is Privacy Worth? J. Legal Stud. **42**(2) (2013). Article no. 1
3. Barnes, S.B.: A privacy paradox: social networking in the United States (2006)
4. Brandimarte, L., Acquisti, A., Loewenstein, G.: Misplaced confidences privacy and the control paradox. Soc. Psychol. Pers. Sci. **4**, 340–347 (2013)
5. Kehr, F., Wentzel, D., Kowatsch, T.: Privacy paradox revised: pre-existing attitudes, psychological ownership, and actual disclosure. In: Proceedings of the Thirty Fifth International Conference on Information Systems, Auckland, New Zealand (2014)
6. Lee, H., Park, H., Kim, J.: Why do people share their context information on Social Network Services? A qualitative study and an experimental study on users' behavior of balancing perceived benefit and risk. Int. J. Hum. Comput. Stud. **71**, 862–877 (2013). doi:10.1016/j. ijhcs.2013.01.005
7. MIT: Sensitive Data: Your Money AND Your Life (2008)
8. Hellenic Data Protection Authority (HDPA). http://www.dpa.gr
9. Norberg, P.A., Horne, D.R., Horne, D.A.: The privacy paradox: personal information disclosure intentions versus behaviors. J. Consum. Aff. **41**(1), 100–126 (2007)
10. Tzourakis, M.: The health care industry and data quality (1996)
11. Google Forms. https://goo.gl/forms/ZXuMRHGeZJA7UJA92
12. Google Forms. https://goo.gl/forms/upq6BlwSh2SG9VO03
13. Google Forms. https://goo.gl/forms/3lxKzMI6oiGCraVY2
14. Web Privacy Wiki. http://web-privacy.wikia.com/wiki/Privacy_Paradox
15. Kehr, F., Kowatsch, T., Wentzel, D., Fleisch, E.: Blissfully ignorant: the effects of general privacy concerns, general institutional trust, and affect in the privacy calculus. Inf. Syst. J. **25**, 607–635 (2015). doi:10.1111/isj.12062
16. Holland H.B.: Privacy Paradox 2.0 (2010)

IT Governance

IT Value Management Capability Enabled with COBIT 5 Framework

Cristiano Pereira[1(✉)], Carlos Ferreira[2], and Luis Amaral[3]

[1] STIC, Universidade de Aveiro, Aveiro, Portugal
cristiano@ua.pt
[2] DEGEIT/IEETA, Universidade de Aveiro, Aveiro, Portugal
carlosf@ua.pt
[3] DSI, Universidade do Minho, Guimarães, Portugal
amaral@dsi.uminho.pt

Abstract. One of the most common dilemmas faced today by organizations and their leaders is how to guarantee value from the high level IT investments, *i.e.* how organizations ensure expected benefits from this growth in IT investments. A superior understanding of how to deliver value to the business from IT initiatives is critical. Value should not be view only as a financial return, but also as other strategic factors that affect the business. This paper adopts a resource-based theory perspective to identify and propose a set of competences, resources, and practices, which contribute to develop and conceptualize an IT Value Management Capability Model, with an oriented practical perspective to existing IT Value Management professional frameworks, namely; COBIT 5 and Val IT 2.0. Based on literature findings, our model supports managers on developing resources, practices, and competences that contribute to business value and competitive advantage.

Keywords: IT Value Management Capability · IT governance · IT value management · Resource based view · COBIT 5

1 Introduction

Organizations have been facing in the last years with an increasing dependence on Information Technologies (IT), essential for the sustainability of their business. The growth in IT investments increases the concern of organizations to ensure the expected benefits [1], which point to cases of failure with investments made in IT [2]. Knowledgeable about this reality, organizations seek solutions in IT Governance (ITG) frameworks proposed by professional community [3–5] or choose to design its own models, adapted to their organizational reality. Organizations should adopt a series of practices, which constitute an organizational IT Value Management (ITVM) capability that answer to the challenge of obtain value from IT investments.

Schryen [6], reveals the need to investigate how practices influence each other and how IT capabilities change over time. With this call, and inspired by [9–11] this paper adopts Resource Based View (RBV) theory to identify and propose, through a literature review, a set of competences, resources and practices which contribute to an ITVM

© Springer International Publishing AG 2017
M. Themistocleous and V. Morabito (Eds.): EMCIS 2017, LNBIP 299, pp. 431–446, 2017.
DOI: 10.1007/978-3-319-65930-5_35

capability, enabled by a set of resources and other organizational elements grounded on COBIT 5 enablers, [4], categories.

The remainder of the paper is organized as follow: Sect. 2 outline the background of the study with a brief review of literature related to Enterprise Governance of IT (EGIT), ITVM, and conclude with the concept of capabilities, resources and practices. Section 3 describes the literature review approach. Section 4 discusses the proposed ITVM capability model components. Section 5 describes the literature findings that complement the model. Finally, some conclusions, along with future research proposals outlined in Sect. 6.

2 Background and Theory

2.1 Enterprise Governance of IT, Evolution and Value for Organizations

Initially, ITG was used to describe how the board of directors and executive management consider IT in their supervision, monitoring, control and direction of organizations [10]. In 2004, Weill and Ross [11], proposed one of the most referenced definitions of ITG and argue that *"effective IT governance is the single most important predictor of the value an organization generates from IT"*. [12] highlights the transversal characteristic of ITG across all organization (Business and IT); this view is sustained by several authors [13–15], pointing to a more broad IT Governance concept called Enterprise Governance of IT (EGIT). De Haes and Van Grembergen [14], consider it encompasses an organizational capacity, and the outcomes it enables, specifically business/IT alignment and in the end more value creation out of IT-enabled investments. A question emerges from this perspective:- What constitutes value creation from IT investments?

The difficulty in answer this question is one of the main reasons that make it hard to obtain value for the business from IT enabled investments, *i.e.* the ambiguity in identifying what is value for the business. This misperception and complexity about value creation are mentioned by [15]. In line with [16] more than concentrate solely on IT management aspects, we argue that an organizational capability like ITVM will give a more strategical and aligned view of value across all the organization.

2.2 A Resource Based View Outlook

Resource-based theory view has been one of the most well-known, powerful theories for understanding organizations over the past two decades [17]. In the IS field research RBV provides a valuable way for researchers think about how IT/IS relate to firm strategy and performance [18]. The RBV claims that resources (including IT resources) enable them to achieve competitive advantage [19], but not all resources create sustainable competitive advantage, only a subset with specific characteristics known as VRIN (valuable, rare, imperfectly imitable and non-substitutable).

Barney, [19], consider resource as *"all assets, capabilities, organizational processes, firm attributes, information, knowledge, etc. controlled by a firm that enable the firm to conceive of and implement strategies that improve its efficiency and effectiveness*". Afterword, Barney [20], in line with Amit & Schoemaker, [21] contradicts itself and refers to "resources and capabilities" as separate things [22]. A firm's resource includes not only its physical assets such as plant and location (tangible resources) but also human resources, knowledge resources, and relationship resources (intangible ones). These intangible resources, are unique in an organization, and cannot be imitated easily, thus satisfying the VRIN conditions of RBV theory [17]. In contrast to resources, [21] define capabilities, as "*the firm's capacity to deploy resources, usually in combination, using organizational processes, to effect a desired end*". This concept is particularly pertinent in IT resources, because, in isolation, this resources rarely have a direct impact in business performance [23]. The influence of IT resources is usually verified only when they have a complementary relationship with other assets and IT capabilities.

Commonly, an organizational capability (by inherence IT capability) is what you can do, with what you know. Here, what you know should be seen as a new concept that emerges from related RBV literature, known as competences. Strategy literature has been debated the distinction between "competence" and "capability". The closeness of the concepts could be perceived when compare the definition of competence and capability proposed respectively by [8] and [21]. To elucidate the concept of competence, [8] stated that competences consist of a combination of personal skills, knowledge and experience, organizational roles and processes, called practices[1] to operationalize the organizational competences.

To support the adoption of RBV in this paper, we recall [19], who classify firm's resources in: (i) physical capital resources, (ii) human capital resources and (iii) organizational capital resources. All of them are necessary for an effective ITVM, that requires, boards of directors and top management teams to participate in managing and governance (organizational capital resources) of IT enabled investments (physical capital resources). This involvement requires skills, knowledge, experience and insights from individual managers and workers of the firm (human capital resources). From this perspective, we assume that organizational competences are a subset of organizational capabilities, and comprise a blend of personal skills, knowledge and experience, roles, organizational structures and processes.

3 Review Methodology

The review process was supported in each of the phases of [24] framework. The first phase of the process aims to frame and define the scope of the literature review. In the second phase, the model was developed (Sect. 4) mapping the key concepts under study, and simultaneously exposing the existing gaps, thus opening the opportunity for this study. In the third stage, a literature search process was carried out with reference to RBV, Resources, Competences and IT Value Management practices. After

---

[1] A set of socially defined ways of doing things, in a specific domain, to achieve a defined and generally measurable outcome.

collecting the literature on the subject under investigation this should be analyzed and synthesized, deriving the results (Sect. 5).

Our literature review focuses mainly on academic and professional publications with the objective of identifying a set of competences, resources and COBIT 5 enablers related to ITVM practices in a neutral or independent perspective. The search has been performed in multiple e-databases (EBSCO, JSTOR, WILEY and ScienceDirect) for scholarly peer-reviewed journal publications without any date range restriction mentioning "value management", "IT investment", "capability", "RBV Theory" and "information systems" or "information technology" in the 'full text'. Each paper was then examined through qualitative content analysis to interpret the context and application of ITVM competences and practices, and the data collected were organized according to the conceptual model in Sect. 4. The obtained results are considered of particular interest to academics and professionals who carry out their activity in this area of knowledge.

4 The IT Value Management Capability Model

A capability-based model should focus on integrating; configure internal and external organizational skills; resources and functional competences concerning the transformational environment that in these days is subjacent to any organization. A model based on capabilities is a more stable concept due to the fact that capabilities are independent of the business processes that materialize them and are independent of the required resources, making capabilities stable over time [25]. Based on [8, 26] viewpoints, a model can be constructed to represent the constituents of the ITVM Capability. The model comprises a set of competences; each competence developed by a restricted and orchestrated use of enablers or practices, supported by firm´s resources (Fig. 1). We restrict the practices to the enabler's categories, described in COBIT 5 framework.

To reflect the ITVM Capability to extend across all levels of organization, and influenced by [7, 27] we stratify our model in three vertical layers; from bottom to top, Resource Level, Organizational Level and Business Level. Resource Level include resource components, defined as enterprise resources in COBIT 5, supported by the other enablers and are key ingredients to ITVM competences. The Organizational Level deals with, processes, organizational structures, principles and policies enablers, and the way they orchestrate resources to develop or create the ITVM competences, which are also part of the Organizational Level. Given the meta-level feature of the Capability concept, that integrates multiple competences across all organization, is only at the business level that ITVM capability arises [7].

4.1 IT Value Management Capability

The notion of Capability is not uniformly understood in the literature [28]. Multiple authors (from strategy literature to IS scholars) have proposed definitions for capabilities, *e.g.* are [21, 29, 41]. Other authors present more concrete IS/IT Capabilities

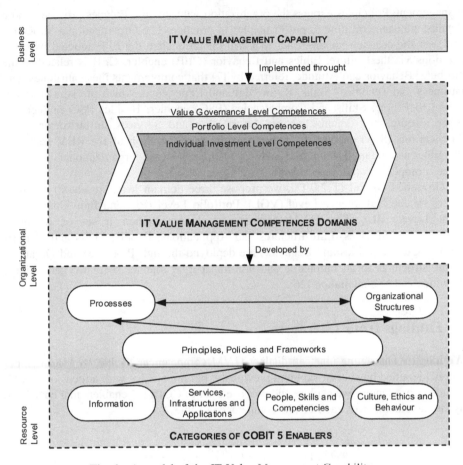

Fig. 1. A model of the IT Value Management Capability

types; Information Management Capability [30], Value Management Capability [26]. In fact, IS/IT is a commodity available to all, what organizations require is competences to leverage value from their IT investments. Drawing on past research on IS/IT capabilities we conceptualize the IT Value Management Capability as the organizational ability to systematically perform a set of transversal and multilevel competences through the orchestrated use of enablers that support and contribute to business and strategic value of IT.

4.2 From COBIT 5 Enablers to Competences

Following the approach of [8], we use COBIT 5 enablers as "practices", to add some granularity to ITVM competences. The first category, Principles, Policies and Frameworks (PPF), is the central hub, that influences all remaining enablers [31], being the vehicle to translate the desired behavior into practical guidance for day-to-day management. Principles express the organization values while Policies provide a more detailed guidance on how to put Principles in practice. The Organizational Structures (OS) set the decision-making roles that will operationalize the PPF adopted by organizations via the Culture, Ethics and Behavior (CEB) enabler. CEB is related to the attribute, Behavior and Attitude, referred by [7] that complements these attributes with Business and Technical Skills, Knowledge and Experience, which are similar in our model to People, Skills and Competencies (PSC) enabler. Besides PSC enabler, the other resources of organization are Information and Services, Infrastructures and Applications (SIA) enabler. In addition to the representation of the RBV resources (tangible and intangible), these enablers also present the organizational processes where competences are embedded.

Grounded on Val IT 2.0 [3], we propose three domain levels to classify competences; Value Governance Level (VGL), Portfolio Level (PL) and Individual Investment Level (IIL). The ITVM competences emerge when a set of resources (Information; Services, Infrastructures and Applications; People, Skills and Competencies; Culture, Ethics and Behavior) are deployed through Processes and Organizational Structures in an inimitable and non-transparent manner so that organizational activities can be undertaken [26].

5   Findings from Literature

We identify and define nineteen distinct ITVM competences (Table 1). Most competences identified have been explored previously, but not always with a focus on ITVM. Beside competences, we explore and propose a set of COBIT 5 enablers that contribute to the development of each ITVM competence (Table 2).

Table 1. IT Value Management competences

	Competence	The ability to...	References
Value governance level	VGL1 – establish ITVG framework	... Create an appropriate IT governance framework consistent with the overall enterprise governance environment, define policies, roles, responsibilities with appropriated organizational structures established, and ensure IT related decisions are made in line with the enterprise's strategies and objectives	[3, 33, 36–38]
	VGL2 – Build informed business/IT relationship	... integrate and align IT function with other functional areas or departments of the firm, bridge the culture gap between IT and business domains ensuring IT development plans are integrated with organizational strategic plans and goals	[3, 7, 17, 38, 39]
	VGL3 – establish external IT relationships	... ability to manage linkages between the IT function and stakeholders (suppliers, customers, and partners) outside the firm, understanding technology and other external factors (legal, regulatory and contractual obligations), identify	[32, 36, 39–41]

(continued)

Table 1. (*continued*)

	Competence	The ability to...	References
		opportunities that may be inherent in redefinition of business processes and use of new IT	
	VGL4 – identify IT innovation	... recognize business opportunities from current and emerging technologies, incorporate their potential in long term business development and understand their responsibilities for value to be realized from those opportunities	[3, 7, 33, 41, 42]
	VGL5 – structuring portfolio and investment criteria	... shaping portfolio and investment types, categories, relative weightings and establish appropriate criteria for decision making on investments in information, systems and technology	[3, 7, 41, 43]
	VGL6 – financial planning	... establish and monitor practices for financial planning in order to effectively manage IT assets, costs of acquisition and ownership to provide input to business cases for new investments in IT	[3, 7, 39, 41]
	VGL7 – establish value governance monitoring	... control strategies and monitor critical performance variables, identify and define key goals, metrics and techniques for reporting performance, assigning specific business value to expected initiative outcomes from IT enabled investments	[3, 41, 44]
	VGL8 – business performance improvement	... identify the knowledge and information needed to deliver strategic objectives through improved value governance, portfolio and investment management practices ensuring that appropriate corrective actions are initiated and controlled to ensure that value is optimized	[7, 33, 41, 45]
Portfolio level	PL1 – establish business strategic direction	... review and ensure that business strategy incorporates new opportunities for IT influence or support the strategy, i.e., business and IT strategies are aligned, including IT budget or a willingness to invest in IT	[3, 7, 33]
	PL2 – prioritize	... assess business cases, evaluating the characteristics of the investments portfolio, with an investment mix based on costs, financial measures over the full lifecycle, risk and alignment that reflect the business strategy and produce the maximum return from resources	[3, 7, 41, 46]
	PL3 – IS/IT staff development	... optimize human resources utilization and capabilities continuously and balanced across the needs of the investment portfolio, and business as usual, train and deploy appropriate staff and ensure technical, business and personal skills	[3, 7, 17, 41]
	PL4 – managing change	... make the business and organizational changes required to maximize the benefits of IT adoption, re-evaluate and reprioritize the portfolio to ensure that it is aligned with the business strategy when business cases are updated to reflect changes in requirements or performance	[3, 7, 39, 47]

(*continued*)

Table 1. (*continued*)

	Competence	The ability to...	References
Individual investment level	IIL1 – idea generation	... recognize new ideas and opportunities for IT investments to create value in support of business strategy from stakeholders and capture, collect and categorize ideas with respect to investment portfolio categories to integrate into an initial concept business case	[3, 16]
	IIL2 – business case	... define and develop the concept document business case with outcomes, benefits, assumptions, costs and risks, and other content that enables efficient and expedient decision making	[3, 16, 33, 48]
	IIL3 – programme planning	... define and document all projects, including those that are needed to bring about changes to the business. Specify required resources, including project managers and project teams as well as business resources. Specify funding, timing and interdependencies of multiple projects. Specify the basis for acquiring and assigning competent staff members and/or contractors to the projects	[3]
	IIL4 –benefits planning	... explicitly identify and plan to realize the benefits from IS investments, understand how business benefits will be realized and who will realize them, costs and risks to be mitigated and/or managed	[3, 7, 8]
	IIL5 –service management	... define service arrangements and performance criteria to match the business requirements including project management, project scope, resources and time, through planning, organizing and controlling, identify deviations from the plan and take timely remedial action when required	[3, 7, 33]
	IIL6 – benefit delivery	... ensure that planned benefits always have owners and are achieved, sustained and optimized, i.e. benefits derived from IT investments are monitor, measure and evaluated	[3, 7, 8]
	IIL7 - post implementation evaluation	... to monitor the performance of the overall programme, IT investment implementation and the projects including the business and the IT functions contributions to the projects, and report to executives in a timely, complete and accurate fashion	[3, 16]

5.1 ITVM Competences in Value Governance Level

Our research identified eight competences, in VGL, that a firm needs to have in order to guarantee the incorporation of an adequate value governance approach in the organization. VGL competences are de glue of any ITVM program; these abilities ensure optimal value from its IT investments throughout their full life cycle [3].

Table 2. COBIT 5 enablers that contribute to develop ITVM competences

(↑) – Enabler that contribute to deploy ITVM Competence

Legend for competence columns: **Value Governance Level** = VG1–VG8; **Portfolio Level** = PL1–PL4; **Individual Investment Level** = IIL1–IIL7 (all under "IT Value Management Competences").

Category	IT Value Management Enablers (COBIT5)	VG1	VG2	VG3	VG4	VG5	VG6	VG7	VG8	PL1	PL2	PL3	PL4	IIL1	IIL2	IIL3	IIL4	IIL5	IIL6	IIL7
Processes	EDM01 - Ensure Governance Framework	↑	↑	↑	↑															
	EDM02 - Ensure Benefits Delivery					↑		↑	↑											
	APO01 - Manage the IT Management Framework	↑			↑															
	APO02 - Manage Strategy		↑																	
	APO04 - Manage Innovation				↑															
	APO05 - Manage Portfolio									↑	↑			↑	↑	↑				
	APO06 - Manage Budget and Costs						↑													
	APO07 - Manage Human Resources											↑								
	APO10 - Manage Suppliers			↑																
	BAI01 - Manage Programmes and Projects														↑	↑	↑	↑	↑	↑
	BAI05 - Manage Organizational Change																			↑
Organizational Structures	Board	↑	↑	↑	↑	↑		↑	↑	↑	↑				↑	↑				
	Strategy executive committee	↑	↑	↑	↑						↑	↑								
	Value Management Office					↑						↑			↑	↑				↑
	Project Management Office														↑	↑	↑	↑	↑	↑
	Steering committee (programs/projects)														↑	↑	↑	↑	↑	↑
	Audit and compliance																			↑
	C-suite executive [CEO;CIO;CFO;COO;CRO]	↑	↑	↑	↑	↑	↑	↑	↑	↑	↑	↑	↑		↑	↑				↑
Culture, Ethics & Behavior	Communication (enforcement and rules)	↑	↑	↑						↑		↑								↑
	Incentives and Rewards									↑		↑								
	Awareness of desired behavior	↑								↑	↑									
Principles, Policies and Frameworks	Enterprise Governance Principles	↑																		
	Value Governance and Management Policy	↑																		
	Principles for allocation of resources		↑									↑			↑					
	Vendor Management Policy			↑																
	Portfolio Management Policy				↑							↑	↑		↑	↑	↑	↑	↑	↑
	IT Investment Management Policy										↑									
	Human Resources Policy										↑									
	Business Case Management Policy										↑				↑					
	Value Monitoring and Reporting Policy								↑										↑	↑
People, Skills and competencies	Knowledge of frameworks for governance of IT	↑	↑	↑	↑															
	IT strategy and IT policy formulation	↑	↑		↑															
	Supplier management skills			↑																
	Knowledge of innovation				↑															
	Knowledge of portfolio and project management									↑	↑			↑	↑	↑	↑	↑	↑	↑
	Knowledge of HR management											↑								
	Knowledge of financial management						↑													
Information	Constitution/bylaws/statutes of organization			↑	↑															
	Investment types and criteria						↑			↑	↑									
	Feedback on portfolio performance									↑	↑			↑						↑
	IT Budget and plan						↑													
	Business case assessment and value proposition										↑			↑						↑
	Skills and competences development plan	↑		↑								↑				↑	↑			
	Business trends and Innovation plans		↑	↑																
	Service level agreements												↑							
	Evaluation of Investments and services portfolio											↑	↑							
	Benefit results and related communications																	↑	↑	↑
Services, Infrastructures & Applications	Buy or Build Decisions			↑								↑		↑	↑		↑			
	Inventory of Information (Systems and Data)								↑	↑	↑			↑				↑		↑

The ability to establish an appropriate ITVG Framework (VGL1) is fundamental to devise organizational arrangements (structures, processes, staffing). The relationships between IT and business (VGL2) and among the IT function and external stakeholders (suppliers, customers and partners) (VGL3) strongly influence the ability of an organization to use IT for strategic objectives [17, 32].

The success of IT adoption arises normally from organizations that constantly look at new IT technologies, and how to incorporate them into the business [33]. Thus, it is fundamental to search and identify IT innovations (VGL4). Most companies use simple and straightforward financial models to make investment decisions. For these companies, the IT portfolio management is incomplete [34]. To overcome such issues, the ability to structuring portfolio and investments criteria (VGL5) for decision-making is fundamental. To build Business Case for IT investments, an important input is financial information. The ability to establish and monitor practices for financial planning (costs, frequency of reporting, cross-sharing provisions), (VGL6) should be integrated with other enterprise practices [3]. From a value governance monitoring (VGL7) point of view, evaluating the business value of IT investments is of high importance [35], in order to control strategies and monitor critical performance variables, key goals and metrics. In ITVM is always a need to keep focus on performance parameters, and the ability to identify knowledge and information necessary to improve practices and corrective actions to ensure optimized business performance (VGL8) from IT investments.

5.2   ITVM Competences in Portfolio Level

At PL, we identified four competences to ensure enterprise ability to analyze, prioritize, select and manage a portfolio of IT investments. Without proper alignment between business and IT strategies (PL1), firms may be misguided in their efforts to generate positive returns from their IT investments [49]. The ability to evaluate and prioritize the IT investments (PL2) to decision-making and moving selected investments to portfolio for execution, based on portfolio and investments criteria defined previously, is fundamental to produce the maximum return from available resources. Besides financial resources, IT human resources development (PL3) is one of the major contributors to the strategic value of IT [17]. To survive in highly dynamic environment, organizations need to balance on a duality of change and stability [50]. With that in mind, we propose Managing Change competence (PL4), which is able to make the necessary changes in order to maximize the benefits of IT adoption.

5.3   ITVM Competences in Individual Investment Level

At IIL, we got seven competences to ensure that individual IT investments contribute to optimizing value, through identification of business requirements, detailed business cases that include IT Value throughout the full economic life cycle of investment (expected and realized) and monitor and report projects performance. The first step is to identify good opportunities to invest. The ability to identify new ideas and opportunities (IIL1) is a competence related to (VGL4), through a formal process, generate and report new ideas and opportunities for IT investments. A second competence is the ability to document the proposals in a structured way through business case (IIL2), to allow an efficient and expedient decision-making. Organizations need a Programme Planning competence (IIL3) to get a complete picture of how it is going the work and monitor the dynamics of the relationships between each project (documentation, resources, structures, quality of execution, risks, communications). To show what the benefits are, how they are going to be realized, the expected delivery schedule and who will realize them along with an explanation of the risks to be mitigated is essential a benefits planning competence (IIL4). Service management competence (IIL5) is the ability to manage programme performance *e.g.* scope, schedule, quality, cost and risk, and identify deviations from plan in order to take timely remedial actions if required. Benefits should be structured within benefits realization plan that will be evaluated on benefits delivery performance (IIL6) to ensure that planned benefits have owners and are achieved. Learning, from past implementations is an important element for that a post implementation evaluation competence (IIL7) is present.

5.4   Organizational Resources – COBIT 5 Enablers

Principles, Policies and Frameworks

Value Governance and Management Policy are good practice that helps organizations to clarify the meaning of value to the organization in terms of benefits, costs and risk [51]. A Portfolio Management Policy establishes requirements to categorize evaluate risk and select investments using agreed-on criteria. The diversity of investments (*e.g.* risk, cost, benefits) should be evaluated and managed differently, which implies that organizations state an IT Investment Management Policy. Given the importance of Business Case document [51, 52] we propose Business Case Management Policy to grant that IT investments will be managed through their full economic life cycle. To establish a formal monitoring and reporting of benefits realization through processes and procedures, a Value Monitoring and Reporting Policy is proposed. Organizations that wish to improve their performance should align individual's performance with the corporate goals and objectives, to achieve this, top management should ensure an adequate Human Resource Policy. To contribute to the deployment of VGL3 competence, we identify Vendor Management Policy, dedicated to the sourcing and management of external IT relationships.

Culture, Ethics and Behavior

Culture, Ethics and Behavior enabler, refers to the set of collective and individual behaviors within organization. Good practices for creating and maintaining desired behavior through the organization identified are: Communication of the desired behavior and the values of firm, awareness of desired behavior, reinforced by example from senior management, and incentives and rewards that encourage the desired behavior. Those in charge of the organization must sponsor this enabler from the outset, conveying by example, a culture of participation and commitment to the desired behaviors.

People, Skills and Competencies

Skills and competences in organization are related to the education and qualification levels, technical skills, experience. COBIT 5 identifies the People involved in ITVM, in profiles defined in Organizational Structures enabler. The main skills identified for ITVM are related knowledge necessary to deploy IT Value Governance Level Competences, *e.g.* Knowledge of ITG Frameworks, IT Strategy, innovation, supplier skills management and financial management knowledge fundamental for IT value management. A useful tool for identifying and defining skills is the Skills for Information Age Framework (SFIA), from SFIA Foundation[2].

Services, Infrastructures and Applications

The delivery of IT services is the responsibility of the IT department; even if a certain service is contracted to third parties, the decision must be made between organization and IT. The integration of cloud, mobile applications, and other emerging technologies into the organizations, are pushing enterprises to make changes to their policies and IT Infrastructures. We identify two type of aspects that directly affect the build of ITVM competences; the aspects related to architecture (inventory) and aspects such vendor relationship and staff skill resourcing that are changing the way organizations are consuming technology services (buy or build).

Information

The recognition that information is an asset or resource that generates benefits to organizations justifies the amount of resources found in the Information category. Each information resource identified in this study supports other enabler categories, in particular as inputs to processes or as data to use in communication practices. Information has no intrinsic value, it is only through putting information into action that value can be generated [51]. To reinforce the importance of this enabler category, ISACA develops the COBIT 5 Information Model [53], where a more detailed relationship with other enablers is described.

Processes

Procedures and practices to accommodate processes are fundamental to deploy organizational resources in order to create business value [26]. The COBIT 5 process enablers that we identify with relevant practices to ITVM competence are, from Governance Domain; the processes EDM01 – Ensure Governance Framework and EDM02 – Ensure Benefits Delivery. From IT management domain; APO01 – Manage the IT Management Framework, APO02 – Manage Strategy, APO04 – Manage Innovation, APO05 – Manage Portfolio, APO06 – Manage Budget and Costs, APO07 – Manage Human Resources, APO10 – Manage Suppliers, BAI01 – Manage Programmes and Projects and BAI05 – Manage Organizational Change Enablement. Each of these processes are detailed in [41].

---

[2] For more information on the Skills Framework for the Information Age (SFIA), refer to www.sfia-online.org.

Organizational Structures

The organizational structures are the key decision-making entities in the organization. COBIT 5, propose a set of common structures to enterprises, trough RACI charts with responsibilities and roles for each enabler process. The identified organizational structures and roles, consider the Accountabilities and Responsibilities, from each of the RACI charts in [41] related to each of the ITVM Processes identified previously. To EDM processes, accountabilities and responsibilities are mainly at the board and other C-suite executives (CxO). The Strategy Executive Committee is responsible for stimulating awareness amongst the board of directors on the potential value and viability of proven and emerging technologies [26]. Value Management Office (VMO) is the function that acts as the secretariat for managing investment and service portfolios, assessing and advising on investments opportunities and business cases. For manage programmes, projects and organizational changes a Steering Committee and Project Management Office are pointed out. Audit and Compliance, are responsible for risk management and control of IT investments and should grant that investments are guided by legal, regulatory and contractual compliance.

6   Conclusions and Future Work

The focus of this research work was on what main components constitute ITVM Capability, grounded on RBV theory. Building on previous research and in the literature review, we propose Organizational Structures, Processes, Principles, Competences and a set of other resources; Information, Services, Infrastructures and Applications and People, Skills and Competencies, from COBIT 5 enablers, that we consider suitable to develop competences that implement and support IT Value Management Capability.

Three enablers categories deserve to be highlighted, namely; Principles, Policies and Frameworks, that we view as resources that operate as translators of the desired behavior (enterprise and individual) and are a simplified guidance for day-to-day management of organizations. For the authors, develop an Organizational IT Value Management Capability should be viewed as an organizational change; with this in mind, the Culture, Ethics and Behavior enablers, are extremely important to maximize the benefits of IT adoption, reevaluate an reprioritize investment portfolio. Numerous studies have seen organizational culture as an important factor that may explain significant variations in IT business value [53]. CEB enabler is fundamental in the challenge to change organizational culture, once this change involves people´s value, attitudes and behaviours. Attracting, retaining and motivating staff are important practices to motivate people to behavior modification in response to organizational change. Finally, from Process enabler we verify the presence of six ITVM Processes in the list of top 10 most important processes of COBIT 5 identified by Bartens et al., [30]. This supports the importance that must be given from the first moment to the component of ITVM in the practical implementation of EGIT.

With this approach, we provide a more practical insight to organizations that are in a phase of adoption or review of EGIT, on how to develop the main ITVM

Competences, supported on industry best practices. Obviously, the proposed practices, and competences should be closely related to the reality of the firm, as noted by [14], a "silver bullet approach" does not exist. Each organization has to select its own set of enterprise governance of IT practices suitable for their sector, size, culture, etc.

Of course, this research has some limitations. The proposed model should be complemented with a practical validation. Thus, it is our intention to submit the current model to a panel of experts, through a Delphi Study, to validate and potentially identify other competences and resources from COBIT 5, or eventually from other industry best practices, that help to build a more robust ITVM Capability model. Identify which resources and/or competences are more appropriate in different organizational contexts, like, public or private, profit or non-profit institutions is also interesting to explore in a future research.

References

1. Kelly, S.C.: Failings in management and governance: report of the independent review into the events leading to the Co-operative Bank's capital shortfall (2014)
2. Wilkin, C., Campbell, J., Moore, S., Van Grembergen, W.: Co-creating value from IT in a contracted public sector service environment: perspectives on COBIT and Val IT. J. Inf. Syst. **27**(1), 283–306 (2013)
3. ITGI, Enterprise Value: Governance of IT Investments, The Val IT Framework 2.0. (2008)
4. ISACA, COBIT: A Business Framework for the Governance and Management of Enterprise IT. (2012)
5. ISO/IEC: ISO/IEC 38500:2015 - Information technology - Governance of IT for the organization (2015)
6. Schryen, G.: Revisiting IS business value research: what we already know, what we still need to know, and how we can get there. Eur. J. Inf. Syst. **22**(2), 139–169 (2013)
7. Peppard, J., Ward, J.: Beyond strategic information systems: towards an IS capability. J. Strateg. Inf. Syst. **13**(2), 167–194 (2004)
8. Ashurst, C., Doherty, N., Peppard, J.: Improving the impact of IT development projects: the benefits realization capability model. Eur. J. Inf. Syst. **17**(4), 352–370 (2008)
9. Pereira, C., Ferreira, C.: Identification of IT value management practices and resources in COBIT 5. RISTI – Rev. Ibérica Sist. e Tecnol. Informação **2015**(15), 17–33 (2015)
10. ITGI: Board Briefing on IT Governance, 2nd edn. IT Governance Institute (2003)
11. Weill, P., Ross, J.W.: IT Governance: How Top Performers Manage IT Decision Rights for Superior Results. Harvard Business School Press, Boston (2004)
12. Van Grembergen, W., De Haes, S., Guldentops, E.:Structures, processes and relational mechanisms for IT governance. In: Strategies for Information Technology Governance, IGI Global, pp. 1–36 (2004)
13. Peppard, J.: Unlocking the performance of the Chief Information Officer (CIO). Calif. Manag. Rev. **52**(4), 73–99 (2010)
14. De Haes, S., Van Grembergen, W.: Enterprise Governance of Information Technology. Springer International Publishing, Cham (2015)
15. Laursen, M., Svejvig, P.: Taking stock of project value creation: a structured literature review with future directions for research and practice. Int. J. Proj. Manag. **34**(4), 736–747 (2016)

16. Ali, S., Green, P., Robb, A.: Information technology investment governance: what is it and does it matter? Int. J. Acc. Inf. Syst. **18**, 1–25 (2015)
17. Ashrafi, R., Mueller, J.: Delineating IT resources and capabilities to obtain competitive advantage and improve firm performance. Inf. Syst. Manag. **32**(1), 15–38 (2015)
18. Chuang, S.H., Lin, H.N.: Performance implications of information-value offering in e-service systems: examining the resource-based perspective and innovation strategy. J. Strateg. Inf. Syst. **26**(1), 22–38 (2017)
19. Barney, J.: Firm resources and sustained competitive advantage. J. Manag. **17**(1), 99–120 (1991)
20. Barney, J.B.: Looking inside for competitive advantage. Acad. Manag. Exec. **9**(4), 49–61 (1995)
21. Amit, R., Schoemaker, P.J.H.: Strategic assets and organizational rent. Strateg. Manag. J. **14**(1), 33–46 (1993)
22. Seddon, P.B.: Implications for strategic IS research of the resource-based theory of the firm: a reflection. J. Strateg. Inf. Syst. **23**(4), 257–269 (2014)
23. Lunardi, G.L., Maçada, A.C.G., Becker, J.L., Van Grembergen, W.: Antecedents of IT governance effectiveness: an empirical examination in Brazilian firms. J. Inf. Syst. **31**(1), 41–57 (2017)
24. vom Brocke, J., Simons, A., Niehaves, B., Riemer, K., Plattfaut, R., Cleven, A.: Reconstructing the giant: on the importance of rigour in documenting the literature search process. In: 17th European Conference on Information Systems, pp. 2206–2217 (2009)
25. Amiri, A.K., Cavusoglu, H., Benbasat, I.: Enhancing strategic IT alignment through common language: using the terminology of the resource-based view or the capability-based view? In: International Conference on Information Systems, pp. 1–12 (2015)
26. Maes, K., De Haes, S., Van Grembergen, W.: Developing a value management capability: a literature study and exploratory case study. Inf. Syst. Manag. **32**(2), 82–104 (2015)
27. Caldeira, M.: Understanding the Adoption and Use of Information Systems/Information Technology In Small And Medium-Sized Manufacturing Enterprises: A Study In Portuguese Industry. Cranfield University, Cranfield (1998)
28. Wójcik, P.: Exploring links between dynamic capabilities perspective and resource-based view: a literature overview. Int. J. Manag. Econ. **45**(1), 83–107 (2015)
29. Wang, N., Liang, H., Zhong, W., Xue, Y., Xiao, J.: Resource structuring or capability building? an empirical study of the business value of information technology. J. Manag. Inf. Syst. **29**(2), 325–367 (2012)
30. Mithas, S., Ramasubbu, N., Sambamurthy, V.: How information managent capability influences firm performance. MIS Q. **35**(1), 237–256 (2011)
31. Bartens, Y., De Haes, S., Lamoen, Y., Schulte, F., Voss, S.: On the way to a minimum baseline in IT governance: using expert views for selective implementation of COBIT 5. In: 2015 48th Hawaii International Conference on System Sciences, pp. 4554–4563, March 2015
32. Khani, N., Nor, K.M., Samani, M.B., Hakimpoor, H.: An empirical investigation of capability factors affecting strategic information system planning success. Int. J. Strateg. Inf. Technol. Appl. **3**(2), 1–17 (2012)
33. Cragg, P., Caldeira, M., Ward, J.: Organizational information systems competences in small and medium-sized enterprises. Inf. Manag. **48**(8), 353–363 (2011)
34. Maizlish, B., Handler, R.: IT Portfolio Management Step-by-Step: Unlocking the Business Value of Technology. Wiley, Hoboken (2005)
35. Pajić, A., Pantelić, O., Stanojević, B.: Representing IT performance management as metamodel. Int. J. Comput. Commun. Control **9**(6), 758–767 (2014)

36. Webb, P., Pollard, C., Ridley, G.: Attempting to define IT governance: wisdom or folly? In: Proceedings of the Annual Hawaii International Conference on System Sciences, vol. 8, no. C, pp. 1–10 (2006)
37. Willcocks, L., Feeny, D., Olson, N.: Implementing core IS capabilities: feeny-willcocks IT governance and management framework revisited. Eur. Manag. J. 24(1), 28–37 (2006)
38. Feeny, D.F., Willcocks, L.P.: Re-designing the IS function around core capabilities. Long Range Plan. 31(3), 354–367 (1998)
39. Wade, M., Hulland, J.: Review: the resource-based view and information systems research: review, extension, and suggestions for future research. MIS Q. 28(1), 107–142 (2004)
40. Feeny, D.F., Willcocks, L.P.: Core IS capabilities for explointing information technology. Sloan Manag. Rev. 39(3), 9–21 (1998)
41. ISACA: Enabling Processes (2012)
42. Kim, M., Song, J., Triche, J.: Toward an integrated framework for innovation in service: a resource-based view and dynamic capabilities approach. Inf. Syst. Front. 17(3), 533–546 (2015)
43. Rungi, M.: How lifecycle influences capabilities and their development. Int. J. Manag. Proj. Bus. 8(1), 133–153 (2015)
44. Lentz, C.M.A., Gogan, J.L., Henderson, J.C.: A comprehensive and cohesive IT value management capability: case studies in the North Amercian life insurance industry. In: Proceedings of the 35th Annual Hawaii International Conference on System Sciences, vol. 0, no. c, pp. 1–10 (2002)
45. ITGI: Measuring and demonstrating the value of IT, pp. 1–25 (2005)
46. Ward, J.L., Daniel, E.: Benefits Management: Delivering Value from IS & IT Investments. Wiley, Hoboken (2006)
47. Peppard, J., Lambert, R., Edwards, C.: Whose job is it anyway?: organizational information competencies for value creation. Inf. Syst. J. 10(1), 291–322 (2000)
48. Pereira, C., Ferreira, C., Amaral, L.: Shape a business case process: an IT governance and IT value management practices viewpoint with COBIT 5.0. In: 17.ª Conferência da Associação Portuguesa de Sistemas de Informação (CAPSI'2017), pp. 60–75 (2017)
49. Masli, A., Richardson, V.J., Sanchez, J.M., Smith, R.E.: The business value of IT: a synthesis and framework of archival research. J. Inf. Syst. 25(2), 81–116 (2011)
50. Maes, K., De Bruyn, P., Oorts, G., Huysmans, P.: On the need for evolvability assessment in value management. In: Proceedings of the Annual Hawaii International Conference on System Sciences. pp. 4406–4415 (2014)
51. ISACA: COBIT 5 for Business Benefits Realization (2016)
52. Maes, K., Van Grembergen, W., De Haes, S.: Identifying multiple dimensions of a business case: a systematic literature review. Electron. J. Inf. Syst. Eval. 17(1), 47–59 (2014)
53. ISACA: Enabling Information (2013)

Assessing IT Governance Processes Using a COBIT5 Model

Gonçalo Rodrigues Cadete[(⊠)] and Miguel Mira da Silva

Instituto Superior Técnico, Universidade de Lisboa, Lisbon, Portugal
{goncalo.cadete,mms}@tecnico.ulisboa.pt

Abstract. COBIT5 process assessments are conducted to support process improvement and thus enable business-IT alignment. For providing an assessment, assessors engage in planning, data collection, and data validation activities. Typically, these activities are assisted by spreadsheet-like artifacts, that are used for recording the COBIT5 Process Assessment Model (PAM) entities, as well as the corresponding assessment evidence. However, spreadsheet-like artifacts are not an optimal solution for assisting assessment activities, since they do not form part of an integrated and up-to-date enterprise architecture (EA) repository. Contradicting the COBIT5 recommendation of using EA – namely for improving alignment, increase agility, and generate potential cost savings – COBIT5 assessors often do not practice what they preach, i.e. that they do not use EA in their own business activities. Some EA tool vendors provide solutions for addressing this gap, but they are proprietary and lack scientific validation, thus presenting interoperability and adoption barriers. In this paper, we propose a set of COBIT5 viewpoints, based on standard Archi-Mate extensions to enable interoperability and ease adoption. We designed and tested the solution using a Design Science Research Methodology process model. We demonstrated the EA proposal in two public sector organizations, and evaluated its efficacy, consistency, and structural quality.

Keywords: COBIT · Enterprise architecture · ArchiMate · Enterprise governance of IT · Business and IT alignment

1 Introduction

The COBIT5 framework [1] provides a good practice approach for implementing Enterprise Governance of IT (EGIT) initiatives. EGIT is instrumental for improving business-IT alignment, maximizing the value from IT investments, manage IT-related risks and achieve compliance [2].

COBIT5 promotes the enabling process *APO03 Manage Enterprise Architecture* as key for enabling EGIT initiatives [3]. EA is relevant for providing holistic organizational self-awareness [4], including relevant entities, relationships, and principles [5], cutting across business domains, as well as bridging business and technology divides [6]. Therefore, EA provides shared inter-domain knowledge and shared viewpoints [7], which enable different stakeholder to conduct effective conversations for engaging in compliance assessments [8], risk management [9], and change initiatives [10, 11].

© Springer International Publishing AG 2017
M. Themistocleous and V. Morabito (Eds.): EMCIS 2017, LNBIP 299, pp. 447–460, 2017.
DOI: 10.1007/978-3-319-65930-5_36

Given the importance of EA for EGIT initiatives, is it unfortunate that auditors and consultants still face EA adoption barriers. Although standard EA languages like ArchiMate help address the interoperability challenges, the lack of freely available EA reference models for COBIT5 hinders easy and widespread adoption. Some EA tool vendors provide solutions for addressing this gap, but they are either proprietary or lack scientific validation, thus hampering adoption. Also, the COBIT5 documentation does not provide such EA artifacts, instead relying on natural language descriptions, tables, and informal diagrams.

In this paper, we validated the assertion that ArchiMate models facilitate COBIT5 process assessments, and provide a set of standards-based viewpoints that may be used in COBIT5 process assessment initiatives. The viewpoints were developed and tested incrementally, using three iterations of a Design Science Research Methodology (DSRM) process model (see Fig. 1). The EA artifacts presented in this paper are the outcomes of the final (i.e. third) DSRM iteration. We demonstrated the EA solution in two public sector organizations. For evaluating the artifacts, we used the demonstrations, evaluation questionnaires, as well as group sessions, and measured the goal efficacy, the environment consistency, and the structural quality of the proposed model.

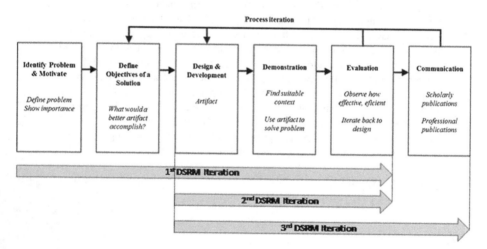

Fig. 1. The adopted DSRM process model, taken from [13].

2 Design Science Research Methodology (DSRM)

Scientific validation is important for promoting adoption. Scientific methods also offer rigorously defined languages, procedures, and criteria that are important for comparing and discussing experimental results, thus facilitating continual improvement of the artifacts. Accordingly, we have adopted a DSRM process model [13], for guiding the construction and evaluation of the architectural artifacts.

DSRM incorporates principles, practices, and process models which are adequate [21] to conduct design science research in applied research disciplines, such as engineering research and – more recently – information systems research, whose cultures value incrementally effective solutions. The design science paradigm seeks to create and evaluate "what is effective" [22] in the problem space. The DSRM process model includes an evaluation activity (see Fig. 1). The evaluation taxonomy we adopted in this work – in all but the first (i.e. the prototype) DSRM iteration – was based on the approach described in [23] (see Fig. 2).

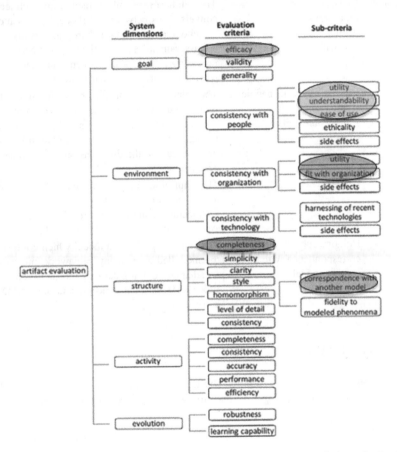

Fig. 2. Evaluation criteria, taken from [23]. For this paper, we used the criteria that are highlighted in the figure (with ellipses).

Accordingly, the evaluators were provided with evaluation forms (questionnaires) with a set of five questions for measuring the goal efficacy, the environment consistency, and the structural quality of the proposed artifacts (see Table 1). For rating the agreement level with each question, the following options were provided: "strongly disagree", "disagree", "agree", and "strongly agree".

Group sessions were also conducted, to elicit informal feedback regarding the solution's usefulness claims.

Table 1. Questions used for evaluating the solution's usefulness claims, regarding the goal efficacy, the environment consistency, and the structural quality.

Dimension	Evaluation criteria	Questions: does the evaluator agree with the following usefulness claims?
Goal	Efficacy	The solution is useful for improving the effectiveness of Process Assessment initiatives
Environment	Consistency with organization: - utility - fit with organization	The solution is useful for facilitating architectural conversations between the Process Assessment stakeholders, enabling a shared understanding of the assessment rationale (i.e. "why") and providing a link for the system implementation representations (i.e. "what" and "how")
	Consistency with people: - utility - understandability - ease of use	The solution is useful for providing architectural representations of the Process Assessment rationale and for providing a link to the system implementation. The graphical notation is easy to understand and the diagrams are easy to use in practice
Structure	Completeness	The solution is complete, meaning that it provides a template for representing all the key architectural concepts required for Process Assessment initiatives using COBIT5
	Homomorphism: correspondence with another model	The solution provides a model which conforms to the modelled assessment framework, presenting an adequate ontological mapping between the Process Assessment concepts and the ArchiMate constructs

3   Research Problem

In this section, we review the state-of-the-art regarding EGIT and EA, as well the relation between EGIT and EA, to formulate the research problem in the correct context.

3.1   COBIT5

In the introduction section, we recalled that the COBIT5 process *APO03 Manage Enterprise Architecture* is proposed as a key process for helping to guide the creation and maintenance of governance and management enablers [3]. The purpose of the *APO03* process is to "represent the different building blocks that make up the enterprise and their inter-relationships as well as the principles guiding their design and evolution over time, enabling a standard, responsive and efficient delivery of operational and strategic objectives" [14]. Accordingly, the EA solution presented in this paper represents concepts from both the business and IT domains, as well as their inter-relationships – note that these are especially important for enabling business-IT alignment.

3.2   TOGAF

The *APO03* COBIT5 process promotes TOGAF as related guidance for providing an EA framework. The TOGAF Architecture Development Method (ADM), as well as other TOGAF components, can be easily mapped to the *APO03* process key management practices [14]. The current TOGAF 9.1 specification recognizes the need for a tailored approach for COBIT, by stating that TOGAF tailoring "may include adopting elements from other architecture frameworks, or integrating TOGAF methods with other standard frameworks, such as ITIL, CMMI, COBIT, PRINCE2, PMBOK, and MSP" [15]. However, the standard TOGAF approach is generic, in the sense that it is not specifically tailored to the COBIT5 rationale – note that the latest TOGAF version was released in 2011, hence before the COBIT5 framework was released.

Therefore, we may conclude that the TOGAF recommendations, by themselves, are not sufficient to guide the design of effective EA solutions for assisting COBIT5 process assessment initiatives. An effective EA solution for assisting COBIT5 EGIT initiatives, such as the one sought in this paper, goes beyond TOGAF's scope.

3.3   ArchiMate

The TOGAF and ArchiMate specifications are both developed by The Open Group, and their development efforts are becoming increasingly coordinated [16, 17]. The ArchiMate standard is also supported by several modelling tool providers, which makes the language an attractive option for easy adoption. The language provides extension mechanisms to extend the core language, through adding attributes to ArchiMate concepts and relationships, as well as specialization of concepts and relationships. Extending the ArchiMate language can be useful for optimizing the ontological fit of the architectural representations. However, in this work we based our artefacts on constructs that are an integral part of the ArchiMate specification, to facilitate the use of popular ArchiMate-compatible modelling tools – thus enabling easy and widespread adoption. Other proposals have used the same standards-based modelling approach, namely for modelling ITIL [18], ITIL process assessments [19], as well as security and risk concepts [20].

3.4   Research Problem and Proposed Solution

Since 2014, the authors have been engaged in EGIT initiatives using COBIT5 that required assessing the capability levels of COBIT5 enabling processes. The practical experience thus gained confirmed, as stated in the COBIT5 framework, that EA can be useful for EGIT initiatives – when compared with the exclusive use of spreadsheet-like artifacts. Indeed, the lack of up-to-date organizational self-awareness capabilities and interactive synchronization tools forces the assessment stakeholders to engage a wasteful amount of resources, especially during the evidence collection and evidence validation iterations. Besides the efficiency penalty, resource-constrained initiatives will also be ineffective, because fewer resources will be made available for performing the actual assessment, as well as other auditing and consulting activities such as documenting exceptions and gaps, communicating the assessment conclusions, and optimizing recommendations and roadmaps [16].

However, the current TOGAF standard does not provide specific architectural building blocks for COBIT5.

Also, existing EA tool vendors' COBIT5 templates present interoperability and adoption barriers, because they are either proprietary or lack scientific validation.

Finally, the COBIT5 documentation does not provide such formal EA artifacts, instead relying on natural languages descriptions, tables, and informal diagrams. To address these EA capability gaps, in this paper we sought to answer the question of whether the use of standard ArchiMate models may facilitate COBIT5 process assessments, as well as provide a set of EA artifacts that may be easily adopted in EGIT process assessment initiatives.

4 Design and Development

Since COBIT5 is a significantly large holistic body of knowledge, we took advantage of the iterative nature of DSRM to split the problem complexity in several stages, as presented in Table 2.

Table 2. Iterative DSRM design and testing plan, showing the increasing design scope, as well as the alternating demonstration settings, evaluation strategies, and evaluator groups.

DSRM iteration	Scope (PAM)	Demonstration setting	Evaluation strategy	Evaluator profile
1st	Capability level 1 (performance)	Academia	*ex ante*	EA academic
2nd	Capability level 1 (performance)	Large public organization (military)	*ex ante*	EA practitioner
3rd	Capability levels 1–5 (all levels)	Large public organization (non-military)	*ex post*	EA practitioner

Using this agile approach, we got frequent feedback that allowed for improving the proposals, controlling the risks, gradually increasing the design scope, as well as alternating between *ex ante/ex post* demonstrations and academic/practitioner evaluations.

The proposed viewpoints (i.e. designed in the 3rd DSRM iteration) are presented in Figs. 3, 4, 5, and 6. To support COBIT5 process assessment initiatives, the proposed artifacts cover the following areas:

- Goals cascade (Fig. 3): entities relevant for translating high-level stakeholder needs and enterprise goals into manageable, specific, IT-related goals, as well as mapping these to specific governance processes.

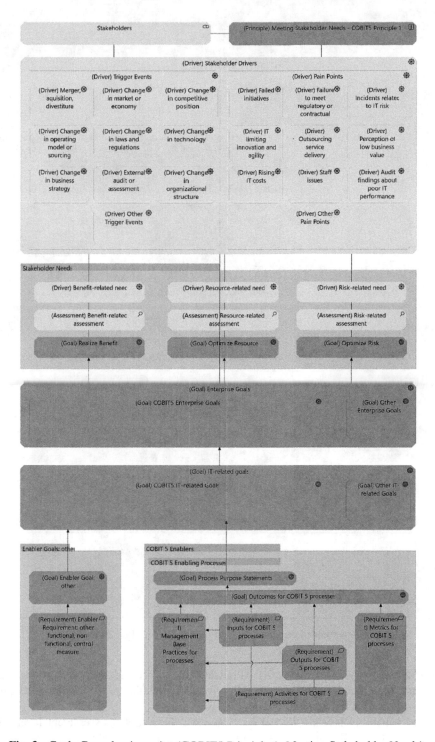

Fig. 3. Goals Cascade viewpoint (COBIT5 Principle 1: Meeting Stakeholder Needs).

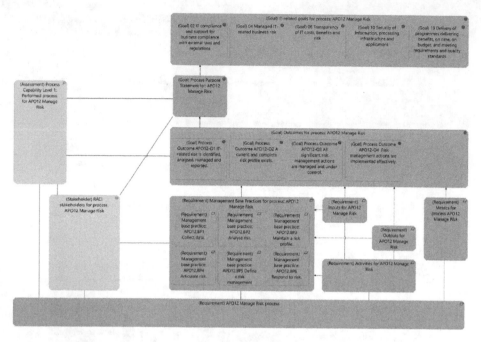

Fig. 4. Viewpoint for process capability level 1 (for APO12 manage risk).

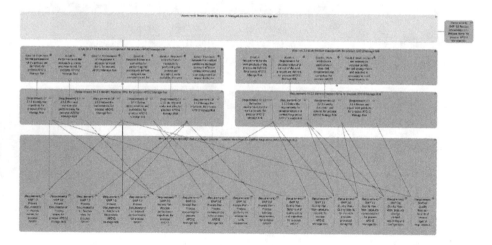

Fig. 5. Viewpoint for process capability level 2 (for APO12 manage risk).

- PAM capability level 1 (Fig. 4): IT-related goals, process purpose, process outcomes, base practices and corresponding RACI stakeholders, work products (inputs and outputs), activities, and metrics;
- PAM capability levels 2 to 5 (Figs. 5 and 6): these higher assessment levels are structurally similar, and include the entities: process attributes (PA), generic

practices (GP), and generic work products (GWP). Figure 6 illustrates the structural similarities of the viewpoints for the process capability levels 2, 3, 4, and 5. Also, note the structural differences between the viewpoint for capability level 1 (Fig. 4) and the viewpoints for higher capability levels (Figs. 5 and 6).

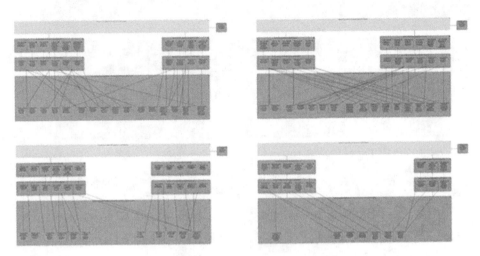

Fig. 6. Structural similarities of the viewpoints for process capability levels 2, 3, 4, and 5 (for APO12 manage risk).

Since the COBIT5 PAM [12] prescribes a fundamental assessment structure that is process-independent, in this paper we show the artifacts for just one such process (APO12 Manage Risk). However, during the demonstration sessions, we modelled several different enabling processes, precisely to illustrate their structural similarity.

For designing the viewpoints, we have used the following design criteria:

- As stated in the ArchiMate specification, motivational elements are related to the core elements via the requirement or constraint concepts.
- **COBIT5 concepts** are modelled using the **ArchiMate Motivation Extension**.
- **Assessment evidence** entities are modelled using **core ArchiMate concepts** (see the demonstration artifact in Fig. 7, yellow and blue graphical elements).

These design criteria help to understand and represent COBIT5 as a holistic body of principles, goals, and requirements for best practice (thus modelled using motivational concepts), that actual IT implementations (modelled using core concepts) should be aligned with.

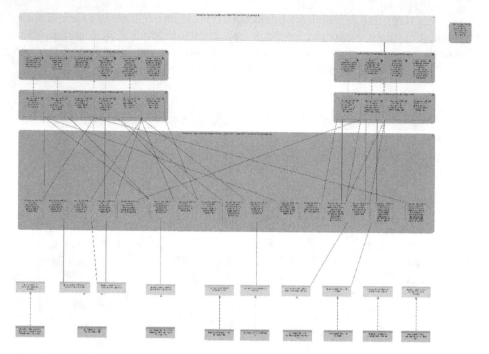

Fig. 7. *Ex post* demonstration: the assessment evidence – i.e. business objects (yellow elements) and data objects (blue elements) – are linked to the COBIT5 assessment requirements (COBIT5 generic work products). (Color figure online)

5   Demonstration

As shown in Table 2, we performed demonstrations with academic specialists in EGIT and EA (1st DSRM iteration, academic setting), as well as with practitioners from two large public sector organizations (2nd and 3rd DSRM iterations). When the demonstrations were performed, these two large organizations had low EA maturity levels, and were actively pursuing enhanced EA capabilities. The evaluators were engaged in EGIT and EA education, training, as well as procurement of advanced EA tools. In the 2nd DSRM iteration, ten evaluators from a military IS/IT department provided the formal (questionnaires) and informal feedback (group sessions). The final (3rd) DSRM evaluation was performed by four information systems professionals in a large (non-military) public organization. In this last iteration, the demonstration included hands-on modelling sessions using an open-source ArchiMate modelling tool.

Figure 7 presents one of the artifacts that was used for demonstration purposes. This *ex post* example is useful for illustrating how to use these artifacts may be used in practice, as well as their utility for assisting COBIT5 process assessments:

- How to use the model in practice: the procedure for constructing the demonstration artifact in Fig. 7 was the following:

(a) The viewpoint for process capability 2 (see Fig. 5) was copied into the EA repository.
(b) The relevant **assessment evidence** was included in the viewpoint, using core ArchiMate constructs, which for this real-world scenario were represented as business objects (yellow elements) and data objects (blue elements).
(c) Relationships (links) were created, connecting **assessment evidence** elements (business and data objects, in yellow and blue) to the corresponding **COBIT5 assessment requirements** (generic work products, lavender elements).

- Utility: using this viewpoint, it becomes trivial to see the influence of each evidence element (yellow and blue elements), all the way up to the assessment result (pink element), through the complex web composed of process attributes, attribute goals, generic practices, and generic work products (lavender elements). This complex web would be difficult to grasp from the COBIT5 PAM documentation alone – which is based on tables and natural language descriptions.

More importantly, using EA artifacts, these information entities and their relationships may easily become part of an integrated, managed, and thus consistent repository. This would not be necessarily the case, if spreadsheet-like artifacts were used for recording the same informational entities.

6   Evaluation

The 1st DSRM evaluation was performed by academic IS/EA expert evaluators, in the context of a graduate's thesis, mainly to validate the prototype artifacts, ensure scientific correctness, as well provide as state-of-the-art guidance regarding design and testing methodologies. For the 2nd and 3rd DSRM iterations, we used demonstrations (as described in the previous section), as well as formal evaluation questionnaires based on the questions presented in Table 1. Group sessions were also conducted to elicit informal and spontaneous feedback, and help make sense of the formal feedback.

Complementing the questions presented in Table 1 – which were used for evaluating the ArchiMate artifacts' usefulness – a set of three additional claims was included in the questionnaires:

- EA facilitates Assessments;
- The Assessment criteria should be included in the architectural representations;
- ArchiMate is useful for providing architectural diagrams.

For rating the agreement level with each of the above claims, we used the same options that were used to evaluate the artifact's usefulness: "strongly disagree", "disagree", "agree", and "strongly agree". The questionnaire results are presented in Fig. 8 (for the 3rd, i.e. final, DSRM iteration).

Regarding the evaluator's ratings for the usefulness of EA and ArchiMate (left chart on Fig. 8), note that the results for EA are better than those for ArchiMate. This fact is consistent with the ratings regarding homomorphism (right charts), as well as with the informal feedback received during the group sessions, where the evaluators expressed that the ontological fit between COBIT5 concepts and ArchiMate constructs

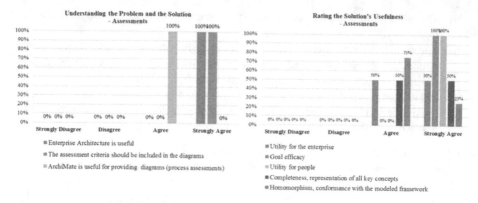

Fig. 8. Evaluation results for the 3ʳᵈ DSRM iteration.

is not perfect. These evaluation results confirm that there are ontological issues, with practical relevance, when trying to map COBIT5 concepts and ArchiMate concepts.

During the informal groups sessions, the evaluators highlighted the following benefits of using the COBIT5 EA model for assessing IT governance processes:

- The visual representations help reduce the complexity and ambiguity of the communication activities required for providing an assessment;
- The diagrams help to illustrate and simplify communication between diverse business areas, as well as diverse management levels;
- COBIT5 concepts are communicated in a simple, structured, and intuitive manner.

Overall, the results are consistent with the hypothesis that EA models facilitate COBIT5 process assessments. However, the evaluators also commented that sufficient education and training must be provided to all EA stakeholders, to ensure successful usage and maintenance of the EA repositories. Also, leadership and commitment from top management are required for EA to reach adequate maturity levels.

7 Conclusion

In this paper, we have demonstrated and evaluated EA models for assisting COBIT5 process assessments, using three iterations of a DSRM process model. These artifacts may be integrated in enterprise architecture (EA) repositories, and reused in real-world assessment initiatives. The main contributions of this research are:

- The validation of the hypothesis that EA models may be used to facilitate COBIT5 process assessments, by improving the effectiveness of the assessment activities;
- A set of COBIT5 viewpoints based on standard ArchiMate constructs, that:
  - may be reused in real-world assessments;
  - may be further developed in future research (e.g. by performing additional DSRM iterations, in adequate demonstration and evaluation settings);
  - address interoperability and adoption barriers, such as those found in proprietary and non-standard solutions.

Regarding limitations, note that the ArchiMate viewpoints were demonstrated and evaluated in large public organizations with relatively low EA maturity levels. In different settings, the results and conclusions may be different. However, in organizations with higher EA maturity levels, one may expect that enhanced EA expertise will provide even better leverage for assisting COBIT5 initiatives – such as process assessments.

Also, the authors do not claim that the proposed artifacts represent the best way to model COBIT5 concepts, but simply that it is possible to produce effective ArchiMate models for assisting COBIT5 process assessment activities. In fact, some evaluators commented that the ontological fit between COBIT5 concepts and ArchiMate language constructs is not perfect. Future work may address these ontological mapping deficiencies, namely to provide insights and recommendations for helping develop the next versions of COBIT and ArchiMate. Future work may also address different settings e.g. private enterprises, small and medium enterprises, as well as multi-enterprise settings such as public-private partnerships.

COBIT5 guidance promotes EA as instrumental to improve business-IT alignment. The authors hope that this work will facilitate and encourage more widespread adoption of EA among COBIT5 practitioners.

References

1. ISACA, COBIT 5: A Business Framework for the Governance and Management of Enterprise IT. ISACA, Rolling Meadows (2012)
2. Van Grembergen, W., De Haes, S.: Enterprise Governance of IT: Achieving Strategic Alignment and Value. Springer Science+Business Media, New York (2009)
3. ISACA. COBIT 5 Implementation. ISACA, Rolling Meadows (2012)
4. Tribolet, J., Pombinho, J., Aveiro, D.: Organizational self-awareness: a matter of value. In: Magalhães, R. (ed.) Organization Design and Engineering: Coexistence, Cooperation or Integration. Palgrave Macmillan, Basingstoke (2014)
5. Greefhorst, D., Proper, E.: Architecture Principles: The Cornerstones of Enterprise Architecture. Springer, Heidelberg (2011)
6. Cameron, B., Malik, N.: A Common Perspective on Enterprise Architecture. In: The Federation of Enterprise Architecture Professional Organizations (FEAPO) (2013)
7. Lankhorst, M.: Enterprise Architecture at Work: Modelling, Communication and Analysis, 3rd edn. Springer, Heidelberg (2013)
8. ISACA. COBIT 5 for Assurance. ISACA, Rolling Meadows (2013)
9. NIST. Joint Task Force Transformation Initiative - Managing Information Security Risk: Organization, Mission, and Information System View, NIST Special Publication 800-39. NIST - National Institute of Standards and Technology (2011)
10. Uhl, A., Gollenia, L.A.: Business Transformation Essentials, Gower (2013)
11. Uhl, A., Gollenia, L.A.: Business Transformation Management Methodology, Gower (2012)
12. ISACA. COBIT Process Assessment Model (PAM): Using COBIT 5. ISACA, Rolling Meadows (2013)
13. Peffers, K., Tuunanen, T., Rothenberger, M.A., Chatterjee, S.: A design science research methodology for information systems research. J. Manag. Inf. Syst. **24**(3), 45–77 (2007)
14. ISACA. COBIT 5: Enabling Processes. ISACA, Rolling Meadows (2012)

15. The Open Group. TOGAF Version 9.1 (2011)
16. ISACA, COBIT Assessor Guide: Using COBIT 5, Rolling Meadows, IL, USA: ISACA, 2013
17. Estrem, W., Gonzalez, S., Thorn, S.: A Practitioner's Guide to Using the TOGAF Framework and the ArchiMate Language, The Open Group (2014)
18. Vicente, M., Gama, N., da Silva, M.M.: Using ArchiMate to represent ITIL metamodel. In: CBI 2013 Proceedings of the 2013 IEEE 15th Conference on Business Informatics, pp. 270–275. IEEE Computer Society, Washington, D.C. (2013)
19. Silva, N., da Silva, M.M., Barafort, B.., Vicente, M., Sousa, P.: Using ArchiMate to model a process assessment framework. In: SAC 2015 Proceedings of the 30th Annual ACM Symposium on Applied Computing, pp. 1189–1194. ACM, New York (2015)
20. Wierda, G.: Mastering ArchiMate, 2nd edn. R&A, Amsterdam (2014)
21. Hevner, A., Chatterjee, S.: Design Research in Information Systems. Springer, New York (2010)
22. Hevner, A., March, S., Park, J., Ram, S.: Design science in information systems research. MIS Q. **28**(1), 75–105 (2004)
23. Prat, N., Comyn-Wattiau, I., Akoka, J.: Artifact evaluation in information systems design-science research - a holistic view. In: PACIS 2014 Proceedings, Paper 23 (2014)

The Influence of Shadow IT Systems on Enterprise Architecture Management Concerns

Melanie Huber[1]([✉]), Stephan Zimmermann[2], Christopher Rentrop[1], and Carsten Felden[3]

[1] Konstanz University of Applied Sciences, Konstanz, Germany
{melanie.huber, christopher.rentrop}@htwg-konstanz.de
[2] BITCO³ GmbH, Konstanz, Germany
stephan.zimmermann@bitco3.com
[3] TU Bergakademie Freiberg, Freiberg, Germany
carsten.felden@bwl.tu-freiberg.de

Abstract. Shadow information technology systems (SITS) coexist with formal enterprise systems in organisations. SITS pose risks but also increase flexibility of business units. Practice shows that SITS emerge, despite that Enterprise Architecture Management (EAM) aims at controling all IT systems in an organization. Studies acknowledge this problem in general. However, they neither show the specific influencing areas of SITS nor provide approaches to address them. To close this gap, we use a literature review to analyse examples of practical SITS and their interference with EAM concerns. Thus, we find that they hinder especially transparency, reduction of EA complexity and governance. Research has focused on achieving transparency, governing the evolution of the EA but lacks strategies for reducing complexity. This study contributes to research and practice by uncovering the main influencing areas of SITS on EAM, as well as by laying a foundation for future research on this topic.

Keywords: Shadow IT systems · Shadow IT · Enterprise architecture

1 Introduction

Three-fourths of Information Technology (IT) deciders acknowledge in surveys that shadow IT systems (SITS) support business processes in their organisation [1]. SITS are "deployed autonomously within business departments" [2]. They are "involved neither technically nor strategically in the IT service management of the organization" [2]. SITS coexist with formal IT systems provided by the organisations [3, 4]. Business units implement them in situations of misalignment of business and IT [2, 5] to overcome limitations of formal systems [6] or fill gaps in their IT landscape [7]. Thereby, e.g., business units develop solutions based on end user computing platforms [8], install applications locally [9], or employ cloud services [10]. The use of SITS in addition to formal IT contributes to the already increasing complexity of IT landscapes [6, 11]. Business unit often do not consider the existing enterprise architecture

© Springer International Publishing AG 2017
M. Themistocleous and V. Morabito (Eds.): EMCIS 2017, LNBIP 299, pp. 461–477, 2017.
DOI: 10.1007/978-3-319-65930-5_37

(EA) [12], which leads to an uncontrolled development of the IT landscape. EA Management (EAM) aims at controlling this development and improving the business–IT alignment [13]. EAM provides methods to deal with complexity [14]. and enables organisations to govern the evolution of their IT [11], as well as their business landscape [15]. Buckl et al. [16] and Khosroshahi et al. [17] identified typical EAM stakeholder concerns, which serve as entry points for EAM activities [18]. While one would assume that an existing EAM controls all IT systems in an organization, case studies show that SITS nevertheless exist and are often not part of EAM tasks [19]. Studies acknowledge a general influence of SITS on EAM but do not show how they interfere with individual EAM tasks. Therefore, this paper aims to systemise this influence on the level of granular EAM concerns and current research focuses. This study may be used as a basis for future research directions.

Research recognises that SITS influence the EA in general [7, 20, 21]. Some studies describe specific effects on EA complexity, maintenance costs, or possible break-downs [6] as users do not consider the EA when implementing SITS [12, 22]. Research has also demonstrated that SITS affect EAM success [19]. Integrating them with other elements of the EA is challenging [23, 24]. SITS influence the EA and vice versa. Changes in the EA can lead to new SITS and new enterprise values can result in perceiving SITS as an opportunity [19]. Studies describe these individual effects briefly. However, no research has focused explicitly on these influences and provided a systematic overview of the various effects that SITS have on the EA. In a structured literature analysis we identify examples of SITS and map them to their effect on EAM concerns. The examples shown prevent transparency, reduction of complexity and IT governance. Research has focused on transparency and governance but lacks approaches in reducing unnecessary EA complexity. This illustration contributes to the discourse in EAM by showing research focuses. It additionally highlights the importance of including SITS in EAM activities in practice.

2 Enterprise Architecture Management and Shadow IT Systems

EA is a concept in information system research and practice describing the basic structure of an organisation [15]. It comprises business and technology elements, such as data, hardware, software [25], processes, strategic initiatives, as well as their interrelations [26]. EAM is the maintenance and planned development of the EA [15] and includes the management tasks of identification, analysis, planning, implementation, and control [11, 27]. EAM is a complex topic that research investigates from different perspectives such as EA standards [28] or success factors [14]. A proceeding approach fragments EAM into granular EAM concerns [18], and further EAM studies base their analyses on this approach [29–31]. Concerns in general represent "aspects that are critical […] to one or more stakeholders" [32]. EAM concerns represent entry points for management activities [18] and can thus be regarded as initial steps in achieving EAM goals [33]. These EA concerns are identified in cooperation with stakeholders from various organizations [16, 17], which is crucial given that stakeholders highly influence EAM realization [11]. Therefore, the identified concerns are useful to organize the complex topic of EAM.

Alongside formal systems, other systems support business processes in the EA. Business departments and users deploy these so-called SITS autonomously in their business processes for individual as well as collective use and, although part of the EA, they are not involved in the IT service management (ITSM) of the organization [2]. SITS can emerge in the form of local applications [9], spreadsheets [8], end devices [34], cloud services [10, 35], databases or combined solutions [36]. Given that SITS are not part of related ITSM processes, implementation and maintenance is often less professional [2, 37]. This exclusion poses a risk for disrupting business processes [5] and causes inefficiencies [34, 38]. However, SITS support users to increase performance and are innovative and flexible [20, 21].

Besides being excluded from ITSM tasks, organizations often exclude SITS from EAM [2, 19]. This is problematic because SITS can be an important part of it [6, 7]. Often, SITS do not fit [21] and users don't consider the EA when implementing SITS [7, 12]. Such disregard can increase EA complexity, maintenance costs and shape its evolution [6]. Excluding SITS from EAM means not benefiting from their innovative potential and revealed user needs [19]. As a result, identifying and controlling SITS can decrease the mentioned problems and converts them into business-located IT [20, 39]. Yet, integrating these now transparent systems with other EA elements can still create conflicts with current technologies and standards [23].

Based on the described literature, SITS, as part of the EA, influence its current state and evolution. However, most research only mentions individual effects briefly [7, 12, 22, 23], determines the relevance of SITS for the overall EA [6] or relegates to the handling of SITS as part of EA policies [19]. Prior studies neither systemize the variety of influences of SITS on EAM nor provide methods to address them. Such an overview highlights research focuses as well as gaps and opens a new discussion on SITS from an EAM point of view. Therefore, we pose the following research question: *How does existing literature address the influences of SITS on EAM?*

3   Research Method

To answer our research question, we conducted a literature review, based on the established procedure of Cooper [40], a common method in EAM [27, 41] and appropriate as SITS research used it previously to find real SITS examples in literature [2]. Examples describe the phenomenon in an organizational context [42] from a technical and a business perspective. Including all information given in their description, e.g. details on their emergence, evolution and interaction, we use them to deviate the influences of SITS on EAM concerns. The identified influences serve as a basis to discuss existing methods addressing them. Hereafter, we explain our method, consisting of the search for examples, as well as their analysis based on concerns.

3.1 Literature Search and Results

We conducted a literature search to identify relevant literature in the research field of SITS in February 2017[1], including all preceding years. We documented this search rigorously to increase the transparency of our procedure [43], illustrated in Table 1. In the search process, we removed duplicate and irrelevant publications that did not deal with the phenomenon of SITS and identified examples of SITS in the full text[2]. We conducted a backward and forward search [44] to complement existing examples and find additional publications not situated in the field of SITS itself. To include an example, the publication where it appeared had to mention it in the context of the term SITS or a synonym. Additionally, the definition that the authors used had to correspond to the SITS definition, given in the prior chapter. At last, the given information had to be enough to be able to map it to at least one EAM concern.

Table 1. Literature search

Found publications	Screened publications	Relevant publications	Publications with examples	Number of examples	Number of unique examples
817 (43 Duplicates)	774	105	45	113 (20 Duplicates)	93

3.2 Literature Analysis

As a result, we identified 93 examples of SITS. We used the concept matrix of Webster and Watson [44] to analyze them. This is appropriate because it focuses on concepts rather than authors and we wanted to draw general conclusions for the examples throughout the publications. However, our concepts did not emerge from the examples. We used EAM concerns to structure the topic of EAM. As explained in the previous section, they cover various important aspects of EAM and are a useful basis for analyzing the influence of SITS on granular EAM activities. We analyzed the 43 EAM concerns included in EAM Pattern Catalogue version 1 [16] and the additional 18 of version 2 [17], which various organizations and EAM stakeholders identified. This leads to a broad representation of 63 important aspects of EAM and a useful basis for analyzing the influence of SITS on granular EAM activities.

---

[1] To identify relevant literature, we based on existing literature [2] and detected keywords and databases that have proven relevant in the context of SITS. As a result, we conducted a full-text literature search using the keywords (*shadow/feral*) *systems*, (*shadow/grey/hidden/rogue*) *IT* joined with *information* (*technology/services/systems/security*) and querying several academic databases (Ebscohost, ScienceDirect, ProQuest, IEEE Xplore, ACM Digital Library, AISeL and Jstor), IS journals and proceedings of established IS conferences.

[2] Applying the four-eye principle we removed publications by screening the title, abstract and keywords that did not fit our topic: Some use the term *SITS* in a different context, e.g. in the field of databases. Other examples are publications, where *it* does not mean information technology but rather just the term *it*.

The pattern catalogue version 1 assigns each of the concerns to one of seven EAM topics [16]. We maintained this classification to structure our analysis and assigned the new 18 concerns to the EAM topics based on their description [17]. We then adapted the concept matrix (Appendix A) and used a qualitative approach for analysis. We examined the effect of each example on each concern. Once we found an effect, we noticed the relevant text part of the example. For further analysis, we used open coding [45] to find similar patterns in the effect. Using axial coding we clustered the effects that the SITS had on the concerns to derive general conclusions. One researcher conducted this analysis, consulting a second author to apply a four-eye principle and ensure the appropriateness of the procedure. We then discussed the results with the other authors to enhance the validity and scientific correctness. These steps led to a structured representation of the influence of SITS on EAM concerns.

4   Findings

We mapped 43 of 61 concerns to the identified SITS examples. The following describes our findings regarding the seven EAM topics. Each section clusters and discusses influences of the examples on the mapped concerns.

4.1   Technology Homogeneity

The first topic that the EAM Pattern Catalogue version 1 mentions is *Technology Homogeneity* (TH). In this area, the main focus is "a homogeneous set of technologies and architectures" [16]. In total, it contains 12 concerns. The examples of SITS influence eleven concerns in this topic (Table 2, based on [16, 17]).

Table 2. Technology Homogeneity concerns

Concern	Concern description excerpt	Examples
C-2	Architectural blueprints/standards used? Areas of breaching?	48
C-4	Replace/keep which technologies are used in the IT landscape?	13
C-5	Activities (reduce heterogeneity) to increase conformance?	35
C-8	Reduce usage of individual software, replace with standard	53
C-9	Reorganize the IT landscape with respect to used technologies	4
C-19	Applications correspond to architectural blueprints/solutions/standards?	37
C-46	Available knowledge about specific subjects?	3
C-100	Individual and standard software in the application landscape	78
C-101	Activities (modify applications) to improve conformance?	35
C-124	Reduce application landscape complexity	10
C-141	Obtain transparency about IT costs	5

Numerous SITS influence the transparency of the use of individual and standard software (C-100). The minority of these examples are standard software such as known solutions [34, 46] or undescribed external software [39, 47, 48]. In contrast, a majority of the examples are self-developed systems [6, 34, 49–52] or explicitly described as end user computing (EUC) [53–55]. These examples also influence the concern to reduce individual software (C-8).

Besides transparency, many SITS influence the area of architectural standards. More than half of the examples show areas where standards are breached (C-2), and many provide reasons for this breach (C-19), which might help improve EA conformance (C-101, C-5). These reasons are mainly deficient formal systems [49, 53] caused by a lack of functionality [37, 56–58] or inefficiencies [47, 53, 59].

Organizations include some identified SITS into their decisions and they influence the technologies in the IT landscape (C-4). In several cases, formal systems thereby replaced business-located IT [60]. Other examples describe decommission [61], the rebuilding or keeping of former SITS [20] or the decision against new technology [56]. One example [22] led to new knowledge about specific technology (C-46).

Other examples mention SITS increasing the complexity of the EA [6, 50] by addressing a high number of users [61] or by introducing parallel instances [38]. Their control could therefore help reduce complexity (C-124), which also enables better cost control. Some of the detected examples can lead to a reorganization of the EA with regard to technology and especially cost reduction (C-9) and transparency (C-141) [5, 61]. The analyzed SITS do not describe an influence on the creation of an architectural blueprint or solution (C-50) as the latter seems to be independent from it.

4.2 Business Processes

The second topic in the EAM Pattern Catalogue is the area of *Business Processes* (BP). It focuses on "analyzing the interaction of business applications, business processes, and related entities" [16]. We did not find any SITS influencing the outsourcing of business processes (C-55), the definition of core competencies (C-56) or whether the processes consider the organizational environment (C-54). However, SITS might have an indirect influence on these decisions not specifically captured in our examples.

4.3 Applications Landscape Planning

Application Landscape Planning (ALP) is the third management topic of EAM concerns. It contains 21 concerns that target the planning and analysis of "the structure and evolution of the application landscape, focusing on current, planned, and target landscapes" [16]. We could map 14 to our examples (Table 3 based on [16, 17]).

Similar to TH, unidentified SITS in this topic influence transparency and prevent a complete picture of the EA. All examples influence the determination of the current EA (C-35) and the prediction of its development (C-88). Several examples report supported business processes (C-87) and different business units using or supporting the application (C-33, C-172). This information stays unknown in the case of SITS. A few

Table 3. Application Landscape Planning concerns

Concern	Concern description excerpt	Examples
C-33	Which applications are used by which organizational units?	38
C-35	Look of the application landscape at a specific date?	93
C-36	Dependencies between applications and effect by projects?	16
C-44	Reduce operating expenses/maintenance costs (redundancy)?	38
C-86	Business applications hosted by which organizational unit?	7
C-87	Which application supports which business processes?	65
C-88	How will the application landscape evolve over time? Differences?	93
C-89	Business applications that will be affected by projects soon?	14
C-90	Phase of lifecycle of an application at a certain time point?	4
C-110	Analyze failure propagation of application landscape	16
C-119	Definition of target application landscape	19
C-127	Integrate business application in application landscape	8
C-157	Detection of consolidation potentials	35

examples also influence knowledge about the hosting of applications (C-86). Business units conduct this by themselves [37], or source externally [7]. Additionally, four examples describe the effect that SITS have on determining their lifecycle phase (C-90) as they would either disappear or a new system would arise [61].

Besides elements in the EA, some SITS preclude transparency of the dependencies between applications (C-36) and about the failure propagation (C-110) that depends on them. The examples describe systems generally connected to numerous elements in the IT landscape [6, 62]. Others mention the upload of data into [37] or the download of data from the enterprise system (ES) into EUC tools [39, 61, 63, 64].

Various examples additionally influence the detection of consolidation potential in the IT landscape (C-157) to reduce costs (C-44). We found systems that replicate ES [7, 37, 61, 65], use the same data [47, 53], or replicate functionality [7, 62, 66].

Few SITS also influence the target state of the IT landscape (C-119). Technological changes are included in C-4 in the last section. Further examples show organizational changes as detected SITS move to the central IT unit [20, 37, 61] or were reengineered [34] by considering an EA integration (C-127).

Finally, some examples influence applications in projects (C-89). SITS emerge as a result of the implementation of formal systems [67], or identified SITS are the target of IT replacement projects [7, 51]. Projects decommission, reengineer [68] detected SITS, or embed them in ITSM [39]. We did not find examples that influence the long-term vision of the application landscape (C-34), the visualization of an application (C-108), the evaluation of the alignment between the landscape and the business model (C-122), or the assessment of change requests (C-169). However, these decisions might indirectly depend on SITS. Furthermore, no example influences the removal of monolithic applications (C-12) as SITS are rarely monolithic. Additionally, we mapped no example to the influence on the IT landscapes merger (C-147) as our dataset does not include such cases.

4.4   Support of Business Processes

The fourth topic of the Pattern Catalogue is *Support of Business Processes* (SBP). It analyses, "how a specific business process is supported by IT" [16]. We found a match in the examples for all of the five concerns (Table 4, based on [16, 17]).

Table 4. Support of Business Processes concerns

Concern	Concern description excerpt	Examples
C-78	Support of processes by applications? Extent of manual support?	66
C-80	IT supports flexibility of processes? Flexibility put at risk?	13
C-132	Evaluate business capabilities on strategy conformity	27
C-142	Map business applications to business capabilities	73
C-171	Which business capabilities are supported by applications?	70

As already mentioned in the topics of TH and ALP, the examples have a strong negative influence on the transparency of the application landscape itself. The missing transparency also very often influences the mapping of applications to capabilities (C-142) and capabilities to applications (C-171).

Several SITS prevent the concern of extending an automated IT support (C-78). These examples influence the mapping to the processes that they support. Some of them prevent an automated support as they describe a manual [53, 61, 64, 69, 70] or partially manual [62] support. However, in some cases, SITS also positively extend the automated support [37, 71].

Besides, some of the examples prevent the evaluation of capabilities to strategy conformity (C-132). Capabilities that mostly do not conform to strategy are data collection and reporting [49, 51, 53, 56, 60, 70].

Other SITS examples preclude the identification of flexibility potential in the EA (C-80). Yet, they influence this potential positively as they are quick [9], and flexible [7, 38]. We did not find an example influencing the realization of a more continuous IT support of the business processes (C-95), although an influence might be possible.

4.5   Project Portfolio Management

The next management topic of the Pattern Catalogue is *Project Portfolio Management* (PPM). It includes concerns dealing with "managing the portfolio of projects changing the application landscape" [16]. Our examples influence two of the given concerns (Table 5, based on [16, 17]).

Some of the detected SITS lead to projects for a modification of the EA to align with needs defined by the underlying strategies (C-91). Organizations acknowledge this need [38], and moved SITS to the IT division [37, 61], or the IT unit developed a new system that fits their strategic requirements [20].

Table 5. Project Portfolio Management concerns

Concern	Concern description excerpt	Examples
C-91	Modification of IT landscape aligned to needs and specified by strategies Consider environment of the organization	12
C-179	Determine regulatory issues	2

Few examples influence regulatory issues (C-179) in which SITS pose a risk to the regulatory environment such as in the banking sector [68]. We did not find SITS that influence the explicit assignment of IT budget to IT projects (C-29) as they are excluded from ITSM [2] and therefore from this activity. However, the decision of business units or users to implement SITS affects the use of IT costs.

4.6   Infrastructure Management

The next management topic is *Infrastructure Management* (IM) which "analyses the technical infrastructure [...] and what impacts this infrastructure can have on the support of the business applications" [16]. We mapped both concerns to our examples (Table 6, based on [16, 17]).

Table 6. Infrastructure Management concerns

Concern	Concern description excerpt	Examples
C-41	Infrastructure software used by the business applications?	7
C-98	Effect of the shut-down of an infrastructure element?	7

Few of the examples influence transparency about infrastructure software (C-5) and therefore about the effect of their shutdown (C-98). The examples report the use of shadow databases [49, 53, 62], a shadow web server [2] or external hosting [39].

4.7   Interface, Business Object and Service Management

The final management topic is *Interface, Business Object and Service Management (IBOSM)*. It covers "services in the context of service-oriented architectures" [16]. The examples influence nine of 14 concerns (Table 7, based on [16, 17]).

As evident in the topics of TH, ALP and SBP, excluding SITS from the concerns leads to a non-transparent EA. This scenario is also important in terms of services (C-65, C-66), interfaces (C-67, C-70, C-99), type of interface (C-68) and business objects (C-61, C-51). We identified examples to upload into [37] or download data from [39] the ES, as well as interfaces between SITS [62]. We found examples with manual interface [62] and one automatic interface [53]. Some examples describe the

Table 7. Interface, Business Object and Service Management concerns

Concern	Concern description excerpt	Examples
C-51	Exchange/use of business objects by applications or services?	6
C-61	Which business objects are exchanged over which interfaces?	6
C-65	Which services are offered by which business application?	70
C-66	Which business processes are supported by which services?	67
C-67	Which interfaces are offered/used by which application?	9
C-68	Type of specific interface? Implementation/capabilities?	5
C-70	Effect of shut-down of interface on which applications?	9
C-71	How does the lifecycle of a service look like?	4

exchanged business objects [37] with the ES [37, 47] or between SITS [62]. Four examples also affect the definition of the lifecycle, described in ALP. We did not discover examples describing dependencies between business objects (C-52) and EAM concerns such as the communication of added value (C-128), the definition of services (C-64) and domains (C-62) because of their independence from SITS.

5 Discussion

In summary, our research confirms a general influence of SITS on EAM [2, 6, 19]. Furthermore, we identified the different influences. Based on our analysis across the described EAM topics, we now further cluster the influences into three main areas: transparency, complexity, and governance. We will summarize and discuss our findings with regard to existing research in these areas in the following sections.

5.1 Transparency

Systems that are still in the shadows lead to a non-transparent EA. The SITS in our examples are unmanaged [2] and therefore not part of EAM. As a result of the exclusion from EAM, SITS also hinder subsequent EA analyses [11, 27]. This limitation leads to examples of undetected compliance issues such as non-conformance to standards in numerous cases and the harm of regulatory demands in some. Besides, some examples preclude the analyses of flexibility that SITS provide and the knowledge of the business units about technologies inherent in SITS. Thus, organizations cannot benefit from these positive effects. Literature provides several methods to include SITS in EAM tasks. These methods aim to achieve a transparent EA including the formal systems and identified SITS [6, 20, 29, 72, 73]. Through the detection of SITS and their inclusion in EAM, organizations can conduct correct EA analyses and further manage the EA.

5.2 Complexity

The described examples confirm that SITS often preclude an important goal of EAM, namely, the reduction of unnecessary EA complexity [28]. Numerous examples increase EA complexity [6] as they are connected to or exist parallel to formal systems. Some of the SITS replicate data or functionality of formal systems [47], which suggests, that this redundancy is unnecessary. A number of examples do not fulfil the standardization and integration requirements of organizations [36]. They often do not confirm to architectural standards as business units use them instead of formal systems. As a result, some examples hinder automation. Given that the reduction of complexity is a major target of EAM, one expects a variety of approaches to reduce complexity that doesn't add value to the organization with regard to SITS. Yet, only few contributions have targeted this issue. One paper measures the role of business-located systems in the EA based on its centrality and derives management prioritization [6], others use standardized cloud applications [34]. The last paper that we found concerning this issue analyses the general integration potential of single SITS with ERP systems [47]. No approaches target the topics of standardization or automation within the research field of SITS.

5.3 Governance

After organizations identify SITS, they have to set responsibilities of now transparent business-located systems. In a number of examples, responsibility for these systems was retained at least partly in the business department. In other cases, organizations reengineered or turned them over to the central IT [39]. Thus, the examples influence IT flexibility and IT efficiency, which are major EAM goals [11]. Additionally, several of the examples harm the conformance to standards, regulatory requirements and the improvement of business–IT alignment [13]. Research focuses on solving these governance issues by providing methods to allocate responsibilities between business and IT for business-located IT [7, 20, 37, 39]. Therefore, they aim to preserve the flexibility of former SITS [20, 37] and balance the positive effects with existing problems. Methods first target a risk reduction [20, 72, 73]. Several control approaches follow regulatory requirements [34, 35] and regulatory risks of SITS [8] that might also influence outsourcing decisions [10]. Efficiency is another important focus to consider [2, 6, 10, 34] because the decision of business units to implement a system by themselves affects overall IT costs. Overall, balancing these measures contributes to an alignment of strategic goals [5, 10, 21].

6 Conclusion

To determine how existing literature addresses the influences of SITS on EAM, we identified and analyzed practical examples of SITS and their influence on various EAM concerns. They then served as a basis to discuss how research addresses these effects.

Thus, we can cluster our examples in three areas of influence: transparency, complexity and governance. Numerous approaches target the influence on transparency by providing methods to identify the EA. Less research has focused on decreasing unnecessary EA complexity. Research has examined the issue of governance, especially from the perspective of task allocation, risk and costs. Studies have also highlighted regulatory requirements and achieving business–IT alignment.

Practitioners can focus on the three main areas of influence when dealing with SITS in their EAM efforts. Subsequently, they can use the presented methods to address these areas and adapt them to their specific needs. Moreover, this study shows that organizations must consider SITS in their digitization efforts, as they often lack integration and automated tasks. Our study contributes to the scientific discussion on SITS as it explicitly confirms the relation of SITS and EAM. Moreover, we identified the influence and provided an overview about existing approaches.

However, there are some limitations to our study. We mapped the examples to the concerns based on the coding of their description that did not explicitly mention the concerns. Although we conducted the analysis methodically and applied a four-eyed principle, subjectivity might not be completely eliminated. Additionally, we chose EAM concerns as they cover a variety of influences that SITS have on EAM. Other approaches, such as EAM success factors might present a different perspective. Lastly, our analysis of the effect of SITS on EAM solely focuses on literature. Future research attempts with practical approaches would complement our findings.

Further methods can use our contribution and develop methods to deal with SITS in terms of EAM. First, future research attempts should focus on the reduction of unnecessary complexity that SITS cause. Thus, approaches for standardization and application integration could be a research focus. This increase in integrated and standardized applications would also benefit automation and digitization in general. Additionally, it can contribute to the reduction of costs and increase efficiency of EAM. Second, EAM measures in terms of SITS should always strive for a balance between positive and negative effects. Therefore, a holistic approach that considers risk, costs, flexibility and innovative potential should be the aim of research attempts. Third, an investigation of a subset of the examples, such as SITS for collective or personal use, cloud applications or EUC, could be interesting to point out their different influence on EAM. Fourth, the reasons for the breaching of standards of SITS could be an interesting starting point for analyzing how organizations can improve single systems. Therefore, they could further improve the whole EA.

Appendix A – Coding of Examples

			TH											ALP														SBP					PPM		IM		IBOSM								
			Concern*																																										
No.	Name	Reference	002	004	005	008	009	019	046	100	101	124	141	033	035	036	044	086	087	088	089	090	110	119	127	157	172	078	080	132	142	171	091	179	041	098	051	061	065	066	067	068	070	071	099
E1	Webfuse/MyInfocom	[5, 37]	X		X	X	X	X		X	X	X	X	X	X	X	X	X	X	X			X	X	X	X	X	X	X		X	X	X				X	X	X	X	X		X		X
E2	Pete's tracker	[3]													X					X											X	X							X	X					
E3	University Reporting	[38]			X	X		X		X	X	X		X	X		X		X	X						X		X	X		X	X	X						X	X					
E4	Tracking Goods	[55, 63]	X		X	X		X		X	X			X	X				X	X								X	X		X	X							X	X					
E5	Inventory	[55, 63]				X				X				X	X	X			X	X			X					X			X	X			X	X	X	X	X	X	X		X		X
E6	Contract System	[55, 63]	X		X	X		X		X	X			X	X	X			X	X			X					X		X	X	X							X	X	X	X	X		X
E7	Client Lists	[71]				X				X				X	X		X		X	X						X	X	X	X		X	X							X	X					
E8	Work Statistics	[71]												X	X				X	X							X	X			X	X							X	X					
E9	Blogs	[22]							X	X					X					X				X	X				X		X	X	X						X	X					
E10	Metrics Database	[53]	X		X	X	X	X		X	X		X		X	X	X		X	X			X			X		X		X	X	X					X	X	X		X	X	X		X
E11	Data Collection	[53]	X		X	X		X		X	X			X	X	X	X	X	X	X						X	X	X		X	X	X			X	X	X	X	X	X	X	X	X		X
E12	Data Mashing	[53]	X			X				X					X	X	X		X	X			X			X		X		X	X	X							X	X					
E13	Profit & Loss	[53]				X				X					X				X	X								X			X	X							X	X					
E14	Travel Expenses	[73]	X			X				X					X		X		X	X						X				X	X	X							X	X					
E15	Google Apps	[67]								X					X					X																									
E16	Human Resources	[67]								X					X					X							X																	
E17	Salesforce	[67]								X					X		X			X						X																		
E18	Dropbox	[34]	X		X			X		X	X				X				X	X				X				X			X	X	X						X	X					
E19	App Distribution	[34]				X				X		X		X	X				X	X				X				X			X	X	X						X	X					
E20	ProjectReporting	[51]	X	X	X	X		X		X	X			X	X		X		X	X	X			X		X		X		X	X	X							X	X					
E21	Bolt Ons	[51, 64]	X												X	X			X	X			X					X			X	X							X	X					
E22	Reporting	[60]	X	X	X	X		X		X	X				X		X		X	X	X			X		X		X		X	X								X	X					
E23	HR System	[48]								X					X					X							X																	
E24	BI Systems	[52]	X			X				X					X				X	X								X		X	X	X							X	X					
E25	CRM	[52]	X							X					X		X		X	X						X	X	X			X	X							X	X					
E26	Additional Systems	[68]	X	X	X	X		X		X	X				X					X	X			X																					
E27	PI Report	[59]			X			X			X				X		X		X	X						X		X			X	X							X	X					
E28	Data Warehouse	[62]			X	X		X		X	X	X			X				X	X							X	X			X	X							X	X					
E29	Reporting	[66]	X			X				X					X		X			X						X		X		X	X	X							X	X					
E30	Health Care Reporting	[66]				X				X					X				X	X								X			X	X							X	X					
E31	Data and Reporting	[57]	X		X	X		X		X	X				X				X	X								X			X	X							X	X					
E32	CSD	[56]	X	X	X	X		X		X	X			X	X				X	X	X			X				X		X	X	X							X	X					
E33	Consumer Device	[72]				X				X				X	X				X	X							X	X																	
E34	Budget Tracking	[54]	X			X				X					X		X		X	X						X		X		X	X	X							X	X					
E35	University Budget	[65]	X		X	X		X		X	X			X	X	X			X	X								X	X	X	X	X					X	X	X	X	X	X	X		X
E36	Reporting	[58]	X		X	X		X		X	X			X	X				X	X								X			X	X							X	X					
E37	Several Systems	[49]	X		X	X		X		X	X				X				X	X								X		X	X	X							X	X					
E38	Student System	[49]	X		X	X		X		X	X			X	X					X							X	X		X	X	X													
E39	Scholarship	[49]				X				X				X	X				X	X							X	X	X		X	X			X	X			X	X					
E40	Electronic Time Clock	[49]	X		X	X		X		X	X				X					X							X			X	X														
E41	Mobile Collection	[7]				X				X					X				X	X								X			X	X							X	X					
E42	Reagent Expiry	[7]				X				X					X				X	X								X			X	X							X	X					
E43	Access Database	[7]				X				X					X					X																X								
E44	Orthopedics Audit	[7]				X				X					X				X	X								X			X	X							X	X					
E45	Organizational Chart	[7]		X		X				X					X		X			X	X			X		X																		
E46	Moodle	[7]	X		X			X		X	X				X		X			X						X																		
E47	Unrestricted Access	[7]	X		X			X			X				X			X		X									X															
E48	Tour planning	[6]	X			X				X		X		X	X	X			X	X			X				X	X		X	X	X							X	X					
E49	RANO (ERP)	[6]	X		X	X		X		X	X			X	X					X												X							X	X					
E50	Dashboard	[6]													X				X	X								X			X	X							X	X					
E51	Product Data	[70]	X		X	X		X		X	X				X				X	X								X		X	X	X							X	X					
E52	Silo Applications	[50]	X		X	X		X		X	X	X			X					X																									
E53	Mobile Devices	[9]								X					X					X											X	X							X	X					
E54	Small Apps	[9]								X					X					X									X															
E55	Open Source	[9]								X					X					X																								
E56	Portable Skype	[9]	X							X					X					X																								
E57	Offer Creation	[47]	X			X				X				X	X		X		X	X						X		X			X	X			X	X			X	X					
E58	Online Product Portal	[47]	X											X	X		X		X	X						X		X			X								X	X					
E59	Inventory Control	[47]	X											X	X		X		X	X						X					X	X							X	X					
E60	Content Management	[47]				X				X				X	X				X	X								X			X	X							X	X					
E61	Editing	[47]				X				X				X	X				X	X								X			X	X							X	X					
E62	Web Application	[47]												X	X				X	X								X			X	X							X	X					
E63	Stamp Drawings	[47]				X				X				X	X	X	X		X	X			X			X		X			X	X					X	X	X	X	X		X		X
E64	Online Shop	[47]												X	X	X	X		X	X			X			X		X			X	X													
E65	Insurance Reporting	[47]	X		X	X		X		X	X			X	X	X	X		X	X			X			X		X			X	X							X	X					
E66	Order Web Application	[47]	X											X	X	X	X		X	X			X			X		X			X	X							X	X					
E67	ASP Archiving	[47]				X				X				X	X	X	X		X	X			X			X		X			X	X							X	X					
E68	Projects Languages	[47]												X	X				X	X								X			X	X							X	X					
E69	Pricing Tool	[47]												X	X		X		X	X						X		X			X	X							X	X					
E70	Social Media	[46]	X		X			X		X	X				X		X		X	X						X		X		X	X	X							X	X					
E71	Google Apps	[46]	X							X					X		X		X	X						X		X	X	X	X	X							X	X					
E72	Microsoft Cloud	[46]	X							X					X		X		X	X						X		X			X	X							X	X					
E73	WhatsApp #1	[46]	X		X			X		X	X				X		X		X	X						X		X		X	X	X							X	X					
E74	WhatsApp/Facebook	[46]	X		X			X		X	X				X		X		X	X						X		X			X	X							X	X					
E75	Skype	[46]	X		X			X		X	X				X		X		X	X						X		X		X	X	X							X	X					
E76	WhatsApp/Telegram	[46]	X		X			X		X	X				X		X		X	X						X		X			X	X							X	X					
E77	WhatsApp/Skype	[46]	X		X			X		X	X				X		X		X	X						X		X			X	X							X	X					
E78	WhatsApp #2	[46]	X		X			X		X	X				X		X		X	X						X		X			X	X							X	X					
E79	Pidgin	[46]	X		X			X		X	X				X		X		X	X						X		X		X	X	X							X	X					
E80	Salesforce and Others	[46]								X					X					X																								
E81	Mobile Devices	[46]								X					X				X	X								X		X	X	X							X	X					
E82	Customer Issues	[6]		X		X				X				X	X				X	X	X			X	X			X			X	X							X	X					
E83	Web Product Database	[6]		X		X				X				X	X			X	X	X	X			X	X		X	X			X	X	X		X	X									
E84	Small Solutions	[6]		X		X	X			X			X		X					X				X	X								X												
E85	Order Book Tool	[61, 69]	X		X	X		X	X	X	X			X	X		X		X	X	X	X		X	X		X	X	X	X	X	X	X	X					X	X				X	
E86	Servers	[61]	X												X		X	X		X																								
E87	Trading System	[61, 69]		X		X		X	X	X		X		X	X			X		X	X	X	X	X			X		X			X	X	X					X	X				X	
E88	Workflow System	[39]		X		X				X				X	X					X	X			X				X			X	X							X	X					
E89	Further Instances	[39]		X									X		X					X																								
E90	Order Processing	[39]		X		X				X		X		X	X			X	X	X	X		X	X	X		X			X	X	X	X		X	X			X	X					
E91	Event Management	[39]		X						X				X	X	X			X	X	X		X	X	X		X	X			X		X								X		X		X
E92	Board-Computer	[69]			X	X	X	X		X	X	X	X		X	X		X		X	X	X	X	X			X	X	X	X	X	X	X		X				X		X	X	X	X	X
E93	File Service	[69]	X			X		X		X		X			X		X	X		X	X	X		X			X			X	X								X					X	
			48	13	35	53	04	37	03	78	35	10	05	38	93	16	38	09	65	93	14	04	16	19	08	35	20	66	13	27	73	70	12	02	07	07	06	06	70	67	09	05	09	04	09

*No examples mapped to concerns C-29,C-34,C-50,C-52,C-54,C-55,C-56,C-62,C-64,C-92,C-95,C-108,C-120,C-122, C-128,C-129,C-147,C-169

References

1. NTT Communications: The People Vs The Ministry of No (2016). http://www.ministryofno.com/
2. Zimmermann, S., Rentrop, C.: On the emergence of shadow IT - a transaction cost-based approach. In: 22nd ECIS, pp. 1–17 (2014)
3. Behrens, S., Sedera, W.: Why do shadow systems exist after an ERP implementation? Lessons from a case study. In: 8th PACIS, Paper 136 (2004)
4. Silva, L., Fulk, H.K.: From disruptions to struggles: theorizing power in ERP implementation projects. Inf. Organ. **22**, 227–251 (2012)
5. Jones, D., Behrens, S., Jamieson, K., Tansley, E.: The rise and fall of a shadow system: lessons for enterprise system implementation. In: 15th ACIS, Paper 96 (2004)
6. Fürstenau, D., Rothe, H.: Shadow IT systems: discerning the good and the evil. In: 22nd ECIS, pp. 1–14 (2014)
7. Chua, C., Storey, V., Chen, L.: Central IT or shadow IT? Factors shaping users' decision to go rogue with IT. In: 35th ICIS, pp. 1–14 (2014)
8. Panko, R.R., Port, D.N.: End user computing: the dark matter (and dark energy) of corporate IT. In: 45th HICSS, pp. 4603–4612 (2012)
9. Silic, M., Back, A.: Shadow IT – a view from behind the curtain. Comput. Secur. **45**, 274–283 (2014)
10. Gozman, D., Willcocks, L.: Crocodiles in the regulatory swamp: navigating the dangers of outsourcing, SaaS and shadow IT. In: 36th ICIS, pp. 1–20 (2015)
11. Schmidt, C., Buxmann, P.: Outcomes and success factors of enterprise IT architecture management: empirical insight from the international financial services industry. Eur. J. Inf. Syst. **20**, 168–185 (2011)
12. Müller, S., Holm, S., Søndergaard, J.: Benefits of cloud computing: literature review in a maturity model perspective. Commun. Assoc. IS **37**, 851–878 (2015)
13. Luftman, J., Zadeh, H.S., Derksen, B., Santana, M., Rigoni, E.H., Huang, Z.D.: Key information technology and management issues 2011–2012: an international study. J. Inf. Technol. **27**, 198–212 (2012)
14. Lange, M., Mendling, J., Recker, J.: An empirical analysis of the factors and measures of enterprise architecture management success. EJIS **25**, 411–431 (2015)
15. Aier, S., Gleichauf, B., Winter, R.: Understanding enterprise architecture management design-an empirical analysis. In: 10th International Conference on Wirtschaftsinformatik, pp. 645–654 (2011)
16. Buckl, S., Ernst, A.M., Lankes, J., Matthes, F.: Enterprise Architecture Management Pattern Catalog. Release 1.0 (2008). http://sho.rtlink.de/EAMPatternCatalogV1
17. Khosroshahi, P.A., Hauder, M., Schneider, A.W., Matthes, F.: Enterprise Architecture Management Pattern Catalog. V. 2.0 (2016). http://sho.rtlink.de/EAMPatternCatalogV2
18. Buckl, S., Ernst, A.M., Lankes, J., Matthes, F., Schweda, C.M.: Enterprise architecture management patterns–exemplifying the approach. In: 12th IEEE EDOC (2008)
19. Tambo, T., Baekgaard, L.: Dilemmas in enterprise architecture research and practice from a perspective of feral information systems. In: 17th IEEE EDOCW, pp. 289–295 (2013)
20. Zimmermann, S., Rentrop, C., Felden, C.: Managing shadow IT instances – a method to control autonomous IT solutions in the business departments. In: 20th AMCIS (2014)
21. Györy, A., Cleven, A., Uebernickel, F., Brenner, W.: Exploring the shadows: IT governance approaches to user-driven innovation. In: 20th ECIS, Paper 222 (2012)
22. Shumarova, E., Swatman, P.A.: Informal eCollaboration channels: shedding light on "Shadow CIT". In: 21st BLED eConference eCollaboration, pp. 371–394 (2008)

23. Koch, H., Zhang, S., Giddens, L., Milic, N., Yan, K., Curry, P.: Consumerization and IT department conflict. In: 35th ICIS, pp. 1–15 (2014)
24. Hetzenecker, J., Sprenger, S., Kammerer, S., Amberg, M.: The unperceived boon and bane of cloud computing: end-user computing vs. integration. In: 18th AMCIS (2012)
25. Richardson, G.L., Jackson, B.M., Dickson, G.W.: A principles-based enterprise architecture: lessons from texaco and star enterprise. MIS Q. **14**, 385–403 (1990)
26. Bernard, S.A.: Using enterprise architecture to integrate business, technology, and business planning. J. Enterp. Archit. **2**, 11–28 (2006)
27. Simon, D., Fischbach, K., Schoder, D.: An exploration of enterprise architecture research. Commun. Assoc. Inf. Syst. **32**, 1–72 (2013)
28. Boh, W.F., Yellin, D.: Using enterprise architecture standards in managing information technology. J. Manag. Inf. Syst. **23**, 163–207 (2007)
29. Buckl, S., Ernst, A., Lankes, J., Schneider, K., Schweda, C.: A pattern based approach for constructing enterprise architecture management information models. In: Wirtschaftsinformatik Proceedings, pp. 145–162 (2007)
30. Balabko, P., Wegmann, A.: Systemic classification of concern-based design methods in the context of enterprise architecture. Inf. Syst. Front. **8**, 115–131 (2006)
31. Khosroshahi, P.A., Hauder, M., Matthes, F.: Analyzing the evolution and usage of enterprise architecture management patterns. In: 22nd AMCIS (2016)
32. Software Engineering StandardsCommittee: IEEE Recommended Practice for Architectural Description of Software Intensive Systems. The Institute of Electrical and Electronics Engineers Inc, New York (2000)
33. Engelsman, W., Quartel, D., Jonkers, H., van Sinderen, M.: Extending enterprise architecture modelling with business goals and requirements. EIS **5**, 9–36 (2011)
34. Beimborn, D., Palitza, M.: Enterprise app stores for mobile applications - development of a benefits framework. In: 19th AMCIS, pp. 1–11 (2013)
35. Haag, S.: Appearance of Dark Clouds? - An Empirical Analysis of Users' Shadow Sourcing of Cloud Services. In: Wirtschaftsinformatik Proceedings 2015, Paper 96 (2015)
36. Themistocleous, M., Irani, Z., O'Keefe, R.M., Paul, R.: ERP problems and application integration issues: an empirical survey. In: 34th HICSS, pp. 1–10 (2001)
37. Behrens, S.: Shadow systems: the good, the bad and the ugly. Commun. ACM **52**, 124–129 (2009)
38. Graham, I.J.W.: Constructing a student data warehouse. In: Proceedings of the 30th Annual ACM SIGUCCS Conference on User Services, pp. 174–175 (2002)
39. Zimmermann, S., Rentrop, C., Felden, C.: Governing identified shadow IT by allocating IT task responsibilities. In: 22nd AMCIS, pp. 1–10 (2016)
40. Cooper, H.M.: Synthesizing Research: A Guide for Literature Reviews. Sage, Thousand Oaks (1998)
41. Lucke, C., Krell, S., Lechner, U.: Critical issues in enterprise architecting–a literature review. In: 16th AMCIS, Paper 305 (2010)
42. Gupta, R.K., Awasthy, R.: Qualitative Research in Management. Methods and Experiences. SAGE Response, SAGE Publications, New Delhi, Thousand Oaks (2015)
43. Vom Brocke, J., Simons, A., Niehaves, B., Riemer, K., Plattfaut, R., Cleven, A.: Reconstructing the giant: on the importance of rigour in documenting the literature search process. In: 17th ECIS, pp. 2206–2217 (2009)
44. Webster, J., Watson, R.T.: Analyzing the past to prepare for the future: writing a literature review. Manag. Inf. Syst. Q. **26**, 8–13 (2002)
45. Corbin, J., Strauss, A.: Basics of Qualitative Research: Techniques and Procedures for Developing Grounded Theory. Sage Publications, Thousand Oaks (2014)

46. Mallmann, G.L., Maçada, A.C.G., Oliveira, M.: Can shadow IT facilitate knowledge sharing in organizations? An exploratory study. In: 17th ECKM, pp. 1–10 (2016)
47. Huber, M., Zimmermann, S., Rentrop, C., Felden, C.: The relation of shadow systems and ERP systems-insights from a multiple-case study. Syst. – Spec. Issue: ERP Syst. **4**, 1–13 (2016)
48. Craig, R.: Laurier enterprise system upgrade. In: 20th ICIS, pp. 654–662 (1999)
49. Singh, H.: Emergence and Consequences of Drift in Organizational Information Systems. In: 19th PACIS, Paper 202 (2015)
50. Kretzer, M., Maedche, A., Gass, O.: Barriers to BI&A generativity: which factors impede stable BI&A platforms from enabling organizational agility? In: 20th AMCIS (2014)
51. Berente, N., Yoo, Y., Lyytinen, K.: Alignment or Drift? Loose Coupling over Time in NASA's ERP Implementation. In: 29th ICIS, Paper 180 (2009)
52. Houghton, L., Mackrell, D.: The impact of individual, collective and structural sensemaking on the usefulness of business intelligence data. In: MCIS (2012)
53. Spierings, A., Kerr, D., Houghton, L.: What drives the end user to build a feral information system? In: 23rd ACIS, pp. 1–10, Geelong, Australia (2012)
54. Ritchie, W.J., Drew, S.A., Srite, M., Andrews, P., Carter, J.E.: Application of a learning management system for knowledge management: adoption and cross-cultural factors. Knowl. Proc. Manag. **18**, 75–84 (2011)
55. Kerr, D., Houghton, L., Burgess, K.: Power relationships that lead to the development of feral systems. Australas. J. Inf. Syst. **14**, 141–152 (2007)
56. Morton, P.: Using critical realism to explain strategic information systems planning. J. Inf. Technol. Theory and Appl (JITTA) **8**, 1–20 (2006)
57. McElheran, K.: Economic and business dimensions: decentralization versus centralization in IT governance. Commun. ACM **55**, 28–30 (2012)
58. Wagner, E., Newell, S.: Making software work: producing social order via problem solving in a troubled ERP implementation. In: ICIS 2005 Proceedings, pp. 446–458 (2005)
59. Kutar, M., Light, B.: Exploring cultural issues in the packaged software industry: a usability perspective. In: ECIS 2005 Proceedings (2005)
60. Boudreau, M.-C., Robey, D.: Enacting integrated information technology: a human agency perspective. Organ. Sci. **16**, 3–18 (2005)
61. Fürstenau, D., Sandner, M., Anapliotis, D.: Why do Shadow Systems Fail? An Expert Study on Determinants of Discontinuation. In: 24th ECIS, Paper 157 (2016)
62. Houghton, L., Kerr, D.V.: A study into the creation of feral information systems as a response to an ERP implementation within the supply chain of a large government-owned corporation. Int. J. Internet Enterp. Manag. **4**, 135–147 (2006)
63. Bob-Jones, B., Newman, M., Lyytinen, K.: Picking up the pieces after a "successful" implementation: networks, coalitions and ERP systems. In: 14th AMCIS, p. 373 (2008)
64. Scott, S.V., Wagner, E.L.: Networks, negotiations, and new times: the implementation of enterprise resource planning into an academic administration. Inf. Organ. **13**, 285–313 (2003)
65. McAlearney, A.S., Robbins, J., Hirsch, A., Jorina, M., Harrop, J.P.: Perceived efficiency impacts following electronic health record implementation: an exploratory study of an urban community health center network. Int. J. Med. Inform. **79**, 807–816 (2010)
66. Walters, R.: Bringing IT out of the shadows. Netw. Secur. **2013**, 5–11 (2013)
67. Kumar, V., Maheshwari, B., Kumar, U.: An investigation of critical management issues in ERP implementation: empirical evidence from Canadian organizations. Technovation **23**, 793–807 (2003)
68. Fürstenau, D., Rothe, H., Sandner, M., Anapliotis, D.: Shadow IT, risk, and shifting power relations in organizations. In: 22nd AMCIS, pp. 1–10 (2016)

69. Huuskonen, S., Vakkari, P.: "I did it my way": social workers as secondary designers of a client information system. Inf. Proc. Manag. **49**, 380–391 (2013)
70. Röder, N., Wiesche, M., Schermann, M.: A situational perspective on workarounds in IT-enabled business processes: a multiple case study. In: 22nd ECIS, pp. 1–15 (2014)
71. Niehaves, B., Köffer, S., Ortbach, K.: IT consumerization under more difficult conditions. In: 14th Annual International Conference on Digital Government Research, pp. 205–213 (2013)
72. Rentrop, C., Zimmermann, S.: Shadow IT evaluation model. In: FedCis (2012)
73. Rentrop, C., Zimmermann, S.: Shadow IT - management and control of unofficial IT. In: 6th International Conference on Digital Society, pp. 98–102, Valencia, Spain (2012)

Towards Conceptual Meta-Modeling of ITIL and COBIT 5

Inês Percheiro[1](✉), Rafael Almeida[1], Pedro Linares Pinto[2],
and Miguel Mira da Silva[1]

[1] Instituto Superior Técnico, Av. Rovisco Pais, 1049-001 Lisbon, Portugal
{ines.percheiro,rafael.d.almeida,mms}@tecnico.ulisboa.pt
[2] INOV, Rua Alves Redol, 9, 1000-029 Lisbon, Portugal
pedro.pinto@tecnico.ulisboa.pt

Abstract. Enterprise Governance of Information Technology (EGIT) promotes the alignment of business with Information Technology (IT), ambitioning to create value from IT-enabled business investments. EGIT offers practices, mainly frameworks and standards, to support the organization's business strategy regarding IT. Although many enterprises have recognized the importance of EGIT practices many have yet to adopt them. Every practice has its limitations and specific application, therefore the implementation of a combination of practices can benefit the enterprise enabling features that would be unavailable through their isolated use. A lack of formal consensus on EGIT practices terminology originates conflicts on the integration of multiple practices. Thus, the main goal of this research is to reduce the perceived complexity of EGIT practices integration in order to help organizations better understand how to use them simultaneously. Accordingly, we present the well-known practices COBIT and ITIL meta-modeled and integrated in a conceptual way and, thereby, representing the underlying logical and semantically rich structures. We propose to use ArchiMate as the modeling language.

Keywords: COBIT 5 · ITIL · Meta-modeling · ArchiMate · Enterprise Governance of IT

1 Introduction

One of the main focuses of an organization is to align its Information Technology (IT) strategy with its business strategy in order to create value [1]. Enterprise Governance of IT (EGIT) as defined by Van Grembergen and De Haes [2] addresses the definition and implementation of processes, structures and relational mechanisms that enable both business and IT people to execute their responsibilities, enabling the alignment of business with IT with the ambition of creating value from IT-enabled business investments. For this, EGIT proposes a set of practices, mainly frameworks and standards, that improve, within a set of measures, the relationship between business and IT.

© Springer International Publishing AG 2017
M. Themistocleous and V. Morabito (Eds.): EMCIS 2017, LNBIP 299, pp. 478–491, 2017.
DOI: 10.1007/978-3-319-65930-5_38

Within the wide range of standards and frameworks we can highlight the most relevant according to Goeken and Alter [3]: Control Objectives for Information and Related Technology (COBIT) and IT Infrastructure Library (ITIL).

COBIT 5 is composed of 37 processes, 5 governance processes and 32 management processes, in which each process is defined through inputs and outputs of work products, process activities and practices, goals and metrics to measure the achievement of the process. The COBIT 5 Process Reference Model (PRM) is a model that represents all the processes related to IT that are usually found in enterprises. COBIT also presents RACI charts to document stakeholder responsibility [4,5].

COBIT 5 defends that the main function of an enterprise is to create value for their stakeholders, through benefit realization and optimal resource cost while optimizing risk, and to translate these needs into an actionable strategy COBIT provides the Goals Cascade. The goals cascade maps stakeholder need into enterprise goals, IT-related goals and enabler goals [5].

In turn, ITIL is a set of five comprehensive publications: Service Strategy, Service Design, Service Transition, Service Operation and Continual Service Improvement, that provide descriptive guidance on management of IT processes, functions, roles, and responsibilities related to IT service management [6] providing a wide range of prescriptive information, indicating "what should be done" instead of "how it should be done" [7].

A proper implementation of ITIL contributes to EGIT by developing synergies between business and IT process [7].

COBIT and ITIL, often classified as complementary [8], attempt to give a holistic representation of all the processes and tasks of an organization. On one hand, ITIL is considered the de facto standard for IT Service Management [9–11] but lacks in structure [3], on the other hand COBIT is referred as the "Integrator" because it has the capacity to cover the universe of several frameworks [12].

Organizations use practices for different applications and with different motivations [13]. According to Nicho and Muamaar [14], organizations are being forced to implement multiple practices to comply with the demands of the industry. There is a need to integrate EGIT practices to align and support the enterprise in a balanced way [15].

Each practice has its limitations and since practices often overlap, because they share the same application area [16], they cannot be perceived as mutually exclusive. When combined they provide powerful EGIT [17] and it is important to use them as a whole to assert full IT management and governance [16].

The adoption of several frameworks simultaneously is for IT Managers a matter of legal compliance and for others a risk management strategy, a cost saving measure or a mean to satisfy customers more effectively [17].

There is a set of challenges allied to the integration of EGIT practices [14], but through the integration of EGIT practices, it is possible to foster the development of features that would not be possible when using practices individually [16–18]. Users need more guidance on how to integrate the leading global EGIT practices to realize value from IT investments and services [12].

This paper follows the following structure: Sect. 2 addresses the Problem of this research, Sect. 3 exposes the literature that supports it. On Sect. 4 the research methodology used on our research is presented; The proposal is presented on Sect. 5 and on Sect. 6 we evaluate it. Conclusions and future work is presented on the last Section.

2   Problem

Organizations still battle with EGIT complexity even if there is an increase in the use of these practices [8]. ITIL and COBIT 5 help organizations meet the challenge of transitioning to IT management through business perspective [8]. The adoption of these practices by organizations that are implementing EGIT for the first time claims to be the most effective approach [19].

Many enterprises recognize the importance of EGIT practices, but many have yet to adopt them, according to a recent study from ISACA [20]. EGIT practices are still perceived as confusing and there are still problems regarding its correct use [8]. Organizations still face challenges to adopt these practices and opt for avoiding adopting them even before acknowledging their value.

Winniford et al. [21] point out that less than half of the US companies surveyed had adopted any type of IT service management practices.

In turn, COBIT 5 adoption is challenging because of its level of complexity [22], lack of detail [23] and abundance of generality making it require expert knowledge [24]. ITIL adoption requires a lot of time spent to perceive process diagrams and to achieve results due to lack of work instructions and failure to maintain momentum [25].

According to Othman et al. [26] and Othman and Chan [27], the adoption of EGIT practices includes different inhibitors such as lack of top management support, communication, slack resources, formalization and regulatory environment, compatibility with existing practices and complexity of understanding and using these practices and politics.

Moreover, organizations are being forced to adopt and integrate multiple EGIT practices to comply with the increasing demands of the industry coupled with compliance requirements [14] but struggle with the complexity and difficulty of understanding and adopting several practices at the same time [28], because each practice defines its own scope, definitions and terminologies.

Using practices independently prevents organizations from achieving the full benefits of EGIT because every practice has its limitations on its application to specific IT areas and all these practices overlap [16]. Due to interrelationships between multiple practices is difficult to have an optimal sequence for implementing the processes within each practice [29].

Due to the lack of formal consensus on the terminology used, inconsistencies and terminology, conflicts arise when an organization decides to use multiple practices, making it difficult to understand the main concepts involved [28,30].

Wasting resources by having different departments handling different EGIT practices independently is counterintuitive since organizations seek efficiency and effectiveness [31]. Choosing how to integrate practices is a major challenge when integrating EGIT in organizations [14], Gehrmann [16] defends that EGIT must comprise a combination of two sets of frameworks.

There are many benefits resulting from the integration of EGIT practices. The primary one is the enabling of features that would be unavailable through the use of practices individually, leading to a more comprehensive and efficient approach on EGIT [16–18]. Challenges arise when integrating COBIT 5 and ITIL [14]. To sum up, this paper recognizes that the integration of EGIT practices is challenging for organizations but the benefits of implementing multiple frameworks exceeds the obstacles.

3   State of the Art

3.1   ArchiMate

ArchiMate [32] is a graphical modeling language used to describe enterprise architectures that aims to provide well-defined relationships between concepts in different architectures [33]. ArchiMate provides a set of concepts and relationships to model its three main layers of "Service orientation", the Business Layer, the Application Layer and the Technology Layer. These layers expose functionality in the form of a service to the above layer [32,33].

3.2   Meta-Modeling

Enterprise modeling is more than analyzing and designing information systems—and it is certainly much more than drawing "bubbles and arrows" [1]. Enterprise modeling is about conceptualizing an important part of the world—as it actually is and as it might be [1]. Since the objects of research are mainly models, and not reality, then we first need to create models of models, which, are called meta-models.

A meta-model provides all concepts, properties, operations and relations between concepts necessary for designing any kind of models [34] and so, a meta-model makes it possible to map multiple models into a single model [34].

Schutte and Rotthowe [35] proposed a set of Guidelines of Modelling (GoM) divided in principles as an assessment framework to evaluate information models. The next points give a brief explanation of each one.

- *Principle of Construction Adequacy:* The first principle, the principle of Construction Adequacy refers that the quality of a model depends on the representation of the reality and on the viewpoint from the designer, but mostly on the context of modeling. This principle requires a consensus regarding the problem to be constructed and the definition of the type of construction. The minimalism criteria mention that a model is minimal, if it is its most simplified version where no information objects can be removed [36,37].

- *Principle of Language Adequacy:* The principle of Language Adequacy describes the relationship between the model to be constructed and the language used and how it should be sustainable and appropriate. The language syntax must be correctly applied for this principle to be put into practice.
- *Principle of Economic Efficiency:* The Economic Efficiency principle alludes to economic efficiency and how in modeling information resources can be conserved. It is crucial that the cost of the design process is covered by the savings.
- *Principle of Clarity:* For the comprehension and the explicit definition of a model, the principle of clarity is applied. Models are arranged according to the structural guidelines with reference to various levels of abstraction and the information objects are displayed complying with rules to support lucidity.
- *Principle of Systematic Design:* The systematic design principle relates to the different modeling views and how they can be accepted. The models show how an information system behaves and what is its logical structure.
- *Principle of Comparability:* The last and sixth principle, the principle of comparability, intents to compare two models based on their semantic similarity. This comparison requires analysis of the language and grammar of the two models to assess compatibility.

3.3   ArchiMate and COBIT 5

There is already a proposal for integration of COBIT 5 and ArchiMate by Almeida et al. [38], in an attempt to integrate, map and model COBIT 5 and COSO frameworks. ArchiMate was used by Almeida et al. [38] as a standard language to support organizations on the integration of these frameworks, providing a visual representation.

Another proposal was performed by Cadete [39] that intended to integrate COBIT 5 in Enterprise Architecture (EA) representations to improve the outcomes of COBIT 5 process assessment and process improvement initiatives. [39] also proposed an Ontological mapping between COBIT 5 process performance assessment concepts and the ArchiMate constructs.

4   Research Methodology

Our research follows the Design Science Research Methodology (DSRM). It involves a rigorous process to design artifacts to solve observed problems, to make research contributions, to evaluate the designs, and to communicate the results to appropriate audiences [40].

As advisable in [41] this research methodology is applied according to the two processes of DSR in Information Systems: Build and Evaluate. Building is the process of constructing an artifact for a specific purpose; evaluation is the process of determining how well the artifact performs [41]. The Build process is in this research composed by two stages (Constructs Definitions and Model Construction) whereas the Evaluate process is comprised by only one (Evaluation) (Table 1).

Table 1. DSR research methodology

Build		Evaluate
Constructs definitions	Model construction	Evaluation
- Domain definition - ITIL to ArchiMate ontological mapping	- ITIL meta-model - A viewpoint of ITIL meta-model using ArchiMate - Integration of COBIT 5 and ITIL meta-models	- Österle principles - Critical analysis

5   Proposal

In order to obtain a meta-model for integration between ITIL and COBIT it is necessary to first propose a meta-model for ITIL that satisfactorily represents the purpose and context of this research problem, achieving the objectives of this paper. In Fig. 1 is presented a meta-model for ITIL. The chosen view is justified by the need for the integration between ITIL and COBIT at the process level. In Fig. 2 is presented a view of the same ITIL meta-model that is used for integration with COBIT meta-model, only the concepts relevant to the integration with COBIT 5 are present in this meta-model.

To fundament the decisions regarding the proposed meta-models for ITIL (Fig. 1) and ITIL Integration with COBIT 5 (Fig. 2), we took into consideration the general guidelines of modeling (GoM) proposed by [35]. Table 2 explains the presence of the informational objects present on this models.

– *Principle of Construction Adequacy:* In this principle, it is judged the level of correctness of a model based on its comparison with reality. The construction of a quality model relies essentially on the consensus about its representation. The proposed meta-model needs an explicit definition of its context so that its appropriateness can be clear and the minimalism criteria are evidenced. The minimalism criteria show that all the information presented in the meta-model is essential and that the last stage of reduction of information was achieved without importance to the future user. There are core concepts of the ITIL meta-model that can undoubtedly be mapped into the COBIT meta-model because of their straight definition. This results in their presence being justified on its own. These concepts are Process, Activity, Input/Output, Purpose, and Process Description. The Key Performance Indicators concept, refers to the metrics used to evaluate Critical Success Factors in ITIL and therefore perform an assessment.

COBIT provides a RACI Chart that comprises all the information with respect to the roles performed by the stakeholders describing their typical responsibilities for each practice. In ITIL there is no direct picture showing what are the accountabilities of each position in an organization, however throughout the ITIL books it is possible to note that there are several references to expected behavior from certain positions, for example, the service

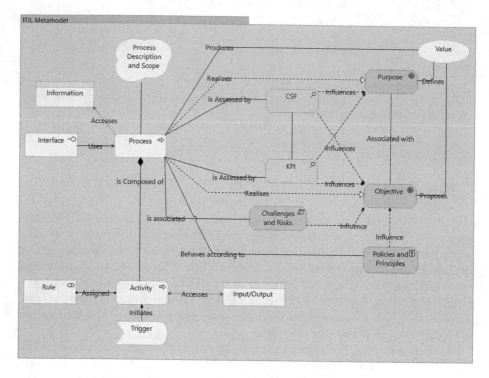

Fig. 1. ITIL meta-model in ArchiMate language.

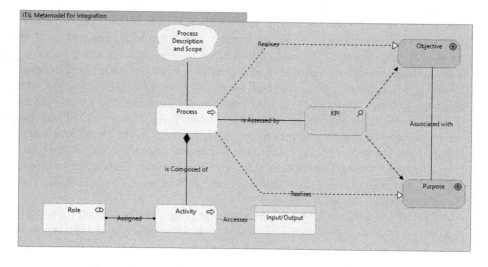

Fig. 2. A view of the ITIL meta-model using ArchiMate.

Table 2. ITIL and ArchiMate ontological mapping

ITIL Concept	ITIL Concept Description [42]	ArchiMate Notation	ArchiMate Concept Description [43]	ArchiMate Representation
Process	A structured set of activities designed to accomplish a speci c objective. A process takes one or more de ned inputs and turns them into de ned outputs.	Business Process	Behavior element that groups behavior based on an ordering of activities	
Activity	A set of actions designed to achieve a particular result. Activities are usually de ned as part of processes or plans, and are documented in procedures.	Business Process	Behavior element that groups behavior based on an ordering of activities	
Role	A set of responsibilities, activities and authorities assigned to a person or team. A role is de ned in a process or function.	Business Role	A business role is de ned as the responsibility for performing speci c behavior, to which an actor can be assigned.	
Input/ Output	Structured sets of data that make the process work and that are expected to be produced by the process. Inputs are gradually converted into outputs.	Business Object	A business object is de ned as a passive element that has relevance from a business perspective.	
Process Description and Scope	Contextual description of the process in question and the boundary or extent to which a process, procedure, certi cation, contract etc. applies.	Meaning	Meaning is de ned as the knowledge or expertise present in a business object or its representation, given a particular context.	
Information	Information on which the process is dependent.	Meaning	A business object represents a concept used within a particular business domain.	
Interface	Determines which inputs each process receives from other processes, and which outputs it must produce so that subsequent processes are able to function.	Business Interface	A business interface is a point of access where a business service is made available to the environment	
Trigger	Event that requires the activities de ned in a particular process to be called into action	Business Event	A business event is a business behavior element that denotes an organizational state change. It may originate from and be resolved inside or outside the organization.	
Value	Value is generated through exchange of knowledge, information, goods or services.	Value	Value represents the relative worth, utility, or importance of a core element or an outcome	
CSF	Something that must happen if an IT service, process, plan, project or other activity is to succeed	Assessment	An assessment is de ned as the outcome of some analysis of some driver.	
KPI	A metric that is used to help manage an IT service, process, plan, project or other activity. Key performance indicators are used to measure the achievement of critical success factors.	Assessment	An assessment is de ned as the outcome of some analysis of some driver.	
Objective	The outcomes required from a process, activity or organization in order to ensure that its purpose will be ful lled	Goal	A goal is de ned as an end state that a stakeholder intends to achieve.	
Purpose	De nition of the motivation for the Process.	Goal	A goal is de ned as an end state that a stakeholder intends to achieve.	
Challenges & Risks	A possible event that could cause harm or loss, or a ect the ability to achieve objectives.	Constrain	A factor that prevents or obstructs the realization of goals.	
Policies & Principles	Formally documented management expectations and intentions.	Principle	A principle represents a qualitative statement of intent that should be met by the architecture.	

owner, process owner, or process manager. Therefore, we can conclude that there is the performance of a role, and with this we justify the presence of the Role concept on the ITIL meta-model.

Both ITIL and COBIT have a section with a short introduction to the process, the Process Description that succinctly describes it and that is relevant to the problem because provides a holistic view of the process and allows a direct parallelism between the two processes from a low-level perspective.

While ITIL only has one defined and structured section for its objectives, in COBIT they are sparsely divided into governance objectives, enterprise goals, IT-related Goals and Process goals, and therefore to map this concept between the two frameworks a one-to-many relationship is obtained.

- *Principle of Language Adequacy:* According to Roux-Rouquié and Soto [34] a modeling language is "a language that contains all the elements with which a model can be described. It is a set of symbols and rules used to specify concepts and constructs for any kind of system; they may be textual and/or visual, structural and/or behavioral. Modeling languages are true languages and have syntax and semantics."

  Lankhorst [44] enumerates several languages for modeling IT and business such as IDEF, BPMN, ARIS and UML. However, Lankhorst [44] also identifies common issues, among them all, like poorly defined relations between domains, models not integrated, weak formal basis and lack of clearly defined semantics, and most of them miss the overall architecture vision being confined to either business or application and technology domains.

  In turn, ArchiMate provides a uniform representation for diagrams that describe EAs and offers an architectural approach that describes and visualizes the different architecture domains and their underlying relations and dependencies [32]. In that way we choose ArchiMate since it fulfills the main objective of this research.

- *Principle of Economic Efficiency:* This principle has a maxim that in the process of information modeling is applicable economic efficiency. In our proposition this principle manifests itself on the integration of the two EGIT frameworks. Once the main goal of this principle is defining the modeling intensity as a restriction for the maximization of profit when promoting the joint use of these frameworks that in some aspects overlap and can be used more efficiently together than apart.

  If an enterprise already puts into practice one of these frameworks and wants to complement its EGIT strategy, it can easily save resources by comparing which processes are already implemented between the two frameworks and how structurally one framework can be mapped into the other. This not only saves time, because there is no need of implementing both frameworks as separate, but also the existing resources can be reused.

- *Principle of Clarity:* The model design must be perceived by its users has a lucid model that has the information objects arranged with clarity and respecting structural guidelines to allow comprehensibility.

  The proposed meta-model, supported by the ArchiMate language, has its informational objects explained in Table 2.

- *Principle of Systematic Design:* In relation to the proposed meta-model, this principle reveals itself on the choice of the modeling view, in this specific case the ArchiMate viewpoint. To predict the behavior of a system, it is significant to sustain the appointed structure and assure consistency.

  An analysis was made regarding the most appropriate ArchiMate viewpoint for our approach, however, from the various existing viewpoints it was not possible to choose one that fully suited our problem. It was then necessary to fuse two ArchiMate Viewpoints, Product Viewpoint Description and Information Structure Viewpoint, to obtain the appropriate elements and concerns.

  It is necessary to make a note that the elements related to the application layer were not used since our problem is based on the motivational and business layer.
- *Principle of Comparability:* This principle is what underlies all the research regarding the integration of the two meta-models. Comparing the ITIL meta-model with the COBIT meta-model is comparing the problem that both propose. Since they are two EGIT practices that exist in the same context and that use semantically compatible aspects and similar language and grammar, it is possible to compare both meta-models and construct a relationship model for the integration of the EGIT practices.

Finally, we present in Fig. 3 an association between the proposed ITIL meta-model and the COBIT meta-model proposed by Almeida et al. [38].

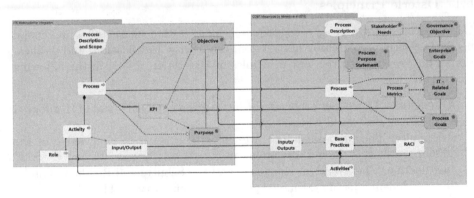

Fig. 3. ITIL and COBIT integration meta-model in ArchiMate language.

Almeida et al. [38] shows the relationship between COBIT 5 PAM and TIPA for ITIL by demonstrating the mapping between COBIT 5 Manage Service Request and Incidents process and ITIL Incident Management and Request Fulfillment, focusing on the concepts of Process Purpose, Base Practices and Outputs. To demonstrate how the mapping proposed in Fig. 3 can be applied, the same processes used by Almeida et al. [38] are now mapped in Fig. 4 regarding ITIL and COBIT 5 objectives. Due to space limitations the authors only present this demonstration.

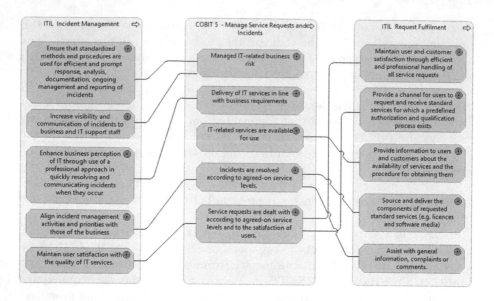

Fig. 4. Objectives mapping in ArchiMate language.

6 Evaluation

6.1 Österle Principles

Österle et al. [45] proposes a set of four principles that characterize in general needs a scientific research by abstraction, originality, justification, and publication. Thus, we fulfilled the four principles of Österle et al. [45] in the following manner, if:

- Abstraction: This paper proposes the integration of COBIT and ITIL adopted by organisations world-wide [3].
- Originality: The proposal is not present in the Book of Knowledge of the domain, as fare as the authors are aware.
- Justification: The ArchiMate language and the Schutte and Rotthowe Guidelines of Modeling justify the model proposal for integration of these two framework meta-models.
- Benefit: To demonstrate that meta-modeling is a useful technique to gain a theoretical foundation on the one hand, and to analyze, compare, and integrate them on the other. Also, by offering visual models of EGIT Practices we believe we are reducing the high perceived complexity of EGIT Practices. Through the process of meta-modeling it is possible to perfect the research on EGIT practices by assessing a comparison between different practices and the completeness of each one, as concluded by Goeken and Alter [3].

7   Conclusion

In this paper we discussed and presented a way to represent the ITIL meta-model in ArchiMate, as well as its integration with the COBIT meta-model. Also we present an approach for integrating these EGIT practices in a conceptual way.

Following the approach proposed by [3] the intention was to demonstrate that meta-modeling is a useful technique to gain a theoretical foundation on the one hand, and to analyze, compare, and integrate them on the other. Also, by offering visual models of EGIT Practices we believe we are reducing the high perceived complexity of EGIT Practices.

In order to achieve this goal, we propose to model ITIL and associate it with COBIT using ArchiMate. However, this paper has also some limitations: Enterprise Architecture (EA) models size, level of detail and complexity can make its analysis a hard task [33].

In these respects, it is important to note that meta-modeling is not a final goal but an interface between data from the real world and models. Therefore, it is important to use an Enterprise Architecture tool that will allow us to promote Governance, Risk and Compliance initiatives aligned with the unique enterprise structure and behavior. Simply put, an EA tool will allow stakeholders to gather evidences that may be spread out through the different EA layers (e.g. application, technology) but that are related to a process from a given EGIT practice, which may facilitate audits. Furthermore, in the future we intend to perform more demonstrations in different organizations and with more EGIT Practices in order to prove the suitability, applicability and benefits of the proposed models.

Finally, EGIT Practices models should be validated through different ways: for example, by interviewing experts and practitioners and by using focus group, we aim as a proposition of future work to validate the integration of the this two frameworks with the use of semantic analysis to compare the content of both documents and map ITIL processes on COBIT processes.

References

1. Frank, U.: Multi-perspective enterprise modeling: foundational concepts, prospects and future research challenges. Softw. Syst. Model. **13**(3), 941–962 (2014)
2. Van Grembergen, W., De Haes, S.: A research journey into enterprise governance of IT, business/IT alignment and value creation. In: Business Strategy and Applications in Enterprise IT Governance, pp. 1–13 (2012)
3. Goeken, M., Alter, S.: Towards conceptual metamodeling of IT governance frameworks approach - use - benefits. In: 42nd Hawaii International Conference on System Sciences, pp. 1–10 (2009)
4. ISACA: COBIT 5: A Business Framework for the Governance and Management of Enterprise IT (2012)
5. ISACA: COBIT 5: Enabling Processes (2012)
6. Lema, L., Calvo-Manzano, J.A., Colomo-Palacios, R., Arcilla, M.: ITIL in small to medium-sized enterprises software companies: towards an implementation sequence. J. Softw. Evol. Process **27**(8), 528–538 (2015)

7. dos Santos, J.H., Todesco, J.L., Fileto, R.: ITIL ontology-based model for IT governance: a prototype demonstration
8. Hill, P., Turbitt, K.: Combine ITIL and COBIT to meet business challenges. BMC Software (2006)
9. Vicente, M., Gama, N., da Silva, M.M.: Using archimate to represent ITIL metamodel. In: IEEE International Conference on Business Informatics (2013)
10. Hanna, A., Windebank, J., Adams, S., Sowerby, J., Rance, S., Cartlidge, A.: ITIL V3 Foundation Handbook (2008)
11. Hochstein, A., Zarnekow, R., Brenner, W.: ITIL as common practice reference model for IT service management: formal assessment and implications for practice. In: IEEE International Conference on eTechnology eCommerce and eService, Nagoya, Japan, vol. 21, pp. 704–710 (2005)
12. Nastase, P., Nastase, F., Ionescu, C.: Challenges generated by the implementation of the IT standards COBIT 4.1, ITIL v3 and ISO/IEC 27002 in enterprises. Econ. Comput. Econ. Cybern. Stud. Res. **43**(1), 16 (2009)
13. Looso, S., Goeken, M.: Application of best-practice reference models of IT governance. In: ECIS (Paper 129) (2010)
14. Nicho, M., Muamaar, S.: Towards a taxonomy of challenges in an integrated IT governance framework implementation. J. Int. Technol. Inf. Manag. **25**(2), 2 (2016)
15. Wessels, E., Loggerenberg, J.: IT governance: theory and practice. In: Conference on Information Technology in Tertiary Education, Pretoria, South Africa (2006)
16. Gehrmann, M.: Combining ITIL COBIT and ISO/IEC 27002 for structuring comprehensive information technology for management in organizations. Navus-Rev. Gestão Tecnol. **2**(2), 66–77 (2012)
17. Cater-Steel, A., Tan, W.G., Toleman, M.: Challenges of adopting multiple process improvement frameworks. In: 14th European Conference on Information Systems (ECIS) (2006)
18. Ula, M., Ismail, Z., Sidek, Z.M.: A framework for the governance of information security in banking system. J. Inf. Assur. Cyber Secur. **2011**, 1–12 (2011)
19. Willson, P., Pollard, C.: Exploring IT governance in theory and practice in a large multinational organisation in Australia. Inf. Syst. Manag. **26**(2), 98–109 (2009)
20. IT Governance Institute: Global Status Report on the Governance of Enterprise IT (GEIT) (2011)
21. Winniford, M., Conger, S., Erickson-Harris, L.: Confusion in the ranks: IT service management practice and terminology. Inf. Syst. Manag. **26**(2), 153–163 (2009)
22. De Haes, S., Van Grembergen, W., Debreceny, R.S.: COBIT 5 and enterprise governance of information technology: building blocks and research opportunities. J. Inf. Syst. **27**(1), 307–324 (2013)
23. Mataracioglu, T., Ozkan, S.: Governing information security in conjunction with COBIT and ISO 27001. Int. J. Netw. Secur. Appl. **3**, 4 (2011)
24. Pereira, R., da Silva, M.M.: A literature review: guidelines and contingency factors for IT governance. In: 16th IEEE International EDOC Conference on Enterprise Distributed Object Computing (2012)
25. Sharifi, M., Ayat, M., Rahman, A.A., Sahibudin, S.: Lessons learned in ITIL implementation failure. In: Information Technology (ITSim) (2008)
26. Othman, M.F.I., Chan, T., Foo, E., Nelson, K., Timbrell, G.: Barriers to information technology governance adoption: a preliminary empirical investigation. In: 15th International Business Information Management Association Conference, Cairo, Egypt, pp. 1771–1787. Queensland University of Technology (2011)

27. Othman, M.F.I., Chan, T.: Lessons learned in ITIL implementation failure. In: 46th Hawaii International Conference on System Sciences, Wailea, HI, USA, pp. 4415–4424 (2013)
28. Pardo, C., Pino, F.J., García, F., Piattini, M., Baldassarre, M.T.: An ontology for the hamonization of multiple standards and models. Comput. Stand. Interfaces **34**, 48–59 (2012)
29. Mendel, T., Parker, A.: Not all ITIL processes are created equal. Network World, 16 March 2005
30. Pardo, C., García, F., Piattini, M., Pino, F., Baldassarre, T.: A 360-degree process improvement approach based on multiple models. Rev. Fac. Ing. Univ. Antioq. **77**, 95–104 (2015)
31. Vicente, M., Gama, N., da Silva, M.M.: Using ArchiMate and TOGAF to understand the enterprise architecture and ITIL relationship. In: Franch, X., Soffer, P. (eds.) CAiSE 2013. LNBIP, vol. 148, pp. 134–145. Springer, Heidelberg (2013). doi:10.1007/978-3-642-38490-5_11
32. Open Group Standard and the Open Group (2013)
33. Lankhorst, M.: Enterprise Architecture at Work: Modelling, Communication and Analysis. The Enterprise Engineering Series, 2nd edn. Springer, Heidelberg (2009)
34. Roux-Rouquié, M., Soto, M.: Virtualization in systems biology: metamodels and modeling languages for semantic data integration. In: Priami, C. (ed.) Transactions on Computational Systems Biology I. LNCS, vol. 3380, pp. 28–43. Springer, Heidelberg (2005). doi:10.1007/978-3-540-32126-2_3
35. Schuette, R., Rotthowe, T.: The guidelines of modeling – an approach to enhance the quality in information models. In: Ling, T.-W., Ram, S., Lee, M. (eds.) ER 1998. LNCS, vol. 1507, pp. 240–254. Springer, Heidelberg (1998). doi:10.1007/978-3-540-49524-6_20
36. Batini, C., Ceri, S., Navathe, S.B.: Conceptual Database Design: An Entity-Relationship-Approach. Benjamin/Cummings Publishing Company, Redwood City (1992)
37. McMenamim, S.M., Palmer, J.F.: Essential Systems Analysis. New York (1984)
38. Almeida, R., Pinto, P.L., da Silva, M.M.: Using ArchiMate to integrate COBIT and COSO metamodels. In: European, Mediterranean and Middle Eastern Conference on Information Systems, Krakow, Poland, p. 5 (2016)
39. Cadete, G.R.: Using enterprise architecture for COBIT 5 process assessment and process improvement (2015)
40. Hevner, A.R., March, S.T., Park, J., Ram, S.: Design science in information systems research. MIS Q. **28**(1), 75–105 (2004)
41. March, S.T., Smith, G.F.: Design and natural science research on information technology. Decis. Support Syst. **15**(4), 251–266 (1995)
42. Axelos: ITIL® Glossary of Terms (2011)
43. The Open Group: ArchiMate 3.0 Specification (2016)
44. Lankhorst, M.: Enterprise Architecture at Work, vol. 10, 3rd edn. Springer, Heidelberg (2013)
45. Österle, H., Becker, J., Frank, U., Hess, T., Karagiannis, D., Krcmar, H., Loos, P., Mertens, P., Oberweis, A., Sinz, E.J.: Memorandum on design oriented information systems research. Eur. J. Inf. Syst. **20**, 7–10 (2011)

Mapping of Enterprise Governance of IT Practices Metamodels

Renato Lourinho[1(✉)], Rafael Almeida[1], Miguel Mira da Silva[1],
Pedro Pinto[2], and Beatrix Barafort[3]

[1] Instituto Superior Técnico, Lisbon, Portugal
{renato.lourinho, rafael.d.almeida,
mms}@tecnico.ulisboa.pt
[2] INESC-INOV, Lisbon, Portugal
pedro.linares.pinto@gmail.com
[3] Luxembourg Institute of Science and Technology, Esch-sur-Alzette,
Luxembourg
beatrix.barafort@list.lu

Abstract. The paper proposes a metamodel for ISO 27001 and its mapping with COBIT 5 using ArchiMate, an Enterprise Architecture (EA) modeling language. The metamodel's purpose is to reduce the perceived complexity of implementing these Enterprise Governance of IT (EGIT) practices simultaneously. For the ontological mapping to be complete, the metamodel is extended with the ISO Technical Specification 33052 and 33072 which propose a Process Reference Model and a Process Assessment Model respectively, specifying Base Practices and Information Items from the ISO TS 33072 – composing the ISO TS 33052 processes - mapped to ISO 27001 controls. By applying best-known metamodeling techniques and modeling principles in conjunction with the use of EA models we further simplify the understanding of different EGIT practices by providing a standard based visualization on how these practices work together. Furthermore, we present the mapping and modeling of a COBIT 5 process and respective ISO 27001 controls as an example. The paper concludes by summarizing the considerations and techniques used in this research, as well as discussing limitations and future work in this domain.

Keywords: ArchiMate · COBIT 5 · Enterprise Architecture · Enterprise Governance of Information Technology · ISO 27001 · ISO TS 33072

1 Introduction

IT has the potential to support both existing and new business strategies, and as such it has moved from being a commodity service to be a strategic asset within today's digital enterprises [1]. Given this relatively new-found importance of IT, Enterprise Governance of IT (EGIT) has also gained new focus [1].

EGIT can be defined as "an integral part of corporate governance and addresses the definition and implementation of processes, structures and relational mechanisms in the organization that enable both business and IT people to execute their responsibilities in

© Springer International Publishing AG 2017
M. Themistocleous and V. Morabito (Eds.): EMCIS 2017, LNBIP 299, pp. 492–505, 2017.
DOI: 10.1007/978-3-319-65930-5_39

support of business/IT alignment and the creation of business value from IT-enabled business investments" [1].

Examples of process mechanisms are EGIT frameworks, Best Practices and ISO standards. The term EGIT Practices is used throughout this paper to refer to all standards and frameworks described.

While there is no single, complete, off-the-shelf EGIT Practice, there are several EGIT Practices available that can serve as useful starting points for developing a governance model [2]. Researchers agree that COBIT, ITIL, and ISO 27000 family are the most valuable and popular practices currently being adopted [3–5].

However, a recent study highlighted the fact that, despite acknowledging the importance of adopting EGIT Practices, many organizations have not adopted them [6]. Also, Winniford et al. points out that less than half of the US companies surveyed had implemented any type of IT service management practice [7].

Thus, the main goal of this research is to reduce the complexity of COBIT 5 and ISO 27001 by designing visual models of these EGIT Practices, facilitating in this way their understanding. Therefore, we propose to use ArchiMate, as the Enterprise Architecture (EA) language, to model COBIT 5 and ISO 27001 metamodels, enabling in this way the mapping of these EGIT Practices.

To enable this mapping, we also present a modeled extension of ISO 27001 with the recent ISO Technical Specification (TS) 33052 – Process Reference Model (PRM) and ISO TS 33072 – Process Assessment Model (PAM) as these documents define relevant concepts that may be matched to COBIT 5 concepts.

This paper is structured as follows: Sect. 2 states the problem in-depth; Sect. 3 presents the related literature which describe theories and other approaches that inspired this research; Sect. 4 presents the proposal; Sect. 5 demonstrates the proposal with a mapping example; Finally, Sect. 6 contains final conclusions, limitations and discussion of future work related to the topic.

2 Problem

IT organizations are facing the challenging, but necessary, transition to manage IT based on business priorities. They are looking for EGIT mechanisms, such as ISO 27001 and COBIT 5, to help them meet the challenge [8]. In fact, their adoption and practice is argued to be the most effective approach and guidance for organizations first considering proper implementation of EGIT [9].

Moore [10] identified approximately 315 standards, guides, handbooks, and other prescriptive documents which were taken as reference models and maintained by 46 different organizations. This number of models has now increased, as have their application areas which provide best practices for different needs, e.g. Information Security Management System (ISMS) such as ISO 27001 Information Technology Governance Processes (IT Governance) and Services Management such as COBIT, ITIL, among others.

Organizations can benefit from the numerous models and standards when assessing and institutionalizing new or improved processes, thus becoming more competitive and producing high quality products [11]. Independently of the model to be used, its implementation requires specific experience and knowledge, along with a high degree of effort and investment, as key factors for it to be successful. All this signifies that the task is not easy and there is a significant risk of failure [12].

For example, there is no fully complete EGIT Practice to be used as a comprehensive off-the-shelf solution to ensure the alignment between service management and the organization's concepts and artifacts [13]. In fact, different EGIT Practices are often used as complementary and, most of the times, simultaneously too. Parallel projects imply a duplication of investments and costs, and even with shared infrastructures we cannot avoid a duplication of data repositories, procedures and human resources, being hard to define a way for teams not to compete or maintain different efforts aligned [13].

Since many EGIT Practices overlap, using them independently prevents organizations from asserting full IT management and governance because each practice has limitations in its application to the management of specific IT areas [14].

The implementation of EGIT Practices should be consistent with the enterprise's risk management and control framework, appropriate for the enterprise, and integrated with other methods and practices that are being used [15]. Therefore, management and staff must understand what to do, how to do it and why it is important to do it [16]. However, there seems to be some confusion regarding EGIT Practices and how best to use them [8].

The heterogeneity of models is positive for organizations as it allows them to choose the best models with which to satisfy their goals. However, each model defines its own characteristics such as: structure, terminology, scope, approach and level of abstraction or detail, domain and size of the organization [16]. This situation has led to certain problems in the use of the models, e.g. ambiguity, instability, subjectivity, incompatibility, amongst others [17].

Individually, it has been stated that COBIT cannot work alone as it is not very detailed, and shows what to do but not how to do [18]. Moreover, its implementation was found to be difficult as it is too generic, and thus requires expert knowledge [19].

Regarding ISO 27001 many organizations find it difficult and challenging to implement this practice along with other information security management practices [20]. Being employed as a standalone guide and not being integrated into a wider practice for EGIT makes it difficult for organizations that adopt ISO 27000 family standards to implement other EGIT Practices [21]. Therefore, in a time when organizations strive to be efficient and effective, it seems counter-intuitive to be wasting resources by having different organizational departments handling both approaches independently [22].

The adoption of COBIT 5 in organizations is widely described as challenging due to the high perceived complexity of COBIT 5 [1]. In contrast to objectively measurable complexity, perceived complexity results from the distinctions made by a subjective observer [23].

In sum, the problem that this research intends to help solve is that organizations struggle with the perceived complexity and difficulty of understanding different EGIT Practices, and thus adopting these practices simultaneously needs large investments, that may even prevent their adoption altogether.

3 Related Work

3.1 COBIT 5

COBIT 5 is based on five principles: meeting stakeholder needs; covering the enter-prise end-to-end; applying a single, integrated framework; enabling a holistic approach; and separating governance from management [24]. Together these principles enable enterprises to assemble and deploy an effective EGIT and management framework and thus support striking balance between benefits realization, risk management and resources [24].

COBIT 5 evolution unified ISACA's three frameworks: Val IT, a value delivery focused framework; Risk IT, a risk management focused framework and previous COBIT versions. Hence this allowed COBIT 5 to cover the lifecycle of governance and management within the scope of enterprise IT [1] COBIT 5 also introduced a new process-reference model, new processes, updated and expanded goals and metrics, and alignment with the ISO 15504 process-capability-assessment model [24].

3.2 ISO 27001 and ISO TS 33052/33072

ISO 27001 provides requirements for implementing, maintaining and improving ISMS [25, 26]. Organizations implement this standard to address security requirements in a consistent, repeatable and auditable manner [27]. ISMS provide risk management processes such that it preserves confidentiality, integrity and availability of information. It is of importance that this risk management process is integrated with the organiza-tion's processes and information security is included in a holistic manner within the scope of process design, information systems and controls.

Published in 2016, ISO TS 33072 [28] is an International Standard Technical Specification that proposes a Process Assessment Model (PAM) enabling the assess-ment of processes based on the ISO 27001 requirements statements. To be able to perform an assessment, ISO TS 33072 presents Base Practices and Information Items/Outcomes which compose the processes defined in ISO TS 33052 [29]. Con-ceptually, these Base Practices and Information Items are similar to COBIT 5 own Practices and Inputs/Outputs and can be shown to be related as COBIT's holistic nature provides coverage over the domain of information security.

3.3 Mapping EGIT Practices

Alignment between COBIT and ISO 27001 has been approached by several researches [5, 12, 30, 31] but these researches either map EGIT Practices at a very abstract level, matching process similarity criteria [27, 31], or have mapped previous versions that have been superseded such as COBIT 4.1 and ISO 27001:2005 [5, 12, 30].

Despite these obstacles, such researches provide valuable guidance regarding the alignment of the current versions of COBIT and ISO 27001. One such case is the choice of which process is most adequate for a mapping demonstration. A sought for trait is a high level of similarity between corresponding processes described by the

EGIT Practices. For this issue, [31] provides a process level method for analysis between ISO 27001 and COBIT.

3.4   ArchiMate

The objective of the ArchiMate language is to provide well-defined relationships between concepts in different architectures, the detailed modeling of which may be done using other, standard or proprietary modeling languages. Concepts in the ArchiMate language cover the business, application, and technology layers of an enterprise and provide an extended layer that represents the motivation. Services offered by one layer to another play an important role in relating the layers [32].

ArchiMate provides a uniform representation for diagrams that describe EAs and offers an architectural approach that describes and visualizes the different architecture domains and their underlying relations and dependencies [33].

ArchiMate is used in this research as it provides a uniform representation for diagrams that describe EA. It offers an approach to describe and visualize different architectural domains and their relations and dependencies. It also supports different viewpoints for selected stakeholders [33]. As a visual design language, EA models become more readable and understandable than textual descriptions, thus lowering the user's perceived complexity.

3.5   ArchiMate and EGIT Practices

As far as the authors are aware, there are few approaches that propose to model and integrate EGIT Practices using ArchiMate as the architecture's modeling language, enabling the mapping of these EGIT Practices in a standard-based EA representation. We would like to highlight three of them:

Almeida et al. mapped, modelled and integrated COBIT 5 and COSO in ArchiMate [34].

Another research [35] proposed a model that uses TIPA [36] for the Information Technology Infrastructure Library (ITIL), COBIT PAM and ArchiMate to analyze the impact of ITIL implementation on COBIT processes performance, and vice-versa.

Furthermore, a technical report from the Luxembourg Institute of Science and Technology [37] presents the whole outputs of the conceptual alignment between concepts used to model EA (based on ArchiMate, TOGAF, IAF and DoDAF) and concepts of the Information System Security Risk Management domain model.

These researches were an important contribution to our research. In this research, we decided to use the COBIT 5 metamodel proposed by [35], since the scope of this paper is to map COBIT 5 and ISO 27001.

3.6   Modeling Techniques and Principles

While there are several EGIT Practices well established to support management and Enterprise Governance of IT, there is a lack of theoretical foundation [38], which can contribute to the evolution and adaptation of said EGIT Practices.

To support and enable these evolutions it is common to use models as a form of abstraction from real word scenarios. If the object of research is an abstraction and as such, already a model, then we create models of models, or so-called metamodels as per the defined model stack [39].

Metamodels provide concepts, properties, operations and relations needed to design any kind of model [40], enabling the integration of multiple models into a single model by establishing well-defined relationships.

In order to develop high quality models, [41] proposed the so-called guidelines of modeling, which propose six principles to raise the quality of information modeling. Since our metamodels are models of models, these principles are applicable and we present their fitness evaluation in the proposal.

4   Proposal

The purpose of the proposed mapping is to provide an integrated way for complementary use of COBIT 5 and ISO 27001. COBIT 5 is well structured in domains, processes and other components and, therefore, closed and self-contained. Also, COBIT is holistic and represents (nearly) all tasks and processes an IT organization should carry out. ISO 27001 was chosen because it is a security standard for Information Security Management System (ISMS) that is a highly dynamic and complex task due to constant change in the information technology domain [42].

To achieve this mapping, for each COBIT 5 process we looked for every related ISO 27001 control category. Upon assessing the applicability of each mapping, we mapped each individual ISO 27001 control to all COBIT 5 processes. We found that every control in each category was related to the process, meaning that once we found matching process and control categories, none of its controls were irrelevant. To enable a consistent and comprehensive model, all related concepts must be mapped, including COBIT 5 Practices and Inputs/Outputs. Although the ISO 27001 does not describe equivalent concepts, the recent ISO TS 33052 and ISO TS 33072 do.

As stated, there are in the literature some mappings regarding these EGIT Practices [30, 31]. However, as the authors are aware, none of the researches use the latest version of both COBIT 5 and ISO 27001 (this means that they map older versions of these mechanisms). Moreover, they do not use an EA representation.

Therefore, we propose to use ArchiMate, as the EA language, to model COBIT 5 and ISO 27001 metamodels, enabling in this way the mapping of these EGIT Practices.

4.1   ISO 27001 Metamodel

To develop a metamodel for ISO 27001 using ArchiMate, we first mapped the main ISO 27001 and ArchiMate concepts, as shown in Table 1. As this research extends the

ISO 27001 metamodel with ISO TS 33052 and 33072 concepts, these were also ontologically mapped to ArchiMate in Table 2. Based on these ontological mappings, we propose the metamodel as shown in Fig. 1.

Table 1. ISO 27001 and ArchiMate ontological mapping

ISO 27001 Concept	ISO 27001 Concept Description [25]	Archi-Mate Notation	ArchiMate Concept Description [33]	ArchiMate Representation
Require-ment	Need that is stated, generally implied or obligatory.	Require-ment	A statement of need that must be realized by a system.	Requirement
Control Objective	Statement describing what is to be achieved as a result of implementing controls.	Goal	An end state that a stakeholder intends to achieve.	Goal
Control	Measure that is modifying risk.	Business Process	A behavior element that groups behavior based on an ordering of activities. It is intended to produce a defined set of products of business services.	Business Process
Organi-zation	Person or group of people that has its own functions with responsibilities to achieve its objectives.	Business Actor	An entity that performs behavior in an organization such as business processes or functions.	Business Actor
Top Manage-ment	Person or group of people who directs and controls an organization at the highest level.	Stake-holder	The role of an individual, team, or that represents their interests in, or concerns relative to, the outcome of the architecture.	Stakeholder
Risk Owner	Person or entity with the accountability and authority to manage a risk.	Business Role	A named specific behavior of a business actor participating in a given context.	Business Role
Infor-mation Security Needs	Policies	Driver	A driver is defined as something that creates, motivates, and fuels the change in an organization.	Driver
	Information Needs			
	External Context			

Table 2. ISO TS 33052/33072 and ArchiMate ontological mapping

ISO TS 33052/33072 Concept	ISO TS 33052/3072 Concept Description [28, 29]	ArchiMate Notation	ArchiMate Concept Description [33]	ArchiMate Representation
Process	Set of interrelated or interacting activities which transforms inputs into outputs.	**Business Process**	As per Table 1.	Business ⇨ Process
Base Practice	Activity that, when consistently performed, contributes to achieving a specific process purpose.	**Business Process**	As per Table 1.	Business ⇨ Process
Information Item	Observable result of the successful achievement of the process purpose.	**Business Object**	A business object is defined as a passive element that has relevance from a business perspective.	Business object

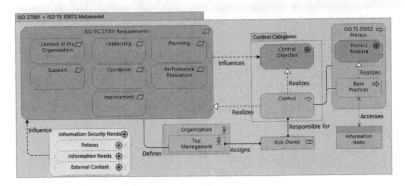

Fig. 1. ISO 27001 and ISO TS 33052/33072 metamodel

This ArchiMate metamodel is adapted from the metamodel developed in [42], an ontology based analysis of the ISO 27001 document by extrapolating the relations of core information security concepts as defined in [25]. This adaptation focuses on the Business Layer and Motivational Extension of the ArchiMate language.

ISO 27001 presents a set of normative requirements, including a set of controls for management and mitigation of the risks associated with the information assets which the organization seeks to protect. This motivation – Driver in the ArchiMate language – influences which requirements the organization should implement, whether they are security, legal or business requirements.

Thus, as the organizational needs influence the general requirements, so does the choice of those general requirements influence the set of controls (or control categories) and other specific requirements to be implemented.

Also, it is important to note that while ISO TS 33075 Information Items/Outcomes are conceptually equivalent to COBIT's Work Products as they follow the same logic and metamodel defined in the ISO 15504 standard (which is the reference model for maturity models such as the ISO TS 33072), we keep the Information Items terminology throughout this paper to distinguish the COBIT and ISO input/output concepts.

4.2 Mapping of EGIT Practices Metamodels Using ArchiMate

In Fig. 2 we propose a metamodel that encompasses COBIT 5, ISO 27001 and ISO TS 33052/33072 using ArchiMate. Some considerations regarding this model: COBIT 5 processes and ISO 27001 controls are related by structural association, meaning they can be mapped from one to another and vice-versa; A COBIT 5 process is composed by one or more ISO 27001 control categories. Each category contains a single control objective and one or more controls.

This model is based on the mapping between COBIT 5 processes and ISO 27001 controls performed by the authors. By semantically assessing the descriptions of both

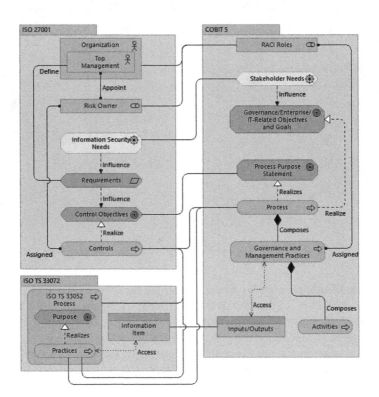

Fig. 2. COBIT 5 – ISO 27001 – ISO TS 33052/33072 metamodel

processes, control objectives and controls, it was found that when a COBIT 5 process matches one or more ISO 27001 control categories, all the controls pertaining to that set are relevant to the COBIT 5 process. Thus, we consider a direct structural association between ISO 27001 controls and a COBIT 5 process due to the more generalized scope of COBIT 5 where its processes, in some way or another, relate to an information asset which can be protected.

Regarding the relationships between Practices, while some ISO 27001 controls map exclusively to some ISO TS 33052 process and vice-versa, and therefore its related Practices, this is not always the case and thus we cannot state that all controls map directly to these processes; but all controls map to one or more Practices in an *ad-hoc* relation. Consequently, the same reasoning applies to COBIT Inputs/Outputs and Information Items.

Regarding responsibility assignments (ArchiMate Business Roles and Actors) and the Motivational Layer (ArchiMate Drivers, Requirements and Goals) influenced by decision makers; while ISO 27001 is not as encompassing as COBIT 5's RACI (Responsible, Accountable, Consulted, Informed) tables, it establishes that the adoption of risk modifying measures (controls) and the motivational rationale behind that adoption must be accountable by certain organizational roles. Thus, it is semantically sound to match ISO 27001 and COBIT 5 at this level.

A similar rationale applies to COBIT 5's stakeholder drivers and ISO 27001 information security needs. As COBIT is the broader EGIT Practice and includes to a certain degree the information security domain, it is also sound to establish a relationship between these motivational concepts.

4.3   Modeling Principles Fitness

1. *Principle of construction adequacy* - This principle judges the adequacy of the model to the reality, the designer's viewpoint and modeling context. Context wise, the models in this proposal are fundamentally theoretical, and as such they have been developed based on the EGIT Practices' documentation and the related literature.

   Accordingly, the designer's (the authors') viewpoint is also theoretical and thus the models are focused on abstracting the concepts and establishing the existing relationships between the determined concepts. As the models strictly follow the architectures described in the EGIT Practices' documentations, we consider the models to fit adequately for the intended purposes.

2. *Principle of language adequacy* - The ArchiMate core language provides the basic concepts and relationships to fulfil the general EA modeling needs. It offers an architectural approach that describes and visualizes the different architecture domains and their underlying relations and dependencies.

   As such, ArchiMate fits the purpose of our metamodels which is to compare, map and integrate different EGIT Practices at a component level. As this principle also includes consistency and completeness, meaning that our models do not include any symbol that is not present in the language metamodel, our ontological mapping between ISO 27001, ISO TS 33052/33072 and COBIT 5 (in the related literature) shows that ArchiMate is adequate as our modeling language.

3. *Principle of economic efficiency* - As our goal is to develop a metamodel from a theoretical perspective, this principle is in a sense, not applicable to the development of our models, although in practice reducing the perceived complexity of mapping COBIT 5 and ISO 27001 promotes economic efficiency within an organization as an outcome of this research.

4. *Principle of clarity* and *systematic design* - This principle is assured by the modeling language ArchiMate, as a visual architectural language, and our ontological mapping between its concepts and the EGIT Practices concepts, since we include all relevant concepts for the scope of this research, thus obtaining a comprehensive metamodel which fulfils the systematic design principle.

5. *Principle of comparability* - As the goal of this research is to compare and map the COBIT 5 and ISO 27001 metamodels, this principle is fulfilled by bridging semantical discrepancies through the ArchiMate metamodel. Moreover, as COBIT is a comprehensive EGIT Practice that also provides coverage of the ISM domain, many concepts are semantically compatible and therefore comparable.

5   Theoretical Demonstration

The demonstration of the mapped COBIT 5 and ISO 27001 metamodels is shown in Fig. 3. We chose the COBIT 5 process Manage Service Requests and Incidents for this instantiation, which per [31] is representative of a strongly related COBIT process to ISO 27001 controls. Note that the colored relations between base practices has no additional semantic value other than to improve readability.

Due to space limitations, we show only the relations mapped to the "16.1 – Management of information security incidents and improvements" ISO 27001 control.

Some COBIT Inputs/Outputs show no relation because either they map to an ISO TS 33072 Information Item from another control or have no relation at all since COBIT is a wider coverage EGIT Practice than the domain of information security management. The COBIT and ISO Base Practices show no relations for readability reasons, but follow the same reasoning as the Work Products relationships.

5.1   Demonstration Benefit Analysis

We attempt to demonstrate the reduced perceived complexity of mapping and connecting different EGIT Practices' concepts. In this subset notice the numerous and complex mapping between different concepts, yet easily traced through visualization. Missing relationships are quickly noticeable, thus enabling focus on the process practices and outcomes relevant to the implementation. In this subset, we can notice how COBIT 5 outcomes such as "Approved service requests" or "Fulfilled service requests" are not relevant to the ISO 27001 control "16.1 – Management of information

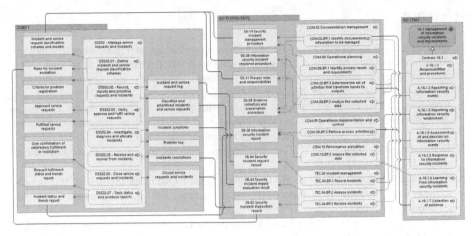

Fig. 3. Instantiation of a COBIT process with ISO controls, practices and information items.

security incidents and improvements" as service continuity is not within the scope of this control.

These EA models are then suitable to establish an overview of the "as-is" situation of organizations and their EGIT Practices implementation as well as facilitating the understanding, planning and communication of the "to-be" future situation.

6   Conclusion

In this paper, we illustrate the modeling and mapping of COBIT 5 and ISO 27001 metamodels using ArchiMate, enabling the mapping of these EGIT Practices and their inclusion into the scope of existing EA techniques.

We identified the problem at hand as the perceived complexity when implementing more than one EGIT Practice in parallel within organizations. While the heterogeneity of EGIT Practices is good for organizations as it presents multiple options for different governance and management needs [16], there is often an overlap of respective domains which cause problems in the organizational models such as ambiguity, instability, subjectivity, incompatibility of concepts and existing artifacts [17].

Thus, we believe the visual representation of COBIT 5 and ISO 27001 metamodels facilitate knowledge sharing, understanding and communication of these EGIT Practices.

Yet, this research also has some limitations. EA models size, level of detail and complexity can make its analysis by human means only a hard task [32]. Moreover, as ArchiMate is a graphical language, it is not prone to automatic analysis.

In the future, we plan to implement our EGIT Practices models into an EA Management software that will allow us to answer to questions such as: "Attending to the allocated resources, within the different EA layers, what is the cost of maintaining a given COBIT 5 process in my organization?" or "How many resources do we have allocated to comply with a given ISO 27001 control?"

References

1. De Haes, S., Van Grembergen, W.: Enterprise Governance of Information Technology: Achieving Strategic Alignment and Value, Featuring COBIT 5. Springer, New York (2015)
2. Symons, C.: IT governance framework: structures, processes and communication. IT Governance Series, Forrester Research (2005)
3. Coleman, T., Chatfield, A.: Promises and successful practice in IT governance: a survey of Australian senior IT managers. In: 15th Pacific Asia Conference on Information Systems: Quality Research in Pacific, PACIS 2011, Queensland, pp. 1–15 (2011)
4. Debreceny, R.S., Gray, G.L.: IT governance and process maturity: a multinational field study. J. Inf. Syst. **27**(1), 157–188 (2011)
5. Sahibudin, S., Sharifi, M., Ayat, M.: Combining ITIL, COBIT and ISO/IEC 27002 in order to design a comprehensive IT framework in organizations. In: Asia International Conference on Modeling (2008)
6. IT Governance Institute: Global Status Report on the Governance of Enterprise IT, ISACA, COBIT 5: Enabling Processes (2011)
7. Winniford, M., Conger, S., Erickson-Harris, L.: Confusion in the ranks: IT service management practice and terminology. Inf. Syst. Manag. **26**(2), 153–163 (2009)
8. Hill, P., Turbitt, K.: Combine ITIL and COBIT to meet business challenges. BMC Softw. (2006)
9. Willson, P., Pollard, C.: Exploring IT governance in theory and practice in a large multinational organization in Australia. Inf. Syst. Manag. **26**(2), 98–109 (2009)
10. Moore, J.W.: An integrated collection of software engineering standards. J. IEEE Softw. **16**(6), 51–57 (1999)
11. Oud, E.J.: The value to IT of using international standards. Inf. Syst. Control J. **3**, 35–39 (2005)
12. Aaen, I.: Software process improvement: blueprints versus recipes. IEEE Softw. J. **20**, 86–93 (2003)
13. Gama, N., Sousa, P., Mira da Silva, M.: Integrating enterprise architecture and IT service management. In: 21st International Conference on Information Systems Development, Italy (2012)
14. Gehrmann, M.: Combining ITIL, COBIT and ISO/IEC 27002 for structuring comprehensive information technology for management in organizations. Navus: Revista de Gestão e Tecnologia **2**(2), 66–77 (2012)
15. Nastase, P., Nastase, F., Ionescu, C.: Challenges generated by the implementation of the IT standards CobiT 4.1, ITIL v3 and ISO/IEC 27002 in enterprises. Econ. Comput. Econ. Cybern. Stud. Res. **43**(3), 1–16 (2009)
16. Biffl, S., Winkler, D., Hörn, R., Wetzel, H.: Software process improvement in Europe: potential of the new V-Model XT and research issues. Softw. Process: Improv. Pract. **3**(3), 229–238 (2006)
17. Liao, L., Qu, Y., Leung, H.K.N.: A software process ontology and its application. In: Proceedings of 4th International Semantic Web Conference (ISWC 2005), Galway, Ireland (2005)
18. Mataracioglu, T., Ozkan, S.: Governing information security in conjunction with COBIT and ISO 27001. arXiv preprint arXiv:1108:2150 (2011)
19. Pereira, R., Mira da Silva, M.: Designing a new integrated IT governance and IT management framework based on both scientific and practitioner viewpoint. Int. J. Enterp. Inf. Syst. **8**(4), 1–43 (2012)
20. Susanto, H., Almunawar, M.N., Tuan, Y.C.: Information security management system standards: a comparative study of the big five. Int. J. Electr. Comput. Sci. **11**(5), 23–29 (2011)

21. Von Solms, B.: Information security governance: COBIT or ISO 17799 or both? Comput. Secur. **24**(2), 99–104 (2005)
22. Vicente, M., Gama, N., Mira da Silva, M.: Using ArchiMate to represent ITIL metamodel. In: IEEE International Conference on Business Informatics, pp. 270–275 (2013)
23. Schlindwein, S.L., Ison, R.: Human knowing and perceived complexity: implications for systems practice. Emerg.: Complex. Organ. **6**, 27–32 (2004)
24. ISACA, COBIT 5: A Business Framework for the Governance and Management of Enterprise IT (2012)
25. Information Technology – Security Techniques – Information Security Management Systems – Overview and Vocabulary, 3rd edn. ISO Standard 27000 (2014)
26. Information Technology – Security Techniques – Information Security Management Systems – Requirements, 2nd edn. ISO Standard 27001 (2013)
27. Nicho, M., Muamaar, S.: Towards a taxonomy of challenges in an integrated IT governance framework implementation. J. Int. Technol. Inf. Manag. **25**, 2 (2016)
28. Information Technology – Process Assessment – Process Capability Assessment Model for Information Security Management – ISO Technical Specification 33072 (2016)
29. Information Technology – Process Assessment – Process Reference Model for Information Security Management, ISO Technical Specification 33052 (2016)
30. Sheikhpour, R., Modiri, N.: An approach to map COBIT processes to ISO/IEC 27001 information security management controls. Int. J. Secur. Appl. **6**(2), 13–28 (2012)
31. Haufe, K., Colomo-Palacios, R., Dzombeta, S., Brandis, K., Stantchev, V.: Security management standards: a mapping. Procedia Comput. Sci. **100**, 755–761 (2016)
32. Lankhorst, M.: Enterprise Architecture at Work: Modeling, Communication and Analysis. The Enterprise Engineering Series, 2nd edn. Springer, Heidelberg (2009)
33. The Open Group: ArchiMate 2.0 Specification (2012)
34. Almeida, R., Pinto, P., Mira da Silva, M.: Using ArchiMate to integrate COBIT 5 and COSO metamodels. In: European, Mediterranean & Middle Eastern Conference on Information Systems, Krakow, Poland (2016A)
35. Almeida, R., Pinto, P., Mira da Silva, M.: Using ArchiMate to assess COBIT 5 and ITIL implementations. In: 25th International Conference on Information Systems Development, Poland (2016B)
36. Luxembourg Institute of Science and Technology: TIPA for ITIL. http://www.tipaonline.org
37. Mayer, N., Aubert, J., Grandry, E., Feltus, C., Goettelmann, E.: An integrated conceptual model for information system security risk management and enterprise architecture management based on TOGAF, ArchiMate, IAF and DoDAF. Luxembourg Institute of Science and Technology Technical report (2016)
38. Goeken, M., Alter, S.: Towards conceptual metamodeling of IT governance frameworks approach-use-benefits. In: 42nd Hawaii International Conference on System Sciences (2009)
39. Hinkelmann, K.: Meta-modeling and Modeling Languages. FHNW School of Business, University of Applied Sciences, Northwestern Switzerland
40. Roux-Rouquié, M., Soto, M.: Virtualizations in systems biology: metamodels and modeling languages for semantic data integration. Trans. Comput. Syst. Biol. I **3380**, 132 (2005)
41. Schütte, R., Rotthowe, T.: The guidelines of modeling – an approach to enhance the quality in information models. In: Ling, T.W., Ram, S. (ed.) Conceptual Modeling ER 98, Singapore, pp. 240–254 (1998)
42. Milicevic, D., Goeken, M.: Ontology-based evaluation of ISO 27001. In: Cellary, W., Estevez, E. (eds.) I3E 2010. IAICT, vol. 341, pp. 93–102. Springer, Heidelberg (2010). doi:10.1007/978-3-642-16283-1_13

Perception of Enterprise Risk Management in Brazilian Higher Education Institutions

Gustavo de Freitas Alves[1([⊠])], Waldemar Lima Neto[1],
Marçal Chagas Coli Jr.[1], Paulo Henrique de Souza Bermejo[1],
Tomás Dias Sant' Ana[2], and Eduardo Gomes Salgado[3]

[1] Post-Graduate Administration Program, Faculty of Economics,
Administration and Accounting, Campus Darcy Ribeiro,
University of Brasília, Brasília 70910-900, Brazil
{gustavo.alves,marcal.chagas}@next.unb.br,
waldemarlimaneto@gmail.com, paulobermejo@unb.br
[2] Pro-Rector of Planning, Budget, and Institutional Development, Federal
University of Alfenas, Rua Gabriel Monteiro da Silva, 714, Centro, Alfenas,
Minas Gerais 37130-000, Brazil
tomas@bcc.unifal-mg.edu.br
[3] Exact Sciences Institute, Federal University of Alfenas, Rua Gabriel Monteiro
da Silva, 700, Centro, Alfenas, Minas Gerais 37130-000, Brazil
eduardo.salgado@unifal-mg.edu.br

Abstract. Enterprise risk management (ERM) has been used as a methodology
for evaluating the risks to an organization's achievement of its objectives. In
private companies, plenty of initiatives and related works have addressed this
topic, but studies in the public sector are lacking. Studying how the top managers of public sector federal universities perceive risk management can provide
insight into developmental gaps in ERM and related software. This work used a
survey to capture public servants' perceptions of risk management using both
quantitative and qualitative questions. The main findings can be used to inform
decision making and ERM implementation as well as to promote related
objectives, planned responses, efficient resource usage, error reduction, and
organizational self-knowledge. The top difficulties perceived were related to
lack of training, limited staff, absence of risk culture, undefined department
structure, lack of interest shown by public servants, reduced budget, and low
engagement of top management with ERM. Management practices such as
ERM could improve public sector managers' work performance, enabling them
to provide better service to society.

Keywords: Enterprise risk management · Public sector · Risk perception ·
Higher education institutions · Public governance

1 Introduction

The Brazilian National Forum of Pro-Rectors of Planning and Administration (FOR-
PLAD) is a workgroup with the responsibility of studying problems related to the
planning and administration of higher education institutions (HEIs) as well as proposing

© Springer International Publishing AG 2017
M. Themistocleous and V. Morabito (Eds.): EMCIS 2017, LNBIP 299, pp. 506–512, 2017.
DOI: 10.1007/978-3-319-65930-5_40

solutions to these problems. This board is composed of federal public servants that work with more than 60 different universities and educational institutions in Brazil. Recently, the Brazilian government created a specific risk management law known as Normative Instructions 01/2016 [1] to develop systematic risk management practices in public organizations. As HEIs are part of public sector, they must develop a risk management strategic plan to control possible risks to the institution. Following this initiative, risk management must be carried out at the organizational level. Risk management software is crucial for assessing multiple risks and enabling consistent communication and monitoring. In this respect, information systems can enable risk management at HEIs based on systematic and rational practices, as determined by software.

Enterprise risk management (ERM) is also a methodology that can provide managers with a new perspective for monitoring and achieving organizational objectives. Some research has shown consistent benefits from ERM [2]. Other research has suggested that risk governance be compared cross-nationally to understand differences between nations given unique regulatory contexts [3]. Finally, risk management dimensions have not yet been clearly or definitively defined, and the benefits to organizations are not clear [4]. Public and private organizations have a proven need for ERM in order to assess risks that could affect their objectives, although ERM studies on the public sector are largely absent. This lack of studies compromises the effectiveness of information systems software. Therefore, risk management in the public sector must be studied in order to adequately develop related management software and to achieve the best risk management results.

Improving the practice of public management can enhance the social benefits provided by public institutions as well as the quantity and quality of services delivered to citizens, thereby improving a nation's development and welfare. The objective of this study is to identify how public servants at HEIs perceive risk management in order to define the main requirements for an ERM software in this environment. To achieve this goal, a survey was created to evaluate the risk management perception of both non-specialists and risk managers. This would allow the correct requisites for assessing risk to be identified and incorporated in information systems, thereby enabling the development of software to subsequently aid managers in ERM practices.

2   Background

In this section, we will present some background on modern theories of public governance and new public management based on a rational approach for public sector management. The main enterprise risk management methodologies will be presented in sequence. Finally, the implementation of risk management in the Brazilian public sector will be detailed.

2.1   Public Governance and New Public Management

In Brazil, starting in the 1990s, a series of attempts at administrative reforms focused on creating a more efficient Brazilian state with greater management capacity. Similar to other countries, the scope of these attempted transformations was framed within a new management paradigm called "new public management," which represents the adoption of managerial ideas and practices arising from the private sector in the public management domain, with a results-oriented and citizen-oriented focus. This involves downsizing, privatization of companies and activities, decentralization of activities to subnational governments, outsourcing of public services, regulation of activities conducted by the private sector, and the establishment of mechanisms for measuring costs and evaluating results [5–7].

The concept of public governance is sometimes used in an abstract way and subjected to different interpretations [8, 9]. However, much research on public governance has emphasized that, currently, traditional mechanisms of political management and control are no longer effective. For this reason, it makes sense to speak of government but also of governance, referring to the participation of different sectors of society in the government and their involvement in the formation and management of public policies.

Risk management occupies an important place in both new public management and in public governance, although numerous approaches exist. The core set of principles typically recommended by new public management approaches are, for example, total quality management, service management, productivity compensation, and risk management. Notwithstanding the "managerial fads" typical of new public management, interest in risk management has had a strong temporal correlation with the advent of new public management.

2.2   Risk Management Methodologies

The Committee of Sponsoring Organizations of the Treadway Commission (COSO®) issued the Enterprise Risk Management – Integrated Framework in 2004. Later, this framework was simply known as ERM Cube ® or COSO II ® [10]. The objective of this framework is to help business managers and other entities to better deal with risk in achieving their objectives.

ISO 31000® 2009, Risk Management – Principles and guidelines, provides a framework for risk management and assessment that can be adopted by different organizations during strategic decision making, business processes, and projects as well as during functional and service activities [11]. It can be applied to different types of risks, regardless of their nature, with potentially positive or negative impacts.

The M_o_R® 2010 (Management of Risk) framework, developed by OGC®, is a guide designed to assist organizations in making decisions about risks that may affect their achievement of strategic, program, project, or operational objectives [12]. It is a methodology that addresses principles, approaches, and processes in a set of interrelated steps, considering various dimensions of risk management and techniques for determining risk in organizations. Notes and references exist for ISO 31000®. Both ISO 31000® and M_o_R® are not competitors, but complementary to risk management.

Brazil, as most developing countries, has adopted international standards and methodologies to help managers and to teach them best practices and guidelines, for example ISO 21500, PMI®, and Prince2® for project management as well as ISO 31000® [11], ERM Cube ® [10], and M_o_R® [12] for risk management. The use of abroad disciplines and guidelines can be a good start, but these sometimes are not reflective of a local culture or values and may therefore give results that are not apt at the local level.

3 Research Method and Data Collection

This work is an exploratory study. Quantitative and qualitative data were collected from a survey of top managers' at Brazilian HEIs in order to understand their perception of risk management.

The survey was developed by a risk management specialist, a public administration consultant, and an independent management consultant. The questionnaire has three different sections: an initial risk management assessment with two objective questions, two open questions on risk management perception, and five non-obligatory questions that include respondent identification and the option to request future feedback or analyzed data from the research project via e-mail (for those respondents who are interested).

The questionnaire was distributed by an internet link and e-mail using the Survey Monkey® platform.

During the pre-test analysis [13], the survey was evaluated by two risk management experts. The 20 initial results were used to collect respondent feedback and to adjust the survey questions. After receiving these 20 responses, some adjustments were made to the questionnaire; these first 20 answers were not considered in the analysis.

The number of questionnaires distributed to HEI managers was 180, and a total of 86 questionnaires were answered. After validating the responses, 67 questionnaires were considered for the analysis.

4 Results

At following, the survey questions and results are presented. Table 1 shows the results for the two initial objective questions: (1) Has your institution already defined a committee and/or those responsible for risk management? (2) Are you a member of this committee and/or are you responsible for risk management at your institution?

The next two questions were open ended, as the respondents were able to write out their response or perception.

Table 1. Survey objective questions 1 and 2

		Question 1			
		I do not know	No	Yes	**Total**
Question 2	No	5	18	15	**38**
	Yes	0	1	28	**29**
	Total	**5**	**19**	**43**	**67**

Question 3 was as follows: Why is risk management important for your institution to achieve its results? The main answers were grouped into the following reasons: objective goal achievement, informed decision making, uncertainty reduction, resources use optimization, planned responses, error reduction, compliancy and organizational self-knowledge.

Question 4 was as follows: According to your perception, what are the main challenges, difficulties, and limitations for effective implementation and practice of risk management at your institution? The main responses dealt with lack of training, limited staff, absence of risk management culture, undefined department structures, lack of interest shown by public servants, reduced budget, complexity of risk management, and low engagement of top management with ERM.

The other five questions were non-obligatory (personal) and will not be shown.

5 Discussion and Final Considerations

In a related study, we were able to verify that some public organizations were struggling to define an ERM methodology for their organization [14, 15]. Difficulties were also identified in the implementation of ERM tools and their analysis as well as in the application of techniques for internal auditing [16, 17]. In addition, the perception of top public sector managers of ERM was analyzed [18].

Brazilian higher educational institutions are facing similar difficulties in implementing ERM. We noticed that respondents responsible for risk management at their institutions provided answers that referenced the main risk management frameworks, such as ISO 31000 ® [11] and ERM Cube ® [10]; apparently, the benefits of adopting risk management are known by these respondents. Those who are not responsible for ERM appear to also notice the importance of risk management even though they lack knowledge on its related methodologies or other aspects. Both those who were responsible and non-responsible for ERM perceived a lack of staff dedicated to ERM and minimal training on this subject.

After analyzing the answers, the main requirements for ERM software are related to the need for a simple and objective methodology that can aid risk management through the use of gamification or other tools. Such software can increase the engagement of actors and top managers, allowing for risks to be easily monitored and aligned to objectives. As a result, the working time necessary for controlling risks can be reduced. Some additional functions may include e-mail reminders for important dates and simple risk management tutorials and videos for guiding users on risk assessment and software usage. Furthermore, the development of an open software would be beneficial due to financial restrictions.

Some limitations in this study were related to the small number of top managers available to participate in the survey. The low number of participants who fully completed the survey meant a reduced amount of opinions and perceptions. A second survey application would increase the number of participants and enable comparison with the first survey. In future works, we recommend measuring the perception of specific variables related to the performance of the public sector in implementing ERM. Also, a review of the main methodologies that have been adopted could help managers decide which framework to use. Finally, ERM maturity could be measured and compared among similar organizations.

Risk management skills could improve the work efficiency and efficacy of public sector servants, enabling them to provide better quality service to citizens and to implement more efficient spending in the public budget.

Acknowledgements. We would like to acknowledge the support of Fundação de Apoio à Cultura, Ensino, Pesquisa e Extensão de Alfenas – FACEPE (Support Foundation for Culture, Education, Research and Extension of Alfenas) for this project titled "Gestão de Riscos nas Universidades Federais: Elaboração de Modelo de Referência e Implantação de Sistema (Risk Management in Federal Universities: Development of a Reference Model and System Implementation)."

References

1. Ministério do Planejamento Orçamento e Gestão, Controladoria Geral da União, B.: Instrução Normativa N 01/2016, Brasília, DF (2016)
2. Bromiley, P., McShane, M., Nair, A., Rustambekov, E.: Enterprise risk management: review, critique, and research directions. Long Range Plann. **48**, 265–276 (2015)
3. Stein, V., Wiedemann, A.: Risk governance: conceptualization, tasks, and research agenda. J. Bus. Econ. **86**, 813–836 (2016)
4. Hillson, D.: The Risk Management Handbook: A Practical Guide to Managing the Multiple Dimensions of Risk. KoganPage, London (2016)
5. Osborne, D., Gaebler, T.: Reinventing government: how the entrepreneurial spirit is transforming government. Adison Wesley, Boston (1992)
6. Kettl, D.F.: The Global Public Management Revolution. Brookings Institution, Washington, DC (2005)
7. Pollitt, C., Bouckaert, G.: Public Management Reform: A Comparative Analysis. Oxford, USA (2011)
8. Denhardt, R.B., Catlaw, T.J.: Theories of Public Organization. Cengage Learning, Stamford (2015)
9. Marini, C., Martins, H.F.: Governança Pública Contemporânea: uma tentativa de dissecação conceitual. Revista TCU (2014)
10. Committee of Sponsoring Organizations of the Treadway Commission - COSO: Enterprise Risk Management: Integrated Framework (2004)
11. International Electrotechnical Commission - IEC. International Organization for Standardization - ISO: ISO 31000. Risk Management - Principles and guidelines. (2009)
12. Office of Government Commerce - OGC: Management of Risk : Guidance for Practitioners. Axelos, London (2010)
13. Hair, J.F., Black, W.C., Babin, B.J., Anderson, R.E., Tatham, R.L.: others: Multivariate data analysis. Prentice hall, Upper Saddle River (1998)
14. Leung, F., Isaacs, F.: Risk management in public sector research: approach and lessons learned at a national research organization. R D Manag. **38**, 510–519 (2008)
15. Oulasvirta, L., Anttiroiko, A.-V.: Adoption of comprehensive risk management in local government. Local Gov. Stud. **43**, 451–474 (2017)

16. Vinnari, E., Skaebaek, P.: The uncertainties of risk management A field study on risk management internal audit practices in a finnish municipality. Account. Audit. Account. J. **27**, 489–526 (2014)
17. Schiller, F., Prpich, G.: Learning to organise risk management in organisations: what future for enterprise risk management? J. Risk Res. **17**, 999–1017 (2014)
18. Coetzee, P.: Contribution of internal auditing to risk management Perceptions of public sector senior management. Int. J. Publ Sect. Manag. **29**, 348–364 (2016)

Management and Organizational Issues in Information Systems

A Practice-Based Methodology to Enlighten Strategic Alignment Research

James Holohan[1]([⊠]) and Joe McDonagh[2]

[1] Department of Information Technology, Limerick Institute of Technology,
Moylish Park, Limerick, Ireland
jim.holohan@lit.ie
[2] School of Business, Trinity College Dublin, Dublin 2, Ireland
jmcdongh@tcd.ie

Abstract. Strategic alignment (SA) has been researched extensively since it first emerged in the early 1990s as a theme within the information systems (IS) strategy domain. The current research trajectory for SA is limited given the predominantly static focus placed on the concept to date, hence we know relatively little about what practitioners do in their day-to-day activity to achieve SA. We therefore propose a practice-based research methodology to help close this knowledge gap. More specifically, we propose joining together, the three pillars of, a Strategy-as-Practice lens, the constructivist grounded theory coding method and case study research design. This practice-based methodology will aid construction of mid-range theory that will have practical relevance in terms of how SA is practiced by practitioners. We recommend applying the methodology to two new research avenues, as a means to build mid-range theory to help us better understand what practitioners actually do, in their day-to-day activity to achieve SA.

Keywords: Strategic alignment · Practice-based methodology · Strategy-as-Practice · Mid-Range theory

1 Introduction

The strategic value of information systems (IS) to organisations is reflected in the stream of research consisting of three closely related sets of literature developed since the late 1970s within the IS strategy domain (Chan and Huff 1992; Ward 2012). These sets comprise; IS for competitive advantage, strategic information systems planning, and strategic alignment (SA) (Chen et al. 2010; Merali et al. 2012). Each literature set is centred around the key concept of IS strategy, which is in turn a major element of corporate strategy (Pyburn 1983).

Since the early 1990s, perspectives on SA have evolved due to its complexity and the ever changing business environment it seeks to address (Chan 2002; Merali et al. 2012). Today's challenge is not to achieve SA only when plans are devised, rather it is to continuously align IS and business goals by periodically addressing all major aspects of related IS planning (Salmela and Spil 2002). Undoubtedly, SA's evolution has contributed to the numerous definitions put forward in the literature, thus making it

M. Themistocleous and V. Morabito (Eds.): EMCIS 2017, LNBIP 299, pp. 515–530, 2017.
DOI: 10.1007/978-3-319-65930-5_41

difficult to determine what exactly SA is (Holohan and McDonagh 2014b). Avison et al. (2004:225) note that although a clear and agreed definition for SA does not exist, in all cases, *"it concerns the integration of strategies relating to the business and IS"*, and this is a view fully supported within the nascent SA literature (Cumps et al. 2009; Al-Hatmi and Hales 2010; Leonard and Seddon 2012; Renaud et al. 2016).

From our review of the SA literature (Holohan and McDonagh 2014a, b) we found the main focus of attention within the SA literature to be on the intellectual dimension of SA and within this focus a strong emphasis exists on measuring organisation level SA in the pursuit of increased organisation performance. With approximately 24% of the literature devoted to models/measurement that provide a mechanistic view, as distinct from practical relevance, these studies provide little value in aiding our understanding of issue that are relational (people) based (Henderson and Venkatraman 1993; Palmer and Markus 2000; Gartlan and Shanks 2007; Leonard and Seddon 2012). The current research trajectory for SA is limited given the predominantly static focus placed on the concept to date (Campbell 2005). The literature also suggests SA results more from relationships between people than from any methodological analysis or business strategy (Reich and Benbasat 1996; Campbell et al. 2005; Ghosh and Scott 2009; Gast and Zanini 2012). Therefore, research into SA would significantly benefit from a greater focus on the practices carried out by people, as they endeavour to achieve SA (Coltman et al. 2015; Karpovsky and Galliers 2015).

In Sect. 2 we provide the reasons why we believe the time is ripe for an alternative methodology to research SA. A brief history of practice research is provided in Sect. 3, followed by Sect. 4 wherein the rationale as to why a practice-based approach is the most appropriate is elucidated. Section 5 describes our proposed practice-based research methodology, which comprises the three pillars of Whittington's (2006b) integrative framework for Strategy-as-Practice (SaP), the constructivist grounded theory coding method espoused by Charmaz (2014) and case study research design. In Sect. 6 our proposed practice-based research methodology is positioned relative to the extant SA literature. Section 7 proposes two new SA research avenues to which the methodology can be applied, to enlighten our understanding as to what practitioners actually do, in their day-to-day activity to achieve SA. Section 8 brings the paper to a close.

2 The Requirement for an Alternative Methodology to Research SA

The current research trajectory for SA is limited given the predominantly static focus placed on the concept to date, hence we know relatively little about what people do in their day-to-day activity to achieve SA (Campbell 2005). Therefore, research into SA would benefit greatly from a greater focus on the practices carried out by people, as they endeavour to achieve SA (Coltman et al. 2015; Karpovsky and Galliers 2015). One such benefit would be the development of mid-range theory that would extend our understanding as to how SA is enacted in practice, rather than relying on the rationality of current prescriptive models (Karpovsky and Galliers 2015; Renaud et al. 2016).

A review of the SA literature undertaken by Holohan and McDonagh (2014a) found, that despite decades of research into SA, we still haven't deciphered how exactly strategists construct SA in practice. As social scientists our primary role is to make theoretical progress (Aguinis and Vandenberg 2014) and hence the alternative methodology that we propose to research SA, is focused on helping to develop mid-range theory as to how organisation actors practice SA. Given this objective and the calls of Ciborra (1997), Renaud and Walsh (2010), and Hiekkanen et al. (2013) for work with strong links to practice, suggests we need practice based studies rooted in our everyday experiences, to aid our understanding as to what practitioners do to help achieve SA.

3   Brief History of Practice Research

Practice is an umbrella term under which particular views are taken. A practice perspective on strategy builds on the work of seminal scholars in social theory (Jarzabkowski et al. 2013). Philosophers such as Wittgenstein (1998) are of the view that practice draws attention to tacit knowledge and illustrates intelligence. Social and organisational theorists such as Giddens (1993) and Bourdieu (1990) see practice as a means of liberation from social structures and systems, to question activities and their influence in shaping social phenomena. Cultural theorists such as Foucault (1980) see practice as routinised interpretations, as a means to oppose structures and systems (Rasche and Chia 2009). The various theorists, particularly those in philosophy and social sciences, see human activity at the core of practice, and acknowledge the social, historical and structural contexts in which knowledge and meaning are generated (Corradi et al. 2010; Nicolini 2012). Given these different perspectives, it is evident a unified theory of practice does not exist.

At the forefront of all practice theories is an emphasis on activity and the requirement to uncover work that makes up this activity, therefore placing the focus on practices carried out by practitioners as they engage in their work (Jarzabkowski and Seidl 2008; Nicolini 2012).

A significant contribution from the practice perspective is that the practice lens enables the researcher to develop an account of the practices that take place (Kaplan 2007). Reality is shaped by practices and therefore our understanding of organisational phenomena is enhanced by treating practices as the focal lens through which to study such phenomena (Orlikowski 2010).

4   Rationale for a Practice-Based Approach

Within the field of IS we study human actions and our questions are of the;

"How do you do something type?" (Niederman et al. 2009:649)

Such actions are practices, hence the focus of attention to help understand organisational phenomena is placed on the practices themselves (e.g. the doing of strategy) rather than on the practitioners who execute them (Reckwitz 2002; Nicolini 2012).

With our focus on practices, we place human action at the centre, rather than the performance of the organisation.

Practice theory, with its focus on dynamics, human relations and enactment, is therefore particularly receptive to the application of our proposed practice-based research methodology, as the methodology can help us understand the day-to-day practices undertaken by practitioners in their quest to achieve SA, whether or not those who carry out the actions are tasked with a formal strategy role (Hendry and Seidl 2003; Whittington 2014). From an ontological perspective, this firmly places practitioners and processes subordinate to practices, thus instilling practices as the primary unit of analysis (Jarzabkowski et al. 2007). In addition, while the "practice turn" focuses on the practices of the practitioner(s), it also focuses on how such practices are rooted in and driven by general organisational and institutional accepted practices, (Bartunek et al. 2011). This linkage between lower level entities and higher level organisational practices, has been researched by social scientists for well over a century (Kozlowski et al. 2013).

5   An Alternative Methodology to Research SA

Three specific research streams have given rise to the current movement in practice-based organisational studies. They are the study of learning and knowing phenomena as situated practices, the study of technology as practice, and the study of SaP (Kaplan 2007; Corradi et al. 2010).

We advocate joining together, within the interpretive paradigm, the three research pillars of; Whittington's (2006) integrative framework for SaP, the constructivist grounded theory coding method espoused by Charmaz (2014) and case study research design. This inter-relationship, comprises our proposed practice-based methodology to research SA and build mid-range theory as to how SA is practiced by practitioners.

5.1   Why We Draw upon a SaP Lens

While the SaP stream has its theoretical roots in sociological theories of practice (Regnér 2008), it also has a strong history with closely observed studies of strategy in an organisational context. SaP is concerned with;

> "the doing of strategy; who does it, what they do, how they do it, what they use and what implications this has for shaping strategy." (Jarzabkowski and Spee 2009:69)

Therefore, it brings practitioners, their actions, their interactions and context, into the heart of strategy research, thus extending the range of outcomes from traditional strategy research beyond that of economic performance to the performance of individuals, groups and organisations (Rousseau 1985; Paroutis et al. 2016).

In line with the broader practice approach, SaP conceptualises strategizing as consisting of three interrelated routines i.e. practitioners, practices and praxis (Suddaby et al. 2013). Practitioners are those who do the work of strategy (Jarzabkowski and Spee 2009), practices represent the done thing i.e. routinised types of behaviour and praxis represents what is actually done i.e. the whole of human action (Reckwitz 2002). The difference between practice and praxis is the difference between the routine that guides activity and the actual activity carried out. Therefore, while praxis is informed and guided by practices, it is unique in that it exists only in the present and includes the routine as well as the non-routine. SaP scholars see strategy as something people do and not something organisations possess, and by studying how people engage in the doing of strategy, we come to understand the nature of strategising and the practices drawn upon to do so (van Wessel et al. 2011; Jarzabkowski et al. 2013). Therefore, the value offered by SaP is in the moving away from what strategy should be or how it should be done, to how it is actually done in practice by people as they construct, implement and realise their strategic objectives.

SaP scholars place a high importance on understanding micro phenomena not in isolation, but rather within the wider social context from which they draw (Jarzabkowski et al. 2007), therefore enabling a better understanding of micro activity and how it impacts on the stability and instability of organisations. SaP offers an alternative to focusing on organisation performance alone, which is so prevalent within the SA literature. Such dominance contributes significantly to researchers paying little attention to the day-to-day practices that can help shape SA (Jarzabkowski et al. 2007; Peppard et al. 2014). By going beyond rational strategy analysis, SaP draws attention to practices engaged in by practitioners, which includes the ability to connect micro level analysis with micro (e.g. department), meso (e.g. organisation) and macro (e.g. organisation's external environment) level considerations (Regnér 2008). Therefore, with a focus on the doing of strategy work, SaP offers the opportunity to obtain a deep level of understanding as to how SA is practiced by practitioners.

5.2 Integrative Framework for SaP

We reviewed the three most prominent SaP theories put forward in the literature. They being; the conceptual framework for analysing SaP by Jarzabkowski et al. (2007), the activity theory framework by Jarzabkowski (2003) and the integrative framework for SaP by Whittington (2006b). Of the three, Whittington's (2006) integrative framework for SaP was the most appropriate theory to;

i. enable identification of practices engaged in by practitioners at the micro, meso and macro levels of activity; and
ii. facilitate the development of mid-range theory in terms of how the alignment of business and IS strategies is practiced by practitioners.

The integrative framework for SaP (refer Fig. 1) developed by Whittington (2006b) elucidates the mutual dependency and interconnectedness of practitioners, practice and praxis. In addition, the integrated nature of Whittington's framework helps to explain the importance of connecting micro activity with the micro, meso (intra-organisational field) and macro levels (extra-organisational field), therefore allowing for the recognition of multilevel implications.

Whittington's (2006b) integrative framework for SaP integrates practitioners, praxis and practices into a mutually dependent whole (refer Fig. 1). At the base are a

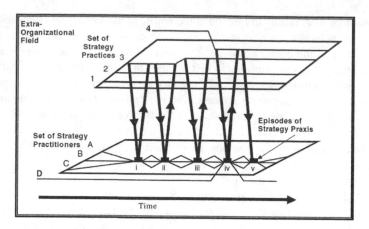

Fig. 1. Integrative framework for SaP (Whittington 2006b).

set of practitioners e.g. IS managers, finance managers and marketing managers (A, B, and C) who at particular intervals engage in various strategising activities e.g. meetings, writing reports, presentations, seminars and form filling or otherwise known as praxis (i, ii, iii, iv and v). As the practitioners engage in strategising, they employ an assorted array of strategy practices e.g. Porter's five force model, Andrew's SWOT, Boston Consultancy Grid (1, 2 and 3).

If an organisation was to rely solely on its existing inventory of practices, then in turn, praxis can end up ensnared by such routine. Figure 1 illustrates how strategy praxis can be developed by not allowing an organisation rely solely on its existing practices. Focusing on practitioners A, B, and C, they utilise a range of practices 1, 2 and 3 (at the micro and meso levels), and in so doing they become more and more reliant on these practices unless the practices are further developed. Such development can occur when one or more of the practitioners learn from being involved in episode (s) of strategy praxis (e.g. exchange information at seminars in the extra-organisational field) and as a consequence, develop what becomes accepted as a new practice (e.g. an amalgamation of the SWOT and Boston Consultancy Grid – practice 2 and 3) resulting in the emergence of practice 4 (brainstorming). Alternatively, practitioner D (a consultant) could enter the organisation from the extra-organisational field at interval iv and bring with him/her a new practice (e.g. brainstorming). This new practice of brainstorming then becomes an established practice within the organisation, thus refreshing the current set of practices at the meso and micro levels. In addition, this practice of brainstorming may well be further developed while being utilised in the organisation and be brought by practitioner D (the consultant) outside the organisation to be applied elsewhere at the macro level.

Referring to Fig. 1, the upward arrows denote the strategy practices e.g. SA practices typically carried out by practitioners. The downward arrows indicate the influence of practices on praxis, and the opportunity for change, particularly change shaped by practices that come from outside the organisation. Therefore, Whittington's (2006b) integrative framework for SaP, with its multilevel design, facilitates designating the practices as the unit of analysis through which to understand how SA is practiced by practitioners, through the collection of qualitative data, which are central to a SaP study. The integrated nature of Whittington's (2006b) integrative framework for SaP helps to explain how practitioners, practice and praxis bind at the micro, meso and macro levels and how such binding can subsequently shape what takes place at the various levels. This provides the framework with a clear mechanism to link the micro activity with the micro, meso and macro levels. Therefore, the framework is very well positioned to support qualitative multilevel research and hence aid the development of mid-range theory in terms of how SA is practiced by practitioners.

Practice research provides knowledge about practices, denoting the empirical field as an array of practices. This requires theory be built based on the knowledge gained from what is taking place within practices, hence the lived experiences of the practitioners including their thoughts and activities, constitute the building blocks of such theory (Chia and MacKay 2007). As Whittington's (2006b) integrative framework for SaP is consistent with our view to studying practice, we advocate applying the framework to guide data collection and analysis.

There are challenges faced by the researcher in analysing data when undertaking practice research. Practice research lacks a formalised and codified set of research techniques, which can lead to the accusation it lacks standards and risks developing an invalid and unreliable knowledge base. Codification is therefore necessary to help the researcher work out what to do and to keep track of what he/she has done. This is particularly true if practice research is to take place in large-scale settings and deal with large amounts of empirical data, so as to help ensure all those involved in the research are adhering to an agreed set of procedures. Codification demonstrates how the research was undertaken, how data was collected, how data and theory are linked, and how conclusions are drawn. Although practice research lacks a formalised and codified set of research techniques, it is similar to other interpretive research approaches insofar it is an art rather than a science thus requiring creative insight, laden with intuition, judgement and tacit knowledge (Pettigrew 1997). For this reason it can draw on procedures and rules from other interpretive research methodologies to overcome these shortcomings (Langley and Tsoukas 2010).

5.3 Grounded Theory - The Constructivist Approach

Whether grounded theory is discovered (positivist tradition) or whether it is constructed (interpretive tradition), the basic premise is the innovative theory is developed from empirical data systematically obtained by researching social phenomena, where the researcher engages with the actors and their contexts (Urquhart et al. 2010; Glaser and Strauss 1967).

The main difference between the constructivist approach and the earlier approaches of grounded theory, is the tools and guidelines of grounded theory are adopted while the positivist assumptions are not (Fendt and Sachs 2008; Charmaz 2014). Unlike the view put forward by Glaser and Strauss (1967) and Corbin and Strauss (2008), whereby they assume an external reality exists awaiting discovery by an unbiased observer who records facts leading to theory discovered from systematically obtaining and analysing data, the constructivist approach does not take the view data is waiting to be discovered and nor does it take the view the researcher should enter the field without prior knowledge from the literature (Bryant 2002; Seidel and Urquhart 2013). The constructivist is of the belief that a qualitative approach cannot depend on deduction alone and the researcher needs to construct what he/she interprets as data, based on his/her engagement in the field (Mills et al. 2006; Seidel and Urquhart 2013).

5.3.1   The Constructivist Grounded Theory Coding Method

Many studies in IS do not use the full set of grounded theory methods with a view to generating theory, but they do avail of a grounded theory coding method as a qualitative data analysis tool to supplement other qualitative research methodologies (Urquhart et al. 2010; Seidel and Urquhart 2013). While this may appear limited, it is perfectly suited to cases where theory building via the full application of grounded theory methods is not the primary purpose. Therefore, to guard against practice research's lack of a formalised and codified set of research techniques, our proposed practice-based methodology will draw on the constructivist grounded theory coding method as a means to analyse data.

The coding process selects, separates and sorts the data gathered, thus enabling the researcher to begin analysis. Charmaz (2014) describes coding as the pivotal link between collecting data and developing an emergent theory to explain these data. Her constructivist grounded theory coding method involves four sets of coding namely; initial, focused, axial and theoretical (refer Table 1). To help initial coding stay close to the data, the researcher should try to code with gerunds thus helping to stifle the tendency towards conceptual leaps prior to undertaking analysis (Charmaz 2014). Coding with gerunds will help us place the emphasis on practices and actions rather than on explaining, thus supporting the SaP approach. The resultant theory, while acknowledging subjectivity, will be tied closely to the empirical data gathered.

Table 1. Constructivist grounded theory coding method by Charmaz (2014).

Constructivist grounded theory coding method
Initial coding
Focused coding
Axial coding
Theoretical coding

Focused coding involves using the most significant and frequent codes from initial coding to sift through the large amounts of data the researcher has gathered. Charmaz (2014) differs from Corbin and Strauss (2008) when it comes to axial coding, as she does not advocate adhering rigidly to their formal and time consuming procedures that aid verification, because it depends too heavily on preconceived prescriptions and introduces a needless level of complexity, which together introduce the threat of forcing the data to fit the theory. Theoretical coding follows closely the codes already selected during focused coding by specifying relationships between the categories developed during focused coding, and the theory begins to take shape.

Our approach to coding, is to build the code syntax based on the primary elements within Whittington's (2006) integrative framework for SaP. Therefore the syntax for initial codes is as follows:

Gerund-level-practitioner-prct/prx-int/ext

The gerund reflects the practice. The level is the level at which the practice takes place (micro, meso or macro). The practitioner is the person who carries out the practice. The prct/prx reflects whether it is a routinised practice or not, and the int/ent reflects whether the practice has been established within the organisation or has come from outside.

5.4 Case Study

The view, whereby a case study is an intensive study of one unit with the aspiration to generalise/shed light on a question concerning a larger set of units (Eisenhardt 1989; Gerring 2004), supports a case study approach to answering research questions focused on understanding real-life phenomenon, in-depth, within organisations.

Although case research in IS has been dominated by the positivist tradition, there now exists a notable increase in interpretive case research to study IS issues (Myers and Liu 2009). A major difference between interpretive case study research and positivist case study research is interpretive case study research places a strong emphasis on close interaction between researcher(s) and participant(s) throughout the case study process, therefore considering case members as participants in the building of the case narrative (Bygstad and Munkvold 2011). In fact, case study is the most widely used design for the qualitative approach to research in IS and is ideally suited to understanding the interactions between IS related issues and organisational contexts (Dubé and Paré 2003).

5.4.1 Case Study as a Research Design

The basic framework for undertaking research, is the research design itself. A research design is;

> *"a logical plan for getting from here to there, where here may be defined as the initial set of questions to be answered, and there is some set of conclusions (answers) about these questions." Yin (2014:28)*

A number of steps have to be undertaken to get from the "here" to the "there" and the principal purpose of research design is to help the researcher ensure evidence he/she seeks and obtains, addresses the research question(s). According to Yin (2014) and supported by Dubé and Paré (2003), five components of research design are specifically important for case studies:

- *A study's question.* As our proposed practice-based research methodology is focused on exploring practices, thus lending itself to questions of the "how do you do something" type, it is very suited to incorporating an exploratory (as distinct from explanatory and descriptive) case study research design.

- *A study's propositions, if any.* As our proposed practice-based research methodology is focused on developing mid-range theory, it therefore has a very clear purpose.
- *Unit(s) of analysis.* The main unit(s) of analysis in our proposed practice-based research methodology are the practices engaged in by practitioners.
- *The logic linking the data to the purpose.* Our proposed practice-based research methodology will uncover the logic, by utilising Whittington's (2006b) integrative framework for SaP to guide data collection and analysis.
- *Criteria for interpreting the findings.* Our proposed practice-based research methodology will avail of the constructivist grounded theory coding method for analysing data.

As current state of the art for case study research design does not provide detailed guidance on how to carry out steps 4 and 5 (Yin 2014) i.e. there is no standard for data collection and analysis, greater demands are placed on the researcher during these two phases. However, Whittington's (2006b) integrative framework for SaP coupled with the constructivist grounded theory coding method, will help overcome these shortcomings.

5.4.2   Methods of Data Collection

Yin (2014) identifies six of the most common sources of evidence referenced during case study research as; documentation, archival records, interviews, direct observation, participant observation and physical artefacts. Each source has its strengths and weaknesses, but taken together the sources complement one another by providing a rich set of data that support the research findings, hence the researcher using case study as a research design should strive to use as many of the sources as possible (triangulation) to limit his/her biases.

Documentation is very useful in case study research for corroborating and augmenting evidence from alternative sources (Miles et al. 2014; Yin 2014). Archival records and their usefulness vary from case to case. Not unlike documentation, the researcher should determine the conditions under which the records were produced and their level of accuracy (Paré 2004; Yin 2014). Interviews are the primary data source for the researcher who acts as an outside observer, as it is the most appropriate method for accessing participant's interpretations (Walsham 1995; Paré 2004). The interview should be guided rather than structured, to take advantage of the information that can be gathered within a face-to-face real-time setting. By taking place in the natural setting and as long as the phenomena of interest are not entirely historical, case studies provide the opportunity to undertake direct (real-time) observation. Direct observation can include sitting in on meetings, observing a teacher in the classroom or watching how factory floor work takes place. A participant observer can take on a number of roles within a case study and may even participate in the events under study (Paré 2004; Yin 2014). More suited to the ethnographic methodology, collecting physical artefacts can provide additional data (Paré 2004; Yin 2014).

5.4.3 Executing Data Analysis

A major criticism levelled at case study is related to the analysis of large amounts of qualitative data, whereby there is no standard analysis approach (Darke et al. 1998). Eisenhardt (1989) maintains that analysing data is at the heart of building theory from case studies, and that this part of the process is the most difficult and least codified. In introducing his chapter on analysing case study evidence, Yin (2014) proffers that analysis of case study evidence is one of the most difficult and least developed aspects of doing case studies. He holds the view that the researcher, as part of his/her research design, should develop a data analysis strategy, because without one, the data will remain as raw data without a story to tell.

When data analysis is guided by Whittington's (2006b) integrative framework for SaP and executed via the constructivist grounded theory coding method, the full benefits of applying case study as a research design can be realised.

6 The Proposed Practice-Based Research Methodology Relative to Extant SA Literature

Apart from the TSAM model developed by Kalika and Walsh (2010) aimed at having practical meaning, the main area of difference between the SA models presented in the literature and the focus of our proposed practice-based research methodology, is that the models are mechanistic and take no account of human action in shaping SA. Therefore they are not very suitable for today's dynamic world of IS and strategy.

Within the SA literature, strategic planning and SA are closely associated, with numerous calls for the highest ranking IS executive to participate in business planning and for business executives to participate in IS planning (Pyburn 1983; Teo and King 1996; Kearns and Sabherwal 2007). Our proposed practice-based research methodology, when utilised, will aid the development of mid-range theory in terms of how SA is practiced by practitioners.

Unfortunately, the vast majority of the SA literature neglects the social dimension, which is an essential dimension for practitioners at the micro, meso and macro levels (Renaud and Walsh 2010; Schlosser 2012). These relationships are fostered through the sharing of knowledge between business and IS executives, resulting in narrowing the knowledge gap that can impede SA (Ghosh and Scott 2009; Enns and McDonagh 2012). Our proposed practice-based research methodology contributes towards answering the calls of Ciborra (1997), Hiekkanen et al. (2013) and others, for research into the social dimension of SA, rooted in our everyday experiences that include both the formal and informal role of practitioners.

There is a dearth of practice-based studies within the SA literature, to aid our understanding as to how people in organisations practice SA (Holohan and McDonagh 2014b). Our proposed practice-based research methodology facilitates the extension of prior concepts reported in the SA literature, by enabling a holistic view as to how practitioners practice SA at the micro, meso and macro levels (Karpovsky and Galliers 2015).

7   Application of the Proposed Practice-Based Research Methodology Within the SA Domain

Our proposed practice-based research methodology will aid the construction of mid-range theory that will have practical relevance in terms of how SA is practiced by practitioners. Therefore, the methodology has the potential to contribute towards reducing the theory-practice gap that currently exists within IS research and help improve IS strategic management practice, of which SA is a major component.

The review of the SA literature undertaken by Holohan and McDonagh (2014a) revealed that studies into SA occur predominately at the meso level, with a concentration on activities undertaken by groups of practitioners from within the organisation. The review also revealed that the current research trajectory for SA is limited, given the predominantly static focus placed on the concept to date. Having identified these gaps, Holohan and McDonagh (2014b) propose two new SA research avenues to which our proposed practice-based research methodology can be applied. The first of these avenues is to identify SA practices carried out by individual practitioners, both internal and external to the organisation, at all three levels of praxis (i.e. micro, meso and macro). The second avenue is to identify SA practices carried out by groups of practitioners, both internal and external to the organisation, principally at the micro and macro levels of praxis. Fulfilling this program of research will contribute towards enlightening our understanding as to what practitioners actually do, in their day-to-day activity to achieve SA. The resultant mid-range theory can be availed of by others, to explore this phenomenon further.

8   Conclusion

Our proposed practice-based research methodology is concentrated on facilitating inquiry into how SA is practiced by practitioners in an organisational context. To guard against attributing data from one level to another level, we require a guiding theory that takes issues of level into account. We therefore chose Whittington's (2006b) integrative framework for SaP, as it is the only SaP theory we found that takes such issues into account.

Although SaP research lacks a formalised and codified set of research techniques, it can draw on procedures and rules from other interpretive research methodologies to overcome these shortcomings. We have chosen the constructivist grounded theory coding method, as it complements the Whittington (2006b) integrative framework for SaP as a data analysis tool, and is based on a qualitative approach to research within the interpretive paradigm.

Our choice of research design is case study, a design that has made valuable contributions to the field of IS theory and practice. The case study design is very suitable when the research is focused on real-life phenomena where little, if any, theoretical knowledge exists and where the phenomena cannot be studied outside the context in which it occurs. Although case study is an appropriate design for our proposed practice-based research methodology, it lacks guidance for data analysis, giving rise to practical limitations in terms of rigor and effectiveness. However, we are satisfied that when data analysis is guided by Whittington's (2006b) integrative framework for SaP and executed via the constructivist grounded theory coding method, the full benefits of applying case study as a research design can be realised. This makes case study a very suitable research design for our proposed practice-based research methodology and will help ensure findings and their validity, stand up to rigorous scrutiny.

Together, these three pillars comprise our proposed practice-based research methodology, which will aid the development of mid-range theory as to how practitioners practice SA. This proposed practice-based research methodology contributes to answering the calls of Ciborra (1997), Renaud and Walsh (2010), Hiekkanen et al. (2013), Karpovsky and Galliers (2015) and Holohan and McDonagh (2017a) for practice-based studies, to ameliorate our understanding as to how practitioners practice SA.

References

Aguinis, H., Vandenberg, R.J.: An ounce of prevention is worth a pound of cure: improving research quality before data collection. Ann. Rev. Organ. Psychol. Organ. Behav. **1**, 569–595 (2014)

Al-Hatmi, A., Hales, K.: Strategic alignment and IT projects in public sector organisation: challenges and solutions. In: European and Mediterranean Conference on Information Systems, Abu Dhabi, 12–13 April 2010

Avison, D., Jones, J., Powell, P., Wilson, D.: Using and validating the strategic alignment model. J. Strateg. Inf. Syst. **13**, 223–246 (2004)

Bartunek, J.M., Balogun, J., Do, B.: Considering planned change anew: stretching large group interventions strategically, emotionally, and meaningfully. Acad. Manag. Ann. **5**, 1–52 (2011)

Bourdieu, P.: The Logic of Practice. Polity Press, Cambridge (1990)

Bryant, A.: Re-grounding grounded theory. J. Inf. Technol. Theory Appl. **4**, 25–42 (2002)

Bygstad, B., Munkvold, B.E.: Exploring the role of informants in interpretive case study research in IS. J. Inf. Technol. **26**, 32–45 (2011)

Campbell, B.: Alignment: resolving ambiguity within bounded choices. In: 9th Pacific Asia Conference on Information Systems, 7–10 July 2005, Bangkok, Thailand, pp. 656–669 (2005)

Campbell, B., Kay, R., Avison, D.: Strategic alignment: a practitioner's perspective. J. Enterp. Inf. Manag. **18**, 653–664 (2005)

Cavaye, A.L.M.: Case study research: a multi-faceted research approach for IS. Inf. Syst. J. **6**, 227–242 (1996)

Chan, Y.E.: Why haven't we mastered alignment?: the importance of the informal organisation structure. MIS Q. Exec. **1**, 97–112 (2002)

Chan, Y.E., Huff, S.L.: Strategy: an information systems research perspective. J. Strateg. Inf. Syst. **1**, 191–204 (1992)

Charmaz, K.: Constructing Grounded Theory. Sage, Thousand Oaks (2006)

Charmaz, K.: Constructing Grounded Theory. Sage, Thousand Oaks (2014)

Chen, D., Mocker, M., Preston, D., Teubner, A.: Information systems strategy: reconceptualization, measurement, and implications. MIS Q. **34**, 233–259 (2010)

Chia, R., Mackay, B.: Post-processual challenges for the emerging strategy-as-practice perspective: discovering strategy in the logic of practice. Hum. Relat. **60**, 217–242 (2007)

Ciborra, C.: De Profundis? Deconstructing the concept of strategic alignment. Scand. J. Inf. Syst. **9**, 67–82 (1997)

Coltman, T., Tallon, P., Sharma, R., Queiroz, M.: Strategic IT alignment: twenty-five years on. J. Inf. Technol. **30**, 91–100 (2015)

Corbin, J., Strauss, A.L.: Basics of Qualitative Research: Techniques and Procedures for Developing Grounded Theory. Sage Publications, Thousand Oaks (2008)

Corradi, G., Gherardi, S., Verzelloni, L.: Through the practice lens: where is the bandwagon of practice-based studies heading? Manag. Learn. **41**, 265–283 (2010)

Cumps, B., Martens, D., Debacker, M., Haesen, R., Viaene, S., Dedene, G., Baesens, B., Snoeck, M.: Inferring comprehensible business/IT alignment rules. Inf. Manag. **46**, 116–124 (2009)

Darke, P., Shanks, G., Broadbent, M.: Successfully completing case study research: combining rigour, relevance and pragmatism. Inf. Syst. J. **8**, 273–289 (1998)

Dubé, L., Paré, G.: Rigor in information systems positivist case research: current practices, trends, and recommendations. MIS Q. **27**, 597–636 (2003)

Eisenhardt, K.M.: Building theories from case study research. Acad. Manag. Rev. **14**, 532–550 (1989)

Enns, H., McDonagh, J.: Irish CIOs' influence on technology innovation and IT-business alignment. Commun. Assoc. Inf. Syst. **30**, 1–11 (2012)

Fendt, J., Sachs, W.: Grounded theory method in management research: users' perspectives. Organ. Res. Methods **11**, 430–455 (2008)

Foucault, M.: Power/Knowledge: Selected Interviews and Other Writings 1972–1977. Pantheon, New York (1980)

Gartlan, J., Shanks, G.: The alignment of business and information technology strategy in Australia. Aust. J. Inf. Syst. **14**, 113–139 (2007)

Gast, A., Zanini, M.: The social side of strategy. McKinsey Q. **2**, 82–93 (2012)

Gerring, J.: What is a case study and what is it good for? Am. Polit. Sci. Rev. **98**, 341–354 (2004)

Ghosh, B., Scott, J.E.: Relational alignment in offshore IS outsourcing. MIS Q. Exec. **8**, 19–30 (2009)

Giddens, A.: New Rules of Sociological Method: A Positive Critique of Interpretative Sociologies. Polity Press, Cambridge (1993)

Glaser, B.G., Strauss, A.L.: The Discovery of Grounded Theory: Strategies for Qualitative Research. Aldine de Gruyter, New York (1967)

Hassan, N., Lowry, P.: Seeking middle-range theories in information systems research. In: 36th International Conference on Information Systems, 13–16 December 2015 Forth Worth, Texas (2015)

Heath, H., Cowley, S.: Developing a grounded theory approach: a comparison of Glaser and Strauss. Int. J. Nurs. Stud. **41**, 141–150 (2004)

Henderson, J.C., Venkatraman, N.: Strategic alignment: leveraging information technology for transforming organisations. IBM Syst. J. **32**, 4–16 (1993)

Hendry, J., Seidl, D.: The structure and significance of strategic episodes: social systems theory and the routine practices of strategic change. J. Manag. Stud. **40**, 175–196 (2003)

Hiekkanen, K., Helenius, M., Korhonen, J., Patricio, E.: Aligning alignment with strategic context: a literature review. In: Benghozi, P.-J., Krob, D., Rowe, F. (eds.) DED&M 2013. AISC, vol. 205, pp. 81–98. Springer, Heidelberg (2013). doi:10.1007/978-3-642-37317-6_8

Holohan, J., Mcdonagh, J.: Towards a systematic approach to reviewing the strategic alignment literature. In: 19th UKAIS Conference on Information Systems, 7–9 April 2014, St. Catherine's College, University of Oxford, England (2014a)

Holohan, J., McDonagh, J.: Reimagining strategic alignment research: a strategy as practice perspective. In: 28th Annual Conference of the Britsish Academy of Management, 9–11 September 2014, University Business School, University of Ulster, Belfast, Northern Ireland (2014b)

Holohan, J., McDonagh, J.: How information systems managers align business and information systems strategies in public service organisations: a practice-based taxonomy. In: 17th European Academy of Management Conference, 21–24 June 2017, University of Strathclyde, Glasgow, Scotland (2017a)

Jarzabkowski, P.: Strategic practices: an activity theory perspective on continuity and change. J. Manag. Stud. **40**, 23–55 (2003)

Jarzabkowski, P., Balogun, J., Seidl, D.: Strategizing: the challenges of a practice perspective. Hum. Relat. **60**, 5–27 (2007)

Jarzabkowski, P., Seidl, D.: The role of meetings in the social practice of strategy. Organ. Stud. **29**, 1391–1426 (2008)

Jarzabkowski, P., Spee, A.: Strategy-as-practice: a review and future directions for the field. Int. J. Manag. Rev. **11**, 69–95 (2009)

Jarzabkowski, P., Spee, A.P., Smets, M.: Material artifacts: practices for doing strategy with 'stuff'. Eur. Manag. J. **31**, 41–54 (2013)

Kalika, M., Walsh, I.: Re-conceptualising IS strategic alignment: the translated strategic alignment model (TSAM). In: 16th Americas Conference on Information Systems, 12–15 August 2010, Lima, Peru (2010)

Kaplan, S.: Book review - strategy as practice: an activity based approach. Acad. Manag. Rev. **32**, 986–990 (2007)

Karpovsky, A., Galliers, R.D.: Aligning in practice: from current cases to a new agenda. J. Inf. Technol. **30**, 136–160 (2015)

Kearns, G.S., Sabherwal, R.: Strategic alignment between business and information technology: a knowledge based view of behaviours, outcome and consequences. J. Manag. Inf. Syst. **23**, 129–162 (2007)

Kozlowski, S.W.J., Chao, G.T., Grand, J.A., Braun, M.T., Kuljanin, G.: Advancing multilevel research design: capturing the dynamics of emergence. Organ. Res. Methods **16**, 581–615 (2013)

Langley, A., Tsoukas, H.: Introducing "perspectives on process organization studies". In: Maitlis, T.H.A.S. (ed.) Process, Sensemaking, and Organizing. Oxford University Press, New York (2010)

Leonard, J., Seddon, P.: A meta-model of alignment. Commun. Assoc. Inf. Syst. **31**, 231–258 (2012)

Merali, Y., Papadopoulos, T., Nadkarni, T.: Information systems strategy: past, present, future? J. Strateg. Inf. Syst. **21**, 125–153 (2012)

Miles, M.B., Huberman, M., Saldana, J.: Qualitative Data Analysis: A Methods Sourcebook. Sage Publications, Thousand Oaks (2014)

Mills, J., Bonner, A., Francis, K.: Adopting a constructivist approach to grounded theory: implications for research design [corrected] [published erratum appears in Int. J. Nurs. Pract. **12**(2):119, April 2006]. Int. J. Nurs. Pract. **12**, 8–13 (2006)

Myers, M.D., Liu, F.: What does the best IS research look like? An analysis of the AIS basket of top journals. In: Pacific Asia Conference on Information Systems, Hyderabad, India (2009)

Nicolini, D.: Practice Theory, Work & Organization. Oxford University Press, Oxford (2012)

Niederman, F., Gregor, S., Grover, V., Lyytinen, K., Saunders, C.: ICIS 2008 panel report: IS has outgrown the need for reference disciplines theories, or has it? Commun. Assoc. Inf. Syst. **24**, 637–657 (2009)

Orlikowski, W.J.: Practice in Research: Phenomenon, Perspective and Philosophy. Cambridge University Press, New York (2010)

Palmer, J., Markus, M.: The performance impacts of quick response and strategic alignment in speciality retailing. Inf. Syst. Res. **11**, 241–259 (2000)

Paré, G.: Investigating information systems with positivist case study research. Commun. Assoc. Inf. Syst. **13**, 233–264 (2004)

Paroutis, S., Loizos, H., Angwin, D.: Practicing Strategy: Text and Cases. Sage, London (2016)

Peppard, J., Galliers, R.D., Thorogood, A.: Information systems strategy as practice: micro strategy and strategizing for IS. J. Strateg. Inf. Syst. **23**, 1–10 (2014)

Pettigrew, A.M.: What is a processual analysis? Scand. J. Manag. **13**, 337–348 (1997)

Pyburn, P.J.: Linking the MIS plan with corporate strategy: an exploratory study. MIS Q. **7**, 1–14 (1983)

Rasche, A., Chia, R.: Researching strategy practices: a genealogical social theory perspective. Organ. Stud. **30**, 713–734 (2009)

Reckwitz, A.: Toward a theory of social practices: a development in culturalist theorizing. Eur. J. Soc. Theory **5**, 243–263 (2002)

Regnér, P.: Strategy-as-practice and dynamic capabilities: steps towards a dynamic view of strategy. Hum. Relat. **61**, 565–588 (2008)

Reich, B.H., Benbasat, I.: Measuring the linkage between business and information technology objectives. MIS Q. **20**, 55–81 (1996)

Renaud, A., Walsh, I.: The lost dimension of strategic alignment. In: 5th Mediterannean Conference on Information Systems, 12–14 September 2010, Tel Aviv, Israel (2010)

Renaud, A., Walsh, I., Kalika, M.: Is SAM still alive? A bibliometric and interpretive mapping of the strategic alignment research field. J. Strateg. Inf. Syst. **25**, 75–103 (2016)

Rousseau, D.: Issues of level in organizational research: multi-level and cross-level perspectives. Res. Organ. Behav. **7**, 1–37 (1985)

Salmela, H., Spil, T.A.M.: Dynamic and emergent information systems strategy formulation and implementation. Int. J. Inf. Manag. **22**, 441–460 (2002)

Schlosser, F.: Mastering the social IT/business alignment challenge. In: 18th Americas Conference on Information Systems, 9–11 August 2012, Seattle, Washington, USA (2012)

Seidel, S., Urquhart, C.: On emergence and forcing in information systems grounded theory studies: the case of Strauss and Corbin. J. Inf. Technol. **28**, 237–260 (2013)

Suddaby, R., Seidl, D., Lê, J.K.: Strategy-as-practice meets neo-institutional theory. Strateg. Organ. **11**, 329–344 (2013)

Teo, T.S.H., King, W.R.: Assessing the impact of integrating business planning and IS planning. Inf. Manag. **30**, 309–321 (1996)

Urquhart, C., Lehmann, H., Myers, M.D.: Putting the "theory" back into grounded theory: guidelines for grounded theory studies in information systems. Inf. Syst. J. **20**, 357–381 (2010)

van Wessel, M., van Buuren, R., van Woerkum, C.: Changing planning by changing practice: how water managers innovate through action. Int. Publ. Manag. J. **14**, 262–283 (2011)

Walsham, G.: Interpretive case studies in IS research: nature and method. Eur. J. Inf. Syst. **4**(2), 74–81 (1995)

Ward, J.M.: Information systems strategy: Quo vadis? J. Strateg. Inf. Syst. **21**, 165–171 (2012)

Whittington, R.: Completing the practice turn in strategy research. Organ. Stud. **27**, 613–634 (2006)

Whittington, R.: Information systems strategy and strategy-as-practice: a joint agenda. J. Strateg. Inf. Syst. **23**, 87–91 (2014)

Wittgenstein, L.: Culture and Value. Blackwell, Oxford (1998)

Yin, R.K.: Case Study Research: Design and Methods. Sage Publications, Thousand Oaks (2009)

Yin, R.K.: Case Study Research: Design and Methods. Sage Publications, Thousand Oaks (2014)

Focus Area Maturity Models:
A Comparative Review

Felix Sanchez-Puchol[1,2(✉)] ⓘ and Joan A. Pastor-Collado[1,3] ⓘ

[1] Open University of Catalunya (UOC), Barcelona, Spain
{fsanchezpu, jpastorc}@uoc.edu
[2] Seidor SBS Learning Services, Barcelona, Spain
[3] Technical University of Catalonia (UPC), Barcelona, Spain

Abstract. Focus area maturity models (FAMMs) have been presented as a good alternative to traditional approaches of continuous or staged maturity models (MMs). FAMMs differ from previous approaches by defining a specific number of maturity levels for a set of focus areas, which embrace concrete capabilities to be developed in order to achieve maturity in a targeted domain. Due to the uninterrupted emergence of new MMs, several literature reviews have been conducted to increase transparency and facilitate understandability on existing MMs. However, none of them has been directly focused on FAMMs. Therefore, the purpose of this paper is to provide a first structured review on such specific models. A total of 16 different FAMMs are identified and compared from a structural point of view. Results suggest that FAMMs are still an under-researched topic. Suggestions for future research are provided in order to foster recognition and acceptance of such MMs.

Keywords: Maturity · Literature review · Focus area maturity models

1 Introduction

Maturity Models (MMs) can currently be considered as an accepted process management approach to systematically asses and gradually improve organizational capabilities with the aim of reaching a determined goal [1, 2]. Recently, Focus Area MMs (FAMMs) have been posed as an alternative to traditional approaches of continuous or staged MMs [3, 4]. FAMMs differ from previous models by defining a specific number maturity levels for a concrete set of focus areas, which in turn, embrace the capabilities to be developed in order to achieve maturity in a targeted functional domain [5, 6]. Such fine-grained and flexible structural configuration postulate FAMMs as an excellent tool for providing incremental process improvement advice, as they can be easily particularized and configured for a wide range of particular domains. This can be specially relevant for (relatively) young disciplines or functional domains, where professional practices and principles still have to be well–established [5].

Over the last decades, we have been witnessing a continuous and uninterrupted emergence of new MMs developed by both practitioners and scholars [7–10], triggering "the risk of losing track" [2, p. 329] due to the constantly growing number of existing exemplars. In order to increase the transparency, understandability and

© Springer International Publishing AG 2017
M. Themistocleous and V. Morabito (Eds.): EMCIS 2017, LNBIP 299, pp. 531–544, 2017.
DOI: 10.1007/978-3-319-65930-5_42

reusability of such a plethora of MMs, several literature reviews have been conducted, partially dismissing the aforesaid risk [2, 7–12]. But despite the existence of a great variety of literature reviews with different focus and scope, none of them has been focused on the topic of FAMMs. Moreover, and from a more pragmatic perspective, a first review on existing FAMMs could be useful for managers as a decision support tool. For example, they could help them in the identification of existing MMs suitable to be tailored (or adapted) to concrete business domain requirements, avoiding thus, unnecessary development expenditures and time-wasting costs [9, 13].

Against this background, the main goal of this paper is to provide an introductory review on FAMMs. Applying basic guidelines for conducting structured literature reviews from the Information Systems (IS) discipline, we examine prior literature in order to (1) identify constructed FAMMs, and (2) to develop a structural comparative analysis between them. Furthermore, and on the basis of the previous analysis, suggestions for future research are also derived. In so doing, we also view the present work as an instrument to create awareness among IS and Management researchers on the fact that FAMMs are still an under-researched topic.

The remainder of the paper is structured as follows. First, we conceptualize MMs and FAMMs in order to provide readers with a basic understanding of the research topic, as well as previous existing research. In Sect. 3, we define the research questions and the methodological details of the review. Next, the main research findings are outlined and discussed, to be subsequently used in Sect. 5 for deriving brief suggestions for further research. Finally, a conclusion section is provided, which also discusses the main limitations of the research.

2   Background

To better understand the main research topic of the paper, in the following paragraphs we introduce the main characteristics of both MMs and FAMMs.

2.1   Maturity Models and Focus Area Maturity Models

Several definitions for MMs can be found in the literature. From a general perspective, the term maturity has been characterized as "the state of being complete, perfect or ready" [9]. Maturity implies an "evolutionary progress in the demonstration of a specific ability or in the accomplishment of a target from an initial to a desired or normally occurring end stage" [9, p. 83]. Thus, MMs can be seen as tools for guiding and facilitating such transformation process [15]. A generic and well-accepted definition for MMs has been proposed by Becker et al. [16, p. 213], who state that MMs "consists of a sequence of maturity levels for a class of objects. It represents an anticipated, desired, or typical evolution path of these objects shaped as discrete stages". Conceptually, MMs are an organizational set of constructs describing certain aspects of maturity of a design domain and how they evolve over time [17, 18]. The entities or class of objects describing the maturity of a certain domain can be diverse, ranging from a person, a process, an object or even a social system [12, 14]. From a practical point of view MMs can be seen as a concrete type of conceptual models capturing organizational change "insofar as they represent an instrument for decision-makers to assess an organization's actual state, derive actions for improvement, and evaluate these actions afterwards in terms of their effectiveness and efficiency" [19, pp. 4–5].

From a structural perspective, several classifications of MMs can be found in the literature. For example, MMs can be generically classified into *staged* or *continuous models* [5, 15, 19]. *Staged models* present an ideal path specifying of development of a targeted domain through a set of concrete stages. *Continuous models* facilitate the review of certain quality features of a design domain at regular intervals. Whilst the latter propose the scoring of individual capabilities or requirements specified at different levels or dimensions of maturity, the former require compliance with all capabilities or requirements defined. Other authors [14, 20] also distinguish among *maturity grids* – which just define textual descriptions for several levels of maturity –; *likert-scale questionnaires* – allowing practical assessment and scoring of maturity on the basis of "good practices" –; and *CMMI-like models*, based "on a [more] formal architecture, which specifies goals and practices to reach a maturity level" [20, p. 1].

FAMMs can be viewed as an extension of previously typologies of MMs. They were first proposed by Koomen and Pol [21, 22] but further formalized by van Steenbergen et al. [3–5]. They pointed out that FAMMs differ from previous MMs by (1) departing from the common practice of considering only five generic maturity levels; and by (2) defining the targeted maturity domain through a set of focus areas encompassing the capabilities to be implemented, having each their own specific number of maturity levels. Hence, the overall maturity of the assessed domain is obtained through the combination of the maturity levels achieved by each one the focus areas considered. Besides, FAMMs define the logical order in which capabilities should be implemented (including their interdependencies). FAMMs can be represented through a *maturity matrix*, as depicted in Fig. 1.

		MS$_1$			...			MS$_n$	
		ML$_1$	ML$_2$	ML$_3$	ML$_4$	ML$_5$	ML$_6$	...	ML$_n$
FAG$_1$	FA$_1$	A$_1$			B$_1$		C$_1$		
	FA$_2$		A$_2$			B$_2$			
...	FA$_3$	A$_3$		B$_3$		C$_3$	D$_3$		E$_3$
FAG$_n$	...								
	FA$_n$		A$_n$			B$_n$			N$_n$

Fig. 1. A focus area oriented maturity model

The basic working principles of FAMMs [4, 6], can be summarized as follows:

- A FAMM is used to establish the *maturity* of an organization in a specific functional domain, which is defined by a set of *focus areas* (FA$_1$ .. FA$_n$) that configure it. They can be grouped into *focus area groups* (FAG$_1$ .. FAG$_n$).
- Each focus area has a set of associated *capabilities* (A$_1$ .. N$_n$), which are juxtaposed relative to each other in the *maturity matrix*. Hence, a *capability* is positioned to the right of another capability if is to be implemented after that capability because it is dependent on its realization.

- Based on the positioning of the capabilities in the *maturity matrix* a number of *maturity levels* (ML_n) can be distinguished, which in turn, can be grouped into *maturity stages* (MS_n). In both cases, an informative *descriptor* can also be specified.
- The *maturity* of an organization in the functional domain is established through an *assessment instrument*, which is based on a set of *assessment questions,* linked to capabilities. To guide the organization in incremental development of the functional domain, *improvement actions* are also associated with the capabilities.

We assume this characterization as our working definition for the review purposes.

2.2   Previous Work

Before conducting the review, and with the aim of getting engaged with prior literature [23] we manually searched for existing reviews on MMs. On the one hand, we found reviews on MMs conducted from a broad IS disciplinary scope (Becker et al. [10], Poeppelbuss et al. [8], Lasrado et al. [24]). Also with a broad scope, Wendler [9] presented a systematic mapping overview of MMs research topics, reporting the use/development of 241 MMs in more than 20 concrete application domains. Literature reviews focussed on business processes MMs can also be easily identified (Rosemann and vom Brocke [25], Röglinger et al. [2], Van Looy [11], Tarhan et al. [7]). Additionally, reviews for concrete types of MMs have also been developed, as von Wangenheim et al. [26] on software process improvement MMs or Albliwi et al. [27] critical comparative study on MMs for business process excellence, to cite a few. But despite the existence of such corpus of reviews, none explicitly refers to FAMMs.

In fact, in our search we only were able to identify a couple of reviews referring explicitly to FAMMs. On the one hand, in a recent systematic literature review conducted in 2016, Saavedra et al. [28] inferred a classification of organizational MMs architectural styles, based on their internal structure and characteristics. The authors suggest *"progression focus area oriented models"* as one of such architectural styles, providing evidence of 5 studies reporting examples of such MMs. On the other hand, García-Mireles et al. [29] conducted a systematic literature review on theoretically-based procedure methods for developing MMs. In their work, the authors reference van Steenbergen et al. [30] procedure for building FAMMs, and roughly mention two exemplars of FAMMs developed using such methodology.

In sum, all these findings clearly reveal the need for conducting a more in-depth focused review devoted to analyse and compare existing FAMMs. In this vein, this paper complements and extends the existing body of knowledge providing improved understanding and greater transparency on what we know about FAMMs. We do so by using a basic framework facilitating their identification, structural analysis and comparison. Also, our study encompasses a wider scope with respect to extant studies, covering both a longest time horizon as well as a greater number of information sources. To the best of our knowledge, there still has been no attempt to conduct a review addressing the concrete topic of FAMMs to date.

3   Research Design

In terms of recently emerged taxonomies of IS literature reviews (Paré et al. [31], Rowe [32]) we characterize or work as rather a descriptive review. From an epistemological point of view [33], with the present review we expect to contribute to a better understanding of FAMMs by (i) summarizing the existing knowledge on FAMMs and their structural characteristics, and by (ii) deriving a set of suggestions for further research. To do so, we followed Webster and Watson [34] recommendations of applying a concept-centric framework for organizing and classifying the literature. We use existing MMs meta-models [13, 30] for such purposes (see Sect. 3.3 for further details).

3.1   Research Questions and Methodology

According to the main review's goal, this paper seeks to specifically answer the following retrospective research questions:

- RQ1 - How many FAMMs have already been developed?
- RQ2 - What are the structural characteristics of the existing FAMMs?

Answering the previous questions will allow us to provide an initial version of a FAMMs catalogue, built through the compilation of existing heterogeneous instances of such MMs. On the basis of the analysis of the structural characteristics of the identified FAMMs, suggestions for plausible research will be outlined.

Finally, and from a methodological process perspective, we proceed by applying Bandara et al. [35] four-step approach as the guiding research structure for the paper, since it provides and adequate support for our review purposes: (i) extraction of relevant literature, (ii) organization and preparation for the analysis, (iii) coding and analysis of the literature, and (iv) writing-up and presentation of results.

3.2   Extraction of Relevant Literature

According to vom Brocke et al. [36], the literature search process has to be comprehensibly documented and be made as transparent as possible in order to proof credibility and increase methodological rigor. Adhering to such principle, we structure and design our search strategy following the procedural stages suggested by the aforementioned authors: (i) database and journal search; (ii) keyword search; and (iii) back-ward and forward search. In addition, and as an ongoing process for evaluating the retrieved sources we also include a two-round screening process.

The strategy was based primarily on a search on multidisciplinary digital databases (covering both peer-reviewed as well as professional sources) as we expected to detect MMs endorsed both by academics or practitioners. The selection of the concrete sources was partly informed by previous literature reviews on MMs. We paid special attention in covering disciplines as IS, Management, Computer Science or Operations Research. To identify an initial sample of highly relevant literature for our research, we use the expression *"focus area"* *<AND>* *"maturity model"* as the keyword string for searching in title, keyword and abstract of documents in database sources. For affordability purposes, in cases like Google Scholar or Springer Link where too many initial results were obtained, the complete keyword string *"focus area maturity model"* was used applying a full-text search. The initial database search was performed in February 2017 and yielded a total set of 123 potentially relevant items for the study. To ensure that only relevant documents were included in the final review an screening process was performed using criteria illustrated in Table 1.

Table 1. Exclusion criteria applied in the review

Criteria type	Criteria description
Language	Studies not written in English
Accessibility	Studies with full text version not available electronically
Publication type	Short papers, less than 2-page length (i.e., posters, editorials, comments, etc.)
Publication content	Duplicate studies (by tittle and/or content)
	Studies not containing relevant information for answering research questions

As we were interested in identifying and collecting information of as many existing FAMMs as possible, no temporal restriction was applied. A first round of content-related screening based on title, keyword and abstract reading of retrieved documents were performed first. However, and with the aim of avoiding the risk of losing important information for the final analysis, doubtful records were also kept in this first round of content-related screening in order to be further analyzed at the latter step of the search process. Overall, after the first screening round, a subset of 36 references were retained for further review.

With the aim of broadening the coverage of our literature search, we proceed to apply a backward and forward snowballing strategy [34, 36]. We used Scopus and Google Scholar services as support tools for forward search. After applying the exclusion criteria, a further set of 5 additional documents (3 identified by backward search and 2 identified by forward search) was collected, which led us led us to a total subset of 41 documents for the final in-depth full-text review. At this point, we explicitly decided to exclude references [37–39] documenting MMs considered as *"progression-focus area MMs"* in Saavedra et al. [28] review, since we considered that they do not strictly follow our established working definition of FAMMs.

To conclude with the literature search process, we performed the second round screening process based on an in-depth full text reading of the remaining documents. We excluded 16 documents that either not appeared to be relevant for the purposes of our review or were considered as content duplicate documents (as for example, a couple of doctoral thesis [40, 41]). Our structured search process resulted in final set of 25 items relevant for our FAMMs review. Finally, to complete the final set of documents to be analysed in the final review, we further added four additional references:

- We knew about the existence of a book by Koomen and Pol [21] documenting a Test Improvement FAMM. As we were not able to get a printed version of it, we manually search on the Internet for alternative references. Hence, two additional working reports [22, 42] were added to the final set of reviewed documents.
- We were also aware of the existence of a social media related FAMMs [43]. This reference also was added to the final set of documents reviewed.
- Finally, we also identified a Wikipedia page including information referring to an Implementation FAMM [44]. We considered the information provided by such source as sufficiently relevant to be included in our final set of documents.

All in all, a total amount of 29 documents were retained for final analysis.

3.3   Preparing for Analysis and Coding Process

To perform the final analysis of the collection of selected documents, we created a Zotero reference database with all the citation data as well as all the retrieved electronic version of the documents. We also used it for the writing-up of the paper and as a support tool for citing and referencing purposes. In addition, and given the relatively manageable amount of papers to analyze, we lastly decided to create a Ms. Excel database for coding the documents alongside the categories of interest for our review.

To give answer to the posed research questions, we applied a concept-centric approach [34] for structuring and coding the literature. For such aim we used as a reference framework the FAMM's meta-model stated by van Steenbergen et al. [6], which in turn, is fully compatible with the more generic MM's meta-model presented by Patas et al. [13]. For each FAMM identified we collected the structural information presented in Sect. 2.1. The coding and classification procedure was as follows: (i) first, we identified all the FAMMs documented in the whole literature retrieved; (ii) next, we associated each individual information source with respect to the each one of the identified FAMMs. This association was made on the basis of a content analysis of each one of the literature sources found in the search process. We found both *direct references* – content and focus of the source mainly devoted to the release, development or exemplary use of an identified FAMM – as well as *indirect references* – the source only provided a simple reference or a brief description to a concrete FAMM – (iii), finally, we extracted the information relevant of each source in order to complete the sub-categories of the FAMMs meta-model used as reference for the analysis.

4   Major Findings and Discussion of Results

In this section, we present and discuss the major findings and results of our analysis.

4.1   Descriptive Analysis

Information extracted from sources retrieved indicate that although the origins of FAMMs can traced back to the late 1990s, the level of publications has been very scarce until the year 2010, coinciding with the publication of van Steenbergen and colleagues' article [30] formalizing this type of MMs. Hence, the temporal distribution of publications clearly shows a marked increase since year 2010, which, in turn, represents the pick of the series. In this vein, 25 of the 29 identified documents correspond to the temporal period between 2010 and 2016, representing the 86.2% of the documents retrieved. Besides, and on the basis of the screening process undertook for the review, we learned that many conference papers – which represent the 41% of retrieved analyzed documents – are in fact published versions of the research results achieved in PhD and Master Dissertations, as well as technical reports. Together, these tree types of publications represent the 75.86% of the overall set of documents retrieved, whilst journal articles only represent the 13.79% of the total.

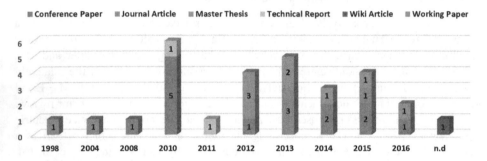

Fig. 2. Timeline (segmented per type of contribution) of documents analyzed

In our opinion, such percentages suggest that there is much emphasis on the need to target FAMMs-related contributions on journal articles, in order to these models to become a more internationally and widely accepted research topic. In such vein, and given the absolute numbers of documents retrieved, the research topic still has to gain such recognition. Moreover, the influence of van Steenbergen and colleagues as reference authors in the topic is also clearly reflected in the geographical origin of the identified contributions, as the 76% of the total of items retrieved for analysis belong to Dutch authors, who clearly dominate the discussions of FAMMs.

4.2   Identified FAMMs and Main Structural Characteristics

We identified a total of 16 different FAMMs in our analysis. The information collected shows a predominance of FAMMs developed for information systems and technology and software engineering domains. Besides, in general terms, FAMM identified are also quite generic, in the sense that they can be applied/used in a wide variety of industries. Exceptions to the previous considerations can be the SMPI-MM and the GPIS models, which are specifically addressed to healthcare organizations, and the SpicEA model, which is tailored for the automotive industry (Table 2).

Table 2. Identified FAMMs in the retrieved literature

#N	Focus area maturity model	Abbrev.	Sources
#1	Implementation Maturity Model	IMM	[44]
#2	Test Process Improvement (TPI) model	TPI	[22, 42]
#3	DyA (Enterprise) Architecture Maturity Matrix	DyAMM	[3, 6, 28, 30, 45]
#4	Software Product Management Maturity Matrix	SPM-MM	[28, 30, 46–51]
#5	Software Product Portfolio Management Maturity Matrix	SPPM-MM	[52]
#6	Master Data Management Maturity Model	MD3M	[53, 54]
#7	Social Media Maturity Model	SM-MM	[43]
#8	FAMM for Information Use in Organizations	IUO-MM	[55]
#9	FAMM for a State-wide Master Person Index	SMPI-MM	[56]
#10	Software Product Line Management Maturity Matrix	SPLM-MM	[57]
#11	Disaster Risk Management Focus Area Maturity Model	DRM-MM	[58]
#12	IT Maturity Model for (healthcare) general practice IS	GPIS	[59]
#13	IT Govern. Maturity Model for Hard & Soft Governance	ITGOV-MM	[60, 61]
#14	Information Security Focus Area Maturity Model	ISFAM	[62, 63]
#15	IT Carve-Outs Focus Area Maturity Model	ITCVO-MM	[64]
#16	Automotive Spice & EA-based Software Manag. FAMM	SpicEA	[65]

With regards to the structural configuration of the identified FAMMs, the information retrieved reveals that FAMMs building efforts concentrated on 4 structural elements, namely *focus areas, focus area groups, capabilities* and *maturity levels.* However, the concrete configuration of such elements is quite diverse among different instances of available FAMMs. For example, *focus areas* defined usually range from 12 to 20, whilst *focus area group levels* (of aggregation) tend to be from 3 to 6. The number of *capabilities* defined also presents a high level of variability, usually ranging from 30 to 70 *capabilities* (only 4 of the identified FAMMs have less than 30 *capabilities* defined, meanwhile 3 of them have more than 70 capabilities defined) (Table 3).

Finally, the number of *maturity levels* defined by identified FAMM perhaps presents a minor level of volatility, as most of them range from 12 to 15. However, only 4 of them define *maturity stage levels* – i.e. aggregations of maturity levels – through a basic maturity stage level descriptor or tag for such aggregated stage. Finally, and regarding the remaining structural components of a FAMM, the achievements under-taken can be characterized as dissimilar. Hence, the 57% of FAMMs identified include *assessment questions*, which is clearly an important indicator of applicability of the model. However, efforts on developing *improvement actions* have been minimal, as only 2 FAMMs include such items. We found these last finding as rather surprising, given their theoretical suitability as incremental process improvement artefacts.

Table 3. FAMM structural comparative

FAMM	Focus area groups	Focus areas	Capabilities	Maturity levels	Maturity stages	Mat. stage levels descriptor	Mat. stage levels description	Assessment questions	Improvement actions
IMM	5	19	68	15	ND	NO	NO	ND	ND
TPI	ND	20	58	13	3	NO	NO	ND	ND
DyAMM	ND	18	54	13	ND	NO	NO	137	ND
SPM-MM v.1[a]	4	15	68	10	ND	NO	NO	68	ND
SPM-MM v.2[b]	4	16	52	12	ND	NO	NO	ND	ND
SPPM-MM	3	9	28	10	ND	NO	NO	ND	ND
MD3M	5	13	65	5	5	NO	NO	65	ND
SM-MM	3	20	79	20	ND	NO	NO	79	ND
IUO-MM	ND	2	6	13	4	YES	NO	ND	ND
SMPI-MM	ND	5	14	8	ND	NO	NO	ND	ND
SPLM-MM	4	16	84	10	ND	NO	NO	ND	ND
DRM-MM	ND	6	ND	5	ND	NO	NO	ND	ND
GPIS	3	12	53	12	ND	NO	NO	115	59
ITGOV-MM	6[1]	12	20	5	ND	NO	NO	ND	ND
ISFAM[c]	4	13	51	12	4	YES	NO	161	ND
ISFAM 2.0[d]	4	14	55	12	4	YES	NO	176	ND
ITCVO-MM	4	16	49	15	ND	NO	NO	NS	NS
SpicEA	3[1]	15	71	15	ND	NO	NO	225[2]	ND

[a]As stated in [46, 47], [b]as stated in [49], [c]as stated in [62], [d]as stated in [63]
[1]Presents two levels of aggregation, [2]covers only a subset of the of focus areas defined, ND: Not Defined, NS: Not Specified

5   Suggestions for Further Research

Given all the above, we finally propose a set of research suggestions for FAMMs.

- **RS1 - Improve transparency and completeness of FAMMs.** Contributions on FAMMs should clearly reference the structural composition of the proposed artifact. Such recommendation is especially relevant in terms of *assessment questions* and *improvement actions*, as these components tend to be not clearly provided in existing contributions. Moreover, new developed FAMMs should make emphasis in the inclusion of *improvement actions*, in order to achieve better quality artefacts.

- **RS2 - Develop rigorous strategies for validating FAMMs.** Whilst a great number of contributions outlining the development of FAMMs include evaluation activities for the proposed artifact, they are usually centered in showing the utility of such models as assessment instruments. Hence, more formal evaluation strategies should have to be developed in order to validate FAMMs, especially regarding their underlying declared utility (i.e. demonstrate that more levels of maturity achieved imply an improvement regarding the utility function claimed for the MM.

- **RS3 - Expand applicability of FAMMs to new and relevant application domains.** According with van Steembergen et al. [30], examples of focus areas could be "the development and maintenance of certain processes or deliverables, alignment with other disciplines, and training of certain competences". Hence, FAMMs can be a powerful instrument for guiding incremental improvements in new disciplines or current emerging fields of application (i.e. quality management, supply chain management, big data, e-government, etc.).

- **RS4 - Improve actionability of developed FAMMs.** FAMMs have to come with clear action plans on how to and for what purpose can be used for. Complementary web-based and graphical tools with dedicated information visualization and reporting features should increment the practical utility of FAMMs for practitioners.

6 Concluding Remarks and Research Limitations

This work presents a first comparative review on the topic of FAMMs. A total of 16 different FAMMs are detailed and compared from a structural point of view. Based on such analysis, a first tentative set of research suggestions is provided to foster the recognition of such research topic. Further research in terms of more systematic scoping reviews on the topic will help to increase its understanding in the future.

As any other, especially for emerging topics, this piece of research also comes with its own limitations. Hence, and besides the inherent subjectivity that can be attributed to any review, concerns may arise regarding its rigor, especially in terms of the literature search strategy executed or related to the subjective quality appraisal of the sources. However, and despite the aforementioned limitations, and given the main and limited purpose of the paper, we believe that it will proof useful for both researchers and practitioners as an introductory and comparative reference catalog of the current existing FAMMs, including the major domain areas where they have been applied.

Acknowledgments. Industrial Doctorates Plan - Generalitat de Catalunya.

References

1. Van Looy, A., De Backer, M., Poels, G.: Questioning the design of business process maturity models. In: Proceedings of the 6th SIKS Conference on EIS, pp. 51–60, Delft, The Netherlands, 31 October 2011
2. Röglinger, M., Pöppelbuß, J., Becker, J.: Maturity models in business process management. Bus. Process Manag. J. **18**, 328–346 (2012)
3. van Steenbergen, M., van den Berg, M., Brinkkemper, S.: A balanced approach to developing the enterprise architecture practice. In: Filipe, J., Cordeiro, J., Cardoso, J. (eds.) ICEIS 2007. LNBIP, vol. 12, pp. 240–253. Springer, Heidelberg (2008). doi:10.1007/978-3-540-88710-2_19
4. van Steenbergen, M., Bos, R., Brinkkemper, S., van de Weerd, I., Bekkers, W.: The design of focus area maturity models. In: Proceedings of the 5th SIKS/BENAIS Conference on EIS, pp. 17–19, Eindhoven, The Netherlands, 16 November 2010
5. van Steenbergen, M., van den Berg, M., Brinkkemper, S.: An instrument for the development of the enterprise architecture practice. In: Proceedings of the ICEIS 2007, pp. 14–22, Funchal, Portugal, 12–16 June 2007
6. van Steenbergen, M., Bos, R., Brinkkemper, S., van de Weerd, I., Bekkers, W.: Improving IS functions step by step: the use of focus area maturity models. Scand. J. Inf. Syst. **25**, 2 (2013)
7. Tarhan, A., Turetken, O., Reijers, H.A.: Business process maturity models: a systematic literature review. Inf. Softw. Technol. **75**, 122–134 (2016)

8. Poeppelbuss, J., Niehaves, B., Simons, A., Becker, J.: Maturity models in information systems research: literature search and analysis. Comm. Assoc. Inf. Syst. **29**, 505–532 (2011)
9. Wendler, R.: The maturity of maturity model research: a systematic mapping study. Inf. Softw. Technol. **54**, 1317–1339 (2012)
10. Becker, J., Niehaves, B., Pöppelbuß, J., Simons, A.: Maturity models in IS research. In: Proceedings of the ECIS 2010, Pretoria, South Africa, 7–9 June 2010
11. Van Looy, A.: Business Process Maturity. A Comparative Study on a Sample of Business Process Maturity Models. Springer International Publishing, Heidelberg (2014)
12. Kohlegger, M., Maier, R., Thalmann, S.: Understanding maturity models. Results of a structured content analysis. In: Proceedings of I-KNOW 2009 and I-SEMANTICS 2009, pp. 51–61, Graz, Austria, 2–4 September 2009
13. Patas, J., Pöppelbuß, J., Goeken, M.: Cherry picking with meta-models: a systematic approach for the organization-specific configuration of maturity models. In: Brocke, J., Hekkala, R., Ram, S., Rossi, M. (eds.) DESRIST 2013. LNCS, vol. 7939, pp. 353–368. Springer, Heidelberg (2013). doi:10.1007/978-3-642-38827-9_24
14. Mettler, T.: Maturity assessment models: a design science research approach. Int. J. Soc. Syst. Sci. **3**, 81–98 (2011)
15. Lahrmann, G., Marx, F.: Systematization of maturity model extensions. In: Winter, R., Zhao, J.L., Aier, S. (eds.) DESRIST 2010. LNCS, vol. 6105, pp. 522–525. Springer, Heidelberg (2010). doi:10.1007/978-3-642-13335-0_36
16. Becker, J., Knackstedt, R., Pöppelbuß, J.: Developing maturity models for IT management: a procedure model and its application. Bus. Inf. Syst. Eng. **1**, 213–222 (2009)
17. Fraser, P., Moultrie, J., Gregory, M.: The use of maturity models/grids as a tool in assessing product development capability. In: Proceedings of the IEMC 2002 IEEE International Conference, pp. 244–249 (2002)
18. Klimko, G.: Knowledge management and maturity models: building common understanding. In: Proceedings of the ECKM 2002, pp. 269–278, Bled, Slovenia, 8–9 November 2001
19. Ofner, M., Otto, B., Österle, H.: A maturity model for enterprise data quality management. Enterp. Model. Inf. Syst. Archit. **8**, 4–24 (2015)
20. Baars, T., Mijnhardt, F., Vlaanderen, K., Spruit, M.: An analytics approach to adaptive maturity models using organizational characteristics. Decis. Anal. **3**(1), 5 (2016)
21. Koomen, T., Pol, M.: Test Process Improvement: A Practical Step-By-Step Guide to Structured Testing. Addison-Wesley, Harlow (1999)
22. Koomen, T., Pol, M.: Improvement of the Test Process using TPI. Sogeti Nederland B.V., Diemen (1998)
23. Boell, S.K., Cecez-Kecmanovic, D.: A hermeneutic approach for conducting literature reviews and literature searches. Commun. Assoc. Inf. Syst. **34**, 257–286 (2014)
24. Lasrado, L.A., Vatrapu, R., Andersen, K.N.: Maturity models development in IS research. In: Selected Papers of the IRIS, Oulu, Finland (2015)
25. Rosemann, M., vom Brocke, J.: The six core elements of business process management. In: Brocke, J., Rosemann, M. (eds.) Handbook on Business Process Management 1. IHIS, pp. 105–122. Springer, Heidelberg (2015). doi:10.1007/978-3-642-45100-3_5
26. von Wangenheim, C.G., Hauck, J.C.R., Salviano, C.F., von Wangenheim, A.: Systematic literature review of software process capability/maturity models. In: Proceedings of the SPICE 2010, Pisa, Italy (2010)
27. Albliwi, S.A., Antony, J., Arshed, N.: Critical literature review on maturity models for business process excellence. In: IEEM 2014 IEEE International Conference, pp. 79–83, Malaysia, 9–12 December 2014

28. Saavedra, V., Dávila, A., Melendez, K., Pessoa, M.: Organizational maturity models architectures: a systematic literature review. In: Mejia, J., Muñoz, M., Rocha, Á., San Feliu, T., Peña, A. (eds.) Trends and Applications in Software Engineering. AISC, vol. 537, pp. 33–46. Springer, Cham (2017). doi:10.1007/978-3-319-48523-2_4

29. García-Mireles, G.A., Moraga, M.Á., García, F.: Development of maturity models: a systematic literature review. In: Proceedings of the EASE 2012, pp. 279–283. IET, Ciudad Real, 14–15 May 2012

30. van Steenbergen, M., Bos, R., Brinkkemper, S., van de Weerd, I., Bekkers, W.: The design of focus area maturity models. In: Winter, R., Zhao, J.L., Aier, S. (eds.) DESRIST 2010. LNCS, vol. 6105, pp. 317–332. Springer, Heidelberg (2010). doi:10.1007/978-3-642-13335-0_22

31. Paré, G., Trudel, M.-C., Jaana, M., Kitsiou, S.: Synthesizing information systems knowledge: a typology of literature reviews. Inf. Manag. **52**, 183–199 (2015)

32. Rowe, F.: What literature review is not: diversity, boundaries and recommendations. Eur. J. Inf. Syst. **23**, 241–255 (2014)

33. Schryen, G.: Writing qualitative IS literature reviews–guidelines for synthesis, interpretation and guidance of research. Commun. Assoc. Inf. Syst. **37**, 286–325 (2015)

34. Webster, J., Watson, R.T.: Analyzing the past to prepare the future: writing a literature review. MIS Q. **26**, xiii–xxiii (2002)

35. Bandara, W., Furtmueller, E., Gorbacheva, E., Miskon, S., Beekhuyzen, J.: Achieving rigor in literature reviews: insights from qualitative data analysis and tool-support. Commun. Assoc. Inf. Syst. **37**, 879–910 (2015)

36. vom Brocke, J., Simons, A., Niehaves, B., Riemer, K., Plattfaut, R., Cleven, A.: Reconstructing the giant: on the importance of rigour in documenting the literature search process. In: Proceedings of the ECIS 2009, Verona, Italy, 8–10 June 2009

37. Rudolph, S., Krcmar, H.: Maturity model for it service catalogues. An approach to assess the quality of IT service documentation. In: Proceedings of the AMCIS 2009, p. 750, San Francisco, California, 6–9 August 2009

38. Mantovani, R., Meyer Jr., V., Reinehr, S., Malucelli, A.: Progressive outcomes: a framework for maturing in agile software development. J. Syst. Softw. **102**, 88–108 (2015)

39. Salleh, H., Alshawi, M., Sabli, N.A.M., Zolkafli, U.K., Judi, S.S.: Measuring readiness for successful information technology/information system (IT/IS) project implementation: a conceptual model. Afr. J. Bus. Manag. **5**, 9770–9778 (2011)

40. van Steenbergen, M.: Maturity and effectiveness of enterprise architecture (2011). https://dspace.library.uu.nl/handle/1874/205434

41. Bekkers, W.: Situational Process Improvement in Software Product Management (2012). https://dspace.library.uu.nl/handle/1874/256455

42. Andersin, J.: TPI–a Model for Test Process Improvement. University of Helsinki (2004)

43. van de Kerkhof, J.: Social media: towards a social media maturity model (2012). https://dspace.library.uu.nl/handle/1874/242246

44. Wikipedia: Implementation maturity model assessment. https://en.wikipedia.org/wiki/Implementation_maturity_model_assessment

45. van Steenbergen, M., Schipper, J., Bos, R., Brinkkemper, S.: The dynamic architecture maturity matrix: instrument analysis and refinement. In: Dan, A., Gittler, F., Toumani, F. (eds.) ICSOC/ServiceWave -2009. LNCS, vol. 6275, pp. 48–61. Springer, Heidelberg (2010). doi:10.1007/978-3-642-16132-2_5

46. Bekkers, W., Spruit, M.: The Situational Assessment Method put to the test. Improvements based on case studies. In: WSPM 2010 International Workshop, pp. 7–16. IEEE (2010)

47. Bekkers, W., van de Weerd, I.: SPM Maturity Matrix. Department of Information and Computing Sciences, Utrecht University, Utrecht, The Netherlands (2010)

48. Bekkers, W., Spruit, M., van de Weerd, I., van Vliet, R., Mahieu, A.: A situational assessment method for software product management. In: Proceedings of the ECIS 2010, Pretoria, South Africa, 7–9 June 2010
49. van de Weerd, I., Bekkers, W., Brinkkemper, S.: Developing a maturity matrix for software product management. Department of Information and Computing Sciences, Utrecht University, Utrecht, The Netherlands (2010)
50. Bekkers, W., Brinkkemper, S., van den Bemd, L., Mijnhardt, F., Wagner, C., van de Weerd, I.: Evaluating the software product management maturity matrix. In: Requirements Engineering Conference, RE 2012, 20th IEEE International, pp. 51–60 (2012)
51. Slooten, R.: Software release planning: Investigating the use of an advanced assessment instrument and evaluating a novel maturity framework (2012). https://pure.tue.nl/ws/files/47035156/731255-1.pdf
52. Jagroep, E., van de Weerd, I., Brinkkemper, S., Dobbe, T.: Software Product Portfolio Management: Towards improvement of current practice. Department of Information and Computing Sciences, Utrecht University, Utrecht, The Netherlands (2011)
53. Spruit, M., Pietzka, K.: MD3M: The master data management maturity model. Comput. Hum. Behav. **51**, 1068–1076 (2015)
54. Pietzka, K.: MD3M Master Data Management Maturity Model - Developing an Assessment to Evaluate an Organization's MDM Maturity (2012). http://dspace.library.uu.nl/handle/1874/255375
55. Alemão Alves, J.F.: Finding Maturity Evolution Paths for Organisational use of Information. A Moviflor Case Study (2013). https://fenix.tecnico.ulisboa.pt/downloadFile/395145528220/DMEIC-57552-Joana-Alves.pdf
56. Duncan, J., Xu, W., Narus, S.P., Clyde, S., Nangle, B., Thornton, S., Facelli, J.: A focus area maturity model for a statewide master person index. Online J. Public Health Inform. **5**, 210 (2013)
57. Sprockel, Y.H.B.: The impact of Software Product Lines from a Product Management perspective (2013). http://dspace.library.uu.nl/handle/1874/276057
58. Van der Waldt, G.: Disaster risk management: disciplinary status and prospects for a unifying theory: original research. Jamba J. Disaster Risk Stud. **5**, 1–11 (2013)
59. Hermanns, T.I.: Towards an IT maturity model for general practice information systems (2014). http://dspace.library.uu.nl/handle/1874/294630
60. Smits, D., van Hillegersberg, J.: IT Governance maturity: developing a maturity model using the Delphi method. In: 48th Hawaii International Conference on System Sciences (HICSS 2015), pp. 4534–4543. IEEE, Hawaii, 5–8 January 2015
61. Smits, D., van Hillegersberg, J.: The development of an IT governance maturity model for hard and soft governance. In: Proceedings of the 8th European Conference on IS Management and Evaluation (ECIME 2014), pp. 347–355, Ghent, Belgium, 11–12 September 2014
62. Spruit, M., Röling, M.: ISFAM: the information security focus area maturity model. In: Proceedings of the ECIS 2014, Tel Aviv, Israel, 9–11 June 2014
63. Slot, G.C.A.: Towards Rule-based Information Security Maturity. The Next Level (2015). https://dspace.library.uu.nl/handle/1874/315919
64. Pflügler, C., Böhm, M., Krcmar, H.: Coping with IT carve-out projects-towards a maturity model. In: Proceedings der 12, Internationalen Tagung Wirtschaftsinformatik (WI 2015), pp. 1664–1678, Osnabrück, Germany, 4–6 March 2015
65. Willems, T.: Supporting the Enterprise in Implementing a Global Vehicle Software Management Strategy (2016). http://dspace.ou.nl/handle/1820/7189

Aligning Business Models with Requirements Models

Eric Souza[(⊠)], Ana Moreira, and João Araújo

NOVA LINCS, Departamento de Informática,
Faculdade de Ciência e Tecnologia - FCT,
Universidade NOVA de Lisboa - UNL, Caparica, Portugal
er.souza@campus.fct.unl.pt,
{amm,joao.araujo}@fct.unl.pt

Abstract. There is a widespread agreement of the business models' importance to represent the economic point of view of a company in a value model. Such value models are core for the successful alignment of a business and its information systems, and they can be used as input of the software requirements specification process. However, creating requirements models, such as goal-oriented models or use cases, aligned with value models is a time consuming and error-prone task. Automating the construction of more refined models from more abstract ones, can contribute to accelerate the software development cycle and increase the developer's productivity. This paper offers a systematic approach to automatically generate goal-oriented models from value models. This is achieved by both defining conceptual mappings between value models and goal-oriented models and using model-driven techniques to define automatic transformations between both types of models. We have used our approach in several case studies and the results show that our approach reduces the requirements engineer effort to build goal-oriented models and ensure the alignment between business and IT.

Keywords: Business model · Value model · Value-driven · Goal-oriented model

1 Introduction

There is a widespread agreement regarding the importance of business models for a company to express value, be it economic, social, or other [1–3]. On the other hand, the success of a company's impact in the market, also depends on the alignment between its information systems and those value models expressing its economic perspective. To achieve this alignment, the value model can be used as input of the software requirements specification process[1], guiding the software development according to the business economic values.

Among the existing requirements specification techniques, goal-oriented modeling is an approach used to discover software requirements. Its main focus is on analyzing

---

[1] Software requirements specifications are created in the initial stages of information systems' development describing the software system to be developed.

© Springer International Publishing AG 2017
M. Themistocleous and V. Morabito (Eds.): EMCIS 2017, LNBIP 299, pp. 545–558, 2017.
DOI: 10.1007/978-3-319-65930-5_43

the "whys" (or "goals") of the software behavior and associated qualities. In goal-oriented modeling, the requirements engineer is not yet interested in the "operational" details of the processes or system requirements, not yet relevant for the specification of early requirements. Omitting such details at these early stages gives the engineer a strategic view of the system, which is often not captured as the engineer is typically more concerned with issues that can be addressed in the late requirements (or more detailed specification) stage [4]. However, there is a wide gap between value models and goal-oriented models, because: (1) requirements engineers find it difficult to extract knowledge from value models to design information systems [5]; and (2) business specialists do not usually understand commonly used requirements techniques to express system behavior [6]. Therefore, creating requirements models (IS area) aligned with value models (business area) is a difficult task [7], lacking approaches to realize this alignment automatically [8].

This paper describes a systematic approach to automatically generate goal-oriented models from value models. For this, we created a conceptual mapping between the Dynamic Value Description (DVD) [9] language (a concise value model) and goal-oriented models, and show how model-driven techniques can be used to perform systematic transformations between those two types of models. An evaluation of our approach with a case study shows a reduction of the requirements engineer effort during goal-oriented model specification, facilitating the transition from business to early stages of IS development.

The remaining of this paper is organized as follows. Section 2 offers an overview on business values, goal-oriented requirements, and model-driven engineering. Section 3 describes the conceptual mapping created between value model and goal-oriented models. Section 4 evaluates our approach, using the generation strategy to produce models for two different languages (iStar and KAOS). This replication of the study proves that our approach is generic and reusable. Finally, Sect. 5 discusses related work, and Sect. 6 concludes and summarizes directions for further work.

2 Background

This section presents an overview of business value, goal-oriented requirements, and model driven engineering, offering a background on the main concepts and technologies supporting our approach.

2.1 Business Value

A business model is a lightweight, semi-formal and conceptual technique, inspired in business science, requirements engineering and conceptual modeling to model business ideas [1]. It is worth noticing the difference between business modeling and process modeling [2]. The goal of a process model (e.g., BPMN [3]) is to clarify how processes should be carried out, and by whom. In contrast, a business model main goal is to identify who is offering what to whom and expects what in return. The central notion in a business model is the concept of value, to explain the creation, addition, and the exchange of value between stakeholders [2]. Thus, a value model is a type of graphical business model that shows how a business value is created and exchanged in an inter-organizational network, facilitating the discovering business opportunities [4]. Value is the reason why companies and people trade with each other, offering money to get something in return. Therefore, a value model represents a business model from an economic perspective, and must determine the economic value exchanged and their intervenients [4].

The Dynamic Value Description (DVD) [5] language aims at representing and analyzing business values exchange. It provides an environment wherein stakeholders can share their values exchanged views in a semi-structured mind map model[2]. Thus, DVD inherits some of the well-accepted mind map benefits, such as the organization of ideas and concepts, emphasis on the relevant keywords, association between elements in branches, and grouping of ideas [6]. DVD describes the basic concepts found in other value models (e.g., REA [7], and e3value [4]), such as actors, resources, and resource transfer between actors [8], plus three more complementary concepts: who starts the transfer of resources, value port, and value level agreement.

Figure 1 presents an example of DVD model for an abstract shop business. Actors are environment entities economically independent. The focus of the business analyst defines the main actor (central node of the model), and the focus changes along the specification process. As the focus changes, a new model is built and new actors and value exchanges appear. Each time the analyst focuses on one actor (the main actor), identifies its relationship with other environment actors, producing an inter-organizational network. From each such relationship, a value exchange (transfer of resources) is defined, showing economic reciprocity through two value ports (blue arrows[3] connected to value exchange in Fig. 1, which points to value objects such as money, goods, services). Each value exchange has a textual description which is not represented on the visual model. Next, we define who starts the value exchanges through a configuration of arrows (in red) between the main actor and the environment actors, helping understanding the model. For example, Shopper starts the value exchange with Store, by making a payment Money in exchange of a product from the store Good. As the business analyst focuses on one actor, setting the main actor, the supporting tool displays it as the central node of the model, dynamically. Each value exchange requires a value level of agreement between the actors involved, which refers to the minimal business rule agreed among them with no clear-cut criteria to achieve it. In the example, the shopping transaction between Shopper and Store must be secure, leading to add Security to the corresponding value exchange.

2.2 Goal-Oriented Requirements

Goal-oriented requirements engineering uses goals for eliciting, elaborating, structuring, specifying, analyzing, negotiating, documenting, and modifying requirements [9].

---

[2] A mind map is a type of cognitive map used to view, classify and organize concepts, and to generate new ideas in a straightforward and intuitive way [6].

[3] The direction is set based on the environment actor.

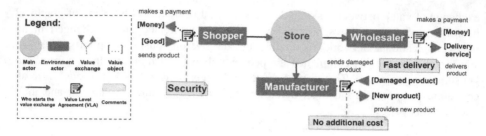

Fig. 1. DVD model for an abstract shop. (Color figure online)

A goal-oriented model uses goal as the concept to provide the rationale (i.e., the why) for the envisioned system [9]. Several goal-oriented approaches exist, each one focusing on different activities of the early stages of information system development, offering a variety of procedures for reasoning about goals (e.g., KAOS [10], EKD [11], BMM [12], i*/Tropos [13], GSN [14], NFR [15], GBRAM [16], Techne [17], and GRL [18]). Detail about all these goal-oriented approaches can be found in [19].

We analyzed nine goal-oriented modeling languages to identify and align business value concepts with goal-oriented concepts (cf. Section 3). This mapping is a fundamental piece in model-driven development to enable the generation of an initial goal-oriented model from a business value model, automating error-prone tasks through model transformation. This reduces the accidental complexity involved in the transition of the business area to technological area.

2.3   Model-Driven Engineering

Model-Driven Engineering (MDE) is an established approach to building complex software systems and has been successfully implemented in many industries, including telecommunication, automotive, aerospace, and business information systems [20]. MDE automates repetitive and error-prone tasks through an automatic processing model aiming at reducing the accidental complexity involved in software development [21]. MDE focuses on the abstraction of the details of a complex problem, concentrating developers on the production of top-level abstraction models to generate complex software artifacts automatically. Hence, MDE uses models as first class entities, with the advantage of increasing productivity, augmenting interoperability, and facilitating communication [22, 23].

Developing software through models requires a rigorous definition of these models [24]. To achieve this, metamodels and automatic transformations of models play a significant role in MDE [25], where metamodels are used to implement model transformations as well as to create Domain Specific Languages (DSLs). The model transformations automatically refine, refactor or re-engineer source models [26] and a DSL is a language designed to be useful for a specific set of tasks and a particular domain [27], realize a particular point of view of a problem, and create rigorous modeling editors. Indeed, metamodels and automatic transformations of models are key concepts in MDE [25]. Hence, abstract models are incrementally refined through model transformations (known as Model to Model, or M2M, and Model to Code, M2C), starting from a problem model until the production of a solution model. Our approach offers a conceptual mapping which enables the use of MDE techniques to generate goal-oriented models from value models (M2M transformation).

3 Aligning a Value Model with Goal-Oriented Models

Figure 2 shows the process to show how a goal-oriented model can be generated from a business value model using MDE. The six activities of the process are discussed next.

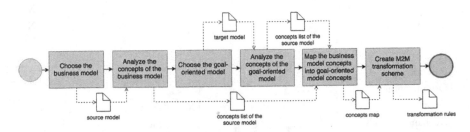

Fig. 2. A generic process to generate goal-oriented models from a value model.

Choose the Business Model. The majority of M2M transformation languages assume two models participating in each transformation: the source model and the target model. This activity selects the business model that will be used as the source of the transformation. A DVD model will be our source model.

Analyze the Concepts of the Business Model. One idea of M2M transformation is to create different views of a problem without losing semantics among the views. So, we need to identify the concepts of the source model that will later be mapped into concepts of the target model. The DVD concepts are: main actor, environment actor, value exchange, value object, and the value level agreement.

Choose the Goal-Oriented Model. Usually, one target model would be chosen for M2M transformations. We will not select one specific goal-oriented model, as want to offer a generic approach able to generate any goal-oriented model from a DVD model. Thus, we analyzed the following nine candidate goal-modeling techniques and their respective models: KAOS [10], EKD [11], BMM [12], i*/Tropos [13], GSN [14], NFR [15], GBRAM [16], Techne [17], and GRL [18].

Analyze the Concepts of the Goal-Oriented Model. Table 1 summarizes the existing concepts of the nine goal-oriented target models.

Map the Business Model Concepts into Goal-Oriented Model Concepts. This activity creates a set of heuristics to map the concepts between the source and target models. Six heuristics are needed to map DVD concepts to goal-oriented concepts. Table 2 describes these heuristics.

Table 1. Goal-oriented concepts

#	Concept	KAOS	EKD	BMM	iStar/Tropos	GSN	NFR	GBRAM	Techne	GRL
1	Goal	X	X	X	X	X		X	X	X
2	Soft goal	X			X		X		X	X
3	Vision			X						
4	Mission			X						
5	Objectives			X						
6	Strategies			X		X				
7	Tactics			X						
8	Operation	X	X				X	X		
9	Task				X				X	X
10	Agent	X			X			X		X
11	Actor				X					X
12	Role				X					X
13	Position				X					X
14	Organization		X	X						
15	Context					X				
16	Domain property	X								
17	Issues		X	X						
18	Obstacles	X						X		
19	Requirements	X						X		
20	Events	X								
21	Resource				X					X
22	Expectation	X								
22	Claim						X			
23	Argument		X							

Create the M2M Transformation Scheme. We selected the two goal-oriented models with more concepts mapped through the heuristics to illustrate the transformation scheme: KAOS (five concepts) and iStar (four concepts)[4]. It is important to notice that iStar has two different types of models: strategic dependency model and strategic rationale model. The iStar dependency model is the one representing dependencies between actors, and hence the most appropriate for this study. Figure 3 shows a visual scheme for a M2M transformation from DVD to KAOS (a) and iStar (b), illustrated with one economic value exchange between Shopper and Store (partial example from Fig. 1). This scheme shows which concepts of the target model the DVD's concepts must be transformed into and the respective heuristic for each transformation. For example, the DVD main actor concept must be transformed into a KAOS software agent and into an iStar actor, using heuristic H3.

---

[4] GRL had the same four concepts mapped, however, as the graphical representation of these concepts is similar to iStar, the same transformation scheme can be used to both goal-oriented models.

Table 2. Heuristics (concepts map)

# DVD concept (source)	Goal oriented concept (target)	Description
H1 value exchange	Goal	A goal is an objective the system under consideration should achieve [9]. Regarding a value exchange, it is an objective the business under consideration should achieve, most likely with the use of a system
H2 value level agreement (VLA)	Softgoal	Softgoal is a goal for which there are no clear-cut criteria for whether the condition is achieved [28]. VLA refers to the minimal business rule agreed among actors with no clear-cut criteria to achieve it. Both concepts are related to the specification of quality attributes and constraints
H3 actors	Agent/Actor	Agents are stakeholders who interact with the system. They are a responsible for achieving requirements and expectations [28]. They are sub-units of a complex social actor, each of which is an actor in a more specialized sense [29]. There are software agents and environment agents. Thus, as the DVD main actor is directly related to the information system under development, it is mapped into a software agent, and the environment actor is mapped into an environment agent
H4 value object	Expectation	An expectation is a type of goal to be achieved by an environment agent [28]. Therefore, when transferring a resource, the expectation is what the actor must provide to receive something in return, and the value object is what the actor provides or receives
H5 value object	Requirement	A requirement is a type of goal to be achieved by a software agent [28]. Therefore, when transferring a resource, the requirement is what the actor must provide to receive something in return, and the value object is what the actor provides or receives
H6 value object	Resource	A resource is what the actor desires to acquire [29], and the value object is what the actor provides or receives

4   Case Study

Our approach to align DVD with goal-oriented models will be described using the waste exportation case study extracted from [30]. Figure 4 shows the DVD model for the waste exportation business. In this case study, waste is traded between an exporter and an importer. In principle, we must distinguish between two different kinds of trading. In general, the exporter has to pay the importer for the waste, handling this as cheaply as possible, for the recovery or disposal of the waste. Accordingly, waste and

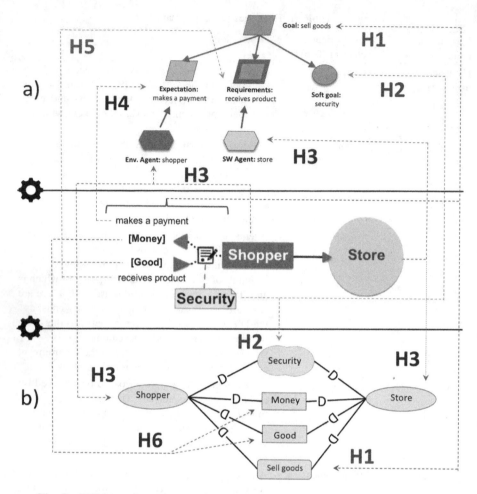

Fig. 3. M2M transformation (a) from DVD to KAOS and (b) from DVD to iStar.

money are given to the importer, and the exporter gets the waste handling service in return (this value exchange between exporter and importer has been represented with identifier "1" on the left side of Fig. 4). In some cases, the waste is traded like a regular good. The trading of recycled paper is a typical example of such a case. This means that the importer has to pay a low price for the waste. In other words, the importer gets the waste, and the exporter receives money in return (this value exchange between exporter and importer has been represented with identifier "2" on the left hand side of Fig. 4). Moreover, the international trading of waste has some legal implications. Competent authorities in the countries of the exporter and the importer control the trading. Accordingly, the exporter has to inform the export authority about a waste transport. The exporter delivers relevant environmental information about the transport and the export authority issues a transport allowance in return. The value of a transport allowance is considered equally to fulfill the legal regulations and, thus, to avoid

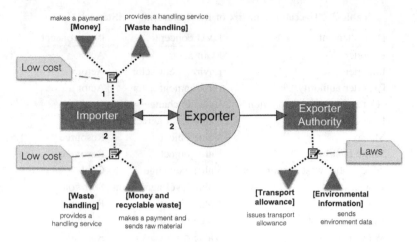

Fig. 4. DVD model for the waste exportation business.

possible fines (this value exchanges between exporter and exporter authority is represented on the right side of Fig. 4).

4.1   Transformation from DVD to iStar

Figure 5 shows the result of the transformation from DVD to iStar by using the set of heuristics described previously. This M2M transformation was performed automatically. Table 3 shows the traceability matrix of the DVD and iStar concepts.

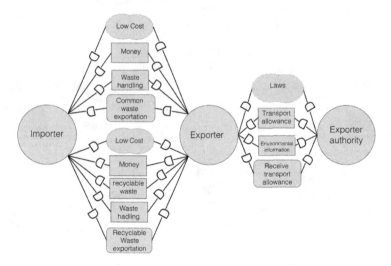

Fig. 5. iStar model generated from the DVD model, for the waste exportation business.

Table 3. Traceability matrix of DVD elements × iStar concepts

DVD element	DVD concept	iStar concept
Exporter	Main actor	Actor
Importer	Environment actor	Actor
Exporter authority	Environment actor	Actor
"Common waste exportation"*	Value exchange	Goal
Low cost	Value level agreement	Soft goal
Money	Value object	Resource
Waste handling	Value object	Resource
"Recyclable waste exportation"*	Value exchange	Goal
Low cost	Value level agreement	Soft goal
Money	Value object	Resource
Recyclable waste	Value object	Resource
Waste handling	Value object	Resource
"Receive transport allowance"*	Value exchange	Goal
Laws	Value level agreement	Soft goal
Transport allowance	Value object	Resource
Environment information	Value object	Resource

* Note that the value exchange description exists in the metamodel, but it
is not shown in the DVD visual model.

4.2   Transformation from DVD to KAOS

Figure 6 shows the result of the M2M transformation from DVD to KAOS, using the
set of heuristics previously established. This transformation was performed fully

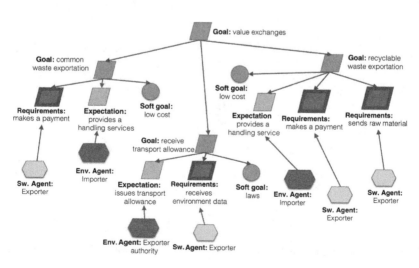

Fig. 6. KAOS model generated from the DVD model, for waste exportation business.

Table 4. Traceability matrix of DVD elements × KAOS concepts

DVD element	DVD concept	KAOS concept
Exporter	Main actor	Software agent
Importer	Environment actor	Environment agent
Exporter authority	Environment actor	Environment agent
"Common waste exportation"*	Value exchange	Goal
Low cost	Value level agreement	Soft goal
"Makes a payment": money	Value object	Requirements
"Provides a handling service": waste handling	Value object	Expectation
"Recyclable waste exportation"*	Value exchange	Goal
Low cost	Value level agreement	Soft goal
"Makes a payment": money	Value object	Requirements
"Sends raw material": recyclable waste	Value object	Requirements
"Provides a handling service": waste handling	Value object	Expectation
"Receive transport allowance"*	Value exchange	Goal
Laws	Value level agreement	Goal
"Issues transport allowance": transport allowance value object "receives environment information"	Value object	Expectation
Environment information	Value object	Requirements

* Note that the value exchange description exists in the metamodel, but it is not shown in the DVD visual model.

automatically; in others words, there is no human decision making activity during the generation process. Table 4 shows the traceability matrix between DVD and KAOS.

4.3 Discussion

From the case study, we can observe that it is possible to create an automatic alignment between DVD and a goal-oriented model in a way that reduces the effort during the goal-modeling specification process. We repeated the process for two different goal-oriented languages, showing that the approach is generic and reusable. Also, once our approach was applied to the waste exportation case study, we identified some important points of improvement in the implemented M2M transformations. In particular, regarding the conversion from DVD to iStar, we detected that we could simplify the resulting iStar model (by reducing the number of dependencies), by not transforming the value exchange to goal (heuristic 1). The reason is that the goal produced in iStar is increasing the size of model, as the other transformed elements (e.g., softgoal and resources) could be seen as decompositions of this goal. This decomposition structure is evident in the KAOS model generated from the same DVD model, since KAOS shows these elements structured in a tree.

Regarding the conversion from DVD to KAOS, we believe that the structure of the KAOS model generated helps the modularization of the requirements specification for IS based on economic value exchanges. Also, this KAOS structuring helps to prioritize the activities of the next stages of IS development. Considering points of improvement in the M2M transformations implemented, we can observe that some elements are repeated (e.g., software agent: exporter). Removing these repeated elements makes the model more concise semantically but maybe also more confusing because the number of relationships to the same visual element will increase and the graphical disposition of the elements will become more difficult (most likely some graphical elements will overlap others). Thus, we decided that our approach should support both situations through a previous configuration before the transformation, giving the modeler the freedom to choose.

5   Related Work

The closest researches on the alignment of both the business and the information technology areas use the business process model as the key business artifact (source) to reduce the gap between both areas [31–33]. However, there is a significant difference between business modeling and business process modeling [2]. Our work uses a business model that represents economic business value (value model) as the essential artifact of this alignment. We are aware of only two approaches to business/IT alignment using a value model [30, 34]. Huemer et al. [30] propose a mapping from an e3value model to a UML profile. Gordijn et al. [34] aim to analyze the economic feasibility of a radio station. For this, they made an e-service design using iStar and e3value models, offering a conceptual mapping between these models. Our work shows how to do any transformation from a business model to a goal-oriented model and also uses model-driven techniques for automatic generation of the target model. We develop a tool to support the automation of the transformations[5].

6   Conclusion

In this paper, we describe a systematic approach for an automatic generation of goal-oriented models from a value model (Dynamic Value Description – DVD).

The approach is generic and has been used to generate two types of goal models: iStar and KAOS. To accomplish this, we created a conceptual mapping between the

---

[5] The proof-of-concept tool can be found at https://goo.gl/9paig7.

DVD and nine goal-oriented models. Then, we show how we can use model-driven techniques to allow automatic transformations between the models. Finally, we evaluated our approach through a waste importation case study. The results show that our approach systematize the automatic generation of the any goal models from any business model using model-driven techniques in order to ensure the alignment between business and IT.

For future work, we plan to integrate our method with existing services identification methods (particularly service-oriented architecture approaches) that use goal models as an input of a top-down analysis (problem decomposition until reaching the services' level). This will be one step forward towards integrating requirements and software architecture, increasing further the alignment between the business and the software services [35].

Acknowledgments. This research is supported by the NOVA LINCS Research Laboratory (Ref. UID/CEC/04516/2013), and Programa Ciência sem Fronteiras - CAPES (Ref. 99999.009047/2013-01).

References

1. Rasiwasia, A.: Meta model for business model design: designing a meta model for E3 value model based on MOF. Master's thesis, Stockholm, Sweden (2013)
2. Gordijn, J., Akkermans, H., van Vliet, H.: Business modelling is not process modelling. In: Liddle, S.W., Mayr, H.C., Thalheim, B. (eds.) ER 2000. LNCS, vol. 1921, pp. 40–51. Springer, Heidelberg (2000). doi:10.1007/3-540-45394-6_5
3. White, S.A.: Introduction to BPMN. IBM Cooperation (2004)
4. Gordijn, J.: Value-based requirements engineering. Ph.D. thesis, May 2002
5. Souza, E., Abrahao, S., Moreira, A., Araújo, J., Insfran, E.: Comparing value-driven methods: an experiment design. In: HuFaMo 2016 Held in MODELS, Saint Malo, France (2016)
6. Buzan, T., Buzan, B.: The Mind Map Book (1996)
7. McCarthy, W.E.: The REA accounting model: a generalized framework for account-ing systems in a shared data environment. Account. Rev. **57**, 554–578 (1982)
8. Andersson, B., et al.: Towards a reference ontology for business models. In: Embley, D.W., Olivé, A., Ram, S. (eds.) ER 2006. LNCS, vol. 4215, pp. 482–496. Springer, Heidelberg (2006). doi:10.1007/11901181_36
9. van Lamsweerde, A.: Goal-oriented requirements engineering: a guided tour. In: Requirements Engineering Conference – RE 2001 (2001)
10. Dardenne, A., van Lamsweerde, A., Fickas, S.: Goal-directed requirements acquisition. Sci. Comput. Program. **20**(1–2), 3–50 (1993)
11. Loucopoulos, P., Kavakli, V., Prekas, N., Rolland, C., Grosz, G., Nurcan, S.: Using the EKD approach: the modelling component (1997)
12. OMG: Business Motivation Model (BMM) (2010)
13. Yu, E., Giorgini, P., Maiden, N., Mylopoulos, J.: Social Modeling for Requirements Engineering. MIT Press, Cambridge (2011)
14. Kelly, T., Weaver, R.: The goal structuring notation – a safety argument notation. In: Dependable Systems and Networks Workshop on Assurance Cases (2004)

15. Chung, L., Nixon, B.A., Yu, E.S.K., Mylopoulos, J.: Non-functional Requirements in Software Engineering. Springer, Boston (1999)
16. Anton, A.I.: Goal-based requirements analysis. In: RE 1996, pp. 136–144. IEEE (1996)
17. Borgida, A., Ernst, N., Jureta, I., Lapouchnian, A.: Techne: a(nother) requirements modeling language. In: ICSE 2009 (2009)
18. Amyot, D., Horkoff, J., Gross, D., Mussbacher, G.: A lightweight GRL profile for i* modeling. In: Heuser, C.A., Pernul, G. (eds.) ER 2009. LNCS, vol. 5833, pp. 254–264. Springer, Heidelberg (2009). doi:10.1007/978-3-642-04947-7_31
19. Fayoumi, A., Kavakli, E., Loucopoulos, P.: Towards a unified meta-model for goal oriented modelling. In: EMCIS 2015 (2015)
20. Mussbacher, G., et al.: The relevance of model-driven engineering thirty years from now. In: Dingel, J., Schulte, W., Ramos, I., Abrahão, S., Insfran, E. (eds.) MODELS 2014. LNCS, vol. 8767, pp. 183–200. Springer, Cham (2014). doi:10.1007/978-3-319-11653-2_12
21. Kolovos, D.S., Matragkas, N., Williams, J.R., Paige, R.F.: Model driven grant proposal engineering. In: Dingel, J., Schulte, W., Ramos, I., Abrahão, S., Insfran, E. (eds.) MODELS 2014. LNCS, vol. 8767, pp. 420–432. Springer, Cham (2014). doi:10.1007/978-3-319-11653-2_26
22. Schmidt, D.C.: Guest editor's introduction: model-driven engineering. IEEE Comput. 39(2), 25–31 (2006)
23. Almeida, J.P.A.: Model-driven design of distributed applications. In: Meersman, R., Tari, Z., Corsaro, A. (eds.) OTM 2004. LNCS, vol. 3292, pp. 854–865. Springer, Heidelberg (2004). doi:10.1007/978-3-540-30470-8_99
24. Singh, Y., Sood, M.: Model driven architecture: a perspective. In: Advance Computing Conference (2009)
25. Kleppe, A.G., Warmer, J.B., Bast, W.: MDA Explained: The Model Driven Architecture: Practice and Promise. Addison-Wesley Professional, Reading (2003)
26. Wanderley, F., Araújo, J.: Generating goal-oriented models from creative requirements using model driven engineering. In: MoDRE 2013 (2013)
27. Gronback, R.C.: Eclipse Modeling Project: A Domain-Specific Language (DSL) Toolkit. Addison-Wesley, Reading (2009)
28. RespectIT: A KAOS Tutorial. v1.0 edn, October 2007
29. Horkoff, J., Yu, E.: I-Star Wiki (2017). http://istar.rwth-aachen.de
30. Huemer, C., Schmidt, A., Werthner, H.: A UML profile for the e3-value e-business model ontology. In: BUSITAL 2008 Held in CAiSE 2008 Conference (2008)
31. Dwivedi, V., Kulkarni, N.: A model driven service identification approach for process centric systems. In: IEEE Congress on Services Part 2 (2008)
32. Inaganti, S., Behara, G.K.: Service Identification: BPM and SOA Handshake (2007)
33. Koliadis, G., Ghose, A.: Relating business process models to goal-oriented requirements models in KAOS. In: Hoffmann, A., Kang, B.-h., Richards, D., Tsumoto, S. (eds.) PKAW 2006. LNCS, vol. 4303, pp. 25–39. Springer, Heidelberg (2006). doi:10.1007/11961239_3
34. Gordijn, J., Yu, E.S.K., van der Raadt, B.: e-Service design using i* and e3value modeling. IEEE Softw. 23(3), 26–33 (2006)
35. Weigand, H., Johannesson, P., Andersson, B., Bergholtz, M.: Value-based service modeling and design: toward a unified view of services. In: Eck, P., Gordijn, J., Wieringa, R. (eds.) CAiSE 2009. LNCS, vol. 5565, pp. 410–424. Springer, Heidelberg (2009). doi:10.1007/978-3-642-02144-2_33

A BPMN Extension for Integrating Knowledge Dimension in Sensitive Business Process Models

Mariam Ben Hassen[✉], Mohamed Turki, and Faïez Gargouri

ISIMS, MIRACL Laboratory, University of Sfax, B.P. 242, 3021 Sfax, Tunisia
{mariem.benhassen, faiez.gargouri}@isims.usf.tn,
mohamed.turki@isetsf.rnu.tn

Abstract. While importance of knowledge dimension is well recognized, there is no clear theoretical background and successful practical experiments of inclusion and integration of this dimension in BP meta-models and BPM formalisms. This paper proposes to extend the well-known Business Process Modeling Notation (BPMN 2.0) with the knowledge dimension so that using a common modeling formalism it would be possible to relate the various types of knowledge, information and data to the business process model. This extension, called «BPMN4KM», is designed methodically by application of the extension mechanisms of BPMN 2.0. We aim at incorporating relevant issues at the intersection of knowledge management (KM) and business process modeling (BPM) in order to enrich the graphical representation of sensitive Business processes (SBPs) and improve the localization and identification of crucial knowledge mobilized and created by these processes. Besides, we evaluate the relevance of BPMN4KM concepts through a real SBP scenario from medical domain.

Keywords: Knowledge management · Knowledge identification · Sensitive business process modeling · BPM4KI meta-model · Ontologies · BPMN 2.0 · Extension mechanism

1 Introduction

Sensitive Business process (SBP) modeling has become primary concern for any successful organization to improve the management of their individual and collective knowledge. A Sensitive Business Process is a BP which comprises a high number of critical organizational activities (individual/collective) with intensive acquisition, sharing, storage and (re)use of very specific knowledge «crucial knowledge». It mobilizes a large diversity of information and knowledge sources, consigning a great amount of heterogeneous knowledge. Moreover, an SBP requires a high dynamic conversion of knowledge and a high degree of collaboration and interaction (intra/inter-organizational) among participants. Its execution involves many external agents and the assistance of many experts, who apply, create and share a great amount of very important tacit organizational knowledge, in order to achieve collective objectives and create value. In addition, SBP are typically an unstructured or semi-structured

© Springer International Publishing AG 2017
M. Themistocleous and V. Morabito (Eds.): EMCIS 2017, LNBIP 299, pp. 559–578, 2017.
DOI: 10.1007/978-3-319-65930-5_44

organizational actions, requires substantial flexibility, encompassing a highly dynamic complexity [1, 2]. Due to those characteristics, modeling and organizing the knowledge involved in SBP is relatively critical.

In order to enrich and improve the SBP modeling, we have proposed, in previous work, a conceptual specification of SBP organized in new multi-perspective meta-model, entitled «BPM4KI: Business Process Meta-Model for Knowledge Identification» [1–3]. BPM4KI explicit and organize the key concepts and relationships that characterize an SBP. It integrates all relevant perspectives/dimensions relating to BPM-KM, i.e. the functional, the organizational, the behavioral, the informational, the intentional and the knowledge perspectives. In this research work, we focus more on the «Knowledge Dimension» which is not yet explicited, fully supported and integrated within BPs models and BPM approaches and formalisms.

However, while importance of knowledge dimension is well recognized, there is no clear theoretical background and successful practical experiments of inclusion and implementation of this dimension in BP/SBP models. In such languages as IDEF0, IDEF3, GRAPES BM in GRADE tool, EPC diagrams in ARIS tool [3], UML 2.0 activity diagram [4] and BPMN 2.0 [5], data, information and material flows are often represented in BP models by the same symbols/artifacts and without any unambiguous definitions of the concepts. At the same time knowledge has poor or no modeling capabilities in these formalisms. On the other hand, knowledge modeling languages (KMDL [6, 7], PROMOTE [8] and NKIP [9]) have shortcomings concerning their ability to explicitly and fully include the knowledge dimension within BPs models as well as relevant issues at the intersection of KM and BPM. They have limited process perspective representation, i.e. they do not address process logic to full extent and thus there is no possibility to represent data and information. To address this research gap, we propose to extend one of the best known modeling formalism, the Business Process Modeling Notation (BPMN) [5], with the knowledge dimension in order to explicitly incorporate all relevant aspects related to KM within BPs models, and on the other hand, to enrich the graphical representation of SBPs and improve the localization and identification of crucial knowledge mobilized and created by these processes. In fact, BPMN 2.0 was selected as the most suitable BPM notations for SBP representation, because addresses the highest representation coverage of the set of BPM4KI concepts and incorporates requirements for SBP modeling better than other formalisms [10, 11]. Nevertheless, the main weaknesses identified in this specification regards the knowledge dimension modeling.

In this research work we present BPMN4KM: a BPMN 2.0 extension, including all relevant aspects related to knowledge dimension in SBP modeling. The proposed extension is developed using the extensibility mechanisms of BPMN [5]. Furthermore, we develop a specific plug-in based on the Eclipse platform, called K4BPMN Modeler, implementing and supporting BPMN4KM.

The rest of the paper is structured as follows: Sect. 2 presents BPMN 2.0 and related works relevant to the research problem. Section 3 presents the central concepts that describe the knowledge dimension of SBP modeling. Section 4 presents the proposed approach for extending BPMN 2.0 with the knowledge dimension. Section 5 illustrates the application and the relevance of some BPMN4KM concepts, based on a real case study. Section 6 concludes the paper and underlines some future research topics.

2   Background and Related Work

This section presents background research: Sect. 2.1 describes BPMN as one of the most suitable BPM notations; Sect. 2.2 briefly present related works relevant to the research problem.

2.1   Business Process Model and Notation (BPMN 2.0)

BPMN 2.0.2 stands for Business Process Model and Notation [5]. It is a graphical representation for specifying BPs in a BP model, and a standard for BP modeling notations. BPMN is initiated as a standard BPM language for conventional business, B2B and services process modeling. It can be used within many methodologies and for many purposes, from high-level descriptive modeling to detailed modeling intended for process execution providing a standardized bridge for the gap between BP design and its implementation. BPMN considers notational elements grouped in five basic categories (Flow Objects, Data, Connecting Objects, Swimlanes and Artifacts). Besides, it has the capabilities of handling B2B BP concepts, such as public, private, collaboration processes and choreographies, as well as advanced modeling concepts, such as exception handling and transaction compensation in addition to the traditional BP.

Several surveys have evaluated the adequacy of BPMN for BPM. From our point of view, BPMN has six main advantages [10, 11]:

- It is the BPM standard backed up by OMG, which is based upon a meta-model [5] built with UML, the notation which is the de facto standard for modeling software engineering artifacts [4].
- It is very simple, easy to use, readily understandable and accessible by all business stakeholders.
- BPMN is one of the most recent and expressive BPM notations, grounded on the experience of earlier BPM formalisms, which ontologically makes it one of the most complete BPM formalisms [12].
- It is appropriate for modeling collaborative BPs actors that display complex flows with high degree of interactions among process' actors and high degree of information exchanged, developed and shared among participants.
- It is currently the BP notation most used among process modeling practitioners, with more BPM tools support available.
- BPMN is extensible. BPMN 2.0 defines an extensibility mechanism for both process model extensions and graphical extensions.
- Finally, BPMN 2.0 presents the broadest coverage of the set of BPM4KI meta-model concepts (except the knowledge dimension) [11].

Based in the previous assessments, BPMN 2.0 is taken as a basis for the representation of SBP models.

2.2   BPMN 2.0 Shortcomings

BPMN stresses the process view representation, offering a number of symbols for modeling various decision points, process, activity and event types. BPMN constructs emphasize mainly the support of the control-flow and data perspective when expressing processes' orchestration and collaboration. As other BPM formalisms, BPMN constructs have a shallow coverage of informational, organizational and intentional aspects of BPM. Moreover, BPMN focuses entirely on the functional and behavioral aspects of the BP model. Nevertheless, the main weaknesses identified in this specification regards the knowledge dimension modeling which represents the core and relevant dimension in SBP models (exploring the collaboration and interaction aspects). Currently, from the point of view of various ways how knowledge (including data and information) are used in organizations, the following issues are not yet fully supported in BPMN 2.0 (neither in any of the above-mentioned BPM and knowledge modeling formalisms):

- Opportunity to clearly distinguish between data, information and knowledge in the representation of flows between SBP activities. The information and data exchange constitutes the basis for knowledge dissemination and generation. Note that, BPMN provides opportunity to model only information and data flow using the same symbols/artifacts and without any unambiguous definitions of the concepts.
- Opportunity to identify the different owners/sources of knowledge involved (used, generated and/or modified) in the BP activities and location where knowledge can be obtained and can be clearly stated.
- Opportunity to consider the roles of humans in BP activities, be it as humans (single persons), teams, or communities of practice who bears the internal/tacit knowledge.
- Opportunity to integrate and separate the different types/kinds of knowledge (tacit/explicit dimension, internal/external dimension, individual/collective dimension, etc.).
- Opportunity to integrate and separate the different nature of knowledge (like experience, basic knowledge, scientific/technical knowledge, general knowledge, etc.).
- Possibility to illustrate knowledge flows between sources and among activities.
- Possibility to represent the dynamic of acquisition, preservation, transfer, sharing, development, and (re) use of individual and organizational knowledge within and between BPs activities.
- Ability to specify more than two opportunities of knowledge conversions (between knowledge types) taking place in single SBP activity.
- Opportunity to enable modeling the critical/knowledge intensity dimensions of organizational activities which are important to determine the crucial knowledge mobilized and created by these activities.
- Opportunity to accurately represent collaborative aspects and specify how do interactions occur (information and knowledge exchange) in SBPs. These aspects are useful to characterize the SBPs, due to, for instance, the high degree of knowledge exchanged and developed and shared among agents through intra/inter-organizational collaboration, and its dynamic nature. In fact, BPMN 2.0 provides a specific choreography model which allows to concentrate only on conversation between performers. However, this model does not show how performer's knowledge changes during the conversation and communication.

To sum up, BPMN 2.0 diagrams are not adequate for the new SBP modeling requirements. So, to overcoming the discussed shortcomings, BPMN 2.0 will be adapted and extended to be convenient for a rich and expressive representation of SBPs, including all or at least most of the relevant issues at the intersection of KM and BPM.

2.3   Related Work

The integration of KM into BPs has rapidly become the most promising practical and theoretical task in KM. In this context, there have been several attempts to integrate the knowledge concept/dimension in BP models as well as in BPM and knowledge modeling formalisms, e.g., [6, 8, 10, 13–16]. However, none of the proposed knowledge oriented BPM approaches and formalisms adequately and fully support and represent all relevant aspects of knowledge dimension within BPs models (e.g., differentiation between tacit and explicit knowledge, the different types of knowledge conversion, the dynamic aspects of knowledge, the different sources of knowledge, etc.). At the same time, BPM is challenging - these notations are weak in representing logic/control flow of the BP and the process perspectives as a whole (i.e., the structural, behavioral, organizational and informational dimensions).

Besides, while importance of knowledge dimension is well recognized, there is no clear theoretical background and successful practical experiments of inclusion of this dimension in the well-known BPM standard. In particular, there are only a few initiatives in the BPM-KM area, which use the BPMN as core formalism and systematically enhance its capabilities and extend it by KM specific aspects [2, 10, 15, 17]. Ammann [17] defined an extension of BPMN 1.1 [18] for knowledge-related BPM, called BPMN-KEC (KEC stands for knowledge, employees, and communities). In this work different objects were used: objects for knowledge and information, for knowledge conversions, for associations and for persons. Nevertheless, the proposal has not the necessary expressivity and features to represent the relevant SBP elements, including the knowledge aspect. Another work by Supulniece et al. [13], proposed an extension of BPMN which roots in concepts implemented in knowledge-oriented modeling language (KMDL) (such as an information object, knowledge object, type of knowledge conversion) [6] with few additions and changes in graphical representation. However, experiments with the integrated notation revealed that the relationship between the phenomena behind the symbols is somewhat unclear in the BPM. Moreover, the relevant aspects of knowledge dimension do not fully supported and represented (like the different types of knowledge mobilized and created by each BP activity, the knowledge flow between particular collective members, the different sources/supports of knowledge, etc.).

To date, to the best of our knowledge, there is a lack of works providing systematic approaches for the development of extensions to the BPMN 2.0 meta-model to consider the knowledge aspect in BPM. However, there are previous works providing approaches to extend BPMN 2.0 to represent their domain specific requirements. Some interesting extension proposals are presented in [19–22]. The differences between the different research works unveil the need for a unified method for the conceptual modeling of extensions and their representation in terms of the BPMN extension mechanism.

In this paper, we aim to solve the discussed shortcomings and address the gap between BPM and KM. Precisely, this research work presents a rigorous scientific approach to extend BPMN 2.0 for KM. This extension must consider and incorporate all relevant aspects of SBP modeling, including the knowledge dimension, in order to allow a rich and expressive representation of SBPs and improve the localization and identification of crucial knowledge mobilized and created by these processes.

3   Knowledge Dimension in SBP Models

In order to develop a rich and expressive graphical representation of SBPs, we proposed a semantically rich conceptualization for specifying a SBP organized in a new generic Business Process Meta-model for Knowledge Identification (BPM4KI). The current version of BPM4KI offers a referential of generic concepts and relationships relevant to the BPM-KM domain semantically rich and well-based on «core» domain ontologies[1] [23–25], which are based on top of the DOLCE foundational ontology [26]. BPM4KI were categorized in six perspectives (or dimensions), namely, the functional, the organizational, the behavioral, the informational, the intentional and the knowledge perspectives. The different dimensions are crucial for a complete understanding, characterization and representation of an SBP [1–3].

In this research work, we focus more on the description and analysis of the knowledge dimension which represents the most relevant aspects of SBP modeling, exploring the KM aspect, the collaboration and interaction and all relevant SBP elements. We point out that the knowledge dimension (supporting the new SBP modeling requirements) is not yet, however, not yet explicited, fully supported and integrated within BPs models and BPM formalisms [2, 3]. So, we aim at obtaining new knowledge helpful for developing BPM formalisms that could adequately support above-mentioned issues in BP/SBP modeling.

Concretely, our approach for modeling the new extended knowledge dimension is based on the reuse and the specification of central generic concepts (and the relationships between them) defined in different ontological modules of the global and consistent ontology OntoSpec [23–25]: *Capacity-OS*, *Action-OS*, *Action of Organization-OS*, *COOP*, *Partcipation-role-OS*, *Agentive Entity-OS*, *Organization-OS*, and *I&DA-OS (Information and Discourse Acts)*. Besides, several features of Knowledge have been further described in the literature and this leads us to classify it according to some dimensions (see Fig. 1). In this paper, we retain the following dimensions: (a) *affiliation of knowledge/agentive entity*, (b) *source of knowledge*, (c) *organizational coverage of knowledge*, (d) *nature of knowledge*, (e) *strategic*, which seem useful and relevant to the context of our research work, *the localization and identification of knowledge*. With respect to the limited space of this paper, a comprehensive description of the different knowledge dimensions and concepts cannot be presented. The new extended concepts related to the knowledge dimension are marked in red (see Fig. 1).

---

[1]   http://home.mis.u-picardie.fr/ ~ site-ic/site/spip.php?article53.

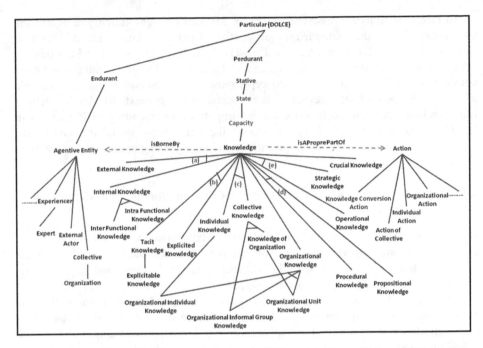

Fig. 1. Extension of the ontological module capacity- OS for modeling the knowledge dimension of SBP

It is important that an appropriate BPM formalism provides explicit representation of the different issues related to the knowledge dimensions in BPM. In this context, the SBPs can be graphically represented, using the well-known standard for BPM, BPMN 2.0 [5], in order to localize and identify the knowledge that is mobilized and created by these processes.

4 BPMN4KM: A BPMN Extension for Including the Knowledge Dimension in SBP Modeling

At the root of the success of modeling, design, reengineering, and running BPs/SBPs is effective use and support of organizational knowledge. Knowledge must be considered as one of the BP dimensions, because knowledge is related to action, it is implemented in the action, and is essential to its development. Knowledge is used to perform a process, it is created as a result of process execution, and it is distributed among process participants. Then, the integration of knowledge dimension and BPs is indispensable. Benefits provided by extending BP models with knowledge dimension are described in Ben Hassen et al. [3]. In this section, we define an extension of the BPMN specification [5], called BPMN4KM, which introduces the knowledge dimension aspects and provides a rich and expressive representation of SBPs to identify and localize the crucial knowledge mobilized by these BPs.

In fact, despite its expressiveness, BPMN 2.0 does not yet explicitly represent the key concepts of the Knowledge perspective (such as Individual Tacit Knowledge, Collective Tacit Knowledge, Expert, Explicit Knowledge, Socialization, Internalization, etc.). To overcoming the shortcomings of BPMN 2.0, some of its concepts must be adapted and extended to include all or at least most of the relevant SBP elements. In this context, BPMN 2.0 defines four standard extension mechanisms that are important for extending SBP model with knowledge dimension. We have introduced the main concepts of the knowledge dimension into BPMN with a some additions and changes in graphical representation.

4.1 Equivalence Check: Analysis of BPMN Support for the Knowledge Dimension Concepts

Based on both SBP elements and requirements, the comparison with standard BPMN is conducted in order to identify a reasonable need for extension. According to the presented domain ontology (see Fig. 1), each concept is examined regarding its semantically equivalence with standard elements.

Therefore, the respective element descriptions, rules and explanations within the BPMN specification [5] were analyzed in-depth. This leads implicitly to the derivation of the BPMN4KM meta-model and its stereotypes. According to Braun et al. [27], the following rules are defined for the equivalence check (correspondence between concepts of the knowledge perspective concepts (extract of BPM4KI) and the BPMN mata-model):

- *Equivalence*: There is a semantically equivalent construct in the BPMN in the sense of a permitted combination of elements or just a single element. In this case, no extension is necessary and the domain concept is represented as BPMN concept.
- *Conditional equivalence*: There is no obvious semantic matching with standard elements, but rather situational discussion is necessary in order to provide arguments for a possible mapping or to explain why it is not feasible. This situation is caused by the partial under specification of BPMN elements [5]. Consequently, the concept is either treated as equivalent concept or as non-equivalent concept.
- *No equivalence*: There is no equivalence to any standard element for three reasons: First, the entire concept is missing. In this case, the domain concept is represented as Extension Concept in the BPMN4KM meta-model. Second, a relation between two concepts is missing. Therefore, an association between the affected concepts is constructed in the BPMN4KM meta-model. Third, properties of a concept are missing. Then, an owned property is assigned to the element in the extended model.

Table 1 provides the conducted equivalence check and its implications for the extended BPMN meta-model. As shown in this table, BPMN lacks support for several concepts of the knowledge aspect (with relevant inter aspects relationships, namely the functional and organizational dimensions). As result of the correspondence check, the concepts of the BPMN4KM meta-model are classified/characterized as *BPMN Concepts* (are those that match with some concept of the BPMN meta-model) or as *Extension Concepts* (are those defined in the domain of the extension).

Table 1. Analysis of the BPMN support for the knowledge dimension concepts (with relevant inter aspects relationships) and derivation of concepts for the BPMN meta-model of the extension.

BPM4KI concepts		Equivalence check/ BPMN concept	Support level	Extended BPMN meta-model
Knowledge perspective	Knowledge	No equivalence (no appropriate marker)	–	Extension concept
	Crucial knowledge	No equivalence		Extension concept
	Internal knowledge	No equivalence	–	Extension concept
	Inter functional knowledge	No equivalence	–	Extension concept
	Intra functional knowledge	No equivalence	–	Extension concept
	External knowledge	No equivalence	–	Extension concept
	Explicited knowledge	No equivalence	–	Extension concept
	Tacit knowledge	No equivalence	–	Extension concept
	Explicitable knowledge	No equivalence	–	Extension concept
	Individual knowledge	No equivalence	–	Extension concept
	Organizational individual knowledge	No equivalence	–	Extension concept
	Collective knowledge	No equivalence	–	Extension concept
	Knowledge of organization	No equivalence	–	Extension concept
	Informal group knowledge	No equivalence		
	Organizational knowledge	No equivalence	–	Extension concept
	Organizational unit knowledge	No equivalence	–	Extension concept
	Procedural knowledge	No equivalence	–	Extension concept
	Propositional knowledge	No equivalence	–	Extension concept
	Strategic knowledge	No equivalence	–	Extension concept
	Operational knowledge	No equivalence	–	Extension concept
	Physical knowledge support	No equivalence	–	Extension concept

(*continued*)

Table 1. (*continued*)

BPM4KI concepts		Equivalence check/ BPMN concept	Support level	Extended BPMN meta-model
Functional perspective	Action of collective	Conditional equivalence[a] → Process	+	Extension concept
	Organizational activity	Equivalence → Activity, task, sub process	+	BPMN concept
	Deliberate action	Conditional equivalence → Activity	Partly	Extension concept
	Discourse act	Conditional equivalence → Activity, task	Partly	Extension concept
	Critical organizational activity	Conditional equivalence → Activity	Partly	Extension concept
	Collaborative organizational activity	Conditional equivalence → Activity, choreography activity	Partly	Extension concept
	Knowledge intensive activity	Conditional equivalence → Activity	Partly	Extension concept
	Communication	Conditional equivalence → Activity, choreography, collaboration, conversation	Partly	Extension concept
	Knowledge conversion action	Conditional equivalence → Activity, choreography activity	–	Extension concept
	Socialization	No equivalence	–	Extension concept
	Internalization	No equivalence	–	Extension concept
	Explicitation	No equivalence	–	Extension concept
	Externalization	No equivalence	–	Extension concept
	Combination	No equivalence	–	Extension concept
Organizational perspective	Agentive entity	Conditional equivalence → Resource role/performer, participant	Partly	Extension concept
	Collective	Conditional equivalence → Resource role/performer, participant	Partly	Extension concept
	Organization	Conditional equivalence → Resource role	Partly	Extension concept
	Organizational unit	Conditional equivalence → Resource role, participant (partner/role entity)	Partly	Extension concept

(*continued*)

Table 1. (*continued*)

BPM4KI concepts		Equivalence check/ BPMN concept	Support level	Extended BPMN meta-model
	Informal group	No equivalence	–	Extension concept
	External actor	Conditional equivalence → Resource role, partner/role entity (with exchange of message flow)	Partly	Extension concept
	Human	Equivalence → Resource role, human performer	+	BPMN concept
	Experiencer	Conditional equivalence → Human performer, potential owner	Partly	Extension concept
	Expert	No equivalence	–	Extension concept

[a] Process only define the Action of Organization (Business Process) which is an Action of Collective performed by a group of individuals affiliated with the organization [24]. However, Process cannot be used to specify the actions that can be carried out collectively by the individuals making up the Collective.

The following section shows the developed BPMN meta-model extension using the BPMN 2.0 extensibility mechanisms.

4.2   The Extended BPMN4KM Meta-model

The BPMN meta-model [5] can be extended by integrating new domain-specific concepts to standard and predefined BPMN elements. This is supported by a standard extension mechanism consisting of four elements:

- `ExtensionDefinition`- specifies a named group of new attributes, that can be used by standard BPMN elements. Thus, both new concepts and new additional attributes can be defined (jointly added/attached to the original BPMN elements).
- `ExtensionAttributeDefinition`- defines new/particular attributes that can be specified for an `ExtensionDefinition` element.
- `ExtensionAttributeValue` - contains the value assigned to an extension attribute of a BPMN element.
- `Extension`- binds/imports the entire `ExtensionDefinition` element and its attributes to a BPMN model definition in order to make them technically accessible.

Figure 2 presents the class diagram of BPMN extension. By associating a BPMN element with an `ExtensionDefinition`, every BPMN element which subclasses the BPMN `BaseElement` can be extended with additional attributes. Therefore, BPMN 2.0 with their different extension mechanisms appear to provide the most complete coverage of the concepts and constructs needed for analyzing and modeling most of the SBP characteristics.

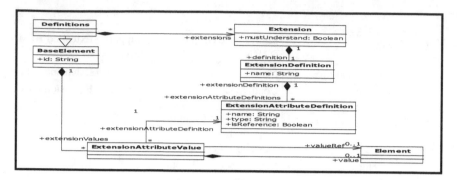

Fig. 2. Representation of the BPMN extension mechanism

Despite the fact that BPMN offers a well-defined extension interface, only very few BPMN extensions make use of it [22], what hampers comprehensibility, comparability between developed extensions and impedes the straightforward integration of extensions in modeling tools. We suppose that the missing procedure model for extension building in BPMN causes this lack of rigor.

Based on the model transformation rules stated in Stroppi et al. [20], we define the BPMN4KM extension model (BPMN + X model). Figure 3 below presents the resulting extended BPMN meta-model. In this figure only the relevant standard BPMN classes are shown in white. The BPMN4KM concepts are shown in grey. We associate `Knowledge` concept with the `RootElement` of the BPMN specification. The semantics and the abstract syntax of the BPMN4KM elements are based on the specification of the BPMN extension mechanism [5]. Figure 4 shows the stereotypes of the extended BPMN profile and the UML metaclasses that it specializes [20].

ExtensionModel is the topmost container of all the elements defining a BPMN extension. *BPMNElement* allows representing an original element of the BPMN meta-model. *ExtensionElement* allows representing a new element in the extension model which is not defined in the BPMN meta-model (such as `Knowledge`, `InternalKnowledge`, `TacitKnowledge`, `ExplicitedKnowledge`, `ExplicitableKnowledge`, `ProceduralKnowledge`, `ExternalKnowlede`, `InternalKnowledge`, `PhysicalKnowledgeSupport`, `Information`, `DistalIntention`, `Combination`, `Socialization`, `Internalization`, `Externalization` and `Explicitation`). *ExtensionDefinition* allows specifying a named group of attributes which are jointly added to the original BPMN elements (such as `KnowledgeFlow`, `Experiencer`, `Collective`, `KnowledgeConversionAction`, `Knowledge Intensive Activity`, `Critical Organizational Activity`, `Collaborative Organizational Activity`, and `Sensitive Business Process`). *ExtensionDefinition* has the same meaning than the `ExtensionDefinition` element of the BPMN metamodel. The semantics defined by the *ExtensionAttributeDefinition* element of the BPMN meta-model is captured by the Property metaclass of the UML metamodel. Thus, ExtensionAttributeDefinition is represented in BPMN4KM models by UML properties, either owned by the ExtensionDefinition elements or navigable from them through associations. The properties of

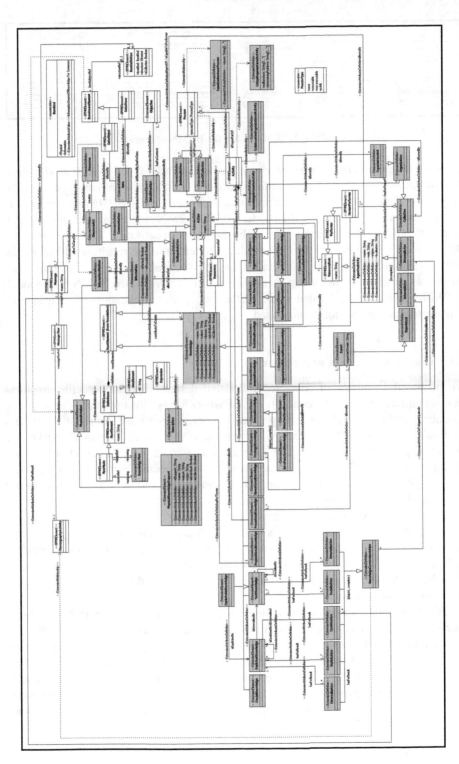

Fig. 3. Abstract syntax of the BPMN extension: the extended BPMN4KM meta-model

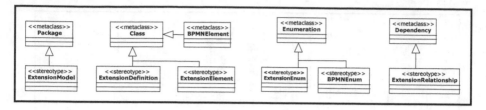

Fig. 4. BPMN + X UML profile for BPMN extension models

ExtensionDefinition and *ExtensionElement* elements can be typed as a *BPMNElement*, *ExtensionElement*, *BPMNEnum*, *ExtensionEnum* or UML primitive type. Finally, *ExtensionRelationship* specifies a conceptual link between a *BPMNElement* and an *ExtensionDefinition* element aimed to extend it. The BPMN extension mechanism cannot express the BPMN element to be extended by an extension definition. Thus, the definition of an *ExtensionRelationship* does not produce any effect in the resulting BPMN extension. *ExtensionRelationship* is provided to help conceptualizing extensions since extensions are generally defined to customize certain elements of the BPMN meta-model.

With respect to the limited space of this paper, the application of each applied transformation rule cannot be presented.

4.3   Concrete Syntaxes and Editor

We proposed an advanced concrete syntax that defines new and specific graphical representation for the new concepts of BPMN4KM as illustrated in Table 2. For instance, the `Action` element is specified by new markers for representing `Individual Action`, `Collective Action`, `Critical Organizational Activity` and `Collaborative Organizational Activity`. Furthermore, we have incorporated new notational elements with specific properties for `Knowledge` typologies (tacit/explicit dimension, declarative/procedural dimension, etc.) and `Information`, for knowledge conversions, `Knowledge Flows` (between knowledge, activities and agentives entities), `Physical Knowledge Supports` and `Agentive Entities` (`Expert` and `Collective`). The concrete syntax of the extended `Knowledge` concept is depicted in Fig. 5. Knowledge elements are marked with source, nature, organizational value and organizational coverage information according to the four knowledge dimensions [2] as introduced in Sect. 3, see Fig. 5 for the notational details.

We have implemented an editor supporting this syntax as shown in Fig. 5. More precisely, we have developed a specific Eclipse plug-in, entitled «K4BPMN: Knowledge for Business Process Modeling Notation», to integrate and represent all relevant aspects related to the knowledge dimension in SBP models (to improve the localization of crucial knowledge that is mobilized and created by these processes). This plug-in extends the open source editor Eclipse BPMN2 Modeler plug-in [28]: it completes this later by integrating new attributes, properties, elements and specific icons for introduce new SBP semantics.

Table 2. Concrete syntax of BPMN4KM

Elements	Modeling Notation	Elements	Modeling Notation
Internal Knowledge (Individual/Collective)	Individual Internal Knowledge Collective Internal Knowledge	Organizational Critical Activity (Individual/Collective)	Individual Critical Task Collective Critical Task
Tacit Knowledge	Tacit Knowledge [T]		Collective Critical Sub-Process
Conscious Knowledge (Strategic/Collective)	Conscious Knowledge [C] Collective Strategic Conscious Knowledge [S]	Physical Knowledge Support (Individual/Collective)	Individual Physical Knowledge Support Collective Physical Knowledge Support
Explicit Knowledge (Individual/Collective, Strategic/procedural)	Individual Strategic Explicit Knowledge [S] Collective Explicit Knowledge Collective Procedural Explicit Knowledge [D]	Expert	
		Collective	
External Knowledge (Individual/Collective, Procedural/Propositional)	External Knowledge Collective Procedural External Knowledge Ind. Are Propositional External Knowledge [P]	Knowledge conversion/creation Internalization (Flow)	
Information	Information	Knowledge conversion: Explicitation	
Knowledge Flow		Knowledge conversion: Externalization	
Socialization	[S]	Knowledge conversion: Combination	

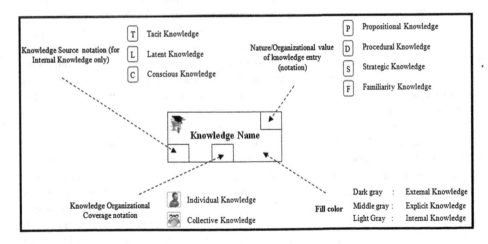

Knowledge Source notation (for Internal Knowledge only)
- [T] Tacit Knowledge
- [L] Latent Knowledge
- [C] Conscious Knowledge

Nature/Organizational value of knowledge entry (notation)
- [P] Propositional Knowledge
- [D] Procedural Knowledge
- [S] Strategic Knowledge
- [F] Familiarity Knowledge

Knowledge Name

Knowledge Organizational Coverage notation
- Individual Knowledge
- Collective Knowledge

Fill color
- Dark gray : External Knowledge
- Middle gray : Explicit Knowledge
- Light Gray : Internal Knowledge

Fig. 5. Graphical notation of knowledge concept

The icons in this figure indicate the different types of activities (e.g. differential diagnosis of neurological abnormalities is a collaborative critical organizational activity), the different types of knowledge (e.g. A_4K_{u1} related to the ability to master the neurological examination is an internal knowledge which is conscious and strategic), the agentive entity (e.g. medical community that bears the internal knowledge) and the different knowledge physical support (icon represents the organizational structure type). The attributes of the new concepts are accessible via the property sheet once the concept is selected.

The following section illustrates the practical applicability of BPMN4KM considering a real SBP scenario from medical domain.

5   Illustrative Example of the Use of Extended BPMN4KM

The research project presented in this paper has been done in the context of the Association of Protection of the Motor-disabled of Sfax-Tunisia (ASHMS) [11]. This organization is characterized by highly dynamic, unpredictable, complex and highly intensive knowledge processes. We intend to apply some concepts proposed by BPMN4KM meta-model to evaluate their practical utility and suitability in providing an adequate and expressive representation of a SBP, to improve the localization and the identification of crucial medical knowledge. Particularly, we are interested in the early care of the disabled children with cerebral palsy (CP). An depth analysis of this care has been made by Ben Hassen et al. [11]. In fact, the medical care process is very complex. The mass of medical knowledge mobilized and produced during this process is very important, heterogeneous and recorded on various scattered sources. One part of this knowledge is embodied in the mind of health professionals. Another part, is preserved in the organizational memory (as reports, medical records, data bases, therapeutic protocols and clinical practice guidelines). The created knowledge stems from the interaction of a large number of multidisciplinary healthcare professionals with heterogeneous skills, expertise and specialties (such as neonatology, neuro-pediatrics, physical therapy, orthopedics, psychiatry, physiotherapy, speech therapy, and occupational therapy) and located on geographically remote sites. Most of them are volunteers and they come from different organizations (e.g. the university hospital of Sfax, the medicine faculty, the health graduate school, etc.). The raised problem concerns on the one hand, the insufficiency and the difficulty to localize and understand the medical knowledge that is necessary for decision-making, and on the other hand, the loss of knowledge held by these experts during their scattering or their departure at the end of the treatment. Thus, the ASHMS risks losing the acquired know-how for good and transferring this knowledge to new novices if ever no capitalization action is considered. Our main purpose is to locate, identify, make visible, know how to coordinate, preserve, share, generate and integrate the different types and modalities of crucial medical knowledge necessary for performing the medical care process of children with CP, in the suitable time and in the suitable context, in order to improve the care quality and help healthcare professionals to make the right clinical decisions in complex circumstances.

The global care process of the disabled children with CP (i.e. the most important BP in the ASHMS organization), which is a SBP is made up of several sub-processes. It consists of a succession of many actions in the form of medical and paramedical examinations and evaluations of children with cerebral palsy in different specialties. The different sub-processes (like process related to neonatology care, process related to neuro-pediatric care, process related to physiotherapy, global evaluation process, etc.) require certain medical information as well as certain medical knowledge (results of clinical exams, hospitalization reports, patient-specific knowledge recorded in the medical case file or practice guidelines).

In this study, we take into consideration the results of experimentation of the multi-criteria Sensitive Organization's Process Identification Methodology (SOPIM) proposed by Turki et al. [29] which was validated in the ASHMS and aims at evaluating and identifying SBPs for knowledge localization. We have opted for the SBP «*Process of initial neuro-motor evaluation of a child with CP*». In Fig. 6, we illustrate an SBP model extract of the initial evaluation process using BPMN 2.0, extended and enriched with some extended concepts related to the knowledge dimension (according to BPMN4KM meta-model). The resulting BPMN model is the result of many individual meetings/interviews for evaluation and validation conducted with each of the two stakeholders implied in the SBP realization: the Neonatologist and the Neuro-pediatrician. During our

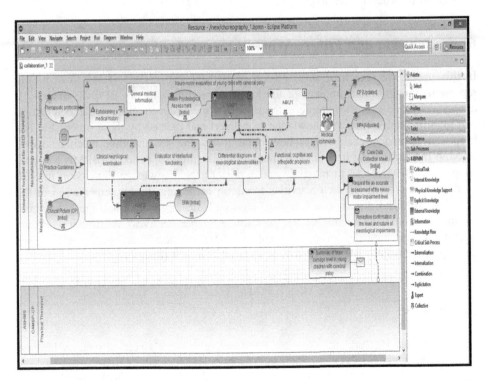

Fig. 6. Fragment of SBP model related to the initial neuro-motor evaluation of a child with CP using extended BPMN modeler

experimentation, we have identified different types of medical knowledge mobilized and created by each critical activity related to the SBP.

For instance, the knowledge A_2K_{p2} related to «Synthesis of neurological abnormalities related to motor, somatic and sensory development of the young children with cerebral palsy, as well as the different clinical signs» is produced by the critical activity A_2 «Clinical neurological examination». Note that this materialized/externalized knowledge is created as a result of the activity execution by the medical community (the Neuropediatric and the Neonatologist), during which they interact with information (i.e. source of knowledge information) related to the child with CP (based on their competences, skills and previous experiences) to generate and communicate their own knowledge. A_2K_{p2} is stored in the following physical media: the neurological and neuro-motor assessment sheets (BNM). These physical media of knowledge are located internally within the Neonatology service in the University Hospital Hedi Chaker, precisely in the various archives drawers or patients' directories. A_2K_{p2} is of a scientific, technical and measure nature which is related to patients. It represents an external and propositional knowledge which is collective. This knowledge is imperfect (general, incomplete and uncertain). A_2K_{p2} is mobilized by the activity A_4 «Differential diagnosis of neurological abnormalities».

It is important to mention that not all Knowledge Perspective concepts are applicable and must be instantiated in every SBP scenario. The graphical representation of SBP is in its experimental stage.

6   Conclusion and Future Work

In this paper, we presented an extension of BPMN 2.0 «BPMN4KM» to explicitly represent, integrate and implement the knowledge dimension in BP/SBP models. It allows a rich and expressive representation of SBPs in order to improve the localization and identification of crucial knowledge mobilized and created by these processes. The proposed approach extension is developed using the extensibility mechanisms of BPMN. In addition, we have implemented a Eclipse plug-in, called K4BPMN Modeler supporting BPMN4KM to integrate and represent the knowledge dimension in SBP models. Our current research activities focus on achieving the implementation of the different BPM4KI dimensions (i.e. the functional, the organizational, the behavioral, the informational, the intentional and the knowledge perspectives). As further work, we intend to automatically generate graphical SBP models and specify executable descriptions of these BPs to enhance the knowledge identification.

References

1. Ben Hassen, M., Turki, M., Gargouri, F.: Modeling dynamic aspects of sensitive business processes for knowledge localization. In: International Conference on Knowledge Based and Intelligent Information and Engineering Systems, KES 2017, Marseille, France (2017)

2. Ben Hassen, M., Turki, M., Gargouri, F.: A proposal to model knowledge dimension in sensitive business processes. In: Madureira, A.M., Abraham, A., Gamboa, D., Novais, P. (eds.) ISDA 2016. AISC, vol. 557, pp. 1015–1030. Springer, Cham (2017). doi:10.1007/978-3-319-53480-0_100

3. Ben Hassen, M., Turki, M., Gargouri, F.: Towards extending business process modeling formalisms with information and knowledge dimensions. In: Benferhat, S., Tabia, K., Ali, M. (eds.) IEA/AIE 2017. LNCS, vol. 10350, pp. 407–425. Springer, Cham (2017). doi:10.1007/978-3-319-60042-0_45

4. OMG: Unified Modeling Language (UML). Version 2.0 (2007). http://www.uml.org/

5. OMG: Business Process Model and Notation (BPMN), Version 2.0.2 (2013). http://www.omg.org/spec/BPMN/2.0.2/pdf/

6. Gronau, N., Korf, R., Müller, C.: KMDL-capturing, analysing and improving knowledge-intensive business processes. J. Univ. Comput. Sci. **11**(4), 452–472 (2005)

7. Arbeitsbericht: KMDL® v2.2 (2009). http://www.kmdl.de/

8. Woitsch, R., Karagiannis, D.: Process oriented knowledge management: a service based approach. J. Univ. Comput. Sci. **11**(4), 565–588 (2005)

9. Netto, J.M., Franca, J.B.S., Baião, F.A., Santoro, F.M.: A notation for knowledge-intensive processes. In: IEEE 17th International Conference on Computer Supported Cooperative Work in Design, vol. 1, pp. 1–6 (2013)

10. Ben Hassen, M., Turki, M., Gargouri, F.: Choosing a sensitive business process modeling formalism for knowledge identification. Proced. Comput. Sci. **100**, 1002–1015 (2016)

11. Ben Hassen, M., Turki, M., Gargouri, F.: Sensitive business processes representation: a multi-dimensional comparative analysis of business process modeling formalisms. In: Shishkov, B. (ed.) BMSD 2016. LNBIP, vol. 275, pp. 83–118. Springer, Cham (2017). doi:10.1007/978-3-319-57222-2_5

12. Recker, J., Rosemann, M., Indulska, M., Green, P.: Business process modeling: a comparative analysis. J. Assoc. Inf. **10**, 333–363 (2009)

13. Supulniece, I., Businska, L., Kirikova, M.: Towards extending BPMN with the knowledge dimension. In: Bider, I., Halpin, T., Krogstie, J., Nurcan, S., Proper, E., Schmidt, R., Ukor, R. (eds.) BPMDS/EMMSAD -2010. LNBIP, vol. 50, pp. 69–81. Springer, Heidelberg (2010). doi:10.1007/978-3-642-13051-9_7

14. Businska, L., Kirikova, M.: Knowledge dimension in business process modeling. In: Nurcan, S. (ed.) CAiSE Forum 2011. LNBIP, vol. 107, pp. 186–201. Springer, Heidelberg (2012). doi:10.1007/978-3-642-29749-6_13

15. Ammann, E.M.: Modeling of knowledge-intensive business processes. World Acad. Sci. Eng. Technol., Int. J. Soc. Behav. Educ. Bus. Ind. Eng. **6**(11), 3144–3150 (2012)

16. dos Santos França, J.B., Netto, J.M., do ES Carvalho, J., Santoro, F.M., Baião, F.A., Pimentel, M.: KIPO: the knowledge-intensive process ontology. Softw. Syst. Model. **14**(3), 1127–1157 (2015)

17. Ammann, E.: BPMN-KEC – an extension of BPMN for knowledge-related business process modeling. Internal Report, Reutlingen University (2008)

18. OMG Final Adopted Specification. Business Process Modeling Notation Specification (2008). http://www.omg.org/spec/BPMN/1.1/

19. Charfi, A., Turki, S.H., Chaabane, A., Witteborg, H., Bouaziz, R.: A model-driven approach to developing web service compositions based on BPMN4SOA. Int. J. Reason.-Based Intell. Syst. **3**(3–4), 194–204 (2011)

20. Stroppi, L.J.R., Chiotti, O., Villarreal, P.D.: Extending BPMN 2.0: method and tool support. In: Dijkman, R., Hofstetter, J., Koehler, J. (eds.) BPMN 2011. LNBIP, vol. 95, pp. 59–73. Springer, Heidelberg (2011). doi:10.1007/978-3-642-25160-3_5

21. Jankovic, M., Ljubicic, M., Anicic, N., Marjanovic, Z.: Enhancing BPMN 2.0 informational perspective to support interoperability for cross-organizational business processes. Comput. Sci. Inf. Syst. **12**(3), 1101–1120 (2015)
22. Braun, R., Esswein, W.: Classification of domain-specific BPMN extensions. In: Frank, U., Loucopoulos, P., Pastor, Ó., Petrounias, I. (eds.) PoEM 2014. LNBIP, vol. 197, pp. 42–57. Springer, Heidelberg (2014). doi:10.1007/978-3-662-45501-2_4
23. Kassel, G.: Integration of the DOLCE top-level ontology into the OntoSpec methodology (2005)
24. Kassel, G., Turki, M., Saad, I., Gargouri, F.: From collective actions to actions of organizations: an ontological analysis. In: Symposium Understanding and Modelling Collective Phenomena (UMoCop), University of Birmingham, Birmingham, England (2012)
25. Turki, M., Kassel, G., Saad, I., Gargouri, F.: A core ontology of business processes based on DOLCE. J. Data Semant. **5**(3), 165–177 (2016)
26. Masolo, C., Vieu, L., Bottazzi, E., Catenacci, C., Ferrario, R., Gangemi, A., Guarino, N.: Social roles and their descriptions. In: Dubois, D., Welty, C. (eds.) Proceedings of the Ninth International Conference on the Principles of Knowledge Representation and Reasoning, pp. 267–277 (2004)
27. Braun, R., Schlieter, H., Burwitz, M., Esswein, W.: Extending a business process modeling language for domain-specific adaptation in healthcare. In: Wirtschaftsinformatik, pp. 468–481(2015)
28. BPMN2 Modeler. http://www.eclipse.org/bpmn2-modeler/
29. Turki, M., Saad, I., Gargouri, F., Kassel, G.: A business process evaluation methodology for knowledge management based on multi-criteria decision making approach. In: Information Systems for Knowledge Management. Wiley-ISTE (2014). ISBN 978-1-84821-664-8

Towards Managing Key Performance Indicators for Measuring Business Process Performance

Emna Ammar El Hadj Amor[1(✉)] and Sonia Ayachi Ghannouchi[2]

[1] ISITCom Hammam Sousse/RIADI Laboratory,
ENSI Manouba, Manouba, Tunisia
emnahouda@yahoo.fr

[2] ISG Sousse/RIADI Laboratory, ENSI Manouba, Manouba, Tunisia

Abstract. Organizations always need to continually improve and review their critical business processes (BP), especially in the healthcare field. This improvement requires an efficient mean to support the management and the analysis of healthcare processes, to collect all relevant indicators designed for both effective management and process improvement and to understand all interesting results based on data instance logs that reflect the performance of business processes. In order to meet these challenges, we propose a novel approach for managing business process performance enabling the evaluation and optimization of BPs. This approach is illustrated through a real case study in the emergency department of "Farhat Hached" hospital in Sousse (Tunisia).

Keywords: Business process · Business process management (BPM) · Performance measurement · Key performance indicators (KPI) · Ontology · Association rules · Data mining · Emergency department · Health care process

1 Introduction

The evaluation of performance stays a topical question to assess how far the organization's goals are achieved. For this reason, BPM technology has become an important instrument for supporting complex coordination scenarios and for improving business process performance [1]. This approach has become a valuable asset in the healthcare domain [2]. It includes methods, techniques, and tools to support the design, enactment, management and analysis of operational business processes involving humans, organizations, applications, documents and other sources of information understanding, task design, and relevant result interpretation of organization's performance [3]. It can be defined also as a structured method of understanding, documenting, modeling, analyzing, simulating, executing, and continuously changing end-to-end business processes and all relevant resources in relation to an organization's ability to add value to the business [4]. In addition, it is crucial not only to understand the actual situation in the emergency department (ED) and to remodel the business processes, if necessary but also to continue the improvement of healthcare business processes based on comprehensive measurement of organization's performance. Indeed, Key Performance Indicators (KPIs) provide critical information to the organization for monitoring and predicting business performance in accordance with strategic objectives [6].

© Springer International Publishing AG 2017
M. Themistocleous and V. Morabito (Eds.): EMCIS 2017, LNBIP 299, pp. 579–591, 2017.
DOI: 10.1007/978-3-319-65930-5_45

Performing business process analysis in healthcare organizations is particularly difficult due to the highly dynamic, complex, ad-hoc, and multi-disciplinary nature of healthcare processes [5].

Like it is the case for any other business process, continuous performance monitoring of healthcare process is important to assess how far the emergency department goals are achieved while promoting the patient satisfaction and improving the business process. However, in practice, performance measurement activity in health organizations has several practical issues, such as unavailability of performance measurement system, distribution of data in multiple locations and performance indicators that are expressed informally in natural language, incompleteness, lack of traceability of the BP etc.; or they define them from the technical view which becomes hardly understandable to non-technical users. In addition, in many organizations including healthcare sector, there is several missing relevant information related to performance qualitative aspect (e.g. KPIs related to patient satisfaction level in the organization) where this kind of performance indicators is in general expressed in natural language. The problem above in such context is how to perform BP in more effective and efficient way without ignoring patient satisfaction aspect. We argue that it is interesting when we associate qualitative measures to quantitative measures and establish the possible links between them.

In this work, we are especially interesting in the improvement of the healthcare process. we can say that there are various challenges related to healthcare performance measurement data which may play serious hurdles in the improvement of health care process. This improvement heavily depends on both key performance indicators and knowledge. Furthermore, we need to know more about the real meaning and the implications of performance measurement value and to understand why an indicator is needed and what it is measuring and what decisions they support. Also, we need to analyze all relevant data in order to make proper decisions. These involve a clear need for approaches that facilitate the understanding of context when implementing healthcare processes. This challenge can be stated by the following three research questions: Is it possible to define a great variety of KPIs so that they are amenable to process evaluation? How can we present their mutual relationships in a way that facilitates decision making? Which KPIs are valuable for the analysis and the evaluation of the BP and how can they be identified?

The remainder of this paper is organized as follow: Sect. 2 introduces related work. Section 3 gives an overview of the proposed approach. Section 4 describes the healthcare process. Section 5 deals with the various performance measurements. Section 6 describes the proposed semantic representation (ontology) and the possible relationships that hold between different concepts. Section 7 focuses on building association rules using data mining technique, to describe the relationships that hold between different KPIs. The last section gives a brief conclusion.

2   Related Works

In [8], Ortega establishes that, in practice, KPIs are informally defined usually in ad-hoc, natural language, with its well-known problems or they are defined from an implementation perspective, hardly understandable to not– technical people. In order to solve this problem, the authors propose an approach to improve the definition of PPIs using templates and linguistic patterns. In [16], the authors introduce a methodology for the application of process mining techniques that leads to the identification of regular behavior, process variants, and exceptional medical cases.

Some authors [18–21] have pointed out the vagueness, imprecision of KPIs values that they are intended to represent, and the lack of an explicit representation of their semantics. According to [21], the major obstacles for effective design and management of Performance Indicators (PI) monitoring systems are related to the fact that PIs are complex objects with an aggregate/compound nature. This often leads to unawareness of indicator semantics as well as of dependencies among indicators. So, the authors in [21] propose to enrich the data cube model with the formal description of the structure of an indicator given in terms of its algebraic formula and aggregation function.

Despite the difficulties in measuring performance indicators in the current health care process, another important aspect is how to cultivate the existing information into useful practices. To solve this issue, Data mining has a great potential to enable healthcare systems to use data more efficiently and effectively [9]. The ability to use a data in databases in order to extract useful information for quality health care is a key of success of healthcare institutions [10]. Data mining techniques provide better medical services to the patients and help the healthcare organizations in various medical management decisions. Some of the services to which apply the data mining techniques in healthcare are: number of days of stay in a hospital, ranking of hospitals, better effective treatments, fraud insurance claims by patients as well as by providers, readmission of patients. They contribute to better identify treatment methods for a particular group of patients and to the construction of an effective drug recommendation systems, etc. [9].

Many works applied [11–14] Apriori algorithm in the healthcare domain. For example, in [11] the author used this algorithm in order to find out the associations between diagnosis and treatments in medical billing data. In [13] this algorithm was applied to discover frequent diseases in medical data in particular geographical locations at a particular period of time.

3 Proposed Approach

Actually, BPM tools are not yet implemented in the emergency department, and consequently, the data related to measuring each activity in the BP is missing and the data related to measuring patient satisfaction is not supported. Consequently, it will result in high error rate and more effort is necessary in order to develop measuring performance which employs several quantitative and qualitative key performance indicators. Those indicators can be derived from guidelines, either from observation and conversation with experts or communication with patients, to get all available data about the tasks in this process and then to build the BP and acquire all KPI values. We try to collect as much data and information as possible. This data will enable us to put together a comprehensive picture of the BP and its KPIs.

In this paper, in order to support the decision, several qualitative and quantitative measurements are retained, and a set of association rules organized for overall improvement purposes. In this view, to measure the performance of an Emergency Department (ED), a list of KPIs is gathered from healthcare process and domain expert and is used for testing. In this work, we are based on our observation, conversation with experts and questionnaire with patients to get all available data to acquire all KPI values. This data will enable us to put together a comprehensive picture of the BP.

Hence, selecting KPIs requires a high cooperation between its quantitative indicators and their related qualitative indicators. For this reason, for quantitative indicators, we are based essentially on the availability of jBPMlog to extract all relevant data. Indeed, the data collected during the business process execution play a key role in improving the overall performance of the business process itself. And for qualitative indicators which are more difficult to measure because they are related to patient experience in the ED, we use a Likert scale to record the level of satisfaction of the patient toward the ED. In this case, all KPIs must be consistently organized for overall improvement purposes.

Furthermore, we aim at developing a KPI ontology that fulfills two conditions. On the one hand, it includes relationships between indicators and process tasks. On the other hand, these relationships must help to reduce the existing visual gap between different kinds of KPIs allowing a comprehensive view of both assets (KPIs and tasks).

After that, the analysis of relationships between indicators can be interpreted into knowledge using data mining techniques which are crucial for making useful decisions. In this use case, since the formats of data are different from the quantitative to the qualitative measurements, the analysis of data may take longer time than usual. Due to that, we inserted all persistent data in a new KPI database. The measurement data is very useful in order to extract the meaningful information from it for improving the BP and the satisfaction of patients toward the ED.

4 Emergency Health Care Process

In this section, we provide a description of the healthcare process and we present its corresponding process model and dashboard and the history logs based on jBPM software. JBPM is an open source BPMS. A BPM system is defined as a software system which extends the functionality of traditional workflow management system, not only limited to work routing, but also to all activities of BPM [22]. It features a robust management console and development tools, with user support during the business process lifecycle including development, deployment, and versioning [15]. This is important to make sure that the BPM system is able to facilitate performance measurement for the adopted business process.

4.1 Description the Healthcare Process

The design phase involves us to look at the big picture of the emergency department which initiates with the patient registration and payment and ends with her/his discharge of the ED. So, we need to focus on what components are involved in the BP.

First, at the beginning of this process, every patient has to pass by Registration activity. After that, in order to arrive to a preliminary conclusion about the status of the patient, sorting activity represents the second task in the BP and the first point of contact with medical staff. This activity consists in recording the preliminary observations and prioritizing the patients, according to their degree of urgency. The following tasks depend on the status of the patient. We find various cases of consultations such as consultation in the delayed emergency sectors in the case of non-urgent patients, consultation in the box, it can be a simple consultation or surgical consultation, consultation in the crash room the patient is of serious harm which requires immediate medical attention and finally the last case Consultation in the supervision room if the condition of the patient is not stable.

At the end of this process, three possible cases exist: the patient is treated and leaves the emergency department, the patient is sent to another service (hospitalization or specialized consultation) to ensure continuity of care or the situation of the patient requires long treatments.

4.2 Execution and Monitoring of the Healthcare Process

The proposed business process model is deployed with the use of jBPM software, where 100 instances have been created through the execution of the process which can be considered as statistically significant and consequently it forms the basis for continuous process optimization. The jBPM core engine stores the process and task history and provides APIs to perform the Business Activity Monitoring (BAM) operations. Further, the jBPM tooling includes the dashboard builder, which enables its users to create and customize dashboards from the business process history [15]. Measuring performance indicator is basically passing queries to the data source and retrieving the query result.

For example, Table 1 and Fig. 2 represents a graphical representation of KPIs in KIE Workbench environment.

Table 1. Task duration related to process instance id 180

process_id	start_date	end_date	taskname	createddate	enddate	status	task duration	duration
180	05/29/16 11:40	05/29/16 12:10	Registration	05/29/16 11:40	05/29/16 11:47	Completed	0h 3m 8s	0h 29m 59s
180	05/29/16 11:40	05/29/16 12:10	Sorting	05/29/16 11:47	05/29/16 11:58	Completed	0h 5m 25s	0h 29m 59s
180	05/29/16 11:40	05/29/16 12:10	Consultation in b	05/29/16 11:58	05/29/16 12:10	Completed	0h 12m 25s	0h 29m 59s

In Table 1 we can find all quantitative indicators values related to task duration details. This dashboard shows all tasks that have been executed.

Figure 1 displays the number of patients manipulated during process executions.

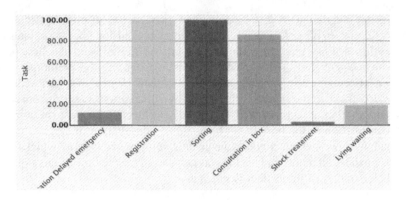

Fig. 1. Number of patients in each activity

5 Emergency Health Indicators

There are many measurements that could be used in the evaluation of healthcare process. The healthcare process requires a good understanding of what is important to the Emergency Department. To acquire a complete knowledge about the data involved in the business process improvement, the patient's experience and the level of satisfaction seem very interesting. After discussion with experts, we arrange a list of relevant indicator where 32 quantitative indicators (including process indicators, aggregated indicators, and administrative indicators) and 19 qualitative indicators are selected for this study. In the next paragraph, we are especially interested in the quantitative indicators that are related to the process execution and the qualitative indicators according to patient satisfaction level.

5.1 Process Indicators

By default, jBPM deals with an H2 database. In our work, to persist jBPM historical data, we configure PostgreSQL database system to store the history log through its persistence module. PostgreSQL is an open source object-relational database system [13]. An overview of the Bamtasksummary table in jBPM data base is presented in Table 2.

Quantitative KPIs are derived from a close look at the activities involved in the BP. For more details, we define 6 indicators (Quanti_KPI7 The duration of registration by patient, Quanti_KPI8 The duration of sorting by patient, Quanti_KPI9 The duration of consultation by patient, Quanti_KPI10 The duration in supervision room by patient, Quanti_KPI11 The duration of delayed emergency by patient, Quanti_KPI12 The duration in crash room by patient) related to the time that the process actor actually spends doing each activity. We also define another indicator related to the duration of all the activities by the patient (Quanti_KPI13 The sum of all previous activities

Table 2. An overview of Bamtasksummary table

createddate timestamp without time zone	duration bigint	enddate timestamp without time zone	processinstanceid bigint	startdate timestamp without time zone
2016-05-26 14:32:23.699	42695	2016-05-26 14:33:13.467	153	2016-05-26 14:32:30.772
2016-05-26 14:33:13.464	364083	2016-05-26 14:41:30.896	153	2016-05-26 14:35:26.813
2016-05-26 14:41:30.892	246660	2016-05-26 15:35:54.844	153	2016-05-26 15:31:48.184
2016-05-26 11:44:25.83	134152	2016-05-26 11:46:49.535	155	2016-05-26 11:44:35.383
2016-05-26 11:46:49.524	218437	2016-05-26 11:50:42.52	155	2016-05-26 11:47:04.083
2016-05-26 11:50:42.515	197293	2016-05-26 12:03:22.647	155	2016-05-26 12:00:05.354
2016-05-26 12:03:22.643	9322161	2016-05-26 15:00:57.104	155	2016-05-26 12:25:34.943
2016-05-26 12:33:44.993	73510	2016-05-26 12:35:09.739	156	2016-05-26 12:33:56.229
2016-05-26 12:35:09.736	269042	2016-05-26 12:41:52.786	156	2016-05-26 12:37:23.744
2016-05-26 12:41:52.779	1050390	2016-05-26 13:00:35.476	156	2016-05-26 12:43:05.086
2016-05-26 09:37:12.317	79784	2016-05-26 09:38:48.809	157	2016-05-26 09:37:29.025
2016-05-26 09:38:48.806	82116	2016-05-26 10:03:26.327	157	2016-05-26 10:02:04.211
2016-05-26 10:03:26.322	332223	2016-05-26 10:56:15.236	157	2016-05-26 10:50:43.013

duration). The value for this indicator corresponds to the duration column in the bam task summary table. We also define 6 indicators related to the time the process waits for the activity in question to be done, from the initial request to the eventual delivery. The value corresponding to these indicators represent the waiting time of an activity. Thus, such indicators are important to see for example, how long a patient waits for a consultation. It is defined as the time difference between created date of a consultation task and its actual start date. In jBPM logs, we don't have a column for this duration so we need to create SQL queries to retrieve this value. Another interesting value that can be derived from this measure is Quanti_KPI23 (The waiting time per patient in all activities) which represent the sum of waiting time per patient in all activities. Finally, we define Quanti_KPI27 (The total time spent in the Emergency Department by the patient in all activities). Its value is retrieved from the duration column in process instance log table.

From the available data in jBPM logs, we conclude that not all KPI can be directly retrieved from these logs and the representation of KPI values especially in the calculation of waiting time and duration in milliseconds can be misunderstood by the decision maker. For this reason, we create another data table which contains all real values related to our research. An overview of this table is presented in Table 3.

Table 3. An example of real values of quantitative indicators table

	quanti_kpi7 character varying(255)	quanti_kpi8 character varying(255)	quanti_kpi9 character varying(255)
22	00:00:18	00:01:22	00:04:17
23	00:01:08	00:00:53	00:05:14
24	00:03:13	00:00:10	00:01:26
25	00:01:04	00:02:16	
26	00:01:16	00:01:33	00:24:31
27	00:02:11	00:02:59	01:35:20
28	00:02:11	00:03:23	00:15:17
29	00:03:11	00:03:13	
30	00:00:37	00:03:57	00:01:11

5.2   Qualitative Indicators

By addressing patient queries, the qualitative aspect will be in a better position to improve satisfaction toward the healthcare processes. This qualitative inquiry is necessary to get close enough to the patient and capture the level of his/her satisfaction and to examine the quality of care provided to patients attending the ED. Hence, at the end of the process, each patient was invited to provide feedback about his satisfaction.

Those indicators concern the qualitative aspects of care in the ED, such as staff attitudes towards patients and the quality of care but they also concern some quantitative aspects such as paramedical staff availability, the overall waiting time before treatment by paramedical personnel and the regularity of doctor visits. This is due for example to the fact that sometimes the nurse is available but the patients who are in less urgent state are not given sufficient attention by nursing staff. So, in order to provide a higher level of quality and consistency with quantitative measurement and with the processes in place, we record the patient's level of satisfaction with those indicators. The aim of this questionnaire is to determine why patients are unsatisfied and what we can do to make them better satisfied.

In order to feed the qualitative database, we are based on the responses of the patient, and consequently, a set of qualitative results is recorded by using a Likert scale. A new table in the database is then created in which respective responses of the patient are inserted as the values of KPIs. Since the exactly followed paths in the process are different from an instance to another, the number of questions asked to each specific patient varies and then some columns in the database have a null value. Table 4 show some data values in the qualitative table.

Table 4. Example from qualitative database (Quali_KPI1 to Quali_KPI6 column)

	qualikpi1 character varying(255)	qualikpi2 character varying(255)	qualikpi3 character varying(255)	qualikpi4 character varying(255)	qualikpi5 character varying(255)	qualikpi6 character varying(255)
17	very satisfied	very satisfied	very satisfied	very satisfied	very satisfied	very satisfied
18	very satisfied	very satisfied	very satisfied	very satisfied	dissatisfied	very satisfied
19	very satisfied	very satisfied	very satisfied	very satisfied	very satisfied	very satisfied
20	satisfied	satisfied	very satisfied	very satisfied	very satisfied	very satisfied

5.3   Integration View of Quantitative and Qualitative KPIs Values

KPIs data mainly contain all the quantitative and qualitative measurements regarding patients. The storage of such type of data is variously depending on the period of analysis and the evaluation of organization performance.

Due to the continuous increasing of the size of measurement healthcare data, we focus on the analysis of the KPIs in the observation period.

Recording real/estimated values of indicators of both quantitative and qualitative aspect offers a better comprehension of the situation. So, we insert all qualitative indicator values from the questionnaire into the new KPI table where we take into account the consistency of several quantitative and qualitative measurements based on the process instance identifier.

At this stage, we track the healthcare process and we identify and extract the appropriate KPI. The data collected during business process execution is used for deriving the key performance indicators (KPI). In addition, we record all real values of indicators of both quantitative and qualitative aspect, which offer a better comprehension of the situation.

The final table contains all information related to quantitative process indicators and qualitative indicators related to the same process instance id. In order to give more sense to ensure that the health care process is fulfilling the expectations of experts' domain and patients, this data need to be checked and evaluated to detect if KPI values are reaching the desired results. As a result, we create another table which contains all estimated values related to the patient instances. This table contains two values ("Ok"codes the KPI value is tolerant and accepted by ED and "Not ok" is the KPI value is not acceptable). Table 5 displays some data from this table.

Table 5. Data from Estimated KPI table

	quali_kpi16 character va	quali_kpi17 character va	quali_kpi18 character va	quali_kpi19 character va	quanti_kpi7 character va	quanti_kpi8 character va	quanti_kpi9 character varying(255)
1	ok	ok			ok	Not ok	ok
2	ok	ok	ok		ok	ok	ok
3	ok	ok			ok	ok	Not ok
4	Not ok	Not ok			ok	ok	
5	ok	ok	ok		ok	ok	ok
6	ok	ok			ok	ok	ok
7	ok	ok			ok	Not ok	ok
8	ok	ok			ok	ok	ok

The main advantage of this table is that it provides the basic information to the decision maker to execute corrective actions in case of important deviations. So, in order to extract the meaningful information from 100 KPIs healthcare rows data, Data mining is beneficial in such a situation.

6 KPI Ontology

The main benefit of this ontology is to offer the opportunity to continuously improve the BP by manipulating relationships and later to identify the reason of bottlenecks. The input in this phase is process tasks and qualitative and quantitative indicators, thus facilitating their future improvement. In this section, we present in Fig. 2 our ontology for the representation of KPIs and process activities. So, in order to make more informed decisions, we collected as much data and information as possible about the possible relations.

Ontology is defined as a set of terms used to describe a given domain and derive inferences from it [7]. OWL ontology is composed of individuals, properties, and classes. The tool used for this ontology is protégé editor. Individuals represent all KPIs and all activities in which we are interested in our use case. In order to represent in one

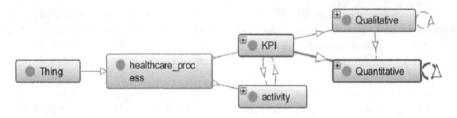

Fig. 2. The main classes of our ontology

hand, the relationship between an activity and the attached KPI, and in another hand, the relationship between indicators, we define a set of object properties. Therefore, we created three object properties related to our KPIs. For example, the property "related_with" might link the possible individuals from the qualitative class (KPI related to a patient satisfaction) to the related individuals in the quantitative class. Also, in order to represent the relationships between indicators in the same category, the owl model implements the properties "Depend quantitative" and "Depend_qualitative" which indicate the links between quantitative/qualitative indicators, their relationships and mainly represent the need to share data between them. Furthermore, properties can have inverses, transitive or symmetric. For example, the inverse of "has_activity" is "has_indicator". Those definitions of relationships and the relative individuals need to be highlighted and modeled in order to aid the monitoring of KPIs that contribute to performance improvement. As well, we envision a collection of Datatype properties to describe relationships between individuals and data values.

7   Extracted Knowledge from KPIs Data

The data mining module uses a well-known data mining algorithm to extract association rules from the given data. With the help of raw data in KPI database, we use SIPINA data mining tool where the Apriori algorithm for data mining is applied.

Data Mining mainly extracts the meaningful KPIs data which were previously recorded in the Estimated KPIs table and derived from the event logs and the qualitative inquiry. This performance measurement can be then interpreted and translated into knowledge where discovering interesting decisions become possible.

In fact, to make any decision, it is important to analyze all the relevant row data of our performance measurement. For this purpose, we use data mining, especially association rule learning as a research method in this stage.

7.1   Association Rule Mining Algorithm: Apriori

Agarwal and his colleagues at IBM Almaden Research Center introduced a novel association rule algorithm called [17] where the association mining can be applied to real databases to extract association rules.

The Apriori algorithm requires two user parameters configuration: the first one is support and second is confidence. Such parameters are used to significantly limit the search for frequent item sets.

Apriori algorithm is interested in finding all such rules having high enough support and confidence. It has two steps:

(1) finding frequent itemsets, that is, those which have enough support,
(2) converting them to rules with enough confidence, by splitting the items into two, as items in the antecedent and items in the consequent [16].

7.2   Association Rule Mining Algorithm: Experimentation Results

Based on Estimated KPI value table, the objective of this step is to understand KPIs interaction. By using Apriori algorithm we aim to find frequent associations and correlations among sets of items. If a KPI value is not frequent, no association rules related to the KPIs are generated. Association rule mining algorithm needs to be configured before learning. So, we give appropriate values for the parameters in advance. Figures 3 and 4 show some results obtained based on Apriori algorithm to predict the occurrence of an item based on the occurrences of other items in the transaction.

Id	Antecedent	Consequent	Length	Support	Confidence	Recall	F-measure	Lift	Conviction
1	quali_kpi17=ok	quali_kpi16=ok	2	0.6200	0.9118	0.9841	0.9466	1.4472	3.7255
2	quali_kpi16=ok	quali_kpi17=ok	2	0.6200	0.9841	0.9118	0.9466	1.4472	10.5147

Fig. 3. Example 1 of association rules

For example, for Rule 1 (Id = 1) the fraction of transactions that contain both quail_KPI17 (The overall waiting time before treatment by medical staff) and quail KPI16 (The overall waiting time before treatment by paramedical personnel) is 0.62 and the confidence for this rule is 0.81. This value measures how often items in quail KPI16 appear in transactions that contain quail_KPI17.

We can see in Fig. 4 that quail_KPI2 is consequent in the rule form, which can be used to determine another measurement that should be associated with it to have a high level of satisfaction.

5	quali_kpi1=ok	quali_kpi2=ok	2	0.6700	0.9853	0.9054	0.9437	1.3315	9.2647
6	quali_kpi3=ok	quali_kpi2=ok	2	0.6600	0.8800	0.8919	0.8859	1.1892	2.0382
7	quali_kpi4=ok	quali_kpi2=ok	2	0.6700	0.8816	0.9054	0.8933	1.1913	2.0647
8	quali_kpi5=ok	quali_kpi2=ok	2	0.6300	0.8400	0.8514	0.8456	1.1351	1.5679
9	quali_kpi6=ok	quali_kpi2=ok	2	0.7100	0.8161	0.9595	0.8820	1.1028	1.3858
10	quali_kpi7=ok	quali_kpi2=ok	2	0.6200	0.9118	0.8378	0.8732	1.2321	2.6471
11	quali_kpi8=ok	quali_kpi2=ok	2	0.6400	0.9143	0.8649	0.8889	1.2355	2.7227

Fig. 4. Example 2 of association rules

Those rules can be used to see what other KPIs should be taken into account to promote a high satisfaction with the quail_KPI2.

This analysis of frequent items aims to find all interesting rules that correlate the presence of one set of items with that of another set of items.

8 Conclusion

Regarding the research questions addressed in this work, not only we track the process behavior and we derive qualitative and quantitative key performance indicators but we also understand all necessary concepts involved in the BP and incorporate domain knowledge of the field. We also, extract implicit information from them and their relationships with other indicators. This information can assist process analysis in the evaluation of KPIs, as well as in the optimization of the associated BPs. An example of implementation of our proposed contribution as well as its validation on a real case study in the healthcare domain is presented.

References

1. Pual Puah K.Y., Nelson Tang K.H.: Business process management, a consolidation of BPR and TQM. In: ICMIT 2000: Proceeding of the 2000 IEEE International Conference on Management of Innovation and Technology, vol. 110, no. 5. IEEE (2000)
2. Stefanelli, M.: Knowledge and process management in health care organizations. Methods Inf. Med. **43**(5), 525–535 (2004)
3. Park, J.H.J., et al. (eds.): Information Technology Convergence, Secure and Trust Computing, and Data Management: ITCS 2012 & STA 2012, vol. 180. Springer Science & Business Media, Heidelberg (2012)
4. De Bruin, T., Rosemann, M.: Application of a holistic model for determining BPM maturity. In: Akoka, J., Comyn-Wattiau, I., Favier, M. (eds.) Proceedings of the 3rd Pre-ICIS Workshop on Process Management and Information Systems. Washington DC, USA: BPTrends, February 2005
5. Rebuge, Á., Ferreira, D.R.: Business process analysis in healthcare environments: a methodology based on process mining, Inf. Syst. **37**(2), 99–116
6. Andrikopoulos, V., Benbernou, S., Bitsaki, M., Danylevych, O., Hacid, M., van den Heuvel, W., Karastoyanova, D., Kratz, B., Leymann, F., Mancioppi, M., Mokhtari, K., Nikolaou, C., Papazoglou, M., Wetzstein, B.: Survey on business process management, July 2008
7. Yadav, U., Narula, S.H., Duhan, N., Jain, V., Murthy, B.K.: Development and visualization of domain specific ontology using protege. Indian J. Sci. Technol. **9**(16) (2016)
8. Del Río-Ortega, A., Resinas, A., Durán, A., et al.: Using templates and linguistic patterns to define process performance indicators. Enterp. Inf. Syst. **10**(2), 159–192 (2016)
9. Ahmad, P., Qamar, S., Rizvi, S.Q.A.: Techniques of data mining in healthcare: a review. Int. J. Comput. Appl. **120**(15) (2015)
10. Abdullah, U., Ahmad, J., Ahmed, A.: Analysis of effectiveness of apriori algorithm in medical billing data mining. In: 2008 International Conference on Emerging Technologies, IEEE-ICET 2008, Rawalpindi, Pakistan, October 18–19 (2008)
11. Patil, M., Joshi, R.C., Toshniwal, D.: Association rule for classification of type-2 diabetic patients. In: Second International Conference on Machine Learning and Computing (2010)

12. Ilayaraja, M., Meyyappan, T.: Mining medical data to identify frequent diseases using apriori algorithm. In: Proceedings of the 2013 International Conference on Pattern Recognition, Informatics and Mobile Engineering (2013)
13. Eapen, A.G.: Application of Data mining in Medical Applications. University of Waterloo, Waterloo (2004)
14. Kai, E., et al.: Empowering the healthcare worker using the portable health clinic. In: IEEE Transactions (2014). doi:10.1109/AINA.2014.108
15. Fiorini, S., Gopalakrishnan, A.V.: Mastering jBPM6. Packt Publishing, Birmingham (2015)
16. Rebuge, Á., Ferreira, D.R.: Business process analysis in healthcare environments: a methodology based on process mining. Inf. Syst. 37(2), 99–116 (2012)
17. Agrawal, R., Srikant, R.: Fast algorithms for mining association rules. In: Proceedings of the 20th International Conference on Very Large Data Bases (VLDB 1994), pp. 487–499. Morgan Kaufmann, Santiago (1994)
18. Pitzos, G., Matsas, M., Chryssolouris, G.: Defining manufacturing performance indicators using semantic ontology representation. In: Proceedings of CIRP 3 2012, Athens, Greece, vol. 3, pp. 8–13 (2012)
19. Opoku-Anokye, S., Tang, Y.: The design of a semantic-oriented organizational performance measurement system. In: 14th International Conference on Informatics and Semiotics in Organisation (ICISO), Stockholm, Sweden, pp. 45–49 (2013). http://centaur.reading.ac.uk/31975. Accessed 10 Mar 2014
20. Shen, Y., Ruan, D., Hermans, E.: Modeling qualitative data in data envelopment analysis for composite indicators. Int. J. Syst. Assur. Eng. 2(1), 21–30 (2011)
21. Diamantini, C., Potena, D., Storti, E.: Extended drill-down operator: digging into the structure of performance indicators. Concurr. Comput. Pract. Exp. 28, 3948–3968 (2015)
22. Dumas, M., et al.: Fundamentals of Business Process Management, vol. 1. Springer, Heidelberg (2013)

Business Process Management Systems in Support of Corporate Governance: Applying Orlikowski's Theoretical Lens

Henk Pretorius[1]([⊠]), Alta van der Merwe[1], and Knut Hinkelmann[1,2]

[1] Department of Informatics, University of Pretoria, Pretoria, South Africa
{henk.pretorius,alta}@up.ac.za,
knut.hinkelmann@fhnw.ch
[2] School of Business,
University of Applied Sciences Northwestern Switzerland FHNW,
Olten, Switzerland

Abstract. The value of corporate governance has received attention in the last decade after a number of incidents where fraudulent activities resulted in close down of organizations. Many countries responded with stricter regulations. In this article we argue that the correct use of Business Process Management Systems (BPMS) in an organization may support good governance in an organization. Theories such as the theory of technologies from Orlikowski [15] play a central role in IS research to provide a web of meaning about a phenomenon under investigation. We discuss the potential of BPMS through the lense of Orlikowski's theory of technologies. We also provide some insights on the use of BPMS to support good governance from a case study conducted. The findings suggest that automated business processes can result in improved corporate governance, as well as business value.

Keywords: Business Process Management Systems · Corporate governance · Theory of technologies

1 Introduction

Corporate governance is described as the system by which companies are directed and controlled [1, 2] to the failures of companies such as Enron, WorldCom, Tyco, Adelphia and Global Crossing [3]. There is evidence that US industries lose about USD 400 billion a year from unethical and criminal behaviour [4]. The resignation and arrest of top US managers suggests that there is an increasing level of managerial negligence and corporate irresponsibility that erode domestic and global trust in those firms [5].

Europe's biggest corporate failure was Parmalat [6, 7] where the company collapsed in 2003 with an EU 14 billion shortfall in its accounts. Calisto Tanzani, Chief Executive Officer (CEO) of Parmalat, was detained hours after the firm was declared insolvent, charged with financial fraud and money laundering and sentenced to 10 years in prison.

In South Africa (the biggest economy in Africa) the corporate governance situation is very similar to that in Europe and the US. Amongst many cases of lapses in corporate

© Springer International Publishing AG 2017
M. Themistocleous and V. Morabito (Eds.): EMCIS 2017, LNBIP 299, pp. 592–605, 2017.
DOI: 10.1007/978-3-319-65930-5_46

governance, information technology (IT) vendors are often accused of offering bribes to government employees [8], such as the State Information Technology Agency (SITA). In 10 years of existence of SITA (2002–2012), it spent approximately ZAR 10 billion on ICT [9]. There have been reports of various forms of corruption and fraud within SITA, at the cost of service delivery to South African citizens. The dimensions of corruption and fraud at SITA include bribery, embezzlement, extortion, nepotism, favoritism, collusion, split purchases, abuse of power, conflict of interest and over- or under- invoicing [9].

In response to the many corporate failures around the globe, legislative changes (e.g. the Sarbanes-Oxley Act of 2002) and regulatory changes (e.g. governance guidelines for the NYSE and NASDAQ) were introduced in various countries [3, 10, 11]. In the US, the purpose of the Sarbanes-Oxley Act of 2002 (SOX) was to build and restore confidence in US and international capital markets [3]. However, many sceptics argue that compliance with legislative and regulatory acts are time consuming, costly and cause overregulation [10, 12]. Furthermore, these efforts do not always provide business value to organisations. There is also no guarantee that adherence to these measures can be enforced, indeed, in the first three years of SOX, this was at best regarded as an overreaction to Enron and at worst ineffective and unnecessary [10, 14].

Business process management (BPM) has the potential to support governance processes within organizations [30, 31]. A business process is defined as a collection of activities that takes one or more kinds of input and creates an output that is of value to a customer [13].

Theories such as the theory of technologies from Orlikowski [15] play a central role in IS research to provide a web of meaning about a phenomenon under investigation [16]. Theory guides the sense making processes of complicated real-world phenomena [17]. The main contribution of this study is the extension of Orlkowski's [15] applied theoretical framework, namely "technologies-in-practice", by adding a situated forces component, giving a new perspective on how a Business Process Management System (BPMS), which is used to automate process logic and business logic, can be used in support of corporate governance.

The applied framework was created and tested by conducting a BPMS vendor case study and seven BPMS client case studies. In understanding the components and requirements for the applied theoretical perspective by following a BPMS approach, the following was found from the perspective of agency, influences and structuration.

The paper consists of an overview on corporate governance and business process models in Sect. 2. Section 3 follows with an overview on Orlikowski's theoretical lens and Sect. 4 consists of the research approach followed. Section 5 provides a discussion on the findings.

2 Background

2.1 Corporate Governance

The corporate governance problems identified in the introduction (see Sect. 1) are in accordance with the findings of King [12], who has done extensive work in the field of corporate governance in South Africa. King's work includes three groundbreaking reports: The King Reports on Corporate Governance, which are called King I [18], King II [1] and King III [10]. These guidelines strive to improve the quality of governance in South African and international firms operating in South Africa [3, 12]. The King Reports do consider how good corporate governance could be promoted by IT, but do not present detailed guidelines for corporations and practitioners to achieve this.

In the three King Reports, the King Committee identifies nine principles of good corporate governance [1, 3, 4, 12, 18]. These may be summarised as adhering to the good corporate governance principles of fairness, accountability, responsibility, transparency, discipline, independence, social responsibility, leadership and sustainability. Most importantly, the foundation of these concepts is intellectual honesty, acting in good faith and acting in the best interests of the company.

2.2   Business Process Management Sytems for Process Automation

A Business Process Management System (BPMS) is defined as a generic software system that is designed to manage and automate operational business processes [19]. A BPMS gives an organisation the ability to rapidly make changes to business processes in the real-time business environment. This reduces the risk of firms losing a competitive advantage, further minimises business process complexity for the user and contributes to the strategic alignment of business processes with business objectives [20]. A BPMS makes business processes visible to process owners, users and auditors who are directly affected by the regularity and conformance pressures of acts e.g. the Sarbanes-Oxley Act [21] in America.

A BPMS consists of a number of architectural components [21–23], including BPM Repository and Database, Process Modelling, Business Rules, BPM Engine, Software Integration Engine, and Monitoring.

3   Orlikowski's Theory of Technologies

In her early work, Orlikowski [24] depicts the relationships between technology, humans (agents) and the organisation. People (agents) design information systems and information systems change the way in which people work. The way in which agents work further changes the characteristics (including social behaviour and social action) of the organisation, referred to as structuration in a continuous process.

Orlikowski [15] advances the structuration perspective by explaining that social structures are not and cannot be embedded in material artefacts, such as technology. Structuration, the enforcement or change in social behaviour and action, can only be achieved through recursive, ongoing technology use when users of a technological artefact interact with certain properties of the technology.

Typically, the properties of the technological artefact are designed and developed by technology designers and developers for a specific organisational purpose, while properties are also added by users. The inscription process of technology [15] explains that when people use technology, they draw on the inscribed properties of the technological artefact – those that were inscribed by the designers and those added by the users. In the process of use, the users also draw on their own interpretive schemes (skills, power, knowledge, emotional abilities, intellectual abilities, other), norms and other facilities (hardware and software) in a specific organisational setting as indicated in in Fig. 1.

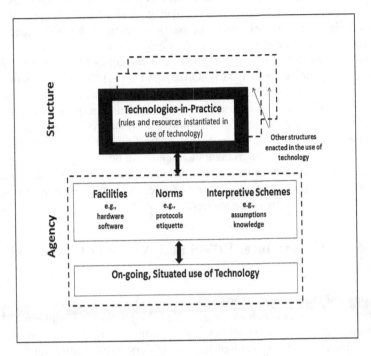

Fig. 1.   Source: Orlikowski [15]

In an organisational setting, a community of users with similar work practices is required to use the technological artefact with its inscribed properties in a similar way; repeated use leads to institutionalisation in the organisation. Orlikowski [15] refers to this institutionalised process of similar technology use as "technology-in-practice", as indicated in Fig. 1. At this stage she argues, institutionalised and similar technology use for a community of users become firm prescriptions for social action that may impede change or reinforcement. Recurrent use of technology may simultaneously enact multiple structures, as indicated in Fig. 1. Over time and as contexts change, different structures will emerge. However, in change lies the possibility and potential for innovation and learning [15]. In the case of this research, to improve corporate governance.

4   Research Methodology

In this research we adopted an interpretive research paradigm approach using qualitative methods. The research context involves case studies of a BPMS vendor company in South Africa and seven BPMS user companies of varying sizes and from different industry sectors (banking, technology, manufacturing, energy and petro-chemical).

The case studies were documented through interviews and surveys between June 2010 and January 2011. At the BPMS vendor company, data was collected from 12 managers (24%), 14 business analysts (29%), 12 developers (24%), eight trainers (16%) and three other positions (7%). At the BPMS user companies, data was collected from six IT managers (24%), two general managers (8%) and 17 business analysts (68%). In all eight case studies, the participants represent different language groups, social backgrounds and genders.

Data was systematically coded into themes and categories (thematic analysis) using the constant comparative method [25]. Orlikowski's [15] "technologies-in-practice" theory was used to synthesise the themes and categories that emerged informing how a BPMS can be used in support of corporate governance. During the research process, Orlikowski's [15] current "technologies-in-practice" framework was extended.

Triangulation was used to increase the credibility and validity of the research results [26, 27]. In this study, the triangulation approach brings together data from eight perspectives, namely the BPMS vendor company and the seven BPMS user companies, to gain a richer and more plausible account of a research phenomenon than would be possible with only one or two case studies.

5   Corporate Governance, BPMS and Applying Orlikowski's Lens

Organisations and technology go through dramatic changes in form and function [15]. On the other hand, researchers have long studied the relationship between technology and organisational processes, structures and outcomes [15]. Notions of innovation, learning and improvement were often used to understand the role, implications and influence of new technologies on the organisation.

Orlikowski [15] designed a theoretical framework for studying technology in organisations that builds on earlier articulated research of structuration introduced by Anthony Giddens [29]. This framework of Orlikowski [15] advances the view that structures are not located in organisations or technology, but are enacted by users of the technology in the organisation.

Orlikowski and Iacono [28] state that any analysis and usage of an IT artefact must acknowledge that the IT artefact is shaped by the interest, values and assumptions of designers and users is embedded in a historical context and emerges from on-going social, political and economic practice. Therefore consideration must be given to the cultural aspects of the implementation journey, composed of a multiplicity of fragile and fragmentary components.

All of these aspects form part of various points of discussion when applying the theoretical lens of Orlikowski [15] to the use of a BPMS in support of corporate governance. The theory of Orlikowski [15] is applied to the research phenomenon, which is using a BPMS in support of corporate governance, as illustrated in Fig. 2. The enactment of corporate governance is discussed from the Agency, Forces (added from the discussion of Orlikowski and Iacono [28]) and Structure Perspective in the following sections.

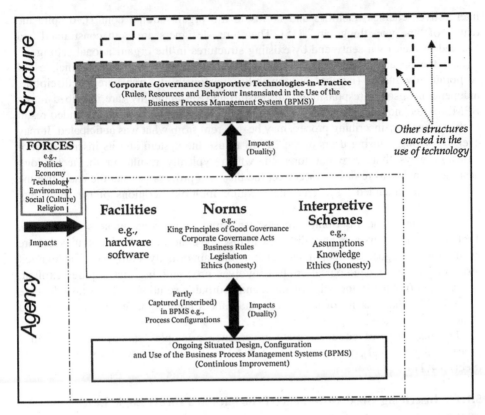

Fig. 2. The enactment of corporate governance supportive technologies-in-practice (adapted from Orlikowski [15])

5.1   Agency: Using a BPMS Approach

In revision of Orlikowski's [15] framework, a technology is constructed and inscribed, with the developers' assumptions and knowledge about the world or context at that specific time. According to Orlikowski [24], a technology only comes into existence through creative human action and is sustained by human action through the on-going maintenance and adaption of technology.

As specified earlier, Orlikowski and Iacono [28] state that any analysis and usage of an IT artefact must acknowledge that the IT artefact is composed of a multiplicity of fragile and fragmentary components. A typical BPMS consists of several architectural components, as indicated in Sect. 2.2. The components are: the BPM Engine, the Process Modeller, the Business Rule Engine, the Software Integration Engine, Monitoring (Reporting Engine) and the BPM Repository and Database.

According to Orlikowski [15], the inscription process is typically conducted by human agents namely the technology designers of a technology, with their own assumptions and knowledge about the world at that time. The designers proactively play a role in bringing forth their own realities of "how things are", through their own interpretive schemes, facilities and norms in the organisational context, despite the reality of "the way things are" [15]. Therefore, designers are both constrained and enabled in their own sense and by existing structures in the organisational setting.

On the negative side, when a new set of norms, such as King's principles of corporate governance (fairness, accountability, responsibility, transparency, discipline, independence, social responsibility, leadership and sustainability) are inscribed into a BPMS by technology designers during the technology design time, the intended result of the technology inscribing process may be different from what was anticipated. It may even be wrong, causing users of the BPMS to use the system and its inscribed properties in ways that were not foreseen, which typically results in the technology designers having to re-design or correct the business process, so that the users use the system in its intended way. This may require multiple iterations of the inscribing process.

To continue, the inscribing process itself may have consequences of its own. Technology designers may have difficulty to inscribe concepts or business rules that are unstructured, vague, ad-hoc, unpredictable and abnormally complex. The explicit inscribing of any technological artefact involves programming that requires explicit, externalised (not tacit), logical, structured and codifiable solutions. This is also the case for a BPMS that has its own program and modelling languages, conventions and standards.

The inscribing of good corporate governance principles can be viewed from a process logic level perspective (Sect. 5.1.1) or an activity level perspective (Sect. 5.1.2).

5.1.1   Inscribing Good Governance Principles at Process Logic Level

The Business Process Modeller component of a BPMS (the part in which a designer designs business processes) is the most prominent component that is used to further good governance at process logic level.

The Business Process Modeller is used to assign specific user roles and user groups to process activities which enforce accountability and responsibility in the organisation [22]. The process modelling tool may further allow for the simulation of real-time execution of business processes. Therefore, owners and designers of processes have the ability to review the impact of business process changes, identify potential bottlenecks and review the time and cost impacts of process changes [22], which supports the good governance principle of sustainability (see Sect. 6).

Process designers e.g. with the necessary access rights may make changes to processes by checking processes in and out of the system. Nobody else is able to access a specific process until it is checked in again by the designer (cf. [22]). The process of checking in and checking out processes is enforced by the BPMS. Various corporate governance principles e.g. accountability, responsibility, discipline and transparency are enforced in this way by the BPMS.

The Process Modeller can also assist the organisation to determine the fairness (a good governance principle) of work distribution between certain resources. This can for example halt possible discrimination between certain gender groups, social groups, work groups and groups of specific ethnical origins.

Finally, the Process Modeller is also used in the design of processes that furthers the principle of independence, which includes processes with independent check-points (also called control points) and audit-points. This is called segregation of duties. Therefore, the process modeller can be used to inscribe and further principles of good corporate governance, such as the principles from the King reports (Sect. 6).

5.1.2   Inscribing Good Governance Principles at Activity Level

Principles of good governance can also be inscribed at activity level, typically where the business rules of a company are implemented.

At activity level typically during an audit, or when some form of mismanagement occurred during corruption, the Monitoring component can be used to trace at activity level what occurred and who was responsible for what exactly happened. The Monitoring component allows this high level of transparency in the organisation allowing everyone to see who was responsible and accountable for a given activity in the specific process instance. The Monitoring component also assists to identify those employees with "vested interests" in the organisation by studying audit trails. Transparency, independence, accountability and responsibility are all principles of good governance from the King reports (Sect. 6).

The BPMS also allows for the ability to set and handle time constraints (business rules set around time) on activity level. Those accountable and responsible for certain tasks within a business process within an organisation, only have a certain time to complete an activity. Therefore, when an activity is not completed in time, a time constraint (business rule) fires and escalation and notification rules execute which informs someone of a higher authority, typically a manager, to take the required action. (cf. [22]).

All the governance principles as suggested (Sect. 6) are supported by various kinds of reports at activity level. One can report on various performance measures and indicators and if duties were performed on time. Managers can use the report data to make better decisions for the survival and improvement of their organisation.

On activity level, the Business Rules Engine is an extension of the Process Modelling capability of the BPMS to configure business rules within a process activity. The rule engine helps to handle and manage business exceptions without human intervention (cf. [22]). This removes human biasness from business rules and decisions, but also supports good governance within processes to curb possible instances of corruption and mismanagement. In some BPMSs, the rule engine consists of a fully programmable programming language and these days' artificial intelligence components are also included in the rule engine component that allows for machine learning in various business situations.

The literature and arguments presented indicate that all King's principles of good corporate governance can be inscribed into a BPMS and its architectural components on process logic level and activity logic level. However, there are positive and negative implications of when inscribing the King principles of good governance into a BPMS. This is discussed later. Next, the contextual forces that influence corporate governance are discussed.

5.2 Situated Forces Influencing Agency

When people use a BPMS, they draw on the properties of the BPMS, those that were inscribed by the designers and those that were later added by the users [15]. Users also draw on their own abilities, assumptions, experiences, ethical frameworks, skills, knowledge and expectations associated with the technology, referred to as the interpretive schemes of the users [15]. However, facilities, norms and interpretive schemes are influenced by forces inside and outside the organisation, for example politics, culture and religion that influence agency.

As indicated by Orlikowski and Iacono [28], any analysis and usage of an IT artefact must acknowledge:

- That the IT artefact is embedded in a historical context and therefore consideration must be given to the cultural aspects of the implementation journey.
- That the IT artefact emerges from on-going social, political and economic practice.

This supports our research, which suggests that there are on-going forces that influence the usage and design of an IT artefact. In other words, if a BPMS that is inscribed with principles of good governance is used in support of corporate governance, it implicates that there are on-going forces that influence corporate governance in the organisation and other social structures [15, 28].

A number of checklists have been developed as way of analysing the contextual forces that might affect an organisation. One such checklist, is a PEST checklist that categorizes contextual and environmental influences as political, economic, social and technological. Sometimes two additional factors, environmental and legal, are added to form a PESTEL checklist, but these themes can easily be subsumed in the others [11].

Different industry sectors have different laws, legislations, acts and regulations that organisations have to adhere to and which influence the way in how actors in these organisations may implement or apply corporate governance. The results of a PEST analysis may be used in a specific business and strategic context to improve corporate governance or to identify corporate governance threats [11]. These situated forces also contribute to the way in which an organisation is governed and shaped, which partly defines the organisational culture [4].

According to Kreitner and Kinicki [4], an organisation's culture is passed on to new employees in the organisation through a process of socialisation (values, norms and required behaviours are learned) and mentoring. The organisational culture influences organisational structures, practices, group and social processes, employee attitudes and employee behaviour that in turn influence organisational outcomes.

Furthermore, social, legal, political, technological, economic and a combination of cultural factors within a country may impact the way in which human actors of an organisation govern the organisation and do business with the organisation [4].

With regards to corporate governance, there should be no absolute definition or understanding of corporate governance because it is situated. Over different contexts there are similarities, but also differences in corporate governance, which is dependent on similarities and differences in the forces that play out inside and outside the organisation which influence facilities, norms and interpretive schemes in the organisation.

5.3   Situated Corporate Governance Supporting Structuration

When people use a technology, they draw on the properties of the artefact, those that were inscribed by the designers and those that were later added by the users [15]. Earlier, it was shown at process and activity levels how principles of good governance could be inscribed into a BPMS and its architectural components, to achieve better corporate governance.

Consequently, the users of a BPMS draw on the inscribed principles of corporate governance when they use the BPMS. However, users also draw on their own abilities, power, assumptions, previous experiences, training, skills, knowledge and expectations associated with the technology [15]. Lastly, users draw on their knowledge and experiences within specific institutional contexts in which they live and work, and the social and cultural conventions associated with these contexts [15]. In this way, the people's use of technology (in this case a BPMS inscribed with Kings' principles of corporate governance) becomes structured by their experiences, knowledge, norms, habits, meanings and technological artefacts [15].

According to Orlikowski [15], the on-going situated use of a technology-in-practice reinforces that technology in the organisation. It becomes regularised and routinised through habitual and repeated use of the technology [15]. Re-enactment of the same technology-in-practice occurs through habitual use of the technology, thus further reinforcing it in the organisation that it becomes taken for granted in the organisation [15]. The technology-in-practice becomes or serves as a behavioural and interpretive template for people who use the technology [15].

From a corporate governance point of view (after inscribing principles of good governance in to the BPMS), one can argue that specific BPMS use serves as behavioural template to improve behaviour that supports better corporate governance. The author termed this behavioural template of technology-use in support of corporate governance: *corporate governance supportive technologies-in-practice.*

Orlikowski and Iacono [28] state that any analysis and usage of an IT artefact must acknowledge that an IT artefact emerges from on-going social, political and economic practice. The framework of Orlikowski [15] also caters for continuous improvisation and change as designers reconfigure the BPMS, or as users alter their habits of use [15]. Users deliberately or inadvertently use a technology in ways that was not anticipated by the developers of the technology [15]. There are no exceptions in the case of a BPMS. Users, typically ignore, alter or work around the inscribed properties of a technology. They then might modify the technological artefact so that it suites their particular interests or requirements [15]. In other words, the change in behaviour may be different from that which was originally anticipated, because users may deliberately or inadvertently use the BPMS in ways that was not anticipated by the developers and designers of the BPMS.

The physical properties of artefacts ensure that there are always boundary conditions in how the technology is used. Many employees use organisation politics to further their own interests, but the more a particular technological artefact is integrated into a larger social system or network or technological configuration, the narrower the range of alternative uses that may be crafted with it by users or designers of the technology [15].

According to Orlikowski [15], users with similar work practices enact similar technologies-in-practice. Similar technology work practices enact similar behaviour templates, also in the case of a BPMS. Similar technology work practices occur through common training sessions, shared socialisation, comparable job experiences, and mutual coordination and storytelling [15]. Over time and through repeated enforcement the technologies-in-practice may become institutionalised in the organisation [15]. At the point of institutionalisation the technologies-in-practice become predetermined and firm inscriptions for social action and change [15].

Continuous improvisation and change is in essence part of BPM and the BPM life-cycle, therefore, when the BPMS is altered to improve corporate governance with BPMS use, or when users alter their BPMS use habits, they enact different, typically improved corporate governance supportive technologies-in-practice, which will become predetermined and frim inscriptions for social action and change.

6 Discussion

This study proposes a theoretical perspective to improve corporate governance by drawing on the premises of Orlikowski's [15] theoretical framework, namely "technologies-in-practice". The applied framework was created and tested by conducting a BPMS vendor case study and seven BPMS client case studies.

In understanding the components and requirements for the applied theoretical perspective by following a BPMS approach, the following was found from the perspective of agency, influences and structuration:

- **Agency:** This research project showed how situated corporate governance properties can be inscribed in a BPMS at process logic level and activity level. According to Orlikowski [15], the inscription process is typically conducted by human agents namely designers of a technology with their own assumptions and knowledge about the world at that time. Users of a BPMS draw on the inscribed and situated corporate governance properties. However, the intended result of the technology inscribing process may play out to be different from what was anticipated. This may require multiple iterations of the inscribing process, until the intended result is achieved. The inscribing process itself may have consequences. Technology designers may have difficulty to inscribe concepts or business rules that are unstructured, vague, ad-hoc, unpredictable and abnormally complex.
- **Situated forces that influences agency:** Users of a BPMS draw on the properties of the BPMS, also those properties that were inscribed to improve corporate governance. Users also draw on their own abilities, assumptions, experiences, ethical frameworks, skills, knowledge and expectations associated with the technology, referred to as the interpretive schemes of the users [15]. However, facilities, norms and interpretive schemes are influenced by forces inside and outside the organisation, for example politics, culture and religion that influence agency. Several frameworks can be used to analyse the influence and interplay of situated forces. One such framework is a PESTEL analysis. With regards to corporate governance this means there should be no absolute definition or understanding of corporate

governance because it is situated. Corporate governance is dependent on similarities and differences in the forces that play out inside and outside the organisation which influence facilities, norms and interpretive schemes in the organisation.

- **A structuration component:** Users of a BPMS draw on the inscribed principles of corporate governance when they use the BPMS. Users further draw on their own abilities, power, assumptions, previous experiences, training, skills, knowledge and expectations that is associated with the BPMS [15]. In this way, the use of technology becomes structured by these experiences, knowledge, norms, habits, meanings and technological artefacts [15]. The on-going situated use of a technology-in-practice reinforces that technology in the organisation [15]. It becomes regularised and routinised through habitual and repeated use of the technology [15]. The technology-in-practice becomes or serves as a behavioural and interpretive template for people who use the technology [15], in the case of this research a BPMS serves as a behavioural and interpretive template for improved corporate governance. Therefore, organisational behaviour is changing in support of corporate governance. BPMS-use now serves as a behavioural template in support of corporate governance. Over time and through repeated enforcement the technologies-in-practice may become institutionalised in the organisation [15].
- **An influence component:** The use of a BPMS in an organisation and the changes in organisational behaviour that it causes may influence other overlapping social systems. It may cause a ripple-effect of corporate governance supportive behaviours inside and outside the organisation.

7 Conclusion

This study proposes a theoretical perspective to improve corporate governance by using a BPMS, drawing on the premises of Orlikowski's [15] theoretical framework. The applied framework was created and tested by conducting a BPMS vendor case study and seven BPMS client case studies. The various components (agency, situated forces, structuration and influence) and requirements for the applied theoretical perspective are explained. The findings suggest that automated business processes can result in improved corporate governance as well as business value. BPMSs applications can result in better risk management and lower organisational risk. On the contrary, where business processes are performed manually, there tends to be less compliance, decreased observation and visibility, reduced monitoring and control, poorer risk management, less corporate governance supportive behaviour, resulting in poor corporate governance practices.

Future research may investigate how other theoretical lenses and perspectives can complement and contribute towards Orlikowski's [15] theoretical perspective when using a BPMS to improve corporate governance.

References

1. IODSA: The King report on corporate governance for South Africa (King II). Institute of Directors of Southern Africa (IODSA) and King Committee on Corporate Governance, Johannesburg (2002)
2. Vu, N.H., Nguyen, T.: Impacts of corporate governance on firm performance, Master's program in International Strategic Management (2017). http://lup.lub.lu.se/luur/download?func=downloadFile&recordOId=8917364&fileOId=8917365. Accessed June 2017
3. Hough, J., Thompson, A., Strickland, A., Gamble, J.: Crafting and Executing Strategy (South African Edition). McGraw-Hill, Berkshire (2009)
4. Kreitner, R., Kinicki, A.: Organisational Behaviour, 6th edn. McGraw-Hill, New York (2004)
5. Michell, L.: Corporate Irresponsibility: America's Newest Export. Yale University Press, New Haven (2002)
6. BBC News: Parmalat in bankruptcy protection (2003). http://news.bbc.co.uk/2/hi/business/3345735.stm. Accessed 14 May 2010
7. Gumber, P.: How it all went so sour. Time Magazine, 23 November 2004
8. Jarvis, K.: IT vendors take rap for corruption. GovernmentIT 1(3), 13 (2009)
9. Mtimunye, M.: SITA lays fraud and corruption ghosts to rest. GovernmentIT 1(3), 15–16 (2009)
10. IODSA: King code of corporate governance in South Africa 2009 and the King report of corporate governance in South Africa 2009 (King III). Institute of Directors in Southern Africa (IODSA) and King Committee on Corporate Governance, Johannesburg (2009)
11. Pearlson, K., Saunders, C.: Strategic Management Information Systems, 4th edn. Wiley, New York (2009)
12. King, M.: The Corporate Citizen: Governance for All Entities. Penguin Books, Johannesburg (2006)
13. Hammer, M., Champy, J.: Reengineering the Corporation: A Manifesto for Business Revolution. Harper Business, New York (1993)
14. Richardson, C.: Process governance best practices: building a BPM center of excellence. Project Performance Corporation (2006). http://www.bptrends.com/publicationfiles/09-06-ART-ProcessGovernanceBestPractices-Richardson1.pdf. Accessed 15 Oct 2012
15. Orlikowski, W.J.: Using technology and constituting structures: a practice lens for studying technology in organisations. Organ. Sci. 11(4), 404–428 (2000)
16. Neuman, W.L.: Social Research Methods: Qualitative and Quantitative Approaches. Allyn and Bacon, Boston (1991)
17. Truex, D., Holmström, J., Keil, M.: Theorizing in information systems research: a reflexive analysis of the adaptation of theory in information systems research. J. Assoc. Inf. Syst. 7 (12), 797–821 (2006)
18. IODSA: The King report on corporate governance for South Africa (King I). Institute of Directors of Southern Africa (IODSA) and King Committee on Corporate Governance, Johannesburg (1994)
19. Haffer, R.: Development of a model of business performance measurement system for organizational self-assessment. The case of Poland. J. Posit. Manag. 7(3), 20–46 (2016)
20. McGoveran, D.: Enterprise integrity: BPMS concepts, part 8. Aternative Technol. 3(8), 1–2 (2001)
21. Palmer, N.: BPM 2003 market milestone report. A Dephi Group White Paper (2003). http://www.dephigroup.com. Accessed Oct 2004
22. Miers, D., Harmon, P.: The 2005 BPM suites report. Business process trends: Version 1, March 2005

23. Miers, D., Harmon, P., Hall, C.: The 2007 BPM suites report (2007). http://www.bptrends. com/reports_landing.cfm. Accessed 20 May 2010
24. Orlikowski, W.J.: The duality of technology: rethinking the concept of technology in organisations. Organ. Sci. 3(3), 298–427 (1992)
25. Strauss, A., Corbin, J.: Basics of Qualitative Research: Grounded Theory Procedures and Techniques, 2nd edn. Sage Publications, California (1998)
26. Kennedy, P.: How to combine multiple research methods: practical triangulation (2009). http://johnnyholland.org/2009/08/practical-triangulation/. Accessed Apr 2012
27. Olsen, W.: Triangulation in social research: qualitative and quantitative methods can really be mixed. In: Holbron, M. (ed.) Developments in Sociology. Causeway Press, Ormskirk (2004)
28. Orlikowski, W.J., Iacono, C.S.: Research commentary: desperately seeking the "IT" in IT research—a call to theorizing the IT artifact. Inf. Syst. Res. 12(2), 121–134 (2001)
29. Giddens, A.: The Constitution of Society. University of California Press, Berkeley (1984)
30. Pretorius, H.W.: Towards a theoretical framework to support corporate governance through the use of a business process management system: a south african perspective. Ph.D. thesis, University of Pretoria (2014)
31. Pretorius, H.W., Leonard, A.C., Strydom, I.: Towards an electronic monitoring, observation and compliance framework for corporate governance using business process management systems. Afr. J. Inf. Commun. (AJIC) 13, 62–75 (2013)

Latent Factor Model Applied to Recommender System: Realization, Steps and Algorithm

Maryam Jallouli[1(✉)], Sonia Lajmi[1,2], and Ikram Amous[1]

[1] Miracl Laboratory, Sfax University, Sfax, Tunisia
jallouli.maryam@gmail.com, slajmi@bu.edu.sa,
ikram.amous@isecs.rnu.tn
[2] Al Baha University, Al Bahah, Saudi Arabia

Abstract. Nowadays, internet has offer an overabundance of available information. In social networks, users confront gigantic number of items. To overcome this phenomenon, known as information overload, recommender systems are intended to filter information and help users to make their choice. Many models based collaborative filtering have been used in the literature to solve the problem of recommendation. Among these models, latent factor model has become the most popular due to his performed results of accuracy. This work is part of research into Recommender System domain and aims to present a detailed explication on works based latent factor model. We first describe a general view of this model. Its realization in field of recommendation is next presented. A detailed study on different steps is then exposed. The most important works that have been developed are then presented. To the author's knowledge, there has been no work that tries to explain in detail how latent factor model is applied to Recommender Systems.

Keywords: Recommender system · Collaborative filtering · Latent factor model

1 Introduction

In our days, recommender systems (RS) have been emerged as a way to offer for a specific user the most appropriate item to meet the following challenges: In one hand, assist users to find the most relevant items. In the other hand, facilitate sales for e-commerce sites.

Latent factor model is among the most successful model based collaborative filtering in the field of recommendation. According to this model, input is presented in a form of a matrix, as demonstrated in Fig. 1.

This $m * n$ matrix, named user-item rating matrix stores interactions of users with items. It consists of m users and n items, where each entry corresponds to the evaluation (rating) of user u for item i. The challenge of the recommender System (RS) is to predict the missing evaluations, in order to propose recommendations that satisfy each user.

In a previous work [1], the general structure of a recommender system is presented by three steps: (1) an input, (2) a recommender system algorithm and (3) an output.

© Springer International Publishing AG 2017
M. Themistocleous and V. Morabito (Eds.): EMCIS 2017, LNBIP 299, pp. 606–618, 2017.
DOI: 10.1007/978-3-319-65930-5_47

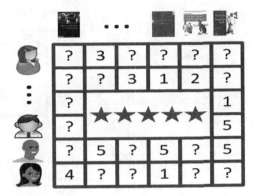

Fig. 1. Structure of evaluation matrix (Source: Weike, 2015)

According to [1], the first step is well detailed whereas steps two and three are not explained. This work can be considered as a continuity of [1] to present the second step about the recommendation algorithm.

This paper is intended to expose several approaches in the state of the art that are relevant to latent factor model in the field of Recommender System. Thus, we present a basic idea behind the latent factor model. Then, its realization is detailed by focusing on the "matrix factorization" and "tensor factorization" models and describing the baseline. By the following, we recapitulate the general steps and algorithm. Existing works that have extending this model and adding other information are then described according to these steps. Finally, we summarize the paper and provide some future works.

2 Basic Idea of Latent Factor Model

Latent factor model is at the basis a mathematical model. Its usage in the field of recommendation consist of explaining the rating by characterising users and items in a latent factor space, and then, try to predict the unknown rating from that space. A simplified example can be represented in Fig. 2. Assume that two factors are done characterised as female versus male and serious versus escapist. The figure shows the localisations of movies and users based upon these two factors. For this model, the predicted rating of a user for a movie is equals to the dot product of the movie's and user's locations on that space.

For example, what we could guess about the user "Gus"? He like "Dumb and Dumber" and "Independence Day" and hate the film "The Color Purple".

In general, an item's location on the graph is a vector that explains his ownership to the selected factors. For a user, his location on the graph is a vector that explains his preference to the same factors. These vectors are called latent factor vector. Then, users and items are mapped into a latent factor space, which define their locations on the graph according to their latent factor vector.

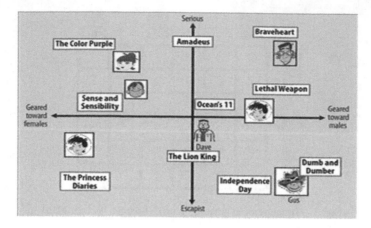

Fig. 2. Simplified illustration of the latent factor model (Source: Yehuda et al. [2])

The question here is, given a user-item matrix, how to obtain latent factor vectors of both users and items in order to predict preference of users for items. In other terms, what is the process allowing us to pass from a rating matrix to a representation in a latent factor space.

Among the most successful achievements of the latent factor model is based on matrix and tensor factorization. These two types of factorization are presented in the next section.

3   Realization of Latent Factor Model

3.1   Matrix Factorization

Matrix factorization consists of factorizing a matrix into two matrices such that the multiplication of these matrices will recover the original matrix [2]. The use of this technique in our context allows us to obtain from an uncomplete rating matrix, a matrix of items, that represents the ownership of all items to factors and a matrix of users, that represents the preferences of all users for the factors. In this way, each element (user and item) is associated with a latent factor vector. It should be noted that, in contrast with the example explained earlier, this technique infers characteristics without exactly knowing each feature. Then, the multiplication of matrix of users and matrix of items construct a complete matrix, very close to the original one. An example of matrix factorization is explained in Fig. 3.

According to it, we consider:

- d: the number of latent factors that "explain" the vote. According to a number n, we have d_1, d_2, d_f, where each d_f refers to a specific factor.
- P: represents the matrix of preference of the user to the factors (d_1, d_2, etc.), of size ($m \times d$)
- Q^T: represents the matrix of items belonging to the factors, of size (n × d)

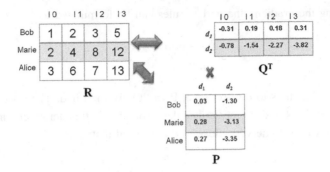

Fig. 3. Simplified example of matrix factorization

- *R*: represents the original matrix where each row represents a user, each column represents an item, and the intersection of a row and a column represents the user's rating value given to an item, of size $(m \times n)$.

$$\hat{r}_{00} = -0,03 . - 0,31 + - 1,30 . - 0,78$$
$$\hat{r}_{00} = 1,01 \tag{1}$$

Once the system computes the latent factor vectors p_u and q_j of each user and item, it is relatively simple task for it to predict the rating a user will give to a given item by using Eq. (2). So, the major challenge here is how to obtain matrices P and Q?

In fact, matrices P and Q are obtained once the error between the original matrix and the predicted matrix (obtained by the product of P and Q is very minimal. We are talking about a function that tries to minimize this error, called objective function. This error is calculated by the sum of the errors between each value $\hat{r}_{uj}$ and ruj as follow:

$$\mathcal{L} = \sum_{u} \sum_{j \in I_u} (p_u \cdot q_j^T - r_{uj})^2 \tag{2}$$

To obtain P and Q, we first start by initializing it with small random values, Then, we compute the product of these two matrices, to obtain a new predicted matrix, called $\hat{R}$. The difference (error) between matrices R and $\hat{R}$ is calculated as defined in [1]. For that, we loop over all the ratings r_{uj} of matrix R, and for each case, the error between r_{uj} and $\hat{r}_{uj}$ is calculated. If this difference is still significant, we must update matrices P and Q to minimise the error between these two matrices. So, we have to know in each direction we have to modify the values, that's mean we need to know the gradient at the current values.

$$\nabla_{p_u} \mathcal{L} = 2 \left(p_u^T q_j - r_{uj} \right) (q_j)$$
$$\nabla_{q_j} \mathcal{L} = 2 \left(p_u^T q_j - r_{uj} \right) (p_u) \tag{3}$$

After obtaining the gradient, the update rules can be formulated as:

$$p_u = p_u + \nabla_{p_u} \mathcal{L}$$
$$q_j = q_j + \nabla_{q_j} \mathcal{L} \qquad (4)$$

Applying these rules to our example, P and Q are modified. Then, we recalculate the predicted matrix $\hat{R}$ and we recalculate the error. We do this iteratively until the new predicted matrix $\hat{R}$ become very close to the original matrix R.

3.2 Tensor Factorization

Context [3] is an important source of information. It can be, for example, domain, time, location, etc. In fact, users change preferences depending on the context. The rating of a given user changes with context. For example; the rating of a given user for ice cream in summer is not the same in winter.

The model of the matrix factorization does not provide a simple way for the integration of further information, such as context. To embed this information, recent works make use of tensor factorization [4], a generalization of matrix factorization, that allows flexible and generic contextual information integration by modelling data of a tensor with 3 dimensions user-item-context instead of the traditional two-dimensional matrix.

At this level, we have seen how matrix and tensor factorisation techniques can be used in the field of latent factor model. But, what is the relation between these methods and the problem of recommendation? To answer this question, we need to see how the work of [2, 5] have proceeded to use matrix factorization techniques.

4 Baselines Latent Factor Model in the Field of Recommendation

As explained in [1], any recommendation system contains three parts: an input data, a theoretical framework and an output data. According to the works of [2, 6], which are denoted the baselines of latent factor model in the field of RS, these three steps of these two works are well explained as follow:

4.1 Define the Input

In general, input is presented in the form of $R \in \mathbb{R}^{m \times n}$, to represent am uncomplete rating-matrix, with two dimensions, user and item. Each value in this matrix, denoted by r_{uj}, represents the rate of user u, about item j. m and n denotes respectively the number of users and items. Since this matrix is uncomplete, it includes unknowns rating values of users towards items.

4.2   Define the Output

The output consists of obtaining a matrix $\hat{R} \in \mathbb{R}^{m*n}$. This matrix is complete and contains two sorts of values: values that are very close to the original one, and the other values represents the predicted values of items that are not yet appreciated by users.

4.3   Theoretical Framework of the Approach

After defining the input, we can understand that the intuition behind solving the problem of recommendation using matrix factorization is to factorize an incomplete (contains unknown ratings) user-item matrix into two matrices P and Q^T which represents respectively user- matrix and item matrix that can adequately approximate R. This approximation enables us to construct a complete matrix and consequently resolve the problem of recommendation. Let us now see how this model can be applied for an incomplete matrix.

A question might have become to your mind by now is: what about unknown values, if we replace them by zero, the predicted values will be zero too. So, the solution is to ignore these values and to consider only the known values.

As explained previously, our objective now is to obtain a complete matrix $\hat{R}$ that is very close to the original Matrix R. To achieve this goal, we must find P and Q.

Formally, a system can learn the user and item latent factor vectors by minimizing the error between these two matrices. To resolve this problem, we use the following function, named objective function, that either stem from regularized loss functions or probabilistic models. In these two cases, the objective functions are of the form:

$$\mathcal{L}\left(R, \hat{R}\right) + \Omega(\hat{R}) \tag{5}$$

where $\mathcal{L}\left(R, \hat{R}\right)$ is a loss function. This term measures the error between the rating matrix R which contain unknown values, and the matrix $\hat{R}$ (output), approximation of R. $\Omega(\hat{R})$ is a regularization term allowing the generalization of the model. So, the difference here between function (5) and the objective function used in function (2) to calculate the error is the adding of a regularized term to avoid over fitting. It should be noted that it is defined only on the set of observed values of the matrix.

According to [6], the objective function is defined as follow:

$$\mathcal{L} = \frac{1}{2}\sum_{u}\sum_{j \in I_u}(p_u^T q_j - r_{uj})^2 + \frac{\lambda}{2}(\sum_{u}||p_u||_F^2 + \sum_{j}||q_j||_F^2) \tag{6}$$

where $||.||_F$ denotes the Frobenius norm, and λ is a parameter to control the extent of regularization. So, how to minimize this equation?

From this work [6], various and massive approaches spring up. To explain them, in the next section, we will summary the steps followed by authors that have been used this model.

5 Organizing General Steps

In this section, we summarize essential steps allowed by recommender system approach based on latent factor model. Our proposition is to decompose the process of this model into four steps, as presented in Fig. 4: decomposition model, decision function, objective function and solving technique. In the following, we will detail each of these steps.

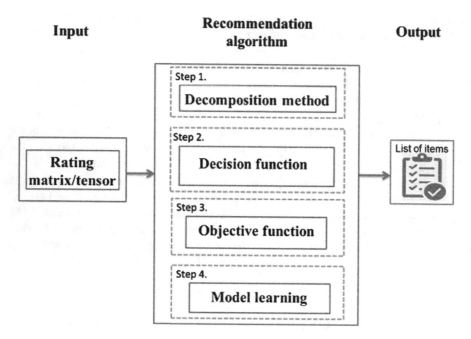

Fig. 4. General structure of a recommender system.

5.1 Step 1: Define Decomposition Model

Different decomposition methods such as matrix factorization [2], tensor factorization [4], are used in the literature. In a previous example described in [2] we have a matrix decomposition expressed by the following equation:

$$R \leftarrow P * Q \tag{7}$$

According to Eq. (7), rating matrix R is decomposed into two matrices P and Q. Each row in matrix P represents preferences of users according to a predefined number of factors. Each row in Q shows belonging of items into the same factors. In case we want to consider context, matrix factorization seems to be not able to include this important source of information. To overcome this limit, tensor factorization is used to include context. In general, tensor factorization is presented in the following way:

$$R \leftarrow P * Q * C \tag{8}$$

where matrix R is decomposed into three matrices P, Q and a context matrix C which represents

5.2 Step 2: Define Decision Function

Decision function consists of defining the rating prediction function of user u for item j. It is represented by a simple multiplication of vectors p_u and q_j^T, and it can be modified in order to include other additional information such as social or contextual information.

5.3 Step 3: Define Objective Function

The function $\mathcal{L}$ measures the error (difference) between the original Matrix R and the predicted matrix $\hat{R}$. In general, the objective function is convex on its variables p and q. So, there is no quick solution to $\mathcal{L}$ (for the minimization of the objective function). Thus, most of the work of the prior state of the art follows an optimization process for the objective function of Eq. (6). This process includes updating alternately at each iteration the user and item latent feature vector p_u. and q_j. until minimising the objective function. For this purpose, numerical methods as the gradient descent are used to solve this problem.

5.4 Step 4: Define Solving Technique

The stochastic gradient descent (SGD) is a widely used method to achieve a minimum of the objective function given by Eq. 6. It consists of updating iteratively each variable until minimizing the error $\mathcal{L}$. Specifically, we follow these steps:

a. Vector Initialization: Initialize vector p_u. and q_j with small random values.
b. Initialize Gradient of $\nabla_{p_u} \mathcal{L}$ and $\nabla_{q_j} \mathcal{L}$: To minimize the error, we have to know in which direction we have to modify the values of p_u and q_j. In other words; we need to know the gradient at the current values.
c. Vector updating: formulate the update rules for both p_{uk} and q_{kj}.
d. Condition criteria: Using the above update rules, we can then iteratively perform the operation until the predicted matrix $\hat{R}$ become very close to the original matrix. That's mean the error converges to its minimum. So, to check this condition, we calculate the overall error using Equation $\mathcal{L}$ as defined in Eq. (3) and determine when we should stop the process. For example, in [7] the condition criteria are set as:

$$\frac{\mathcal{L}_{i+1} + \mathcal{L}_i}{\mathcal{L}_i} \leq 10^{-10}$$

$$\text{Number of iterations } i \leq 10^5 \tag{9}$$

6   Recapitulation of General Algorithm

At this state, we can write the general algorithm of recommendation in the field of latent factor method.

```
Input: Incomplete matrix R ∈ ℝ^{m*n}
Output: Complete matrix R̂ ∈ ℝ^{m*n}
Theoretical framework of the approach:
Initialization:
Step 1: Define decomposition method R̂
Step 2: Define decision function r
Step 3: Define objective function L
Step 4:
        (a) Initialize  p_u and q_i with small random values
        (b) Initialize  ∇_{p_u}L and ∇_{q_j}L
begin

        repeat    (Step 4.c):
                For each user   u ∈ P Do
                    Update   P_u = P_u + α ∇_{p_u}L
                End

                For each item   i ∈ Q Do
                    Update   q_i = q_i + α ∇_{q_i}L
                End

                Computes R̂;
                Computes L ;

        Until (Step 4.e): convergence
    Return   R̂;
end.
```

In the next sections, we first explain how RS based latent factor model have to modify decision function to integrate additional, social and contextual information into SVD model.

7 Factorization in Approaches Including Additional Information

SVD++ model [5] is a top of a state-of-the-art model. In this work, author said that it would be unwise to explain the rating by a simple interaction of the form $q_j^T p_u$. Thus, they formulate the rating prediction function by considering user/item biases and the influence of implicitly rated items other than user/item-specific vectors. Formally, the rating prediction function for user u on item j is as follow:

$$\hat{r}_{uj} = b_u + b_j + \mu + q_j^T \left(p_u + |I_u|^{-\frac{1}{2}} \sum_{i \in I_u} y_i \right) \tag{10}$$

- b_u and b_j represents the user and item biases, respectively. They indicate the observed deviations of user u and item i, respectively, from the average.
- μ denote the global erage rating over all items.
- I_u denote the set of items for which user u expressed an implicit preference
- Each item i is associated with vector y_i

Due to this formulation, we can solve the problem of cold start users. In fact, when we dispose few information about a user, we can use, due to this new rating prediction function, the features of implicitly rated item, associated with a vector y_i, as they are the characteristic of the user, and that's why, instead of using a simple vector p_u, they have considered that $p_u = p_u + |I_u|^{-\frac{1}{2}} \sum_{i \in I_u} y_i$.

8 Factorization in Approaches Including Social Information

In the literature, many approaches differ by one or more steps defined in Sect. 5. In our survey, we explain how approaches have include social information into SVD++ [2] by focusing in step 2. In fact, these approaches have used to modify the decision function to integrate other information such as additional, social or environmental.

8.1 Social Trust Ensemble (STE)

In this work [8], the users' tastes and their trusted friends' tastes were fuses together as follow:

$$\hat{r}_{uj} = \alpha p_u q_j^T + (1 - \alpha) \sum_{v \in N_u} t_{uv} p_u q_i^T \tag{11}$$

where α controls the effect of neighbours on the estimated rating, t_{uv} represents the degree of trust between users u and v and N_u is the set of trusted friends by user u.

8.2   Social Matrix Factorization (SocialMF)

The behavior of a user u is affected by his direct neighbours N_u. In other words, the latent feature vector of user u is dependent on the latent feature vectors of all his direct neighbors $v \in N_u$. On this basis, the latent feature of a user is calculated in [9] as follow:

$$p_u = \sum_{v \in \mathcal{F}_u} t_{uv} p_v \qquad (12)$$

And the predicted rating of user u on item j does not change $\hat{r}_{uj} = p_u q_j^T$

- t_{uv}: measures normalised trust value between users u and v
- $\mathcal{F}_u$: friends of user u

8.3   Socially Enabled Collaborative Filtering (SECOFI)

According to [7], the predicted rating of user u for item j is:

$$\hat{r}_{uj} = p_u q_j^T + \sum_{v \in \mathcal{F}_u} \frac{\alpha_{uv}}{|\mathcal{F}_u|} \cdot p_v q_j^T \qquad (13)$$

where α_{uv} is a weight parameter that weights the level of influence of a friend u in a user v and $\mathcal{F}_u$ is the set of friends of user u.

In this work, and according to Eq. (11), authors [7] has include social influence into matrix factorisation model. They have modelled the predicted value $\hat{r}_{uj}$ as the combination of his own preferences, presented by $p_u q_j^T$, and preferences of his friends, shows by $\sum_{v \in \mathcal{F}_u} \frac{\alpha_{uv}}{|\mathcal{F}_u|} \cdot p_v q_j^T$.

8.4   TrustSVD

TrustSVD [10] takes into consideration the implicit effect of trusted users on rating prediction in the same manner as Koren [5] who employs the influence of rated items. Formally, the rating for user u on item j is predicted by:

$$\hat{r}_{ujk} = b_u + b_j + \mu + q_j^T \left(p_u + |I_u|^{-\frac{1}{2}} \sum_{i \in I_u} y_i + |T_u|^{-\frac{1}{2}} \sum_{v \in T_u} w_v \right) \qquad (14)$$

where I_u is the set of items preferred by user u and y_i denotes latent factor vector of item i, with $i \in I_u$. T_u is the set of users trusted by u and w_v is the user-specific latent feature vector of user v trusted by user u, with $v \in T_u$.

9 Factorization in Approaches Including Contextual Information

In this section, we describe for each approach, step 1 and step 2, that's mean, define decomposition model and decision function. Tensor factorization is the decomposition model applied in order to integrate contextual information in the field of recommendation. The two most widely used tensor decomposition techniques are the Hight Order Singular Value Decomposition technique (HOSVD) [11] and Canonical Decomposition (CANDECOMP) [12] also known as Parallel Factor Analysis (PARAFAC) jointly abbreviated CP.

9.1 Multiverse Recommendation

HOSVD is a generalization of SVD [13] for matrices. This technique decomposes an initial tensor with N dimensions in N matrices and a tensor smaller than the original one.

By analogy, authors in [14] have used this method to factorize a tensor $R \in \mathbb{R}^{n*m*c}$ with 3 dimensions into 3 matrices $P \in \mathbb{R}^{m*d}, Q \in \mathbb{R}^{n*d}$, $C \in \mathbb{R}^{k*d}$ and a central tensor $S \in \mathbb{R}^{m*n*d}$ smaller than the original one.

Following step 2, the decision function for a user u, item j, context k in [14] becomes:

$$\hat{r}_{ujk} = S * P_{u*} * Q_{j*} * C_{k*} \tag{15}$$

9.2 TFMAP

PARAFAC tensor factorisation is a technique that decomposes a tensor into 3 matrices. In [15], authors have used this method to factorise a tensor $R \in \mathbb{R}^{n*m*c}$ with 3 dimensions into 3 matrices $P \in \mathbb{R}^{m*d}, Q \in \mathbb{R}^{n*d}$, $C \in \mathbb{R}^{k*d}$.

The decision function for a user u, on item j, in context k is:

$$\hat{r}_{ujk} = P_{u*} * Q_{j*} * C_{k*} \tag{16}$$

10 Conclusion

In this paper, we have analyzed the most important and recent works in the field of recommender system. Special interest is allowed to works based on Latent factor model. We observed that the commonly adopted gait to propose a recommender system based latent factor model is still unclear. In an attempt to unveil ambiguity, we have proposed to analyze and more precisely to separate the general latent factor model into four steps to facilitate its modelization. In future work, we plan to follow these steps to propose a novel approach of recommendation based on latent factor model.

References

1. Jallouli, M., Lajmi, S., Amous, I.: Similarity and trust metrics used in recommender systems: a survey. In: Madureira, A.M., Abraham, A., Gamboa, D., Novais, P. (eds.) ISDA 2016. AISC, vol. 557, pp. 1041–1050. Springer, Cham (2017). doi:10.1007/978-3-319-53480-0_102
2. Yehuda, K., Robert, B., Volinsky, C.: Matrix factorisation techniques for recommender systems. IEEE Computer Society (2009)
- 3. Anind, K.D., Gregory, D.: Towards a better understanding of context and context-awareness. In: Proceedings of the 1st International Symposium on Handheld and Ubiquitous Computing, Karlsruhe, Germany (1999)
4. Weijia, S.: Tensor Completion. Saarland University (2012)
5. Koren, Y.: Factorization meets the neighborhood: a multifaceted collaborative filtering model. In: Proceedings of the 14th ACM SIGKDD International Conference on Knowledge Discovery and Data Mining, pp. 426–434 (2008)
6. Salakhutdinov, R., Mnih, A.: Probabilistic matrix factorization. In: Advances in Neural Information Processing Systems (NIPS) (2008)
7. Delporte, J., Karatzoglou, A., Matuszczyk, T., Canu, S.: Socially enabled preference learning from implicit feedback data. In: Proceedings, Part II, of the European Conference on Machine Learning and Knowledge Discovery in Databases (ECML/PKDD), pp. 145–160 (2013)
8. Ma, H., King, I., Lyu, M.R.: Learning to recommend with social trust ensemble. In: proceeding of the 32nd International ACM SIGIR Conference on Research and Development in Information Retrieval (SIGIR), pp. 203–210 (2009)
9. Jamal, M., Ester, M.: A matrix factorization technique with trust propagation for recommendation in social. In: Proceedings of the 2010 ACM Conference on Recommender Systems (RecSys), pp. 135–142 (2010)
10. Guo, G., Zhang, J., Yorke-Smith, N.: TrustSVD: collaborative filtering with both the explicit and implicit influence of user trust and of item ratings. In: Proceedings of the 29th AAAI Conference on Artificial Intelligence (AAAI), pp. 123–129 (2015)
11. Lathauwer, L.D., Moor, B.D., Vandewalle, J.: A multilinear singular value decomposition. SIAM J. Matrix Anal. 21(4), 1253–1278 (2000)
12. Harshman, R.A.: Foundations of the PARAFAC Procedure: Models and Conditions for an "explanatory" Multi-modal Factor Analysis, vol. 1, p. 16. University of california, Los Angelos (1970)
13. Datta, B.: Numerical Linear Algebra and Application. Brooks/Cole Publishing Company, Pacific Grove (1995)
14. Karatzoglou, A., Amatriain, X., Baltrunas, L.: Multiverse recommendation: N-dimensional tensor factorization for context-aware collaborative filtering. In: Recommender Systems 2010, Barcelona, Spain (2010)
15. Shi, Y., Karatzoglou, A., Baltrunas, L., Larson, M., Hanjalic, A., Oliver, N.: TFMAP: optimizing MAP for Top-N context-aware recommendation. In: Proceedings of the 35th International ACM SIGIR Conference on Research and Development in Information Retrieval (SIGIR), pp. 155–164 (2012)

View Integration of Business Process Models

João Colaço[1](✉) and Pedro Sousa[1,2]

[1] Instituto Superior Técnico, University of Lisbon, Lisbon, Portugal
{joao.p.colaco,pedro.manuel.sousa}@tecnico.ulisboa.pt,
pedro.sousa@linkconsulting.com
[2] Link Consulting SA, Lisbon, Portugal

Abstract. Business processes usually have a group of organizational stakeholders with contrasting concerns. Considering the multitude of concerns around a given process, different modeling teams tend to obtain distinct diagrams when engaging in business process design. Since managing multiple diagrams of the same business process poses challenges to organizations, we propose the use of an incremental approach to consolidate diverse process diagrams into a single model. This approach consists in using a repository, which contains a taxonomy and a set of dimensions, to support the proposed view integration method. As an outcome of adopting our method, we expect a decrease in the complexity of managing business processes, by providing the organizations with a systematic way for the stakeholders to represent and consolidate their concerns.

Keywords: Business process modeling · View integration · Repository · Enterprise architecture

1 Introduction

Business processes are designed to achieve specific goals and the task of business process modeling is expected to improve the understanding and communication across the different stakeholder groups [1]. However, these goals are difficult to achieve in most cases, where multiple and possibly conflicting diagrams are found for the same process. We argue that these issues arise due to two main reasons [2–6].

On the one hand, business processes often cross multiple organizational units and also tend to cross inter-organizational boundaries. Therefore, processes are often shared among different stakeholders, which have contrasting concerns and focus on distinct perspectives of the same business process, such as performance, auditing, information systems, people or compliance.

On the other hand, a business process model is a representation of the modeler's perspective regarding a given process. Thus, a business process model depends on the team that designed it, and different teams will most likely achieve different specifications for the same process.

© Springer International Publishing AG 2017
M. Themistocleous and V. Morabito (Eds.): EMCIS 2017, LNBIP 299, pp. 619–632, 2017.
DOI: 10.1007/978-3-319-65930-5_48

The purpose of this research work was proposing a standardized way of addressing these issues, through the definition of a method for merging distinct business process diagrams into a single model. The present work will focus on applying view integration to business process modeling.

The problem of view integration was previously addressed in the context of database design [7–11]. Here, the goal of view integration is to produce a global conceptual description of a proposed database. Analogously, the goal of view integration in business process design is to produce a global model of a given business process.

To our knowledge, solely the research work presented by Mendling and Simon [12] provides a method to address the problem of view integration in the context of business process design. Moreover, we argue that there is a gap in knowledge regarding the application of view integration methods to models specified in Business Process Model and Notation (BPMN) 2.0 [13].

Our approach uses a repository that takes business process views (i.e. different process models of the same process) as input and guides the modeler in the classification of the view elements into an organizational taxonomy. Afterwards, this information can be used to generate a consolidated business process model.

In terms of contributions, we expect both to provide a method for the integration of different business process views specified in BPMN 2.0 and also to support the stakeholders in the task of business process design, by proposing a systematic way for representing their concerns.

This work is organized as follows: in Sect. 2 we contextualize the problem and present its key concepts and challenges. Section 3 presents the related research works. In Sect. 4 we describe the architecture of our approach. In Sect. 5 we apply this integration method to a use case. Section 6 concludes the paper.

2   Problem Definition

2.1   Motivation

Organizations often have to manage multiple process diagrams that represent the same business process, which can lead to several inconsistencies, such as heterogeneous schemes for naming its activities and entities, usage of different modeling styles and process hierarchies with arbitrary depth and level of detail [14]. For instance, the processing of job applications can be captured from the perspective of the human resources (HR) unit, where the process model guides the work of HR employees and clarifies hand-overs; or from the perspective of the IT department, where the purpose of the model is to emphasize the interactions of HR employees with an IT-system [15].

A recent survey [16], shows significant research efforts in the past decade regarding business process variability modeling. These research works address the problem of managing families of process variants when it is clear that the process variations exist in the "real world" (e.g. multiple sales processes for different products or multiple bookkeeping processes for different countries).

However, variations and inconsistencies between process models are inherent to the task of business process modeling, for the aforementioned reasons.

We argue that the problem of managing multiple business process views, should be given more importance in terms of research efforts. We also argue that the task of business process modeling must cope with the multiple views and goals of the different organizational stakeholders, while capturing the complex relationships between information, people, goals and systems, as well as the underlying control and data flows [14].

2.2   Problem Description

This section describes the research problem by presenting the illustrative example of an auto repair company, with the purpose of facilitating the readers' understanding of the problem.

Before proceeding to the example, we must clarify the scope of the business process view integration problem. Business process view integration is the process of combining multiple business process views into a single business process model, which will be referred as **consolidated model**. A business process view is a model expressing the architecture of a given organizational business process, from the perspective of specific stakeholder concerns.

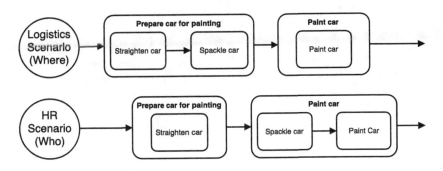

Fig. 1. Logistics and HR teams' concerns regarding the car repair process.

The goal of this work is to define a method for the integration of business process views. The proposed method will be applied to the following example:

- Considering an auto repair company, part of the process of repairing a car is performed by two participants: a painter and a mechanic. The repairing process starts when the damaged car arrives at the mechanical workshop. At this point, the mechanic is responsible for straightening the damaged car plates. Once the straightening is finished, the car plate is spackled by the painter and the car is moved to the paint shop for painting.
- The car repair process is modeled by two different organizational units: the HR team and the logistics team. The main concern of the HR team is regarding the different process participants - *who* dimension - in order to measure

their performance. The main concern of the logistics team is regarding the different locations where the activities are performed - *where* dimension - in order to measure the time participants spend changing location, during the process execution.

The concerns of both departments are illustrated in Fig. 1. Here, regarding the logistics team scenario, the activity "Prepare car for Painting" contains the tasks performed at the mechanical workshop (*straighten car* and *spackle car*) and the "Paint car" activity contains the task performed at the paint shop (*paint car*). Considering the HR team scenario, the "Prepare car for painting" activity contains the task performed by the mechanic (*straighten car*) and the "Paint car" activity contains the tasks performed by the painter (*spackle car* and *paint car*).

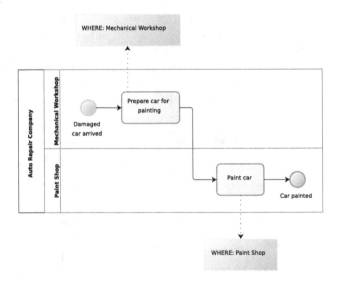

Fig. 2. Logistics team car repair process view.

The diagrams specified in BPMN 2.0 representing the views of the logistics and the HR teams are illustrated in Figs. 2 and 3, respectively. In Fig. 2, since the *where* dimension is the main concern of the logistics department, it is used as a split criterion of the activities. In Fig. 3, since the *who* dimension is the main concern of the HR department, it is used as a split criterion of the activities.

In the scope of our problem, the question that now arises is: "How to integrate the previous views into a single business process model?". This integration must be achieved knowing that the resulting model must address the concerns of both teams. Therefore, to address the concerns of both stakeholders, the resulting model must consider both splitting criteria, regarding performers (*who* dimension) and locations (*where* dimension). Ideally and considering this example, the goal of our method is to produce a consolidated model similar to the

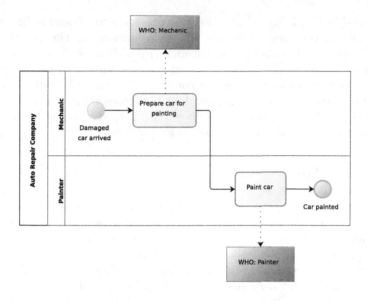

Fig. 3. HR team car repair process view.

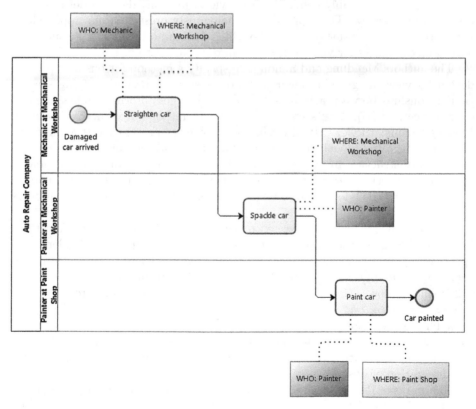

Fig. 4. Consolidated model that meets the concerns of both teams.

one presented in Fig. 4. For graphical representation and understanding reasons, the semantics of the *Lane* element ([13]) is broadened for this solution to be valid. We achieved this integration without any technique due to the simplicity of this specific example. Figure 4 explicitly shows that the splitting criterion of the process activities addresses the concerns of both stakeholders, since the lanes contain formal information regarding the process participants and their locations. Both *who* and *where* dimensions were merged and represented in each lane to consolidate the concerns of both stakeholders.

3   Related Work

As mentioned in Sect. 1, the problem of view integration was firstly addressed in the field of database design. The works of Navathe et al. [7–9] present a conceptual framework for logical database design. Here, the authors [7] describe the main four phases of logical database design, where the first two are "View Modeling" and "View Integration". The "View Modeling" phase consists of extracting and representing each user's perspective of the real world (user view), using requirement specifications as input. The "View Integration" phase integrates the multiple and possibly conflicting user views into one data model that represents a global view. These phases are analogously present in our approach to view integration in business process modeling. The same authors [8,9] contribute with a methodology for view integration in database design.

The authors Mendling and Simon [12], specify a method for business process design by view integration. Their method consists of: first identifying semantic relationships between activities of different business process models, namely equivalence and sequence; second, defining a merge operator for Event-driven Process Chains (EPCs) that also takes two business process views as input to build an integrated model; third, defining a set of restructuring rules to remove unnecessary structure from the integrated model. In our approach, our goal is to propose a method for business process view integration using BPMN models, contributing to the adoption of view integration in the field of business process design.

Sousa et al. [2] highlight the existence of conflicting process specifications for the same organizational process, depending on the distinct stakeholders' perspectives and on the modelers' view regarding that particular process. Moreover, their approach applies the six contextual dimensions to business process modeling, defined in the Zachman Framework [17,18]. These dimensions are *what*, *how*, *where*, *who*, *when*, *why* and each one is used as a criteria for activity decomposition. These criteria for activity decomposition support business process modeling by facilitating the task of different stakeholders consistently modeling the same process. Our approach is also based on these dimensions to define criteria for the integration of business process views.

The works of Pereira et al. [3,4], continue in the same direction of Sousa et al. [2] and add important contributions towards facilitating the consistent modeling of a business process by different stakeholders. For instance, a correspondence between business process modeling notations' elements and the aforementioned dimensions is presented, together with a set of principles that define equivalence between business processes [3]. In [4], an organizational taxonomy is defined together with a definition of business process based in the previously mentioned six dimensions (Fig. 5). By creating a taxonomy for each dimension and classifying each business process element using it, a controlled vocabulary is defined across distinct modeling perspectives. As a result, the gap between stakeholders and process designers is reduced, regarding the task of business process modeling. Our approach is grounded on the organizational taxonomy defined by Pereira et al. as it is used to support the proposed method for business process view integration.

Our approach fills the gap in knowledge existing in the works of Pereira et al. [3–5,19] and Caetano et al. [19], since their focus is on the definition of an organizational taxonomy and on the generation of business process views, without detailing structured ways for the stakeholders to classify their concerns and provide their views as input.

Fig. 5. Definition of business process used in our approach (from [4]).

4   The Approach

4.1   Core Concepts

The approach is based on a business process repository. This repository incrementally receives as input a set of business process views, together with the stakeholders' classification of the process elements and generates a consolidated business process model. As a starting point of our work, we present the definition of business process proposed by Pereira et al. [4,5] and Caetano et al. [19], which is illustrated in Fig. 5. Here, a business process is defined as follows:

- **How.** A business process is defined as a set of connected activities which consumes (inputs) and produces (outputs) tangible or intangible information entities, is performed by someone, contributes to achieve business goals, takes place in a specific location and occurs during a given period of time. A business process can be functionally decomposed in as many levels of detail as required.

- **Who**. An actor plays one or more roles. Actors may represent people, systems and organizations, including departments. An actor may play multiple roles and the same role may be played by different actors.
- **What**. An information entity represents the information about an artifact, person, place, concept, or event that has meaning in the business context.
- **Where**. An organizational unit defines the organizational structure. It may describe a logical unit (e.g. department) or a physical or geographical location.
- **When**. Corresponds to events that are relevant to an organization and its business.
- **Why**. A business goal is an objective that may be decomposed and each one can be associated with the business process that contributes to its achievement.

Given this definition, we consider that organizations and organizational units can be classified, in the context of a specific business process, simultaneously in the *where* and *who* dimensions. For instance, a pool representing an organization which has two lanes representing actors, we consider valid to classify that organization into the *who* dimension.

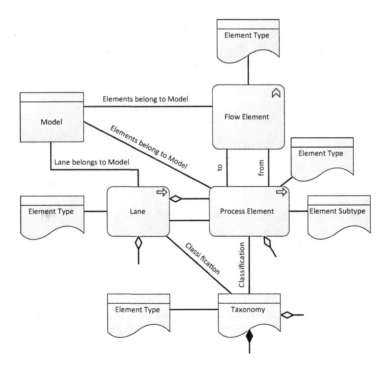

Fig. 6. Repository meta-model.

4.2   Meta-Model

We present the meta-model of the repository that will support the process of view integration represented in Fig. 6. Here, the ArchiMate [20] notation was adapted to represent a class diagram. For instance, the "loose" aggregation and composition connectors represent self-aggregation and self-composition. Each instance of the classes represented in the meta-model has its own life cycle, enabling a timeline of changes in the business process models. Here, we embed the "time factor" into business process models since organizations suffer multiple changes in time.

The meta-model is valid for any business process modeling notation because the element types are delegated into other classes. This conceptual design choice increases flexibility and simplicity. Flexibility in extending the meta-model with more types of the existing classes. Simplicity because if we chose inheritance, the number of classes in the meta-model would increase, decreasing its legibility. The *Taxonomy* class can aggregate an arbitrary number of dimensions and each dimension is composed of taxonomy nodes, i.e. instances of the class taxonomy that represent dimensions and process elements in the taxonomy tree. The *Process Element* class represents all elements of a process and the *Flow Element* class represents all notation elements which imply flow (e.g. data, sequence and message). All the instances of the classes present in this meta-model have a *Model* instance as a parent. The *Model* class allows us to differentiate between input, consolidated and output models. The remaining classes and relationships are self-explanatory.

The types of lanes considered in the repository meta-model are: **System**; **Actor**; **Role**; **Application**; **Organization**; **Organizational Unit**. If a lane is classified as an organization or organizational unit, according to what was mentioned before, it is classified in the dimensions taxonomy as belonging to *who* and *where* simultaneously.

4.3   View Integration Method

The proposed view integration method is represented in Fig. 7. Here, the lanes *Stakeholder A* and *Stakeholder B* can represent any organizational unit, modeling team or stakeholder. The lane *Repository* represents the business process repository supporting our method. The view integration process starts with the identification of a modeling need, followed by the activity of modeling a view of a given process. The respective view is then uploaded into the repository. At this point, the classification of the view elements is performed by the stakeholder, by giving input information to the repository (detailed in the next section). These activities are repeated in the context of another view of the same process, highlighting the incremental aspect of the proposed method. In this case, we assume there are only two views to be integrated, but there could be more. The process ends with the generation of the consolidated model by the repository, based on the information introduced as input by the stakeholders. The details regarding this activity will not be addressed in this work, since generating a model from a populated taxonomy was already addressed in previous works [6,19].

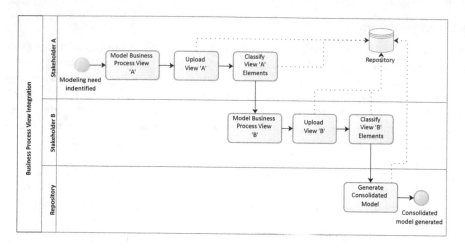

Fig. 7. Diagram of the view integration process.

5   Proof of Concept

This section will describe the view integration method illustrated in Fig. 7, focusing on the details of classifying the elements of each view. As it was previously mentioned, the generation of the consolidated model will not be addressed in this work. Considering that the repository is initially empty and that the logistics team view presented in Fig. 2 was already uploaded, the classification of its elements in the repository taxonomy is performed as follows:

1. **Lane: Auto Repair Company**. The question "From the following list, how do you classify this lane?" is answered by the stakeholder. The logic answer is "Organization" and a new taxonomy node is created in both dimensions *who* and *where*.
2. **Lanes: Paint Shop & Mechanical Workshop**. The same question is now answered regarding both lanes. The answer is the same for both, but in this case there are two valid answers: "Location" and "Organizational Unit". Here we considered "Location", resulting in the assignment of both lanes to the *where* dimension. If we chose the alternative answer, the final consolidated model would be the same. As a result, two child nodes having the Auto Repair Company as a parent are created in the *where* taxonomy.
3. **Events: Paint Shop & Mechanical Workshop**. In this case, the events are automatically assigned as nodes of the *when* taxonomy.
4. **Activities: Prepare for painting & Paint car**. Here it is asked to the stakeholder if he can drill down each activity in further detail. This is possible in the case of the "Prepare for painting" activity. As a result, the more detailed activity node "Straighten car" is selected and the activity "Spackle car" is added as a child node of the "Paint car" activity in the *how* taxonomy.

The state of the repository after uploading and classifying the elements of the logistics team view is presented in Fig. 8. At this point, the repository already has

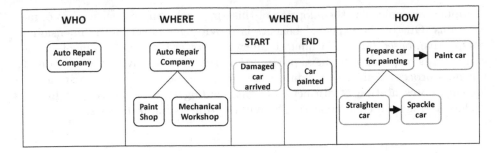

Fig. 8. State of the repository after uploading and classifying the logistics team view.

information regarding the logistics team view. We will now describe the activity responsible for the classification of the elements belonging to the HR team view. The steps for classifying these elements in the taxonomy are the following:

1. **Lane: Auto Repair Company.** Now that the repository is not empty, based on the name of the element, it is suggested to the stakeholder that the lane "Auto Repair Company" already exists. Considering that the stakeholder agrees with the classification of this lane as "Organization", he selects the already existing nodes for this lane in the *who* and *where* dimensions.
2. **Lanes: Painter & Mechanic.** This step is analogous to the one described for the classification of the logistics team view. The difference lies in the classification given to the lanes, which in this case is "Actor". The result of this step is adding both lanes as child nodes of the "Auto Repair Company" lane in the *who* dimension.
3. **Events: Paint Shop & Mechanical Workshop.** Now that the same events are already present in the repository taxonomy, the stakeholder selects them and the taxonomy remains unchanged.
4. **Activities: Prepare for painting & Paint car.** In this step, as previously described, it is asked to the stakeholder if he can drill down each activity in further detail. Regarding the "Prepare for painting" activity, it already exists in the repository, therefore it is selected by the stakeholder. Regarding the "Paint car" activity it is first selected in the taxonomy and then it is possible to add more detail into it. As a result, the more detailed activities "Straighten car" and "Spackle car", including their sequence, are added as child nodes of the "Prepare for painting" activity in the *how* taxonomy.

After uploading and classifying the elements of the HR team view, we reach the final state of the repository presented in Fig. 9. This final result shows that the repository contains the necessary information for the generation of the consolidated model. We state so because all the elements present in both views are classified in the taxonomy and the meta-model (Fig. 6) evidences that the relationships between the elements are sufficient to produce the consolidated model (Fig. 4). Moreover, by proposing a structured way of classifying and giving views as input, we argue that the adoption of this approach can be beneficial to organizations. One of the expected benefits is providing a novel multi-dimensional perspective, based on the Zachman Framework, towards the management, modeling and perception of business processes in organizations. With the division of business process elements into dimensions, organizations can adopt a standard for the representation of distinct concerns, leading to an enhancement in the understanding and communication among the different stakeholders.

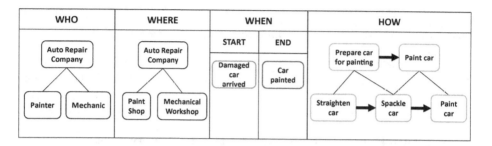

Fig. 9. State of the repository after uploading and classifying the HR team view.

6   Conclusions and Future Work

This work served the purpose of exposing the problem of consistent modeling by multiple stakeholders to the research community. Our contribution to this problem is grounded on applying view integration to business process modeling. The goal of the proposed view integration method is to define the process of uploading and classifying multiple stakeholder views, with the support of a business process repository, enabling the generation of a consolidated model. This work followed the direction of previous works towards helping stakeholders and organizations managing their approach towards business process management. More specifically, by using an organizational taxonomy together with the proposed dimensions, we expect to aid stakeholders and organizations expressing their concerns when engaging in business process design.

As a work in progress, the main limitation of this proposal is the lack of diversity in the examples and limited complexity of the use case. However, it provides a first contribution towards applying view integration to the problem of consistent modeling in BPMN 2.0. Furthermore, this method differentiates itself from other proposals, as it consists of an incremental approach that can adapt to the growth of organizations and their businesses, by embedding time into the business process models. As future work, we intend to detail further the generation of the consolidated model, based on the information of the repository. We also aim to obtain tool support for our approach, in order to provide results with a larger diversity of use cases in terms of complexity, structure and number of elements.

Acknowledgments. This research was supported by the Enterprise Architecture unit at Link Consulting, SA (www.linkconsulting.com) under the EAMS research project.

References

1. Indulska, M., Green, P., Recker, J., Rosemann, M.: Business process modeling: perceived benefits. In: Laender, A.H.F., Castano, S., Dayal, U., Casati, F., Oliveira, J.P.M. (eds.) ER 2009. LNCS, vol. 5829, pp. 458–471. Springer, Heidelberg (2009). doi:10.1007/978-3-642-04840-1_34
2. Sousa, P., Pereira, C., Vendeirinho, R., Caetano, A., Tribolet, J.: Applying the Zachman framework dimensions to support business process modeling. In: Cunha, P.F., Maropoulos, P.G. (eds.) Digital Enterprise Technology: Perspectives and Future Challenges, pp. 359–366. Springer, Boston (2007). doi:10.1007/978-0-387-49864-5_42
3. Pereira, C.M., Sousa, P.: Business process modelling through equivalence of activity properties. In: Proceedings of the Tenth International Conference on Enterprise Information Systems, ICEIS, vol. 3, pp. 137–146. INSTICC, ScitePress, Barcelona, Spain (2008)
4. Pereira, C.M., Caetano, A., Sousa, P.: Ontology-driven business process design. In: Skersys, T., Butleris, R., Nemuraite, L., Suomi, R. (eds.) I3E 2011. IAICT, vol. 353, pp. 153–162. Springer, Heidelberg (2011). doi:10.1007/978-3-642-27260-8_12
5. Marques Pereira, C., Caetano, A., Sousa, P.: Using a controlled vocabulary to support business process design. In: Barjis, J., Eldabi, T., Gupta, A. (eds.) EOMAS 2011. LNBIP, vol. 88, pp. 74–84. Springer, Heidelberg (2011). doi:10.1007/978-3-642-24175-8_6
6. Pereira, C.: Using an organizational taxonomy to support business process design. Ph.D. thesis, Instituto Superior Técnico (2011)
7. Navathe, S.B., Schkolnick, M.: View representation in logical database design. In: Proceedings of the 1978 ACM SIGMOD International Conference on Management of Data - SIGMOD 1978, p. 144. ACM Press, New York (1978)
8. Navathe, S.B., Gadgil, S.G.: A methodology for view inegration in logical database design. In: Proceedings of Eigth International Conference on Very Large Data Bases, Mexico City, Mexico, 8–10 September 1982, pp. 142–164. Morgan Kaufmann (1982)
9. Navathe, S.B., Elmasri, R., Larson, J.: Integrating user views in database design. Computer **19**(1), 50–62 (1986)
10. Batini, C., Lenzerini, M.: A methodology for data schema integration in the entity relationship model. IEEE Trans. Softw. Eng. **SE–10**(6), 650–664 (1984)
11. Batini, C., Lenzerini, M., Navathe, S.B.: A comparative analysis of methodologies for database schema integration. ACM Comput. Surv. **18**(4), 323–364 (1986)
12. Mendling, J., Simon, C.: Business process design by view integration. In: Eder, J., Dustdar, S. (eds.) BPM 2006. LNCS, vol. 4103, pp. 55–64. Springer, Heidelberg (2006). doi:10.1007/11837862_7
13. International Organization Of Standardization: ISO/IEC 19510: 2013 - Information technology - Object Management Group Business Process Model and Notation (2013)
14. Caetano, A., Silva, A.R., Tribolet, J.: Business process decomposition an approach based on the principle of separation of concerns. Enterp. Model. Inf. Syst. Archit.-Int. J. **5**(1), 44–57 (2010)

15. Weidlich, M., Mendling, J.: Perceived consistency between process models. Inf. Syst. **37**(2), 80–98 (2012)
16. La Rosa, M., Van Der Aalst, W.M.P., Dumas, M., Milani, F.P.: Business process variability modeling: a survey. ACM Comput. Surv. **50**(1), 1–45 (2017)
17. Zachman, J.A.: A framework for information systems architecture. IBM Syst. J. **26**(3), 454–470 (1987)
18. Sowa, J.F., Zachman, J.A.: Extending and formalizing the framework for information systems architecture. IBM Syst. J. **31**(3), 590–616 (1992)
19. Caetano, A., Pereira, C., Sousa, P.: Generation of business process model views. Procedia Technol. **5**, 378–387 (2012)
20. The Open Group: ArchiMate® 2.1 Specification. The Open Group series, van Haren Publishing (2013)

A Balanced Scorecard Approach for Evaluating the Utility of a Data Warehousing System

Itilde Martins and Orlando Belo$^{(\boxtimes)}$

Department of Informatics, School of Engineering, ALGORITMI R&D Centre,
University of Minho, Campus de Gualtar, 4710-057 Braga, Portugal
obelo@di.uminho.pt

Abstract. From time to time, on a regular basis, companies often plan and implement internal measurement processes of quality of service and information of their computer systems. These processes are very important for ensuring user confidence in the operation of their systems and consequently on the information they extracted and used from them. In all activities related to decision-making, it is important to ensure that systems supporting such activities, such as data warehousing systems, guarantee a good quality of service and provide data accordingly the needs of decision makers. Most company managers are concerned how to monitor and ensure that their business objectives, strategies and methods will be achieved adopting and using a balanced scorecard solution. In this work, we present and discuss how we can use such kind of approach for monitoring and optimizing the utility of a data warehousing system. We approached the problem defining some of the most conventional perspectives used on balanced scorecards solutions, namely financial, customer, internal processes, and learning and growth, applying them with the goal to measure and optimize the degree of utility of a specific data warehousing system.

Keywords: Decision support systems · Data warehousing systems · Evaluation of services and data structures · Data warehousing quality methodologies · Balanced scorecards

1 Introduction

Today we live in a globalized world. Data is easy to gather even from disparate resources. Companies use and transform it to get competitive advantages in the markets where they are players. However, data will be useful for companies only if they have a well-defined strategic plan supporting their strategic decisions and business actions. Usually, enterprise resource planning systems, and external sources as well, supply large amounts of data for regular business and decision-making processes, making the company consuming significant efforts and doing important investments for using appropriated tools and methodologies, such as balanced scorecard, to optimize, monitor and follow the utility of the systems they have installed.

In many aspects, balanced scorecards [1, 2] aim to improve the concept of a control system beyond of traditional financial indicators, including information such as

© Springer International Publishing AG 2017
M. Themistocleous and V. Morabito (Eds.): EMCIS 2017, LNBIP 299, pp. 633–645, 2017.
DOI: 10.1007/978-3-319-65930-5_49

financial and non-financial, internal and external, business performance or involving current results. Their goals go far beyond of what we can extract in a simple set of business indicators, enabling organizational transformations towards the specialization of processes. Balanced scorecards can be viewed as an interactive construction performed by organization managers, and guided by experienced consultants. All the perspectives must support each other, and the weakness of one perspective may bring a negative influence on the overall company strategies implementation. For many years, databases were the only information source that companies used to support their decision-making activities. But, over the years, it was noted that their managers asked for more, before taking any kind of decision – they required that decisions must be always well supported.

Data warehousing systems [3] have emerged as valuable instruments for supporting decision-making activities, providing a high-specialized source containing historical integrated and non-volatile data, representing faithfully the state of a company and being today one of the most sophisticated technology-based structure for decision making on company management. Yet, the utility of a data warehousing system continues to be quite unmeasured in a lot of practical cases. The cause of this non-measurement is that information technology managers do not have adequate resources to evaluate the benefits of a data warehouse and the information it maintains. Most company managers are concerned how to monitor and ensure that their business objectives and strategies will be achieved by implementing a balanced scorecard methodology, having particular concerns about how to confirm that the strategy set was correctly implemented, or how to check if it is not necessary to make some kind of correction on business settings already defined.

In the context of this work, balanced scorecards were used as a support tool to optimize and monitoring the progress of decisions taken for evaluating the utility of a data warehousing system, and indirectly its performance. The main objective of this work was to evaluate the usefulness of a data warehousing system as a management tool for performance and usefulness using balanced scorecards, as already referred. We approached the problem defining some of the usual perspectives of balanced scorecards namely financial, customer, internal processes, and learning and growth, with the goal to measure and optimize the degree of utility of a particular data warehousing system.

In this paper we will expose and explain the process we carried out to evaluate the utility of a data warehousing system, taking a large diversity of aspects that usually are related with the most important system development stages, with particular focus on data exploration activities. Then, combining all these aspects using a balanced scorecard approach, we reached a interesting platform that allows for us to get a picture of the importance and the utility of a data warehousing system for its users and consequently for the company where it is installed. In the following sections we will expose and discuss some related work (Sect. 2), present the evaluation process we implemented to measure the utility of a data warehouse (Sect. 3), and finally a brief set of conclusions, pointing out some working lines for future work (Sect. 4).

2 Related Work

For a long time business managers evaluated their performance based exclusively on financial indicators such as profit, cash flow, profitability, or return of investment, among others [4]. Although, in the 50's years business managers realized they had to go a little bit beyond such kind of indicators. Even so, the "boom" of awakening only happened in the 80's years with a profound revelation in the areas of business models, economy and company management. Today, non-financial indicators, such as the value of a brand, customer relationships, or organizational cultures, are recognized as very important aspects to take into consideration for evaluating the business performance of a company. However, managers have some difficulties to deal with such non-financial indicators and their inherent complexity, and answering appropriately and timely to know how to follow their changes and what kind of methodologies should be used. For that, they want simple but robust consistent models, so they can handle changes easily and react appropriately without difficulties.

Today, there is a large diversity of evaluation solutions in the market especially oriented for helping business managers in performance evaluation processes. Many of these solutions use to cover a lot of aspects related to performance evaluation and modeling, using a great diversity of techniques and methods, ranging from simulation, monitoring, analytical processing, benchmarking, or even profiling for sustaining effective models of appraising. Performance evaluation of any system should be started immediately at the design stage as an effort to identify and eliminate any kind of design tradeoff. One of the most used techniques to do this is to create a software model and simulating the execution of a set of benchmarks, in order to evaluate functionality of the several components of a system and the way they interact with each other. However, systems continue to growth in dimension and complexity. Now, we need new techniques beyond the most conventional ones for developing more effective evaluation mechanisms, having advanced measuring means with the ability to deal with complex issues, like flexibility, accuracy or speed, and represent their correlation using a simple but meaningful number [5].

In the field of software engineering many approaches were followed by a large diversity of researchers, covering many areas, ranging from system specification and modeling to databases systems [6]. Besides, we can find other works in the area that evaluate the performance of software in its initial state of development, as is the case when we are discussing and analyzing software architectural aspects. See, for example, the works presented in [7] or [8], which address, respectively, some quite pertinent and diversified issues related to the development and maintenance of software components for distributed systems, and a simplified way to evaluate software projects in general. In other works, such as [9], the authors have dedicated their time in the establishment of a method for predicting performance, rather than focusing exclusively on their evaluation. But it is when we read [10] that, in fact, we see the number of works and areas that have been worked on in the field of software performance analysis and recognize the importance of performance evaluation.

This paper illustrates quite well such relevance. At the Internet level, we can find also some other software performance evaluation works, such as [11], in which the authors present a very interesting work for evaluating service differentiation performance of Web servers [12]. In this way we can see that much has been done in the field of performance evaluation, using a great diversity of means, methods and techniques. Therefore, the references in the literature are in great number. Despite this, we did not find many references to works that used balanced scorecards in systems performance or utility evaluation processes, unless the work presented in [13], which presents and discuss some of the most relevant aspects for measuring the performance of a data warehousing system, providing a concrete operational view about data operations. However, we can also highlight some other curious works that we have found, not in the field of performance evaluation, but in performance management, see for example [14, 15], or more recently [16]. However, when we focused on the field of data warehousing systems, we did not find anything related to the evaluation of their performance, whether using any of the techniques or methods mentioned in this section or using balanced scorecards as we intend to do. Thus, as far we know, balanced scorecards have never been used in performance appraisal or in determining the utility of a data warehousing system – the goal of this work. In the next section, we will begin the presentation and discussion of how we have approached the evaluation of the usefulness of a data warehousing system with the use of balanced scorecards.

3   Evaluating the Utility of a Data Warehousing System

3.1   The Use of Balanced Scorecards

The evaluation of the utility of a data warehousing system can be understood as a systematic procedure requiring to answer to a set of specific questions about the value and importance of a particular domain, theme, propose or project. Any evaluating process of a system faces a large variety of meanings, since it can be used in different contexts and in a large set of human activities [17]. In this work we designed and developed a controlling instrument for data warehousing systems based on principles and rules usually used in any balanced scorecard approach. The motivation behind using a balanced scorecard is to make sure that a data warehousing system will be efficient and stable, presenting a highly performance level in its operational tasks.

The performance of a data warehousing system can be measured and evaluated using some of the most important criteria aspects that usually are referred by its architects and engineers, which were analyzed and discussed in a very interesting way by Rahman in [13]. In our point of a view, this is a very relevant work about the application of balanced scorecards to the world of data warehousing systems. Our work goes in the same direction of this work, adopting some of the issues exposed by Rahman, but instead of measuring the performance of a data warehousing system we were concerned about how to evaluate its utility. Additionally, we want to provide a way for ensuring a high use of the information contained in a data warehouse, improve the expertize and skills of the company's staff, redirect the data warehousing system to new information sources and areas of competence, and, finally, improve the way we select the data we stored inside the data warehouse. Next, we will see how we designed and built a balanced scorecard approach especially oriented to satisfy such a goal.

Generally, speaking, the implementation of a balanced scorecard approach in a data warehousing system environment, like in other application area, aims to increase some of the most significant aspects of the system, like its strategic vision definition and consensus, the definition of the work team, the resource allocation and strategic initiatives, just to name a few. Following the ideas of Headley [18], after getting the agreement between the strategic objectives and indicators we want to reach, someone must be selected to lead the entire balanced scorecard project implementation: the data warehousing system administrator. However, others must also help him in the coordination and definition of the strategic indicators – e.g. ou information systems managers, business analysts, or decision-makers. The data warehousing system administrator should lead all the processes, as well as structuring the working program, and collect the necessary documentation for all project tasks.

3.2   The Balanced Scorecard Approach

As in other application cases, the first steps to implement a balanced scorecard approach in a data warehousing system environment must be based on methodological processes, and in a clear consensus about the process of translating the organizational mission and strategy into objectives and indicators [19]. According to Giollo [20] the process of implementing a balanced scorecard solution must start with the creation of a concrete vision and a well-defined strategy, in order to design the agreement on the objectives to be achieved. During the development process of a balanced scorecard there must be what we called as an "architect", which function is to facilitate the process and also to be able to understand the information gained as well as having a full involvement of companies' top managers [19].

On the basis of the technological aspects of a data warehousing system, balanced scorecards perspectives are developed and populated with relevant objectives and measures especially defined to sustain its success. These perspectives are integrated into a consistent data warehouse scorecard. A data warehousing system utility evaluation process approaches a group of very specific factors that allow for checking if a data warehouse helps or not a company (and their managers) in the support of its decision-making processes. It is a process that takes into account several factors such as the frequency of use, and the quality of information stored under the user's perspective, regarding tools and the information itself. Using 42 distinct variables framed in some of the most relevant aspects of our application context, such as usability and user profiling, quality of information, productivity and frequency of use, or strategic implementation, just to name a few, we was able to define and clarify the business strategic vision of the target data warehousing system, establish a more effective communication and link among the different business perspectives, evaluate all the strategic goals' metrics for each business perspective integrated in the system, and, of course, identify all the areas that revealed some kind of negative issues. Basically, all this was set up in order to provide company managers with a better platform to support and develop their decision-making processes.

The construction of a balanced scorecard solution can be influenced by many factors, such as the complexity of its structure and formation. Things like these can influence strongly how the evaluation project should be implemented. Therefore, this process diverges from system to system, and obviously from company to company. Based on the experience of Kaplan and Norton [19], we can structure the balanced scorecard solution in four main development stages (Fig. 1), namely:

(1) Define the evaluation of the system, which involves selecting the system unit, the identification of relationships between the system and the business unit.
(2) Achieving a consensus on strategic objectives, where we define how to conduct interviews and summarize the conclusions, defining meetings with top users – decision makers and business analysts.
(3) Select and draw indicators, this is the stage where we establish the meetings of subgroups, a second series of meetings with top users.
(4) Developing the implementation plan, where we define the construction of its implementation, promoting a third series of meetings with top users, and, finally, complete the implementation plan.

Fig. 1. The development stages of the balanced scorecard approach.

The definition of a strategic plan is one of most critical and important step to build and implement a good balanced scorecard. In this critical task, it is important to focus on the strategic foundation structure, and finish it with a set of strategic targets or maps. The strategic targets, measurements, grids, and programming are the four indispensable key components that allow for building a balanced scorecard solution. The adapted strategy is a mixed strategy (adaptive and offensive) where the principle focus is taking advantage of opportunities through leveraging the strengths - using the data warehouse's information in decision-making processes - and overcoming weaknesses - technical staff and end-users' qualifications.

3.3   Implementing the Solution

The strategic map (Fig. 2) defined for the balanced scorecard solution presented in this work, identify the cause-effect relationship existing in all perspectives, as well as the strategic behavior vectors defined their objectives, and according to them, define the measures indicators necessary to evaluate the accomplishment of these goals. However, in the balanced scorecard building process we will not followed the original principles and methodologies proposed in [19]. This means that instead of using four distinct perspectives, in this work we only used three, namely: (1) exploration (customers); (2) internal processes; and (3) learning and growth. Thus, next we will focus our discussion in these three perspectives, as well as in their strategic objectives and mediation indicators.

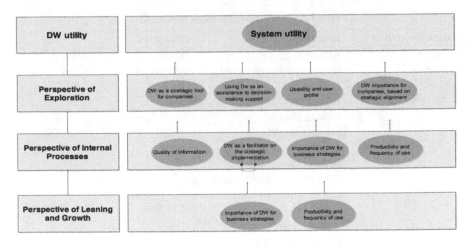

Fig. 2. The strategic map of the balance scorecard solution.

Lets see each one of the referred perspectives, with some more detailed. In the exploration perspective we defined four strategic objectives (the data warehouse as a strategic tool for companies, using the data warehouse as an assistant to support decision-making, usability and user profile, the importance of the data warehouse for companies), and their respective indicators – e.g. company impact, data warehouse flexibility, easy of using, company differentiation, or increasing the market share –, and final objectives – e.g. increase the importance of the data warehouse, improve the data warehouse flexibility, or decrease the difficulties of utilization. Basically, the purpose of this perspective is to find out if each one of indicators has been met or not. For the internal processes perspective we defined to evaluate four other strategic objectives, and their indicators and objectives. Table 1 shows a brief view about the characterization of this perspective. Finally, we have the learning and growth perspective, which allows for understanding the point of view of users, and through it evaluate the data warehousing system utility for a company. In this last perspective we defined only two strategic objectives (innovation and technology, and user qualification), and the correspondent indicators – e.g. the rate of new ideas used after the data warehouse implementation, level of new products and services developed after the data warehouse implementation, or the level of experience of users –, and final objectives – e.g. improve the factors in the innovation and technology sector, or increase user qualification.

3.4   Research Instruments and Survey Implementation

In order to develop and apply appropriately the balanced scorecard approach we needed an effective working base that allows for us to evaluate the three perspectives referred before, which has the ability to provide data about some specific data warehousing systems projects and installations. Basically, we needed to gather some answers for satisfying the questions that were defined for this work. For this, we

Table 1. The internal processes perspective.

Strategic objectives	Indicators	Objectives
Quality of information	Reliability and accuracy of the information provided	Increase the quality of the information stored on DW
	Current and timely information	
	Correct level of information detail	
	Acquisition of information in exact time	
	Perceptible and ambiguity free information	
	Merge information from multiple sources without data inconsistency	
	Data relevance	
	Data localization	
The DW as a facilitator on the strategic implementation	Strategic plan based on the successfully implementation of the DW	Increase the rate of the DW importance to the implementation of new business strategies
	Improvements made with the strategic plan implementation	
Importance of the DW for business strategies	Decrease the lead time	
	Definition of the execution time for each task	
Productivity and frequency of use	Improvements made on the productivity sector	
	Frequency of use	

prepare a specific survey, containing all the most pertinent questions we need to gather an answer, distributing it for several data warehousing systems developer teams that were doing work on projects in the area in one of the curricular units of business intelligence at our university. In fact, it is an academic application case, but it covers all the issues and aspects we use and want for validating this work. The Likert scale [21] was selected for regulating this survey, aiming to transform the order points scale into a linear one. This scale represents a systematic and a refined way for developing metrics through independent answers, weighted and summed. In this work, we used a Likert Scale of 5 points, based on the format suggested in [21], starting with the alternative "Disagree", with three intervals, until the last alternative "Totally Agree", which will be, respectively, stored from 1 to 5.

Having all the survey requirements well established, all members of the data warehousing developer teams that agree to participate in the study received the survey. Later, the answers where gathered, treated and analyzed. The analysis process was done accordingly the three perspectives referred previously that were defined especially for this balanced scorecard approach to data warehousing systems. Results were quite interesting, revealing some curious situations about the utility of a data warehousing system. From the entire study we done, we selected only three cases (simply due to limitations of space) to present here, each one corresponding to a specific strategic objective of one of the three perspectives in analysis. Such results can be seen, respectively, in Tables 2, 3 and 4.

Table 2. Perspective of exploration - the data warehousing system as a strategic tool for companies.

Variables	Topics	Overall average of users information	Overall average for each evaluation measure
V1	The DW implementation created positive impacts on company activities performance	4	4
V2	The obtaining of information from the DW is quick, and allows the decision-making support in a timely way	4	
V3	The DW contributed decisively to the Integration of information from the several different strategic areas of the company	5	
V4	The information generate from the DW is important for the decision-making processes	4	

If we look to Table 2, perspective of exploration, for example, based on the users perception about a data warehousing system, when supporting the strategy of the entire business and decision making activities of a company, it can be stated that, as a strategic tool, it meets completely the expectations of the company. It can be also affirmed that the data warehouse system implementation has generated positive impacts for the company (V1), allowing for obtaining information quickly, being capable to support decision-making processes in time (V2). About its contribution to the integration of information from the many different strategic areas of the company (V3), it was said by users that it exceed all the expectations planned. They also said that, the information generated by the data warehousing system is quite satisfactory for supporting company's decision-making processes (V4). The other two tables (Tables 3 and 4) are also very simple to interpret and understand.

3.5   Implementation Tools

In the implementation of the balanced score card solution presented here we used the software Balanced Scorecard Designer PRO [22]. This software simplifies very much the process of creating and implementing balanced scorecards, helping to develop new scorecards with strategic maps, categories, indicators, etc. With this software, it is also

Table 3. Perspective of internal processes - quality of information.

Variables	Topics	Overall average of users information	Overall average for each evaluation measure
V24	The DW provides precise information (correct and reliable)	4	4
V25	The DW provides current information, opportune and timely for the company decision-making processes	4	
V26	The information provides sufficiently detailed information for the company decision-making processes	4	
V27	The DW allows you to access the data timely	4	
V28	The information provided by the DW is dear, easily understandable and free of ambiguities	3	
V29	The DW covers information about different data sources, without generating any inconsistency	4	
V30	The data stored in DW are relevant to the company decision-making processes	4	
V31	It is easy to determine establish what information is available in DW, and how to find	3	

Table 4. Perspective of learning and growth - innovation and technology.

Variables	Topics	Overall average of users information	Overall average for each evaluation measure
V38	The rate of new ideas is higher after the DW implementation	3	3
V39	The level of new products and services developed after the DW implementation is greater	3	

possible to generate a set of key performance indicators, define the association between groups, goals, and identify the importance of each indicator. This software offered a very flexibility way to calculate the utilities values that are depending on the settings of the indicators.

4 Conclusions and Future Work

Nowadays, companies are increasingly looking for surviving in very diversified contexts that are affected by deep and very frequent modifications, facing high levels of competitiveness in the markets where they are positioned. Consequently, their decision support systems, and in particular their data warehousing systems, should be able to provide the information necessary to support all the business processes that should be implemented (or adjusted) in the face of such levels of competitiveness. It is important to ensure that the information they contain and the services they provide are in fact useful.

Balanced scorecards provide a very interesting definition and clarification of how visions and strategies of a data warehousing system can be seen, establishing coherent communication and connection means between all the perspectives, metrics and strategic objectives established by designers, architects or engineers. Additionally, they also provide the means for verifying the differences of the various perspectives used in an organizational model, taking into consideration all the dependences that may exist on the particularities and on the strategic objectives of each business and decision-making application and context.

The investigation we pursued for evaluating the utility of a data warehousing system, and indirectly decision making activities performance, requires recognition and study of how the information can be collected, stored and processed – the populating process of a data warehouse -, and, finally, how it is explored by final users, in order to achieve some gains in performance and ensure system utility and usage. Although we performed this work based on data collected in some data warehousing system development processes in course in our academic environment, we were able to apply in a very concrete way the balanced scorecards models we idealized and developed in all the stage of the utility evaluation process. This allowed for demonstrating and validating the efficiency of the balanced scorecard model prepared for evaluating the utility of a data warehousing system.

At short term, we intend to improve the presented approach perceiving the difficulties that emerge during the identification of the right evaluation keys, so that the evaluation process can be performed based on more appropriate utility indicators, if companies develop business intelligence applications based on the management of intangible assets, especially when using information generated by a decision support system. Additionally, it is also necessary to demonstrated the effectiveness of the balanced scorecard approach for evaluate a real-world data warehousing system, having the intention to validate what we reach in a controlled environment.

Acknowledgments. This work has been supported by COMPETE: POCI-01-0145-FEDER-007043 and FCT – Fundação para a Ciência e Tecnologia within the Project Scope: UID/CEC/00319/2013.

References

1. Kaplan, R., Norton, D.: Using the Balanced Scorecard as a Strategic Management System. Harvard Business Review, Boston (1996)
2. Kaplan, R., Norton, D.: The Balanced Scorecard: Translating Strategy Into Action. Harvard Business School Press, Boston (1996)
3. Kimball, R., Caserta, J.: The Data Warehouse ETL Toolkit: Practical Techniques for Extracting, Cleaning Conforming, and Delivering Data, vol. 1. Wiley, Hoboken (2004)
4. Pinto, F.: Balanced Scorecard – Alinhar Mudança, Estratégica e Performance nos Serviços Públicos. Edições Sílabo, Lisbon (2007)
5. Kurian, J.: Performance evaluation: techniques, tools and benchmarks. Electrical and Computer Engineering Department, The University of Texas at Austin (2017). https://lca. ece.utexas.edu/pubs/john_perfeval.pdf. Accessed 15 May 2017
6. Balsamo, S., Mamprin, R., Marzolla, M.: Performance evaluation of software architectures with queuing network models. In: Bobenau, C. (ed.) Proceedings of the European Simulation and Modeling Conference (ESMc 2004), pp. 206–213. UNESCO, Paris (2004)
7. Balsamo, B., Marzolla, M.: Performance evaluation of UML software architectures with multiclass queueing network models. In: Proceedings of the 5th International Workshop on Software and Performance, 12–14 July 2005, pp. 37–42 (2005)
8. Kähkipuro, P.: UML-based performance modeling framework for component-based distributed systems. In: Dumke, R., Rautenstrauch, C., Scholz, A., Schmietendorf, A. (eds.) GWPESD/WOSP -2000. LNCS, vol. 2047, pp. 167–184. Springer, Heidelberg (2001). doi:10.1007/3-540-45156-0_11
9. Bharathi, B., Kulanthaivel, G.: A tool for architectural design evaluations using simplistic approach. Int. J. Comput. Appl. Spec. Issue Comput. Sci. New Dimens. Perspect. (4), 162–165 (2011)
10. Geetha, D.E., et al.: Predicting performance of software systems during feasibility study of software project management. In: 2007 6th International Conference on Information, Communications & Signal Processing, pp. 1–5 (2007)
11. Merseguer, J., Campos, J., Mena, E.: Performance analysis of internet based software retrieval systems using Petri Nets. In: Meo, M., Dahlberg, T.A., Donatiello, L. (eds.) Proceedings of the 4th ACM International Workshop on Modeling, Analysis and Simulation of Wireless and Mobile Systems (MSWIM 2001), pp. 47–56. ACM, New York (2001)
12. Chen, X., Mohapatra, P.: Performance evaluation of service differentiating internet servers. IEEE Trans. Comput. 51(11), 1368–1375 (2002)
13. Rahman, N.: Measuring performance for data warehouses - a balanced scorecard approach. Int. J. Comput. Inf. Technol. 04(01), 1–7 (2013)
14. Kaplan, R.S., Norton, D.P.: The Balanced Scorecard – Measures that Drive Performance. Harvard Business Review (January–February), pp. 75–85 (1982)
15. Kaplan, R.S., Norton, D.P.: Transforming the balanced scorecard from performance measurement to strategic management: part I. Acc. Horiz. 15(1), 87–104 (2001)
16. Gomes, R.C., Liddle, J.: The balanced scorecard as a performance management tool for third sector organizations: the case of the Arthur Bernardes Foundation, Brazil. Braz. Adm. Rev. 6 (4), 354–366 (2009)
17. Aguilar, M.J., Ander-Egg, E.: Avaliação de programas e serviços sociais. Edition Vozes, Petrópolis (1994)
18. Headley, J.: Aspectos prácticos de la implementación del Cuadro del Mando Integral. Finanzas y Contabilidad, no. 22, pp. 35–41, March/April 1998

19. Kaplan, R., Norton, D.: Cuadro de Mando Integral – The Balanced Scorecard. Ediciones Gestión 2000, SA, Barcelona (1997)
20. Giollo, P.R.: Modelo de avaliação de desempenho fundamentando no Balanced Scorecard – um estudo de caso no URI – Campus Erechim. Master degree dissertation in Administration. Federal University of Rio Grande do Sul, Porto Alegre (2002)
21. Babbie, E.: Métodos de Pesquisa de Survey. Editora UFMG, Belo Horizonte (1999)
22. BSC Designer, Professional Balanced Scorecard Software and Training (2017). http://www. bscdesigner.com. Accessed 1 July 2017

Crowd and Experts' Knowledge: Connection and Value Through the Notion of Prism

Riccardo Bonazzi[1], Gianluigi Viscusi[2(✉)], and Valérie Barbey[1]

[1] HES-SO Valais/Wallis, Sierre, Switzerland
{riccardo.bonazzi,valerie.barbey}@hevs.ch
[2] École Polytechnique Fédérale de Lausanne (EPFL CDM MTEI CSI),
ODY 1 16 (Odyssea) Station 5, 1015 Lausanne, Switzerland
gianluigi.viscusi@epfl.ch

Abstract. Crowdsourcing is an online activity in which an individual, an institution, a non-profit organization, or company proposes to a group of individuals of varying knowledge, heterogeneity, and number, via a flexible open call, the voluntary undertaking of a task. Crowdsourcing has been traditionally considered suitable to provide different types of support to the decision making process, especially in the design phase, through idea generation and co-creation, in the choice phase, through voting, as well as in the intelligence phase to explore or exploit information about the issue to be investigated. This article aims to investigate how to perform scenario planning by exploring ways to use crowdsourcing as a complement to two standard techniques for idea generation and selection: (a) brainstorming and (b) the Delphi method. Then, we question the cost and the effectiveness of combining these methods, and crowdsourcing to perform scenario planning for policy making. To this end, in this article we propose a model to assess the cost and effectiveness of the intersection between crowd and experts in decision-making activities, with a focus on scenario planning, choosing a public sector research site for its evaluation.

Keywords: Crowdsourcing · Decision making · Policy making · Scenario planning · Design science

1 Introduction

This article aims to investigate how to perform scenario planning by exploring ways to use crowdsourcing as a complement to two standard techniques for idea generation and selection: (a) brainstorming and (b) the Delphi method. Then, we question the cost and the effectiveness of combining these methods, and crowdsourcing to perform scenario planning for policy making. Scenario planning is a disciplined method for imagining possible futures that companies have applied to a great range of issues [1]. Brainstorming is defined here as a group or individual creativity technique by which efforts are made to find a conclusion for a specific problem by gathering a list of ideas spontaneously contributed by its member(s), whereas Delphi method is a structured communication technique, originally developed as a systematic, interactive forecasting method which relies on a panel of experts. Although the two techniques can be

© Springer International Publishing AG 2017
M. Themistocleous and V. Morabito (Eds.): EMCIS 2017, LNBIP 299, pp. 646–654, 2017.
DOI: 10.1007/978-3-319-65930-5_50

combined together, they differ in terms of quantity of ideas generated, ease of implementation of the ideas generated and diversity of ideas generated. Indeed, brainstorming allows exploring a greater set of opportunities and initially focuses on quantity instead of quality, whereas the Delphi method relies on experts to obtain a smaller set of feasible options. Notwithstanding the growing research on management of crowdsourcing challenges and review/selection process [2], questioning the value of crowdsourcing compared to other methods for planning and decision making is still worth developing [3]. To this end, in this article we propose a model to assess the cost and effectiveness of the intersection between crowd and experts in decision making activities, with a focus on scenario planning, choosing the a public sector research site for its evaluation. The model encompasses the concept of *crowd crystal* as a small, delimited, and rigid group of agents, strictly delimited and of great constancy, which serve to precipitate the crowd [22]. Due to its nature, a crowd crystal may allow the resulting crowd to use, e.g., its ideas as example, thus enacting a process of refraction similar to the optic one usually associate to a *"prism"*, that is the second key concept making up our model.

The research presented in this article follows Design Science Research (DSR) method [4, 5] through a sequence of research activities that are building, evaluating, theorizing on and justifying artifacts [4]. The work presented in this paper concerns the definition of a model for assessing the intersection between crowd and experts in decision-making activities (*building*), and the discussion of the outcomes of an evaluation on scenario planning activities within the specific context of policy making in the public sector. Thus, although this research and the subsequent discussion are meant to represent the early stages of application of the DSR method, we claim that its outcomes will contribute to knowledge on crowdsourcing in decision making by improving the understanding of its value compared to the involvement by traditional experts. Accordingly, we argue the research contribution can be classified as "improvement", according to the DSR Knowledge Framework by Gregor and Hevner [6].

The paper is structured as it follows. Section 2 introduces background and motivations for this research. Section 3 discusses the model we propose together with its main constructs. Section 4 presents the site used for the evaluation of the model together with preliminary results. Section 5 concludes the paper by discussing its limitations and future directions of investigation.

2   Background and Motivations

Crowdsourcing is a type of online activity in which an individual, an institution, a non-profit organization, or company proposes to a group of individuals of varying knowledge, heterogeneity, and number, via a flexible open call, the voluntary undertaking of a task [7, 8]. Crowdsourcing has been usually considered suitable to provide different types of support to the decision making process, especially (i) in the *design phase*, through *idea generation* and *co-creation*, (ii) in the *choice phase*, through *voting*, as well as (iii) in the *intelligence phase* to explore or exploit information about the issue to be investigated [9]. A stream of research has explored the opportunity to perform brainstorming with a larger set of participants [10–12], by exploiting the notion of crowdsourcing. On the one hand, these techniques increase the participation of users (e.g., customers or citizens, depending on the nature of the context, which could be private or public) and allow increasing their commitment. On the other hand, the effort required to organize this type of brainstorming event is significantly greater. Thus, our research subject concerns the cost and the effectiveness of combining brainstorming, Delphi method, and crowdsourcing to perform scenario planning for decision-making, thus contributing to the current research streams investigating these specific applications of crowdsourcing [13–16].

Scenario planning helps decision-makers understanding changes in their external environment, spotting early warning signals and refining perceptions of existing or emerging problems and corresponding problem-solving strategies [17]. Moreover, considering, for example, policy-making, participatory scenario planning helps mobilizing action by different public and private actors, surfacing and managing conflicts between diverging societal interests and values, and finding common ground for future action, which in turn is a key essence of policy-making [18]. Hence, we recall the work of Volkery and Ribeiro [19], which listed the key success factors for scenarios planning and other future methodologies, among which three elements seems to be enhanced by crowdsourcing: (a) the skills of those carrying out the scenario exercise, (b) the level of involvement of the audience with the exercise, and (c) the skills of those using the scenario outputs ranked among the important factors as well. Afuah and Tucci [20] were among the early scholars in management posing the question on when crowdsourcing may be a better solution for solving problems than the alternatives (either by people internal to a company or by external experts), arguing that the choice depends on (i) the characteristics of the problem (low tacitness and complexity as well as high modularity), (ii) the knowledge required for the solution (high distance from a focal agent's knowledge as well as high tacitness and complexity of the required knowledge), (iii) the crowd *seriality* and goal orientation (for these issues see specifically the further elaboration by Viscusi and Tucci [21, 22]), (iv) the solution to be evaluated and evaluators, and (v) the information technology (IT) adopted. Yet, notwithstanding the high number of papers subsequently citing Afuah and Tucci [20], little when no empirical research on their initial question have been carried out, focusing instead on some of the above mentioned characteristics [23–26]. For example, Majchrzak and Malhotra [25] points out the lack in crowdsourcing innovation of collaborative discourse that leads to generative co-creation when knowledge differences are relevant in groups [27]. As to this issue, information systems have key a role, but according to Majchrzak and Malhotra [25] the requirements for crowdsourced co-creation (diversified engagement, sharing, volunteering) are not easily implementable because of three tensions, that are (i) simultaneously encouraging competition and collaboration, (ii) idea evolution takes time but crowd members spend little time, (iii) creative abrasion requires familiarity with collaborators; yet crowd consists of strangers.

Considering now specifically crowdsourcing for decision making and planning, a stream of literature has already categorized the types of uses of crowdsourcing for the specific case of policy making. For example, Prpić et al. [28] used virtual labor marketplaces such as Amazon Mechanical Turk and Crowdflower to collect feedback of participants over policy measures on environmental issues. Extending the idea of crowdsourcing to other types of citizen inclusion, Linders [29] has described the classification of services for citizen coproduction– whereby citizens perform the role of partner rather than customer in the delivery of public services- by using three over-arching categories: Citizen Sourcing, Government as a Platform, and Do-It-Yourself Government. Taking the above issues into account, we aim to contribute to prior work by questioning the value of crowdsourcing compared to other methods for planning and decision making further investigating the cost and effectiveness of the intersection between crowd and experts. This topic has received little attention since now, espe-cially in the Information Systems research; whereas, in the field of technology man-agement and innovation, contributions such as the one by Hienerth and Riar [3] have questioned the topic of evaluation, differentiating between expert evaluation and evaluations from crowds.

In the next Section we first provide details about the method, before discussing the proposed models and the research site chosen for its evaluation.

3 Model Description

In this Section we present (1) the first order constructs of our model, (2) the rela-tionships among the constructs, and (3) the predictions associated with each construct. In this study we assumes that the total cost of the process depends on: (a) the use of crowd *crystals* to increase the quality of ideas and (b) the use of refraction process to lower the cost of idea selection. A *crowd crystal* is a small, delimited, and rigid group of agents, strictly delimited and of great constancy, which serve to precipitate the crowd [30]. Actually, as pointed out by Viscusi and Tucci [22], a crowd crystal as a basic unit of crowd, but not yet an actual crowd, can lead to either an online com-munity, a closed crowd, or an open crowd. Due to its nature a crowd crystal may allow the resulting crowd to use, e.g., its ideas as example (*Refraction*), thus enacting a process of similar to the optic one usually associate to a "*prism*". Therefore, the possible values of our constructs are:

- Cost $_{\text{crowdsourcing}}$ = [0 = Low; 1 = Medium; 1 = High];
- Crystal = [0 = No; 1 = Yes];
- Refraction = [0 = No; 1 = Yes].

Following what stated in the previous paragraph, we obtain the following set of equations.

$$Equation\ 1 : Cost_{ideas\ selection} = Cost_{experts\ management} * (1 - Refraction_{effect} * Refraction) \quad (1)$$

$$Equation\ 2 : Cost_{participants\ selection} = Cost_{process\ management} * (1 - Prism_{effect} * Refraction * Crystal)$$
$$(2)$$

$$Equation\ 3 : Cost_{crowdsourcing} = Cost_{idea\ collection} + Cost_{idea\ selection} + Cost_{participants\ selection} \quad (3)$$

The predictions associated to each construct are illustrated in Table 1 for the types making up our typology of experts/crowd dynamics. The first row represents the standard solution, which relies only on crowd for ideas generation. The second row describes a solution where no expert (*Crystal*) is explicitly involved in the crowd, but experts are allowed to access the idea challenge before the rest of the crowd to let them add their ideas and allow the crowd to use their ideas as example (*Refraction*).

Table 1. Types of experts and crowd dynamics.

Type/process	Crystal	Refraction	Cost crowdsourcing
Crowd driven	No	No	High
Expert driven	No	Yes	Medium
Expert enabled	Yes	No	Medium
Prism driven	Yes	Yes	Low

The third row describes the case where decision makers can decide to use a subset of experts to increase the productivity of the crowd, but not to generate ideas on their own. Thus, the third row refers to the notion of crowd crystal whereas the fourth row shows predictions for the "prism" process, where experts (the *Crystal*) are allowed to express ideas and the other participants build on the ideas expressed by the previous crowd (*Refraction*).

4 Evaluation

In this Section we consider the case of decision-making in the public sector for evaluating the model presented in Sect. 4. Besides the opportunity to have access to the field there are research motivations, likewise. Actually, openness and inclusion are key challenges and target for public management and policy makers, especially since the growing diffusion of open government initiatives [31] and the renewed debate on public value among public policy and e-government scholars [32–34]. Quite surprisingly these efforts have been loosely connected to the contributions from the academic and managerial literature on open innovation [35]. Yet, looking at this latter field, public sector and societal issues have been only recently introduced as potential research areas of interest for studying open innovation [36]. Therefore, we are going to illustrate a process, which is expected to reduce the cost of an idea challenge for policy making. One conceptual framework to structure the processes of policy-making is the concept of the policy cycle [37], which breaks the policy-making process down into three phases:

(1) The policy maker recognizes that there is a problem and highlights the societal relevance of the problem, while underlining the need for a response from the political system. At this stage, the policy maker can decide to simply reach for a crowd or to use a subset of experts to increase the productivity of the crowd, defined as "*crowd crystal*", according to the model discussed in Sect. 3.

(2) The policy maker checks for the strengths and weaknesses of different problem-solving strategies make a final selection and formulate the concrete shape of the measure. At this stage, the policy maker can choose to use a set of experts

to select the best ideas, or to rely on the crowds to select the best ideas. Moreover, an extension of the notion of crowd crystal would be to allow the crowd crystal to access the idea challenge before the rest of the crowd to let them add their ideas and allow the crowd to use their ideas as example (according to the model discussed in Sect. 3, we call this process *"refraction"*, because it splits the challenge into two parts).

(3) The policy maker puts the measure into practical action and identifies the effects of the policy measure, while evaluating to which degree they deliver according to their objective, and, if necessary, re-design or terminate the measure, which would start a new cycle. Previous research has already pointed out that a subset of participants is more motivated by factors beyond the monetary compensation of open challenges. For example, building on the data collected during the challenge. gov experience, Desouza [38] points that a way to increase the awareness and interest in a challenge is to use an external panel of well-known judges that increases the visibility of the events and the notoriety of the winner. A possible extension of the notions of "crowd crystal" and "refraction" is the idea of adding the winner of an idea challenge to the subset of crowd crystal for the next competition. As said above, we call this process *"prism"* because it uses the crowd crystal to perform refraction.

To test our model we have performed a set of tests using a crowdsourcing platform to allow citizens to express their opinion about how the public and the private sector could improve their well-being. The total number of participants on the selected platform reaches some 10'000 contributors. In one test we asked to the crowd to list a set of future actions that could enable to improve the well-being at the university campus. Before launching the idea challenge, a subset of experts was contacted and they were allowed to post their ideas in advance. A second test asked the same type of question to a different crowd. Before launching the idea challenge, a subset of experts was contacted but they were not allowed to post their ideas in advance. A third test included a larger set of participants and did not use any subset of experts.

4.1   Discussion of the Results

Table 2 shows the results of our tests. In the first test we obtain a limited amount if ideas, but the number of ideas that were supported by a significant set of people was significantly greater than in the other tests.

Table 2. Results of the three tests

	Test 1	Test 2	Control study
Supported ideas	8.0	3.0	10.0
Low support (n < 3 votes)	16.0	3.0	50.0
No support (0 votes)	5.0	8.0	4.0
Average of votes per idea	1.8	1.6	2.9
Variance	2.4	5.6	14.9

Indeed, the analysis of the votes distribution in the second test shows that the ideas of the experts did not inspire the rest of the crowd and simply won more votes (see Fig. 1). Finally, the control test was the one with the larger set of ideas, which were not good enough to be retained.

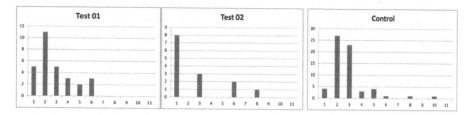

Fig. 1. Votes distributions of the three tests

5   Conclusion

In this paper we have investigated the cost and effectiveness of crowdsourcing for policy making, especially to perform scenario planning. To this end, we have proposed an assessment model and we have identified a typology of dynamics for intersecting experts and crowd together with a set of processes that may contribute to reduce the cost of crowdsourcing for policy making. Yet, our contribution lies in the idea that the crowd and experts can be combined to lower cost and improve generated value. However, it is worth noting here that while in this paper we have focused both the analysis and the experiment on policy making as research domain, the goal is to reach a broader understanding on the cost and effectiveness of the intersection between crowd and experts. Thus, the notion of prism is worth investigating also in business rather than in public sector alone to support managers and decision makers in the right configuration or "blend" of experts and "mobs". In summary, our design guidelines are derived from and extend existing theories and they have been empirically tested. Nonetheless, this is a preliminary study and in future work additional tests will be carried out on the domain considered in this paper as well as on other domains different from public sector to obtain a set of further design guidelines to reduce the cost of crowdsourcing for creative ideas.

References

1. Schoemaker, P.J.: Scenario planning: a tool for strategic thinking. Sloan Manag. Rev. **36**, 25 (1995)
2. Nagar, Y., De Boer, P., Garcia, A.C.B.: Accelerating the review of complex intellectual artifacts in crowdsourced innovation challenges. In: Thirty Seventh International Conference on Information Systems (ICIS2016), Dublin, Ireland, 11–14 December 2016

3. Hienerth, C., Riar, F.: The wisdom of the crowd vs. expert evaluation: a conceptualization of evaluation validity. In: 35th DRUID Celebration Conference, Barcelona, Spain, pp. 17–19, June 2013

4. Hevner, A.R., March, S.T., Park, J., Ram, S.: Design science in information systems research. MIS Q. **28**, 75–105 (2004)

5. March, S., Smith, G.: Design and natural science research on information technology. Decis. Support Syst. **15**(4), 251–266 (1995)

6. Gregor, S., Hevner, A.R.: Positioning and presenting design science research for maximum impact. MIS Q. **37**, 337–355 (2013)

7. Howe, J.: The rise of crowdsourcing. Wired Mag. **14**(6), 1–4 (2006)

8. Estellés-arolas, E., González-ladrón-de-guevara, F.: Towards an integrated crowdsourcing definition. J. Inf. Sci. **38**, 1–14 (2012)

9. Chiu, C.M., Liang, T.P., Turban, E.: What can crowdsourcing do for decision support? Decis. Support Syst. **65**, 40–49 (2014)

10. Vaculin, R., Hull, R., Vukovic, M., Heath, T., Mills, N., Sun, Y.: Supporting collaborative decision processes (2013)

11. AlShehry, M.A., Ferguson, B.W.: A taxonomy of crowdsourcing campaigns. In: Proceedings of the 24th International Conference on World Wide Web, pp. 475–479. ACM, New York (2015)

12. Blohm, I., Leimeister, J.M., Krcmar, H.: Crowdsourcing: how to benefit from (too) many great ideas. MIS Q. Exec. **12**, 199–211 (2013)

13. Kaivo-oja, J., Santonen, T., Myllylä, Y.: The crowdsourcing delphi: combining the delphi methodology and crowdsourcing techniques. In: ISPIM Conference Proceedings, the International Society for Professional Innovation Management (ISPIM), p. 1 (2013)

14. Flostrand, A.: Finding the future: crowdsourcing versus the delphi technique. Bus. Horiz. **60**, 229–236 (2017)

15. Hiltunen, E.: Crowdsourcing the future: the foresight process at finpro. J. Futur. Stud. **16**, 189–196 (2011)

16. Halman, A.: Before and beyond anticipatory intelligence: assessing the potential for crowdsourcing and intelligence studies. J. Strateg. Secur. **8**, 15–24 (2015)

17. Lempert, R.J.: Shaping the Next One Hundred Years: New Methods for Quantitative, Long-Term policy Analysis. Rand Corporation, Santa Monica (2003)

18. Eriksson, E.A., Weber, K.M.: Adaptive foresight: navigating the complex landscape of policy strategies. Technol. Forecast. Soc. Change **75**, 462–482 (2008)

19. Volkery, A., Ribeiro, T.: Scenario planning in public policy: understanding use, impacts and the role of institutional context factors. Technol. Forecast. Soc. Change **76**, 1198–1207 (2009)

20. Afuah, A., Tucci, C.: Crowdsourcing as a solution to distant search. Acad. Manag. Rev. **37**, 355–375 (2012)

21. Tucci, C., Viscusi, G., Gasparetto, F.: Distinguishing crowd dynamics in small teams: a crowdsourcing exercise in higher education. In: Collective Intelligence 2016, New York (2016)

22. Viscusi, G., Tucci, C.: Three's a crowd? In: Tucci, C., Afuah, A., Viscusi, G. (eds.) Creating and Capturing Value Through Crowdsourcing. Oxford University Press, Oxford (2017)

23. Alexy, O., George, G., Salter, A.J.: Cui Bono? The selective revealing of knowledge and its implications for innovative activity. Acad. Manag. Rev. **38**, 270–291 (2013)

24. Henkel, J., Schöberl, S., Alexy, O.: The emergence of openness: how and why firms adopt selective revealing in open innovation. Res. Policy **43**, 879–890 (2014)

25. Majchrzak, A., Malhotra, A.: Towards an information systems perspective and research agenda on crowdsourcing for innovation. J. Strateg. Inf. Syst. **22**, 257–268 (2013)

26. Prpić, J., Shukla, P.P., Kietzmann, J.H., McCarthy, I.P.: How to work a crowd: developing crowd capital through crowdsourcing. Bus. Horiz. **58**, 77–85 (2015)
27. Majchrzak, A., More, P.H.B., Faraj, S.: Transcending knowledge differences in cross-functional teams. Organ. Sci. **23**, 951–970 (2012)
28. Prpić, J., Taeihagh, A., Melton, J.: Experiments on crowdsourcing policy assessment. Oxford Internet Institute, University of Oxford-IPP (2014)
29. Linders, D.: From e-government to we-government: defining a typology for citizen coproduction in the age of social media. Gov. Inf. Q. **29**, 446–454 (2012)
30. Canetti, E.: Crowds and Power. Continuum, New York (1962)
31. Misuraca, G., Viscusi, G.: Is open data enough? E-governance challenges for open government. Int. J. Electron. Gov. Res. **10**, 19–36 (2014)
32. Cordella, A., Bonina, C.M.: A public value perspective for ICT enabled public sector reforms: a theoretical reflection. Gov. Inf. Q. **29**, 512–520 (2012)
33. Harrison, T.M., Guerrero, S., Burke, G.B., Cook, M., Cresswell, A., Helbig, N., Hrdinova, J., Pardo, T.: open government and e-government democratic challenges from a public value perspective. Inf. Polity Int. J. Gov. Democr. Inf. Age **17**, 83–97 (2012)
34. Alford, J.: Public value from co-production by clients. In: Benington, J., Moore, M.H. (eds.) Public Value - Theory and Practice. Palgrave Macmillan, Basingstoke (2011)
35. Viscusi, G., Poulin, D., Tucci, C.: Open innovation research and e-government: clarifying the connections between two fields. In: XII Conference of the Italian Chapter of AIS (itAIS2015). Luiss University Press, Roma (2015)
36. Chesbrough, H., Bogers, M.: Explicating open innovation: clarifying an emerging paradigm for understanding innovation. In: Chesbrough, H., Vanhaverbeke, W., West, J. (eds.) New Frontiers in Open Innovation, pp. 3–28. Oxford University Press, Oxford (2014)
37. Howlett, M., Ramesh, M., Perl, A.: Studying Public Policy: Policy Cycles and Policy Subsystems. Cambridge Univ Press, Cambridge (1995)
38. Desouza, K.: Challenge. gov: using competitions and awards to spur innovation. In: IBM Center for the Business of Government (2012). Accessed 17 Oct 2012

Automation Process for Co-evolution of Enterprise Architecture Meta-Models and Models

Nuno Silva$^{(\boxtimes)}$, Tiago Rechau, Miguel Mira da Silva, and Pedro Sousa

Instituto Superior Tecnico, Av. Rovisco Pais 1, 1049-001 Lisboa, Portugal
{nuno.miguel,tiago.rechau,mms,pedro.manuel.sousa}@tecnico.ulisboa.pt

Abstract. Maintaining up-to-date and accurate enterprise architecture (EA) models is a non-trivial task due to their size and complexity. This becomes more challenging when the meta-model governing these models changes. Current EA maintenance processes are executed manually with little automation, resulting in a time consuming and error prone task. This paper presents a (semi-)automated process for co-evolving EA meta-models and EA models by gathering information from both practitioners and existing literature. This work is one of the steps in the direction of minimizing manual effort for EAM by automation and eliminating modeling errors.

Keywords: Enterprise architecture · Co-evolution · Process · Meta-model · Model

1 Introduction

The practice of Enterprise Architecture Management (EAM) in mid-sized to large organizations aims at capturing the relationships between the elements portraying both the business and the supporting IT-landscape [1]. By documenting the current status of this network of dependencies, the alignment of business and IT can be analyzed and transformed to an architecture that optimally supports business strategy.

In the practice of EA, the corresponding models can grow very large and expose complex relationships between architecture elements. Also, these large and complex models continuously need to be aligned with the real-world enterprise which they represent in order to be most effective [1]. Therefore, maintaining these models is confirmed by EA practitioners as one of the main challenges of EAM in practice [2] since it implies a large effort in synchronizing the model with the organization's reality.

Another aspect that contributes to the difficulty in maintaining EA models is the built-in complexity of EA modeling languages and underlying meta-models. A straightforward adoption of EA modeling languages is most of the times non-trivial since behind each language there is a complex theory defining the structure and semantics of the architectural elements, as well as a specific notation which often raises communication and understanding issues.

© Springer International Publishing AG 2017
M. Themistocleous and V. Morabito (Eds.): EMCIS 2017, LNBIP 299, pp. 655–661, 2017.
DOI: 10.1007/978-3-319-65930-5_51

The necessity of evolving the EA meta-model is perceived as a real issue when applying EAM initiatives. Modeling languages themselves are also subject to change. For example, one can consider the scenario of changing from ArchiMate 2.1 to ArchiMate 3.0. Another scenario can be an organization using parts of existing modeling languages while applying modifications to the meta-model according to their specific needs. A third scenario would be an organization wanting to develop their own meta-model from scratch, enriching the meta-model as needed. Therefore, not only models need to adapt to the organization's needs but also the meta-model governing the rules in which these models are built.

The common practice in updating EA models resumes to a collection of EA information via stakeholder interviews which are then manually entered into an EAM tool [1]. As stated by Farwick et al. *this is a time-consuming manual task, that can only be executed by EA specialists* [3] besides being error-prone when the complexity of changes rises.

In our on-going research, we are working towards (semi-)automated procedures that enable the co-evolution of the EA meta-model and EA model, thus eliminating manual modeling errors and reducing overall manual modeling effort. This paper presents a (semi-)automated process that addresses the co-evolution of both the EA meta-model and conforming EA model. The aim of this process description is to provide the basis for a future technical implementation.

The remainder of this paper is structured as follows. In the next section, we highlight related work on the topic of enterprise architecture maintenance. Section 3 describes the automation process for co-evolving the EA meta-model and EA model. Finally, in Sect. 4 we summarize our approach and point out the future direction of our research.

2   Related Work

EA model maintenance has been addressed in literature both from practice [4–7] and research [1–3,8–15]. Nonetheless, most of the work done in this area falls short of giving concrete approaches and strategies on how to address EA meta-model and EA model co-evolution.

TOGAF's Architecture Development Method (ADM) [4] and other EA frameworks only states that some kind of "monitoring" should be applied to keep the models up-to-date, although no concrete description of such monitoring mechanisms is given nor their implications in practice. Both the Department of Defense Architecture Framework [7] and the Federal Enterprise Architecture Framework [5] are similarly short in giving concrete advice for data collection and maintenance.

On the side of research, Dam et al. emphasized EA model maintenance from a change propagation perspective in which, given a set of primary changes made to the EA model, secondary changes are necessary in order to maintain consistency across multiple levels of the EA [8]. These secondary changes may then lead to further changes. In a large modern organizational structure consisting of many elements, resources, business processes, and infrastructures with complex relationships becomes costly and labor intensive to correctly maintain the EA model. Hence, Dam et al. proposed a (semi-)automatic mechanism of change propagation based on inconsistency management to provide a method for generating repair options from Alloy consistency rules [8]. This propagation mechanism raises the question of whether or not the meta-model should be perceived as an immutable artefact in which the EA model is to be validated against. As argued in Sect. 1 this is not always the case.

Farwick et al. have specified a list of requirements for automated EA model maintenance based on literature review, a survey among EA practitioners and the authors own experience in the field [2]. The outcome of such requirement specification suggested a computer-aided EA maintenance process that would implement these requirements. Such process was proposed by Farwick et al. [1] supported by an EA repository that not only stores, collects, and consolidates EA data, but also has the capability of supporting the data maintenance process with task lists, and a corresponding role system [1].

Despite the body of research addressing the need for a well-defined and (semi-)automated EA maintenance process, no publication considers the changeability of both the EA meta-model and EA model, only focusing on the later. We argue that this vision is overall simplistic, therefore reflecting only a partial reality of the EA maintenance process. The process of maintaining EA should and must regard all architectural description artefacts (EA meta-model, EA model, EA viewpoints, and EA views) as potentially changeable and not only the EA model.

3   The Co-evolution Automation Process

In this section, we present a description of a (semi-)automated co-evolution process. Note the process needs to be understood in the context of a process engine running on top of an EA repository. This means that the EA repository provides both EA meta-data and EA data as input to the process engine. Moreover, the following aspects were considered throughout the process design:

- element dependencies between the meta-model elements, which induces a change propagation through the graph structure representing the meta-model;
- conformance checking of the model's syntactic well-formedness with the new version of the meta-model; followed by
- a model change propagation in order to enforce meta-model conformance.

Hence, the engine should execute the process by applying conformance checking and change propagation mechanisms to both the EA meta-model and EA model with minimal user intervention.

3.1   Process Design Requirements

In addition to the work of [1,8], we conducted a set of semi-structured interviews among EA practitioners to identify relevant aspects that needed to be addressed regarding the meta-model and model co-evolution. It is important to note that, conversely to [1–3], the focus of this research does not consider EA data integration in a federated approach, thus, no data integration from various data sources is expected. The co-evolution of both the EA meta-model and EA model is directly applied to the EA repository. Nevertheless, an *a priori* execution of federated mechanisms does not invalidate *a posteriori* execution of the co-evolution process.

Seven EA practitioners (two EA consultants with fifteen and seven years of experience, four EA researchers with four, three, one, and five years of experience, and one architect with three years of experience) from one private and two public organizations were interviewed. The purpose of these interviews was to identify and assess the necessary requirements that should be initially considered in the co-evolution process design. Therefore, data were collected regarding the description of EA maintenance steps performed by practitioners, mistakes often made during the maintenance process and respective frequency, and the average time spent in changing both the meta-model and model.

After analyzing all data from the interviews, the following requirements were defined:

- The co-evolution process must allow meta-model edition based on a default changes catalog;
- The co-evolution process must allow the user to choose whether or not to execute a change before propagating it to both the meta-model and model in the EA repository;
- The co-evolution process must provide indicators of the degree of impact (like propagation cost, modified model elements, etc.) regarding a specific change.

Based on these requirements and on the change propagation framework introduced by Dam et al., the (semi-)automated process for co-evolution of the EA meta-model and EA model was defined, as shown in Fig. 1.

3.2   Process Activities

The co-evolution process is composed of five activities, with two requiring user intervention. In this section, we detail each activity according to the inputs, outputs, and the mechanisms involved in each one.

Select Change. The *Select Change* activity is the starting activity of the co-evolution process. In this step, the user chooses from an existing *Change Catalog* a change to be performed to the EA meta-model. We categorized each change as one of three types as in [14]:

- *Co-construction* changes are changes that add structure to the EA meta-model. This type of changes does not alter the conforming model since they add expressiveness to the EA modeling language of choice;

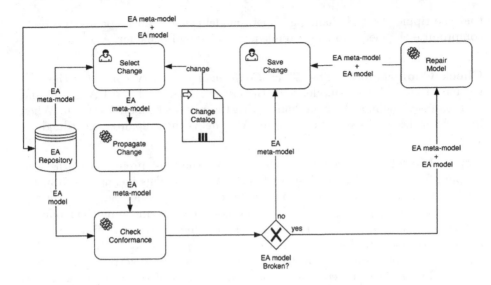

Fig. 1. The (semi-)automated co-evolution process

- *Co-deletion* changes diminish the modeling language's expressiveness by reducing the meta-model's overall structure through element deletion. This type of change can usually compromise the model's well-formedness and respective compliance with the new version of the meta-model. Thus, model repair mechanisms are required to preserve well-formedness;
- *Co-refactoring* changes alter the state of the meta-model elements, meaning they change the defining attributes of the element. This type of change can also cause model inconsistencies.

Propagate Change. The *Propagate Change* activity supports propagation of the effects of changing a meta-model element to other elements that are strongly connected to it, i.e., elements having a single connection to elements they depend on. This step is mandatory only when performing co-deletion type changes, the reason is that a reduction in the meta-model's structure can compromise its expected behavior. For example, assuming a meta-model with concepts A, B, and C. A is related to B (uses relation type), and B to C (uses relation type). By deleting C and not having further elements from which B depends on, the intended behavior of B becomes compromised.

Essentially, the concepts and relations expressing an enterprise network are inter-dependable on one another. If the only dependency between two concepts is erased due to a co-deletion type change to one of the concepts, the remaining concept loses its intended purpose for representing both the structure and behavior of the enterprise. This effect propagates through the entire network of meta-model concepts. Therefore, this step in the co-evolution process ensures that the ripple effect of changing a meta-model element does not unexpectedly compromise the meta-model's intended structure and behavior.

Check Conformance. The *Check Conformance* activity compares the new version of the EA meta-model with the current version of the EA model and then assesses the model's compliance with the new meta-model version. When inconsistencies are found (if any), the EA model is then repaired accordingly.

Repair Model. All three types of changes have the "co-" prefix due to the high coupling between the EA meta-model and EA model. This means that a meta-model change can have repercussions on its conforming model. So, when the EA model becomes broken due to an EA meta-model change, meaning the model no longer conforms to the new version of the meta-model, the model needs to be repaired to ensure well-formedness. Hence, the *Repair Model* activity describes a propagation procedure similar to the *Change Propagation* activity but focused on the EA model while considering the new version of the meta-model, thus guaranteeing meta-model conformance.

Save Change. In this last activity (*Save Change*) of the co-evolution process, the user is asked whether or not to save the change to the EA repository once the EA meta-model and EA model are properly updated. The change and respective propagations made to both the EA meta-model and EA model are discarded if the user opts for not saving them. The user can then choose another change from the *Change Catalog* to be executed.

4   Conclusion

To build and maintain an EA model that reflects the current situation of an enterprise while guaranteeing well-formedness and compliance with the EA meta-model is a challenge that has been identified by both researchers and practitioners. In our research effort on automating the co-evolution of the EA meta-model and conforming EA model, we aim at simplifying the process of maintaining both the meta-model and model by specifying an automation process that reduces manual modeling effort and eliminates manual modeling errors.

The (semi-)automated process involves minimal user interaction. The user is expected only to select a change and to save it after both the EA meta-model and EA model are updated accordingly. Hence, we aim at fully automating the ripple effect of a change to the EA meta-model, the conformance checking between the EA meta-model and the EA model, and consequent EA model repairs to ensure meta-model conformance.

As future efforts in this direction, we will define change propagation mechanisms to be applied to the meta-model and model when well-formedness is compromised. Finally, we will implement the engine supporting the co-evolution process as an extension of a proprietary EAM tool[1]. Validation of both the process and mechanisms will be done in a public organization once the prototype is complete.

References

1. Farwick, M., Agreiter, B., Breu, R., Ryll, S., Voges, K., Hanschke, I.: Automation processes for enterprise architecture management. In: 15th International Enterprise Distributed Object Computing Conference Workshops (EDOCW), pp. 340–349. IEEE (2011)
2. Farwick, M., Agreiter, B., Breu, R., Ryll, S., Voges, K., Hanschke, I.: Requirements for automated enterprise architecture model maintenance. In: 13th International Conference on Enterprise Information Systems (ICEIS) (2011)
3. Farwick, M., Pasquazzo, W., Breu, R., Schweda, C.M., Voges, K., Hanschke, I.: A meta-model for automated enterprise architecture model maintenance. In: 16th International Enterprise Distributed Object Computing Conference (EDOC), pp. 1–10. IEEE (2012)
4. Haren, V.: TOGAF Version 9.1. Van Haren Publishing, Zaltbommel (2011)
5. CIO Council: Federal enterprise architecture framework version 2 (2013). https://obamawhitehouse.archives.gov/sites/default/files/omb/assets/egov_docs/fea_v2.pdf. Accessed 26 May 2017
6. Hanschke, I.: Strategic IT Management: A Toolkit for Enterprise Architecture Management. Springer Science & Business Media, Heidelberg (2009)
7. United States Department of Defense: The DODAF architecture framework version 2.02 (2010)
8. Dam, H.K., Le, L.S., Ghose, A.: Supporting change propagation in the evolution of enterprise architectures. In: 14th International Enterprise Distributed Object Computing Conference (EDOC), pp. 24–33. IEEE (2010)
9. Buckl, S., Matthes, F., Schweda, C.M.: Future research topics in enterprise architecture management – a knowledge management perspective. In: Dan, A., Gittler, F., Toumani, F. (eds.) ICSOC/ServiceWave 2009. LNCS, vol. 6275, pp. 1–11. Springer, Heidelberg (2010). doi:10.1007/978-3-642-16132-2_1
10. Fischer, R., Aier, S., Winter, R.: A federated approach to enterprise architecture model maintenance. Enterp. Model. Inf. Syst. Archit. (EMISA) 2(2), 14–22 (2015)
11. Aier, S., Kurpjuweit, S., Saat, J., Winter, R.: Enterprise architecture design as an engineering discipline. AIS Trans. Enterp. Syst. 1(1), 36–43 (2009)
12. Kaisler, S.H., Armour, F., Valivullah, M.: Enterprise architecting: critical problems. In: Proceedings of the 38th Annual Hawaii International Conference on System Sciences (HICSS), pp. 224b–224b. IEEE (2005)
13. Buckl, S., Matthes, F., Neubert, C., Schweda, C.M.: A lightweight approach to enterprise architecture modeling and documentation. In: Soffer, P., Proper, E. (eds.) CAiSE Forum 2010. LNBIP, vol. 72, pp. 136–149. Springer, Heidelberg (2011). doi:10.1007/978-3-642-17722-4_10
14. Silva, N., Ferreira, F., Sousa, P., da Silva, M.M.: Automating the migration of enterprise architecture models. Int. J. Inf. Syst. Model. Des. (IJISMD) 7(2), 72–90 (2016)
15. Silva, N., Mira da Silva, M., Sousa, P.: Modelling the evolution of enterprise architectures using ontologies. In: 19th Conference on Business Informatics (CBI). IEEE (2017)

---

[1] http://www.linkconsulting.com/eams/.

Collaborative Filtering for Producing Recommendations in the Retail Sector

Dimitrios Poulopoulos$^{(\boxtimes)}$ and Dimosthenis Kyriazis

University of Piraeus, Piraeus, Greece
{james, dimos}@unipi.gr

Abstract. Recommender Systems exploit implicit or explicit user feedback, to create recommendations and provide a personalized user experience. In the case of explicit feedback datasets, the system directly collects the user opinion. On the other hand, to compile implicit feedback datasets the system works passively in the background, tracking different sorts of user behavior, such as browsing activity, watching habits or purchase history. In this work, we focus on implicit feedback recommendation systems. We analyze their unique characteristics and identify their differences to the much more extensively researched explicit feedback systems.

Keywords: Collaborative filtering · Recommender systems · Implicit feedback

1 Introduction

Anderson, in his 2004 article entitled "The Long Tail", said that we are leaving the age of information and entering the age of recommendation [1]. Unless we have a way to filter the information overload that we absorb every day and retain only what is important to us, data reduce to noise. A recommender or recommendation system, aims to analyze the patterns and dynamics of users' behavior, his preferences or dislikes, and utterly provide a personalized experience.

Recommender systems are divided into two categories. Content based approaches aspire to create a profile for each user or product, to capture their character. This information is used to match users and items in a meaningful manner. However, getting this information is not always an easy task.

On the other hand, Collaborative Filtering techniques, a term coined by the developers of the first Recommender System, Tapestry [2], are based on past user behavior or feedback, without requiring previously known user or item features. Collaborative Filtering techniques often produce more accurate results than content-based strategies, while also being more scalable. Be that as it may, they suffer from the cold-start problem, which appears when a new user or item enters the process, or when there are not enough data, e.g. an item has been rated by as small number of users.

Recommender Systems are further divided by the kind of input they depend on. They rely on two distinct types of user feedback; Explicit and implicit. Explicit feedback systems directly collect every user's opinion. For example, Netflix uses a five-star rating system to capture the movie preferences of a user. In the case of implicit

© Springer International Publishing AG 2017
M. Themistocleous and V. Morabito (Eds.): EMCIS 2017, LNBIP 299, pp. 662–669, 2017.
DOI: 10.1007/978-3-319-65930-5_52

feedback, the system works passively in the background, tracking different sorts of user behavior, such as browsing activity, watching habits or purchase history [3]. For instance, sources of implicit feedback that have been explored in the past are time spent reading [4] or URL references in Usenet postings [5].

A direct implication of the latter approach, though, is that we do not have any direct indication of a user's preferences and, specifically, we do not have any considerable evidence on which items they dislike. Thus, implicit feedback recommender systems express confidence instead of preference [6].

The aim of this work is to analyze the unique properties of implicit feedback datasets, especially tailored to fit the retail sector, confronting a vast variety of products, and problems that emerge due to high variance between item prices.

2 Related Work

2.1 Neighborhood Models

Neighborhood models establish the foundations and present the most common approach to collaborative filtering and recommendation systems. They are divided into two sub-categories; user-oriented and item-oriented.

The first approach to neighborhood models is that of user-oriented. This strategy aims to predict the missing ratings, by identifying similar minded users. A good analysis of this method is provided by Herlocker et al. [7].

On the other hand, item-oriented approaches strive to group related items, and fill in the missing ratings by looking at how the same user rates those items [8, 9]. Item-oriented approaches produce more accurate results, while being more scalable [9–11]. In addition, researchers and developers can explain the recommendations clearly and reason about the results in a more precise manner. Thus, the latter approach has become the prominent method for neighborhood models.

As we mentioned before, in implicit feedback systems, the constructed rating, whatever it may be, indicates confidence and not preference. The inability to distinguish between user preference and the confidence we have on it, undermines the performance of item-oriented approaches in such cases.

2.2 Latent Factor Models

Latent factor models aim to uncover hidden or unknown features that could explain the relationships between users and items. Examples of this approach include pLSA [12], neural networks [13], and Latent Dirichlet Allocation [14]. This approach associates every user with a weight vector $x_u \in \mathbb{R}^f$ and each item with a feature vector $y_i \in \mathbb{R}^f$. The weight vector x_u captures the preferences of each user on the corresponding dimension. The feature vector y_i quantifies the value of every item feature, across the dimension space. The prediction is calculated by taking the dot product of those vectors, i.e. $\hat{r}_{ui} = x_u^T y_i$. Thus, the critical part is estimating the parameters. In explicit feedback systems, the model uses only the observed items, e.g. the items that have a rating. Regularization is used, to avoid overfitting.

$$\min_{x_*, y_*} \sum_{r_{\{ui\}} \in K} \left(r_{ui} - x_u^T y_i \right)^2 + \lambda \left(\|x_u\|^2 + \|y_i\|^2 \right)$$

Here, λ is the regularization parameter, and K is the set of (u, i) pairs for which the rating r_{ui} is known. Stochastic gradient descent is usually applied to learn the parameters [11, 15, 16], and the results reported on the Netflix dataset are by far superior to those of Neighborhood models.

Implicit feedback recommendation systems, report impressive results using a Latent Factor Models approach, but they do have some unique characteristics [6]:

1. There is no negative feedback. In explicit feedback datasets, users can directly express their preferences or dislikes. In the case of implicit feedback, we can only infer the user's opinion by looking at attributes like frequency, that indicate how many times a user has interacted with an item. However, we cannot assume that a user dislikes an item if there is no interaction between them, as he might have not been aware of the existence of the product. This has some ramifications. Implicit feedback systems cannot focus only on known interactions. This would produce distorted user profiles, as user-item interactions tend to capture mainly the positive opinions. Hence, it is crucial to consider the empty, unexplored space between users and items as well.
2. Implicit feedback is inherently noisy. Even if a user interacts with an item, we can only assume that this interaction indicates preference. For example, the consumer could have bought an item as a gift. Thus, we treat the numerical value associated with the feedback as a confidence indication instead of preference. If a user has interacted with an item multiple times, e.g. a consumer purchases the same brand in a super market, then we have high confidence that this user prefers this item. This is contrary to explicit feedback, where ratings are directly treated as preference.
3. In implicit feedback recommendation systems, we should consider aspects like the availability of an item, temporal dynamics or competition between items, when we are interpreting and evaluating the results. Having explicit feedback, facilitates the process of evaluation, because every point in the dataset is a clear indication of user preferences.

3 Proposed Approach

In this section, we define the model used in this work. The goal is to build a recommendation system, tailored to fit the needs, and overcome the challenges of the retail sector. In most cases, the retail sector cannot benefit from the use of explicit user feedback. This is because the product catalogue count is in the thousands, and even if the retailer asks users about their opinion, it cannot base its findings on such sparse data.

This means that the users have not expressed any preference for the products they buy, and moreover, we cannot assume anything about the products they have not interacted with yet. Thus, we should work based on implicit feedback, that we extract observing the consumers' behavior.

3.1 Preliminaries

We are considering a set of users X, which consists of user vectors $x_1, \ldots, x_u$, where every vector x_u characterizes the user's u preferences. In other words, each vector captures the weights, which define the preference of each user for a specific item feature. Thus, the user vectors' shape is $D \times 1$, where D is the number of dimensions, i.e. the number of product features. Similarly, the set of products Y consists of item vectors $y_1, \ldots, y_2$, where every vector characterizes the product's features. We consider a third matrix R, which describes the user-item interactions. Thus, r_{ui} is numerical value, that records the interaction between user u and item i.

3.2 The Model

First, for implicit feedback systems, we need to formalize the notion of confidence. In our case, the numerical values that are considered as implicit user feedback, count how many times a user has interacted (e.g. purchased) with a specific item. This indicates our confidence, that this user most likely prefers this item. We should associate low levels of preference with low confidence and work from the bottom up. This does not mean that low levels of confidence indicate dislike. We are just trying to capture the various levels of uncertainty, that exist in the absence of data. Finally, the main notion is that as the numerical value of $r_{u,i}$ raises, the levels of confidence become stronger. Due to the high variance of the values of r_{ui}, we use the following method to capture confidence:

$$c_{ui} = 1 + \alpha \log(1 + r_{ui}/\varepsilon)$$

This method transforms the raw observation r_{ui} into a confidence value c_{ui}. The number ε is just a constant, that can be tuned as a hyperparameter.

To use these confidence levels to our advantage, we need to introduce a new, binary variable, that captures whether a user has interacted with an item or not. Thus, this variable takes the value of zero if there is no interaction and one otherwise:

$$p_{ui} = \begin{cases} 1 \text{ if } r_{u,i} > 0 \\ 0 \text{ if } r_{u,i} = 0 \end{cases}$$

Now, we can assume that if the user u has interacted with the item i, this user has shown a preference for this specific item. On the other hand, if the user has not yet interacted with this item, we have no indication ($p_{ui} = 0$).

Finally, we see that the raw, implicit observations r_{ui}, are now expressed through two, newly created variables, c_{ui} that expresses confidence, and p_{ui} that captures preference. Now, we can associate every user-item interaction with a confidence level, even if there is no interaction at all. In this case, the confidence measure takes its lower value, which is equal to one. The goal is to find the vectors x_u and y_i, in R^D, for every user and item, where D is the number of item features and, consequently, user weights.

Finally, our predictions are given by the dot product of these vectors:

$$\hat{p}_{ui} = x_u^T \cdot y_i$$

The vectors x_u and y_i are called user-factors and item-factors respectively. Thus, the technique we are using is analogous to the matrix factorization techniques used in latent factor models, though there are a couple of differences:

1. We have introduced the notion of confidence to our model
2. We need to take into consideration every data point in our dataset.

Finally, the loss function of the model is shown below, where $\lambda(||x_u||^2 + ||y_i||^2)$ is performing the necessary regularization and λ is the regularization parameter:

$$\min_{x_*,y_*} \sum_{r_{u,i} \text{ is known}} c_{u,i} \left(r_{u,i} - x_u^T y_i \right)^2 + \lambda \left(||x_u||^2 + ||y_i||^2 \right)$$

Solving for x_u and y_i, as we would in the case of a linear model such as linear regression, we have the two equations that calculate the user-factors and item-factors:

$$x_u = \left(Y^T C^u Y + \lambda I \right)^{-1} Y^T C^u p(u)$$
$$y_i = \left(X^T C^i X + \lambda I \right)^{-1} X^T C^i p(i)$$

4   Experimental Study

4.1   Data Description

The dataset consists of customer and product IDs, to define user-item interactions, as well as the quantity purchased and unit price for each item, which was used in the compilation of implicit "ratings". Specifically, the implicit feedback we calculate for any user-item interaction is given as:

$$r_{ui} = quantity * unit\ price$$

Thus, the r_{ui} is a real value, with the number zero indicating no interaction between user u and item i. This implies that the more a user interacts with an item, the more confident we are that this user prefers this item. Furthermore, we should account for the unit price of this item, because items that are priced lower tend to be purchased more. Thus, if the item is for example a couch that has been brought only once, the high price of this item should increase our confidence levels, because the user chose to spend a lot of money on that item. The outcome of the product that computes the r_{ui} value is usually a substantial number. Therefore, we reevaluate our confidence levels in a logarithmic scale, to account for the vast variance between them.

To evaluate the results, a test dataset was used. In this dataset, though, the values for r_{ui} are either one or zero, capturing what is known as the ground truth; a user has either interacted with an item or not.

In the training dataset, a small proportion of interactions has been masked. That means that the real value has been replaced by zero. This is the prominent way to test the algorithms performance in the end.

4.2 Methodology

First, we generate a sorted list of products for each user, using the proposed model, with the top item being the most preferable one. As a base metric, we construct a popularity-based recommendation system, which, for each user, recommends the most popular items that they have not already been purchased. For instance, if the popularity-based recommendation system recommends the top-k most popular items, and the user has already purchased an item in this set, the recommendation system replaces this item with the item in position $k + 1$, and sorts the set again.

In the next step, we compare the recommendations produced by out model against the most popular items. To achieve this, we used the area under the ROC curve (AUC) score, a usual metric for binary classifiers, as out test set consists of zero-one values. We evaluate the results of our model on the user-item interactions that have been masked during the test dataset construction step.

4.3 Results

We used various number of factors, to run our model and we got various results. In every case, our model surpassed the popularity-based recommendation system. For best results, we concentrated on the model with 30 factors, after 100 iterations. We empirically found those hyper parameters to work best, and used a $\lambda = 0.3$ as the regularization factor and an $\alpha = 40$ for building up our confidence levels. We achieved a result of 0.878 AUC score, compared to 0.815 accomplished by the popularity baseline.

5 Conclusion

In this work, we presented collaborative filtering on implicit feedback datasets. We presented notable previous applications on the subject and separated the case of implicit vs explicit feedback datasets. We proposed a model for recommendations, using collaborative filtering for implicit feedback datasets, and we studied the case of retail specifically.

Next steps include ways to improve the accuracy and performance of the algorithm. When we are talking about accuracy, there are a few elements of the algorithm that should be revisited. The way that the confidence matrix is created should consider the high variance of prices in a retail situation. Another part of that process is the evaluation and analysis of the zero values in the user-item interaction matrix. Since those values fill most that matrix, we need to analyze and reevaluate their properties. One simple example is how to treat those zeros when an item is out of stock or if the user selects a competitor.

A major problem of collaborative filtering algorithms is that of high sparsity. The user-item interaction matrix consists mainly of zeros, so, if the sparsity is high, the algorithm yields poor results. Thus, we should consider methods to alleviate this problem.

On the side of the performance, this algorithm will be transformed, to be able to work in parallel and scale linearly. Moreover, part of the ongoing research is to make this algorithm work iteratively, and be able to learn using only delta differences that occur, instead of going through the whole dataset with every change.

References

1. Anderson, C.: The Long Tail: Why the Future of Business is Selling Less of More. Hyperion, New York City (2006)
2. Goldberg, D., Nichols, D., Oki, B.M., Douglas, T.: Using collaborative filtering to weave an information tapestry. Commun. ACM **35**(12), 61–70 (1992)
3. Oard, D.W., Kim, J.: Implicit feedback for recommender systems. In: Proceedings on 5th DELOS Workshop on Filtering and Collaborative Filtering, pp. 31–36 (1998)
4. Morita, M., Shinoda, Y.: Information filtering based on user behavior analysis and best match text retrieval. In: Proceedings on 17th ACM SIGIR Conference on Research and Development in Information Retrieval (1994)
5. Terveen, L., Hill, W., Amento, B., McDonald, D., Creter, J.: PHOAKS: a system for sharing recommendations. Commun. ACM **40**, 59–62 (1997)
6. Yifan, H., Yehuda, K., Chris, V.: Collaborative filtering for implicit feedback datasets. In: Proceedings on IEEE International Conference on Data Mining (ICMD 2008), pp. 263–272 (2008)
7. Herlocker, J.L., Konstan, J.A., Borchers, A., Riedl, J.: An algorithmic framework for performing collaborative filtering. In: Proceedings on 22nd ACM SIGIR Conference on Information Retrieval, pp. 230–237 (1999)
8. Linden, G., Smith, B., York, J.: Amazon.com recommendations: item-to-item collaborative filtering. IEEE Internet Comput. **7**, 76–80 (2003)
9. Sarwar, B., Karypis, G., Konstan, J., Riedl, J.: Item based collaborative filtering recommendation algorithms. In: Proceedings on 10th International Conference on the World Wide Web, pp. 285–295 (2001)
10. Bell, R., Koren, Y.: Scalable collaborative filtering with jointly derived neighborhood interpolation weights. In: IEEE International Conference on Data Mining (ICDM 2007), pp. 43–52 (2007)
11. Takacs, G., Pilaszy, I., Nemeth, B., Tikk, D.: Major components of the gravity recommendation system. SIGKDD Explor. **9**, 80–84 (2007)
12. Hofmann, T.: Latent semantic models for collaborative filtering. ACM Trans. Inf. Syst. **22**, 89–115 (2004)

13. Salakhutdinov, R., Mnih, A., Hinton, G.: Restricted Boltzmann Machines for collaborative filtering. In: Proceedings on 24th Annual International Conference on Machine Learning, pp. 791–798 (2007)
14. Blei, D., Ng, A., Jordan, M.: Latent Dirichlet allocation. J. Mach. Learn. Res. **3**, 993–1022 (2003)
15. Netflix Update: Try This at Home. http://sifter.org/simon/journal/20061211.html. Accessed 27 Apr 2017
16. Paterek, A.: Improving regularized singular value decomposition for collaborative filtering. In: Proceedings on KDD Cup and Workshop, pp. 39–42 (2007)

Technological Readiness of the Czech Republic and the Use of Technology

Libuse Svobodova$^{(\boxtimes)}$ and Martina Hedvicakova

Department of Economics, Faculty of Informatics and Management,
University of Hradec Kralove, Hradec Kralove, Czech Republic
{libuse.svobodova,martina.hedvicakova}@uhk.cz

Abstract. Since 1993, modern technologies have been growing in the Czech Republic like there are in many other European Union countries. This article aims to analyse the current technological readiness of the Czech Republic and compare its current position with other EU countries (especially on the Visegrad Four). The reasons for the growth of spending on ICT investments consists in the expansion of modern technologies, the requirements for speed of information as well as the execution of operations. The ICT boom has brought about not only an increase in overall efficiency and productivity growth, but is also contributing to GDP growth. Using official data obtained from the Czech Statistical Office and statistical methods, the article analyses the situation of technological competitiveness and readiness in the Czech Republic for the years 2008–2015 and evaluates the use of modern technology. Regarding technological readiness the Czech Republic is doing better than the other Visegrad countries.

Keywords: Development · Internet · Technological readiness

1 Introduction

Like businesses, individual states also measure their competitiveness in comparison with other countries. The World Economic Forum has been making comparisons of states all over the world since 2001 [20]. This article describes the technology level of the Czech Republic from 2008 to 2015. It briefly presents the results of the selected and best 10 countries from the last comparison. The previous article by the author was aimed at the results from 2001 to 2006 [18]. The aim was on the sub index (technology index) of the Growth Competitiveness Index (GCI). A new process for making the GCI calculation has been used since 2007 and the technology index was substituted. The Czech Republic was ranked from 19th to 22nd position in the previous evaluation. This topic has also been addressed by Parasuraman [13], Lin et al. [9], Peltier et al. [14], Jong-Wha [8], Moorhouse [11], Richey et al. [16], Scholleová [17], Svobodová [19] and others.

2 Modern Technologies for Shopping on the Internet

The 21st century is marked by a constantly-changing competitive environment and especially by the needs and behaviour of customers, who determine market demand. Modern technologies are increasingly penetrating into the methods of shopping and

© Springer International Publishing AG 2017
M. Themistocleous and V. Morabito (Eds.): EMCIS 2017, LNBIP 299, pp. 670–678, 2017.
DOI: 10.1007/978-3-319-65930-5_53

thus also into marketing strategies. Multichannel retailing is being used more and more. Multichannel retailing, or Omnichannel retailing, is the use of a variety of channels in a customer's shopping experience, including research conducted before a purchase is made [6, 10]. Such channels include retail stores, online stores, mobile stores, mobile app stores, telephone sales and any other method of transacting with a customer [5].

When focusing solely on the situation in the EU, the most important trends and the current situation in the EU include [7]:

- 67% of individuals aged 16–74 in the EU used the Internet on average daily or almost daily in 2015.
- Nearly two-thirds of Internet users in the 12 months prior to the survey (hereafter referred as "Internet users") made online purchases in the same period. Overall, the share of e-shoppers of all Internet users is growing, with the highest proportions being found in the 16–24 and 25–54 age groups (68% in each case). (See Fig. 1).
- The proportion of e-shoppers varied considerably across Member States, ranging from 18% of Internet users in Romania to 87% in the United Kingdom.
- The most popular type of goods and services purchased online in the EU was clothes and sporting goods (60% of e-buyers), followed by travel and holiday accommodation (52%).
- In terms of frequency, the highest proportion of e-shoppers made purchases in the three months prior to the survey only once or twice (39%). In terms of the amount spent, the highest proportion of e-buyers (40%) bought goods or services for a total of €100–499.
- 30% of e-buyers purchased from other EU Member States, compared with 25% in 2012.

In terms of technology and Internet use, the most important current trends in online shopping include the fact that the influence of social networks is inexorably growing.

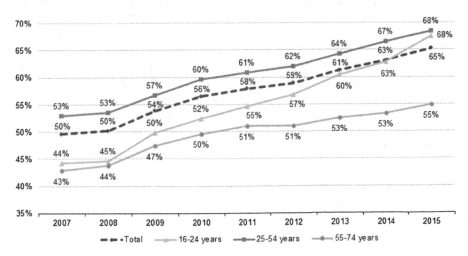

Fig. 1. Internet users who bought or ordered goods or services for private use over the Internet in the previous 12 months by age group, EU 28 (Source: Eurostat (2015a)

Customers share information about products, stores, events, etc. Facebook and Twitter still dominate in Europe but, Pinterest is popular in the US, where consumers use it to share views and experiences with regard to shopping.

Retailers will invest more in digital in-store technologies, especially web-based and mobile technologies. Across the world, this approach already has been named – the "outernet" [15].

In on-line shopping, tradesmen could also use the potential of showrooms to their benefit, as advertising or dispensaries, or charge fees in the future for consulting with vendors and look-over without any subsequent purchase, or ordering directly from the shop. More and more customers are only inspecting products in a showroom and then ordering on-line. Vendors will need to use this trend to their benefit.

Stores are increasingly using advanced information technologies that are able to analyse big data in real time at the individual level and create time- and space-relevant communication and offers for specific customers.

"Brick and mortar" stores can also use touch screens, tablets and other equipment for providing informing about products, user manuals, examples of use, etc.

Digital personalized coupons based on the collection of personal data will soon replace traditional paper coupons.

All of the above show that modern technology plays a key role in shopping. Manufacturers and retailers who want to succeed in today's competitive environment and gain a competitive advantage cannot do without modern technology.

3   The Goal of Article and Methodology

This article aims to compare the competitiveness of individual countries through a sub-index (technology index) of the Growth Competitiveness Index (GCI). It compares the situation regarding the evaluation of the top ten countries in the world. The table will also present the basic results of countries that make up the Visegrad Group (also known as the "Visegrad Four" or "V4"), including the Czech Republic. The Visegrad Group is an association composed of the Czech Republic, the Slovak Republic, Poland and Hungary and focuses on foreign policy activities. The group aims to promote cooperation and stability in the broader region of Central Europe. The article further compares the situation in the Czech Republic since the formation of the new methodology to the latest study, i.e. from 2008/2009 study to the 2015/2016 study.

The result is an analysis of the technological competitiveness and readiness of the Czech Republic for the 2008–2015 period and the use of modern technologies for on-line shopping by the country's population.

The following hypotheses have been established:

- Regarding technological readiness, pursuant to the standing in 2008/2009–2015/2016, the Czech Republic is performing better than the other V4 countries.
- It can be expected that technological readiness is in the Czech Republic continually improved for the each of the past three years - by value.

The article is based on secondary sources. The secondary sources provide information about technological readiness, professional literature and information collected from professional press, web sites, discussions and previous participation at professional seminars and conferences related to the chosen subject. Most of the information was gained from the World Economic Forum and the Czech Statistical Office, the Ministry of Industry and Trade. It was then necessary to select, classify and update accessible relevant information from the numerous published materials that provide the basic knowledge about the selected topic.

The next section analyses the technological readiness and Internet use in the Czech Republic and selected countries.

4 The Global Competitiveness Report and Technological Readiness

The World Economic Forum (WEF) annually prepares the Global Competitiveness Report, which in the last study evaluates the competitiveness of 140 countries.

In today's globalized world, technology is increasingly essential for firms to compete and prosper. The technological readiness pillar measures the agility with which an economy adopts existing technologies to enhance the productivity of its industries, with specific emphasis on its capacity to fully leverage information and communication technologies (ICTs) in daily activities and production processes for increased efficiency and enabling innovation for competitiveness. ICTs have evolved into the "general purpose technology" of our time, given their critical spillovers to other economic sectors and their role as industry-wide enabling infrastructure. Therefore, ICT access and usage are key enablers of a country's overall technological readiness.

Table 1 presents countries that are on the highest positions in the evaluation of technological readiness. Also compared are similar countries such as the Czech Republic and countries from the Visegrad Group, esp. the Slovak Republic, Poland and Hungary. The positions change each year. In the last three comparisons, the same countries occupied the first ten places. The Czech Republic was ranked 29th, the best of the countries that make up the Visegrad Group. Poland, the Slovak Republic and Hungary are closely positioned between 41st and 48th place. The position of countries from the Visegrad Group has improved when comparing 2008/2009 and 2015/2016. The ratings got better in all countries except the Slovak Republic. On the other hand, the Czech Republic remains stagnant at 29th and Czech businesses - although doing comparatively well in a regional context - are less sophisticated and innovative than other economies in the European Union (World Economic Forum, The Global Competitiveness Report 2015–2016).

The subjective feelings of the evaluators may be one drawback of this evaluation. Some people or nations may be less critical, others more.

The next four points are evaluated on the basis of the official results from official authorities. The official authority and source of the information in the Czech Republic is the Czech Statistical Office.

The first 10 places include 8 countries from the European Union. The greatest progress in the selected group was made by Luxembourg when comparing the first 2008/09 evaluation. On the other hand, the Netherlands saw the highest decrease. Ratings were recorded in different variations, which are in the order of exchanging

Table 1. The ranking of technological readiness – the first places and countries from the Visegrad Group from 2008/09 to 2015/16 (Source: World economic forum)

	2008/09	2009/10	2010/11	2011/12	2012/13	2013/14	2014/15	2015/16
Luxembourg	12	5	2	9	2	2	1	1
Great Britain	8	8	8	8	7	4	2	3
Sweden	2	1	1	2	1	1	3	4
Norway	4	7	9	–	13	3	4	7
Hong Kong	10	9	5	6	4	6	5	8
Denmark	3	4	6	4	3	5	6	9
Singapore	7	6	11	10	5	7	7	5
Iceland	6	14	4	3	8	10	8	6
Netherlands	1	2	3	5	9	8	9	10
Switzerland	5	3	7	1	6	9	10	2
Czech Republic	33	30	32	31	31	34	36	29
Poland	46	44	47	48	42	43	48	41
Slovak Republic	36	33	34	37	45	52	52	44
Hungary	40	40	37	36	49	46	50	48

positions. There is not a significant change. In the first comparison, Canada was ranked 9th. In the last evaluation, it was ranked 18th.

Based on the results shown in Table 1, it is possible to accept the hypothesis that the Czech Republic, pursuant to its ranking in 2008/2009–2015/2016, is performing better than the other V4 countries.

Table 2 presents the overall results of the technological readiness of the Czech Republic from 2008/2009 to 2015/2016. A detailed evaluation of the Czech Republic will be presented in Tables 3 and 4.

The technological readiness of the Czech Republic improved not only in the value of technological readiness, but also in its ranking. For the first time, it placed under thirtieth place in the 2015/16 study. Despite the fact that the overall value got better in 2014/15 compared to 2013/14, our position deteriorated compared to other countries. The reason was that other countries improved more than the Czech Republic in the given factors.

Table 2. Technological readiness, total ranking and values of the Czech Republic (Source: World economic forum)

	2008/09	2009/10	2010/11	2011/12	2012/13	2013/14	2014/15	2015/16
Rank	33	30	32	31	31	34	36	29
Value	4.5	4.7	4.5	4.8	5.1	4.9	5.0	5.4

Table 3. The values of technological readiness factors in the Czech Republic (Source: World economic forum)

	2011/12	2012/13	2013/14	2014/15	2015/16
The availability of the latest technologies	5.6	5.5	5.2	5.2	5.6
Firm-level technology absorption	5.2	5.1	4.9	5.0	5.0
FDI and technology transfer	5.3	5.3	5.1	5.0	5.0
Individuals using Internet, as a %	68.8	73.0	75.0	74.1	79.7
Fixed broadband Internet subscriptions/100 pop	14.7	15.7	16.6	17.0	27.6
Int'l Internet bandwidth, kb/s per user	47.7	91.1	101.0	111.2	116.8
Mobile broadband subscriptions/100 pop	-	43.1	44.0	45.3	62.8

Table 4. The ranking of technological readiness factors of the Czech Republic (Source: World economic forum)

	2008/09	2009/10	2010/11	2011/12	2012/13	2013/14	2014/15	2015/16
The availability of the latest technologies	49	48	46	40	43	53	51	32
Firm-level technology absorption	38	35	36	45	49	54	50	48
FDI and technology transfer	13	14	15	15	18	27	36	22
Individuals using the Internet, as a %	40	38	30	30	28	28	31	27
Fixed broadband Internet subscriptions/100 pop	33	32	33	41	38	41	41	24
Int'l Internet bandwidth, kb/s per user	–	–	34	20	16	19	25	25
Mobile broadband subscriptions/100 pop	7	13	–	–	20	35	44	39

The availability of the latest technologies has improved in the last year for nineteen positions. All other factors also improved in the last evaluation with the exception of int'l Internet bandwidth, which remained at the same level. The most problematic rankings were firm-level technology absorption and mobile broadband subscriptions/100 pop.

Table 3 lists the specific values assigned to the Czech Republic. Even though we know the evaluation ranking from the 2008/2009 study, the obtained values were published only from 2011/2012 study.

As was mentioned above, despite the fact that some factors increased (for example, the last three factors in the 2014/15 evaluation), our position deteriorated in an overall comparison with other countries. In the last evaluation, the comparison with the last year's availability of latest technologies, individuals using Internet, fixed broadband Internet subscriptions, Int'l Internet bandwidth and mobile broadband subscriptions all improved. The opinion of people in the evaluation of the availability of the latest technologies was the same value in 2015/16 as in the 2011/12 ranking. The last two criteria stayed the same in comparison with the previous year. On the other hand, they were rated worse than in the 2011/12 and 2012/13 studies.

An increase in individual indicators, however, does not imply a better evaluation in comparison with other countries. Rankings with regard to individual factors is shown in Table 4. In the table above some results in the rankings are missing. The values have not been provided. In making a comparison between 2008/09 and 2015/16, the Czech Republic improved in four criteria. These are:

- The availability of the latest technologies
- Individuals using the Internet, as a %
- Fixed broadband Internet subscriptions/100 pop
- Int'l Internet bandwidth, kb/s per user.

Given the fact that the investigation by the World Bank was exhaustive and joined by all the countries, it is not necessary to use advanced statistical methods. On the basis of obtained and analysed data, it is possible to accept the hypothesis that technological readiness improved in the values in the last three studies completed in the Czech Republic.

The partial results of Global Competitiveness, especially the technological readiness index survey, will be supplemented in the next part by selected results from the Czech Statistical Office [2–4].

5   Conclusion and Discussion

In comparison with the ranking of the other Visegrad countries, the technological readiness of the Czech Republic has occupied the top position over the past 8 studies. Slovakia stands behind the Czech Republic, followed by Poland and Hungary.

Two hypotheses regarding the technological readiness can be confirmed. In comparison with the ranking of the other Visegrad countries, the technological readiness of the Czech Republic has occupied the top position over the past 8 studies. Slovakia stands behind the Czech Republic, followed by Poland and Hungary.

Within the period under review, the Czech Republic moved from the initial position of 33 to position 36 with the penultimate evaluation up to the best standing of 29 in the final evaluation. The Czech Republic has improved and nearly doubled its value in fixed broadband Internet subscriptions, and more than doubly increased its Int'l Internet bandwith between the 2011/12 and 2015/16 studies.

Even though most of the surveyed criteria have improved, the Czech Republic received a worse rating than the developed countries of the European Union and still has room for further improvement. This can be supported by subsidies from the

European Union, which we can get through the end of 2015. However, they can be lost if officials and authorized persons are inactive [12].

In most countries of the European Union there is an aging population. Age group 65+ grows. The question is whether population aging will stop the growing trend in online shopping and social networks. On the other hand, the increasing trend to continually reduce costs and increase labor productivity. This is why Industry 4.0 has emerged. In the framework of further research, the impacts of Industry 4.0 and investment evaluation will be analyzed and investment evaluation.

Acknowledgement. This paper is supported by specific project No. 2103 "Investment evaluation within concept Industry 4.0" at Faculty of Informatics and Management, University of Hradec Kralove, Czech Republic. Thanks to help student Petra Henclová.

References

1. APEK: Association for Electronic Commerce, Vánoce se blíží, v e-shopech začíná hlavní sezona. Češi letos za dárky utratí na Internetu přes 25 miliard Kč (2015). http://www.apek.cz/tiskove-zpravy/vanoce-se-blizi-v-e-shopech-zacina-hlavni-sezona-cesi-letos-za-darky-utrati-na-Internetu-pres-25-miliard-kc/
2. Czech Statistical Office: Information Technologies (2015). https://www.czso.cz/csu/czso/information_technologies
3. Czech Statistical Office: Statistical Yearbook of the Czech Republic – 2015 (2015). https://www.czso.cz/csu/czso/21-information-and-communication-technologies
4. Czech Statistical Office. Information society in figures – 2014 (2015a). https://www.czso.cz/csu/czso/information-society-in-figures-2014-vs2yjvlh5u
5. Deloitte University Press: DeDeloitte's 2014 annual holiday survey2014 (2015). http://dupress.com/articles/holiday-retail-sales-2014/
6. Dholakia, R.R., Zhao, M., Dholakia, N.: Multichannel retailing: a case study of early experiences. J. Interact. Mark. **19**(2), 63 (2005). doi:10.1002/dir.20035
7. Eurostat: Internet users who bought or ordered goods or services for private use over the internet in the previous 12 months by age groups, EU-28, 2007–2015 (% of internet users). png (2015a). http://ec.europa.eu/eurostat/statistics-explained/index.php?title=File:Internet_users_who_bought_or_ordered_goods_or_services_for_private_use_over_the_internet_in_the_previous_12_months_by_age_groups,_EU-28,_2007-2015_(%25_of_internet_users).png&oldid=270018
8. Jong-Wha, L.: Education for technology readiness: prospects for developing countries. J. Hum. Dev. **2**(1), 115–151 (2001)
9. Lin, ChH, Shih, H.Y., Sher, P.J.: Integrating technology readiness into technology acceptance: the TRAM model. Psychol. Mark. **24**(7), 641–657 (2007)
10. McGoldrick, P.J., Collins, N.: Multichannel retailing: profiling the multichannel shopper. Int. Rev. Retail Distrib. Consum. Res. **17**(2), 139 (2007). doi:10.1080/09593960701189937
11. Moorhouse, D.J.: Detailed definitions and guidance for application of technology readiness levels. J. Aircr. **39**(1), 190–192 (2002)
12. Novinky.CZ: Miliardy z EU na Internet ohroženy (2016). http://www.novinky.cz/domaci/392307-miliardy-z-eu-na-Internet-ohrozeny.html
13. Parasuraman, A.: Technology Readiness Index (Tri) a multiple-item scale to measure readiness to embrace new technologies. J. Serv. Res. **2**(4), 307–320 (2000)

14. Peltier, J.W., Zhao, Y., Schibrowksy, J.A.: Technology adoption by small businesses: an exploratory study of the interrelationships of owner and environmental factors. Int. Small Bus. J. **30**, 406–431 (2012)
15. POPAI Central Europe: Retail trendy 2014 - Ještě větší moc zákazníků a on-line! (2014). http://www.popai.cz/d-2-21-307/Retail-trendy-2014Jeste-vetsi-moc-zakazniku-a-on-line!. aspx
16. Richey, R.G., Daugherty, P.J., Roath, A.S.: Firm technological readiness and complementarity capabilities impacting logistics service competency and performance. J. Bus. Logist. **28**(1), 195–228 (2007)
17. Scholleová, H.: Czech Republic innovations evaluated by summary innovation index. In: Hradecké ekonomické dny 2009, pp. 203–210. Universita Hradec Králové, Hradec Králové (2009)
18. Svobodová, L.: Technology readiness of the Czech Republic. In: Hradecké ekonomické dny 2010, pp. 126–130. Universita Hradec Králové, Hradec Králové (2010)
19. Svobodová, L.: Technology readiness of the Czech Republic. In: Proceedings of the 9th International Days of Statistics and Economics (MSED), pp. 1518–1527 (2015)
20. Word Economic Forum: Research (2015). http://www.weforum.org/reports/

Author Index

Printed in the United States
By Bookmasters